ACTIVE NETWORK THEORY

S. S. HAYKIN
Department of Electrical Engineering
McMaster University

ADDISON-WESLEY PUBLISHING COMPANY

READING, MASSACHUSETTS · MENLO PARK, CALIFORNIA · LONDON · DON MILLS, ONTARIO

This book is in the
ADDISON-WESLEY SERIES IN ELECTRICAL ENGINEERING

Consulting Editors:
DAVID K. CHENG
LEONARD A. GOULD
FRED K. MANASSE

 Printed in Great Britain. Library of Congress Catalog Card No. 73-94970.

PREFACE

The objective of this book is to provide a unified and modern treatment of active networks at a level suitable for use by senior undergraduate and graduate electrical engineering students. Wherever possible, physical explanation and meaningful illustrative examples have been included to elucidate the results obtained. Each chapter ends with a list of references and problems, some of which are intended to extend the material of the text.

Chapter 1 reviews the various models used to describe the behaviour of vacuum-tube and semiconductor devices. From topological considerations, systematic procedures are developed in Chapter 2 for formulating the loop, nodal and state-variable equations of linear time-invariant networks. In Chapter 3 we develop the criteria for short-circuit stability, open-circuit stability (with applications to negative resistance amplifiers and oscillators) and passivity of one-port networks. Chapter 4 presents a detailed study of the various parameter sets for the electrical characterization of a two-port network and its equivalent circuits, the behaviour of terminated two-port networks, the indefinite admittance matrix for three-terminal networks and its application to vacuum-tube and transistor amplifiers, the interconnections of two-port networks and the pertinent validity tests (with applications to feedback amplifiers), and the electrical characterization of the ideal transformer, gyrator, negative-impedance converter and negative-impedance inverter. In Chapter 5 the study of two-port networks is continued by developing the criteria for reciprocity, symmetry, stability (with applications to feedback oscillators), potential instability and absolute stability (with applications to tuned amplifiers), passivity, conjugate-image impedances and maximum power gain. Chapter 6 introduces the scattering-matrix method of studying network behaviour.

In Chapter 7 the various integral relationships between the gain and phase components of a minimum phase transfer function are derived using Cauchy's theorem of complex variable theory. In Chapter 8 the conventional transient response analysis using the Laplace transform is considered. Also Elmore's definitions of rise time and delay as applied to low-pass amplifiers with a monotonic transient response are developed; the chapter ends with a discussion of the convolution integral. In Chapter 9 the time response is studied in terms of the state-variable technique, emphasizing the role of the state-transition matrix and the controllability and observability of natural modes.

Chapter 10 introduces the idea of a signal-flow graph, and the evaluation of gain by using a step-by-step reduction of the graph or, alternatively, by using Mason's direct rule. Next, Chapter 11, with the aid of signal-flow graphs, develops the basic feedback concepts of return difference and null return difference, and their relationships to the over-all gain, sensitivity, noise performance and driving-point impedances of single-loop feedback amplifiers. The algebraic method of evaluating the return difference in terms of the circuit determinant, as originally carried out by Bode, is also studied. Procedures are described for the practical evaluation of return difference by breaking the feedback loop and terminating it appropriately, or alternatively by using driving-point impedance measurements. In terms of the return difference and null return difference matrices for a multiplicity of controlled sources, formulae are developed for the over-all gain, sensitivity, and driving-point impedances of multiple-loop feedback circuits. In Chapter 12 the study of feedback amplifiers is completed by examining the stability problem and its design limitations in terms of the closed-loop transient response (and the associated root-locus diagram). Using a conformal transformation, a procedure is described for using the Nyquist diagram to evaluate the dominant oscillatory component of the closed-loop transient response. The Nyquist criterion is next generalized to assess the stability of a multiple-loop feedback circuit. Some of the gain–phase relations of Chapter 7 are used to derive Bode's ideal loop gain characteristic which assures the realization of a specified mid-band return ratio with desired stability margins. A procedure is described for the synthesis of an inter-stage corrective network for achieving such a characteristic.

Chapter 13 is devoted to the synthesis of inductorless filters using negative-impedance converters or gyrators, with particular attention given to the problem of minimizing the sensitivity to parameter changes. The procedures described include Linvill's cascade and Yanagisawa's parallel realizations of a transfer function, Kinariwala's driving-point impedance synthesis and its adaptation to the realization of transfer functions, using negative-impedance converters, and Horowitz's cascade realization using a gyrator. The chapter concludes with gyrator–capacitor adaptations of *LC* ladder filters. The last chapter of the book is devoted to a study of the behaviour of uniform and exponentially tapered distributed *RC* networks in the frequency domain, and some of their applications.

Although the book is largely devoted to a study of linear active networks, the effects of nonlinearity are not completely ignored. Thus, in Chapters 3 and 12 we have included sections on equivalent linearization procedures for evaluating the amplitude of oscillation in nonlinear negative-resistance and feedback oscillators, while in Chapter 4 a procedure is described for evaluating harmonic distortion due to a small degree of nonlinearity in amplifiers. Also, in Chapter 9 the so-called discrete-time approximation for approximate evaluation of the state response is extended to nonlinear networks.

Hamilton, Ontario S.S.H.
September, 1969

CONTENTS

Chapter 5 Further Properties of Two-port Networks

Chapter 6 The Scattering Matrix

Chapter 7 Gain–phase Analysis

Chapter 8 Transient Response

Chapter 9 State-variable Approach

Chapter 10 Signal-flow Graphs

CHAPTER 1

NETWORK ELEMENTS AND MODELS

1.1 INTRODUCTION

The theory of active networks, like so many other disciplines, has undergone some big changes during the past two decades. If we were to single out the one development that has had the greatest and most profound impact on the study of active networks, it must surely be the advent of the transistor and related semiconductor devices which have come to dominate the field of electronics. In order to understand fully the circuit properties and limitations of semiconductor devices and the older vacuum tubes in a unified manner, and be able to deal effectively with the circuit applications of new devices yet to come, it has become more necessary than ever before to emphasize the fundamentals of active network theory. It is with this objective in mind that the present book has been written.

The purpose of this introductory chapter is to review briefly the various types of electric networks and elements, and some of the circuit models used to represent the external behaviour of vacuum tubes and semiconductor devices when operating under small signal conditions.

1.2 CLASSIFICATION OF NETWORKS

Linear and Nonlinear Networks

There are various ways of classifying electric networks and the elements which constitute such networks. For example, we speak of a network element as *linear* or *nonlinear* depending on whether or not its behaviour, within a specified range of operation, is describable by a linear equation. The ordinary resistor, in which the current is proportional to the voltage drop across it, is an example of a linear element.

A network which is made up entirely of linear elements is itself linear and has the important property of *superposition* which can be stated as follows: If a number of independent electrical sources exist in a linear network, the total voltage (or current) in any part of the network is equal to the sum of the voltages (or currents) that would result if the various sources were to act individually. This is known as the *principle of superposition.*

In the category of linear networks we may include a vacuum tube or transistor provided the device is operated under small signal conditions, that is, the deviations from the quiescent operating point are sufficiently small to assure a negligible departure from linear behaviour. Another example is the tunnel diode which can present an essentially linear negative resistance over part of its characteristic curve.

Time-invariant and Time-varying Networks

Another way of classifying networks is by the property of *time invariance*. A network is time invariant if its input–output relations do not change with time, that is, a linear system having the input–output relation

$$r(t) = K\, e(t)$$

is time invariant if and only if

$$r(t - \tau) = K\, e(t - \tau) \tag{1.1}$$

for all $e(t)$ and τ. The network is time varying if it is not time invariant. As illustrated in Fig. 1.1, the definition of Eq. (1.1) implies that the shape of the output signal of a time-invariant network is entirely independent of the time at which the input signal is applied to the network. In this book we shall be only concerned with time-invariant networks.

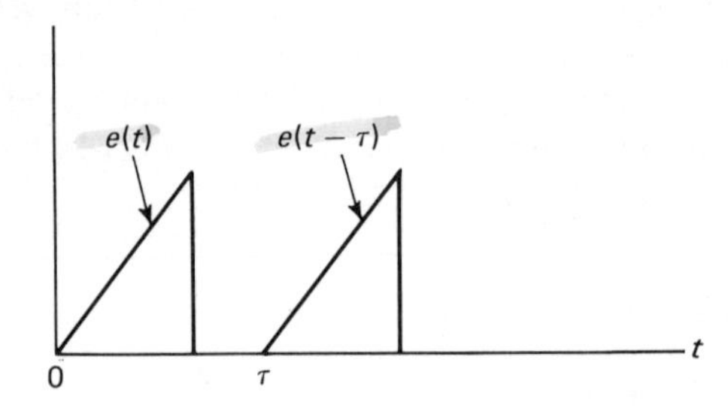

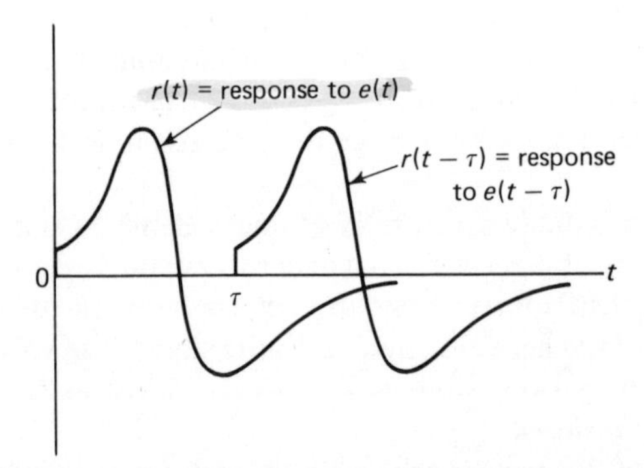

Fig. 1.1. Input and output signals in a time-invariant network.

Passive and Active Networks

We may also classify electric networks into *passive* and *active* networks. Consider an n-port network which is linear and time invariant, with $i_k(t)$ and $v_k(t)$ denoting the instantaneous current and voltage, respectively, at port $k - k'$ (say), as in Fig. 1.2. The total energy into the n-port network is

$$\mathscr{E}(t) = \int_{-\infty}^{t} \sum_{k=1}^{n} v_k(\tau) i_k(\tau)\, d\tau, \tag{1.2}$$

where $v_k(\tau)$ and $i_k(\tau)$ are obtained from $v_k(t)$ and $i_k(t)$, respectively, simply by substituting τ for t. The network is passive if $\mathscr{E}(t) \geq 0$ for all t; otherwise the network is active.

The resistor (which dissipates energy), the inductor (which stores energy by virtue of the current flowing through it), and the capacitor (which stores energy by virtue of the voltage existing across it) are examples of passive-network elements. On the other hand, the negative resistance exhibited by a tunnel diode is an example of an active two-terminal network element. A transistor operated in its linear region is also an example of an active network because the output-signal power which such a device delivers to its load can exceed the input-signal power supplied from an external source. An important characteristic of an active network is that it may, under certain circumstances, become unstable and therefore break into oscillation. This is in direct contrast to a passive network which is inherently stable.

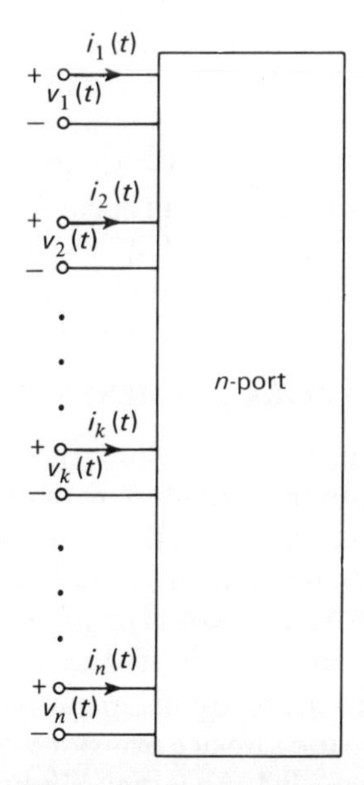

Fig. 1.2. n-Port network.

1.3 LAPLACE TRANSFORM

The *Laplace transform* provides a rather powerful and yet relatively simple tool for the study of linear time-invariant networks. For a function of time, $f(t)$, which is zero for $t < 0$, to be transformable it is sufficient that we have

$$\int_0^\infty |f(t)|\varepsilon^{-\sigma_a t}\, dt < \infty \tag{1.3}$$

for some real, positive σ_a known as the *abscissa of absolute convergence*. This condition is satisfied by functions of time commonly used in the study of electrical systems. The Laplace transform of $f(t)$ is defined as

$$F(s) = \int_0^\infty f(t)\varepsilon^{-st}\, dt = \mathscr{L}[f(t)], \tag{1.4}$$

where $s = \sigma + j\omega$, with both σ and ω being real. The variable s has the dimensions of frequency and is therefore termed the *complex-frequency variable*. Equation (1.4) transforms the function $f(t)$ in the time domain into another function $F(s)$ in the complex-frequency domain. The *inverse Laplace transform* is written as

$$f(t) = \frac{1}{2\pi j}\int_{c-j\infty}^{c+j\infty} F(s)\varepsilon^{st}\, ds = \mathscr{L}^{-1}[F(s)], \tag{1.5}$$

where c is any real number greater than the σ_a (i.e., abscissa of absolute convergence) belonging to the $f(t)$ and its transform $F(s)$, so that the path of integration is parallel to the $j\omega$-axis of the s-plane, passing through the point $s = c + j0$.

The functions $f(t)$ and $F(s)$ constitute a *transform pair* with a one-to-one correspondence. In Table 1.1 we have listed some of the commonly used transform pairs. This table may be used in two ways: to find the direct Laplace transform for a given $f(t)$, or the inverse Laplace transform for a given $F(s)$. Table 1.2 lists some of the important properties of the Laplace transform, the proofs of which follow directly from Eq. (1.4) and can be found elsewhere.[1]

1.4 PASSIVE ELECTRIC NETWORK ELEMENTS

The resistor, inductor and capacitor constitute three network elements the interconnections of which comprise the branch of network theory dealing with *linear, lumped, finite, passive and bilateral* networks. The voltage–current relationships for these elements, in the time domain, are as defined in Table 1.3 depending on whether the current $i(t)$ or the voltage $v(t)$ is regarded as the independent variable. The polarities of $v(t)$ and $i(t)$ are as indicated in Fig. 1.3. In Table 1.3 we have also included the voltage–current relationships in the *complex-frequency domain*; they are obtained by applying the Laplace transform to the pertinent time-domain relationships. We see that provided the initial conditions are zero, that is, for an

Table 1.1. A Short Table of Laplace Transform Pairs

$f(t)$	$F(s)$
$\delta(t)$ (unit-impulse function)*	1
$u(t)$ (unit-step function)†	$\frac{1}{s}$
t	$\frac{1}{s^2}$
$\frac{1}{(n-1)!}t^{n-1} \quad (n = 1, 2, 3, \ldots)$	$\frac{1}{s^n}$
ε^{-at}	$\frac{1}{s+a}$
$\frac{1}{(n-1)!}t^{n-1}\varepsilon^{-at} \quad (n = 1, 2, 3, \ldots)$	$\frac{1}{(s+a)^n}$
$\cos \omega t$	$\frac{s}{s^2+\omega^2}$
$\frac{1}{\omega}\sin \omega t$	$\frac{1}{s^2+\omega^2}$
$\varepsilon^{-at}\cos \omega t$	$\frac{s+a}{(s+a)^2+\omega^2}$
$\frac{1}{\omega}\varepsilon^{-at}\sin \omega t$	$\frac{1}{(s+a)^2+\omega^2}$

* The unit-impulse function $\delta(t)$ is

$$\delta(t) = \begin{cases} \text{undefined} & \text{for } t = 0 \\ 0 & \text{for } t \neq 0. \end{cases}$$

Also

$$\int_{-\infty}^{\infty} \delta(t)\, dt = 1$$

† The unit-step function $u(t)$ is defined as

$$u(t) = \begin{cases} 1 & \text{for } t > 0 \\ 0 & \text{for } t < 0. \end{cases}$$

inductor the initial current $i(0^+) = 0$ and for a capacitor the initial voltage $v(0^+) = 0$, then in the complex-frequency domain a resistor presents an impedance $Z(s) = R$, an inductor presents an impedance $Z(s) = sL$, and a capacitor presents an impedance $Z(s) = 1/sC$, where $Z(s) = V(s)/I(s)$.

In addition to the resistance, inductance and capacitance there exists a fourth network parameter, the mutual inductance which arises when two coils are

Table 1.2. Properties of the Laplace Transform

1. Linearity:

   If
$$F_1(s) = \mathscr{L}[f_1(t)],$$
$$F_2(s) = \mathscr{L}[f_2(t)],$$
   then
$$\mathscr{L}[f_1(t) + f_2(t)] = F_1(s) + F_2(s).$$

2. Scaling:
$$\mathscr{L}[af(t)] = a\mathscr{L}[f(t)] = aF(s).$$

3. Differentiation in the time domain:
$$\mathscr{L}\left[\frac{d}{dt}f(t)\right] = sF(s) - f(0^+),$$
   where $f(0^+)$ denotes the value of $f(t)$ at $t = 0^+$. The + sign indicates that if there is a discontinuity in $f(t)$ at $t = 0$, then $f(0^+)$ is the limit of $f(t)$ as t approaches zero from the positive side.

4. Integration in the time domain:
$$\mathscr{L}\left[\int_{-\infty}^{t} f(t)\,dt\right] = \frac{F(s)}{s} + \frac{f^{-1}(0^+)}{s},$$
   where $f^{-1}(0^+)$ denotes the value of
$$\int_{-\infty}^{0^+} f(t)\,dt.$$

5. Time-displacement or shifting theorem:
$$\mathscr{L}[f(t - a)u(t - a)] = \varepsilon^{-as}F(s),$$
   where $u(t - a)$ is a unit-step function occurring at $t = a$.

6. Initial value theorem:
$$f(0) = \lim_{s\to\infty} sF(s).$$

7. Final value theorem:
$$f(\infty) = \lim_{s\to 0} sF(s).$$

magnetically coupled, as in Fig. 1.4. In this diagram L_1 is the self inductance of the primary winding, and L_2 is the self inductance of the secondary winding. If the currents $i_1(t)$ and $i_2(t)$ are regarded as the independent variables, we obtain the following voltage–current relationships, in the time domain, for the pair of

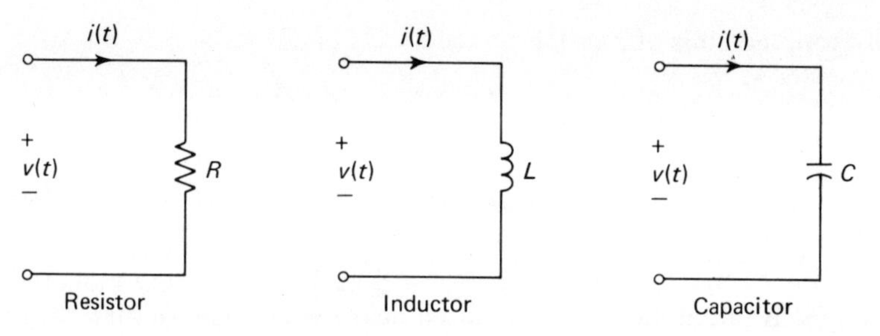

Fig. 1.3. *R*, *L*, and *C*.

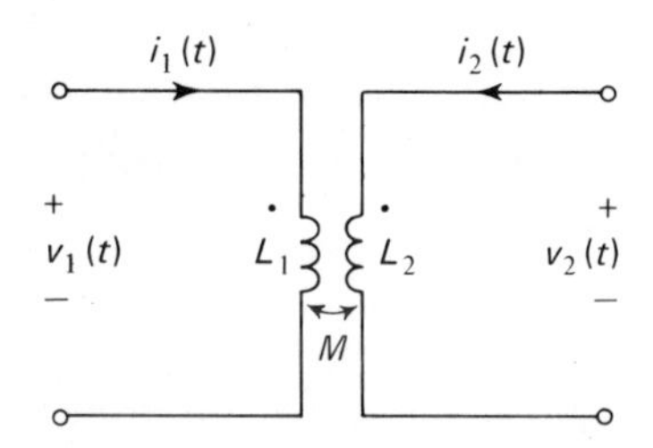

Fig. 1.4. A pair of mutually coupled coils.

Table 1.3. Voltage–Current Relationships for *R*, *L* and *C*

	Time domain	Complex-frequency domain
Resistor R	$v(t) = Ri(t)$ $i(t) = Gv(t)$	$V(s) = RI(s)$ $I(s) = GV(s)$
Inductor L	$v(t) = L\dfrac{di(t)}{dt}$ $i(t) = \dfrac{1}{L}\int_{-\infty}^{t} v(t)\,dt$	$V(s) = sLI(s) - Li(0^+)$ $I(s) = \dfrac{V(s)}{sL} + \dfrac{v^{-1}(0^+)}{sL}$
Capacitor C	$v(t) = \dfrac{1}{C}\int_{-\infty}^{t} i(t)\,dt$ $i(t) = C\dfrac{dv(t)}{dt}$	$V(s) = \dfrac{I(s)}{sC} + \dfrac{i^{-1}(0^+)}{sC}$ $I(s) = sCV(s) - Cv(0^+)$

mutually coupled coils of Fig. 1.4:

$$v_1(t) = L_1\frac{di_1(t)}{dt} + M\frac{di_2(t)}{dt},$$
$$v_2(t) = M\frac{di_1(t)}{dt} + L_2\frac{di_2(t)}{dt}. \tag{1.6}$$

According to the *dot convention* used in Fig. 1.4 the mutual inductance M is always a positive number. The dot placed near one of the terminals for each coil simply specifies that positive current enters that terminal. If the currents $i_1(t)$ and $i_2(t)$ enter by the dots (as in Fig. 1.4) or both leave by the dots, the sign preceding M in the voltage–current relations for the coupled coils is positive, as in Eqs. (1.6). If, however, one current enters and the other leaves by the dots, the sign preceding M is negative.

The mutual inductance M is a measure of the magnetic coupling between the coils; it is related to the self inductances L_1 and L_2 of the two coils by

$$M = k\sqrt{L_1L_2}, \tag{1.7}$$

where k is called the *coefficient of coupling*. Owing to the existence of unavoidable flux leakages, we find that in a practical situation the coefficient of coupling is always less than unity.

In the ideal case when the coefficient of coupling is unity, and at the same time the primary and secondary inductances are infinite but so proportioned that their ratio is finite, the coupled arrangement of Fig. 1.4 is said to constitute an *ideal transformer*. For such a device we have

$$v_2(t) = nv_1(t),$$
$$i_2(t) = \frac{1}{n}i_1(t), \tag{1.8}$$

where n is the *turns ratio*. An ideal transformer is obviously a hypothetical device; however, a practical transformer with an iron core can approximate the conditions of an ideal transformer fairly closely.

1.5 VACUUM TUBES[2]

Under conditions of small signal operations, a vacuum tube or transistor behaves essentially as a linear device in that it may be replaced by an incremental *circuit model* consisting of linear, lumped elements (i.e. resistors, capacitors, and inductors) and ideal sources whose magnitudes are directly proportional to voltages or currents existing in other parts of the model. The parameters of the model are chosen to yield the best approximate fit to the characteristics of the device inside the region of operation.

Consider first the case of a vacuum tube triode having the static characteristic curves of Fig. 1.5. Assuming that the total grid–cathode and plate–cathode voltages, V_{GK} and V_{PK}, are the independent variables, and expressing the total plate current I_P as a function of these variables, we have:

$$I_P = f(V_{GK}, V_{PK}). \tag{1.9}$$

If V_{GK} and V_{PK} are given small increments, the resulting increment in I_P can be expressed, to a first order of approximation, as follows:

$$dI_P = \frac{\partial I_P}{\partial V_{GK}} dV_{GK} + \frac{\partial I_P}{\partial V_{PK}} dV_{PK}. \tag{1.10}$$

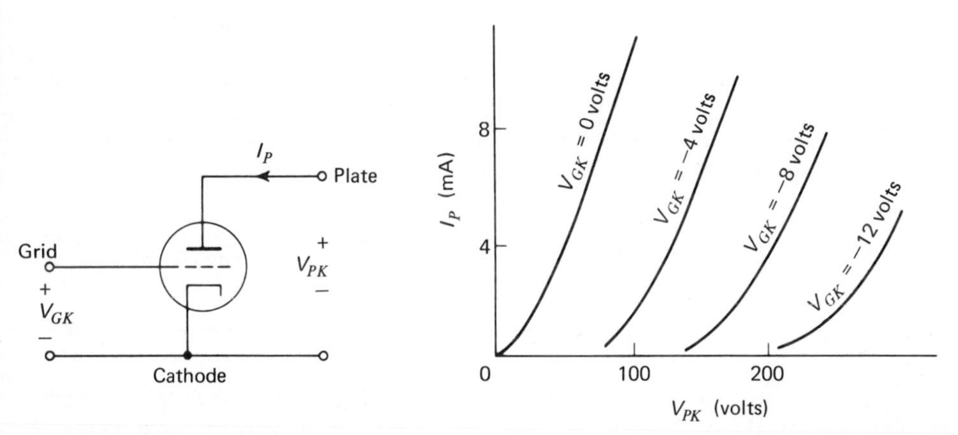

Fig. 1.5. Vacuum-tube triode and its static plate-characteristic curves. (Diagram includes typical values.)

Let

$$g_m = \text{mutual conductance}$$

$$= \left.\frac{\partial I_P}{\partial V_{GK}}\right|_{V_{PK}=\text{constant}} \tag{1.11}$$

$$g_p = \text{incremental plate conductance}$$

$$= \left.\frac{\partial I_P}{\partial V_{PK}}\right|_{V_{GK}=\text{constant}} \tag{1.12}$$

and let the increments in I_P, V_{GK} and V_{PK} be designated i_p, v_{gk}, and v_{pk}, respectively; then Eq. (1.10) becomes

$$i_p = g_m v_{gk} + g_p v_{pk}. \tag{1.13}$$

Equation (1.13) corresponds to the circuit model shown in Fig. 1.6(a). This model

SC. CURRENTS ARE = ⇒ $g_m V_{gk} = \frac{\mu V_{gk}}{r_p}$

 ⇒ $g_m r_p = \mu$

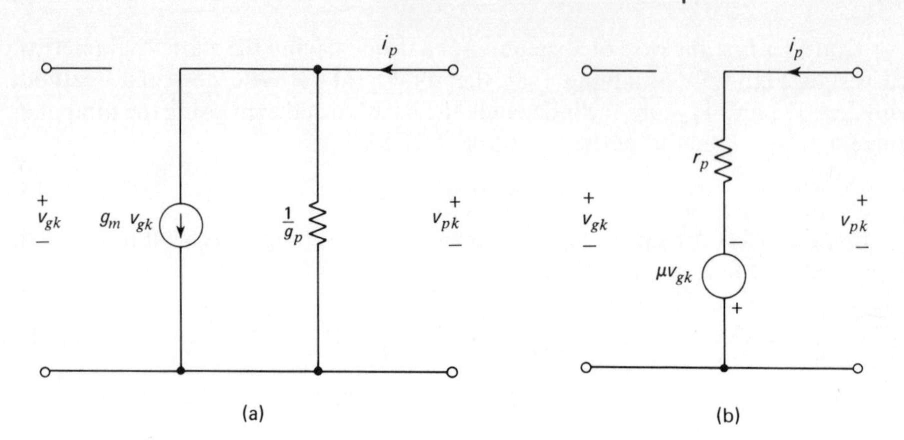

Fig. 1.6. Circuit models of a triode.

may also be transformed into the form of Fig. 1.6(b) where $r_p = 1/g_p$ is the *incremental plate resistance*, and μ is the *amplification factor* defined by

$$-\left.\frac{\partial V_{PK}}{\partial V_{GK}}\right|_{I_P = \text{constant}} = \mu = g_m r_p. \tag{1.14}$$

The minus sign in Eq. (1.14) results from the fact that in order to maintain the plate current I_P constant, the voltages V_{GK} and V_{PK} have to take on opposite increments.

The two models of Fig. 1.6 are both suitable representations for the triode at low frequencies; the choice between them is usually determined by convenience. At high frequencies, however, the effect of the interelectrode capacitances of the tube can become important; to account for them, we add the plate–grid capacitance C_{gp}, the grid–cathode capacitance C_{gk} and the plate–cathode capacitance C_{pk}, as in the model of Fig. 1.7. It should be noted that the actual values of all the parameters in this model depend on the quiescent operating point.

In a vacuum-tube pentode the characteristic curves are as shown in Fig. 1.8 where we see that in the region of normal linear operation of the device, the plate

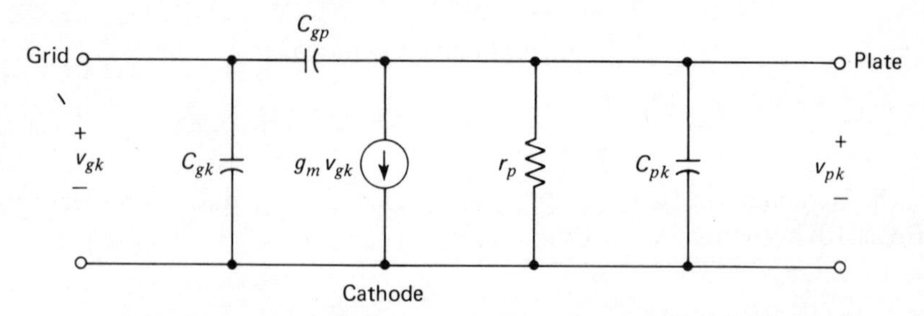

Fig. 1.7. Circuit model including interelectrode capacitances.

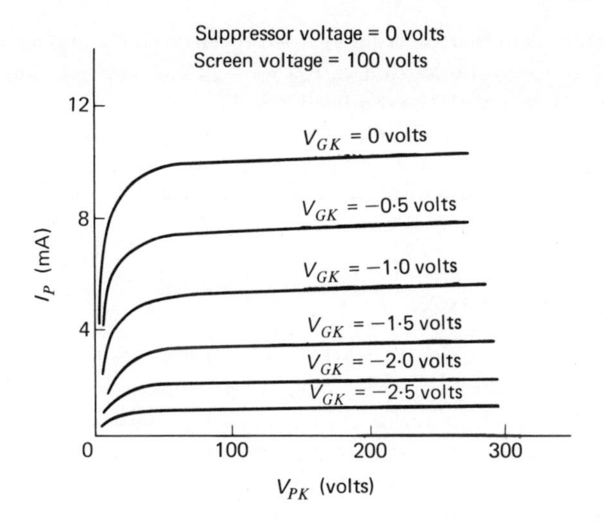

Fig. 1.8. Static plate-characteristic curves of a vacuum-tube pentode. (Diagram includes typical values.)

current is practically independent of the grid–plate voltage. That is to say, the incremental plate resistance r_p is very large (usually of the order of 1 MΩ), and as such, the shunting effect of r_p is often negligible. Furthermore, in a pentode we find that the grid–plate capacitance is much smaller than in a triode.

Biasing Considerations

A commonly used method for establishing the quiescent operating point of a triode amplifier is to include a resistor R_K in the cathode circuit, as shown in Fig. 1.9. Under quiescent conditions, the grid voltage is zero, so that the

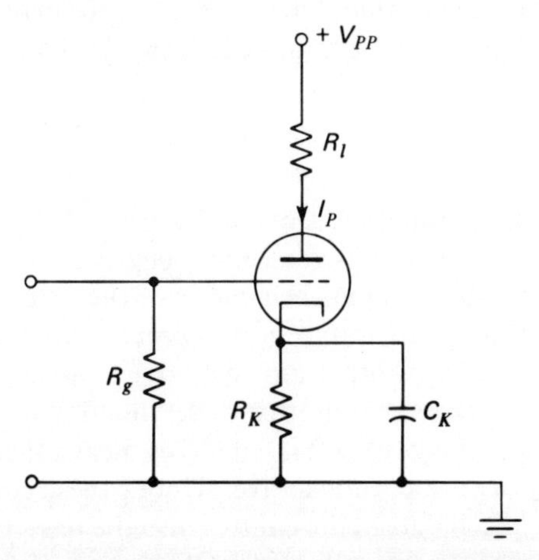

Fig. 1.9. Self-biasing circuit for a vacuum-tube triode.

corresponding quiescent plate current I_P flowing through R_K makes the grid negative with respect to the cathode; that is, $V_{GK} = -I_P R_K$. The cathode resistor thus serves as the means by which the required grid bias is obtained; such a technique is called *self bias*. The capacitor C_K shown connected across R_K in Fig. 1.9 is included so as to provide a low-impedance path for time-varying components of the plate current and thereby to by-pass them around R_K. This usually requires C_K to be fairly large.

1.6 JUNCTION TRANSISTORS[3]

There are two types of junction transistors, *pnp* and *npn*, which are symbolically represented as in Fig. 1.10. Each type consists basically of two *pn*-junction diodes that are fabricated close together in a single crystal of germanium or silicon. We shall only consider a *pnp* transistor, since the operation of an *npn* transistor is identical except that the polarities of biasing voltages are reversed and the roles of holes and electrons are simply interchanged.

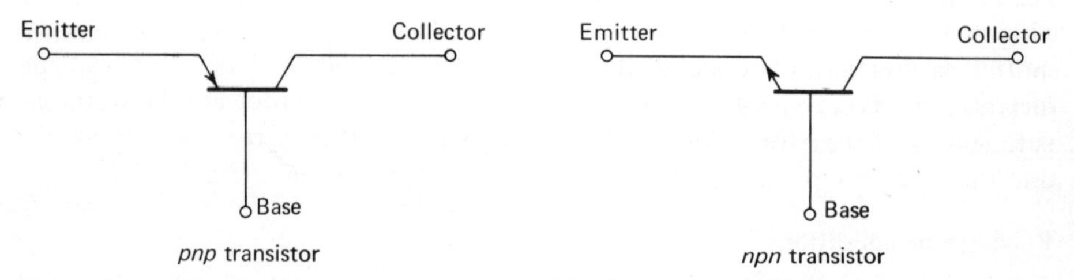

Fig. 1.10. Symbols for *pnp* and *npn* transistors.

The transistor is a three-terminal device, so that we may define three currents and three voltages at its terminals, as indicated in Fig. 1.11, with

$$I_E + I_B + I_C = 0, \tag{1.15}$$

$$V_{EB} + V_{BC} + V_{CE} = 0. \tag{1.16}$$

Hence, only two of the terminal currents and two of the port voltages are independent. That is, we require only two terminal currents and two port voltages to characterize completely the terminal behaviour of the transistor.

When a *pnp* transistor is operated in its *normal active mode* the emitter and collector are maintained at positive and negative potentials, respectively, with respect to the base, so that the emitter–base junction is forward-biased and the collector–base junction is reverse-biased. This means that V_{EB} and I_E are both positive, while V_{CB} and I_C are both negative. Under these conditions we find that nearly all of the holes injected from the emitter into the base are transported across the base and then swept into the collector by the electric field in the space-charge

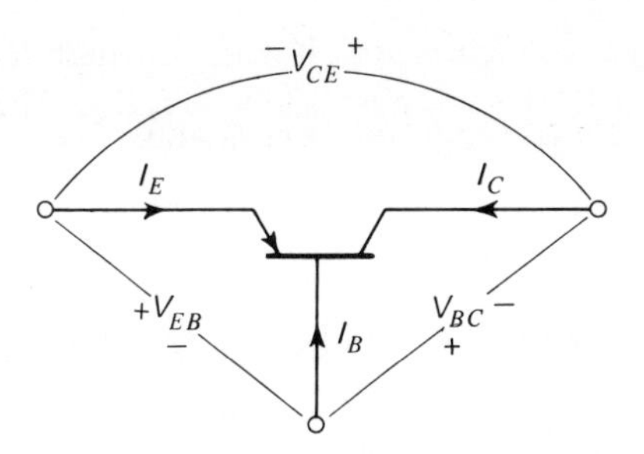

Fig. 1.11. Transistor as a three-terminal device.

layer surrounding the collector–base junction. The resulting collector current I_C is thus nearly equal to the emitter current I_E in magnitude. Moreover, both I_E and I_C are nearly independent of the collector–base voltage V_{CB}.

Circuit Models For the Common-base Connection

The characteristic curves shown in Fig. 1.12 pertain to a *pnp* transistor operated with the base terminal common to the input and output ports. Assuming that the emitter current I_E and collector–base voltage V_{CB} are chosen as the independent variables, we may express the emitter–base voltage V_{EB} and collector current I_C as functions of these variables as follows:

ONE TERMINAL IS A REF. NODE.

$$V_{EB} = f_1(I_E, V_{CB}),$$
$$I_C = f_2(I_E, V_{CB}). \tag{1.17}$$

Suppose I_E and V_{CB} are given small increments; then the corresponding changes in V_{EB} and I_C are obtained as

$$dV_{EB} = \frac{\partial V_{EB}}{\partial I_E} dI_E + \frac{\partial V_{EB}}{\partial V_{CB}} dV_{CB},$$
$$dI_C = \frac{\partial I_C}{\partial I_E} dI_E + \frac{\partial I_C}{\partial V_{CB}} dV_{CB}. \tag{1.18}$$

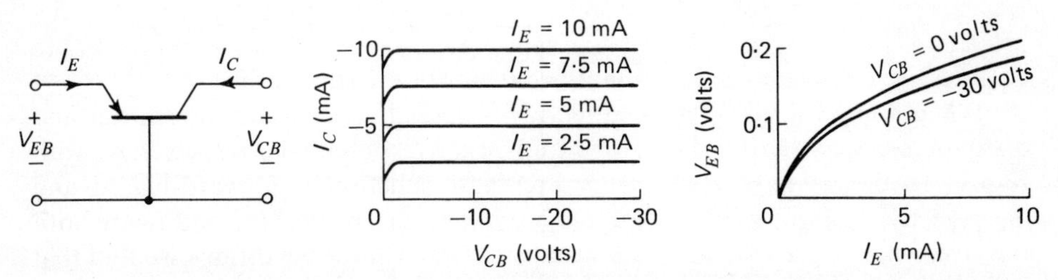

Fig. 1.12. Input and output static characteristic curves for common-base connection. (Diagram includes typical values.)

Let

$$h_{ib} = \text{short-circuit input resistance}$$

$$= \left.\frac{\partial V_{EB}}{\partial I_E}\right|_{V_{CB}=\text{constant}}, \tag{1.19}$$

$$h_{rb} = \text{open-circuit reverse voltage transfer ratio}$$

$$= \left.\frac{\partial V_{EB}}{\partial V_{CB}}\right|_{I_E=\text{constant}}, \tag{1.20}$$

$$h_{fb} = \text{short-circuit forward current transfer ratio}$$

$$= \left.\frac{\partial I_C}{\partial I_E}\right|_{V_{CB}=\text{constant}}, \tag{1.21}$$

$$h_{ob} = \text{open-circuit output conductance}$$

$$= \left.\frac{\partial I_C}{\partial V_{CB}}\right|_{I_E=\text{constant}}, \tag{1.22}$$

where the letter b in the subscripts indicates that these parameters apply to a transistor operated in the *common-base connection.* Also, let the incremental changes in I_E, I_C, V_{EB} and V_{CB} be designated i_e, i_c, v_{eb} and v_{cb} respectively. Then, we may rewrite Eqs. (1.18) in the form:

$$\begin{aligned} v_{eb} &= h_{ib}i_e + h_{rb}v_{cb}, \\ i_c &= h_{fb}i_e + h_{ob}v_{cb}. \end{aligned} \tag{1.23}$$

Equations (1.23) can be represented by the circuit model of Fig. 1.13 which is known as the *common-base hybrid model* because a mixed set of voltages and currents is used as the set of independent variables in Eqs. (1.23). The hybrid model is widely used to characterize the linear behaviour of a transistor, chiefly because the hybrid parameters are all easily measurable. Also, they can be related in a

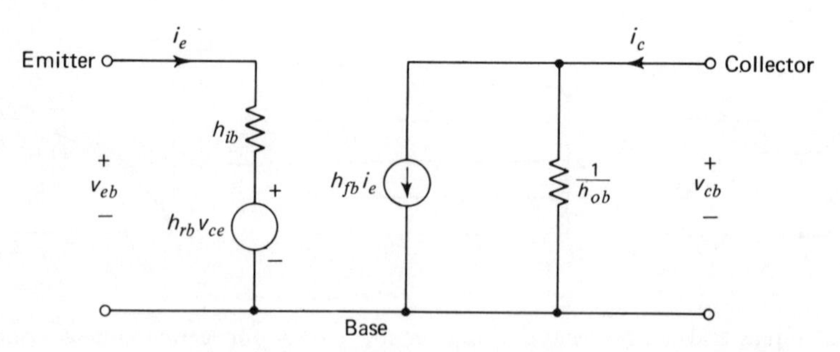

Fig. 1.13. Common-base hybrid model.

simple way to the input and output characteristic curves of the device by using the definitions of Eqs. (1.19)–(1.22).

The hybrid model of Fig. 1.13 is rather similar to Early's model[4] derived from the physical theory of transistors. In the normal active mode, the emitter–base junction of a transistor is forward-biased, while the collector–base junction is reverse-biased, so that under small signal conditions the transistor may be represented to a first order of approximation by the *elementary T-model* shown in Fig. 1.14(a). The resistance r'_e represents the incremental resistance of the forward-biased emitter–base junction, $r_{b'}$ the *base-spreading resistance*, and the current source $\alpha_o i_e$ accounts for the control action exercised by the incremental emitter current i_e upon the collector current. The resistance r'_e is defined by

$$r'_e = \frac{kT}{qI_E}, \tag{1.24}$$

where k = Boltzmann's constant, T = absolute temperature, q = charge of electron, I_E = quiescent emitter current.

The internal node b' in Fig. 1.14(a) represents a node somewhere between the emitter and collector junctions of the transistor.

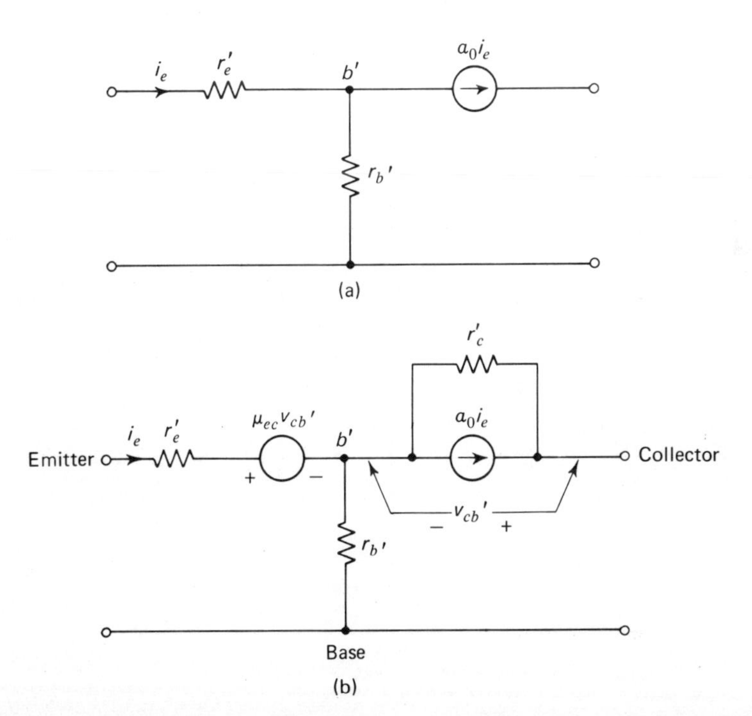

Fig. 1.14. Illustrating the development of Early's *T*-model for the common-base connection.

The model of Fig. 1.14(a) can be expanded to include the effects of *base-width modulation* in the manner indicated in Fig. 1.14(b). Base-width modulation results from the dependence of the effective base width of a transistor upon the collector–base voltage. It produces the finite output conductance $1/r_c'$, while its effect upon the input characteristics is accounted for by the voltage source $\mu_{ec}v_{cb'}$. Clearly, the model of Fig. 1.14(b) is, except for the base-spreading resistance $r_{b'}$, identical with the hybrid model of Fig. 1.13 in form.

The hybrid parameters h_{ib}, h_{rb}, h_{fb}, and h_{ob} of a transistor have real values only at low frequencies. At higher frequencies they all assume complex values which can be explained reasonably well by adding two capacitances $C_{b'e}$ and $C_{b'c}$ to the T-model as indicated in Fig. 1.15(a). The capacitance $C_{b'e}$ appearing in parallel with r_e' consists of two components: the *emitter–base junction capacitance*, and *emitter-diffusion capacitance* that accounts for changes occurring in the charge stored in the base region when the transistor is driven with a time-varying

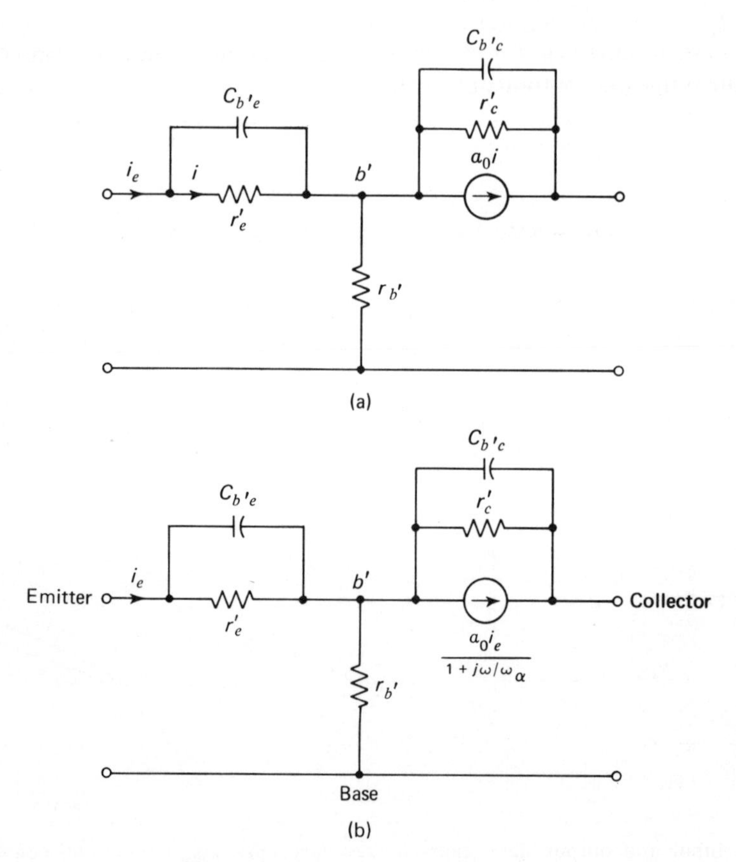

Fig. 1.15. High frequency T-models.

signal. The capacitance $C_{b'c}$ appearing in parallel with r_c' accounts for the *collector–base junction capacitance* of the transistor. In Fig. 1.15(a) we have ignored the source of feedback in the emitter branch, $\mu_{ec}v_{cb'}$, because at high frequencies it becomes overshadowed by the feedback due to $r_{b'}$. Recognizing that

$$i = \frac{i_e}{1 + j\omega/\omega_\alpha}, \tag{1.25}$$

where

$$\omega_\alpha = \frac{1}{r_e' C_{b'e}}, \tag{1.26}$$

we may alternatively use the equivalent model of Fig. 1.15(b) with a frequency-dependent current source.

Circuit Models For the Common-emitter Connection

Consider next the case of a *pnp* transistor operated with the emitter terminal common to the input and output ports; the corresponding characteristic curves are as shown in Fig. 1.16. Thus, with the total base current I_B and collector–emitter voltage V_{CE} as the independent variables, we have

$$\begin{aligned} V_{BE} &= f_1(I_B, V_{CE}), \\ I_C &= f_2(I_B, V_{CE}). \end{aligned} \tag{1.27}$$

Proceeding exactly as was done for the common-base connection, we obtain the following relations between the incremental components of the port voltages and currents:

$$\begin{aligned} v_{be} &= h_{ie}i_b + h_{re}v_{ce}, \\ i_c &= h_{fe}i_b + h_{oe}v_{ce}, \end{aligned} \tag{1.28}$$

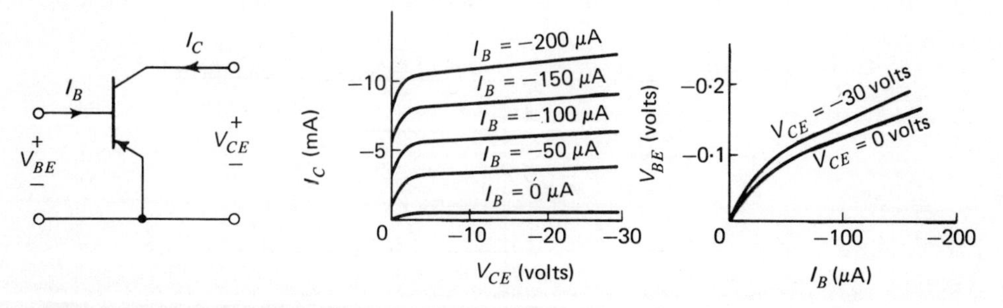

Fig. 1.16. Input and output static characteristic curves for common-emitter connection. (Diagram includes typical values.)

where

$$
\begin{aligned}
h_{ie} &= \left.\frac{\partial V_{BE}}{\partial I_B}\right|_{V_{CE}=\text{constant}}, \\
h_{re} &= \left.\frac{\partial V_{BE}}{\partial V_{CE}}\right|_{I_B=\text{constant}}, \\
h_{fe} &= \left.\frac{\partial I_C}{\partial I_B}\right|_{V_{CE}=\text{constant}}, \\
h_{oe} &= \left.\frac{\partial I_C}{\partial V_{CE}}\right|_{I_B=\text{constant}},
\end{aligned}
\tag{1.29}
$$

The letter e in the subscripts in Eqs. (1.29) signifies that these parameters pertain to a transistor operated in the *common-emitter connection.* Equations (1.28) can be represented by the circuit model of Fig. 1.17 known as the *common-emitter hybrid model.*

Another widely used transistor model is the so-called *hybrid-π model* that can be readily expanded to account for frequency dependence in the transistor characteristics. Consider first Fig. 1.18(a) showing the elementary T-model for the transistor, oriented so that the emitter terminal is common. This model is transformed into the equivalent form shown in Fig. 1.18(b) with

$$
\begin{aligned}
r_{b'e} &= \frac{r_e'}{1 - \alpha_0}, \\
g_m &= \frac{\alpha_0}{r_e'}.
\end{aligned}
\tag{1.30}
$$

The parameter g_m is known as the *mutual conductance* of the transistor. If next we add the resistances $r_{b'c}$ and r_{ce}, as in Fig. 1.18(c), to account for the effects of base-width modulation upon the input and output characteristics of the transistor, respectively, we obtain the low-frequency version of the hybrid-π model. A more general hybrid-π model is obtained by including the capacitances $C_{b'e}$ and $C_{b'c}$,

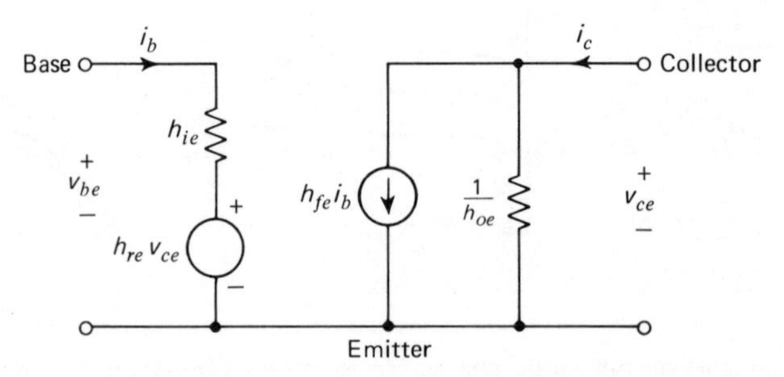

Fig. 1.17. Common-emitter hybrid model.

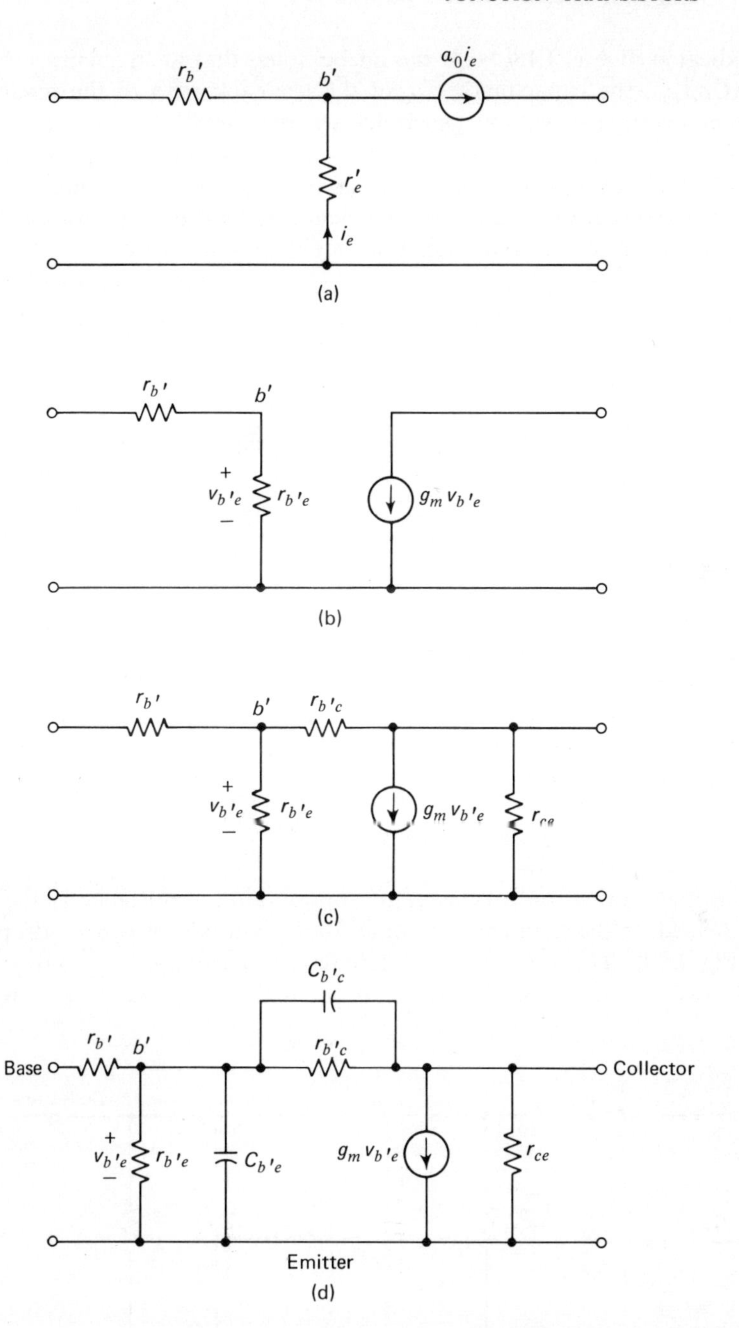

Fig. 1.18. Illustrating the development of the hybrid-π model for the common-emitter connection.

as indicated in Fig. 1.18(d). It should be noted that at frequencies much above $1/(2\pi C_{b'c}r_{b'c})$, the capacitance $C_{b'c}$ in the feedback path of the model becomes dominant compared with $r_{b'c}$, such that we may justifiably ignore $r_{b'c}$.

Suppose a sinusoidal base current is applied to the input port of a common-emitter amplifier with the output port short-circuited, as in the model of Fig. 1.19. Then, we find that the frequency dependence of the short-circuit forward current-transfer ratio of the common-emitter amplifier is given by

$$h_{fe}(j\omega) = \frac{\beta_0}{1 + j\omega/\omega_\beta}, \tag{1.31}$$

where

$$\beta_0 = g_m r_{b'e} = \frac{\alpha_0}{1 - \alpha_0} \tag{1.32}$$

$$\omega_\beta = \frac{1}{r_{b'e}(C_{b'e} + C_{b'c})}. \tag{1.33}$$

In Fig. 1.20 we have sketched the frequency dependence of $|h_{fe}(j\omega)|$ on a logarithmic scale. We see that at $\omega = \omega_\beta$, the magnitude of $h_{fe}(j\omega)$ drops to 0·707 of its zero-frequency value β_0. Thus ω_β serves as a useful measure of the frequency band over which $|h_{fe}(j\omega)|$ remains sensibly constant and nearly equal to its zero-frequency value β_0. For $\omega \gg \omega_\beta$, we find from Eq. (1.31) that $|h_{fe}(j\omega)|$ falls almost linearly with frequency, as expressed by:

$$|h_{fe}(j\omega)|_{\omega \gg \omega_\beta} \simeq \frac{\beta_0 \omega_\beta}{\omega}. \tag{1.34}$$

Another transistor parameter of considerable importance is the frequency ω_T, defined as the frequency at which the magnitude of $h_{fe}(j\omega)$ drops to unity (see Fig. 1.20). This frequency will be referred to as a *figure-of-merit frequency* of the transistor since, from Eq. (1.34), it is equal to the gain–bandwidth product,

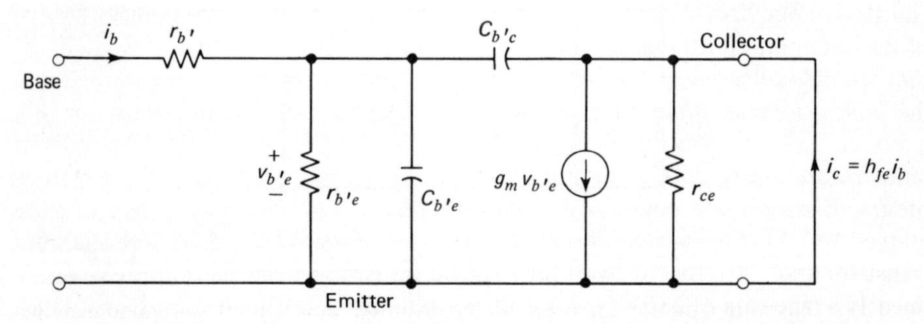

Fig. 1.19. Model of a common-emitter amplifier with the output port short-circuited.

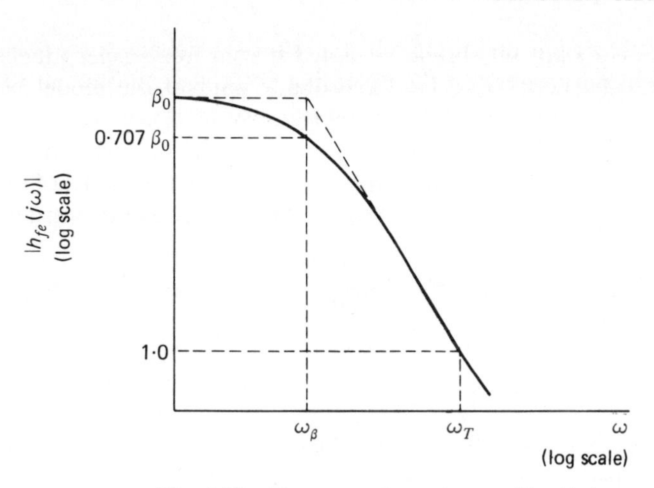

Fig. 1.20. Frequency dependence of $h_{fe}(j\omega)$.

that is:

$$\omega_T = \beta_0\omega_\beta. \tag{1.35}$$

It should be noted that ω_T is essentially independent of β_0, a quantity which is under relatively poor control in that its value varies considerably from sample to sample and is quite sensitive to changes in the quiescent operating point of the transistor.

Biasing Considerations

The transistor is basically a temperature-sensitive device in that a change in temperature has a profound effect on some of the transistor parameters, and hence on the quiescent operating point. This means that the biasing network for a transistor must be so designed that changes in the temperature-sensitive parameters have a minimum effect on the operating point. The temperature-sensitive parameters usually considered are the *collector-saturation current* I_{CO}, base–emitter voltage drop V_{BE}, and the transistor beta β_0 (i.e., the zero-frequency value of the common-emitter short-circuit current-transfer ratio h_{fe}). The current I_{CO} is that special collector current which flows when the emitter is open-circuited and the collector–base junction is reverse biased. In a germanium transistor I_{CO} is about 1 μA at room temperature (25°C), doubling for every 10°C increase in temperature above 25°C, while in a silicon transistor I_{CO} is usually three to four orders of magnitude smaller, doubling for every 7°C increase in temperature above 25°C. At room temperature, V_{BE} is about $-0{\cdot}1$ to $-0{\cdot}2$ volt for a germanium transistor and $-0{\cdot}6$ to $-0{\cdot}7$ volt for a silicon transistor, decreasing approximately linearly at the rate of 2 mV/°C for both germanium and silicon transistors. As for β_0, it usually increases with temperature for most transistors.

In order to determine quantitatively the effects of changes in these temperature-dependent parameters on the operating point, it is convenient to introduce the following *d.c. stability factors* for the collector current:[5]

$$S_1 = \frac{\partial I_C}{\partial I_{CO}}$$
$$S_2 = \frac{\partial I_C}{\partial V_{BE}} \tag{1.36}$$
$$S_3 = \frac{\partial I_C}{\partial \beta_0}$$

where each stability factor is obtained with the other two temperature-dependent parameters maintained constant. For a given temperature change, the total change in collector current is therefore,

$$dI_C = \frac{\partial I_C}{\partial I_{CO}} dI_{CO} + \frac{\partial I_C}{\partial V_{BE}} dV_{BE} + \frac{\partial I_C}{\partial \beta_0} d\beta_0 \tag{1.37}$$

Clearly, the smaller the absolute value of each stability factor, the smaller the change in the collector current.

Example 1.1. *Single-battery biasing network.* As an illustrative example, consider the widely used biasing network of Fig. 1.21 for which we may write, by

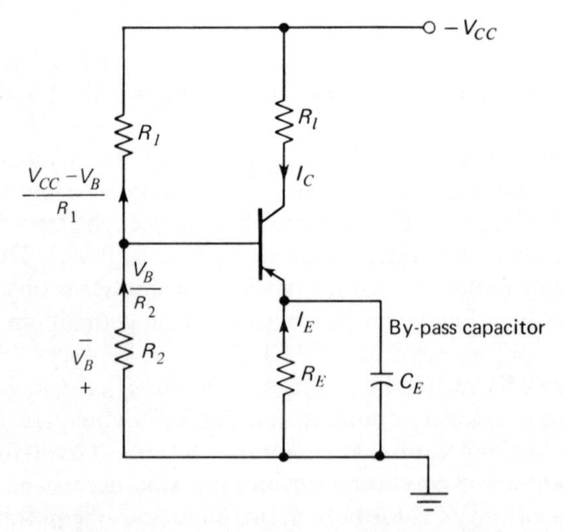

Fig. 1.21. Single-battery biasing network.

inspection,

$$\begin{aligned} V_B &= V_{BE} + I_E R_E \\ \frac{V_B}{R_2} &= \frac{V_{CC} - V_B}{R_1} - (I_C + I_E) \\ I_C &= -(\alpha_0 I_E + I_{CO}). \end{aligned} \tag{1.38}$$

Solving Eqs. (1.38) for the collector current, by eliminating I_E and V_B, we obtain

$$I_C = -\frac{I_{CO}(R_B + R_E) + \alpha_0\left(\frac{R_B}{R_1}V_{CC} - V_{BE}\right)}{R_B(1 - \alpha_0) + R_E},$$

or, since $\beta_0 = \alpha_0/(1 - \alpha_0)$,

$$I_C = -\frac{I_{CO}(1 + \beta_0)(R_B + R_E) + \beta_0\left(\frac{R_B}{R_1}V_{CC} - V_{BE}\right)}{R_B + R_E(1 + \beta_0)}, \tag{1.39}$$

where R_B is the parallel combination of R_1 and R_2,

$$R_B = \frac{R_1 R_2}{R_1 + R_2}. \tag{1.40}$$

Use of Eqs. (1.36) and (1.39) yields:

$$S_1 = \frac{\partial I_C}{\partial I_{CO}} = -\frac{(1 + \beta_0)(R_B + R_E)}{R_B + R_E(1 + \beta_0)}, \tag{1.41}$$

$$S_2 = \frac{\partial I_C}{\partial V_{BE}} = \frac{\beta_0}{R_B + R_E(1 + \beta_0)}, \tag{1.42}$$

and

$$S_3 = \frac{\partial I_C}{\partial \beta_0} = -\frac{(R_B + R_E)\left(R_B I_{CO} + \frac{R_B}{R_1}V_{CC} - V_{BE}\right)}{[R_B + R_E(1 + \beta_0)]^2}. \tag{1.43}$$

We thus see that the choice of a large emitter resistor R_E tends to minimize all three stability factors, and the choice of a small R_B tends to minimize S_1 and S_3 but not S_2, the stability factor related to V_{BE}.

1.7 CONTROLLED SOURCES

The circuit models of Figs. 1.6–1.18 contain a new element called a *dependent* or *controlled source*; thus, for example, in the common-emitter hybrid model of Fig. 1.17, we see that the current source $h_{fe}i_b$ is controlled by the input base current i_b, and the voltage source $h_{re}v_{ce}$ is controlled by the output collector–emitter voltage v_{ce}. Signal transmission in the forward direction (i.e., from input to output) through the transistor is thus controlled by the source $h_{fe}i_b$, while backward transmission (i.e., from output to input) is controlled by the source $h_{re}v_{ce}$. Hence, the transistor is a *non-unilateral* device in that it transmits electric signals in both directions, with the backward transmission being considerably weaker than the forward transmission inside the useful frequency band of the device; a typical value for h_{fe} is 100, while that for h_{re} is 10^{-4}. The vacuum tube, on the other hand, is essentially a *unilateral* device in that at low frequencies the backward transmission is zero, as evidenced by the models of Fig. 1.6.

Types of Controlled Sources

The controlled source differs from an independent voltage or current source in that it is basically a *unilateral two-port element* representing a single constraint between the voltage or current at one pair of terminals and the voltage or current (not necessarily in that order) at the other pair of terminals. Figure 1.22 shows a

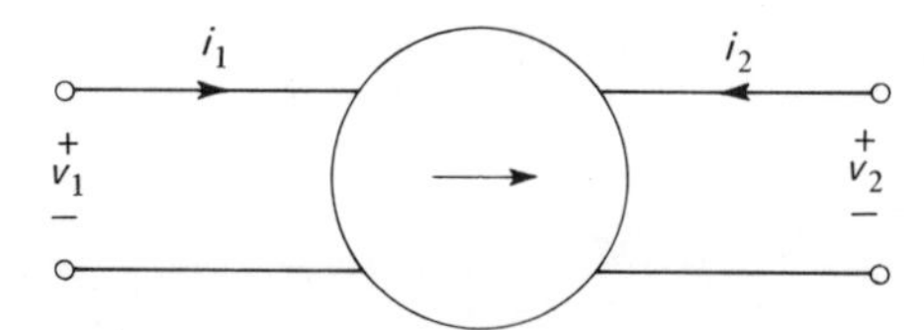

Fig. 1.22. A general diagrammatic representation of a controlled source.

general schematic representation for a controlled source. The current i_1 or voltage v_1 at the input port may constitute the *control signal.* The controlled source may itself be of a voltage or current type equal to v_2 or i_2, respectively. It follows therefore that there are four possible types of controlled sources:

1) A *current-controlled voltage source.* This is shown in Fig. 1.23 where the current i_1 is the control signal, and r_m is the *control parameter* having the dimensions of an impedance. Mathematically, we can define the behaviour of this controlled source by the following matrix relation

$$\begin{bmatrix} v_1 \\ v_2 \end{bmatrix} = \begin{bmatrix} 0 & 0 \\ r_m & 0 \end{bmatrix} \begin{bmatrix} i_1 \\ i_2 \end{bmatrix} \tag{1.44}$$

2) A *voltage-controlled current source.* This is shown in Fig. 1.24 where the voltage v_1 is the control signal; g_m is the control parameter having the dimensions of an admittance. In matrix form the behaviour of this controlled source is defined as follows:

$$\begin{bmatrix} i_1 \\ i_2 \end{bmatrix} = \begin{bmatrix} 0 & 0 \\ g_m & 0 \end{bmatrix} \begin{bmatrix} v_1 \\ v_2 \end{bmatrix} \tag{1.45}$$

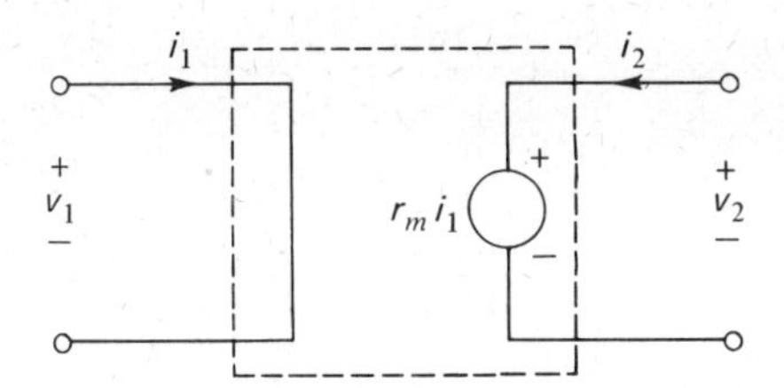

Fig. 1.23. Current-controlled voltage source.

Clearly, a voltage-controlled current source and a current-controlled voltage source are *dual* networks in that their mathematical representations, as defined by Eqs. (1.44) and (1.45), are similar in form with the roles of terminal currents and voltages interchanged.

3) A *current-controlled current source.* This is shown in Fig. 1.25 where the control parameter α is dimensionless, and the behaviour of the controlled source is defined by

$$\begin{bmatrix} v_1 \\ i_2 \end{bmatrix} = \begin{bmatrix} 0 & 0 \\ \alpha & 0 \end{bmatrix} \begin{bmatrix} i_1 \\ v_2 \end{bmatrix} \tag{1.46}$$

4) A *voltage-controlled voltage source.* This is shown in Fig. 1.26 where the control parameter μ is dimensionless, and the behaviour of the controlled source is defined by

$$\begin{bmatrix} i_1 \\ v_2 \end{bmatrix} = \begin{bmatrix} 0 & 0 \\ \mu & 0 \end{bmatrix} \begin{bmatrix} v_1 \\ i_2 \end{bmatrix} \tag{1.47}$$

A voltage-controlled voltage source and a current-controlled current source are, clearly, the duals of each other.

From the circuit representations of Figs. 1.23–1.26 we observe that if a current-controlled voltage source is connected in tandem with a voltage-controlled current source as in Fig. 1.27(a), we obtain a current-controlled current source, while if a voltage-controlled current source is connected in tandem with a current-controlled voltage source, as in Fig. 1.27(b), we obtain a voltage-controlled

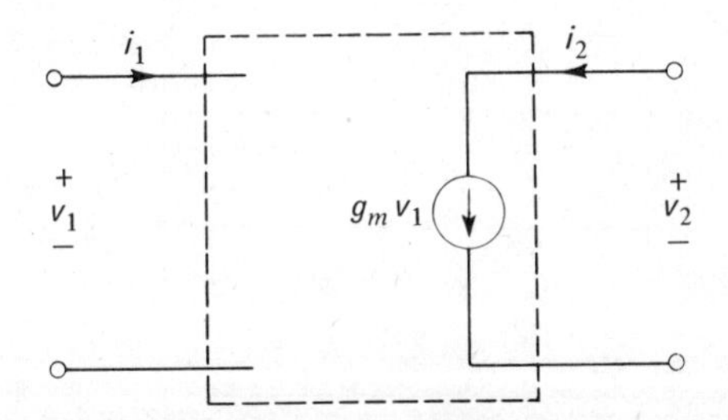

Fig. 1.24. Voltage-controlled current source.

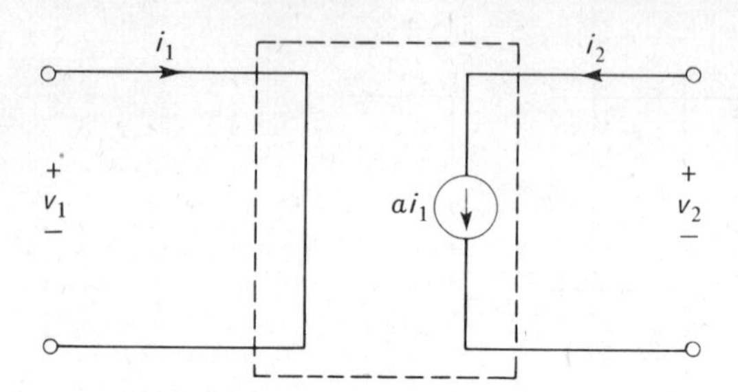

Fig. 1.25. Current-controlled current source.

voltage source. On the other hand, the realization of a voltage-controlled current source and a current-controlled voltage source with the use of a voltage-controlled voltage source and a current-controlled current source is not possible. Therefore, only the voltage-controlled current source and current-controlled voltage source are of a fundamental nature.

1.8 NEGATIVE-RESISTANCE DEVICES

The incremental resistance of a two-terminal circuit element is defined as the ratio of a small increment in the voltage across the element to the resulting increment in the current through it. As such, therefore, the incremental resistance of an element is equal to the reciprocal of the slope of the static current–voltage characteristic of the element at a specified quiescent operating point. The two-terminal element is said to exhibit a negative incremental resistance when the slope of the characteristic curve is negative, that is, when an increment of current flows in opposition to the increment of voltage which produces it.

Negative-resistance characteristics fall into two distinct and important classes: when, as in Fig. 1.28(a), the current is a single-valued function of the voltage, the characteristic and the device which it represents are said to be *voltage controlled.* On the other hand, a device in which the voltage is a single-valued function of the current, as in Fig. 1.29(a), it is said to be *current controlled.* At first glance it might seem there is little difference between these two types of negative-

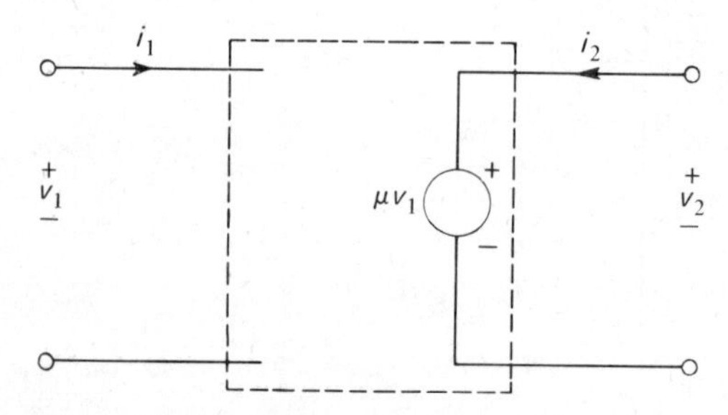

Fig. 1.26. Voltage-controlled voltage source.

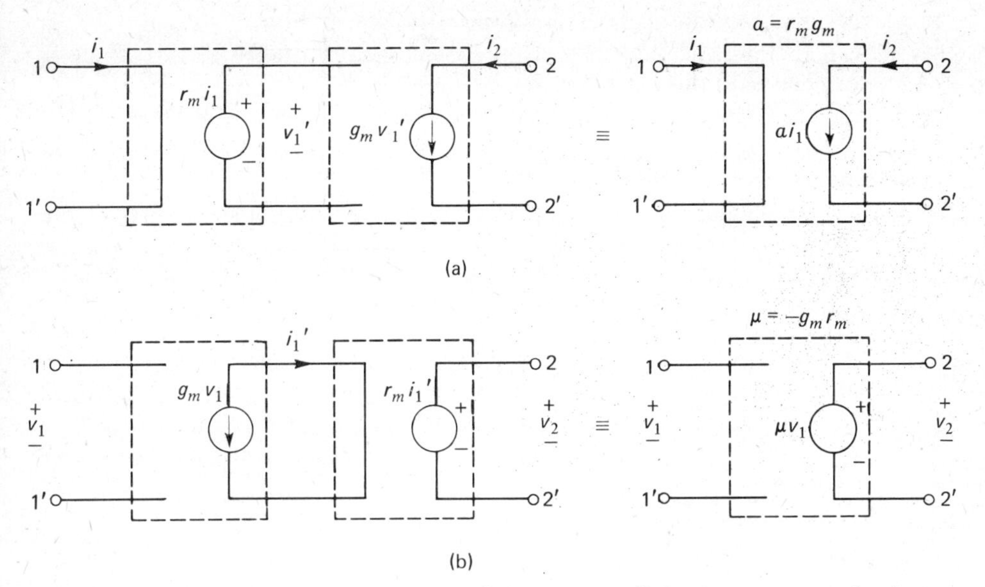

Fig. 1.27. Illustrating fundamental nature of current-controlled voltage source and voltage-controlled current source.

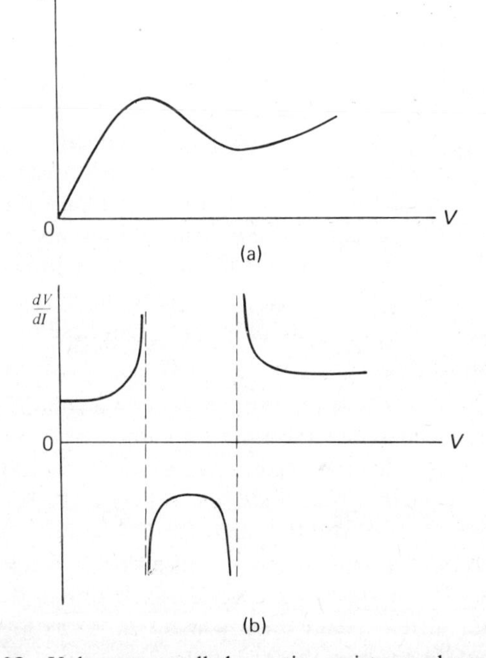

Fig. 1.28. Voltage-controlled negative-resistance characteristic.

resistance characteristics. However, from parts (b) of Figs. 1.28 and 1.29 we see that in making the transition from the positive-resistance to the negative-resistance portion of the characteristic, the incremental resistance passes through infinity in the voltage-controlled case, while in the current-controlled case it passes through zero. Also, outside the useful frequency range of the device, we ordinarily find that a voltage-controlled negative-resistance device departs from the idealized behaviour in such a way that the device is unstable on open-circuit, while a current-controlled negative-resistance device is unstable on short-circuit.

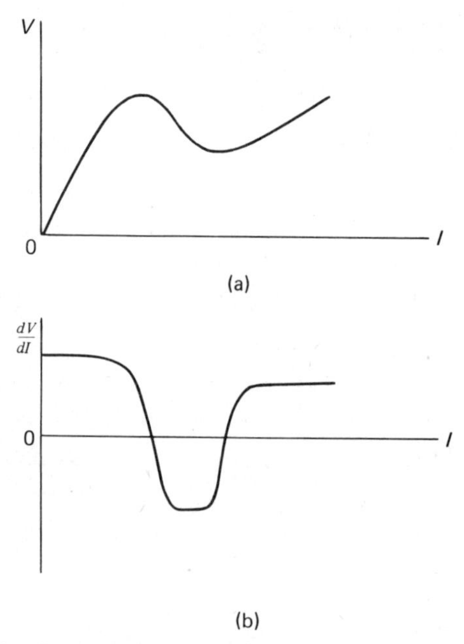

Fig. 1.29. Current-controlled negative-resistance characteristic.

In certain electronic devices such as the *tunnel diode* a negative incremental resistance arises as an intrinsic property of the device. From the static-characteristic curve of Fig. 1.30 we see that the tunnel diode is a voltage-controlled type of negative-resistance device. When the tunnel diode is operated in the negative-resistance part of its characteristic curve, the small-signal behaviour can be represented by the incremental model shown in Fig. 1.31. The negative resistor $-r$ represents the net effect of the tunnelling currents and the ordinary barrier currents at the junction, while r_s represents the ohmic resistance of the semi-conducting material. The frequency response of the tunnel diode is determined principally by the cut-off frequency $1/rC$ which can be as high as 10^{10} rad/sec, thereby allowing amplification in the microwave region. It is on account of the

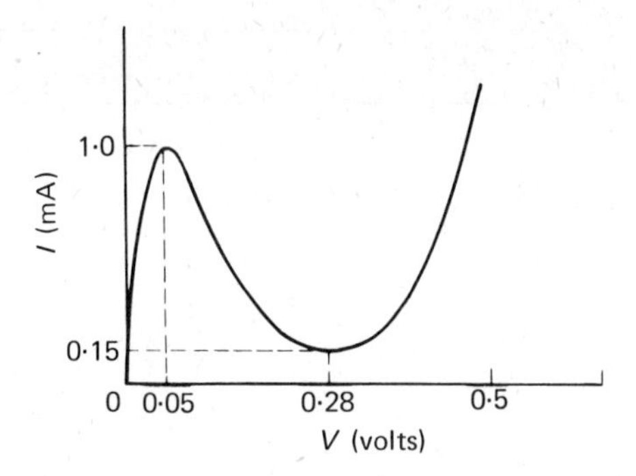

Fig. 1.30. Static-characteristic curve of a tunnel diode (Diagram includes typical values).

very high cut-off frequency of the tunnel diode that we have included the lead inductance L_s in the circuit model.

The voltage-controlled type of negative resistance characteristic can also be produced by artificial means using the feedback arrangement of Fig. 1.32(a) involving a voltage-controlled current source.[6] The input current i is evidently

$$i = -g_m v. \tag{1.48}$$

Looking into port 1–1′ we therefore see a negative conductance $-g_m$.

In a dual manner we find that the current-controlled type of negative-resistance characteristic can be realized artificially using the feedback arrangement of Fig. 1.32(b) involving a current-controlled voltage source. The input voltage v is

$$v = -r_m i, \tag{1.49}$$

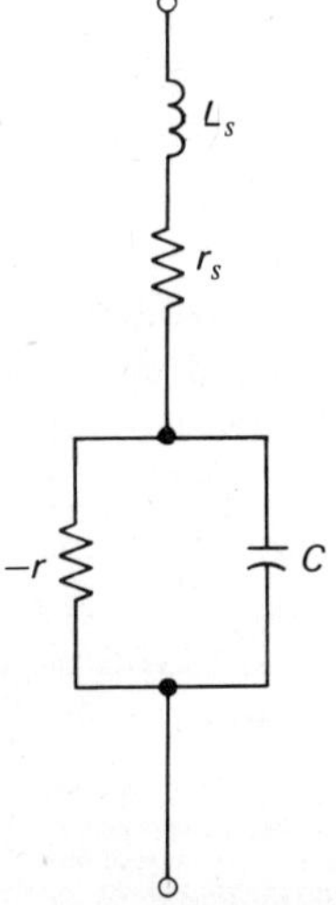

Fig. 1.31. Circuit model of a tunnel diode.

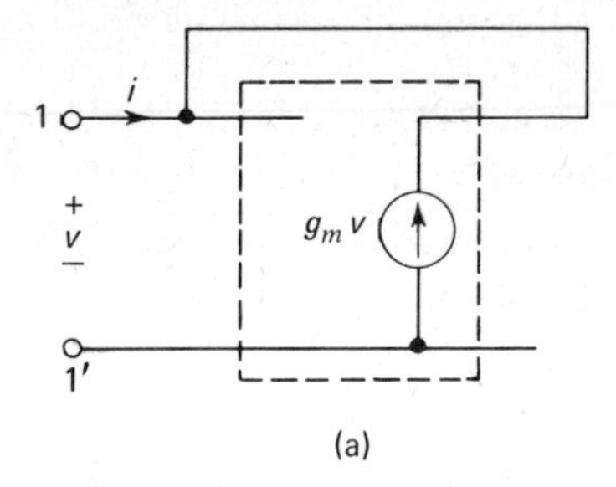

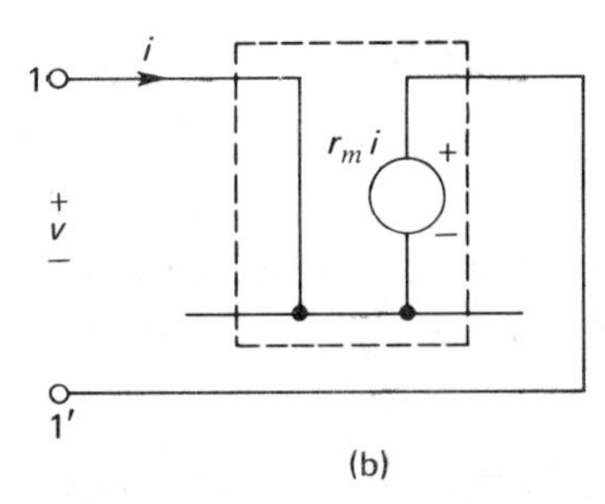

Fig. 1.32. (a) Indirect implementation of a voltage-controlled negative resistance; (b) Indirect implementation of a current-controlled negative resistance.

which represents a negative resistance $-r_m$ looking into port 1–1′. Figure 1.32 further demonstrates the fundamental nature of the voltage-controlled current source and current-controlled voltage source.

Concluding Remarks

To conclude, a physically realizable negative resistance is not the reverse of a positive resistance but comprises a one-port device with an internal source of energy (such as a battery) which is controlled either by the current through or by the voltage across the terminals of the device but not by both. A negative resistance is therefore of two types, current controlled and voltage controlled, which behave differently when inserted in a circuit. The two types are, however, the duals of each other in that the phenomena observed with the resistance in the one case will be observed with the conductance in the other. A basic difference between a positive resistance and a negative resistance is that a positive resistance dissipates energy proportional to the square of the applied voltage or current, whereas a negative resistance generates energy proportional to the square of the applied voltage or current. As a consequence of this latter property a negative resistance may be used for the purpose of amplification or oscillation, as will be demonstrated in Chapters 3 and 6.

REFERENCES

1. D. K. Cheng, *Analysis of Linear Systems*. Addison-Wesley, 1959.
2. T. S. Gray, *Applied Electronics*. Wiley, 1954.
3. J. F. Gibbons, *Semiconductor Electronics*. McGraw-Hill, 1966.
4. J. M. Early, "Effects of Space-charge Layer Widening in Junction Transistors," *Proc. I.R.E.*, **40,** 1401 (1952).
5. L. P. Hunter, *Handbook of Semiconductor Electronics*, pp. 11–70. McGraw-Hill, 1962.
6. G. F. Sharpe, "Axioms on Transactors," *Trans. I.R.E.*, **CT-5,** 189 (1958).

PROBLEMS

1.1 Determine the common-base h-parameters h_{ib}, h_{rb}, h_{fb} and h_{ob} from the model of Fig. 1.14(b).

1.2 Determine the common-emitter h-parameters h_{ie}, h_{re}, h_{fe} and h_{oe} from the hybrid-π model of Fig. 1.18(c).

1.3 The common-emitter h-parameters of a *pnp* transistor operating at $I_C = -2\text{mA}$, $V_{CE} = -6\text{V}$ and room temperature (25°C) have the following low-frequency values:

$$h_{ie} = 560\ \Omega$$

$$h_{re} = 1{\cdot}25 \times 10^{-4}$$

$$h_{fe} = 40$$

$$h_{oe} = 20\ \mu\Omega.$$

Also,

$$C_{b'c} = 10\ \text{pF}$$

$$f_T = \omega_T/2\pi = 20\ \text{MHz}.$$

Evaluate the elements of the hybrid-π model. (Note that Boltzmann's constant $= 1{\cdot}38 \times 10^{-23}$ Joule/°K, charge of electron $= 1{\cdot}6 \times 10^{-19}$ coulomb.)

1.4 Determine the d.c. stability factors S_1, S_2 and S_3 for the transistor amplifier of Fig. P1.4.

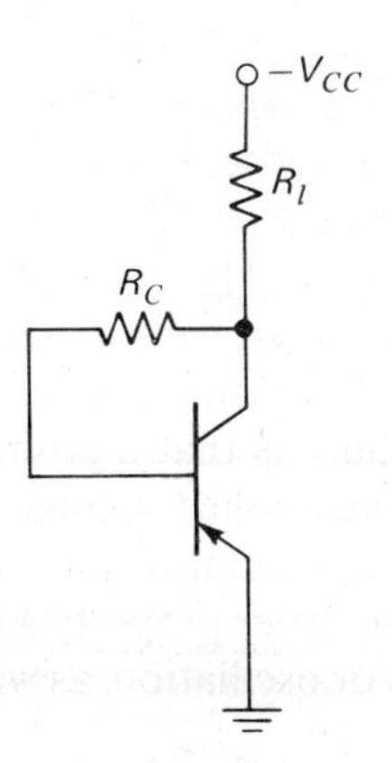

Fig. P1.4.

CHAPTER 2

GENERAL EQUILIBRIUM EQUATIONS

2.1 INTRODUCTION

The object of this chapter is to develop, from topological considerations, the three important methods of network analysis: the loop, nodal, and state-variable approach for the systematic formulation of network equilibrium equations (algebraic or differential) in terms of a set of variables called the *loop currents*, *node voltages*, and *state variables*, respectively. These three procedures complement each other, providing not only a practical means for computing the response of specific networks but also the basis for studying the general properties of networks, a task that we shall undertake in subsequent chapters. In formulating the procedures, we shall find it convenient to use matrix notation for network characterization, and thereby avoid the need for explicitly specifying the number of elements contained in the network and the manner of their interconnection.

2.2 NETWORK TOPOLOGY

In an electrical network made up of the interconnection of elements and sources, the branch currents and voltages are constrained by:

1. The voltage–current relations that characterize the elements.
2. Kirchhoff's current law (KCL) according to which the algebraic sum of the currents entering (or leaving) any node of the network is zero; a node is a connection point at which one or more elements or sources terminate.
3. Kirchhoff's voltage law (KVL) according to which the algebraic sum of voltage drops (or rises) around any closed path of the network is zero.

The voltage–current relations have obviously nothing to do with how the elements are connected in the network. As a result of these relations, however, we may regard either the branch currents or the branch voltages as adequately characterizing the network behaviour. Furthermore, Kirchhoff's current and voltage laws impose additional constraints, so that neither one of these two sets of quantities is independent.

When a network contains a large number of branches, the problem of deciding whether a selected set of current or voltage variables is not only independent but also adequate for the complete characterization of the equilibrium conditions in the network may become quite difficult and complex. It is therefore desirable to develop systematic procedures for selecting the variables, which is rather elegantly accomplished by disregarding the electrical properties of the given network and considering only the manner in which the various elements are interconnected at their junction points. For this purpose, the interconnecting branches may be thought of merely as line segments, so that the network is represented by what topologists call a *linear graph*. For example, consider the network of Fig. 2.1(a); the corresponding graph can be drawn as in part (b) of the diagram where, for convenience, the nodes have been arbitrarily numbered from 1 to 4, and the branches from 1 to 6. Such a graph is said to be *directed* in the sense that each branch has an arbitrary orientation applied to it by means of an arrowhead, defining the direction of the corresponding branch current. It is, therefore, apparent that in constructing the graph we only retain the geometrical features of the network. As such, a voltage source should be graphed as a short-circuit, because an ideal voltage source possesses zero internal impedance. As a dual to this, an ideal current source possesses zero internal admittance, so that it should

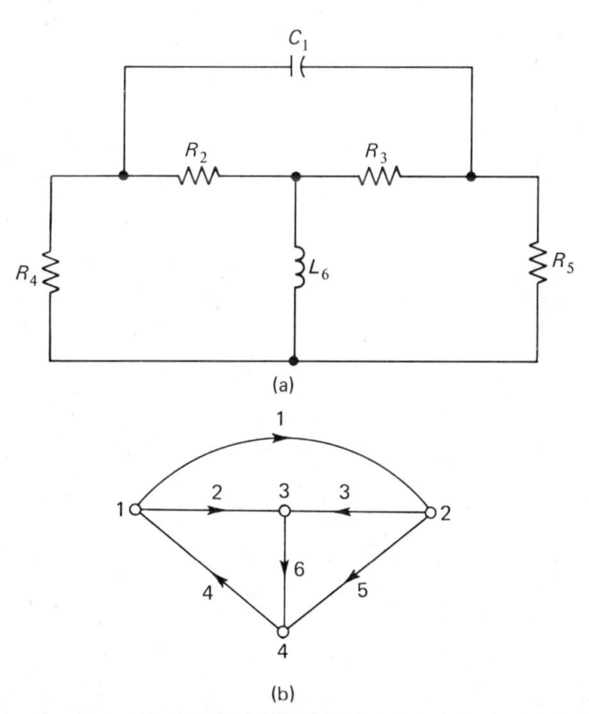

Fig. 2.1. (a) An *RLC* network; (b) Directed graph of the network.

be graphed as an open-circuit. Controlled voltage and current sources are subject to similar treatment.

Separate Parts, Trees and Links

The graph of a network is said to be *connected* if there exists at least one path along branches of the graph between any pair of nodes. When the graph is not connected, then it must consist of a finite number of *separate parts*. Such a situation arises, for instance, when the network contains mutual inductance. Thus, the graph of Fig. 2.2(b), pertaining to a network with two pairs of mutually coupled coils, consists of three parts. However, we may unite these parts by connecting a node of each separate part to a common ground, as shown in Fig. 2.2(c). It is apparent that such a modification leaves all the branch voltages or currents unchanged, and yet it simplifies matters by reducing the number of nodes and the number of separate parts by the same quantity. Therefore, without loss in generality we may, from here on, consider only graphs with one separate part.

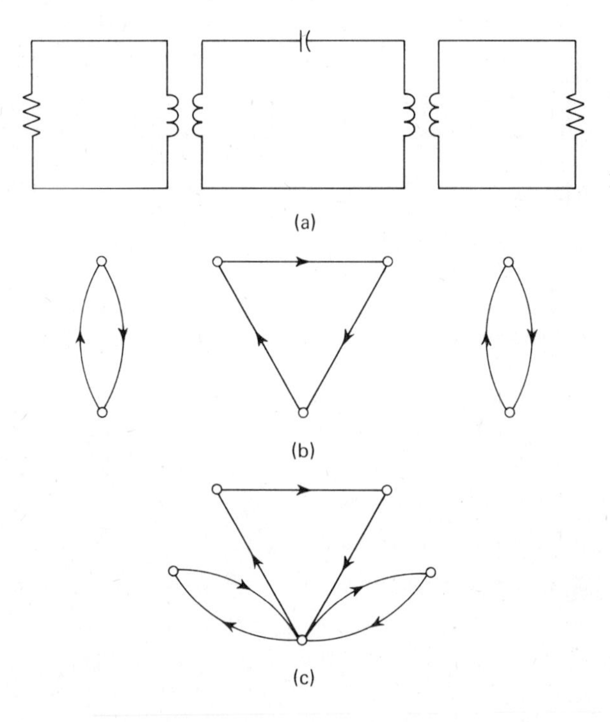

Fig. 2.2. (a) A network with two pairs of mutually coupled coils; (b) Graph with three separate parts; (c) Modified graph with common ground.

A *tree* is defined as any connected set of branches that includes all of the nodes in the given graph, but contains no closed paths; so that there is one and only one path between any two nodes of a tree. In general, a graph will contain several trees; for example, Fig. 2.3 shows three possible trees for the graph of Fig. 2.1(b). Assuming that the graph contains a total number of n nodes and b branches, it is apparent that each tree will divide the b branches into two sets: $n - 1$ branches contained in the tree, and the $b - n + 1$ branches not contained in the tree. These latter $b - n + 1$ branches are called *links* of the tree and are said to constitute a *co-tree* of the graph. For ease of reference, we shall put

$$m = n - 1 \tag{2.1}$$

$$l = b - n + 1 \tag{2.2}$$

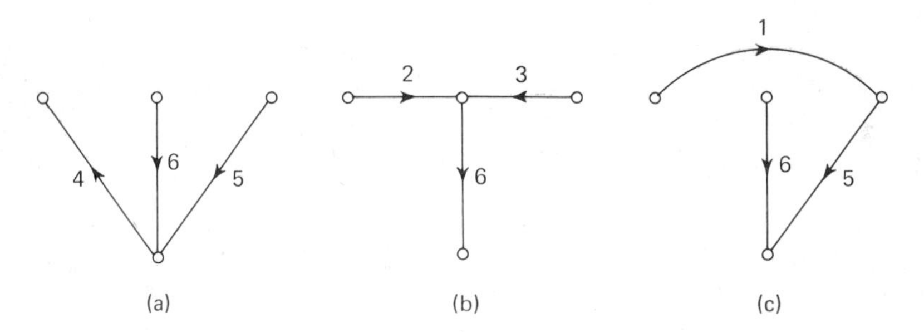

Fig. 2.3. Three possible trees for the graph of Fig. 2.1(b).

The Incidence Matrix

Ordinarily, a graph does not contain any branches which are closed upon themselves. It is then obvious that each branch of the graph has associated with it precisely two nodes. These two nodes and the branch are said to be *incident*. The incidence of the nodes and branches of the graph may be described algebraically by means of an $n \times b$ matrix, called the *incidence matrix*, $\mathscr{A}_a$, whose rows correspond to the nodes of the graph and whose columns correspond to the branches.[1,2] Its elements a_{jk} have the values $+1$, -1 or zero, as shown by

$a_{jk} = +1$, if branch k is incident with node j and oriented away from it.

$a_{jk} = -1$, if branch k is incident with node j and oriented towards it.

$a_{jk} = 0$, if branch k is not incident with node j.

For the graph of Fig. 2.1(b), we have

$$\mathscr{A}_a = \begin{bmatrix} +1 & +1 & 0 & -1 & 0 & 0 \\ -1 & 0 & +1 & 0 & +1 & 0 \\ 0 & -1 & -1 & 0 & 0 & +1 \\ 0 & 0 & 0 & +1 & -1 & -1 \end{bmatrix}$$

Obviously each column of $\mathscr{A}_a$, corresponding to a branch of the graph, contains exactly two non-zero elements, namely, $+1$ and -1. Hence, if we add together all the rows we get a row of zeros, i.e., the rows of $\mathscr{A}_a$ are not linearly independent. The number of linearly independent rows (or columns) of a matrix is called its *rank*. In the case of a connected graph, the rank of $\mathscr{A}_a$ is $m = n - 1$.

The matrix $\mathscr{A}_a$ not only gives all the information about the incidence of branches and nodes but also indicates the branch orientations. The incidence matrix is, therefore, basic for topological synthesis in that any matrix possessing the properties of an incidence matrix may be realized directly in the form of a corresponding graph. A necessary and sufficient condition for an $n \times b$ matrix to be the incidence matrix of a linear graph is that each column of the matrix contain precisely one $+1$ and one -1, all other elements in the column being zero.

In a connected graph, the incidence matrix contains a certain redundancy. Thus, suppose node n is chosen as the *datum node* of the graph, and we form from the $n \times b$ matrix $\mathscr{A}_a$ an $m \times b$ matrix, by deleting the row corresponding to node n. The resultant matrix, denoted by $\mathscr{A}$, is called the *reduced incidence matrix* of the graph. To obtain the reduced incidence matrix of the graph in Fig. 2.1(b), we omit the fourth row from $\mathscr{A}_a$, yielding

$$\mathscr{A} = \begin{bmatrix} +1 & +1 & 0 & -1 & 0 & 0 \\ -1 & 0 & +1 & 0 & +1 & 0 \\ 0 & -1 & -1 & 0 & 0 & +1 \end{bmatrix}.$$

Some of the columns of $\mathscr{A}$ will not now contain both a $+1$ and a -1. Any such column must obviously correspond to a branch incident with the datum node n; the sign of the single non-zero element in that column will indicate the orientation of the pertinent branch. Thus, provided we know the graph in question is connected, the matrix $\mathscr{A}$ contains as much information about the graph as $\mathscr{A}_a$.

The Loop Matrix

A *loop* of a graph is defined as a connected subgraph having precisely two branches incident with each node. In general, the graph will contain many loops, which may be arbitrarily numbered and oriented. For example, Fig. 2.4 shows all the seven possible loops of the graph of Fig. 2.1(b).

If a particular loop contains a particular branch, that loop and that branch are said to be *incident*, so that any loop may be defined in terms of its incidence with the branches of the given graph. This incidence may be described algebraically by means of a matrix, denoted by $\mathscr{B}_a$, containing b columns, one for each branch of the graph, and a finite number of rows, one for each possible loop

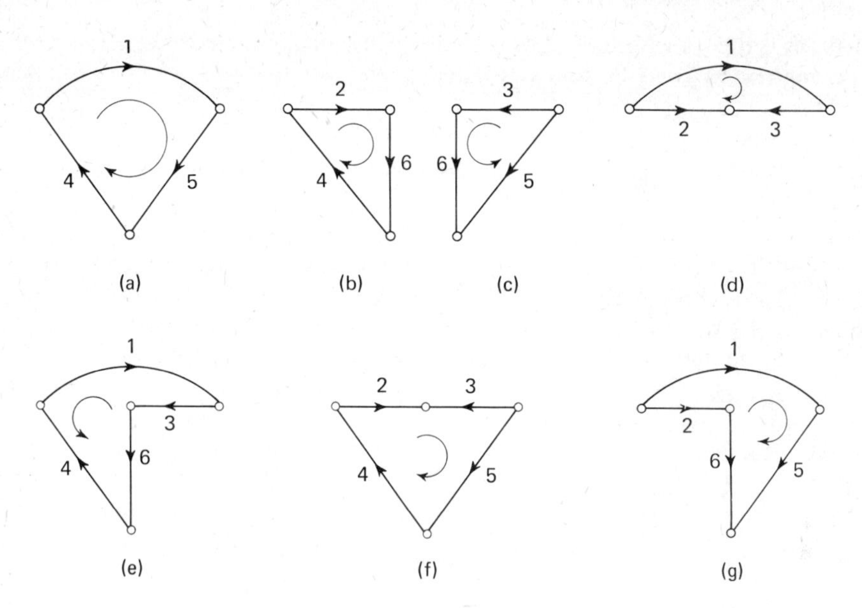

Fig. 2.4. Illustrating the seven possible loops of the graph of Fig. 2.1(b).

of the graph. Its elements b_{jk} have the values $+1$, -1 or zero, as shown by

$b_{jk} = +1$, if branch k is incident with loop j and their orientations coincide there.

$b_{jk} = -1$, if branch k is incident with loop j and their orientations are opposed there.

$b_{jk} = 0$, if branch k and loop j are not incident.

For the arbitrary numbering and orientation of the loops shown in Fig. 2.4 we have

$$\mathscr{B}_a = \begin{bmatrix} +1 & 0 & 0 & +1 & +1 & 0 \\ 0 & +1 & 0 & +1 & 0 & +1 \\ 0 & 0 & +1 & 0 & -1 & +1 \\ +1 & -1 & +1 & 0 & 0 & 0 \\ -1 & 0 & -1 & -1 & 0 & -1 \\ 0 & +1 & -1 & +1 & +1 & 0 \\ +1 & -1 & 0 & 0 & +1 & -1 \end{bmatrix}.$$

If the columns of the matrices $\mathscr{A}_a$ and $\mathscr{B}_a$ are arranged in the same order, then we may write[2]

$$\mathscr{A}_a\mathscr{B}_a^t = 0; \qquad \mathscr{B}_a\mathscr{A}_a^t = 0 \tag{2.3}$$

where $\mathscr{B}_a^t$ is the transpose of $\mathscr{B}_a$, obtained by interchanging the rows and columns of the matrix; $\mathscr{A}_a^t$ may be similarly defined.

There is also an interesting relation between the loops and trees of the graph. Thus, consider any tree; since it is connected, there exists a unique path between any two nodes. Suppose we add the links to the chosen tree, one at a time. The addition of a link between any two nodes of the tree establishes a closed path, which is different for each link. Therefore, each of the l links of the tree defines a loop of the graph. These l loops constitute the set of *fundamental* loops with respect to the chosen tree. Such a set of loops is evidently independent; consequently, the matrix $\mathscr{B}_a$ has a rank of l. It is rather convenient to define the orientations of the fundamental loops so as to be confluent with the corresponding links. Thus, the tree of Fig. 2.3(a) defines the set of three fundamental loops shown in Fig. 2.5(a).

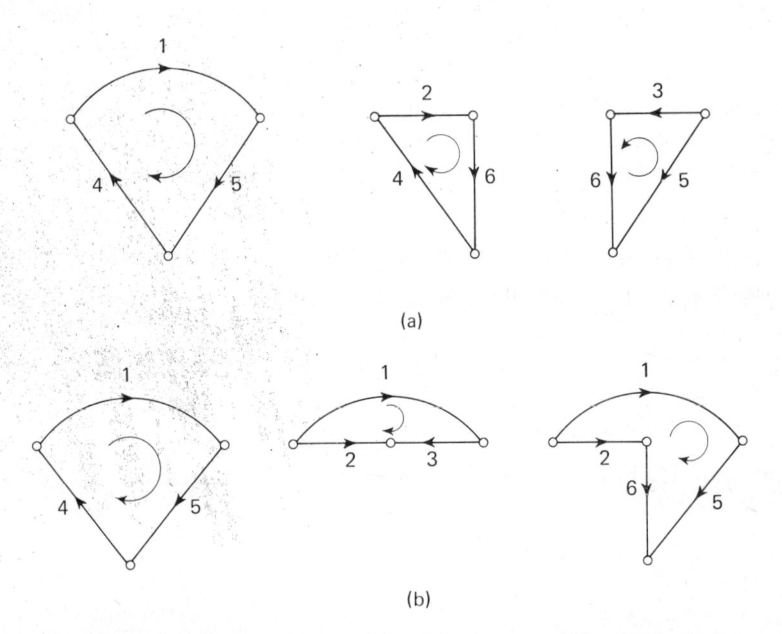

Fig. 2.5. (a) The set of fundamental loops defined by the tree of Fig. 2.3(a); (b) A set of nonfundamental independent loops.

A set of fundamental loops of a graph defines an $l \times b$ submatrix of $\mathscr{B}_a$, called a *fundamental-loop matrix* and denoted by $\mathscr{B}$. Thus, for the set of fundamental loops of Fig. 2.5(a) we have

$$\mathscr{B} = \begin{bmatrix} +1 & 0 & 0 & +1 & +1 & 0 \\ 0 & +1 & 0 & +1 & 0 & +1 \\ 0 & 0 & +1 & 0 & -1 & +1 \end{bmatrix}.$$

Since, by definition, each link of the tree is incident with precisely one of the fundamental loops, we see that with a suitable numbering, the corresponding l columns of $\mathscr{B}$ form a unit matrix of order l.

It should be noted that besides the sets of fundamental loops, there may exist other sets of independent loops. Thus, the rows of any $l \times b$ matrix of $\mathscr{B}_a$ will define a set of independent loops provided it has a rank of l. For example, Fig. 2.5(b) shows a set of three independent loops of the graph of Fig. 2.1(b), which is not a fundamental set.

The Cut-set Matrix

A *cut-set* of a graph is defined as a set of branches whose removal increases the number of separate parts of the graph by one. It is important to note that one of the separate parts may be merely an isolated node. Thus, we may think of a cut-set as a set of branches that divides the graph into two separate parts in such a manner that it is not possible to go from a node of one part to a node of the other part unless we pass through a branch of the cut-set. In general, the graph will contain many cut-sets, which may be arbitrarily numbered and oriented. Figure 2.6 shows all the seven cut-sets of the graph of Fig. 2.1(b).

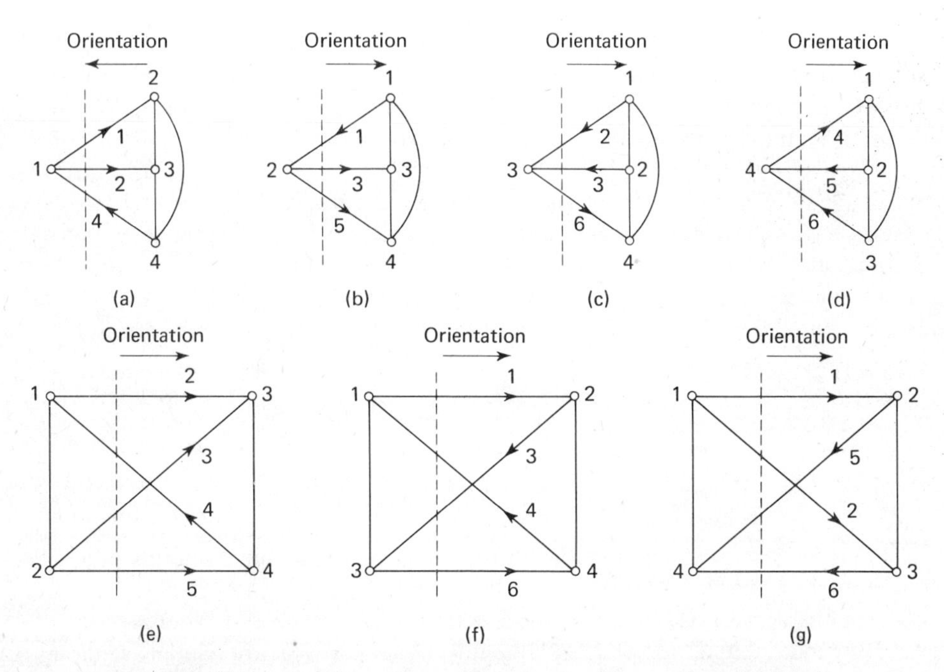

Fig. 2.6. Illustrating the seven possible cut sets of the graph of Fig. 2.1(b).

The incidence between the branches and cut-sets of a given graph may be described algebraically by a matrix, denoted by $\mathcal{Q}_a$, containing b columns, one for each branch of the graph, and a finite number of rows, one for each cut-set of the graph. The elements q_{jk} of $\mathcal{Q}_a$ are $+1$, -1, or zero as follows:

$q_{jk} = +1$, if branch k is incident with cut-set j and has the same orientation.
$q_{jk} = -1$, if branch k is incident with cut-set j and has the opposite orientation.
$q_{jk} = 0$, if branch k and cut-set j are not incident.

For the cut-sets of Fig. 2.6, we thus have

$$\mathcal{Q}_a = \begin{bmatrix} -1 & -1 & 0 & +1 & 0 & 0 \\ -1 & 0 & +1 & 0 & +1 & 0 \\ 0 & -1 & -1 & 0 & 0 & +1 \\ 0 & 0 & 0 & +1 & -1 & -1 \\ 0 & +1 & +1 & -1 & +1 & 0 \\ +1 & 0 & -1 & -1 & 0 & +1 \\ +1 & +1 & 0 & 0 & -1 & -1 \end{bmatrix}.$$

As with loops, trees provide the basis of defining a *fundamental* system of cut-sets as follows. Consider any tree of the graph; each branch of this tree, together with certain of the pertinent links, will define a cut-set of the graph. It may be shown[2] that those links which combine with any one tree branch to form this cut-set are the links whose fundamental loops contain the tree branch. We thus obtain a system of m cut-sets, one from each tree branch. Such a system, called a *fundamental system* of cut-sets with respect to the chosen tree is guaranteed to be independent, so that $\mathcal{Q}_a$ is of rank m in a connected graph. A fundamental system of cut-sets defines an $m \times b$ submatrix of $\mathcal{Q}_a$, called a *fundamental cut-set matrix* and denoted by $\mathcal{Q}$. It is convenient to choose the orientations of the cut-sets to be the same as those of the corresponding tree branches. Thus, the tree of Fig. 2.3(a) defines the fundamental system of cut-sets shown in parts (a), (b) and (c) of Fig. 2.6, for which we obtain

$$\mathcal{Q} = \begin{bmatrix} -1 & -1 & 0 & +1 & 0 & 0 \\ -1 & 0 & +1 & 0 & +1 & 0 \\ 0 & -1 & -1 & 0 & 0 & +1 \end{bmatrix}.$$

Because, by definition, each cut-set of the fundamental system contains precisely one branch of the tree, we see that, with suitable ordering and orientation of the cut-sets, the corresponding columns of $\mathcal{Q}$ form a unit matrix of order m.

The fundamental loop and fundamental cut-set matrices are related as follows: Suppose the numbering and orientations of the cut-sets are such that, for a particular fundamental system, $\mathscr{Q}$ is of the form

$$\mathscr{Q} = [\mathbf{E} \quad \mathbf{1}_m] \tag{2.4}$$

while the loops have been so numbered and oriented that the corresponding matrix $\mathscr{B}$ is of the form

$$\mathscr{B} = [\mathbf{1}_l \quad \mathbf{F}] \tag{2.5}$$

where $\mathbf{1}_m$ and $\mathbf{1}_l$ are unit matrices of orders m and l, respectively. Then we may write

$$\mathbf{E} = -\mathbf{F}^t \tag{2.6}$$

where $\mathbf{F}^t$ is the transpose of $\mathbf{F}$.

2.3 THE EQUILIBRIUM EQUATIONS

Consider an electrical network made up entirely of a finite number of lumped resistors, self inductors and capacitors, with each branch of the network assumed to have an ideal voltage source $v_{sk}(t)$ in series and an ideal current source $i_{sk}(t)$ in parallel, as shown in Fig. 2.7. Since an ideal voltage source is a generalized short-circuit and an ideal current source is a generalized open-circuit, it is clear that so long as the sources appear in the network in the manner illustrated in Fig. 2.7, then their presence does not disturb the network topology in the sense that all matters pertaining to that topology remain unaltered. It is assumed that the values of the sources as functions of time are known, and the unknowns in the network are taken to be the branch currents and voltage drops which exist as a result of the sources. The relative directions of the branch current $i_{bk}(t)$ and voltage drop $v_{bk}(t)$ in each branch, e.g., the kth, are assumed to be as shown in Fig. 2.7.

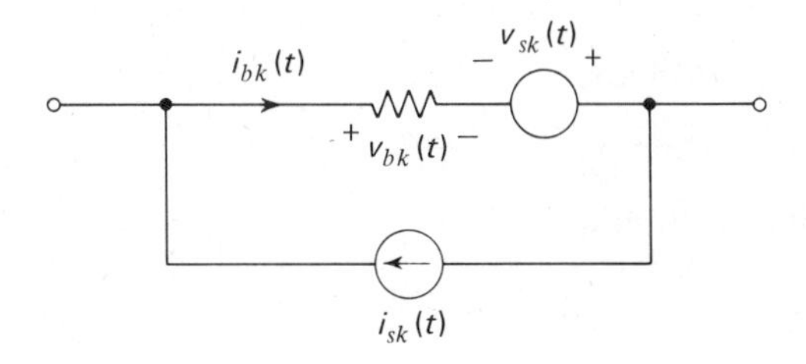

Fig. 2.7. An active branch.

It is rather convenient to formulate the equilibrium equations in the complex frequency domain, so that we may deal straight away with the Laplace transforms of the time-varying currents and voltages in the network. In addition we shall assume that the initial values of the variables are zero, i.e., the network is initially at rest. According to Fig. 2.7, the net voltage drop across, say, the kth oriented active branch is $V_{bk}(s) - V_{sk}(s)$. The b branch variables $V_{b1}(s), V_{b2}(s), \ldots, V_{bb}(s)$ may be arranged in the form of a $b \times 1$ column vector which we denote by $\mathbf{V}_b(s)$. Similarly, we may define the $b \times 1$ column vector $\mathbf{V}_s(s)$. If we choose a total of l independent loops, denoting the corresponding loop matrix by $\mathscr{B}$, and then apply Kirchhoff's voltage law to every loop of the network, there results

$$\mathscr{B}[\mathbf{V}_b(s) - \mathbf{V}_s(s)] = 0$$

or

$$\mathscr{B}\mathbf{V}_b(s) = \mathscr{B}\mathbf{V}_s(s). \tag{2.7}$$

The net current in the kth oriented active branch is $I_{bk}(s) - I_{sk}(s)$. Let the branch current and current-source vectors be denoted by $\mathbf{I}_b(s)$ and $\mathbf{I}_s(s)$, respectively. Then, since the reduced incidence matrix $\mathscr{A}$, with respect to a chosen datum node, specifies the m nodes of the network in terms of their incidence with the branches, we find that the Kirchhoff's current law applied to each of the m nodes yields

$$\mathscr{A}[\mathbf{I}_b(s) - \mathbf{I}_s(s)] = 0$$

or

$$\mathscr{A}\mathbf{I}_b(s) = \mathscr{A}\mathbf{I}_s(s). \tag{2.8}$$

The two Kirchhoff laws, as expressed by Eqs. (2.7) and (2.8) are constraints imposed upon $\mathbf{V}_b(s)$ and $\mathbf{I}_b(s)$ solely by the topology of the network. In addition, $\mathbf{V}_b(s)$ and $\mathbf{I}_b(s)$ are constrained by the voltage–current relations that characterize the elements. Thus, assuming that the network is made up of resistors, self inductors and capacitors, we have:

For a resistive kth branch,

$$V_{bk}(s) = R_k I_{bk}(s). \tag{2.9}$$

For an inductive kth branch,

$$V_{bk}(s) = sL_k I_{bk}(s). \tag{2.10}$$

For a capacitive kth branch,

$$V_{bk}(s) = \frac{1}{sC_k} I_{bk}(s). \tag{2.11}$$

For each of the b branches of the network one of the relations (2.9), (2.10) or (2.11) will hold, giving a total of b voltage–current relations. These may be written in the matrix form

$$\mathbf{V}_b(s) = \mathbf{Z}_b(s)\mathbf{I}_b(s), \tag{2.12}$$

where $\mathbf{Z}_b(s)$ is a non-singular $b \times b$ matrix called the *branch-impedance matrix*.

Suppose the branches of the network are numbered according to the type of element, say capacitors, followed by resistors followed by inductors, as in Fig. 2.1(a). Then we may express $\mathbf{Z}_b(s)$ as

$$\mathbf{Z}_b(s) = \begin{bmatrix} \frac{1}{s}\mathbf{C}_b^{-1} & 0 & 0 \\ 0 & \mathbf{R}_b & 0 \\ 0 & 0 & s\mathbf{L}_b \end{bmatrix}, \tag{2.13}$$

where $\mathbf{L}_b$, $\mathbf{R}_b$ and $\mathbf{C}_b^{-1}$ are respectively the matrices of branch self inductances, resistances, and reciprocal capacitances. The matrices $\mathbf{R}_b$, $\mathbf{L}_b$ and $\mathbf{C}_b$ are diagonal (i.e., with all off-diagonal elements zero).

Since $\mathbf{Z}_b$ is non-singular, we may invert Eq. (2.12) to obtain

$$\mathbf{I}_b(s) = \mathbf{Y}_b(s)\mathbf{V}_b(s), \tag{2.14}$$

where $\mathbf{Y}_b(s)$ is the *branch-admittance matrix* defined by

$$\mathbf{Y}_b(s) = \mathbf{Z}_b^{-1}(s) = \begin{bmatrix} s\mathbf{C}_b & 0 & 0 \\ 0 & \mathbf{R}_b^{-1} & 0 \\ 0 & 0 & \frac{1}{s}\mathbf{L}_b^{-1} \end{bmatrix}. \tag{2.15}$$

Equations (2.7), (2.8) and (2.12) are called the *network branch equations.* They provide a total of $l + m + b = 2b$ independent algebraic equations for the $2b$ unknown variables $I_{bk}(s)$ and $V_{bk}(s)$, with $k = 1, 2, \ldots, b$. However, even for a simple network $2b$ can be a rather large number. Thus, for the network of Fig. 2.1(a), for example, we would need to solve twelve simultaneous equations. This may be avoided by using the well-known *loop* or *nodal* method of analysis. The loop method involves the use of circulating *loop currents,* while the nodal method involves the use of *node voltages.* Although each method can be used when both current and voltage sources are present,[1] the solutions obtained in each case are very much simpler if for loop analysis it is assumed that only voltage sources are present, while for nodal analysis it is assumed that only current sources are present. Hence, to illustrate the methods, we shall assume that for loop analysis $\mathbf{I}_s(s) = 0$, while for nodal analysis $\mathbf{V}_s(s) = 0$.

Loop Method

Suppose, for a chosen set of l independent loops, we define a corresponding set of l loop currents $I_1(s), I_2(s), \ldots, I_l(s)$. If we denote the corresponding loop matrix by $\mathscr{B}$, then from the definition of the loop matrix we see that (with $\mathbf{I}_s(s) = 0$) the branch currents are related to the loop currents by

$$\mathbf{I}_b(s) = \mathscr{B}'\mathbf{I}(s), \tag{2.16}$$

where $\mathbf{I}_b(s)$ and $\mathbf{I}(s)$ denote the $b \times 1$ branch current and $l \times 1$ loop current vectors, respectively, and $\mathscr{B}^t$ is the transpose of $\mathscr{B}$. When a set of fundamental loops is used, we may identify each link current with a loop current.

Consider, next, Eq. (2.7); the product term $\mathscr{B}\mathbf{V}_s(s)$ represents a column vector of l rows. Each element, corresponding to one of the l independent loops in the network, is the algebraic sum of the voltage sources contained in that loop. Thus, we may put

$$\mathscr{B}\mathbf{V}_s(s) = \mathbf{V}_L(s), \tag{2.17}$$

where $\mathbf{V}_L(s)$ is an $l \times 1$ column vector representing the loop voltage sources. Hence, from Eqs. (2.7) and (2.17),

$$\mathbf{V}_L(s) = \mathscr{B}\mathbf{V}_b(s). \tag{2.18}$$

If we substitute Eq. (2.12) into (2.18), we get

$$\mathbf{V}_L(s) = \mathscr{B}\mathbf{Z}_b(s)\mathbf{I}_b(s). \tag{2.19}$$

Next, substitution of Eq. (2.16) into (2.19) yields the loop equations

$$\mathbf{V}_L(s) = \mathscr{B}\mathbf{Z}_b(s)\mathscr{B}^t\mathbf{I}(s),$$

or

$$\mathbf{V}_L(s) = \mathbf{Z}(s)\mathbf{I}(s), \tag{2.20}$$

where $\mathbf{Z}(s)$ is an $l \times l$ matrix called the *loop-impedance matrix* and defined by

$$\mathbf{Z}(s) = \mathscr{B}\mathbf{Z}_b(s)\mathscr{B}^t \tag{2.21}$$

Equation (2.20) represents a system of l loop equations with the loop currents as variables, as shown by the familiar expanded form

$$\begin{bmatrix} V_{L1} \\ V_{L2} \\ \cdot \\ \cdot \\ \cdot \\ V_{Ll} \end{bmatrix} = \begin{bmatrix} Z_{11} & Z_{12} & \cdots & Z_{1l} \\ Z_{21} & Z_{22} & \cdots & Z_{2l} \\ \cdot & \cdot & & \cdot \\ \cdot & \cdot & & \cdot \\ \cdot & \cdot & & \cdot \\ Z_{l1} & Z_{l2} & \cdots & Z_{ll} \end{bmatrix} \begin{bmatrix} I_1 \\ I_2 \\ \cdot \\ \cdot \\ \cdot \\ I_l \end{bmatrix}, \tag{2.22}$$

where the dependences upon s have, for convenience, been omitted. The loop equations are evidently expressions of voltages equilibrium in the network, i.e., they express the fact that the algebraic sum of all the driving voltages in any closed loop of the network is equal to the algebraic sum of voltage drops across all elements in that loop.

A typical coefficient $Z_{jj} = sL_{jj} + R_{jj} + 1/sC_{jj}$ on the principal diagonal of the loop-impedance matrix is equal to the sum of all the branch impedances contained in loop j; Z_{jj} is therefore called the *self impedance* of loop j. The coefficient $Z_{jk} = sL_{jk} + R_{jk} + 1/sC_{jk}$ for $j \neq k$ is the *mutual impedance* of loops j and k; it is equal to the ratio of the voltage drop in loop k to the current flowing in loop j which produces it. That is, if the current in loop k is I_k, and the mutual impedance with loop j is Z_{jk}, then $Z_{jk}I_k$ is the voltage developed in loop j due to I_k. Thus, Z_{jk} is the negative sum of the branch impedances that are common to loops j and k, assuming that all the circulating loop currents are assigned the same direction.

Also, in networks consisting entirely of bilateral elements (i.e., inductors, resistors and capacitors) we have $Z_{jk} = Z_{kj}$ for all k and j, which implies that the loop-impedance matrix is symmetric, that is, $\mathbf{Z} = \mathbf{Z}^t$.

Solving Eqs. (2.22) for the loop current I_k by Cramer's rule, we get

$$I_k = \frac{\Delta_{1k}}{\Delta} V_{L1} + \frac{\Delta_{2k}}{\Delta} V_{L2} + \cdots + \frac{\Delta_{lk}}{\Delta} V_{Ll},$$

or

$$I_k = \frac{1}{\Delta} \sum_{j=1}^{l} \Delta_{jk} V_{Lj}, \tag{2.23}$$

where $\Delta = \det \mathbf{Z}(s)$, this is

$$\Delta = \begin{vmatrix} Z_{11} & Z_{12} & \cdots & Z_{1l} \\ Z_{22} & Z_{22} & \cdots & Z_{2l} \\ \cdot & \cdot & & \cdot \\ \cdot & \cdot & & \cdot \\ \cdot & \cdot & & \cdot \\ Z_{l1} & Z_{l2} & \cdots & Z_{ll} \end{vmatrix} \tag{2.24}$$

Δ is called the *loop-basis circuit determinant*; it is a characteristic of the network depending solely upon the parameters of the elements and the manner of their interconnection. The Δ_{jk} in Eq. (2.23) is called the *cofactor* of Z_{jk} in $\mathbf{Z}(s)$. It is equal to $(-1)^{j+k}$ times the determinant of the $(l-1) \times (l-1)$ submatrix of $\mathbf{Z}(s)$ obtained by deleting row j and column k. The ratio Δ_{jk}/Δ has the dimensions of an admittance.

Having determined all the loop currents, we can then find the branch currents from the loop transformation of Eq. (2.16). Finally, the voltage–current relations of Eq. (2.12) can be used to find the corresponding branch voltages.

Nodal Method

The development of the nodal method proceeds in a manner similar to that used in connection with the loop method. It begins by selecting an independent set of voltage variables capable of representing any distribution of branch voltages that may exist in the network consistent with Kirchhoff's voltage law. The branch voltages of any tree obviously provide such a set, for if all tree branch voltages are forced to be zero (by short-circuiting the tree branches), then by virtue of the fact that the tree branches connect all of the nodes, we find that all branch voltages in the network are reduced to zero. However, a tree that is of particular interest is obtained by selecting one node as a datum node and then constructing lines from this node to each of the other m (i.e., $n - 1$) nodes of the network. If we denote the branch voltages of this particular tree by an $m \times 1$ column vector $\mathbf{V}_b(s)$ and the corresponding reduced incidence matrix by $\mathscr{A}$, then from the definition of the reduced incidence matrix we see that (with no voltage sources present, i.e., $\mathbf{V}_s(s) = 0$) the branch voltages of the network are given by

$$\mathbf{V}_b(s) = \mathscr{A}^t\mathbf{V}(s), \tag{2.25}$$

where $\mathscr{A}^t$ is the transpose of the reduced incidence matrix.

Consider, next, Eq. (2.8); the product term $\mathscr{A}\mathbf{I}_s(s)$ represents a column vector of m rows. Each element corresponds to one of the m nodes of the network and is the algebraic sum of the current sources entering that node. Thus we may put

$$\mathscr{A}\mathbf{I}_s(s) = \mathbf{I}_N(s) \tag{2.26}$$

where $\mathbf{I}_N(s)$ is an $m \times 1$ column vector representing the nodal current sources. Hence, from Eqs. (2.8) and (2.26),

$$\mathbf{I}_N(s) = \mathscr{A}\mathbf{I}_b(s). \tag{2.27}$$

If we substitute Eq. (2.14) in (2.27), we get

$$\mathbf{I}_N(s) = \mathscr{A}\mathbf{Y}_b(s)\mathbf{V}_b(s). \tag{2.28}$$

Next, substitution of Eq. (2.25) in (2.28) yields the nodal equations

$$\mathbf{I}_N(s) = \mathscr{A}\mathbf{Y}_b(s)\mathscr{A}^t\mathbf{V}(s)$$

or

$$\mathbf{I}_N(s) = \mathbf{Y}(s)\mathbf{V}(s) \tag{2.29}$$

where $\mathbf{Y}(s)$ is an $m \times m$ matrix called the *nodal-admittance matrix* and defined by

$$\mathbf{Y}(s) = \mathscr{A}\mathbf{Y}_b(s)\mathscr{A}^t. \tag{2.30}$$

In expanded form, Eq. (2.29) represents the familiar system of nodal equations:

$$\begin{bmatrix} I_{N1} \\ I_{N2} \\ \cdot \\ \cdot \\ \cdot \\ I_{Nm} \end{bmatrix} = \begin{bmatrix} Y_{11} & Y_{12} & \cdots & Y_{1m} \\ Y_{21} & Y_{22} & \cdots & Y_{2m} \\ \cdot & \cdot & & \cdot \\ \cdot & \cdot & & \cdot \\ \cdot & \cdot & & \cdot \\ Y_{m1} & Y_{m2} & \cdots & Y_{mm} \end{bmatrix} \begin{bmatrix} V_1 \\ V_2 \\ \cdot \\ \cdot \\ \cdot \\ V_m \end{bmatrix}. \tag{2.31}$$

The nodal equations are evidently expressions of current equilibrium in the network, i.e., they express the fact that the algebraic sum of all the driving currents flowing into any of the m nodes of the network from outside is equal to the algebraic sum of all currents flowing away from that node into the rest of the network.

A typical coefficient $Y_{jj} = sC_{jj} + G_{jj} + 1/sL_{jj}$ is the sum of all the branch admittances connected directly to node j; Y_{jj} is therefore called the *self admittance* of the jth node. The coefficient $Y_{jk} = sC_{jk} + G_{jk} + 1/sL_{jk}$, for $j \neq k$, is the *mutual admittance* of nodes j and k. Specifically, if the voltage at node k is V_k, and the mutual admittance with node j is Y_{jk}, then $Y_{jk}V_k$ is the resultant current at node j due to V_k. Thus, the admittance Y_{jk} is the negative sum of the branch admittances connecting node j and node k.

Also, in networks containing bilateral elements only, $Y_{jk} = Y_{kj}$ for all k and j, so that the nodal-admittance matrix is symmetric, that is $\mathbf{Y} = \mathbf{Y}^t$.

Solving Eqs. (2.31) for the node voltage V_k by Cramer's rule, we get

$$V_k = \frac{\Delta'_{1k}}{\Delta'} I_{N1} + \frac{\Delta'_{2k}}{\Delta'} I_{N2} + \cdots + \frac{\Delta'_{mk}}{\Delta'} I_{Nm}$$

or

$$V_k = \frac{1}{\Delta'} \sum_{j=1}^{m} \Delta'_{jk} I_{Nj} \tag{2.32}$$

where $\Delta' = \det \mathbf{Y}(s)$, that is,

$$\Delta' = \begin{vmatrix} Y_{11} & Y_{12} & \cdots & Y_{1m} \\ Y_{21} & Y_{22} & \cdots & Y_{2m} \\ \cdot & \cdot & & \cdot \\ \cdot & \cdot & & \cdot \\ \cdot & \cdot & & \cdot \\ Y_{m1} & Y_{m2} & \cdots & Y_{mm} \end{vmatrix}. \tag{2.33}$$

Δ' is the *nodal-basis circuit determinant*; it is a characteristic of the network depending solely upon its topology and the parameters of the elements involved. The Δ'_{jk} is the cofactor of Y_{jk} in $\mathbf{Y}(s)$. The ratio Δ'_{jk}/Δ' has the dimensions of an impedance.

Having determined all the node voltages, we may evaluate the branch voltages using the nodal transformation of Eq. (2.25). Finally, using the voltage–current relations of Eq. (2.14), we can evaluate the branch currents.

Networks Containing Mutual Inductance

Consider, next, a network containing capacitors, resistors and coupled coils. In this case, the branch voltages and branch currents for the coupled coils are related by a matrix of self and mutual inductances, $\mathbf{L}_b$, which is nondiagonal but symmetric that is, $L_{jk} = L_{kj}$ for all j and k. The matrices $\mathbf{C}_b$ and $\mathbf{R}_b$ are, of course, still diagonal. Thus, the branch impedance matrix, $\mathbf{Z}_b(s)$, and its inverse, $\mathbf{Y}_b(s)$, as defined by Eqs. (2.13) and (2.15), respectively, are now no longer diagonal. For the matrix $\mathbf{Y}_b(s)$ to exist, it is necessary that the inductance matrix $\mathbf{L}_b$ be nonsingular, a condition which is always satisfied provided the network contains no perfectly coupled coils.

The loop-impedance matrix, $\mathbf{Z}(s)$, or the nodal-admittance matrix, $\mathbf{Y}(s)$, of the network are both symmetric as before. However, when a network contains mutual inductance, we find that, in general, it is no longer possible to determine all the elements Z_{jk} of the loop-impedance matrix or all the elements Y_{jk} of the nodal-admittance matrix directly by inspection of the network, as previously described. This, therefore, means that in order to characterize the network by a set of nodal equations, say, we have to go back to our systematic procedure, evaluate the reduced incidence matrix $\mathscr{A}$ and branch admittance matrix $\mathbf{Y}_b(s)$ of the network, and then use Eq. (2.30) to evaluate the required nodal-admittance matrix, $\mathbf{Y}(s)$.

Networks Containing Controlled Sources

The networks considered so far have been assumed to consist entirely of bilateral elements. When, however, the network contains controlled sources, the equilibrium equations may be considered to consist of two types:[3]

1. An independent set of l loop or m nodal equations specifying the equilibrium conditions in the network obtained with all sources treated as if they were independent.
2. A set of equations defining the constraints imposed on the controlled sources, consisting of one equation for each controlled source.

In general, the value of each controlled source is proportional to some controlling voltage or current in the network, which is, in turn, always expressible as a linear combination of the loop currents (in the loop method) or node voltages (in the nodal method). Hence, each controlled source can always be expressed as a linear combination of the current or voltage variables used to formulate the first set of equations.

When the controlled source variables are eliminated by substituting each equation of the second set into the first set, we find that the resultant equilibrium equations take on a final form similar to that defined by Eq. (2.22) (for the loop method) or Eq. (2.31) (for the nodal method) with one basic modification, namely the pertinent matrix of coefficients may no longer be symmetric. That is, in a network containing controlled sources we may find that, in general, $Z_{jk} \neq Z_{kj}$ (in the loop method) or $Y_{jk} \neq Y_{kj}$ (in the nodal method).

Example 2.1. *Loop analysis.* As an example illustrating the loop analysis, consider the network of Fig. 2.8(a) which includes a current-controlled voltage

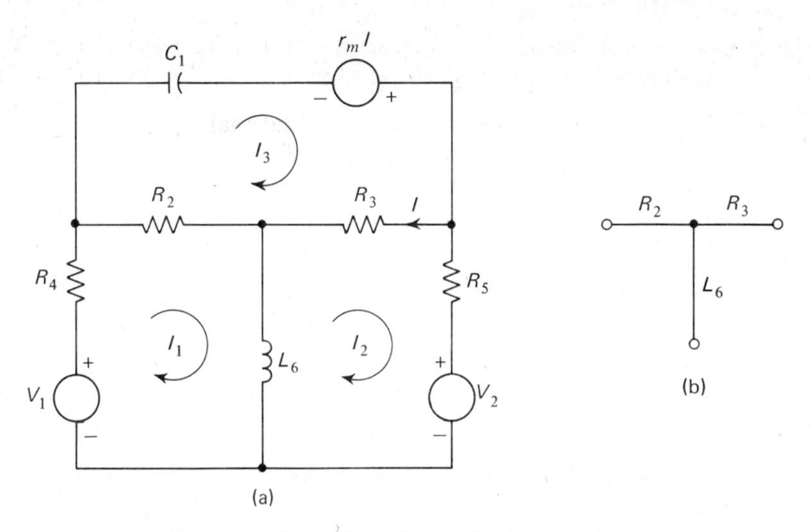

Fig. 2.8. Network for illustrating loop analysis.

source $r_m I$, where I is the branch current flowing through resistor R_3. Treating this controlled voltage source as if it were independent, we find by inspection that the loop equations, in matrix form, are as follows:

$$\begin{bmatrix} V_1 \\ -V_2 \\ r_m I \end{bmatrix} = \begin{bmatrix} R_2 + R_4 + sL_6 & -sL_6 & -R_2 \\ -sL_6 & R_3 + R_5 + sL_6 & -R_3 \\ -R_2 & -R_3 & R_2 + R_3 + \dfrac{1}{sC_1} \end{bmatrix} \begin{bmatrix} I_1 \\ I_2 \\ I_3 \end{bmatrix}, \quad (2.34)$$

where I_1, I_2 and I_3 are the loop currents, assumed to be circulating in a clockwise direction in the fundamental loops defined by the tree of Fig. 2.8(b). The control signal I of the controlled source is equal to $I_3 - I_2$; therefore:

$$r_m I = r_m I_3 - r_m I_2.$$

Substituting this relation into Eq. (2.34) and transposing the unknown currents I_2 and I_3 to the right, we obtain

$$\begin{bmatrix} V_1 \\ -V_2 \\ 0 \end{bmatrix} = \begin{bmatrix} R_2 + R_4 + sL_6 & -sL_6 & -R_2 \\ -sL_6 & R_3 + R_5 + sL_6 & -R_3 \\ -R_2 & r_m - R_3 & R_2 + R_3 + \dfrac{1}{sC_1} - r_m \end{bmatrix} \begin{bmatrix} I_1 \\ I_2 \\ I_3 \end{bmatrix}, \quad (2.35)$$

where we observe that the matrix of coefficients is not symmetric. This is a result of the unidirectional coupling introduced between loops 2 and 3 of the network in Fig. 2.8(a) by the controlled source. The control parameter r_m appears only in the third row of the matrix of coefficients because the controlled source is in the third loop only. Furthermore, r_m appears in columns 2 and 3, since the controlled source depends on both I_2 and I_3.

Example 2.2. *Nodal analysis.* As a second example, illustrating the nodal analysis, consider the network of Fig. 2.9(a) which includes a voltage-controlled current source $g_m V_1$. Treating this controlled current source as if it were independent, we get the following nodal equations, in matrix form, directly from Fig. 2.9(a),

$$\begin{bmatrix} I_1 \\ I_2 \\ -g_m V_1 \end{bmatrix} = \begin{bmatrix} G_2 + G_4 + sC_1 & -sC_1 & -G_2 \\ -sC_1 & G_3 + G_5 + sC_1 & -G_3 \\ -G_2 & -G_3 & G_2 + G_3 + \dfrac{1}{sL_6} \end{bmatrix} \begin{bmatrix} V_1 \\ V_2 \\ V_3 \end{bmatrix} \tag{2.36}$$

where V_1, V_2 and V_3 are the branch voltages pertaining to the tree shown in Fig. 2.9(b). The controlled source $g_m V_1$ is already expressed in terms of the tree-branch voltage V_1; therefore, transposing $g_m V_1$ to the right in Eq. (2.36) yields

$$\begin{bmatrix} I_1 \\ I_2 \\ 0 \end{bmatrix} = \begin{bmatrix} G_2 + G_4 + sC_1 & -sC_1 & -G_2 \\ -sC_1 & G_3 + G_5 + sC_1 & -G_3 \\ g_m - G_2 & -G_3 & G_2 + G_3 + \dfrac{1}{sL_6} \end{bmatrix} \begin{bmatrix} V_1 \\ V_2 \\ V_3 \end{bmatrix} \tag{2.37}$$

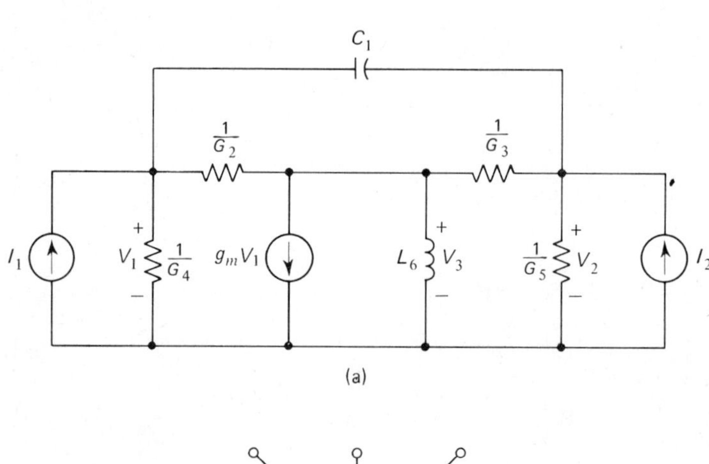

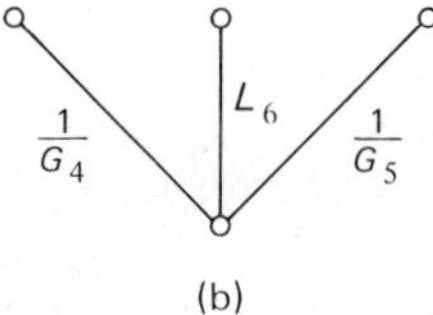

Fig. 2.9. Network for illustrating nodal analysis.

where again we observe that, owing to the presence of a controlled source, the matrix of coefficients is not symmetric. The control parameter g_m appears only in row 3 and column 1 of this matrix because the controlled source is connected to node 3 and is dependent upon V_1.

The Choice Between the Loop and Nodal Methods

For a network containing a total number of b branches and n nodes, the preceding discussion gives the required number of equilibrium equations as $l = b - n + 1$ for the loop system and as $m = n - 1$ for the nodal system. For a given network, the system with the fewer equations is obviously to be preferred. For the same number of equations on either basis, however, the nodal analysis has the advantage that the equations can be more directly correlated with the physical structure of the network than is possible with the loop analysis. Also, the nodal equations can be written down directly, but to use the loop analysis we have to begin by selecting a suitable set of closed loops, which may not be an easy problem in a complicated circuit.

2.4 NETWORK THEOREMS

Thévenin's Theorem

Thévenin's theorem states that, in Fig. 2.10, in so far as the behaviour of the network N with respect to the port k–k' is concerned, the network can be represented by the so-called *Thévenin equivalent network* of Fig. 2.10(b) which consists of a source of voltage V_{oc} connected in series with an impedance Z_{eq}. To prove this theorem and determine V_{oc} and Z_{eq}, suppose that the terminal current I_k is regarded as the independent variable; then if we apply the nodal analysis we find from Eq. (2.32) that the resulting terminal voltage V_k is

$$V_k = \frac{1}{\Delta'} \sum_{\substack{j=1 \\ j \neq k}}^{m} \Delta'_{jk} I_j + \frac{\Delta'_{kk}}{\Delta'} I_k. \tag{2.38}$$

In Eq. (2.38) we have separated the contribution of the current I_k from the other independent current sources inside the network as we are seeking a relation between V_k and I_k. Let

$$V_{oc} = \frac{1}{\Delta'} \sum_{\substack{j=1 \\ j \neq k}}^{m} \Delta'_{jk} I_j \tag{2.39}$$

and

$$Z_{eq} = \frac{\Delta'_{kk}}{\Delta'}. \tag{2.40}$$

Then, substitution of these two definitions in Eq. (2.38) gives

$$V_k = V_{oc} + Z_{eq} I_k. \tag{2.41}$$

This relation is, in essence, a statement of Thévenin's theorem and can be represented as the equivalent network shown in Fig. 2.10(b). The networks shown in parts *a* and *b* of Fig. 2.10 are equivalent in the sense that both may be described by the same voltage–current relation with respect to the port *k*–*k'*.

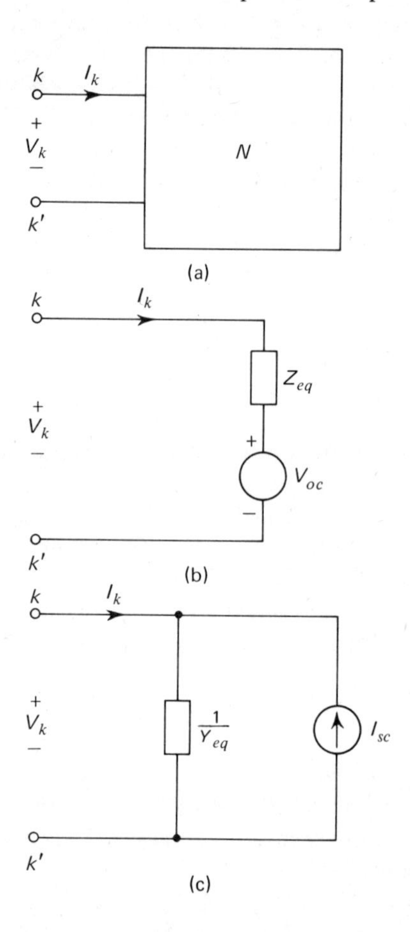

Fig. 2.10. (a) One-port network; (b) Thévenin equivalent network; (c) Norton equivalent network.

We can determine the voltage V_{oc} and the impedance Z_{eq} from the nodal analysis by using Eqs. (2.39) and (2.40). Alternatively, we can use two independent measurements:

1. When I_k is zero, i.e., the port k–k' is open circuited, we see from Eq. (2.41) that V_{oc} is the voltage that appears across the port $k-k'$ of the network N.
2. When V_{oc} is zero, which can be accomplished by reducing all independent sources within the network N to zero, the Thévenin equivalent impedance Z_{eq} is the impedance measured looking into the port k–k'.

It is important to note that if the network contains controlled sources they must not be removed when calculating the impedance Z_{eq}, because a controlled source is by nature dependent on the control signal, so that any change in the control signal produces a corresponding change in the strength of the controlled source.

In Eq. (2.41) the current I_k is the independent variable and the voltage V_k is the dependent variable. If, however, we regard V_k as the independent variable we find from Eq. (2.23) that the loop analysis leads to the following alternative relation between I_k and V_k,

$$I_k = \frac{1}{\Delta} \sum_{\substack{j=1 \\ j \neq k}}^{l} \Delta_{jk} V_j + \frac{\Delta_{kk}}{\Delta} V_k, \tag{2.42}$$

or

$$I_k = -I_{sc} + Y_{eq} V_k \tag{2.43}$$

where

$$I_{sc} = -\frac{1}{\Delta} \sum_{\substack{j=1 \\ j \neq k}}^{l} \Delta_{jk} V_j \tag{2.44}$$

$$Y_{eq} = \frac{\Delta_{kk}}{\Delta}. \tag{2.45}$$

The relation of Eq. (2.43) suggests the alternative representation of Fig. 2.10(c) which is known as the *Norton equivalent network*; it consists of a source of current I_{sc} equal to the short-circuit current, connected in parallel with an admittance Y_{eq} equal to the admittance measured looking into the port k–k' when all independent sources within the network N are set to zero. Equation (2.43) could have also been deduced directly from Eq. (2.41) by solving for I_k in terms of V_k. We thus find that $Y_{eq} = 1/Z_{eq}$ and

$$Z_{eq} = \frac{V_{oc}}{I_{sc}}, \tag{2.46}$$

which states that the Thévenin equivalent impedance Z_{eq} is the ratio of the open-circuit voltage to the short-circuit current at the port k–k'. This provides yet another method of evaluating Z_{eq}.

Example 2.3. *Vacuum-tube triode amplifier.* To illustrate the procedure for evaluating the Thévenin equivalent network, consider the vacuum-tube triode amplifier of Fig. 2.11(a). Replacing the triode with its low-frequency circuit model, we obtain the network of Fig. 2.11(b). When the port 2–2′ is left open-circuited, we have

$$V_s = (R_1 + R_2 + r_p)I_1 - \mu V_{gk} \tag{2.47}$$

and

$$V_{gk} = V_s - R_1 I_1. \tag{2.48}$$

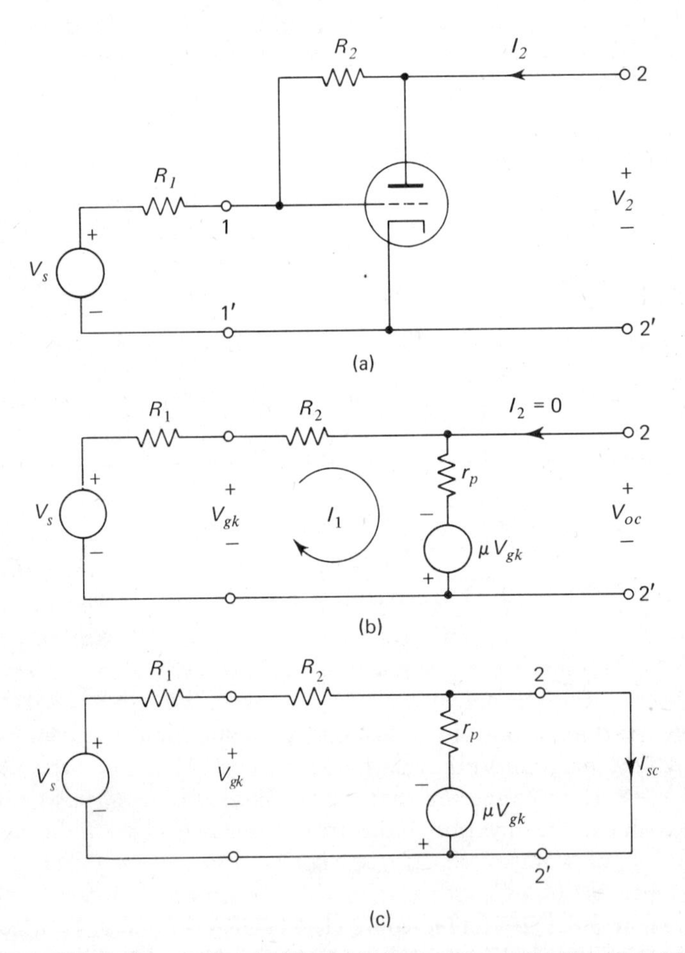

Fig. 2.11. (a) Vacuum-tube triode amplifier; (b) Network for evaluating V_{oc}; (c) Network for evaluating I_{sc}.

Solving Eqs. (2.47) and (2.48) for I_1 by eliminating V_{gk}, we get

$$I_1 = \frac{(1 + \mu)V_s}{R_1(1 + \mu) + R_2 + r_p}. \tag{2.49}$$

Therefore, the open-circuit voltage V_{oc} is

$$V_{oc} = V_s - (R_1 + R_2)I_1,$$

or

$$V_{oc} = \frac{-(\mu R_2 - r_p)V_s}{R_1(1 + \mu) + R_2 + r_p}. \tag{2.50}$$

If next the port 2–2′ is short-circuited, as in Fig. 2.11(c), we find that the short-circuit current I_{sc} is

$$I_{sc} = -\frac{\mu V_{gk}}{r_p} + \frac{V_s}{R_1 + R_2} \tag{2.51}$$

where

$$V_{gk} = \frac{R_2 V_s}{R_1 + R_2}. \tag{2.52}$$

IF 2-2′ IS SHORT-CIRCUITED.

Therefore,

$$I_{sc} = \frac{-(\mu R_2 - r_p)V_s}{r_p(R_1 + R_2)}. \tag{2.53}$$

Substitution of Eqs. (2.50) and (2.53) in Eq. (2.46) gives the Thévenin equivalent impedance Z_{eq} with respect to the port 2–2′ as

$$Z_{eq} = \frac{r_p(R_1 + R_2)}{R_1(1 + \mu) + R_2 + r_p}. \tag{2.54}$$

Substitution Theorem[4]

The *substitution theorem* applies to the network configurations shown in Fig. 2.12. In part (a) of the diagram we have a linear network N and a current-controlled voltage source $r_m I_1$ so connected that the terminal current I_1 flows through the controlled source in the direction of the drop in voltage. Clearly, the currents and voltages in the network N remain unchanged when the controlled source is replaced by a resistance of r_m ohm.

In a dual manner, a voltage-controlled current source $g_m V_1$ which is connected across the voltage V_1, as in Fig. 2.12(b), can be replaced by a conductance of g_m mho without altering the currents and voltages in the network N.

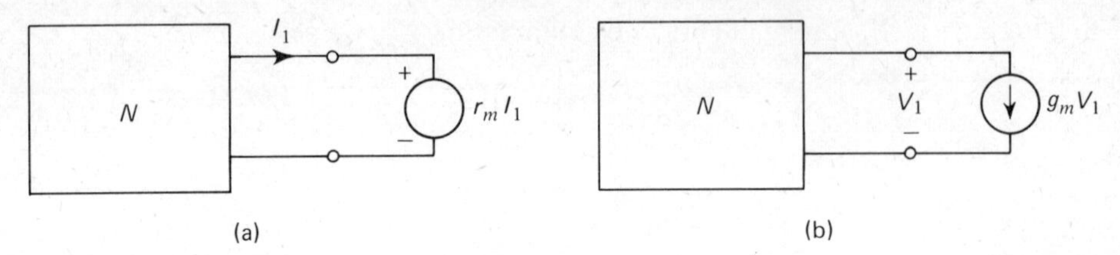

Fig. 2.12. Illustrating the substitution theorem: (a) Network with current-controlled voltage source; (b) Network with voltage-controlled current source.

Reduction Theorem[4]

The *reduction theorem*, pertaining to the network configurations shown in Fig. 2.13, is in effect an extension of the substitution theorem. In Fig. 2.13(a) we have a linear one-port network N and a voltage-controlled voltage source μV_1 connected in series so that the terminal voltage V_1 and μV_1 are additive. The reduction theorem states that all currents in the network N of Fig. 2.13(a) remain unchanged if the controlled source μV_1 is replaced with a short-circuit and if all resistances, inductances, reciprocals of capacitances, and voltage sources in the network N are divided by $1 + \mu$. This follows directly from the form of the loop equations.

In a dual manner, all voltages in the network N of Fig. 2.13(b) remain unchanged if the current-controlled current source αI_1 is replaced with an open-circuit and if all conductances, capacitances, reciprocals of inductances, and current sources in the network N are divided by $(1 + \alpha)$. This follows directly from the nodal equations.

Compensation Theorem[5]

The *compensation theorem* deals with the effect of changes in any one element, passive or active, in a linear network. It is therefore particularly useful in studying the effect of parameter tolerances in network design. To be specific, consider the situation depicted in Fig. 2.14(a) which is assumed to include an independent current source I_1 and a voltage-controlled current source $I_j = g_m V_i$, where the control voltage V_i is developed across the indicated pair of terminals. The prob-

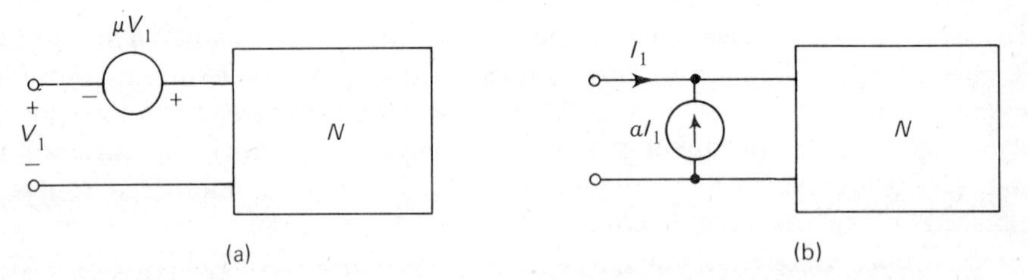

Fig. 2.13. Illustrating the reduction theorem: (a) Network with voltage-controlled voltage source; (b) Network with current-controlled current source.

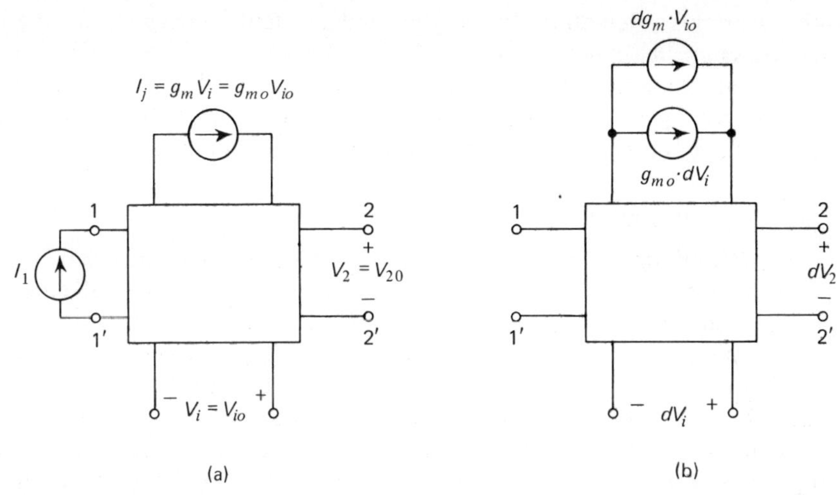

Fig. 2.14. Illustrating the compensation theorem: (a) Network before the change in g_m; (b) Network with perturbation source simulating the change.

lem is to determine the changes produced in any or all branch voltages in the network as a result of a known change in g_m.

Treating the controlled source I_j as if it were independent and setting up the nodal equations for the network, we find that the voltage V_2 developed at some specified port 2–2′ is given by

$$V_2 = \frac{\Delta'_{12}}{\Delta'} I_1 + \frac{\Delta'_{j2}}{\Delta'} (g_m V_i), \tag{2.55}$$

where Δ' is the nodal-basis circuit determinant of the network, excluding the controlled source $g_m V_i$, so that Δ' and the cofactors Δ'_{12} and Δ'_{j2} are all independent of g_m. Suppose g_m changes by a small amount dg_m, with the independent source I_1 and all other parameters of the network maintained constant. To a first order of approximation, the corresponding change in V_2 is therefore

$$dV_2 = \frac{\Delta'_{j2}}{\Delta'} (dg_m V_{io} + g_{mo} dV_i), \tag{2.56}$$

where g_{mo} and V_{io} denote the original values of g_m and V_i before the change. As far as changes in node voltages are concerned, we may, for a known dg_m, treat the term $dg_m V_{io}$ in Eq. (2.56) as a known independent current source simulating the change; we shall refer to it as a *perturbation source.* On the other hand, the term $g_{mo}\, dV_i$ is a controlled source in that it depends on dV_i, which is a quantity we are trying to compute. This suggests that the change in each node voltage is equal to the corresponding response produced by the perturbation source $dg_m V_{io}$ placed in parallel with the controlled current source $g_{mo}\, dV_i$, as in Fig.

2.14(b). For this calculation, the original independent sources (I_1, in the case under study) are set equal to zero.

In a dual manner, we may evaluate the effect of small changes in a controlled voltage source; in this case, the perturbation source is in the form of an independent voltage source placed in series with the changed controlled voltage source.

2.5 STATE-VARIABLE APPROACH

The *state-variable approach* is a method of describing the dynamic behaviour of a network in terms of a set of simultaneous first-order differential equations expressed in matrix form. The characterization is carried out in terms of a minimal set of dynamically independent network variables, called *state variables.* The set is minimal in the sense that no algebraic relations exist between them, and sufficient in that all other variables of the network are uniquely expressible in terms of this set. As a method of analysis, the state-variable approach is well suited to general network studies in that it provides a unifying basis for studying linear, time-varying, and nonlinear networks, although we shall be only concerned with networks that are linear and time invariant. Another advantage is that the system of first-order differential equations obtained from the state-variable approach lends itself readily to programming for numerical solution on a digital computer.

In physical terms the state variables specify the energy stored in a set of independent energy storage elements, so that knowledge of this set of variables implies knowledge of all the energy stored in the network. It is thus natural in electric networks to associate the state variables with currents through inductors and voltages across capacitors. Since the behaviour of an inductor or a capacitor is governed by a first-order differential equation, it follows that we may describe the dynamic behaviour of a linear, time-invariant network by a set of first-order equations expressed in matrix form:

$$\frac{d\mathbf{x}}{dt} = \mathbf{A}\mathbf{x} + \mathbf{B}\mathbf{e}, \tag{2.57}$$

where $\mathbf{x}$ is a column vector representing the state variables; $\mathbf{A}$ and $\mathbf{B}$ are matrices of constant coefficients; and $\mathbf{e}$ is a column vector representing the inputs.

Example 2.4. *A simple RLC network.* As an illustrative example, consider the network shown in Fig. 2.15. Choosing the inductor current $i(t)$ and capacitor voltage $v(t)$ as the state variables, and applying Kirchhoff's voltage law and Kirchhoff's current law to express the inductor voltage $L_2(di/dt)$ and capacitor current $C_3(dv/dt)$ in terms of the state variables and the input $v_s(t)$, we obtain by inspection,

$$L_2 \frac{di}{dt} = v - R_4 i,$$

$$C_3 \frac{dv}{dt} = \frac{1}{R_1}(v_s - v) - i,$$

or after rearrangement,

$$
\begin{aligned}
\frac{di}{dt} &= -\frac{R_4}{L_2} i + \frac{1}{L_2} v, \\
\frac{dv}{dt} &= -\frac{1}{C_3} i - \frac{1}{R_1 C_3} v + \frac{1}{R_1 C_3} v_s.
\end{aligned}
\tag{2.58}
$$

These equations may be written in matrix form as

$$
\begin{bmatrix} \dfrac{di}{dt} \\ \\ \dfrac{dv}{dt} \end{bmatrix} = \begin{bmatrix} -\dfrac{R_4}{L_2} & \dfrac{1}{L_2} \\ \\ -\dfrac{1}{C_3} & -\dfrac{1}{R_1 C_3} \end{bmatrix} \begin{bmatrix} i \\ \\ v \end{bmatrix} + \begin{bmatrix} 0 \\ \\ \dfrac{1}{R_1 C_3} \end{bmatrix} v_s. \tag{2.59}
$$

For the network of Fig. 2.15, we thus have,

$$
\begin{aligned}
\mathbf{x} &= \begin{bmatrix} i \\ v \end{bmatrix} \\
\mathbf{A} &= \begin{bmatrix} -\dfrac{R_4}{L_2} & \dfrac{1}{L_2} \\ -\dfrac{1}{C_3} & -\dfrac{1}{R_1 C_3} \end{bmatrix} \\
\mathbf{B} &= \begin{bmatrix} 0 \\ \dfrac{1}{R_1 C_3} \end{bmatrix} \\
\mathbf{e} &= [v_s].
\end{aligned}
\tag{2.60}
$$

Topological Considerations

The formulation of the state equations for the network of Fig. 2.15 was quite simple and straightforward. However, in order to deal with networks of any degree of complexity we need a more systematic procedure for the general formulation of the state equations. The development of such a procedure, as with the loop or nodal analysis, is based on the topology of the network under study.

Consider a connected network that is made up entirely of linear, time-invariant resistors, self inductors and capacitors (the case of networks containing controlled sources will be considered later). The first step in the state equation formulation procedure is to choose a suitable tree for the network. In the special case of networks that do not contain capacitor-only loops and inductor-only cut-

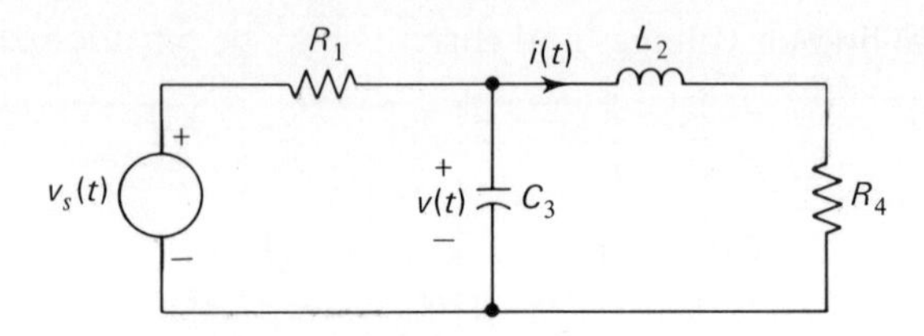

Fig. 2.15. A simple *RLC* network.

sets, it is convenient to choose a tree which contains every capacitor of the network together with resistors, if necessary, to complete the tree. Such a tree is called, after Bashkow,[6] a *proper tree.* All the inductors of the network together with those resistors that are not involved in the construction of a proper tree constitute link elements.

In general, however, an *RLC* network may contain capacitor-only loops and inductor-only cut-sets. Clearly, in such networks it is not possible to construct a proper tree because in a loop containing only capacitors at least one capacitor must be excluded from the tree branches and included in the links, while in a cut-set containing only inductors at least one inductor must be excluded from the links and included in the tree branches. Any capacitors and inductors which prevent the construction of a proper tree will be called *excess.* When a network contains excess elements, we may choose a modified proper tree called the *normal tree* as proposed by Bryant.[7,8] This tree is constructed by designating the capacitive excess elements as links and the inductive excess elements as tree branches. In other words, a normal tree includes the maximum number of capacitors and the minimum number of inductors possible.

In this section we shall follow Bryant's procedure for the general formulation of the state equations. We define a branch of the network to be a single element: a resistor, a self inductor or a capacitor. It is assumed that the network contains a total number of b such branches, n nodes and l fundamental loops. For a chosen normal tree we number the network branches as follows: we number the capacitive links first, the resistive links next, and then the inductive links. We turn now to the tree itself by numbering the capacitive tree branches followed by resistive tree branches, followed by inductive tree branches. The network branches are thus partitioned into six mutually exclusive types α, β, γ, δ, ε and ζ as follows:

Type	
α	capacitive links
β	resistive links
γ	inductive links
δ	capacitive tree branches
ε	resistive tree branches
ζ	inductive tree branches.

Accordingly, the branch voltages and currents may be partitioned as*

$$\mathbf{v} = \begin{bmatrix} \mathbf{v}_\alpha \\ \mathbf{v}_\beta \\ \mathbf{v}_\gamma \\ \cdots \\ \mathbf{v}_\delta \\ \mathbf{v}_\varepsilon \\ \mathbf{v}_\zeta \end{bmatrix} \tag{2.61}$$

and

$$\mathbf{i} = \begin{bmatrix} \mathbf{i}_\alpha \\ \mathbf{i}_\beta \\ \mathbf{i}_\gamma \\ \cdots \\ \mathbf{i}_\delta \\ \mathbf{i}_\varepsilon \\ \mathbf{i}_\zeta \end{bmatrix} \tag{2.62}$$

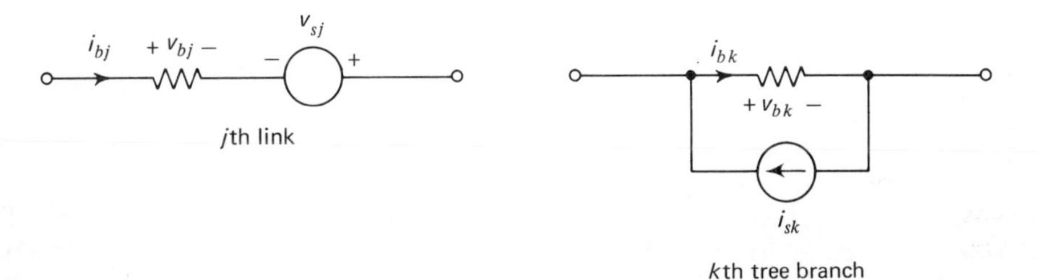

Fig. 2.16. Illustrating an active link and active tree-branch.

For convenience in analysis, we shall assume that the independent sources located in links are voltage sources and the independent sources located in tree branches are current sources, as illustrated in Fig. 2.16. Thus, if $\mathscr{B}$ denotes the fundamental loop matrix for the particular tree chosen, Kirchhoff's voltage law applied to every fundamental loop yields

$$\mathscr{B}\mathbf{v} = \begin{bmatrix} \mathbf{v}_{s\alpha} \\ \mathbf{v}_{s\beta} \\ \mathbf{v}_{s\gamma} \end{bmatrix},$$

* Since the state equations are formulated in the time domain, lower case symbols will be used for voltages and currents throughout this section.

or

$$[\mathbf{1}_l \vdots \mathbf{F}] \begin{bmatrix} \mathbf{v}_\alpha \\ \mathbf{v}_\beta \\ \mathbf{v}_\gamma \\ \cdots \\ \mathbf{v}_\delta \\ \mathbf{v}_\varepsilon \\ \mathbf{v}_\zeta \end{bmatrix} = \begin{bmatrix} \mathbf{v}_{s\alpha} \\ \mathbf{v}_{s\beta} \\ \mathbf{v}_{s\gamma} \end{bmatrix}, \tag{2.63}$$

where $\mathbf{1}_l$ is an $l \times l$ unit matrix; $\mathbf{F}$ is a submatrix expressing the topological relation between links and tree branches; $\mathbf{v}_{s\alpha}$, $\mathbf{v}_{s\beta}$, and $\mathbf{v}_{s\gamma}$ are column vectors representing the total loop voltage sources in the fundamental loops defined by capacitive, resistive and inductive links, respectively. Similarly, if $\mathscr{Q}$ denotes the fundamental cut-set matrix for the chosen tree then Kirchhoff's current law applied to every node yields

$$\mathscr{Q}\mathbf{i} = \begin{bmatrix} \mathbf{i}_{s\delta} \\ \mathbf{i}_{s\varepsilon} \\ \mathbf{i}_{s\zeta} \end{bmatrix}$$

or

$$[-\mathbf{F}^t \vdots \mathbf{1}_m] \begin{bmatrix} \mathbf{i}_\alpha \\ \mathbf{i}_\beta \\ \mathbf{i}_\gamma \\ \cdots \\ \mathbf{i}_\delta \\ \mathbf{i}_\varepsilon \\ \mathbf{i}_\zeta \end{bmatrix} = \begin{bmatrix} \mathbf{i}_{s\delta} \\ \mathbf{i}_{s\varepsilon} \\ \mathbf{i}_{s\zeta} \end{bmatrix}, \tag{2.64}$$

where $\mathbf{1}_m$ is an $m \times m$ unit matrix (with $m = n - 1$); $\mathbf{F}^t$ is the transpose of the matrix $\mathbf{F}$ appearing in Eq. (2.63); $\mathbf{i}_{s\delta}$, $\mathbf{i}_{s\varepsilon}$ and $\mathbf{i}_{s\zeta}$ are column vectors representing the total current sources in the fundamental cut-sets defined by capacitive, resistive and inductive tree branches, respectively.

Equations (2.63) and (2.64) may be re-written as follows, respectively,

$$\begin{bmatrix} \mathbf{v}_\alpha \\ \mathbf{v}_\beta \\ \mathbf{v}_\gamma \end{bmatrix} + \mathbf{F} \begin{bmatrix} \mathbf{v}_\delta \\ \mathbf{v}_\varepsilon \\ \mathbf{v}_\zeta \end{bmatrix} = \begin{bmatrix} \mathbf{v}_{s\alpha} \\ \mathbf{v}_{s\beta} \\ \mathbf{v}_{s\gamma} \end{bmatrix} \tag{2.65}$$

and

$$-\mathbf{F}^t \begin{bmatrix} \mathbf{i}_\alpha \\ \mathbf{i}_\beta \\ \mathbf{i}_\gamma \end{bmatrix} + \begin{bmatrix} \mathbf{i}_\delta \\ \mathbf{i}_\varepsilon \\ \mathbf{i}_\zeta \end{bmatrix} = \begin{bmatrix} \mathbf{i}_{s\delta} \\ \mathbf{i}_{s\varepsilon} \\ \mathbf{i}_{s\zeta} \end{bmatrix}, \tag{2.66}$$

which can be combined to give

$$\begin{bmatrix} \mathbf{v}_\alpha \\ \mathbf{v}_\beta \\ \mathbf{v}_\gamma \\ \cdots \\ \mathbf{i}_\delta \\ \mathbf{i}_\varepsilon \\ \mathbf{i}_\zeta \end{bmatrix} = \begin{bmatrix} \mathbf{0} & \vdots & -\mathbf{F} \\ \cdots & \cdots & \cdots \\ \mathbf{F}^t & \vdots & \mathbf{0} \end{bmatrix} \begin{bmatrix} \mathbf{i}_\alpha \\ \mathbf{i}_\beta \\ \mathbf{i}_\gamma \\ \cdots \\ \mathbf{v}_\delta \\ \mathbf{v}_\varepsilon \\ \mathbf{v}_\zeta \end{bmatrix} + \begin{bmatrix} \mathbf{v}_{s\alpha} \\ \mathbf{v}_{s\beta} \\ \mathbf{v}_{s\gamma} \\ \cdots \\ \mathbf{i}_{s\delta} \\ \mathbf{i}_{s\varepsilon} \\ \mathbf{i}_{s\zeta} \end{bmatrix}. \tag{2.67}$$

In this equation the link voltages are expressed in terms of the independent voltage sources and the tree branch voltages, which are the independent variables used in the nodal analysis; the tree branch currents are expressed in terms of the independent current sources and the link currents, which are the independent variables used in the loop analysis.

It is convenient to partition the matrix **F** according to the nature of the network elements; we may thus write

$$\mathbf{F} = \begin{bmatrix} \mathbf{F}_{\alpha\delta} & \mathbf{F}_{\alpha\varepsilon} & \mathbf{F}_{\alpha\zeta} \\ \mathbf{F}_{\beta\delta} & \mathbf{F}_{\beta\varepsilon} & \mathbf{F}_{\beta\zeta} \\ \mathbf{F}_{\gamma\delta} & \mathbf{F}_{\gamma\varepsilon} & \mathbf{F}_{\gamma\zeta} \end{bmatrix} \tag{2.68}$$

where the matrix $\mathbf{F}_{\alpha\delta}$ expresses the topological relation between capacitive links and capacitive tree branches, and, similarly, $\mathbf{F}_{\beta\delta}$ expresses the topological relation between resistive links and capacitive tree branches, etc. However, from the method of constructing the normal tree the loops defined by the capacitive links must be purely capacitive, so that

$$\mathbf{F}_{\alpha\varepsilon} = \mathbf{0}$$

$$\mathbf{F}_{\alpha\zeta} = \mathbf{0}.$$

Also, the loops defined by the resistive links can contain no inductors, so that

$$\mathbf{F}_{\beta\zeta} = \mathbf{0}.$$

Hence,

$$\mathbf{F} = \begin{bmatrix} \mathbf{F}_{\alpha\delta} & \mathbf{0} & \mathbf{0} \\ \mathbf{F}_{\beta\delta} & \mathbf{F}_{\beta\varepsilon} & \mathbf{0} \\ \mathbf{F}_{\gamma\delta} & \mathbf{F}_{\gamma\varepsilon} & \mathbf{F}_{\gamma\zeta} \end{bmatrix}. \tag{2.69}$$

With this partitioning of $\mathbf{F}$, we find that Eq. (2.67) yields:

$$\begin{aligned}
\mathbf{v}_\alpha &= -\mathbf{F}_{\alpha\delta}\mathbf{v}_\delta + \mathbf{v}_{s\alpha} \\
\mathbf{v}_\beta &= -\mathbf{F}_{\beta\delta}\mathbf{v}_\delta - \mathbf{F}_{\beta\varepsilon}\mathbf{v}_\varepsilon + \mathbf{v}_{s\beta} \\
\mathbf{v}_\gamma &= -\mathbf{F}_{\gamma\delta}\mathbf{v}_\delta - \mathbf{F}_{\gamma\varepsilon}\mathbf{v}_\varepsilon - \mathbf{F}_{\gamma\zeta}\mathbf{v}_\zeta + \mathbf{v}_{s\zeta} \\
\mathbf{i}_\delta &= \mathbf{F}^t_{\alpha\delta}\mathbf{i}_\alpha + \mathbf{F}^t_{\beta\delta}\mathbf{i}_\beta + \mathbf{F}^t_{\gamma\delta}\mathbf{i}_\gamma + \mathbf{i}_{s\delta} \\
\mathbf{i}_\varepsilon &= \mathbf{F}^t_{\beta\varepsilon}\mathbf{i}_\beta + \mathbf{F}^t_{\gamma\varepsilon}\mathbf{i}_\gamma + \mathbf{i}_{s\varepsilon} \\
\mathbf{i}_\zeta &= \mathbf{F}^t_{\gamma\zeta}\mathbf{i}_\gamma + \mathbf{i}_{s\zeta}.
\end{aligned} \tag{2.70}$$

The voltage–current relations for the six types of network branches are defined by

$$\begin{aligned}
\mathbf{i}_\alpha &= \mathbf{C}_{\alpha\alpha}\frac{d\mathbf{v}_\alpha}{dt} \\
\mathbf{v}_\beta &= \mathbf{R}_{\beta\beta}\mathbf{i}_\beta \\
\mathbf{v}_\gamma &= \mathbf{L}_{\gamma\gamma}\frac{d\mathbf{i}_\gamma}{dt} \\
\mathbf{i}_\delta &= \mathbf{C}_{\delta\delta}\frac{d\mathbf{v}_\delta}{dt} \\
\mathbf{v}_\varepsilon &= \mathbf{R}_{\varepsilon\varepsilon}\mathbf{i}_\varepsilon \\
\mathbf{v}_\zeta &= \mathbf{L}_{\zeta\zeta}\frac{d\mathbf{i}_\zeta}{dt},
\end{aligned} \tag{2.71}$$

where $\mathbf{C}_{\alpha\alpha}$, $\mathbf{R}_{\beta\beta}$ and $\mathbf{L}_{\gamma\gamma}$ denote the link-capacitance, resistance and inductance matrices, respectively, while $\mathbf{C}_{\delta\delta}$, $\mathbf{R}_{\varepsilon\varepsilon}$ and $\mathbf{L}_{\zeta\zeta}$ denote the tree-branch capacitance, resistance and inductance matrices, respectively; all of these matrices are diagonal. It is assumed that the network contains no mutual inductance.

In the network under consideration, the inductor currents $\mathbf{i}_\gamma$ and capacitor voltages $\mathbf{v}_\delta$ constitute the desired minimal set of state variables, representing independently specifiable initial conditions. We may thus put

$$\mathbf{x} = \begin{bmatrix} \mathbf{i}_\gamma \\ \mathbf{v}_\delta \end{bmatrix}. \tag{2.72}$$

If, therefore, we eliminate the non-state variables that appear in Eqs. (2.70) and (2.71) by a straightforward algebraic process we finally obtain the normal form for the state equations*

$$\frac{d\mathbf{x}}{dt} = \mathbf{A}\mathbf{x} + \mathbf{B}\mathbf{e},$$

with

$$\mathbf{A} = \begin{bmatrix} \mathscr{L} & \mathbf{0} \\ \mathbf{0} & \mathscr{C} \end{bmatrix}^{-1} \begin{bmatrix} -\mathbf{F}_{\gamma\varepsilon}\mathscr{G}^{-1}\mathbf{F}^t_{\gamma\varepsilon} & -\mathbf{F}_{\gamma\delta} + \mathbf{F}_{\gamma\varepsilon}\mathbf{R}_{\varepsilon\varepsilon}\mathbf{F}^t_{\beta\varepsilon}\mathscr{R}^{-1}\mathbf{F}_{\beta\delta} \\ \mathbf{F}^t_{\gamma\delta} - \mathbf{F}^t_{\beta\delta}\mathbf{R}^{-1}_{\beta\beta}\mathbf{F}_{\beta\varepsilon}\mathscr{G}^{-1}\mathbf{F}^t_{\gamma\varepsilon} & -\mathbf{F}^t_{\beta\delta}\mathscr{R}^{-1}\mathbf{F}_{\beta\delta} \end{bmatrix}, \tag{2.73}$$

$$\mathbf{B} = \begin{bmatrix} \mathscr{L} & \mathbf{0} \\ \mathbf{0} & \mathscr{C} \end{bmatrix}^{-1}$$

$$\times \begin{bmatrix} \mathbf{0} & -\mathbf{F}_{\gamma\varepsilon}\mathbf{R}_{\varepsilon\varepsilon}\mathbf{F}^t_{\beta\varepsilon}\mathscr{R}^{-1} & \mathbf{1} & \mathbf{0} & -\mathbf{F}_{\gamma\varepsilon}\mathscr{G}^{-1} & -\mathbf{F}_{\gamma\zeta}\mathbf{L}_{\zeta\zeta} \\ \mathbf{F}^t_{\alpha\delta}\mathbf{C}_{\alpha\alpha} & \mathbf{F}^t_{\beta\delta}\mathscr{R}^{-1} & \mathbf{0} & \mathbf{1} & \mathbf{F}^t_{\beta\delta}\mathbf{R}^{-1}_{\beta\beta}\mathbf{F}_{\beta\varepsilon}\mathscr{G}^{-1} & \mathbf{0} \end{bmatrix} \tag{2.74}$$

and

$$\mathbf{e} = \begin{bmatrix} d\mathbf{v}_{s\alpha}/dt \\ \mathbf{v}_{s\beta} \\ \mathbf{v}_{s\gamma} \\ \mathbf{i}_{s\delta} \\ \mathbf{i}_{s\varepsilon} \\ d\mathbf{i}_{s\gamma}/dt \end{bmatrix}, \tag{2.75}$$

where

$$\begin{aligned} \mathscr{L} &= \mathbf{L}_{\gamma\gamma} + \mathbf{F}_{\gamma\zeta}\mathbf{L}_{\zeta\zeta}\mathbf{F}^t_{\gamma\zeta} \\ \mathscr{C} &= \mathbf{C}_{\delta\delta} + \mathbf{F}^t_{\alpha\delta}\mathbf{C}_{\alpha\alpha}\mathbf{F}_{\alpha\delta} \\ \mathscr{G} &= \mathbf{R}^{-1}_{\varepsilon\varepsilon} + \mathbf{F}^t_{\beta\varepsilon}\mathbf{R}^{-1}_{\beta\beta}\mathbf{F}_{\beta\varepsilon} \\ \mathscr{R} &= \mathbf{R}_{\beta\beta} + \mathbf{F}_{\beta\varepsilon}\mathbf{R}_{\varepsilon\varepsilon}\mathbf{F}^t_{\beta\varepsilon}. \end{aligned} \tag{2.76}$$

---

* Equations (2.73) to (2.76) may also be used to set up the state equations of a network containing mutual inductance, except that the matrix $\mathscr{L}$ is defined by

$$\mathscr{L} = [\mathbf{1} \quad \mathbf{F}_{\gamma\zeta}] \begin{bmatrix} \mathbf{L}_{\gamma\gamma} & \mathbf{L}_{\gamma\zeta} \\ \mathbf{L}_{\zeta\gamma} & \mathbf{L}_{\zeta\zeta} \end{bmatrix} \begin{bmatrix} \mathbf{1} \\ \mathbf{F}^t_{\gamma\zeta} \end{bmatrix}$$

where the matrix

$$\begin{bmatrix} \mathbf{L}_{\gamma\gamma} & \mathbf{L}_{\gamma\zeta} \\ \mathbf{L}_{\zeta\gamma} & \mathbf{L}_{\zeta\zeta} \end{bmatrix}$$

represents the matrix of self and mutual inductances of the network. For a proof, see P. R. Bryant, "The Explicit Form of Bashkow's *A*-matrix", *Trans. I.R.E.*, **CT-9**, 303 (1962).

The matrix $\mathscr{L}$ is the loop-inductance matrix for those fundamental loops defined by inductive links; $\mathscr{C}$ is the cut-set capacitance matrix for those fundamental cut-sets defined by capacitive tree branches; $\mathscr{G}$ is the cut-set conductance matrix for the fundamental cut-sets defined by resistive tree branches; and $\mathscr{R}$ is the loop-resistance matrix for the fundamental loops defined by resistive links.

Two Special Cases

Case 1. $\mathbf{F}_{\alpha\delta} = \mathbf{0}$ and $\mathbf{F}_{\gamma\zeta} = \mathbf{0}$.

When an *RLC* network does not contain capacitor-only loops there will be no capacitive links, and $\mathbf{F}_{\alpha\delta} = \mathbf{0}$; similarly, when the network does not contain inductor-only cut-sets there will be no inductive tree branches, and $\mathbf{F}_{\gamma\zeta} = \mathbf{0}$. For such networks the normal tree reduces to a proper tree with

$$\mathbf{F} = \begin{bmatrix} \mathbf{F}_{\beta\delta} & \mathbf{F}_{\beta\varepsilon} \\ \mathbf{F}_{\gamma\delta} & \mathbf{F}_{\gamma\varepsilon} \end{bmatrix} \tag{2.77}$$

and so we may simplify Eqs. (2.73)–(2.75) as follows

$$\mathbf{A} = \begin{bmatrix} \mathbf{L}_{\gamma\gamma} & \mathbf{0} \\ \mathbf{0} & \mathbf{C}_{\delta\delta} \end{bmatrix}^{-1} \begin{bmatrix} -\mathbf{F}_{\gamma\varepsilon}\mathscr{G}^{-1}\mathbf{F}^t_{\gamma\varepsilon} & -\mathbf{F}_{\gamma\delta} + \mathbf{F}_{\gamma\varepsilon}\mathbf{R}_{\varepsilon\varepsilon}\mathbf{F}^t_{\beta\varepsilon}\mathscr{R}^{-1}\mathbf{F}_{\beta\delta} \\ \mathbf{F}^t_{\gamma\delta} - \mathbf{F}^t_{\beta\delta}\mathbf{R}^{-1}_{\beta\beta}\mathbf{F}_{\beta\varepsilon}\mathscr{G}^{-1}\mathbf{F}^t_{\gamma\varepsilon} & -\mathbf{F}^t_{\beta\delta}\mathscr{R}^{-1}\mathbf{F}_{\beta\delta} \end{bmatrix} \tag{2.78}$$

$$\mathbf{B} = \begin{bmatrix} \mathbf{L}_{\gamma\gamma} & \mathbf{0} \\ \mathbf{0} & \mathbf{C}_{\delta\delta} \end{bmatrix}^{-1} \begin{bmatrix} -\mathbf{F}_{\gamma\varepsilon}\mathbf{R}_{\varepsilon\varepsilon}\mathbf{F}^t_{\beta\varepsilon}\mathscr{R}^{-1} & \mathbf{1} & \mathbf{0} & -\mathbf{F}_{\gamma\varepsilon}\mathscr{G}^{-1} \\ \mathbf{F}^t_{\beta\delta}\mathscr{R}^{-1} & \mathbf{0} & \mathbf{1} & \mathbf{F}^t_{\beta\delta}\mathbf{R}^{-1}_{\beta\beta}\mathbf{F}_{\beta\varepsilon}\mathscr{G}^{-1} \end{bmatrix} \tag{2.79}$$

$$\mathbf{e} = \begin{bmatrix} \mathbf{v}_{s\beta} \\ \mathbf{v}_{s\gamma} \\ \mathbf{i}_{s\delta} \\ \mathbf{i}_{s\varepsilon} \end{bmatrix}, \tag{2.80}$$

where the matrices $\mathscr{G}$ and $\mathscr{R}$ are defined by Eqs. (2.76) as before.

Case 2. $\mathbf{F}_{\beta\varepsilon} = \mathbf{0}$ (in addition to $\mathbf{F}_{\alpha\delta} = \mathbf{0}$ and $\mathbf{F}_{\gamma\zeta} = \mathbf{0}$).

Consideration of the way in which $+1$, and -1 and 0 enter $\mathbf{F}_{\beta\varepsilon}$ will show that an entry of ± 1 can only arise in $\mathbf{F}_{\beta\varepsilon}$ if the loop formed when a resistive link is inserted into the chosen tree contains a resistive tree branch. This means that $\mathbf{F}_{\beta\varepsilon}$ will be zero if a network formed from the original network by replacing all the capacitors by short-circuits contains no resistor-only loops. Thus for this class of networks $\mathscr{G} = \mathbf{R}^{-1}_{\varepsilon\varepsilon}$ and $\mathscr{R} = \mathbf{R}_{\beta\beta}$, and we may simplify Eqs. (2.78) and (2.79) further as follows[9]

$$\mathbf{A} = \begin{bmatrix} \mathbf{L}_{\gamma\gamma} & \mathbf{0} \\ \mathbf{0} & \mathbf{C}_{\delta\delta} \end{bmatrix}^{-1} \begin{bmatrix} -\mathbf{F}_{\gamma\varepsilon}\mathbf{R}_{\varepsilon\varepsilon}\mathbf{F}^t_{\gamma\varepsilon} & -\mathbf{F}_{\gamma\delta} \\ \mathbf{F}^t_{\gamma\delta} & -\mathbf{F}^t_{\beta\delta}\mathbf{R}^{-1}_{\beta\beta}\mathbf{F}_{\beta\delta} \end{bmatrix}, \tag{2.81}$$

$$\mathbf{B} = \begin{bmatrix} \mathbf{L}_{\gamma\gamma} & \mathbf{0} \\ \mathbf{0} & \mathbf{C}_{\delta\delta} \end{bmatrix}^{-1} \begin{bmatrix} \mathbf{0} & \mathbf{1} & \mathbf{0} & -\mathbf{F}_{\gamma\varepsilon}\mathbf{R}_{\varepsilon\varepsilon} \\ \mathbf{F}^t_{\beta\delta}\mathbf{R}^{-1}_{\beta\beta} & \mathbf{0} & \mathbf{1} & \mathbf{0} \end{bmatrix}. \tag{2.82}$$

The input vector **e** is defined by Eq. (2.80) as in the previous case.

Example 2.5. *Resistively terminated LC ladder network.* Consider Fig. 2.17(a) showing a four-element LC ladder network that is resistively terminated at both ports. This network has no excess elements. We may therefore choose the proper tree shown in Fig. 2.17(b). With the network branches numbered as previously described, the fundamental loop matrix corresponding to this tree is

$$\mathscr{B} = \left[\begin{array}{ccc:ccc} +1 & 0 & 0 & +1 & 0 & 0 \\ 0 & +1 & 0 & -1 & +1 & 0 \\ 0 & 0 & +1 & 0 & -1 & +1 \end{array}\right]. \tag{2.83}$$

Hence,

$$\mathbf{F} = \left[\begin{array}{cc:c} +1 & 0 & 0 \\ \hdashline -1 & +1 & 0 \\ 0 & -1 & +1 \end{array}\right], \tag{2.84}$$

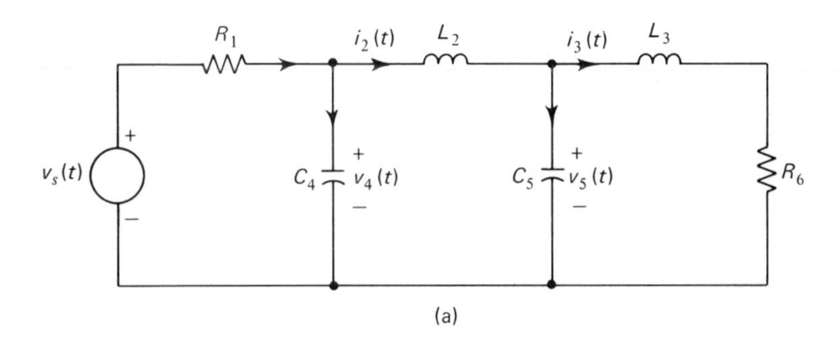

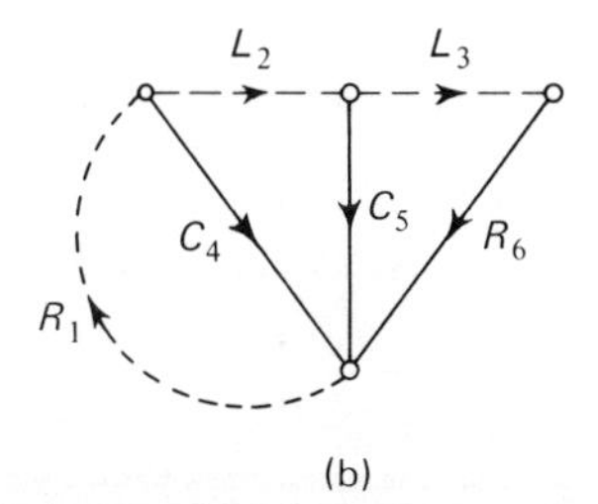

Fig. 2.17. (a) An RLC network; (b) A proper tree of the network.

where the partitioning shows the submatrices of **F**, as in Eq. (2.77). Thus,

$$
\begin{aligned}
\mathbf{F}_{\beta\delta} &= [+1 \quad 0] \\
\mathbf{F}_{\gamma\delta} &= \begin{bmatrix} -1 & +1 \\ 0 & -1 \end{bmatrix} \\
\mathbf{F}_{\beta\varepsilon} &= [0] \\
\mathbf{F}_{\gamma\varepsilon} &= \begin{bmatrix} 0 \\ +1 \end{bmatrix}.
\end{aligned}
\tag{2.85}
$$

The branch element matrices are

$$
\begin{aligned}
\mathbf{R}_{\beta\beta} &= [R_1] \\
\mathbf{L}_{\gamma\gamma} &= \begin{bmatrix} L_2 & 0 \\ 0 & L_3 \end{bmatrix} \\
\mathbf{C}_{\delta\delta} &= \begin{bmatrix} C_4 & 0 \\ 0 & C_5 \end{bmatrix} \\
\mathbf{R}_{\varepsilon\varepsilon} &= [R_6].
\end{aligned}
\tag{2.86}
$$

Since $\mathbf{F}_{\beta\varepsilon}$ is zero, Eqs. (2.81) and (2.82) are applicable to the network under consideration; hence,

$$
\mathbf{A} = \begin{bmatrix} 0 & 0 & L_2^{-1} & -L_2^{-1} \\ 0 & -R_6L_3^{-1} & 0 & L_3^{-1} \\ -C_4^{-1} & 0 & -C_4^{-1}R_1^{-1} & 0 \\ C_5^{-1} & -C_5^{-1} & 0 & 0 \end{bmatrix}, \tag{2.87}
$$

$$
\mathbf{B} = \left[\begin{array}{c:cc:cc:c} 0 & L_2^{-1} & 0 & 0 & 0 & 0 \\ 0 & 0 & L_3^{-1} & 0 & 0 & -R_6L_3^{-1} \\ \hdashline C_4^{-1}R_1^{-1} & 0 & 0 & C_4^{-1} & 0 & 0 \\ 0 & 0 & 0 & 0 & C_5^{-1} & 0 \end{array}\right]. \tag{2.88}
$$

From Eq. (2.80) the input vector is

$$
\mathbf{e} = \begin{bmatrix} v_s \\ \cdots \\ 0 \\ 0 \\ \cdots \\ 0 \\ 0 \\ \cdots \\ 0 \end{bmatrix}. \tag{2.89}
$$

The state equations for the network of Fig. 2.17 are thus given by

$$\frac{d}{dt}\begin{bmatrix} i_2 \\ i_3 \\ v_4 \\ v_5 \end{bmatrix} = \begin{bmatrix} 0 & 0 & L_2^{-1} & -L_2^{-1} \\ 0 & -R_6L_3^{-1} & 0 & L_3^{-1} \\ -C_4^{-1} & 0 & -C_4^{-1}R_1^{-1} & 0 \\ C_5^{-1} & -C_5^{-1} & 0 & 0 \end{bmatrix}\begin{bmatrix} i_2 \\ i_3 \\ v_4 \\ v_5 \end{bmatrix} + \begin{bmatrix} 0 \\ 0 \\ C_4^{-1}R_1^{-1} \\ 0 \end{bmatrix} v_s \tag{2.90}$$

where i_2 and i_3 are the currents through inductors L_2 and L_3, while v_4 and v_5 are the voltages across capacitors C_4 and C_5.

Effect of Controlled Sources

For networks containing controlled sources there are no general explicit formulas defining the state equations as there are for linear, time-invariant *RLC* networks. However, depending upon the type of the controlled source, we may follow certain procedures for formulating the state equations.[8,9] We shall limit ourselves to networks where controlled voltage sources are connected directly in series with link elements or where controlled current sources are connected directly across tree-branch elements. Also, it will be assumed that the control parameters are real quantities. Then, as with the loop or nodal analysis, the formulation is carried out in two steps:

1. The controlled sources are imagined to be independent sources and the state equations are derived as previously described.
2. The control signals are expressed as linear combinations of the state variables. Thus, using the constraints imposed on the controlled sources (one for each source) the effects upon the state variables in question are determined.

This procedure is best illustrated by means of an example.

Example 2.6. *An RC amplifier.* Figure 2.18(a) shows an *RC* network with a voltage-controlled current source $g_m v_1$ and three capacitors C_1, C_4, C_5 which form a loop. This network may represent, for example, the circuit model of a vacuum-tube triode amplifier with all interelectrode capacitances included. The normal tree, as shown in Fig. 2.18(b), excludes the capacitance C_1. The fundamental loop matrix for this choice of tree is therefore

$$\mathscr{B} = \left[\begin{array}{ccc:cc} +1 & 0 & 0 & -1 & +1 \\ 0 & +1 & 0 & +1 & 0 \\ 0 & 0 & +1 & 0 & -1 \end{array}\right]. \tag{2.91}$$

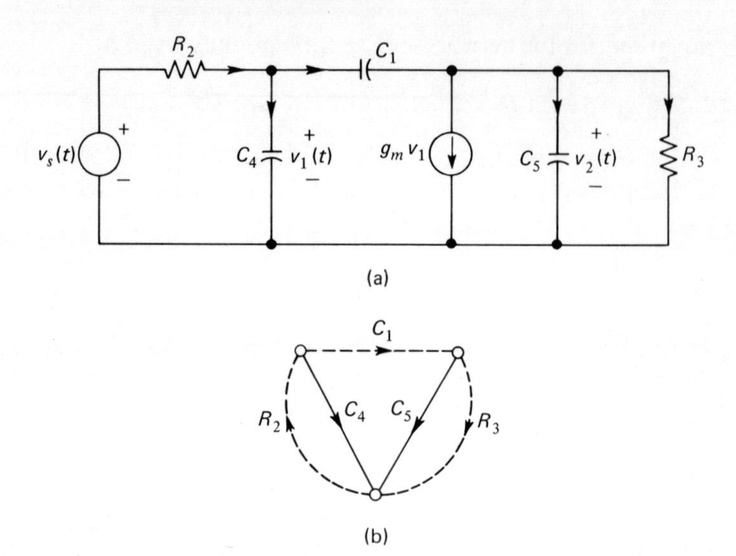

Fig. 2.18. (a) An *RC* amplifier; (b) A normal tree of the network.

Hence,

$$\mathbf{F} = \begin{bmatrix} -1 & +1 \\ \hdashline +1 & 0 \\ 0 & -1 \end{bmatrix}, \tag{2.92}$$

from which we obtain

$$\mathbf{F}_{\alpha\delta} = [-1 \quad +1]$$

$$\mathbf{F}_{\beta\delta} = \begin{bmatrix} +1 & 0 \\ 0 & -1 \end{bmatrix}. \tag{2.93}$$

The branch element matrices are

$$\mathbf{C}_{\alpha\alpha} = [C_1]$$

$$\mathbf{R}_{\beta\beta} = \begin{bmatrix} R_2 & 0 \\ 0 & R_3 \end{bmatrix} \tag{2.94}$$

$$\mathbf{C}_{\delta\delta} = \begin{bmatrix} C_4 & 0 \\ 0 & C_5 \end{bmatrix}.$$

From Eqs. (2.76),

$$\mathscr{C} = \begin{bmatrix} C_1 + C_4 & -C_1 \\ -C_1 & C_1 + C_5 \end{bmatrix}$$

$$\mathscr{R} = \begin{bmatrix} R_2 & 0 \\ 0 & R_3 \end{bmatrix}. \tag{2.95}$$

Use of Eqs. (2.73) and (2.74) yields

$$\mathbf{A} = -\mathscr{C}^{-1}\mathbf{F}^t_{\beta\delta}\mathscr{R}^{-1}\mathbf{F}_{\beta\delta} = -\frac{1}{\Delta_C}\begin{bmatrix}(C_1 + C_5)/R_2 & C_1/R_3 \\ C_1/R_2 & (C_1 + C_4)/R_3\end{bmatrix} \tag{2.96}$$

$$\mathbf{B} = \mathscr{C}^{-1}[\mathbf{F}^t_{\alpha\delta}\mathbf{C}_{\alpha\alpha} \quad \mathbf{F}^t_{\beta\delta}\mathscr{R}^{-1} \quad \mathbf{1}]$$

$$= \frac{1}{\Delta_C}\left[\begin{array}{c|cc|cc} -C_1C_5 & (C_1 + C_5)/R_2 & -C_1/R_3 & C_1 + C_5 & C_1 \\ C_1C_4 & C_1/R_2 & -(C_1 + C_4)/R_3 & C_1 & C_1 + C_4\end{array}\right], \tag{2.97}$$

where

$$\Delta_C = C_1C_4 + C_4C_5 + C_5C_1.$$

With the controlled source $g_m v_1$ treated as if it were independent, Eq. (2.75) gives

$$\mathbf{e} = \begin{bmatrix} 0 \\ \cdots \\ v_s \\ 0 \\ \cdots \\ 0 \\ -g_m v_1 \end{bmatrix}. \tag{2.98}$$

Thus, after collecting terms, the state equations are obtained as

$$\frac{d}{dt}\begin{bmatrix} v_1 \\ v_2 \end{bmatrix} = \frac{1}{\Delta_C}\begin{bmatrix} -g_mC_1 - (C_1 + C_5)/R_2 & -C_1/R_3 \\ -g_m(C_1 + C_4) - C_1/R_2 & -(C_1 + C_4)/R_3\end{bmatrix}\begin{bmatrix} v_1 \\ v_2 \end{bmatrix}$$
$$+ \begin{bmatrix} C_1 + C_5 \\ C_1 \end{bmatrix}\frac{v_s}{R_2\Delta_C}. \tag{2.99}$$

Concluding Remarks

To conclude, the state variable approach yields a set of first-order differential equations called the *normal form.* It proceeds in two stages:

1. selection of a set of state variables **x**, and
2. formulation of the state equations in terms of the variables **x**.

In order that these state equations completely define the behaviour of the network, it is necessary that the state variables **x** be an independent set. For a general *RLC* network containing capacitor-only loops and inductor-only cut-sets, this requirement is satisfied by constructing a normal tree that comprises the maximum number of permissible capacitors, then resistors, and finally, the minimum number of inductors. The voltages across the capacitive tree branches and the currents through the inductive links constitute the required set of state variables.

REFERENCES

1. P. R. BRYANT, "The Algebra and Topology of Electrical Networks," *Proc. I.E.E.*, **108,** Part C, 215 (1961).
2. S. SESHU and M. B. REED, *Linear Graphs and Electrical Networks*. Addison-Wesley, 1961.
3. A. G. BOSE and K. N. STEVENS, *Introductory Network Theory*, p. 268. Harper and Row, 1965.
4. E. J. ANGELO, *Electronic Circuits*, p. 413. McGraw-Hill, 1958.
5. R. D. THORNTON *et al.*, *Multistage Transistor Circuits*. Wiley, 1965.
6. T. R. BASHKOW, "The *A*-matrix, a New Network Description," *Trans. I.R.E.*, **CT-4,** 117 (1957).
7. P. R. BRYANT, "The Order of Complexity of Electrical Networks," *Proc. I.E.E.*, **106C,** 174 (1959).
8. E. S. KUH and R. A. ROHRER, "The State-variable Approach to Network Analysis," *Proc. I.E.E.E.*, **53,** 672 (1965).
9. A. G. J. MACFARLANE, "Notes on the Use of Matrix Theory in the Analysis of Linear Feedback Circuits," *Proc. I.E.E.*, **110,** 139 (1963).
10. H. J. REICH, *Functional Circuits and Oscillators*. Van Nostrand, 1961.

PROBLEMS

2.1 Figure P2.1 shows the circuit diagram of a compensated d.c. amplifier. Using the loop analysis, determine the output voltage developed at the plate of tube Q_1.

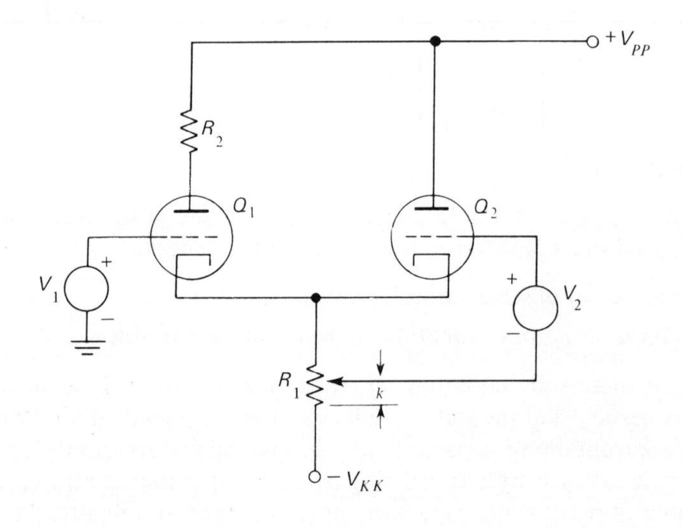

Fig. P2.1.

2.2 The network of Fig. P2.2 represents the model of a transistor operated with its collector terminal common to the input and output ports. Determine the voltage ratio V_{ec}/V_s.

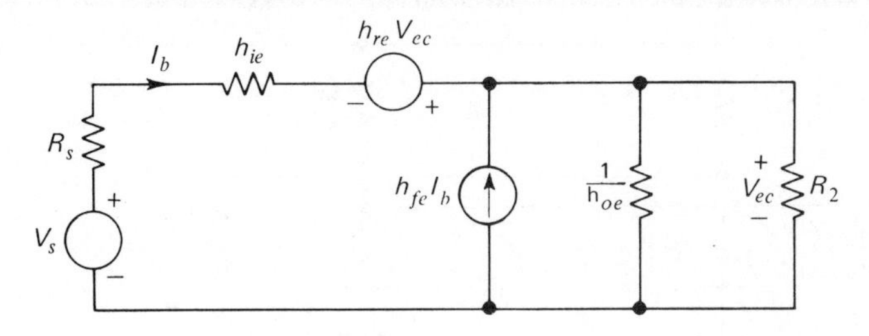

Fig. P2.2.

2.3 Using the nodal analysis, determine the transfer impedance V_2/I_s for the network of Fig. P2.3 which represents the hybrid-π model of a transistor operated with the base terminal common to the input and output ports.

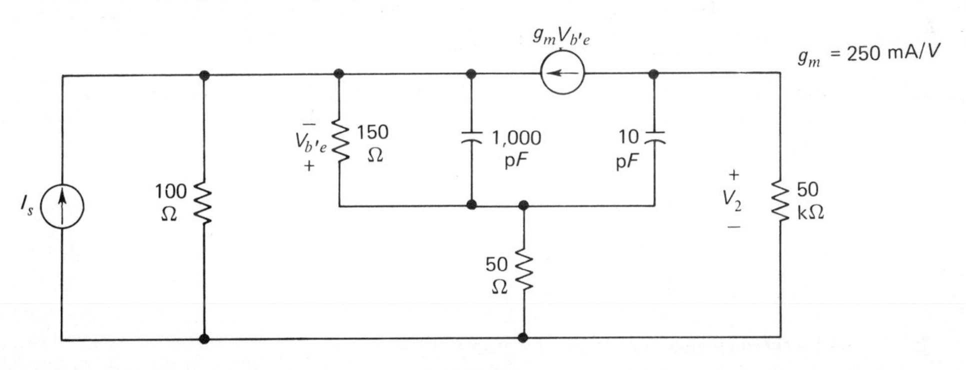

Fig. P2.3.

2.4 Develop the Thévenin and Norton equivalent networks for the circuit of Fig. P2.4 with respect to port 1–1′.

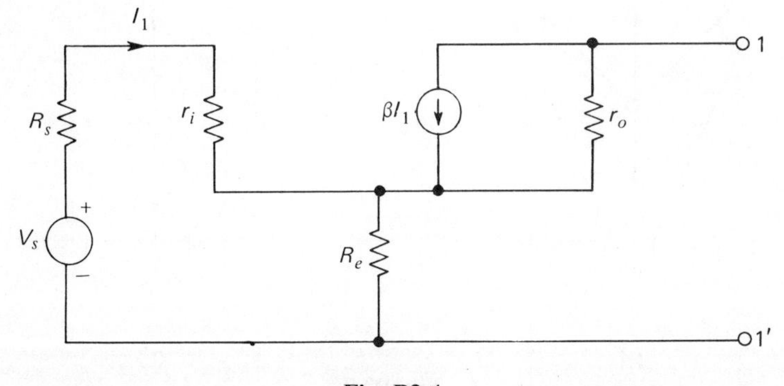

Fig. P2.4.

2.5 Show that the incremental resistance v/i measured between the two cathode terminals of Fig. P2.5 is negative.[10] Determine the magnitude of this resistance, given that the tubes are identical and have the incremental parameters $r_p = 10\,\text{k}\Omega$ and $\mu = 40$.

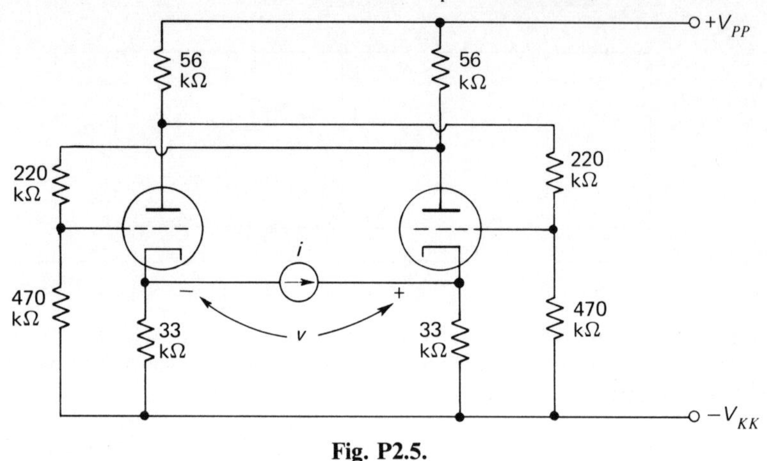

Fig. P2.5.

2.6 Construct the graph of a connected network having the following incidence matrix

$$\mathscr{A}_a = \begin{bmatrix} -1 & -1 & 0 & -1 & 0 & 0 & 0 \\ 0 & +1 & -1 & 0 & 0 & +1 & +1 \\ 0 & 0 & 0 & 0 & -1 & 0 & -1 \\ 0 & 0 & +1 & 0 & +1 & -1 & 0 \\ +1 & 0 & 0 & +1 & 0 & 0 & 0 \end{bmatrix}.$$

2.7 a) Set up the incidence matrix for the graph of Fig. P2.7(a).

b) Set up the fundamental loop matrix and fundamental cut-set matrix as defined by the tree of Fig. P2.7(b).

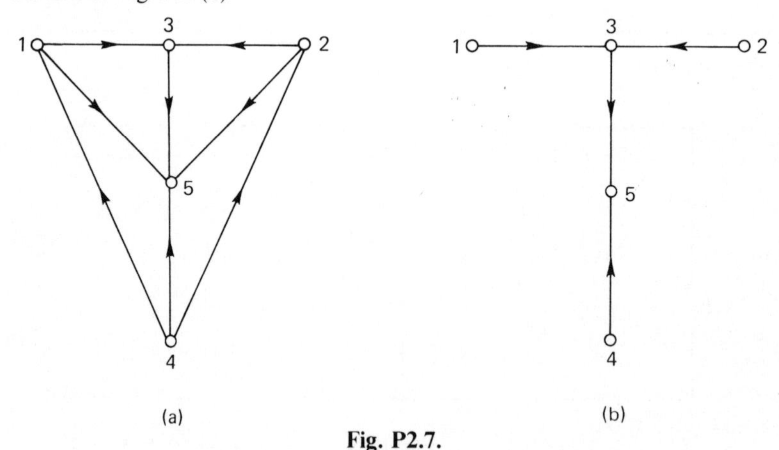

Fig. P2.7.

2.8 Show that for the bridged-T network of Fig. P2.8:

$$\mathbf{A} = \begin{bmatrix} -\dfrac{1}{RC} & 0 \\ 0 & -\dfrac{R}{L} \end{bmatrix}$$

$$\mathbf{B} = \frac{1}{2R}\begin{bmatrix} \dfrac{1}{C} & \dfrac{1}{C} \\ \dfrac{R}{L} & -\dfrac{R}{L} \end{bmatrix}.$$

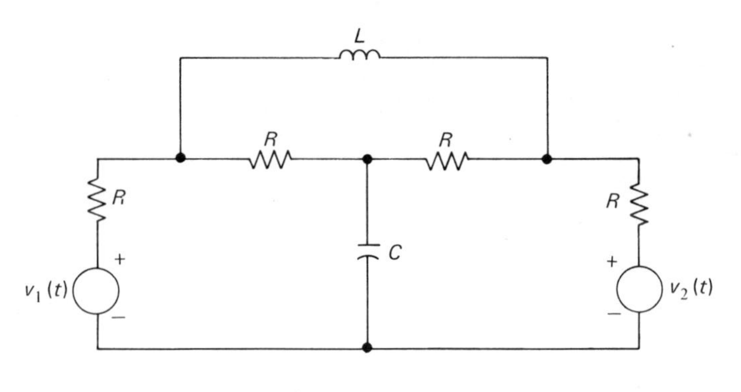

Fig. P2.8.

2.9 Construct a normal tree for the LCR network shown in Fig. P2.9. Hence, set up a system of state equations for the network.

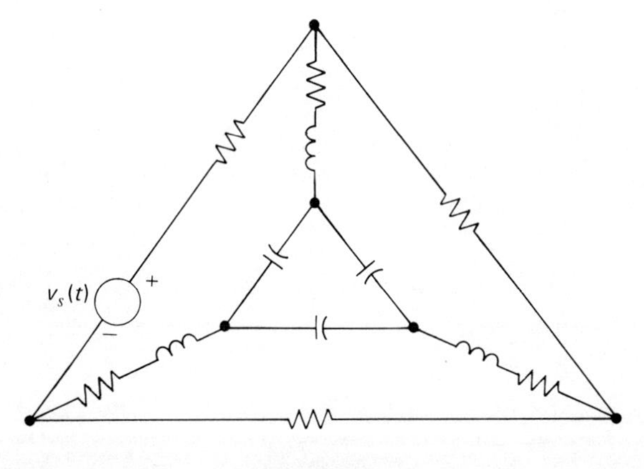

Fig. P2.9.

2.10 The network of Fig. P2.10 contains a voltage-controlled voltage source. Set up the state equations for the network.

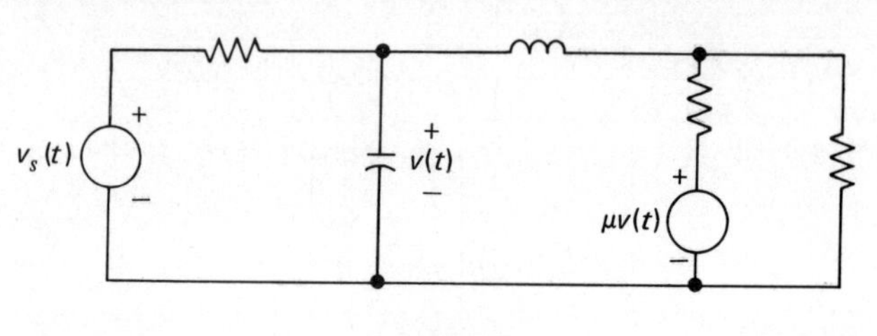

Fig. P2.10.

CHAPTER 3

PROPERTIES OF ONE-PORT NETWORKS

3.1 INTRODUCTION

In a *one-port* network we have a single pair of terminals available for external connection. This chapter is concerned with two important characteristics of linear one-port networks, namely, stability and passivity. A network is said to be *stable* if as a result of subjecting it to an excitation which dies out with time (e.g., an impulse function) we find that the response remains bounded in amplitude as time grows indefinitely; conversely the network is *unstable* if the response is an ever-increasing function of time. In the case of a one-port network we have two aspects of the stability problem to consider, depending on whether the current or the voltage at the terminal pair of interest is regarded as the excitation. As for passivity, the criterion is determined from energy considerations. If the total energy delivered to the network from an externally applied excitation is positive the network is said to be *passive*; the network is *active* if it is not passive. The characteristics of stability and passivity of a one-port network are intimately related to the impedance or admittance measured looking into the network. It is therefore appropriate to begin our study of one-port networks by examining these two *driving-point functions.*

3.2 DRIVING-POINT FUNCTIONS

In Fig. 3.1 the linear network N is assumed to consist of resistors, inductors, capacitors and controlled sources. The presence of independent sources inside the network N is, however, excluded. The behaviour of this network with respect to the port 1–1′ is uniquely described in terms of the *driving-point impedance* $Z(s)$, or its reciprocal, the *driving-point admittance* $Y(s)$. By definition we have

$$Z(s) = \frac{V(s)}{I(s)} \tag{3.1}$$

$$Y(s) = \frac{1}{Z(s)} = \frac{I(s)}{V(s)}, \tag{3.2}$$

where $V(s)$ and $I(s)$ denote the Laplace transforms of the voltage $v(t)$ and current $i(t)$, respectively, existing at the terminal pair 1–1′. The network is assumed to be

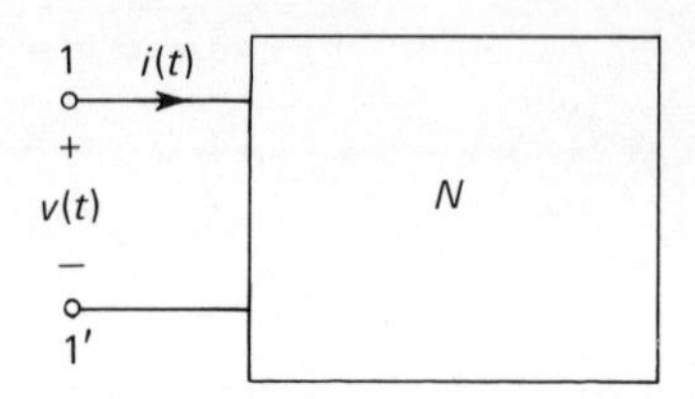

Fig. 3.1. One-port network.

initially at rest. A *system function*, of which $Z(s)$ and $Y(s)$ are specific examples, is defined as the ratio of the response, or output, transform to the excitation, or input transform. It follows, therefore, that according to Eq. (3.1) the current $i(t)$ provides the *excitation* and the voltage $v(t)$ is the resulting *response*, that is, $I(s)$ is the *excitation transform* and $V(s)$ is the *response transform*. On the other hand, according to Eq. (3.2) the voltage $v(t)$ provides the excitation and the current $i(t)$ is the resulting response.

We can determine $Z(s)$ in terms of the elements which make up the network N by treating $i(t)$ as the excitation and applying the nodal analysis. Thus, with one of the input terminals chosen as the reference node, say terminal 1′, we find that

$$Z(s) = \frac{V(s)}{I(s)} = \frac{\Delta'_{11}(s)}{\Delta'(s)}, \tag{3.3}$$

where $\Delta'(s)$ is the nodal-basis circuit determinant, and $\Delta'_{11}(s)$ is the determinant obtained by omitting the first row and the first column of $\Delta'(s)$. A typical element $Y_{jk}(s)$ of these two determinants has the following general form:

$$Y_{jk}(s) = G_{jk} + sC_{jk} + \frac{1}{sL_{jk}}. \tag{3.4}$$

$Y_{jk}(s)$ denotes the self-admittance of a node for $j = k$, or the mutual admittance of two nodes for $j \neq k$.

From Eqs. (3.3) and (3.4) it follows that, in general, the driving-point impedance $Z(s)$ must be the ratio of two polynomials in s, as shown by

$$Z(s) = \frac{p_s(s)}{p_o(s)} = \frac{a_0 s^n + a_1 s^{n-1} + \cdots + a_{n-1} s + a_n}{b_0 s^m + b_1 s^{m-1} + \cdots + b_{m-1} s + b_m}. \tag{3.5}$$

Such an expression is called a *rational function* of s. If each of the numerator and denominator polynomials is expressed in its factored form, we have

$$Z(s) = \frac{a_0(s - s'_1)(s - s'_2)\cdots(s - s'_n)}{b_0(s - s_1)(s - s_2)\cdots(s - s_m)}, \tag{3.6}$$

where a_0/b_0 is a *scale factor*. The constants $s'_1, s'_2, \cdots, s'_n$ are referred to as *zeros* of $Z(s)$; they are roots of the so-called characteristic equation

$$p_s(s) = a_0 s^n + a_1 s^{n-1} + \cdots + a_{n-1}s + a_n = 0, \tag{3.7}$$

which is obtained by setting the numerator polynomial, $p_s(s)$, of $Z(s)$ to zero. The constants $s_1, s_2, \ldots, s_m$ are referred to as *poles* of $Z(s)$; they are roots of a second characteristic equation

$$p_o(s) = b_0 s^m + b_1 s^{m-1} + \cdots + b_{m-1}s + b_m = 0, \tag{3.8}$$

which is obtained by setting the denominator polynomial, $p_0(s)$, of $Z(s)$ to zero. When s has a value equal to any of the zeros, $Z(s)$ is zero; when s has a value equal to any of the poles, $Z(s)$ is infinite. Clearly, the m poles and n zeros specify the impedance function $Z(s)$ completely except for the constant multiplier, a_0/b_0. In a physical system the coefficients $a_0, a_1, \ldots, a_{n-1}, a_n$ and $b_0, b_1, \ldots, b_{m-1}, b_m$ are all real numbers. It follows, therefore, that the poles and zeros of the driving-point impedance of any linear physical system must be real numbers or else occur in pairs of complex-conjugate numbers.

In terms of the constants $s_1', s_2', \ldots, s_n'$ and $s_1, s_2, \ldots, s_m$ we can express the driving-point admittance $Y(s) = 1/Z(s)$ at the port 1–1′ by

$$Y(s) = \frac{b_0(s - s_1)(s - s_2)\cdots(s - s_m)}{a_0(s - s_1')(s - s_2')\cdots(s - s_n')}. \tag{3.9}$$

We see, therefore, that the zeros of the driving-point admittance $Y(s)$ are the same as the poles of the driving-point impedance $Z(s)$, and vice versa.

Open-circuit and Short-circuit Natural Frequencies

An important property of a network is its *natural response*, the form of which depends only on the values of the elements and their interconnection. The natural response is characterized by a set of *natural frequencies* or *natural modes of oscillation,* defined as the frequencies at which signals can exist in the network without the necessity of supplying driving power. A common way of observing the natural response of a network is to store some energy on a capacitor or in an inductor in the network and then to allow the network to behave in its own natural manner. Alternatively, the natural response may be described as the response (for time $t \geq 0^+$) of the network (initially relaxed) to a unit-impulse excitation which will impart to the network a finite amount of energy at time $t = 0$, whereafter the excitation is zero. Because of convenience, we shall only consider the impulse method of determining the natural behaviour of a network.

In a one-port network we have two sets of natural frequencies to consider, depending on whether we examine the natural response of the network with the port in question open-circuited or short-circuited. Consider first Fig. 3.2 where it is assumed that $i(t)$ provides the excitation in the form of a unit-impulse current, and $v(t)$ is the resulting response. In so far as the natural frequencies of the network

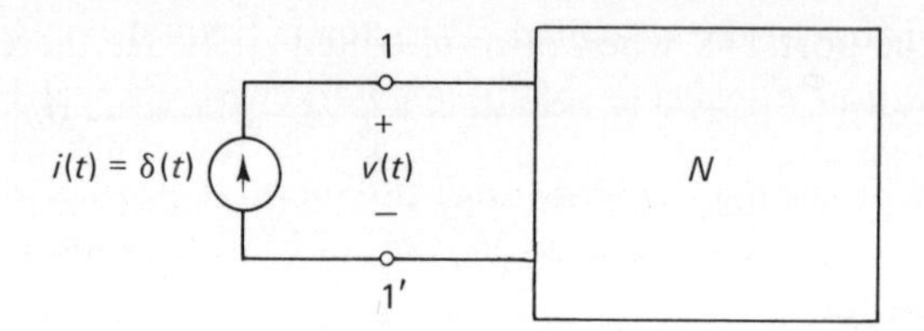

Fig. 3.2. One-port network with unit-impulse current as excitation.

are concerned, the current source represents an open-circuit across the port 1–1′. Since the Laplace transform of a unit-impulse is equal to unity, it follows that

$$\begin{aligned} I(s) &= 1 \\ Z(s) &= V(s). \end{aligned} \tag{3.10}$$

That is, the driving-point impedance of a one-port network is equal to the Laplace transform of the voltage developed across the pertinent port in response to an applied unit-impulse current. Substitution of Eq. (3.10) in Eq. (3.6) yields

$$V(s) = \frac{a_0(s - s_1')(s - s_2')\cdots(s - s_n')}{b_0(s - s_1)(s - s_2)\cdots(s - s_m)}. \tag{3.11}$$

Assuming that all the poles $s_1, s_2, \ldots, s_m$ of $Z(s)$ are simple, we can expand the rational function of Eq. (3.11) in partial fractions as follows:

$$V(s) = \frac{k_1}{s - s_1} + \frac{k_2}{s - s_2} + \cdots + \frac{k_m}{s - s_m}$$

or

$$V(s) = \sum_{\nu=1}^{m} \frac{k_\nu}{s - s_\nu}, \tag{3.12}$$

where k_ν is the *residue* at the pole s_ν and is given by

$$k_\nu = \lim_{s \to s_\nu} (s - s_\nu)Z(s). \tag{3.13}$$

The inverse Laplace transform of $k_\nu/(s - s_\nu)$ is equal to $k_\nu \varepsilon^{s_\nu t}$. Hence, from Eq. (3.12) we find that the voltage $v(t)$ developed across the port 1–1′ in response to an applied unit-impulse current is, for $t \geq 0^+$,

$$v(t) = \sum_{\nu=1}^{m} k_\nu \varepsilon^{s_\nu t}. \tag{3.14}$$

Since the current source in Fig. 3.2 represents an open-circuit across the port 1–1′ as far as the natural response of the network is concerned, we see from Eq. (3.14) that $s_1, s_2, \ldots, s_m$ represent the complex frequencies of the voltage that can exist across the port 1–1′ when open-circuited; they are therefore referred to as *open-circuit natural frequencies* of the network with respect to the port 1–1′. However, the constants $s_1, s_2, \ldots, s_m$ also denote the poles of the impedance $Z(s)$; we conclude, therefore, that the open-circuit natural frequencies of a network with respect to a given port are poles of the driving-point impedance at that port. Equivalently, the open-circuit natural frequencies are zeros of the driving-point admittance, because $Z(s) = 1/Y(s)$.

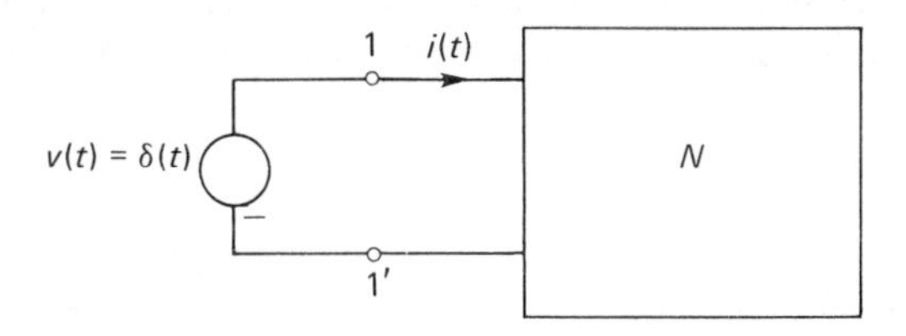

Fig. 3.3. One-port network with unit-impulse voltage as excitation.

Consider next the situation depicted in Fig. 3.3 where it is assumed that a unit-impulse voltage applied across the port 1–1′ provides the excitation, and the terminal current $i(t)$ represents the resulting response. Here we find that

$$\begin{aligned} V(s) &= 1 \\ Y(s) &= I(s). \end{aligned} \tag{3.15}$$

That is, the driving-point admittance is equal to the Laplace transform of the terminal current developed in response to an applied unit-impulse voltage. Substitution of Eq. (3.15) in Eq. (3.9) gives

$$I(s) = \frac{b_0(s - s_1)(s - s_2)\cdots(s - s_m)}{a_0(s - s'_1)(s - s'_2)\cdots(s - s'_n)}. \tag{3.16}$$

Assuming all the poles $s'_1, s'_2, \ldots, s'_n$ of the admittance $Y(s)$ are simple, we can expand Eq. (3.16) in partial fractions as follows:

$$I(s) = \frac{k'_1}{s - s'_1} + \frac{k'_2}{s - s'_2} + \cdots + \frac{k'_n}{s - s'_n}$$

or

$$I(s) = \sum_{\nu=1}^{n} \frac{k'_\nu}{s - s'_\nu}, \tag{3.17}$$

where k'_ν is the residue at the pole s'_ν. The corresponding function $i(t)$ is, for $t \geq 0^+$,

$$i(t) = \sum_{\nu=1}^{n} k'_\nu \varepsilon^{s'_\nu t}. \tag{3.18}$$

In Fig. 3.3 the voltage source represents a short-circuit across the port 1–1′ as far as the natural behaviour of the network is concerned. We see, therefore, from Eq. (3.18) that $s'_1, s'_2, \ldots, s'_n$ represent the complex frequencies of the current that can exist in the port 1–1′ when short-circuited; they are accordingly referred to as the *short-circuit natural frequencies* of the network with respect to the port 1–1′. Since the constants $s'_1, s'_2, \ldots, s'_n$ also denote the zeros of $Z(s)$, we conclude that the short-circuit natural frequencies of a network with respect to a given port are zeros of the driving-point impedance at that port. Equivalently, the short-circuit natural frequencies are poles of the driving-point admittance.

3.3 STABILITY

We have seen that a one-port network has two natural modes of operation, one with the port open-circuited and the other with the port short-circuited. In either mode the stability performance of the network depends on where in the interior of the complex-frequency plane (i.e., the s-plane) the pertinent natural frequencies are located. To establish the relationship between stability and the location of natural frequencies in the s-plane, consider Eq. (3.14) which defines the natural behaviour of a one-port network when the port is open circuited. If the open-circuit natural frequency s_ν, which is a pole of the driving-point impedance at the port in question, is located on the real axis in the left half of the s-plane, as in Fig. 3.4(a), the term $k_\nu \varepsilon^{s_\nu t}$ decreases exponentially and thereby reduces to zero at infinite time, as illustrated in Fig. 3.4(b). If, however, s_ν is a positive real number, that is, it lies on the real axis in the right half of the s-plane, as in Fig. 3.5(a), then the term $k_\nu \varepsilon^{s_\nu t}$ increases exponentially with time, as in Fig. 3.5(b).

If the open-circuit natural frequency s_ν is a complex number the associated multiplying factor k_ν must also be a complex number; so that

$$s_\nu = \sigma_\nu + j\omega_\nu \tag{3.19}$$

$$k_\nu = |k_\nu| \varepsilon^{j\theta_\nu}. \tag{3.20}$$

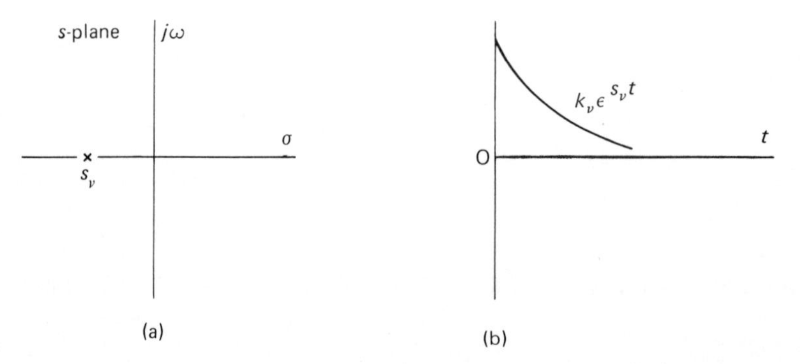

Fig. 3.4. A natural frequency on the negative real axis of the s-plane.

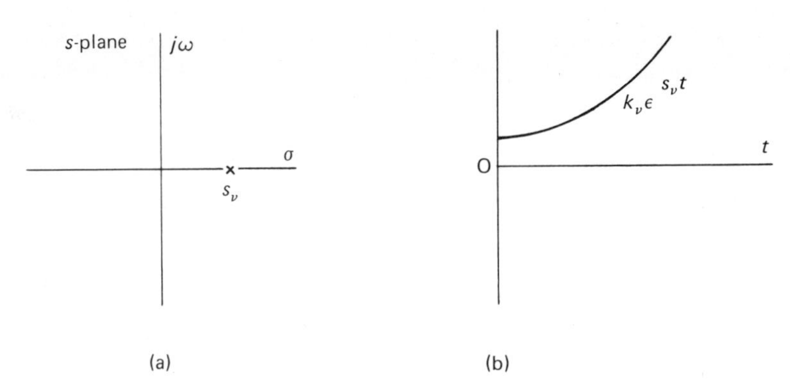

Fig. 3.5. A natural frequency on the positive real axis of the s-plane.

In such a situation we may express $k_v \varepsilon^{s_v t}$ as follows:

$$k_v \varepsilon^{s_v t} = |k_v| \varepsilon^{\sigma_v t} \varepsilon^{j(\omega_v t + \theta_v)},$$

or

$$k_v \varepsilon^{s_v t} = |k_v| \varepsilon^{\sigma_v t} [\cos(\omega_v t + \theta_v) + j \sin(\omega_v t + \theta_v)]. \tag{3.21}$$

Since complex natural frequencies occur in conjugate pairs, it follows that there must be an open-circuit natural frequency at the point $s_v^* = \sigma_v - j\omega_v$, where s_v^* is the complex conjugate of s_v. The transient term having this natural frequency will thus be equal to

$$k_v^* \varepsilon^{s_v^* t} = |k_v| \varepsilon^{\sigma_v t} \varepsilon^{-j(\omega_v t + \theta_v)}$$

or

$$k_v^* \varepsilon^{s_v^* t} = |k_v| \varepsilon^{\sigma_v t} [\cos(\omega_v t + \theta_v) - j \sin(\omega_v t + \theta_v)]. \tag{3.22}$$

Adding Eqs. (3.21) and (3.22) gives the total contribution of a complex-conjugate pair of open-circuit natural frequencies as

$$k_v \varepsilon^{s_v t} + k_v^* \varepsilon^{s_v^* t} = 2|k_v| \varepsilon^{\sigma_v t} \cos(\omega_v t + \theta_v). \tag{3.23}$$

If σ_v is negative, that is, the open-circuit natural frequencies s_v and s_v^* are located in the left half of the s-plane, as in Fig. 3.6(a), then the total contribution of Eq. (3.23) is of an exponentially damped cosine form, as illustrated in Fig. 3.6(b). But if σ_v is positive, that is, s_v and s_v^* lie in the right half plane, as in Fig. 3.7(a), then the total contribution is of a positively damped cosine form which increases with time, as in Fig. 3.7(b). Finally, if σ_v is zero, that is, s_v and s_v^* lie on the imaginary axis of the s-plane, as in Fig. 3.8(a), we have sustained oscillations of constant amplitude, as in Fig. 3.8(b).

We see, therefore, that it is only when all the open-circuit natural frequencies of a network with respect to a given port lie in the left half of the s-plane, or lie

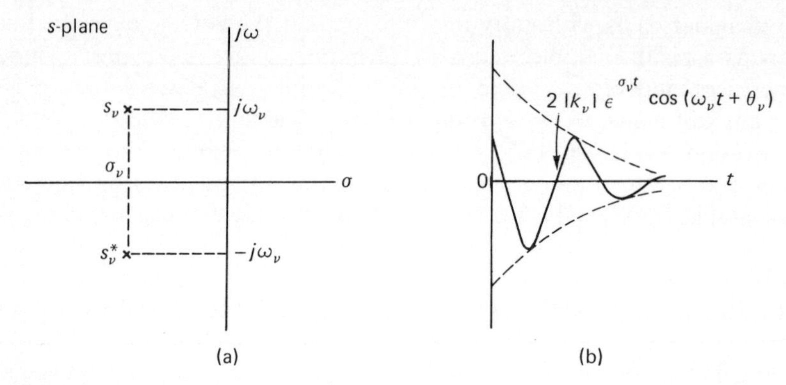

Fig. 3.6. A pair of complex-conjugate natural frequencies in the left half plane.

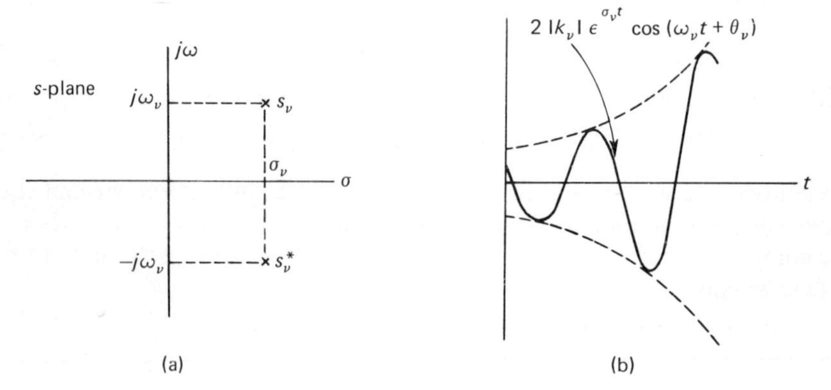

Fig. 3.7. A pair of complex-conjugate natural frequencies in the right half plane.

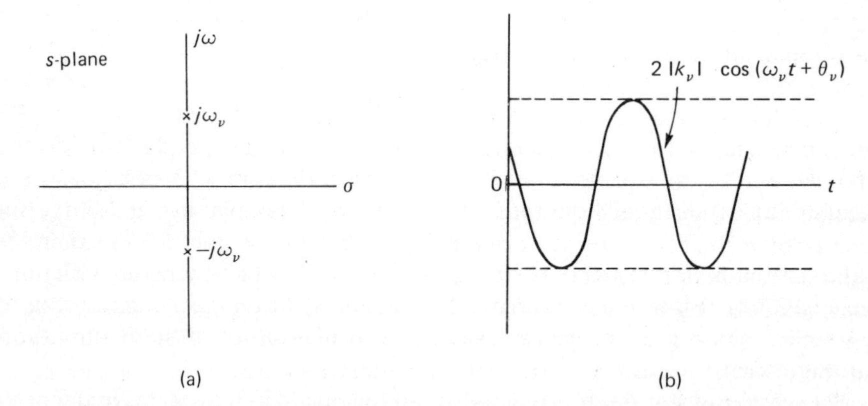

Fig. 3.8. A pair of conjugate natural frequencies on the imaginary axis.

on the imaginary axis with unity multiplicity, that the voltage developed across the port as a result of subjecting it to an impulse current remains bounded as time increases indefinitely. When any of the open-circuit natural frequencies lies in the right half plane, we have a runaway or unstable situation.

In Section 3.2 it was tacitly assumed that the poles of the driving-point impedance $Z(s)$, which are the open-circuit natural frequencies of the network, were all simple. In special cases, however, $Z(s)$ may have multiple poles with the result that the natural response of the network, under open-circuit conditions across the given port, may include exponentials multiplied by powers of t. As an illustration, suppose that the denominator polynomial of $Z(s)$ has a factor of the form $[(s - \sigma_v)^2 + \omega_v^2]^2$, that is, $Z(s)$ has double poles at $s = \sigma_v \pm j\omega_v$. The total contribution of such a pair of double poles to the natural response of the one-port network with the port open-circuited is equal to[1]

$$\frac{1}{2\omega_v^3}\varepsilon^{\sigma_v t}(\sin \omega_v t - \omega_v t \cos \omega_v t).$$

Clearly, if the double poles lie in the left or right half of the s-plane the extra factor t in the second term is of no significance in determining whether the contribution will decrease or increase with time, because it is overpowered by the exponential. If, however, σ_v is zero, that is, the double poles lie on the imaginary axis, then the contribution will increase linearly with time, and we have an unstable situation.

In summary, a one-port network has two natural modes of operation: one with the pertinent port open-circuited and the other with the port short-circuited; corresponding to these modes of operation we have the open-circuit and short-circuit natural frequencies which are, respectively, identical with the poles and zeros of the driving-point impedance function $Z(s)$ at the port in question. The network is open-circuit stable with respect to the given port if and only if the driving-point impedance $Z(s)$ at that port has no poles in the right half of the s-plane, and any poles of $Z(s)$ on the imaginary axis are simple. In a dual manner, the network is short-circuit stable with respect to the given port if and only if the driving-point admittance $Y(s)$ at that port has no poles in the right half of the s-plane, and any poles of $Y(s)$ on the imaginary axis are simple.

Example 3.1. *Tunnel diode stability.* For our first example we shall study the stability performance of a circuit consisting simply of a tunnel diode connected to the terminals of a battery, as in Fig. 3.9. As the negative time constant of a tunnel diode is usually substantially less than 10^{-9} seconds in magnitude, a meaningful analysis must include the finite lead inductance of the battery loop, and the inherent passive resistance of the battery and diode.

The circuit of Fig. 3.9 has two natural frequencies s_1, s_2 which are the zeros of the driving-point impedance $Z(s)$ measured looking away from the battery

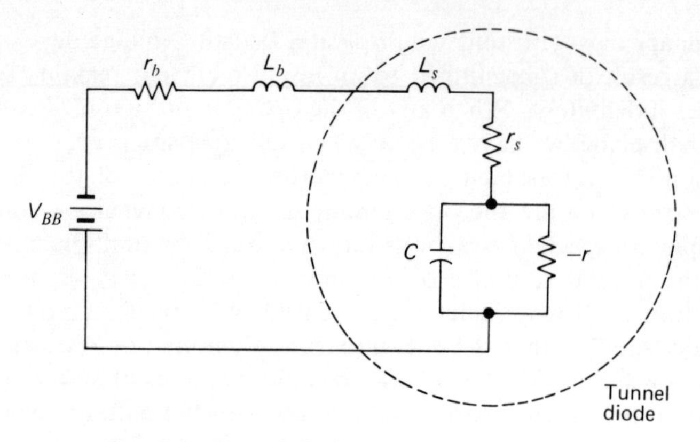

Fig. 3.9. Tunnel diode connected to a battery.

V_{BB}; that is,

$$s_1, s_2 = \frac{1}{2}\left(\frac{1}{rC} - \frac{r_t}{L_t}\right) \pm j\sqrt{\frac{1}{L_t C} - \frac{1}{4}\left(\frac{r_t}{L_t} + \frac{1}{rC}\right)^2} \tag{3.24}$$

where $r_t = r_s + r_b$ and $L_t = L_s + L_b$. For convenience, let us define

$$\omega_0 = \frac{1}{\sqrt{L_t C}} \tag{3.25}$$

and

$$Q_n = \omega_0 r C = r\sqrt{\frac{C}{L_t}}. \tag{3.26}$$

Then, Eq. (3.24) may be re-written in the form

$$s_1, s_2 = \frac{\omega_0 Q_n}{2}\left[\left(\frac{1}{Q_n^2} - \frac{r_t}{r}\right) \pm j\sqrt{\frac{4}{Q_n^2} - \left(\frac{r_t}{r} + \frac{1}{Q_n^2}\right)^2}\,\right]. \tag{3.27}$$

We may identify the following different situations depending upon the choice of circuit parameters Q_n, r_t and r:
a) If

$$\frac{r_t}{r} < \frac{2}{Q_n} - \frac{1}{Q_n^2},$$

then both natural frequencies s_1 and s_2 will be complex. If, further,

$$\frac{r_t}{r} > \frac{1}{Q_n^2},$$

the real parts of s_1 and s_2 will be negative, and an initial disturbance will result in a decaying sinusoid, so that stability of the circuit is ensured. On the other hand, if

$$\frac{r_t}{r} < \frac{1}{Q_n^2},$$

the real parts of s_1 and s_2 will be positive, and the circuit is therefore unstable in that an initial disturbance will result in a growing sinusoid the amplitude of which is limited only by the nonlinearity in the I–V characteristic of the tunnel diode.

b) If

$$\frac{r_t}{r} > \frac{2}{Q_n} - \frac{1}{Q_n^2},$$

then both s_1 and s_2 will be real and an initial disturbance will either decay or grow exponentially. To ensure that both s_1 and s_2 lie in the left half plane, so that the circuit is stable, we must, in addition, require

$$\frac{r_t}{r} < 1.$$

Otherwise, at least one of the real roots will be located in the right half plane, thereby resulting in instability.

We may thus define a regional stability diagram[2] in the $[r_t/r, Q_n]$ plane, as shown in Fig. 3.10. This diagram shows the allowed ranges of the parameters for particular types of transient waves. It indicates that the conditions

$$L_t < r_t rC \qquad \text{and} \qquad r_t < r \tag{3.28}$$

are both necessary and sufficient for stability.

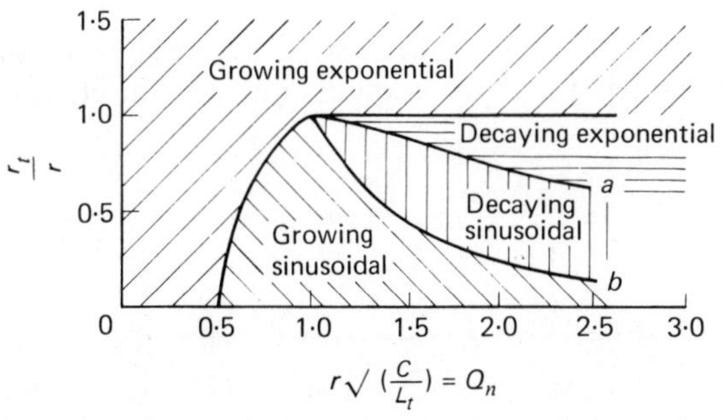

Fig. 3.10. Regional stability diagram for tunnel diode

$\left(\text{For curve } a, \frac{r_t}{r} = \frac{2}{Q_n} - \frac{1}{Q_n^2}; \text{ for curve } b, \frac{r_t}{r} = \frac{1}{Q_n^2}\right)$

It is of interest to note that the second condition of Eq. (3.28) also ensures that the load line with a slope $-1/r_b$ intersects the negative resistance portion of the I–V characteristic of the tunnel diode at a single point, so that this singular quiescent operating point is stable.

Example 3.2. *Tunnel-diode tuned amplifier.* There are two basic types of negative resistance amplifiers: the *transmission* type and *reflection* type.[3] In a transmission-type amplifier the one-port negative resistance device (e.g., tunnel diode) is used to lower the internal resistance of the signal source, increasing its available power; in a reflection-type amplifier, on the other hand, a propagating signal incident on the negative resistance device is reflected at increased amplitude in the same characteristic impedance, some means being provided to separate the input and output signal waves.

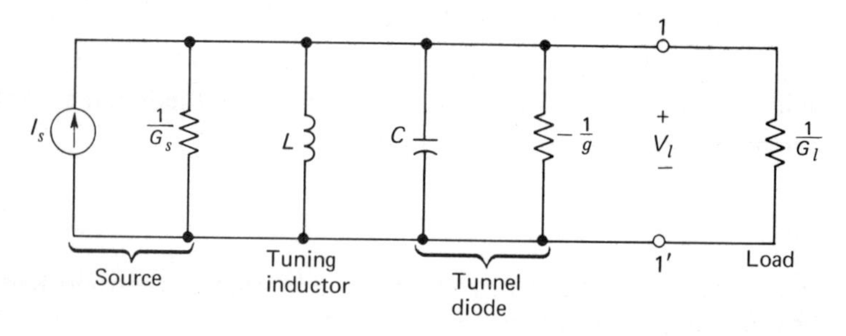

Fig. 3.11. Tunnel-diode tuned amplifier.

In this chapter we will only consider the transmission-type amplifier shown in Fig. 3.11 in which a tunnel diode is connected directly in parallel with a current source of shunt conductance G_s and a load of conductance G_l. To simplify the analysis, the tunnel diode is represented simply as the parallel combination of negative conductance $-g$ and junction capacitance C. The shunt inductance L is chosen to resonate with C at some desired frequency ω_0, so that $L = 1/\omega_0^2 C$. Therefore, at the frequency ω_0 the circuit assumes the purely resistive form shown in Fig. 3.12.

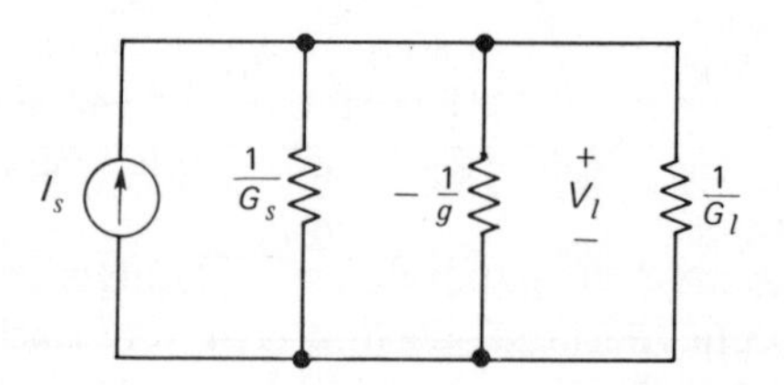

Fig. 3.12. Behaviour of tuned amplifier at resonant frequency.

A useful measure of the performance of a negative-resistance amplifier is the so-called *transducer gain*, G_T, defined as

$$G_T = \frac{P_l}{P_{avs}}, \tag{3.29}$$

where P_l = power delivered to the load

P_{avs} = power available from the source.

At the frequency ω_0, we see from Fig. 3.12 that the load current is $G_l I_s/(G_s - g + G_l)$; hence, the power delivered to the load is

$$P_l = \frac{G_l I_s^2}{(G_s - g + G_l)^2}. \tag{3.30}$$

Also, the power available from the source is

$$P_{avs} = \frac{I_s^2}{4G_s}. \tag{3.31}$$

The transducer gain at the frequency ω_0 is therefore

$$G_T = \frac{4G_s G_l}{(G_s - g + G_l)^2}. \tag{3.32}$$

The gain of Eq. (3.32) is a meaningful quantity only if the complete amplifier of Fig. 3.11 is stable, the necessary condition for which is

$$g < G_s + G_l. \tag{3.33}$$

When the magnitude g of the negative conductance of the tunnel diode exceeds the critical value $G_s + G_l$, the open-circuit natural frequencies of the network with respect to port 1–1′ cross the imaginary axis, into the right half of the s-plane, and oscillation occurs.

Example 3.3. *Wien-bridge oscillator.* As another example, consider the oscillator circuit of Fig. 3.13(a) the key feature of which is the slightly modified Wien bridge serving as both amplitude limiter and frequency selective network. The element r is a thermistor having a negative temperature coefficient of resistance. When the thermistor is cold r is high, while r decreases with increasing voltage across the element owing to heating.

It is assumed that the voltage amplifier is unidirectional with a high input resistance and low output resistance, so that it may be represented by a voltage-controlled voltage source μV_1, as in Fig. 3.13(b). We are interested in evaluating the stability performance of the network with respect to port 1–1′. Analysis on

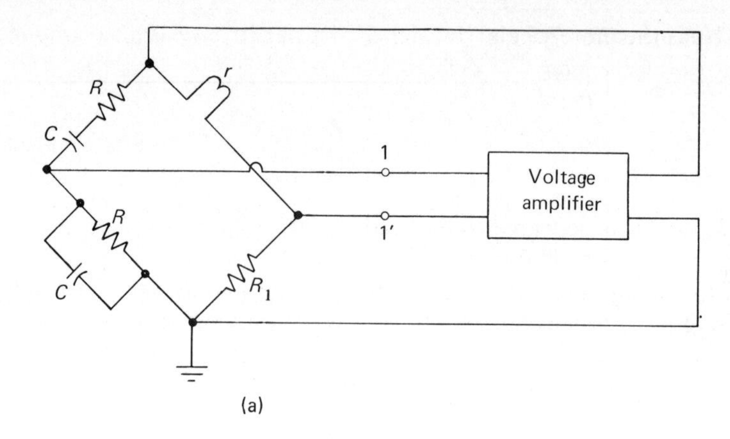

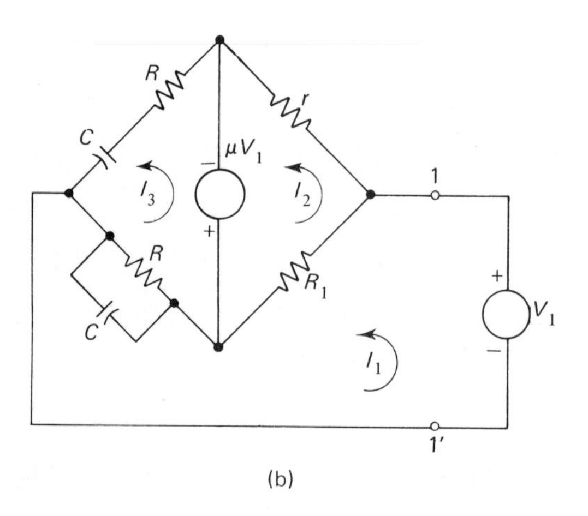

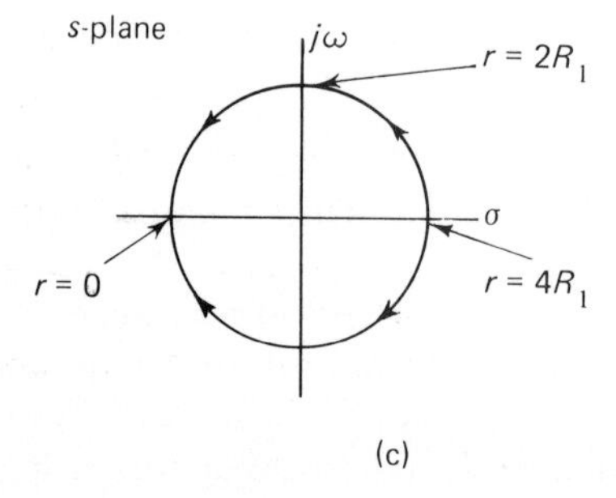

Fig. 3.13. (a) Wien-bridge oscillator; (b) Network for evaluating the driving-point impedance at port 1–1′; (c) Migration of open-circuit natural frequencies inside the s-plane (radius of circle $= 1/CR$).

the loop basis therefore yields

$$\begin{bmatrix} V_1 \\ \mu V_1 \\ -\mu V_1 \end{bmatrix} = \begin{bmatrix} R_1 + \dfrac{R}{1+sCR} & -R_1 & -\dfrac{R}{1+sCR} \\ -R_1 & r + R_1 & 0 \\ -\dfrac{R}{1+sCR} & 0 & R + \dfrac{1}{sC} + \dfrac{R}{1+sCR} \end{bmatrix} \begin{bmatrix} I_1 \\ I_2 \\ I_3 \end{bmatrix}. \quad (3.34)$$

Solving Eq. (3.34) for the driving-point admittance $Y(s) = I_1/V_1$ at the port 1–1′, we obtain

$$Y(s) = \frac{(r + R_1 + \mu R_1)\left[s^2C^2R^2 + \left(3 - \dfrac{\mu(r+R_1)}{r + R_1 + \mu R_1}\right)sCR + 1\right]}{rR_1\left[s^2C^2R^2 + \left(2 + \dfrac{R}{R_1} + \dfrac{R}{r}\right)sCR + \left(1 + \dfrac{R}{R_1} + \dfrac{R}{r}\right)\right]}. \quad (3.35)$$

For all positive values of r the poles of the admittance $Y(s)$, which are the short-circuit natural frequencies with respect to port 1–1′, are located in the left half of the s-plane and the network is therefore short-circuit stable with respect to this port. Also, if the voltage gain μ of the amplifier is very large compared with unity the zeros of $Y(s)$, which are the open-circuit natural frequencies with respect to port 1–1′, would be located at the points

$$s_1, s_2 = \frac{1}{CR}\left[\frac{r}{2R_1} - 1 \pm \sqrt{\frac{r}{2R_1}\left(\frac{r}{2R_1} - 2\right)}\right]. \quad (3.36)$$

Hence, with varying r the open-circuit natural frequencies migrate inside the s-plane, as illustrated in Fig. 3.13(c). When the thermistor is cold, so that $r > 2R_1$, the open-circuit natural frequencies are located in the right half plane. This results in oscillations with increasing amplitude, which in turn will heat up the thermistor, reducing r, and thereby causing the open-circuit natural frequencies to move towards the imaginary axis. Equilibrium is finally obtained when $r = 2R_1$ for which the open-circuit natural frequencies are on the imaginary axis at $s = \pm j/CR$, corresponding with a sinusoidal oscillation of constant amplitude and frequency equal to $1/CR$ radians per second.

Routh's Criterion for Stability

The short-circuit stability, or instability, of a network with respect to a specified port can be determined by establishing whether or not all the roots of the characteristic equation

$$a_0 s^n + a_1 s^{n-1} + \cdots + a_n = 0 \quad (3.37)$$

are located in the left half of the s-plane. This equation is obtained by setting the numerator polynomial of the pertinent driving-point impedance of Eq. (3.5) to zero. The roots of Eq. (3.37) yield the short-circuit natural frequencies which are zeros of $Z(s)$.

Similarly, to investigate the open-circuit stability performance of the network we examine the roots of the characteristic equation

$$b_0 s^m + b_1 s^{m-1} + \cdots + b_{m-1} s + b_m = 0 \tag{3.38}$$

which results from equating the denominator polynomial of $Z(s)$ to zero. The roots of Eq. (3.38) give the open-circuit natural frequencies which are poles of $Z(s)$.

The numerical evaluation of the roots of a given characteristic equation is quite straightforward; but, unless a digital computer is available, it becomes more time-consuming as the order of the highest power of s increases. A simpler method is to apply Routh's criterion* by means of which the number of roots in the right half plane, if any, can be determined from the coefficients of the characteristic equation without actually finding the roots.

Consider the characteristic equation of Eq. (3.37). The coefficients of this equation are arranged in two rows in the following fashion:

Row 1	a_0	a_2	a_4	a_6
Row 2	a_1	a_3	a_5	a_7

where the arrows indicate the order of arrangement. The next step is to use these two rows to construct the following array of terms:

$$\begin{array}{c|cccccc} s^n & a_0 & a_2 & a_4 & a_6 & \cdot & \cdot \\ s^{n-1} & a_1 & a_3 & a_5 & a_7 & \cdot & \cdot \\ s^{n-2} & B_1 & B_3 & B_5 & \cdot & \cdot & \cdot \\ s^{n-3} & C_1 & C_3 & C_5 & \cdot & \cdot & \cdot \\ s^{n-4} & D_1 & D_3 & \cdot & \cdot & \cdot & \cdot \\ s^{n-5} & E_1 & \cdot & \cdot & & & \\ \vdots & \vdots & \vdots & \cdot & & & \end{array} \tag{3.39}$$

* Routh's criterion is also referred to as the "Routh–Hurwitz criterion". A. Hurwitz, who was unaware of E. J. Routh's work in 1877, developed essentially the same result in 1895 but in a somewhat different form. For a proof of the criterion see E. A. Guillemin, *The Mathematics of Circuit Analysis*, p. 395. Wiley, 1949.

where

$$B_1 = \frac{1}{a_1}(a_1 a_2 - a_0 a_3) \tag{3.40}$$

$$B_3 = \frac{1}{a_1}(a_1 a_4 - a_0 a_5) \tag{3.41}$$

$$B_5 = \frac{1}{a_1}(a_1 a_6 - a_0 a_7) \tag{3.42}$$

$$\cdot \quad \cdot \quad \cdot$$

$$C_1 = \frac{1}{B_1}(B_1 a_3 - a_1 B_3) \tag{3.43}$$

$$C_3 = \frac{1}{B_1}(B_1 a_5 - a_1 B_5) \tag{3.44}$$

$$\cdot \quad \cdot \quad \cdot$$

$$D_1 = \frac{1}{C_1}(C_1 B_3 - B_1 C_3) \tag{3.45}$$

$$\cdot \quad \cdot \quad \cdot$$

$$E_1 = \frac{1}{D_1}(D_1 C_3 - C_1 D_3). \tag{3.46}$$

$$\cdot \quad \cdot \quad \cdot$$

Thus, using rows one and two the coefficients of the third row are found by cross multiplication. The second and third rows are next used to form the fourth row, and so on. The process of forming a new row from the preceding two rows is continued until only zeros are obtained. In general, the array will consist of $n + 1$ rows, with the last two rows containing a single element each. It should be remarked that in the course of constructing the array the coefficients of any row may be multiplied or divided by a positive real number without affecting the final result. This is sometimes done so as to simplify the numerical work of finding the coefficients of the succeeding row.

According to Routh's criterion the roots of a given characteristic equation are all located in the left half plane, and the network described by that characteristic equation is therefore stable, if and only if all the elements in the first column of the array constructed in the above manner are positive. If the elements in the first column of such an array are not all of the same sign, then the number of sign changes in that column is equal to the number of those roots of the characteristic equation which are located in the right half plane.

A special case that is found to arise in the study of linear systems is that of an array with a row having zero for all of its elements. This indicates that the characteristic equation has at least one pair of roots equal in magnitude but opposite in sign, that is, located opposite each other and equidistant from the origin. In such a situation the process may be continued by first setting up an *auxiliary polynomial* whose coefficients are the elements in the last non-vanishing row of the array, and then using the coefficients of the derivative of this polynomial in place of the row of zeros.[1] The auxiliary polynomial is an even polynomial (that is, a polynomial is s^2) with its highest power being that of the s indicated at the left of the last non-vanishing row. The equation obtained by setting the auxiliary polynomial to zero gives those root pairs of the original characteristic equation which are equal in magnitude but opposite in sign. When the construction of the array has been completed, after incorporating the modification outlined above, the number of sign changes in the first column gives the number of roots located in the right half plane. If the characteristic equation has roots on the imaginary axis they will be found amongst the roots of the auxiliary polynomial.

Example 3.4. *Colpitts oscillator.* In Fig. 3.14(a) we have the circuit diagram of the so-called *Colpitts oscillator* using a vacuum-tube triode as the active-network component. The problem is to investigate how the open-circuit and short-circuit stability performances of the network with respect to the port 1–1′ are influenced by variations in the mutual conductance g_m of the tube.

Before proceeding with this analysis we shall find it convenient to apply an appropriate impedance and frequency normalization[4] to the network of Fig. 3.14(a). Let R_0 and ω_0 denote the desired impedance- and frequency-normalizing factors, respectively. Each resistor of the network is multiplied by $1/R_0$, each inductor is multiplied by ω_0/R_0, and each capacitor is multiplied by $\omega_0 R_0$. As a result of this normalization process we find that the magnitude of an impedance is scaled from a nominal value Z to Z/R_0, and at the same time the frequency is scaled from a nominal value ω to ω/ω_0. For the network of Fig. 3.14(a) the normalizing factors of $R_0 = 1\ \text{k}\Omega$ and $\omega_0 = 10^5$ rad/sec are considered to be appropriate. Accordingly, using these factors and replacing the tube with its low-frequency circuit model, we obtain the scaled network of Fig. 3.14(b). A current source I_1 has been connected across the port 1–1′ for the purpose of evaluating the driving-point impedance at this port.

Analysing the network of Fig. 3.14(b) on the nodal basis gives

$$\begin{bmatrix} I_1 - g_m V_{gk} \\ g_m V_{gk} \end{bmatrix} = \begin{bmatrix} \left(0{\cdot}5s + 1 + \dfrac{1}{s + 0{\cdot}05}\right) & -(0{\cdot}5s + 1) \\ -(0{\cdot}5s + 1) & (2{\cdot}5s + 1) \end{bmatrix} \begin{bmatrix} V_1 \\ V_2 \end{bmatrix}. \tag{3.47}$$

However, $V_{gk} = -V_2$; therefore, eliminating V_{gk} and collecting terms, we get

$$\begin{bmatrix} I_1 \\ 0 \end{bmatrix} = \begin{bmatrix} \left(0{\cdot}5s + 1 + \dfrac{1}{s + 0{\cdot}05}\right) & -(0{\cdot}5s + 1 + g_m) \\ -(0{\cdot}5s + 1) & (2{\cdot}5s + 1 + g_m) \end{bmatrix} \begin{bmatrix} V_1 \\ V_2 \end{bmatrix}. \tag{3.48}$$

Solving Eqs. (3.48) for V_1, we find that the driving-point impedance at the port

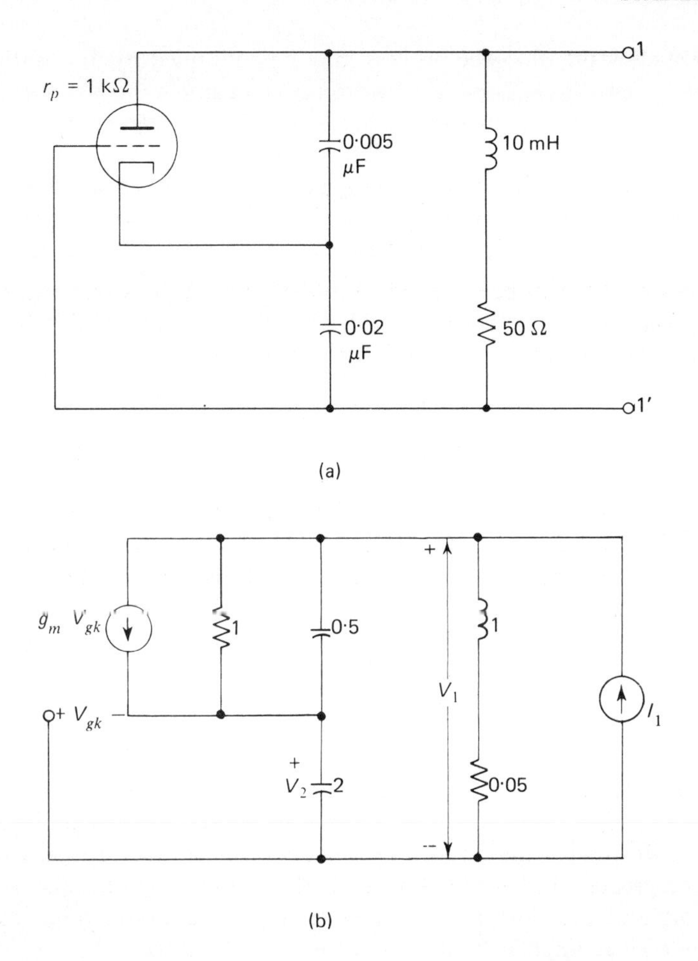

Fig. 3.14. (a) Colpitts oscillator; (b) Scaled network for evaluating the driving-point impedance at port 1–1′.

1–1′ is, after simplification,

$$Z(s) = \frac{V_1}{I_1} = \frac{2{\cdot}5(s + 0{\cdot}05)(s + 0{\cdot}4 + 0{\cdot}4g_m)}{s^3 + 2{\cdot}05s^2 + 2{\cdot}6s + 1 + g_m}. \tag{3.49}$$

We see that $Z(s)$ has two real zeros at $s = -0{\cdot}05$ and $s = -0{\cdot}4(1 + g_m)$. The mutual conductance g_m of a vacuum tube is always positive; hence, both zeros of $Z(s)$ are confined to the left half of the s-plane and the network of Fig. 3.14(a) is short-circuit stable with respect to the port 1–1′ for all positive values of g_m.

To investigate the open-circuit stability performance of the network, we apply Routh's criterion to the denominator polynomial of Eq. (3.49). This leads to the following array of elements:

$$\begin{array}{c|cc} s^3 & 1 & 2{\cdot}6 \\ s^2 & 2{\cdot}05 & 1 + g_m \\ s^1 & 2{\cdot}11 - 0{\cdot}486 g_m & 0 \\ s^0 & 1 + g_m & 0 \end{array} \tag{3.50}$$

All the poles of $Z(s)$ will be confined to the left half of the s-plane, and the network will accordingly be stable, provided all the elements of the first column in the array (3.50) are positive. This, in turn, requires that

$$2{\cdot}11 - 0{\cdot}486 g_m > 0$$

or

$$g_m < 4{\cdot}33. \tag{3.51}$$

If $g_m = 4{\cdot}33$ the third row of the array in (3.50) will vanish, and the network will have a pair of imaginary open-circuit natural frequencies which are roots of the auxiliary equation

$$2{\cdot}05 s^2 + (1 + g_m) = 0$$

or

$$s = \pm j1{\cdot}61. \tag{3.52}$$

With the network of Fig. 3.14(b) having been scaled with respect to $R_0 = 1\ \text{k}\Omega$ and $\omega_0 = 10^5$ rad/sec we may state, in summary, that the original network is open-circuit unstable with respect to the port 1–1′ if the tube has a mutual conductance greater than 4·33 mA/volt. If the mutual conductance is equal to 4·33 mA/volt and the port 1–1′ is open circuited we would obtain sustained oscillations with an angular frequency of $1{\cdot}61 \times 10^5$ rad/sec.

Nyquist's Criterion for Stability[5]

The stability performance of a linear network, with reference to a specified pair of terminals, can also be examined by using the *Nyquist criterion*, which is a graphical method involving a polar plot of the pertinent driving-point impedance $Z(s)$, or admittance $Y(s)$, for $s = j\omega$. The Nyquist criterion not only provides a means of determining the stability status of the network, but also reveals the degree of stability and other useful information. A derivation of the Nyquist criterion can be made with the aid of the following theorem for functions of a complex variable: If a function $F(s)$ is analytic, except for possible poles, within and on a given closed contour C in the s-plane, then

$$\frac{1}{2\pi j}\oint \frac{F'(s)}{F(s)}\,ds = N - P, \tag{3.53}$$

where N is the number of zeros and P is the number of poles of $F(s)$ within the contour, when each zero and pole is counted in accordance with its multiplicity.[6] In Eq. (3.53) we have $F'(s) = dF/ds$ and the contour C is traversed in the positive (counter-clockwise) direction. Since

$$\frac{F'(s)}{F(s)} = \frac{d(\log F)}{ds} \tag{3.54}$$

and if we express the function $F(s)$ in polar form in terms of its magnitude and phase angle,

$$F(s) = |F(s)|\varepsilon^{j \operatorname{ang} F(s)}, \tag{3.55}$$

we find that Eq. (3.53) can be written in the form

$$\frac{1}{2\pi j}\oint d(\log F) = \frac{1}{2\pi j}\oint d(\log|F|) + \frac{1}{2\pi}\oint d(\operatorname{ang} F) = N - P. \tag{3.56}$$

The value of $|F(s)|$ is the same at the beginning as at the end of the closed contour C; therefore, the first integral on the right of Eq. (3.56) vanishes. If, further, $[\operatorname{ang} F(s)]_C$ denotes the change in the phase angle of $F(s)$ as s describes the contour C, we obtain

$$\frac{1}{2\pi}[\operatorname{ang} F(s)]_C = N - P. \tag{3.57}$$

Suppose the function $F(s)$ is represented on a complex plane of its own. Then as the variable s moves around the contour C in the s-plane, we find that $F(s)$ describes a closed curve in the F-plane, as illustrated in Fig. 3.15. Since one revolution corresponds to 2π radians, it follows from Eq. (3.57) that if a function $F(s)$ has a total number of N zeros and P poles inside a closed contour C in the s-plane, then as a result of the variable s moving around the contour C once

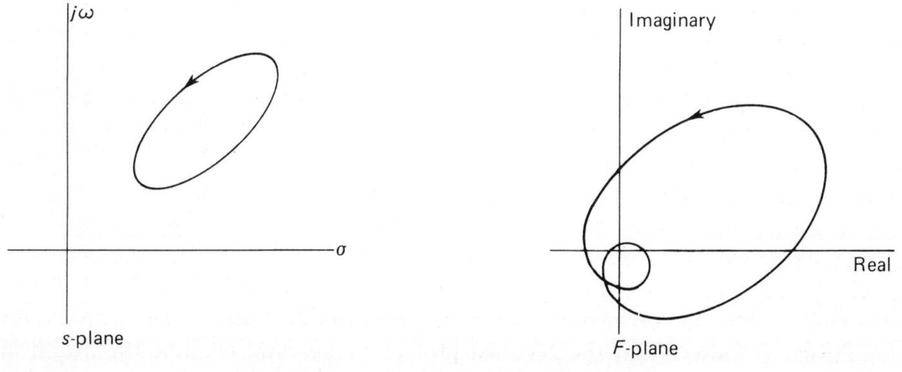

Fig. 3.15. Contours in s- and F-planes.

in the counter-clockwise direction, the plot of $F(s)$ will encircle the origin of the F-plane in the counter-clockwise direction a number of times equal to $N - P$. By encirclement we mean a complete revolution of a radius vector drawn from the origin to a moving point describing the plot of $F(s)$.

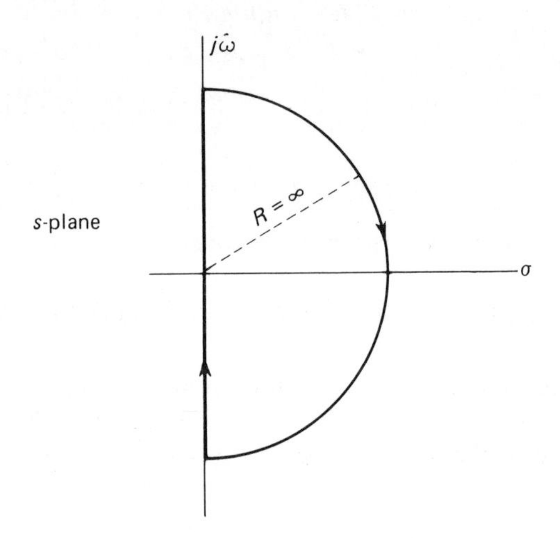

Fig. 3.16. Contour for deriving the Nyquist criterion.

Suppose the function $F(s)$ represents the driving-point impedance $Z(s)$ at the port in question. Also, let the contour C be the path which extends along the imaginary axis and folds back around a semicircle of infinite radius, thereby enclosing the entire right half of the s-plane, as indicated in Fig. 3.16. In this diagram the contour C is shown transversed in the clockwise direction, so that along the imaginary axis it corresponds to increasing frequency. The *Nyquist diagram* is a polar plot of $Z(s)$ for values of s on the contour C of Fig. 3.16. The number of times the plot of $Z(s)$ encircles the origin of the Z-plane in the clockwise direction while s moves along the contour C, is equal to the number of zeros of $Z(s)$ in the right half of the s-plane diminished by the number of poles of $Z(s)$ in the same region. Suppose, however, the network is known to be open-circuit stable; then $Z(s)$ can have no poles in the right half plane. If $Z(s)$ also has no zeros in this region, so that the network is stable under short-circuit conditions, it follows that the Nyquist plot of $Z(s)$ cannot enclose the origin of the Z-plane.

In a dual manner we may state that if the network is known to be short-circuit stable with reference to the specified pair of terminals, then the pertinent driving-point admittance $Y(s)$ has no poles in the right half plane; and if $Y(s)$ also has no zeros in this region, so that the network is stable under open-circuit conditions, it follows that the Nyquist plot of $Y(s)$ cannot encircle the origin of the Y-plane.

The contour C of Fig. 3.16 includes the imaginary axis as well as the large semicircle in the right half plane. If the driving-point function $Z(s)$, or $Y(s)$, being plotted approaches a constant value as s approaches infinity, which is the case when the function has an equal number of poles and zeros, then the semicircular part of the path can be dismissed as it has no effect on the number of times the Nyquist plot will encircle the origin, so that only the values of the function along the imaginary axis are required. If, however, as s approaches infinity, the function behaves as either a positive or a negative power of s, then the Nyquist diagram must be modified to include an arc of a very large or very small circle, respectively, to represent the values assumed by the pertinent function over the semicircular part of the path.[7]

Example 3.5. *Emitter follower with capacitive load.* The emitter follower or common-collector amplifier finds frequent use as a buffer amplifier and as a power driver. In such applications trouble is often encountered with oscillations, the trouble being aggravated by the capacitive load on the output.[8] To illustrate such an effect, consider a transistor having the following hybrid-π parameters: $r_{b'} = 100\,\Omega$, $r_{b'e} = 1\,\text{k}\Omega$, $C_{b'e} = 1000\,\text{pF}$, $g_m = 40\,\text{mA/volt}$. The transistor is operated as an emitter follower, as in Fig. 3.17(a), with a load consisting of a 1 kΩ

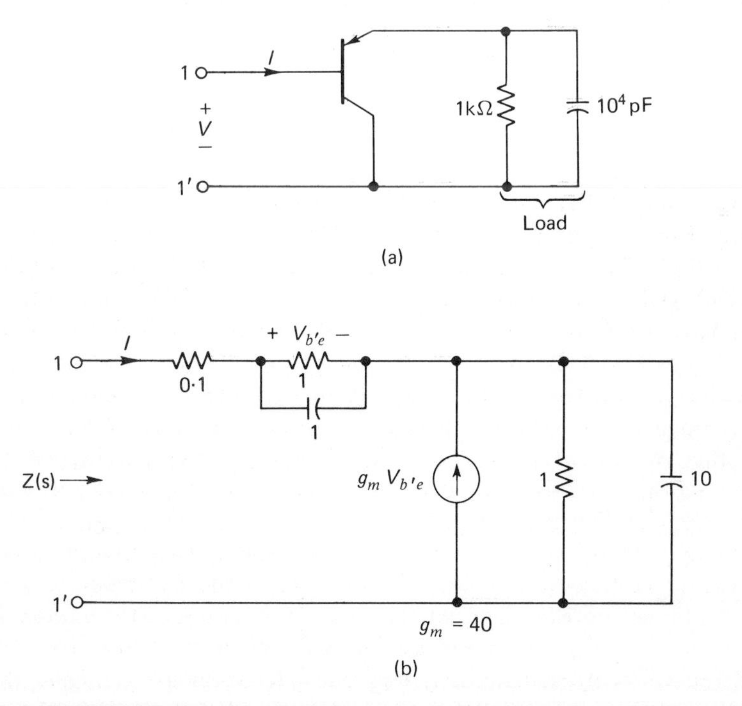

Fig. 3.17. (a) Emitter follower with capacitive load; (b) Network for evaluating the driving-point impedance at port 1–1′.

resistor connected in parallel with a 10^4 pF capacitor. It is required to investigate the stability performance of the stage with respect to the input port 1–1′.

In Fig. 3.17(b) we have replaced the transistor with its simplified hybrid-π model, neglecting the effect of $C_{b'c}$. For convenience, the network is shown scaled with respect to a resistance of 1 kΩ and a frequency of 10^6 rad/sec. Analysing the network on the nodal basis gives the driving-point impedance $Z(s) = V/I$ at port 1–1′ as

$$Z(s) = \frac{s^2 + 12{\cdot}1s + 42{\cdot}1}{10s^2 + 11s + 1}. \tag{3.58}$$

Putting $s = j\omega$, and evaluating the resistive and reactive components of the impedance $Z(j\omega)$, we get

$$R(\omega) = \frac{10\omega^4 - 289\omega^2 + 42{\cdot}1}{100\omega^4 + 101\omega^2 + 1} \tag{3.59}$$

$$X(\omega) = -\frac{\omega(110\omega^2 + 451)}{100\omega^4 + 101\omega^2 + 1}. \tag{3.60}$$

The resistive component $R(\omega)$ becomes negative when

$$10\omega^4 - 289\omega^2 + 42{\cdot}1 < 0$$

or

$$(\omega^2 - 0{\cdot}147)(\omega^2 - 28{\cdot}6) < 0. \tag{3.61}$$

It therefore follows that $R(\omega)$ is negative in the frequency range $5{\cdot}35 > \omega > 0{\cdot}384$. From Eq. (3.60) we find that when $\omega = 0{\cdot}384$, $X(\omega) = -10$ and when $\omega = 5{\cdot}35$, $X(\omega) = -0{\cdot}23$. To find the most negative value of $R(\omega)$, we set $dR(\omega)/d\omega$ to zero, which yields the solution: $R_{min} = -1{\cdot}3$ at $\omega = 0{\cdot}745$. The corresponding value of $X(\omega)$ is $-4{\cdot}34$. From Eqs. (3.59) and (3.60) we also find that $R(\omega) = 42{\cdot}1$, $X(\omega) = 0$ at $\omega = 0$, while at $\omega = \infty$ we have $R(\omega) = 0{\cdot}1$ and $X(\omega) = 0$.

We thus obtain the Nyquist diagram of Fig. 3.18 for $Z(j\omega)$. The network is clearly open-circuit stable with respect to port 1–1′, because if $I = 0$ in Fig. 3.17(b), then we have $V_{b'e} = 0$ and the controlled source $g_m V_{b'e}$ becomes inactivated, so that looking into port 1–1′ we simply see a passive network. Also, since the Nyquist diagram of Fig. 3.18 does not enclose the origin of the Z-plane, it follows that the network is also short-circuit stable with respect to port 1–1′. The network, is, however, potentially unstable in the frequency range $5{\cdot}35 > \omega > 0{\cdot}384$, for which $R(\omega)$ is negative. Thus, if the input of the emitter follower is connected to a source having an internal impedance equal to $R_s + j\omega L_s$, then the circuit will oscillate at a frequency determined by L_s and the input capacitance of the emitter follower, providing this frequency is in the range of $5{\cdot}35 > \omega > 0{\cdot}384$, and the total input loop resistance $R_s + R_{in}(\omega) < 0$.

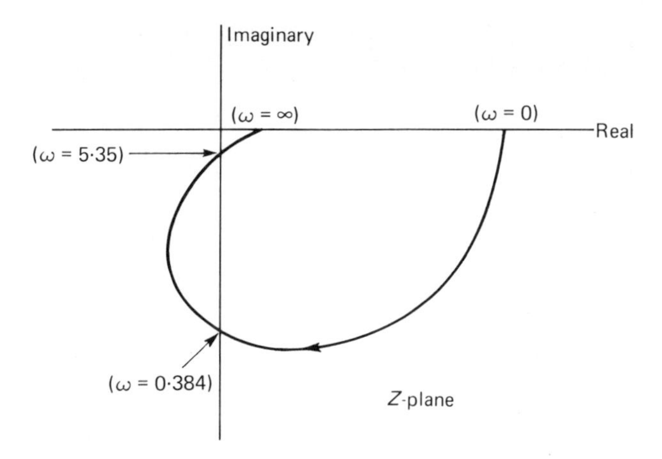

Fig. 3.18. Nyquist locus of the $Z(j\omega)$ of Eq. (3.58).

Amplitude of Oscillation in Nonlinear Negative-resistance Oscillators

The circuit model of a *parallel-tuned negative-resistance oscillator* consists basically of a voltage-controlled negative-resistance device (e.g., tunnel diode), an LC resonator and load connected in parallel, as in Fig. 3.19. The capacitance C is assumed to be the sum of the parasitic capacitance associated with the negative-resistance device and whatever additional circuit capacitance that is required to achieve sinusoidal oscillation at the desired frequency. For the oscillator to be self-starting in the sense that any oscillation, however small (initiated by noise, closing of a switch or some other disturbance), can grow in amplitude it is necessary that the total admittance $Y(s)$ at the port 1–1′ has a pair of complex-conjugate zeros in the right half of the s-plane. Owing to the inherent non-linearities associated with the negative-resistance device, however, we find that the effective conductance presented by the device is a function of amplitude of oscillation, so that as the oscillation builds up the right half plane zeros of $Y(s)$ move to the left until they are located on the imaginary axis, at which point

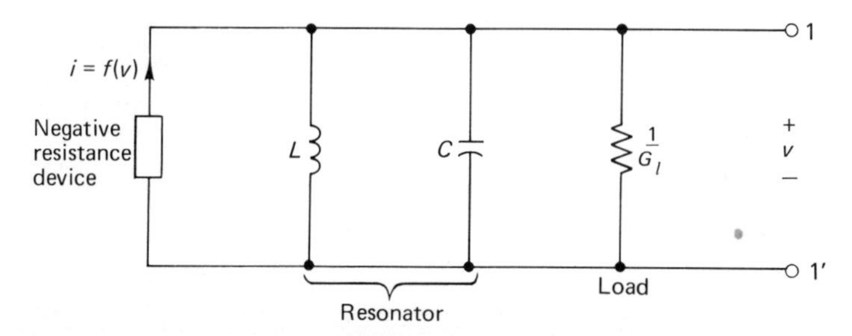

Fig. 3.19. A negative-resistance oscillator.

equilibrium conditions are established. It is clear, therefore, that the system is nonlinear during the build-up period.

To study the build-up performance of the oscillator, we may use the method of *equivalent linearization.*[9] The basic idea of the method is as follows: when a sinusoidal voltage is applied across a nonlinear resistor the resulting current may be resolved by means of the Fourier series into a fundamental component which has the same frequency as the applied voltage, plus harmonic components the frequencies of which are integral multiples of the fundamental. The fundamental component of current is in phase with the voltage and has a magnitude which depends upon the voltage. If the Q-factor of the loaded resonant circuit is fairly high, as is usually the case in a sinusoidal oscillator, the effects of the second and higher order harmonic components may be ignored, and in terms of the fundamental frequency the nonlinear resistance may be replaced by an equivalent linear resistance chosen in such a way that the amplitude of the fundamental component of current is the same for both the linear and nonlinear elements when subjected to the same sinusoidal voltage. The system resulting from this replacement is referred to as *quasi-linear*.

Suppose the nonlinear element is characterized by the voltage–current relation

$$i = f(v) \tag{3.62}$$

and the applied voltage v is sinusoidal, described by

$$v = V_0 \cos \omega_0 t = V_0 \cos \theta. \tag{3.63}$$

Then, the fundamental component of the resulting current will have an amplitude given by

$$I_0 = \frac{1}{\pi} \int_0^{2\pi} f(V_0 \cos \theta) \cos \theta \, d\theta. \tag{3.64}$$

The conductance of the equivalent linear resistor is therefore a function of the amplitude V_0 of the applied voltage, as shown by

$$g_{eq}(V_0) = \frac{I_0}{V_0} = \frac{1}{\pi V_0} \int_0^{2\pi} f(V_0 \cos \theta) \cos \theta \, d\theta. \tag{3.65}$$

Replacing the nonlinear negative-resistance device with the equivalent conductance $g_{eq}(V_0)$, we have for $s = j\omega$

$$Y(j\omega) = g_{eq}(V_0) + G_l + j\left(\omega C - \frac{1}{\omega L}\right). \tag{3.66}$$

When equilibrium is established $Y(j\omega) = 0$, which corresponds to a sustained oscillation with a frequency

$$\omega_0 = \frac{1}{\sqrt{LC}} \tag{3.67}$$

and an amplitude V_0 given by the condition

$$g_{eq}(V_0) + G_l = 0. \tag{3.68}$$

Furthermore for the oscillator to be self starting it is necessary that at zero amplitude,

$$g_{eq}(0) + G_l \leq 0. \tag{3.69}$$

Example 3.6. *Tunnel-diode oscillator.* Various expressions have been derived from theoretical and empirical considerations to describe the I–V characteristic curve of a tunnel diode; the simplest approximation is the classical Van der Pol cubic equation which relates the instantaneous current i and voltage v by[10,11]

$$i = -av + bv^3, \tag{3.70}$$

where a and b are positive constants. In Fig. 3.20 we have sketched a curve corresponding to Eq. (3.70), with the origin shown at the centre of the region of negative slope. The shift in the origin from the position of the actual characteristic of the device merely represents a change in co-ordinates.

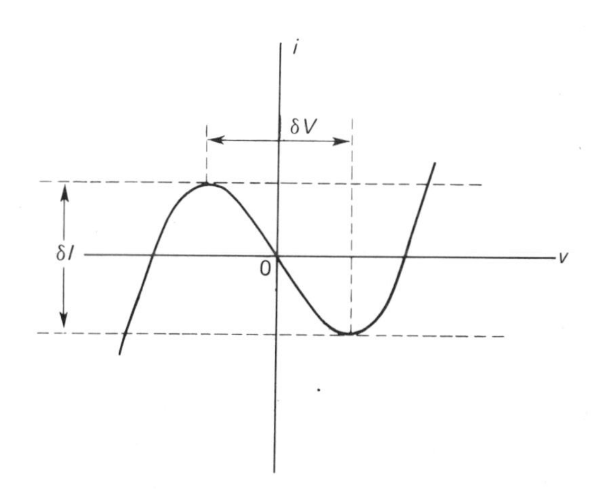

Fig. 3.20. i–v characteristic curve for Van der Pol approximation.

Ignoring the higher harmonics of the voltage across the diode, so that v is nearly sinusoidal, we find from Eqs. (3.65) and (3.70) that the conductance of the equivalent linear element is

$$g_{eq}(V_0) = -a + \tfrac{3}{4}bV_0^2. \tag{3.71}$$

For steady-state oscillation we have

$$g_{eq}(V_0) + G_l = 0$$

or

$$(-a + G_l) + \tfrac{3}{4}bV_0^2 = 0. \tag{3.72}$$

Solving for the voltage amplitude yields

$$V_0 = 2\sqrt{\frac{a - G_l}{3b}}. \tag{3.73}$$

The power delivered to the load is therefore

$$P_l = \tfrac{1}{2}V_0^2 G_l = \frac{2G_l}{3b}(a - G_l), \tag{3.74}$$

which attains the maximum value

$$P_{max} = \frac{a^2}{6b} \quad \text{for} \quad G_l = \frac{a}{2}. \tag{3.75}$$

Referring to Fig. 3.20, we see that the current and voltage swings corresponding to the peak and valley points of the I–V characteristic are related to the parameters a and b as follows

$$a = \frac{3}{2}\frac{\delta I}{\delta V}, \tag{3.76}$$

$$b = 2\frac{\delta I}{(\delta V)^3}, \tag{3.77}$$

in terms of which we may express the maximum power as

$$P_{max} = \tfrac{3}{16}\delta I\, \delta V. \tag{3.78}$$

3.4 PASSIVITY

A one-port network is *passive* if for all excitations the total energy $\mathscr{E}(t)$ delivered to the network is non-negative, the energy being measured starting from $t = -\infty$ when the network is completely at rest with no stored energy. The total energy $\mathscr{E}(t)$ is related to the instantaneous voltage $v(t)$ and the instantaneous current $i(t)$ at the port 1–1′ of the network as follows:

$$\mathscr{E}(t) = \int_{-\infty}^{t} v(\tau)i(\tau)\, d\tau = \mathscr{E}(0) + \int_{0}^{t} v(\tau)i(\tau)\, d\tau, \tag{3.79}$$

where

$$\mathscr{E}(0) = \int_{-\infty}^{0} v(\tau)i(\tau)\, d\tau$$

represents the energy stored in the network at time $t = 0$. The one-port network is passive if $\mathscr{E}(t) \geq 0$ for all t. The network is *active* if it is not passive.

By using the definition of Eq. (3.79) we shall next derive the necessary conditions for the passivity of a one-port network in terms of the driving-point impedance. Suppose that the current $i(t)$ provides the excitation, and is given by

$$i(t) = \varepsilon^{\sigma_0 t} \sin \omega_0 t, \tag{3.80}$$

which, for $\sigma_0 > 0$, represents an exponentially increasing, oscillatory excitation, as in Fig. 3.21. The reason for using this form of excitation is that, with σ_0 positive, the magnitude of the forced (driven) part of the response will increase with time, thereby making it easy to separate the forced response of the network from its natural response. It is assumed that the network is open-circuit stable so as to facilitate its excitation from a current source.

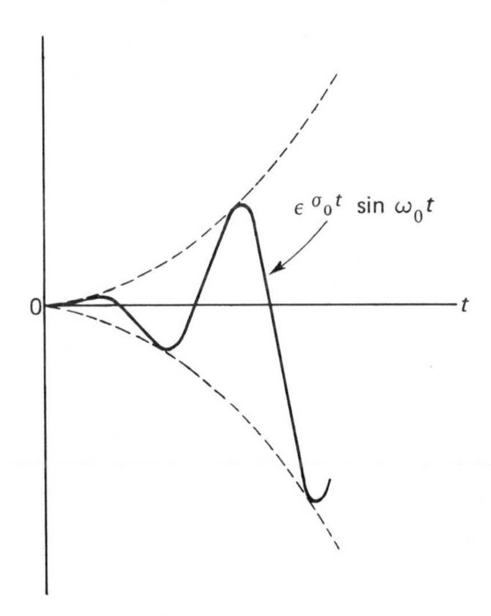

Fig. 3.21. An exponentially growing sinusoidal excitation.

The Laplace transform $I(s)$ of the excitation of Eq. (3.80) is given by

$$I(s) = \frac{\omega_0}{(s - \sigma_0)^2 + \omega_0^2}. \tag{3.81}$$

Therefore, we have a pair of complex-conjugate *forced frequencies* at (see Fig. 3.22)

$$s_0, s_0^* = \sigma_0 \pm j\omega_0 = |s_0| \, \varepsilon^{\pm j\theta_0}. \tag{3.82}$$

The resulting response transform $V(s)$ is

$$V(s) = Z(s)I(s) = \frac{\omega_0 Z(s)}{(s - \sigma_0)^2 + \omega_0^2}, \tag{3.83}$$

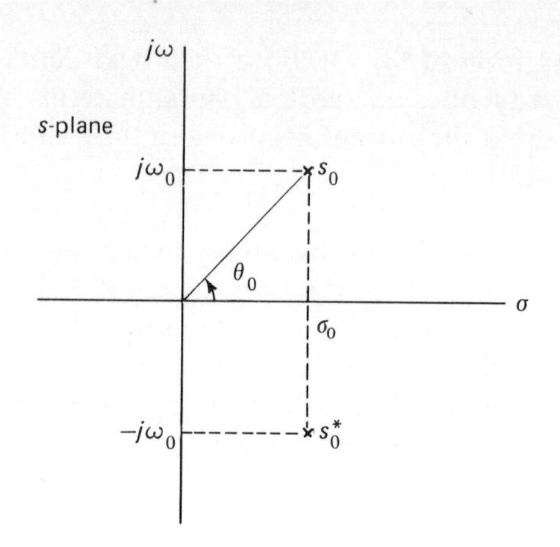

Fig. 3.22. Pair of complex forced frequencies.

where $Z(s)$ is the driving-point impedance at the port 1–1′. Expressing $Z(s)$ in terms of its poles and zeros, as in Eq. (3.6), we have

$$V(s) = \frac{a_0\omega_0(s - s_1')(s - s_2')\cdots(s - s_n')}{b_0[(s - \sigma_0)^2 + \omega_0^2](s - s_1)(s - s_2)\cdots(s - s_m)}. \tag{3.84}$$

As the network is assumed to be open-circuit stable, it follows that the poles $s_1, s_2, \ldots, s_m$ of $Z(s)$ are all located inside the left half plane and are therefore different from the poles of $V(s)$ at $\sigma_0 \pm j\omega_0$ which are due to the externally applied excitation. Expanding the rational function on the right of Eq. (3.84), we get

$$V(s) = \frac{k_0}{s - \sigma_0 - j\omega_0} + \frac{k_0^*}{s - \sigma_0 + j\omega_0} + \sum_{\nu=1}^{m} \frac{k_\nu}{s - s_\nu}. \tag{3.85}$$

From Eq. (3.83) we find that the residue k_0 at the pole $\sigma_0 + j\omega_0$ has the value

$$k_0 = \left.\frac{\omega_0 Z(s)}{s - \sigma_0 + j\omega_0}\right|_{s=\sigma_0+j\omega_0}$$

or

$$k_0 = \frac{1}{2j}|Z_0|\varepsilon^{j\phi_0}, \tag{3.86}$$

where $|Z_0|$ and ϕ_0 are the magnitude and phase angle of $Z(s)$, respectively, evaluated at the point $s = \sigma_0 + j\omega_0$. The residue k_0^* at the pole $\sigma_0 - j\omega_0$ is equal to the complex conjugate of k_0; therefore,

$$k_0^* = -\frac{1}{2j}|Z_0|\varepsilon^{-j\phi_0}. \tag{3.87}$$

The time-domain voltage $v(t)$ developed across the port 1–1′ in response to the current excitation of Eq. (3.80) can be determined by summing the inverse Laplace transforms of the individual terms on the right of Eq. (3.85). The first two terms of this equation give rise to the *forced-response* component of $v(t)$, and the remaining terms, which come from the poles of $Z(s)$, give rise to the natural component of $v(t)$. We may thus express $v(t)$ as

$$v(t) = v_F(t) + v_N(t), \tag{3.88}$$

where

$$\begin{aligned} v_F(t) &= \text{forced-response component} \\ &= |Z_0|\varepsilon^{\sigma_0 t}\left[\frac{\varepsilon^{j(\omega_0 t + \phi_0)} - \varepsilon^{-j(\omega_0 t + \phi_0)}}{2j}\right] \\ &= |Z_0|\varepsilon^{\sigma_0 t}\sin(\omega_0 t + \phi_0), \end{aligned} \tag{3.89}$$

$$\begin{aligned} v_N(t) &= \text{natural response component} \\ &= \sum_{\nu=1}^{m} k_\nu \varepsilon^{s_\nu t}. \end{aligned} \tag{3.90}$$

Since the poles $s_1, s_2, \ldots, s_m$ of $Z(s)$ are all assumed to be located in the left half plane, it follows that the natural response component must decay or at most remain bounded in amplitude. Therefore, after a sufficiently long time the forced component will become large enough to mask the natural component. That is, for large values of t we have

$$v_F(t) \gg v_N(t)$$

and

$$v(t) \simeq v_F(t). \tag{3.91}$$

Thus, ignoring the effect of the natural component, we find that the instantaneous power taken by the network is

$$\begin{aligned} v(t)i(t) &= |Z_0|\varepsilon^{2\sigma_0 t}\sin(\omega_0 t + \phi_0)\sin\omega_0 t \\ &= \tfrac{1}{2}|Z_0|\varepsilon^{2\sigma_0 t}[\cos\phi_0 - \cos(2\omega_0 t + \phi_0)], \end{aligned} \tag{3.92}$$

where we have made use of the trigonometric relation

$$\sin x \sin y = \tfrac{1}{2}[\cos(x - y) - \cos(x + y)]. \tag{3.93}$$

However,

$$|Z_0|\cos\phi_0 = \operatorname{Re} Z_0 \tag{3.94}$$

and

$$|Z_0|\varepsilon^{2\sigma_0 t}\cos(2\omega_0 t + \phi_0) = \operatorname{Re}[Z_0\varepsilon^{2s_0 t}], \tag{3.95}$$

where Re signifies the real part of the expression which follows it. Hence, we may re-write Eq. (3.92) in the form

$$v(t)i(t) = \tfrac{1}{2}\text{Re}[Z_0(\varepsilon^{2\sigma_0 t} - \varepsilon^{2s_0 t})]. \tag{3.96}$$

The energy delivered to the network by the current excitation of Eq. (3.80) is for $t > 0$,

$$\begin{aligned}\mathscr{E}(t) &= \int_0^t v(\tau)i(\tau)\,d\tau \\ &= \int_0^t \tfrac{1}{2}\text{Re}[Z_0(\varepsilon^{2\sigma_0\tau} - \varepsilon^{2s_0\tau})]\,d\tau.\end{aligned} \tag{3.97}$$

Since the operations of Re [ ] and $\int_0^t [\ ]\,d\tau$ are commutative, we find that

$$\mathscr{E}(t) = \frac{1}{4}\varepsilon^{2\sigma_0 t}\text{Re}\left[\frac{Z_0}{\sigma_0} - \frac{Z_0\varepsilon^{j2\omega_0 t}}{s_0}\right] + C_1, \tag{3.98}$$

where C_1 is a constant equal to $\frac{1}{2}\text{Re}[Z_0\{(1/2\sigma_0) - (1/2s_0)\}]$. For large t, the magnitude of the first term on the right of Eq. (3.98) will certainly exceed that of the constant C_1. Further, if the one-port network is to be passive $\mathscr{E}(t)$ must be positive for all t. Therefore, for $\mathscr{E}(t)$ to be positive for large t, it is necessary that

$$\text{Re}\left[\frac{Z_0}{\sigma_0} - \frac{Z_0\varepsilon^{j2\omega_0 t}}{s_0}\right] \geq 0. \tag{3.99}$$

Expanding the expression on the left of (3.99) by finding the real part of each term, and noting that $s_0 = |s_0|\varepsilon^{j\theta_0}$, we get

$$\frac{\text{Re } Z_0}{\sigma_0} - \frac{|Z_0|}{|s_0|}\cos(2\omega_0 t + \phi_0 - \theta_0) \geq 0, \tag{3.100}$$

from which we deduce that there are two requirements to fulfil:

$$\text{Re } Z_0 \geq 0 \tag{3.101}$$

$$\frac{\text{Re } Z_0}{\sigma_0} \geq \frac{|Z_0|}{|s_0|}. \tag{3.102}$$

In other words, the real part of Z_0 must not only be positive but large enough to satisfy the condition of (3.102) if we are to be certain that $\mathscr{E}(t)$ does not become negative during any part of the time cycle of the cosine function $\cos(2\omega_0 t + \phi_0 - \theta_0)$.

Since $|Z_0|$ is the magnitude of $Z(s)$ at $s = \sigma_0 + j\omega_0$ and $\sigma_0 = \text{Re } s_0$ has been chosen to be positive, we can re-state the condition of (3.101) as follows:

$$\text{Re}[Z(s)] \geq 0 \quad \text{for Re } s \geq 0. \tag{3.103}$$

To interpret the second condition of (3.102), we shall first rearrange it in the form

$$\frac{\text{Re } Z_0}{|Z_0|} \geq \frac{\text{Re } s_0}{|s_0|}. \tag{3.104}$$

But

$$\frac{\text{Re } Z_0}{|Z_0|} = \cos \phi_0, \tag{3.105}$$

$$\frac{\text{Re } s_0}{|s_0|} = \cos \theta_0. \tag{3.106}$$

Hence, (3.104) may be re-written as

$$\cos \phi_0 \geq \cos \theta_0$$

or

$$\phi_0 \leq \theta_0 \tag{3.107}$$

which must hold for $|\theta_0| \leq \pi/2$. The inequality of (3.107) which ϕ_0 and θ_0 must satisfy is shown illustrated in Fig. 3.23. Since ϕ_0 is the phase angle of $Z(s)$, and θ_0 is the phase angle of s_0, we may re-state (3.107) in the form

$$|\text{ang } Z(s)| \leq |\text{ang } s| \quad \text{for } 0 < |\text{ang } s| \leq \frac{\pi}{2}. \tag{3.108}$$

Although at first sight it may appear that this condition is stronger than that of (3.103), it is nevertheless contained in the statement $\text{Re}[Z(s)] \geq 0$ for $\text{Re } s \geq 0$. In other words, (3.103) defines the necessary condition for passivity of a one-port network in terms of the real parts of $Z(s)$ and s, while (3.108) defines the necessary condition in an equivalent polar form expressed in terms of the angles of $Z(s)$ and s. A mathematical proof demonstrating the equivalence of Eqs. (3.103) and (3.108) is given elsewhere.[6]

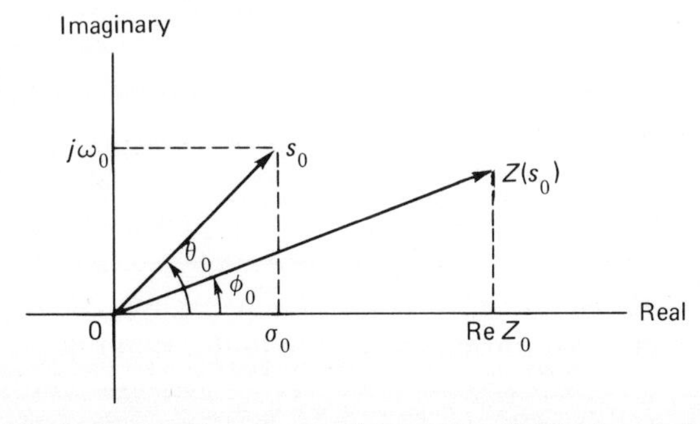

Fig. 3.23. Illustrating the condition of Eq. (3.107) for passivity.

Furthermore, from Section 3.2 it is recalled that $Z(s)$ is a rational function with real coefficients, that is,

$$Z(s) = \frac{a_0 s^n + a_1 s^{n-1} + \cdots + a_{n-1} s + a_n}{b_0 s^m + b_1 s^{m-1} + \cdots + b_{m-1} s + b_m}. \tag{3.109}$$

Clearly, if the variable s is real, then $Z(s)$ is real.

In summary, we have shown that an impedance function $Z(s)$ represents the driving-point impedance of a passive one-port network if and only if

$$Z(s) \text{ is real when } s \text{ is real}$$

and

$$\operatorname{Re}[Z(s)] \geq 0 \quad \text{for} \quad \operatorname{Re} s \geq 0. \tag{3.110}$$

Any function that satisfies the two conditions of (3.110) is called a *positive real function*, which was first introduced into network theory by Brune.[12] Indeed, not only is the positive real requirement necessary, but it is also sufficient for realizability as a passive one-port network.

Following a procedure similar to that outlined above but starting with a voltage excitation, we find that for passivity of a one-port network it is necessary that the driving-point admittance $Y(s)$ is also a positive real function.

We may therefore state that the necessary and sufficient condition for a one-port network to be passive is that its driving-point impedance (or driving-point admittance) be a positive real function. The network is active if it is not passive.

Properties of Positive Real Functions

The first important property of a positive real function is that it cannot have poles or zeros in the right half of the s-plane; poles and zeros along the imaginary axis are permissible only if they are simple. To demonstrate this property, suppose the impedance function $Z(s)$ has a pole of multiplicity k at the point $s = s_a$. Expanding $Z(s)$ as a Laurent series about this point, we have

$$Z(s) = \frac{A_k}{(s - s_a)^k} + \cdots + \frac{A_2}{(s - s_a)^2} + \frac{A_1}{(s - s_a)} + B_0 + B_1(s - s_a) + \cdots. \tag{3.111}$$

However, in the immediate neighbourhood of the point $s = s_a$ we may approximate the series of Eq. (3.111) by its dominant term; so that

$$Z(s) \simeq \frac{A_r}{(s - s_a)^k}. \tag{3.112}$$

For convenience, we shall assume that all points of interest near s_a are within a circle of radius r which is arbitrarily small but finite, as in Fig. 3.24. On this small circle we have

$$s - s_a = r\varepsilon^{j\alpha}. \tag{3.113}$$

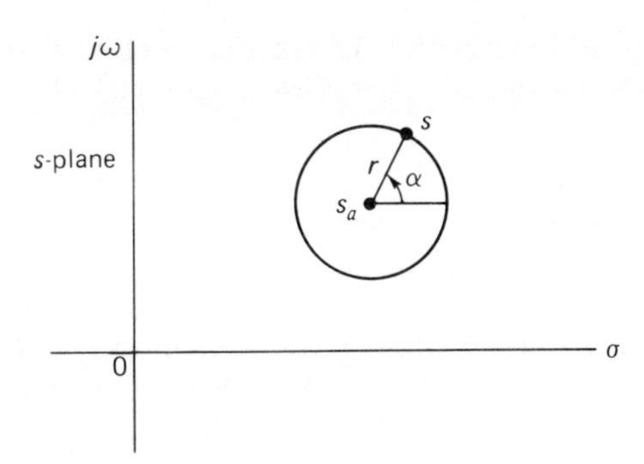

Fig. 3.24. The location of a neighbourhood of a pole s_a in the right half plane.

Expressing the complex coefficient A_k in the polar form

$$A_k = |A_k|\varepsilon^{j\beta}, \tag{3.114}$$

we find that the real part of $Z(s)$ is equal to

$$\mathrm{Re}\left[\frac{|A_k|}{r^k}\varepsilon^{j(\beta - k\alpha)}\right] = \frac{|A_k|}{r^k}\cos(\beta - k\alpha). \tag{3.115}$$

As the variable point s traverses the circular contour once, the angle α changes from 0 to 2π radians, and the real part of $Z(s)$ changes its algebraic sign $2k$ times for a complete traversal around the pole. Accordingly, the only way in which it is possible to fulfil the requirement $\mathrm{Re}[Z(s)] \geq 0$ for $\mathrm{Re}\, s \geq 0$ is to exclude the poles of $Z(s)$ from the right half plane. That is to say, a positive real function is analytic over the entire right half plane as it is not permitted to have poles in this region.

There remains the question of whether a positive real function is permitted to have poles along the imaginary axis. Figure 3.25 shows such a pole with a

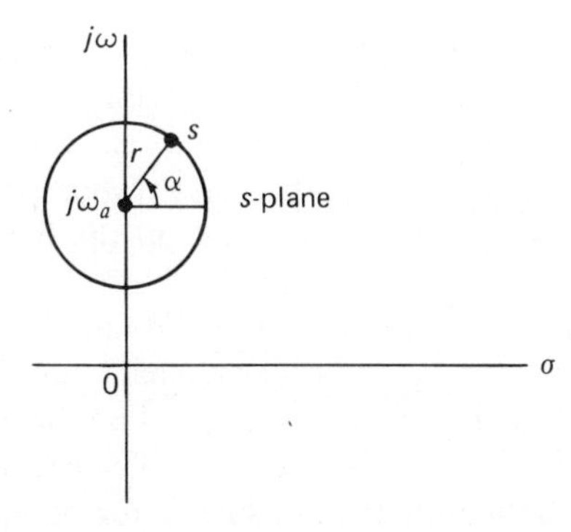

Fig. 3.25. The location of a neighbourhood of a pole $s_a = j\omega_a$ on the imaginary axis.

circle of radius r around the pole. We see that a point in the right half plane corresponds to $-\pi/2 \leq \alpha \leq \pi/2$. Therefore, from Eq. (3.115) it follows that there is but one possible way for a positive real function to have poles along the imaginary axis, and that is if and only if $\beta = 0$ and $k = 1$. In other words, a positive real function is permitted to have poles on the imaginary axis only if they are simple and their residues are real and positive.

If $Z(s)$ is a positive real function, then its reciprocal $Y(s)$ must also be a positive real function, and as such $Y(s)$ cannot have poles in the right half plane and any poles on the imaginary axis must be simple with positive real residues. However, the poles of $Y(s)$ are zeros of $Z(s)$; therefore, a function which is positive real is not permitted to have poles and zeros in the right half plane, and any poles and zeros on the imaginary axis must be simple. This property implies that a passive one-port network is necessarily open-circuit stable and short-circuit stable. However, a stable network is not necessarily passive; stability is but one necessary condition for passivity.

The second property of a positive real function is that the relative degrees of its numerator and denominator polynomials can at most differ by one. Referring to Eq. (3.109) we see that for large values of s the impedance function approximates as

$$Z(s) \simeq \frac{a_0}{b_0} s^{n-m}. \tag{3.116}$$

Since infinity is a point on the imaginary axis, and since imaginary poles and zeros of a positive real function are simple, it follows that if $Z(s)$ is a positive real function, then there are only three possibilities

$$n - m = \begin{cases} 1 \\ 0 \\ -1. \end{cases} \tag{3.117}$$

Furthermore, the scale factor a_0/b_0 must always be positive.

The third property of a positive real function relates to the behaviour of its real part on the imaginary axis of the s-plane. For this evaluation we shall make use of a theorem of complex variable theory[6] which states that if a function of the complex variable is analytic within and on the boundary of a region in the s-plane and has no zero in the region, then the real part of that function attains its maximum and minimum values upon the boundary. The particular part of the s-plane that is of interest to us is the entire right half plane bounded by the imaginary axis which is modified by small semicircles bending the outline to the right around any poles that may be there, as in Fig. 3.26. Clearly a positive real impedance function $Z(s)$ is analytic within and on such a region. Accordingly, the theorem tells us that the minimum value of the real part of $Z(s)$ will be found somewhere on the imaginary axis boundary. In other words, the positiveness of $\text{Re}[Z(j\omega)]$

for all values of ω assures the positiveness of $Z(s)$ over the entire right half plane provided $Z(s)$ is analytic in this region.

We conclude, therefore, that an impedance function $Z(s)$, which is real for real values of s, is a positive real function if:

1. $Z(s)$ has no poles in the right half plane
2. Any poles of $Z(s)$ on the imaginary axis are simple with positive and real residues
3. $\text{Re}[Z(j\omega)] \geq 0$ for all ω. (3.118)

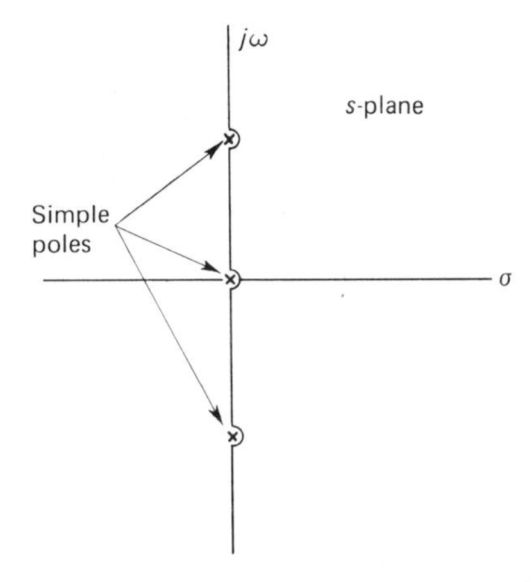

Fig. 3.26. Imaginary axis with small semicircular indentations to avoid simple poles there.

In general, a network containing controlled sources or negative-resistance elements is capable of presenting a driving-point impedance which may violate one or more of the above three conditions necessary for passivity. This, however, does not mean that all one-port networks containing controlled sources or negative-resistance elements are necessarily active. A one-port network containing such elements may be justifiably classified as active only if its driving-point impedance fails to satisfy one or more of the passivity requirements. An example illustrating these observations is given later (see Example 3.7 at the end of next section).

Testing of Passivity Criterion

The first major step in testing a given impedance function $Z(s)$ for passivity is to establish whether it satisfies the stability requirement. To meet this requirement $Z(s)$ must have no poles in the right half plane, and any poles on the imaginary axis must be simple. This we can do readily by applying Routh's criterion to the denominator polynomial of $Z(s)$. In the resulting array we must find that all the elements in the first column are positive. If $Z(s)$ has any imaginary-axis poles we would encounter a vanishing row in the course of constructing the array; the imaginary-axis poles of $Z(s)$ are located by finding the roots of the auxiliary equation formed using the elements in the last non-vanishing row. In this way we can determine the imaginary-axis poles and establish whether or not they are simple. The residues at such poles are next determined to see if they are real and positive. The residue at a simple pole s_a, say, is equal to the limiting value of $(s - s_a)Z(s)$ as s approaches s_a.

There now remains the problem of checking that the real part of $Z(s)$ remains non-negative along the entire imaginary axis. To facilitate evaluation of the real part, the impedance function $Z(s)$ is expressed in the form

$$Z(s) = \frac{m_1 + n_1}{m_2 + n_2}, \tag{3.119}$$

where the m's and n's denote the even and odd parts, respectively, of the numerator and denominator polynomials of $Z(s)$. When $s = j\omega$, and only in this case, we find that the m's are real and the n's are imaginary. To separate $Z(s)$ into its even and odd parts, we multiply the numerator and denominator by $(m_2 - n_2)$ yielding

$$Z(s) = \frac{(m_1 + n_1)(m_2 - n_2)}{(m_2 + n_2)(m_2 - n_2)},$$

or

$$Z(s) = \frac{m_1 m_2 - n_1 n_2}{m_2^2 - n_2^2} + \frac{m_2 n_1 - m_1 n_2}{m_2^2 - n_2^2}. \tag{3.120}$$

The first term on the right of Eq. (3.120) is even, while the second term is odd. When $s = j\omega$, they reduce to the real and imaginary parts of $Z(j\omega)$, respectively. Hence,

$$\text{Re}[Z(j\omega)] = \left. \frac{m_1 m_2 - n_1 n_2}{m_2^2 - n_2^2} \right|_{s = j\omega}. \tag{3.121}$$

Our aim is to determine whether $\text{Re}[Z(j\omega)]$ is non-negative for all ω. In Eq. (3.121) the denominator represents the square of the absolute value of $m_2 + n_2$ for $s = j\omega$. Therefore, $m_2^2 - n_2^2$ must be positive for all ω; so that the condition $\text{Re}[Z(j\omega)] \geq 0$ requires

$$N_0(\omega^2) = [m_1 m_2 - n_1 n_2]_{s = j\omega} \geq 0 \quad \text{for all } \omega. \tag{3.122}$$

$N_0(\omega^2)$ is an even polynomial in ω; expressing it in its factored form, we have

$$N_0(\omega^2) = K(\omega^2 + \delta_1^2)(\omega^2 + \delta_2^2) \cdots (\omega^2 + \delta_r^2). \tag{3.123}$$

We observe that the scale factor K must be positive, otherwise $N_0(\omega^2)$ would become negative for large values of ω. As for a typical factor $\omega^2 + \delta_k^2$ in Eq.

(3.123) we have three possible cases to consider, depending on whether δ_k^2 is positive real, complex, or negative real:

1. If δ_k^2 is a positive real number the corresponding factor $\omega^2 + \delta_k^2$ can never become negative for all real values of ω.
2. If δ_k^2 is a complex number in the factor $\omega^2 + \delta_k^2$, then there must be another factor $\omega^2 + \delta_k^{*2}$ in which δ_k^{*2} is the complex conjugate of δ_k^2. In this case we find that $(\omega^2 + \delta_k^2)(\omega^2 + \delta_k^{*2})$ is the square of an absolute value and cannot therefore become negative for real values of ω either.
3. If δ_k^2 is a negative real number the corresponding factor $\omega^2 + \delta_k^2$ can certainly become negative for some values of ω unless such a factor is of even multiplicity.

We conclude, therefore, that $N_0(\omega^2)$ must not have positive real roots with odd multiplicity if $N_0(\omega^2)$ is to be non-negative for all ω. To find out whether this condition is fulfilled we may determine the various factors of $N_0(\omega^2)$, which becomes more time-consuming as the order of the polynomial $N_0(\omega^2)$ in ω^2 increases. Alternatively, we may use *Sturm's theorem.*[13] To simplify the application of Sturm's theorem, let $\omega^2 = x$. Then, within the interval $-\infty < \omega < \infty$ the new variable x lies in the range $0 < x < \infty$. From Eq. (3.123) we have

$$N_0(x) = \alpha_0 x^r + \alpha_1 x^{r-1} + \cdots \alpha_{r-1} x + \alpha_r, \tag{3.124}$$

where $\alpha_0, \alpha_1, \ldots, \alpha_r$ are real coefficients. The first derivative of $N_0(x)$ is

$$N_1(x) = r\alpha_0 x^{r-1} + \cdots + 2\alpha_{r-2} x + \alpha_{r-1}. \tag{3.125}$$

The polynomials $N_0(x)$ and $N_1(x)$ constitute the first two of a set of so-called *Sturm functions.* To form the next Sturm function, $N_1(x)$ is divided into $N_0(x)$ to give a two-term quotient. The negative of the remainder, which is one degree lower than $N_1(x)$, gives the next Sturm function, $N_2(x)$. The succeeding Sturm function $N_3(x)$ is obtained by dividing $N_2(x)$ into $N_1(x)$; the negative of the remainder is equal to $N_3(x)$. The process is continued till the last remainder is simply a constant, $-N_k$; k is usually equal to (but may be less than) the degree r of the original polynomial $N_0(x)$. We can sum up the process by writing

$$\begin{aligned}
N_0 &= q_1 N_1 - N_2 \\
N_1 &= q_2 N_2 - N_3 \\
&\cdots\cdots\cdots \\
N_{k-2} &= q_{k-1} N_{k-1} - N_k
\end{aligned} \tag{3.126}$$

where $q_1, q_2, \ldots, q_{k-1}$ denote the various two-term quotients.

The use of the Sturm functions in determining whether $N_0(x)$ has any zeros within a specified interval $a < x < b$ is best demonstrated with the aid of Table 3.1. The columns headed $N_0, N_1, N_2, \ldots, N_k$ contain the algebraic signs of these functions when evaluated at $x = a$ and $x = b$. If v_a and v_b denote the number of

Table 3.1

	N_0	N_1	N_2	$\cdots$	N_k	Changes in signs
$x = a$	+	+	−	$\cdots$	+	v_a
$x = b$	−	+	+	$\cdots$	−	v_b

sign changes in the respective rows, then according to Sturm's theorem the absolute value of the difference $v_a - v_b$ gives the number of real zeros of $N_0(x)$ within the interval $a < x < b$, provided the zeros are all simple. In our application the interval of interest is $0 < x < \infty$.

If the procedure terminates prematurely with the last remainder N_k equal to zero, it is an indication that $N_0(x)$ has multiple zeros. This follows from Eq. (3.126) where we see that if $N_k = 0$, then N_{k-1} is a common factor of N_{k-2} and all the preceding Sturm functions right up to N_0. Suppose the original polynomial $N_0(x)$ has a multiple zero of order m; the same zero with a multiplicity $m - 1$ will be possessed by the common factor N_{k-1}. We conclude, therefore, that if N_k is zero N_{k-1} will contain all the multiple zeros of $N_0(x)$ with a multiplicity one less than that contained in $N_0(x)$. The multiple zeros of $N_0(x)$ may thus be determined from N_{k-1} by inspection. If, however, N_{k-1} is found to be a high order polynomial, then we may have to apply Sturm's theorem separately to determine its zeros.

Example 3.7. *Colpitts oscillator* (*continued*). In Example 3.4 we showed that the scaled network of Fig. 3.14(b) is open-circuit unstable with respect to the port 1–1′ if the scaled mutual conductance g_m is greater than 4·33. We shall continue analysis of this network by determining the minimum value of g_m for which the driving-point impedance at the port 1–1′ is active. From Eq. (3.49) we have

$$Z(s) = \frac{2{\cdot}5s^2 + (1{\cdot}13 + g_m)s + 0{\cdot}05(1 + g_m)}{s^3 + 2{\cdot}05s^2 + 2{\cdot}6s + 1 + g_m}. \tag{3.127}$$

To determine the restriction on g_m for $\mathrm{Re}[Z(j\omega)] \geq 0$, we may examine the following polynomial

$$N_0(\omega^2) =$$

$$[(2{\cdot}5s^2 + 0{\cdot}05 + 0{\cdot}05g_m)(2{\cdot}05s^2 + 1 + g_m) - s^2(1{\cdot}13 + g_m)(s^2 + 2{\cdot}6)]_{s=j\omega}.$$

After simplification, we get

$$N_0(\omega^2) = \omega^4(4 - g_m) + \omega^2(0{\cdot}323 - 0{\cdot}0025g_m) + 0{\cdot}05(1 + g_m)^2. \tag{3.128}$$

By inspection, we see that if $g_m > 4$ the coefficient of ω^4 becomes negative and $N_0(\omega^2)$ becomes negative for large values of ω. Summarizing, we may therefore state that the network of Fig. 3.14(a) is

1. passive for $0 < g_m < 4$,
2. active but open-circuit stable for $4 < g_m < 4{\cdot}33$,
3. active and open-circuit unstable for $g_m > 4{\cdot}33$.

Example 3.8. *Maximum frequency of oscillation of a tunnel diode.* A rather useful figure of merit for an active device is the parameter ω_{max} defined as the maximum frequency up to which the device may be made to oscillate. For a tunnel diode ω_{max} is obtained by determining the frequency at which the real part of the driving-point impedance $Z(j\omega)$ ceases to be negative. From the model shown in Fig. 3.27 we see that

$$Z(j\omega) = r_s + j\omega L_s - \frac{r}{1 - j\omega Cr}. \tag{3.129}$$

Hence,

$$\text{Re}[Z(j\omega)] = r_s - \frac{r}{1 + \omega^2 C^2 r^2}, \tag{3.130}$$

which varies with frequency as illustrated in Fig. 3.28. The transition frequency

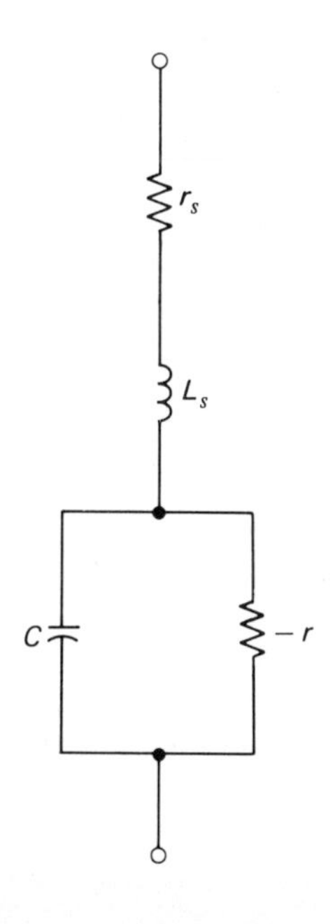

Fig. 3.27. Model of a tunnel diode.

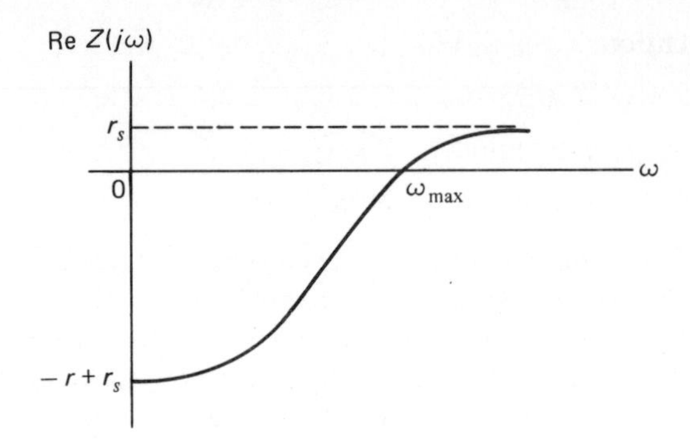

Fig. 3.28. Illustrating frequency dependence of Re $[Z(j\omega)]$ for a tunnel diode.

from a negative to positive real part is therefore

$$\omega_{max} = \frac{1}{rC}\sqrt{\frac{r}{r_s} - 1}. \tag{3.131}$$

Clearly, Eq. (3.131) not only defines the maximum frequency of oscillation but also the upper frequency limit up to which the tunnel diode may be used as an amplifier.

Example 3.9. As a final example, we shall test the driving-point impedance

$$Z(s) = \frac{s^3 + 2s^2 + 2s + 3}{s^3 + s^2 + 2s + 1} \tag{3.132}$$

for passivity. Applying Routh's criterion to the denominator polynomial, to see if the stability requirements are satisfied, we obtain the following array:

$$\begin{array}{c|cc} s^3 & 1 & 2 \\ s^2 & 1 & 1 \\ s^1 & 1 & 0 \\ s^0 & 1 & 0 \end{array}$$

As the elements of the first column are all positive, $Z(s)$ has no poles in the right half plane. There are no poles on the imaginary axis, as no vanishing row was encountered. Therefore, Eq. (3.132) represents the driving-point impedance of an open-circuit stable one-port network.

Next, we have to test whether the condition $\text{Re}[Z(j\omega)] \geq 0$ is satisfied. This condition requires examination of the polynomial

$$N_0(\omega^2) = [(2s^2 + 3)(s^2 + 1) - (s^3 + 2s)^2]_{s=j\omega}$$

or, after simplification,

$$N_0(\omega^2) = \omega^6 - 2\omega^4 - \omega^2 + 3.$$

Putting $\omega^2 = x$, we get

$$N_0(x) = x^3 - 2x^2 - x + 3$$

$$N_1(x) = \frac{d}{dx} N_0(x) = 3x^2 - 4x - 1.$$

To form the next Sturm function we divide $N_1(x)$ into $N_0(x)$; so that

$$\begin{array}{r|l} & \tfrac{1}{3}x - \tfrac{2}{9} \\ \hline 3x^2 - 4x - 1 & x^3 - 2x^2 - x + 3 \\ & x^3 - \tfrac{4}{3}x^2 - \tfrac{1}{3}x \\ \cline{2-2} & -\tfrac{2}{3}x^2 - \tfrac{2}{3}x + 3 \\ & -\tfrac{2}{3}x^2 + \tfrac{8}{9}x + \tfrac{2}{9} \\ \cline{2-2} & -\tfrac{14}{9}x + \tfrac{25}{9} \end{array}$$

Therefore,

$$N_2(x) = 14x/9 - 25/9.$$

Similarly, we divide $N_2(x)$ into $N_1(x)$ yielding

$$N_3(x) = -279/196.$$

Since the process does not terminate prematurely, we deduce that $N_0(x)$ has no multiple roots. Further, using the algebraic signs of the four Sturm functions at $x = 0$ and $x = \infty$, we get the following table:

	N_0	N_1	N_2	N_3	Variations in signs
$x = 0$	+	−	−	−	$v_a = 1$
$x = \infty$	+	+	+	−	$v_b = 1$

We see that $v_a - v_b = 0$; hence, the polynomial $N_0(x)$ has no roots along the entire imaginary axis. That is to say, $\text{Re}[Z(j\omega)] \geq 0$ for all ω.

The various results obtained above, together with the fact that the $Z(s)$ of Eq. (3.132) is real for real s, confirm it to be a positive real function.

Concluding Remarks

To conclude, an impedance function $Z(s)$, which is real for s real, is positive real, and the one-port network in question is therefore passive if

1. $Z(s)$ has no poles in the right half of the s-plane.

2. Any poles of $Z(s)$ on the imaginary axis are simple with positive and real residues.
3. $\text{Re}[Z(j\omega)] \geq 0$ for all ω.

If $Z(s)$ is positive real, then its reciprocal, the admittance function $Y(s) = 1/Z(s)$, is positive real too. Accordingly, a passive one-port network is both open-circuit and short-circuit stable with respect to the given port.

A network is active if it is not passive. An active one-port network may be

a) neither open-circuit nor short-circuit stable but stable for some intermediate resistive termination,

b) open-circuit stable but short-circuit unstable,

c) short-circuit stable but open-circuit unstable, or

d) both open-circuit and short-circuit stable.

Also, a physical active one-port network, having frequency-dependent parameters, may cease to be active above a certain frequency, whereas a passive one-port network stays passive for all frequencies.

REFERENCES

1. M. F. Gardner and J. L. Barnes, *Transients in Linear Systems*, p. 351. Wiley, 1942.
2. M. E. Hines, "High-frequency Negative Resistance Circuit Principles for Esaki Diode Applications," *B.S.T.J.*, **39,** 477 (1960).
3. J. O. Scanlan, *Analysis and Synthesis of Tunnel Diode Circuits*. Wiley, 1966.
4. E. A. Guillemin, *Introductory Circuit Theory*, p. 359. Wiley, 1953.
5. H. Nyquist, "Regeneration Theory," *B.S.T.J.*, **11,** 126, 1932.
6. E. A. Guillemin, *The Mathematics of Circuit Analysis*. Wiley, 1949.
7. H. W. Bode, *Network Analysis and Feedback Amplifier Design*, p. 166. Van Nostrand, 1945.
8. L. P. Hunter, *Handbook of Semiconductor Electronics*, McGraw-Hill, 1962.
9. N. Kryloff and N. Bogoliuboff, *Introduction to Nonlinear Mechanics*. Princeton University Press, 1947.
10. W. J. Cunningham, *Nonlinear Analysis*. McGraw-Hill, 1958.
11. C. S. Kim and A. Brandli, "High-frequency High-power Operation of Tunnel Diodes," *Trans. I.R.E.*, **CT-8,** 416 (1961).
12. O. Brune, "Synthesis of a Finite Two-terminal Network whose Driving-point Impedance is a Prescribed Function of Frequency," *J. Math. and Phys.*, **10,** 191 (1931).
13. J. V. Upensky, *Theory of Equations*. McGraw-Hill, 1948.

PROBLEMS

3.1 Determine the open-circuit and short-circuit natural frequencies of the network shown in Fig. P3.1 with respect to port 1–1′. What is the necessary condition for open-circuit instability?

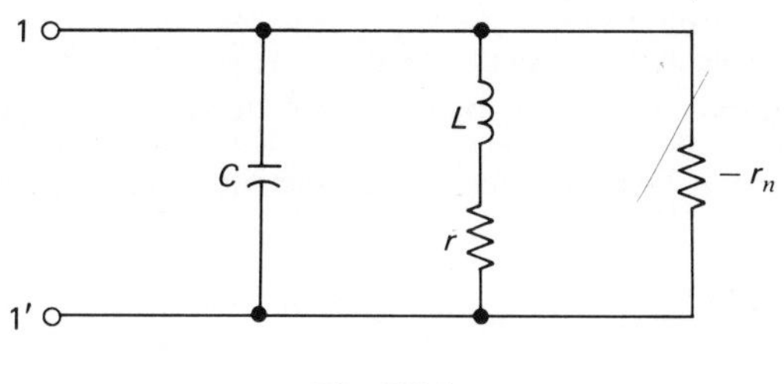

Fig. P3.1.

3.2 Assuming that the multivibrator circuit shown in Fig. P3.2 operates in its linear mode, determine the open-circuit natural frequencies with respect to the port 1–1′. Trace out their loci as the mutual conductance g_m of the pentodes, assumed to be identical, is varied from zero up to 4 mA/volt. The effect of the plate resistances of the tubes may be neglected.

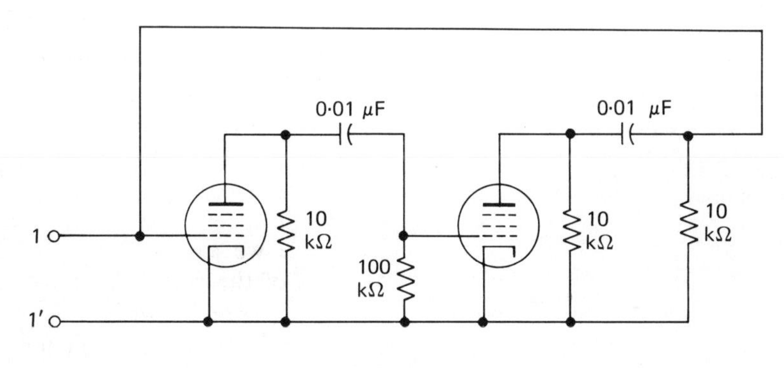

Fig. P3.2.

3.3 Test the following driving-point impedance functions for open-circuit stability, short-circuit stability and passivity:

a) $Z(s) = \dfrac{s^3 + 2s^2 + s + 12}{s^3 + 2s^2 + 2s + 1}$

b) $Z(s) = \dfrac{s^3 + 4s^2 + 5s + 6}{s^3 + s^2 + 2s + 8}$

c) $Z(s) = \dfrac{2s^3 + 3s^2 + 5s + 3}{2s^4 + 4s^3 + 7s^2 + 6s + 2}$

3.4 Fig. P3.4 shows the circuit model of a common-emitter amplifier with an inductive load. Determine the range of values of load inductance L for which the driving-point impedance at port 1–1′ becomes active.

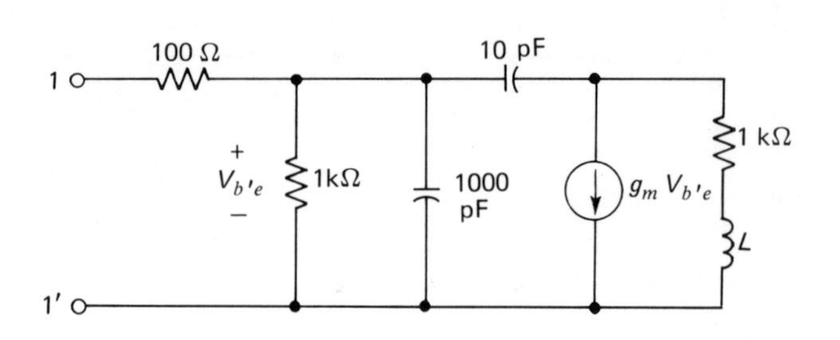

Fig. P3.4.

3.5 Show that[7]

a) The series combination of a short-circuit stable impedance $Z_1(s)$ and an open-circuit stable impedance $Z_2(s)$ will be short-circuit stable if $|Z_1(j\omega)| > |Z_2(j\omega)|$ for all ω.

b) The parallel combination of an open-circuit stable admittance $Y_a(s)$ and a short-circuit stable admittance $Y_b(s)$ will be open-circuit stable if $|Y_a(j\omega)| > |Y_b(j\omega)|$ for all ω.

3.6 For the transistor Colpitts oscillator shown in Fig. P3.6 determine the admittance measured looking into port 1–1′. If a resistor R is connected across this port, what is the minimum value of R for which the circuit can oscillate? It is given that

$$h_{ie} = 1\ \text{k}\Omega$$
$$h_{re} = 10^{-4}$$
$$h_{fe} = 100$$
$$h_{oe} = 25\ \mu\mho$$

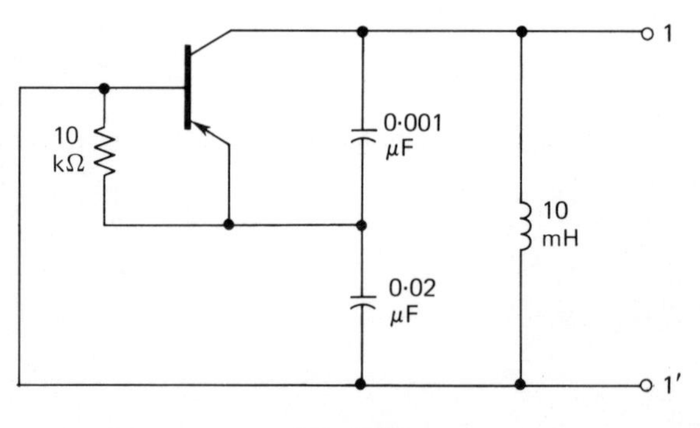

Fig. P3.6.

3.7 Determine the range of values of control parameter r_m for which the network shown in Fig. P3.7 is short-circuit unstable with respect to port 1–1′. Is it possible for the network to become open-circuit unstable? Why?

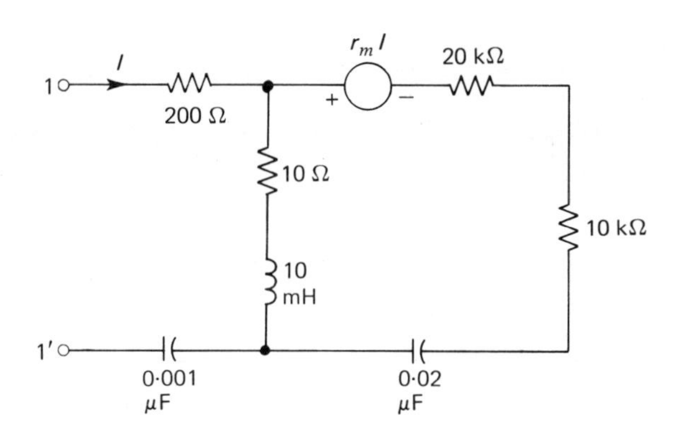

Fig. P3.7.

3.8 Determine the natural frequencies of the network shown in Fig. P3.8 with respect to port 1–1′. Is the network open-circuit stable? Is it short-circuit stable? Assume that the tubes are identical, having the parameters r_p = 10 kΩ and μ = 30.

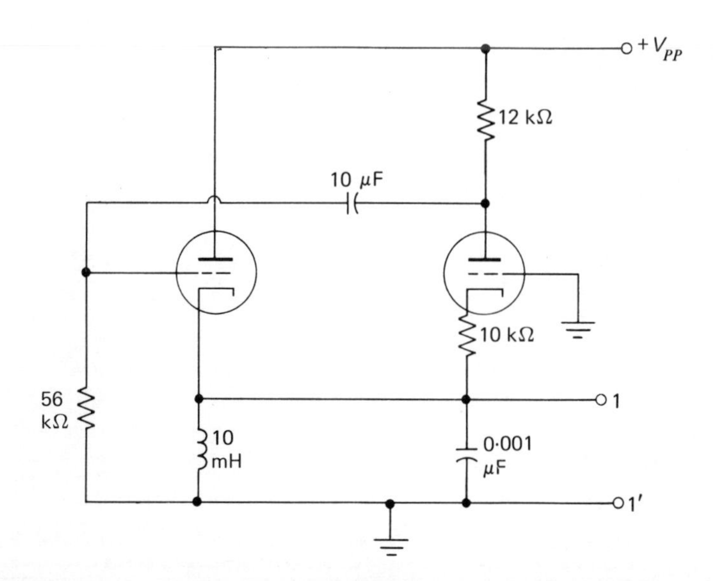

Fig. P3.8.

3.9 Determine the locus of the open-circuit natural frequencies of the network shown in Fig. P3.9 with respect to port 1–1′ as the gain μ of the amplifier is increased from zero up to infinity. What is the critical value of μ for which the network becomes unstable? Assume that the amplifier is unidirectional, and has infinite input impedance and zero output impedance.

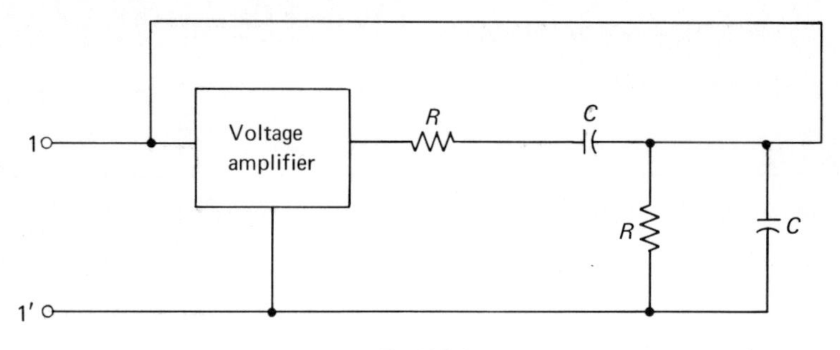

Fig. P3.9.

3.10 A parallel-tuned negative-resistance oscillator uses a tunnel diode having the V–I characteristic of Fig. 1.30, and a junction capacitance of 10 pF. The tuning inductor is 100 μH and has a Q-factor of 100. The load resistance is 200 Ω. Determine

a) the tuning capacitor to produce an oscillation frequency of 1 MHz, and
b) the amplitude of oscillation.

CHAPTER 4

TWO-PORT NETWORKS

4.1 INTRODUCTION

A two-port network is defined as a network with two pairs of terminals available for external connection. Transistor and vacuum-tube amplifiers, filters and transmission lines are important examples of two-port structures. The terminal pairs or ports of the network are usually designated as 1–1′ and 2–2′, as shown in Fig. 4.1. We will refer to the port 1–1′ as the *input port* and to 2–2′ as the *output port*. The external behaviour of the two-port network is completely determined if the current I_1 and voltage V_1 at the input port, and the current I_2 and voltage V_2 at the output port are known. The two-port network may be looked upon as a means of inter-relating these four quantities. In a linear network the inter-relationships will be linear too. Furthermore, the inter-relationships between I_1, V_1, I_2 and V_2 can be expressed in six different ways depending on which two of the four quantities are regarded as the independent variables and which two are regarded as the dependent variables. These six possibilities lead, in turn, to the following six different types of parameters, each with its own particular advantage, for characterizing the performance of a two-port network:

Independent variables	*Dependent variables*	*Parameter type*
I_1, I_2	V_1, V_2	Open-circuit impedances (z)
V_1, V_2	I_1, I_2	Short-circuit admittances (y)
I_1, V_2	V_1, I_2	Hybrid parameters (h)
V_1, I_2	I_1, V_2	Inverse hybrid parameters (g)
V_2, I_2	V_1, I_1	Chain parameters (A, B, C, D)
V_1, I_1	V_2, I_2	Inverse chain parameters ($\mathscr{A}$, $\mathscr{B}, \mathscr{C}, \mathscr{D}$)

4.2 TWO-PORT PARAMETERS

Open-circuit Impedance and Short-circuit Admittance Parameters

Suppose in Fig. 4.1 the current I_1 at the input port and the current I_2 at the output port are regarded as the independent variables. Then assuming that the network N contains no other independent sources, and that it is initially at rest, we find that application of the nodal analysis leads to the following equilibrium

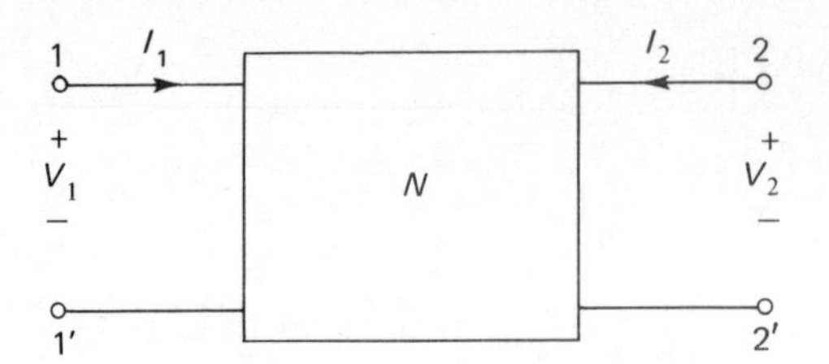

Fig. 4.1. Two-port network.

equation in the complex-frequency domain

$$\begin{bmatrix} I_1 \\ I_2 \\ 0 \\ \cdot \\ \cdot \\ \cdot \\ 0 \end{bmatrix} = \begin{bmatrix} Y_{11} & Y_{12} & Y_{13} & \cdots & Y_{1m} \\ Y_{21} & Y_{22} & Y_{23} & \cdots & Y_{2m} \\ Y_{31} & Y_{32} & Y_{33} & \cdots & Y_{3m} \\ \cdot & \cdot & \cdot & & \cdot \\ \cdot & \cdot & \cdot & & \cdot \\ \cdot & \cdot & \cdot & & \cdot \\ Y_{m1} & Y_{m2} & Y_{m3} & \cdots & Y_{mm} \end{bmatrix} \begin{bmatrix} V_1 \\ V_2 \\ V_3 \\ \cdot \\ \cdot \\ \cdot \\ V_m \end{bmatrix}. \tag{4.1}$$

Solving this equation for the port voltages V_1 and V_2, we get

$$\begin{bmatrix} V_1 \\ V_2 \end{bmatrix} = \begin{bmatrix} \dfrac{\Delta'_{11}}{\Delta'} & \dfrac{\Delta'_{21}}{\Delta'} \\ \dfrac{\Delta'_{12}}{\Delta'} & \dfrac{\Delta'_{22}}{\Delta'} \end{bmatrix} \begin{bmatrix} I_1 \\ I_2 \end{bmatrix} \tag{4.2}$$

where Δ' is the nodal-basis circuit determinant. Clearly, the coefficients on the right of Eq. (4.2) have the dimensions of an impedance. Introducing the definitions

$$\begin{bmatrix} z_{11} & z_{12} \\ z_{21} & z_{22} \end{bmatrix} = \begin{bmatrix} \dfrac{\Delta'_{11}}{\Delta'} & \dfrac{\Delta'_{21}}{\Delta'} \\ \dfrac{\Delta'_{12}}{\Delta'} & \dfrac{\Delta'_{22}}{\Delta'} \end{bmatrix}, \tag{4.3}$$

we may re-write Eq. (4.2) more simply as follows

$$\begin{bmatrix} V_1 \\ V_2 \end{bmatrix} = \begin{bmatrix} z_{11} & z_{12} \\ z_{21} & z_{22} \end{bmatrix} \begin{bmatrix} I_1 \\ I_2 \end{bmatrix}. \tag{4.4}$$

In the expanded form, we have

$$\begin{aligned} V_1 &= z_{11}I_1 + z_{12}I_2 \\ V_2 &= z_{21}I_1 + z_{22}I_2. \end{aligned} \tag{4.5}$$

Each equation in (4.5) expresses the voltage at a port as a linear combination of the port currents I_1 and I_2. The impedances z_{11}, z_{12}, z_{21} and z_{22} are called the *open-circuit impedance parameters* or simply the *z-parameters* of the two-port network. The adjective "open-circuit" signifies the fact that the individual parameters define the relationships between the port voltages and currents when one of the two ports of the network is open-circuited. Thus, suppose that the output port is open-circuited; this would constrain the output current I_2 to be zero. Under this condition, Eqs. (4.5) yield

$$\begin{aligned} z_{11} &= \left.\frac{V_1}{I_1}\right|_{I_2=0} \\ z_{21} &= \left.\frac{V_2}{I_1}\right|_{I_2=0}. \end{aligned} \tag{4.6}$$

Therefore, z_{11} is the *open-circuit input impedance* and z_{21} is the *open-circuit forward transfer impedance* of the two-port network. Similarly, if the input port is open circuited, and the input current I_1 is thereby constrained to be zero, we have from Eqs. (4.5) that

$$\begin{aligned} z_{12} &= \left.\frac{V_1}{I_2}\right|_{I_1=0} \\ z_{22} &= \left.\frac{V_2}{I_2}\right|_{I_1=0}. \end{aligned} \tag{4.7}$$

Therefore, z_{12} is the *open-circuit reverse transfer impedance* and z_{22} is the *open-circuit output impedance* of the two-port network.

If, next, the port voltages V_1 and V_2 are regarded as the independent variables, we may develop the relations expressing the port currents I_1 and I_2 in terms of V_1 and V_2 from the equilibrium equations of the network formulated on a loop basis. Alternatively, we may solve Eqs. (4.5) for I_1 and I_2 obtaining

$$\begin{bmatrix} I_1 \\ I_2 \end{bmatrix} = \begin{bmatrix} \dfrac{z_{22}}{\Delta_z} & \dfrac{-z_{12}}{\Delta_z} \\ \dfrac{-z_{21}}{\Delta_z} & \dfrac{z_{11}}{\Delta_z} \end{bmatrix} \begin{bmatrix} V_1 \\ V_2 \end{bmatrix}, \tag{4.8}$$

where Δ_z is the determinant of the z-matrix:

$$\Delta_z = \begin{vmatrix} z_{11} & z_{12} \\ z_{21} & z_{22} \end{vmatrix} = z_{11}z_{22} - z_{12}z_{21}. \tag{4.9}$$

The coefficients on the right of Eq. (4.8) have the dimensions of an admittance;

we may, therefore, re-write Eq. (4.8) in the simpler form

$$\begin{bmatrix} I_1 \\ I_2 \end{bmatrix} = \begin{bmatrix} y_{11} & y_{12} \\ y_{21} & y_{22} \end{bmatrix} \begin{bmatrix} V_1 \\ V_2 \end{bmatrix}, \tag{4.10}$$

where

$$\begin{bmatrix} y_{11} & y_{12} \\ y_{21} & y_{22} \end{bmatrix} = \begin{bmatrix} \dfrac{z_{22}}{\Delta_z} & \dfrac{-z_{12}}{\Delta_z} \\ \dfrac{-z_{21}}{\Delta_z} & \dfrac{z_{11}}{\Delta_z} \end{bmatrix}. \tag{4.11}$$

The y_{11}, y_{12}, y_{21} and y_{22} are called the *short-circuit admittance parameters* or simply the *y-parameters* of the two-port network. The adjective "short-circuit" is used to signify the fact that the individual parameters define the inter-relationships between the port voltages and currents when one of the two ports of the network is short-circuited. Thus, if the output port is short-circuited, that is $V_2 = 0$, we find from the expanded form of Eq. (4.10) that

$$\begin{aligned} y_{11} &= \left.\frac{I_1}{V_1}\right|_{V_2=0} \\ y_{21} &= \left.\frac{I_2}{V_1}\right|_{V_2=0}. \end{aligned} \tag{4.12}$$

Therefore, y_{11} is the *short-circuit input admittance* and y_{21} is the *short-circuit forward transfer admittance* of the two-port network. If, however, the input port is short-circuited, that is, $V_1 = 0$, we have

$$\begin{aligned} y_{12} &= \left.\frac{I_1}{V_2}\right|_{V_1=0} \\ y_{22} &= \left.\frac{I_2}{V_2}\right|_{V_1=0}, \end{aligned} \tag{4.13}$$

from which we see that y_{12} is the *short-circuit reverse transfer admittance* and y_{22} is the *short-circuit output admittance* of the two-port network.

We could have started our discussion of the two-port network from Eq. (4.10) and used it to express V_1 and V_2 in terms of I_1 and I_2. Thus, solving Eq. (4.10) for V_1 and V_2 leads to the following relation:

$$\begin{bmatrix} V_1 \\ V_2 \end{bmatrix} = \begin{bmatrix} \dfrac{y_{22}}{\Delta_y} & \dfrac{-y_{12}}{\Delta_y} \\ \dfrac{-y_{21}}{\Delta_y} & \dfrac{y_{11}}{\Delta_y} \end{bmatrix} \begin{bmatrix} I_1 \\ I_2 \end{bmatrix}, \tag{4.14}$$

where Δ_y is the determinant of the y-matrix:

$$\Delta_y = \begin{vmatrix} y_{11} & y_{12} \\ y_{21} & y_{22} \end{vmatrix} = y_{11}y_{22} - y_{12}y_{21}. \tag{4.15}$$

Comparing Eqs. (4.4) and (4.14) we observe

$$\begin{bmatrix} z_{11} & z_{12} \\ z_{21} & z_{22} \end{bmatrix} = \begin{bmatrix} \dfrac{y_{22}}{\Delta_y} & \dfrac{-y_{12}}{\Delta_y} \\ \dfrac{-y_{21}}{\Delta_y} & \dfrac{y_{11}}{\Delta_y} \end{bmatrix}. \tag{4.16}$$

That is to say, the y- and z-matrices of a two-port network are the inverse of each other.

Hybrid Parameters

Consider next the choice of I_1 and V_2 as the independent variables; then, by suitably manipulating the expanded form of Eq. (4.10) we find that

$$\begin{bmatrix} V_1 \\ I_2 \end{bmatrix} = \begin{bmatrix} \dfrac{1}{y_{11}} & \dfrac{-y_{12}}{y_{11}} \\ \dfrac{y_{21}}{y_{11}} & \dfrac{\Delta_y}{y_{11}} \end{bmatrix} \begin{bmatrix} I_1 \\ V_2 \end{bmatrix}. \tag{4.17}$$

For convenience, let

$$\begin{bmatrix} h_{11} & h_{12} \\ h_{21} & h_{22} \end{bmatrix} = \begin{bmatrix} \dfrac{1}{y_{11}} & \dfrac{-y_{12}}{y_{11}} \\ \dfrac{y_{21}}{y_{11}} & \dfrac{\Delta_y}{y_{11}} \end{bmatrix}. \tag{4.18}$$

Then, we can re-write Eq. (4.17) more simply in the form

$$\begin{bmatrix} V_1 \\ I_2 \end{bmatrix} = \begin{bmatrix} h_{11} & h_{12} \\ h_{21} & h_{22} \end{bmatrix} \begin{bmatrix} I_1 \\ V_2 \end{bmatrix}. \tag{4.19}$$

From Eq. (4.19) we obtain the following definitions for the h-parameters of a two-port network by letting $I_1 = 0$ or $V_2 = 0$; so that

$$\begin{aligned} h_{11} &= \left.\frac{V_1}{I_1}\right|_{V_2=0} & h_{12} &= \left.\frac{V_1}{V_2}\right|_{I_1=0} \\ h_{21} &= \left.\frac{I_2}{I_1}\right|_{V_2=0} & h_{22} &= \left.\frac{I_2}{V_2}\right|_{I_1=0}. \end{aligned} \tag{4.20}$$

Therefore, h_{11} is the *short-circuit input impedance*, h_{21} is the *short-circuit forward current-transfer ratio*, h_{12} is the *open-circuit reverse voltage-transfer ratio*, and h_{22} is the *open-circuit output admittance* of the two-port network. These parameters are called the *hybrid parameters* or simply *h-parameters* because they are dimensionally mixed and are evaluated under the mixed set of terminal conditions of a short-circuit across the output port and an open-circuit across the input port.

If V_1 and I_2 are regarded as the independent variables we obtain the so-called *inverse hybrid parameters* or *g-parameters*, defined by

$$\begin{bmatrix} I_1 \\ V_2 \end{bmatrix} = \begin{bmatrix} g_{11} & g_{12} \\ g_{21} & g_{22} \end{bmatrix} \begin{bmatrix} V_1 \\ I_2 \end{bmatrix}, \tag{4.21}$$

where

$$\begin{aligned} g_{11} &= \left.\frac{I_1}{V_1}\right|_{I_2=0} & g_{12} &= \left.\frac{I_1}{I_2}\right|_{V_1=0} \\ g_{21} &= \left.\frac{V_2}{V_1}\right|_{I_2=0} & g_{22} &= \left.\frac{V_2}{I_2}\right|_{V_1=0}. \end{aligned} \tag{4.22}$$

Therefore, g_{11} is the *open-circuit input admittance*, g_{21} is the *open-circuit forward voltage-transfer ratio*, g_{12} is the *short-circuit reverse current-transfer ratio*, and g_{22} is the *short-circuit output impedance.* We see that the g-parameters are also dimensionally mixed and are evaluated under the mixed conditions of an open-circuit across the output port and a short-circuit across the input port. Furthermore, from Eqs. (4.19) and (4.21) we see that the h- and g-matrices are the inverse of each other; hence,

$$\begin{bmatrix} g_{11} & g_{12} \\ g_{21} & g_{22} \end{bmatrix} = \frac{1}{\Delta_h} \begin{bmatrix} h_{22} & -h_{12} \\ -h_{21} & h_{11} \end{bmatrix} \tag{4.23}$$

and

$$\begin{bmatrix} h_{11} & h_{12} \\ h_{21} & h_{22} \end{bmatrix} = \frac{1}{\Delta_g} \begin{bmatrix} g_{22} & -g_{12} \\ -g_{21} & g_{11} \end{bmatrix}, \tag{4.24}$$

where Δ_h and Δ_g are determinants of the h- and g-matrices, respectively, that is,

$$\Delta_h = \begin{vmatrix} h_{11} & h_{12} \\ h_{21} & h_{22} \end{vmatrix} = h_{11}h_{22} - h_{12}h_{21} \tag{4.25}$$

$$\Delta_g = \begin{vmatrix} g_{11} & g_{12} \\ g_{21} & g_{22} \end{vmatrix} = g_{11}g_{22} - g_{12}g_{21}. \tag{4.26}$$

Chain Parameters

If we regard V_2 and I_2 as the independent variables, and V_1 and I_1 as the dependent variables, we get

$$\begin{bmatrix} V_1 \\ I_1 \end{bmatrix} = \begin{bmatrix} A & B \\ C & D \end{bmatrix} \begin{bmatrix} V_2 \\ -I_2 \end{bmatrix}. \tag{4.27}$$

This representation is, as we shall see later, useful in the study of a chain of two-port networks viewed from the input end. For this reason the matrix of coefficients on the right of Eq. (4.27) is referred to as the *chain matrix*. Historically, the chain parameters A, B, C and D were the first set of two-port parameters to be used in the study of transmission lines. In such networks if the input current I_1 flows into the line the output current I_2 naturally flows out, that is, the direction of the output current is opposite to that indicated in Fig. 4.1. This accounts for the introduction of the minus sign in Eq. (4.27).

From Eq. (4.27) we find that

$$\begin{aligned} A &= \left.\frac{V_1}{V_2}\right|_{I_2=0} & B &= \left.\frac{V_1}{-I_2}\right|_{V_2=0} \\ C &= \left.\frac{I_1}{V_2}\right|_{I_2=0} & D &= \left.\frac{I_1}{-I_2}\right|_{V_2=0}. \end{aligned} \tag{4.28}$$

Therefore, A is the reciprocal of the open-circuit voltage ratio for signal transmission in the forward direction, and the parameter D is the reciprocal of the short-circuit current ratio for the same direction of signal transmission. The parameters B and C are reciprocals of the short-circuit transfer admittance and the open-circuit transfer impedance, respectively, also for the forward direction of signal transmission.

For the last set of two-port parameters we have V_1 and I_1 as the independent variables as indicated by

$$\begin{bmatrix} V_2 \\ I_2 \end{bmatrix} = \begin{bmatrix} \mathscr{A} & \mathscr{B} \\ \mathscr{C} & \mathscr{D} \end{bmatrix} \begin{bmatrix} V_1 \\ -I_1 \end{bmatrix}. \tag{4.29}$$

This relation defines the output port variables in terms of the input port variables. The matrix of coefficients on the right of Eq. (4.29) is called the *inverse chain matrix*; it is useful in analysing a chain of two-port networks viewed from the output end. From Eq. (4.29) we see that

$$\begin{aligned} \mathscr{A} &= \left.\frac{V_2}{V_1}\right|_{I_1=0} & \mathscr{B} &= \left.\frac{V_2}{-I_1}\right|_{V_1=0} \\ \mathscr{C} &= \left.\frac{I_2}{V_1}\right|_{I_1=0} & \mathscr{D} &= \left.\frac{I_2}{-I_1}\right|_{V_1=0}. \end{aligned} \tag{4.30}$$

Conversions between Parameter Sets

In the study of two-port networks we often find it necessary to convert from one set of parameters to another. For example, Eqs. (4.11) and (4.16) give the relationships between the z- and y-parameters; Eqs. (4.23) and (4.24) give the relationships between the h- and g-parameters, and in Eq. (4.18) we have the h-parameters defined in terms of the y-parameters. Additional relationships between the various sets of parameters can be obtained by using the various sets of voltage–current relationships of the two-port network. In Table 4.1 we have summarized the inter-relationships among the various parameter sets of the two-port network. It is to be noted that the matrices appearing in any one row of Table 4.1 are equivalent.

Example 4.1. *T-section.* Consider the T-section of Fig. 4.2(a). It is convenient to start the characterization of this network by evaluating the z-matrix. Thus applying the definitions of Eqs. (4.6) and (4.7), we obtain the z-matrix

$$\begin{bmatrix} z_{11} & z_{12} \\ z_{21} & z_{22} \end{bmatrix} = \begin{bmatrix} Z_a + Z_b & Z_b \\ Z_b & Z_b + Z_c \end{bmatrix} \tag{4.31}$$

with the determinant

$$\Delta_z = Z_a Z_b + Z_b Z_c + Z_c Z_a \,. \tag{4.32}$$

Substituting Eqs. (4.31) in (4.11), we obtain the y-matrix

$$\begin{bmatrix} y_{11} & y_{12} \\ y_{21} & y_{22} \end{bmatrix} = \frac{1}{\Delta_z} \begin{bmatrix} Z_b + Z_c & -Z_b \\ -Z_b & Z_a + Z_b \end{bmatrix}. \tag{4.33}$$

To evaluate the chain parameters of the network we may use Table 4.1 to convert the z-matrix, as defined by Eq. (4.31), into the equivalent chain matrix, obtaining

$$\begin{bmatrix} A & B \\ C & D \end{bmatrix} = \begin{bmatrix} \dfrac{z_{11}}{z_{21}} & \dfrac{\Delta_z}{z_{21}} \\[2ex] \dfrac{1}{z_{21}} & \dfrac{z_{22}}{z_{21}} \end{bmatrix} = \begin{bmatrix} 1 + Z_a Y_b & Z_a + Z_c + Z_a Y_b Z_c \\ Y_b & 1 + Y_b Z_c \end{bmatrix} \tag{4.34}$$

where $Y_b = 1/Z_b$.

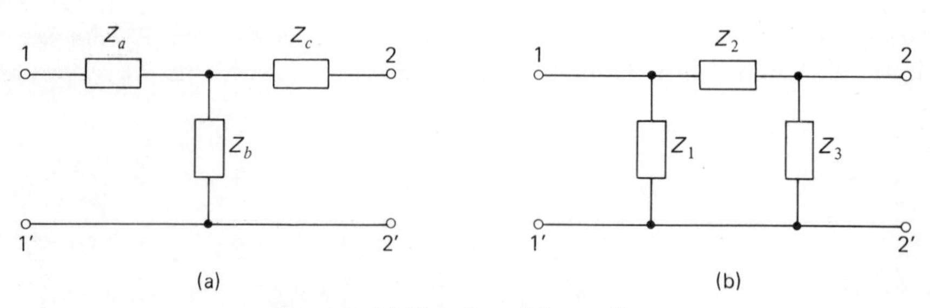

Fig. 4.2. (a) T-section; (b) π-section.

Table 4.1. Matrix Conversion Table

To \ From	z's		y's		h's		g's		A, B, C, D	
z's	z_{11}	z_{12}	$\frac{y_{22}}{\Delta_y}$	$\frac{-y_{12}}{\Delta_y}$	$\frac{\Delta_h}{h_{22}}$	$\frac{h_{12}}{h_{22}}$	$\frac{1}{g_{11}}$	$\frac{-g_{12}}{g_{11}}$	$\frac{A}{C}$	$\frac{AD-BC}{C}$
	z_{21}	z_{22}	$\frac{-y_{21}}{\Delta_y}$	$\frac{y_{11}}{\Delta_y}$	$\frac{-h_{21}}{h_{22}}$	$\frac{1}{h_{22}}$	$\frac{g_{21}}{g_{11}}$	$\frac{\Delta_g}{g_{11}}$	$\frac{1}{C}$	$\frac{D}{C}$
y's	$\frac{z_{22}}{\Delta_z}$	$\frac{-z_{12}}{\Delta_z}$	y_{11}	y_{12}	$\frac{1}{h_{11}}$	$\frac{-h_{12}}{h_{11}}$	$\frac{\Delta_g}{g_{22}}$	$\frac{g_{12}}{g_{22}}$	$\frac{D}{B}$	$\frac{-AD+BC}{B}$
	$\frac{-z_{21}}{\Delta_z}$	$\frac{z_{11}}{\Delta_z}$	y_{21}	y_{22}	$\frac{h_{21}}{h_{11}}$	$\frac{\Delta_h}{h_{11}}$	$\frac{-g_{21}}{g_{22}}$	$\frac{1}{g_{22}}$	$\frac{-1}{B}$	$\frac{A}{B}$
h's	$\frac{\Delta_z}{z_{22}}$	$\frac{z_{12}}{z_{22}}$	$\frac{1}{y_{11}}$	$\frac{-y_{12}}{y_{11}}$	h_{11}	h_{12}	$\frac{g_{22}}{\Delta_g}$	$\frac{-g_{12}}{\Delta_g}$	$\frac{B}{D}$	$\frac{AD-BC}{D}$
	$\frac{-z_{21}}{z_{22}}$	$\frac{1}{z_{22}}$	$\frac{y_{21}}{y_{11}}$	$\frac{\Delta_y}{y_{11}}$	h_{21}	h_{22}	$\frac{-g_{21}}{\Delta_g}$	$\frac{g_{11}}{\Delta_g}$	$\frac{-1}{D}$	$\frac{C}{D}$
g's	$\frac{1}{z_{11}}$	$\frac{-z_{12}}{z_{11}}$	$\frac{\Delta_y}{y_{22}}$	$\frac{y_{12}}{y_{22}}$	$\frac{h_{22}}{\Delta_h}$	$\frac{-h_{12}}{\Delta_h}$	g_{11}	g_{12}	$\frac{C}{A}$	$\frac{-AD+BC}{A}$
	$\frac{z_{21}}{z_{11}}$	$\frac{\Delta_z}{z_{11}}$	$\frac{-y_{21}}{y_{22}}$	$\frac{1}{y_{22}}$	$\frac{-h_{21}}{\Delta_h}$	$\frac{h_{11}}{\Delta_h}$	g_{21}	g_{22}	$\frac{1}{A}$	$\frac{B}{A}$
A, B, C, D	$\frac{z_{11}}{z_{21}}$	$\frac{\Delta_z}{z_{21}}$	$\frac{-y_{22}}{y_{21}}$	$\frac{-1}{y_{21}}$	$\frac{-\Delta_h}{h_{21}}$	$\frac{-h_{11}}{h_{21}}$	$\frac{1}{g_{21}}$	$\frac{g_{22}}{g_{21}}$	A	B
	$\frac{1}{z_{21}}$	$\frac{z_{22}}{z_{21}}$	$\frac{-\Delta_y}{y_{21}}$	$\frac{-y_{11}}{y_{21}}$	$\frac{-h_{22}}{h_{21}}$	$\frac{-1}{h_{21}}$	$\frac{g_{11}}{g_{21}}$	$\frac{\Delta_g}{g_{21}}$	C	D

Example 4.2. *π-section.* Consider, next, the π-section of Fig. 4.2(b). In this case it is convenient to start by evaluating the *y*-matrix. Applying the definitions of Eqs. (4.12) and (4.13) we obtain

$$\begin{bmatrix} y_{11} & y_{12} \\ y_{21} & y_{22} \end{bmatrix} = \begin{bmatrix} Y_1 + Y_2 & -Y_2 \\ -Y_2 & Y_2 + Y_3 \end{bmatrix}, \tag{4.35}$$

where $Y_1 = 1/Z_1$, $Y_2 = 1/Z_2$ and $Y_3 = 1/Z_3$. The determinant of this matrix is

$$\Delta_y = Y_1 Y_2 + Y_2 Y_3 + Y_3 Y_1 . \tag{4.36}$$

Substitution of Eq. (4.35) in (4.16) yields the *z*-matrix

$$\begin{bmatrix} z_{11} & z_{12} \\ z_{21} & z_{22} \end{bmatrix} = \frac{1}{\Delta_y} \begin{bmatrix} Y_2 + Y_3 & Y_2 \\ Y_2 & Y_1 + Y_2 \end{bmatrix}. \tag{4.37}$$

Using Table 4.1 to convert the *y*-matrix, as defined by Eq. (4.35), into the equivalent chain matrix, we obtain

$$\begin{bmatrix} A & B \\ C & D \end{bmatrix} = \begin{bmatrix} \dfrac{-y_{22}}{y_{21}} & \dfrac{-1}{y_{21}} \\ \dfrac{-\Delta_y}{y_{21}} & \dfrac{-y_{11}}{y_{21}} \end{bmatrix} = \begin{bmatrix} 1 + Z_2 Y_3 & Z_2 \\ Y_1 + Y_3 + Y_1 Z_2 Y_3 & 1 + Y_1 Z_2 \end{bmatrix}. \tag{4.38}$$

4.3 EXTERNAL CIRCUIT PROPERTIES OF TERMINATED TWO-PORT NETWORKS

The two-port network is normally used to couple a source to a load; Fig. 4.3 illustrates the general case of a source of voltage V_s and internal impedance Z_s coupled to a load of impedance Z_l. The *input impedance* Z_{in} of the terminated two-port network is defined as the impedance measured looking into the input port when the output port is terminated with Z_l; so that

$$Z_{in} = \frac{V_1}{I_1}. \tag{4.39}$$

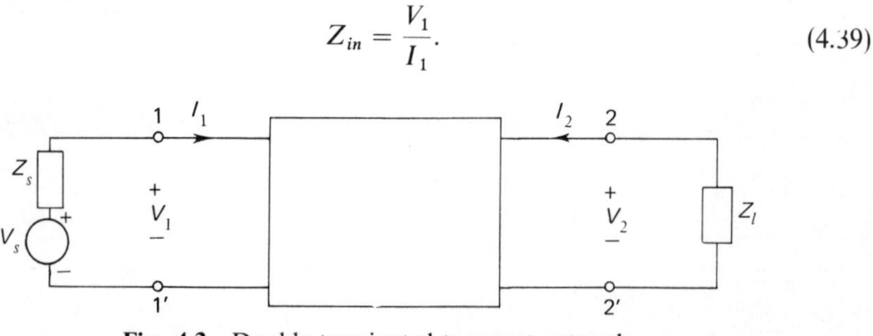

Fig. 4.3. Double-terminated two-port network.

When $Z_s = Z_{in}$ the voltage V_1 developed across the input port is $V_s/2$. We therefore have a second definition for the input impedance of a two-port network in that it is equal to that value of source impedance for which the voltage developed across the input port of the network is half the source voltage.

The *voltage gain* K_v is defined as the ratio of the output voltage V_2, developed across the load, over the input voltage V_1; so that

$$K_v = \frac{V_2}{V_1}. \tag{4.40}$$

The *current gain* K_i is defined as the ratio of output current I_2 over the input current I_1; so that

$$K_i = \frac{I_2}{I_1}. \tag{4.41}$$

When an impedance Z_l is connected across the output port, as in Fig. 4.3, the conditions existing at this port become constrained by the relation

$$V_2 = -I_2 Z_l. \tag{4.42}$$

From Eqs. (4.39)–(4.42) it follows, therefore, that the input impedance, voltage and current gains are related by

$$K_v Z_{in} = -K_i Z_l. \tag{4.43}$$

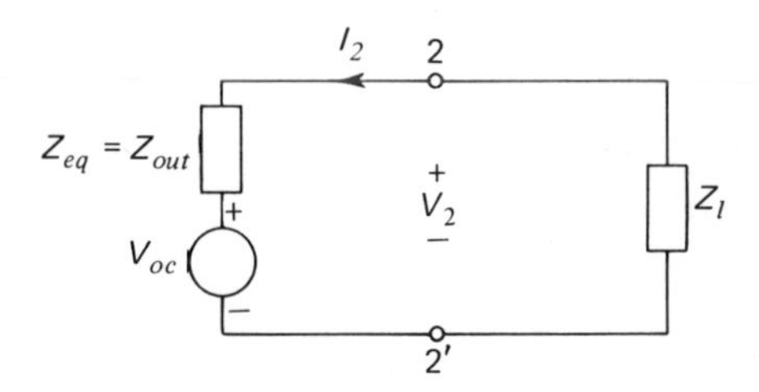

Fig. 4.4. Equivalent network defining the conditions that exist at the output port of a double-terminated two-port network.

The *output impedance* Z_{out} is defined as the impedance measured looking into the output port when the input port is terminated with the source impedance Z_s; so that

$$Z_{out} = \left.\frac{V_2}{I_2}\right|_{V_s=0}. \tag{4.44}$$

Suppose the two-port network, together with V_s and Z_s, is represented by its Thévenin equivalent network with respect to the port 2–2′, as in Fig. 4.4 where

V_{oc} denotes the open-circuit output voltage which is completely independent of the load impedance. Clearly, when $Z_l = Z_{out}$ the voltage V_2 developed across the load is $V_{oc}/2$, so that the output impedance of a two-port network may also be defined as that value of load impedance which causes the output voltage to assume half its open-circuit value.

Example 4.3. *Evaluation in terms of z-parameters.* In terms of the z-parameters, we have for the two-port network

$$\begin{aligned} V_1 &= z_{11}I_1 + z_{12}I_2 \\ V_2 &= z_{21}I_1 + z_{22}I_2. \end{aligned} \tag{4.45}$$

Eliminating V_2 between Eq. (4.42) and the second line of Eqs. (4.45), and then solving for the current gain, we obtain

$$K_i = \frac{-z_{21}}{z_{22} + Z_l}. \tag{4.46}$$

Next, eliminating I_2 between the first line of Eqs. (4.45) and Eq. (4.46), and then solving for the input impedance, we get

$$Z_{in} = z_{11} - \frac{z_{12}z_{21}}{z_{22} + Z_l} \tag{4.47}$$

or

$$Z_{in} = \frac{\Delta_z + z_{11}Z_l}{z_{22} + Z_l}. \tag{4.48}$$

To evaluate the voltage gain we may use Eqs. (4.43), (4.46) and (4.48), obtaining

$$K_v = \frac{z_{21}Z_l}{\Delta_z + z_{11}Z_l}. \tag{4.49}$$

To evaluate the output impedance, consider Fig. 4.3 where the input port has been terminated with the source impedance Z_s; with $V_s = 0$, we find that the input current I_1 and input voltage V_1 are constrained as

$$V_1 = -I_1 Z_s. \tag{4.50}$$

Therefore, solving Eqs. (4.45) and (4.50) for the ratio V_2/I_2 gives the output impedance as

$$Z_{out} = z_{22} - \frac{z_{12}z_{21}}{z_{11} + Z_s}$$

or

$$Z_{out} = \frac{\Delta_z + z_{22}Z_s}{z_{11} + Z_s}. \tag{4.51}$$

In a similar manner we can calculate the input impedance, current gain, voltage gain and output impedance of the terminated two-port network in terms of the other sets of parameters. The results are summarized in Table 4.2.

Table 4.2. Circuit Properties of Terminated Two-Port Network

Expressed in terms of	z's	y's	h's	g's	A, B, C, D
K_v	$\frac{z_{21}Z_l}{\Delta_z + z_{11}Z_l}$	$\frac{-y_{21}Z_l}{1 + y_{22}Z_l}$	$\frac{-h_{21}Z_l}{h_{11} + \Delta_h Z_l}$	$\frac{g_{21}Z_l}{g_{22} + Z_l}$	$\frac{Z_l}{B + AZ_l}$
K_i	$\frac{-z_{21}}{z_{22} + Z_l}$	$\frac{y_{21}}{y_{11} + \Delta_y Z_l}$	$\frac{h_{21}}{1 + h_{22}Z_l}$	$\frac{-g_{21}}{\Delta_g + g_{11}Z_l}$	$\frac{-1}{D + CZ_l}$
Z_{in}	$\frac{\Delta_z + z_{11}Z_l}{z_{22} + Z_l}$	$\frac{1 + y_{22}Z_l}{y_{11} + \Delta_y Z_l}$	$\frac{h_{11} + \Delta_h Z_l}{1 + h_{22}Z_l}$	$\frac{g_{22} + Z_l}{\Delta_g + g_{11}Z_l}$	$\frac{B + AZ_l}{D + CZ_l}$
Z_{out}	$\frac{\Delta_z + z_{22}Z_s}{z_{11} + Z_s}$	$\frac{1 + y_{11}Z_s}{y_{22} + \Delta_y Z_s}$	$\frac{h_{11} + Z_s}{\Delta_h + h_{22}Z_s}$	$\frac{g_{22} + \Delta_g Z_s}{1 + g_{11}Z_s}$	$\frac{B + DZ_s}{A + CZ_s}$

Power Gain, Available Gain and Transducer Gain

In the case of such two-port networks as transistors which require a finite power input, by virtue of a finite input impedance, it is important to have a measure of power flow in the device. The simplest measure of power flow in a two-port network is the *power gain* G defined as follows (see Fig. 4.5)

$$G = \frac{P_{out}}{P_{in}}, \tag{4.52}$$

where P_{out} is the power delivered to the load, and P_{in} is the power supplied to the network. Clearly, the power gain is a function of the two-port parameters and the load impedance; it is, however, independent of the source impedance.

The second measure of power flow in a two-port network is the so-called *available power gain* G_A defined by the formula (see Fig. 4.5)

$$G_A = \frac{P_{avo}}{P_{avs}}, \tag{4.53}$$

where P_{avo} is the power available at the output port of the network, and P_{avs} is the power available from the source under matched conditions. The available power gain G_A is a function of the two-port parameters and the source impedance; it is, however, independent of the load impedance.

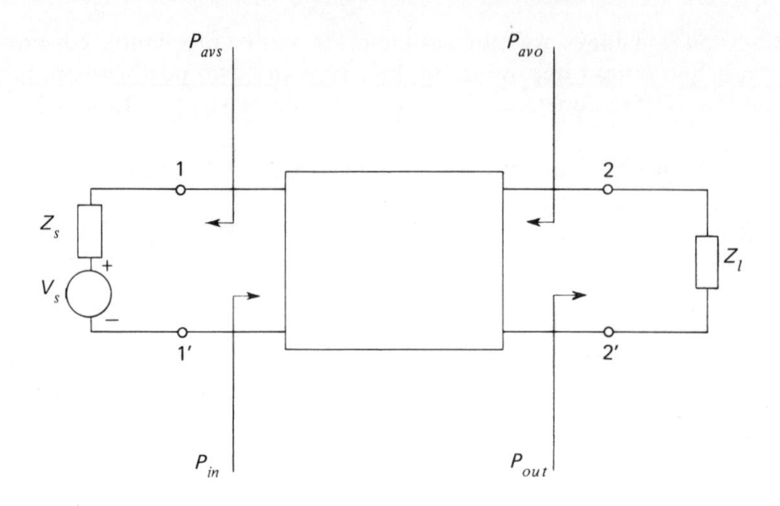

Fig. 4.5. Illustrating the four measures of power flow in a terminated two-port network.

The third measure of power flow is the *transducer power gain* G_T defined by

$$G_T = \frac{P_{out}}{P_{avs}}, \tag{4.54}$$

which is a function of the two-port parameters, the load impedance and the source impedance. The transducer power gain is a particularly important quantity as it compares the power which the two-port network delivers to the load with the power which the source is capable of supplying to the network under optimum conditions.

Since the power gain G does not involve the source impedance Z_s, and the available power gain G_A does not involve the load impedance Z_l, whereas the transducer power gain G_T involves both Z_s and Z_l, it is not possible to formulate general relationships between G, G_A and G_T. We can, however, make the following observations: if the source impedance is conjugately matched to the input impedance of the two-port network, then $P_{in} = P_{avs}$; otherwise, we find that $P_{in} < P_{avs}$. We may, therefore, state that

$$G_T \le G. \tag{4.55}$$

When the load impedance is conjugately matched to the output impedance of the two-port network, then $P_{out} = P_{avo}$; otherwise, we have $P_{out} < P_{avo}$. Therefore,

$$G_T \le G_A. \tag{4.56}$$

In other words, the power gain and available power gain provide upper bounds on the transducer power gain. Furthermore, when the two-port network is conjugately matched to the source and load impedances at its input and output ports, respectively, the gains G, G_T and G_A assume a common maximum value,

$$G_{max} = G_{T,max} = G_{A,max}. \tag{4.57}$$

This relation assumes that the *maximum available power gain* $G_{A,max}$ is finite, which requires the two-port network to be absolutely stable at the frequency at which $G_{A,\max}$ occurs. The criterion for absolute stability and the optimum terminating impedances which result in the maximum available power gain will be developed in the next chapter.

Example 4.4. To illustrate the above definitions for G, G_A and G_T, we shall evaluate them in terms of the z-parameters. The power input and power output are given as follows:

$$P_{in} = |I_1|^2 \operatorname{Re} Z_{in} \tag{4.58}$$

$$P_{out} = |I_2|^2 \operatorname{Re} Z_l. \tag{4.59}$$

Now, the input impedance and current gain of a two-port network are related to the z-parameters by

$$Z_{in} = z_{11} - \frac{z_{12}z_{21}}{z_{22} + Z_l} \tag{4.60}$$

$$K_i = \frac{I_2}{I_1} = \frac{-z_{21}}{z_{22} + Z_l}. \tag{4.61}$$

Therefore, we find that using Eqs. (4.58)–(4.61) gives the power gain to be

$$G = \frac{P_{out}}{P_{in}} = \frac{|z_{21}|^2 \operatorname{Re} Z_l}{|z_{22} + Z_l|^2 \operatorname{Re}[z_{11} - z_{12}z_{21}/(z_{22} + Z_l)]}. \tag{4.62}$$

The power available from the source is

$$P_{avs} = \frac{|V_s|^2}{4 \operatorname{Re} Z_s}. \tag{4.63}$$

The power available from the network may be evaluated from the Thévenin equivalent network of Fig. 4.4 which represents the conditions seen looking back from the load in the terminated network. The open-circuit voltage V_{oc} and Thévenin equivalent impedance $Z_{eq} = Z_{out}$ are given by the following expressions:

$$V_{oc} = \frac{z_{21}V_s}{z_{11} + Z_s} \tag{4.64}$$

$$Z_{eq} = z_{22} - \frac{z_{12}z_{21}}{z_{11} + Z_s}. \tag{4.65}$$

Therefore, the power available from the network results when $Z_l = Z_{eq}^*$, yielding

$$P_{avo} = \frac{|V_{oc}|^2}{4 \text{ Re } Z_{eq}} \tag{4.66}$$

or

$$P_{avo} = \frac{|z_{21}|^2 |V_s|^2}{4|z_{11} + Z_s|^2 \text{ Re}[z_{22} - z_{12}z_{21}/(z_{11} + Z_s)]}. \tag{4.67}$$

Using Eqs. (4.63) and (4.67), we find that the available power gain is

$$G_A = \frac{P_{avo}}{P_{avs}} = \frac{|z_{21}|^2 \text{ Re} Z_s}{|z_{11} + Z_s|^2 \text{ Re}[z_{22} - z_{12}z_{21}/(z_{11} + Z_s)]}. \tag{4.68}$$

From Eqs. (4.59)–(4.63) we find that the transducer power gain is

$$G_T = \frac{P_{out}}{P_{avs}} = \frac{4|z_{21}|^2 \text{ Re } Z_s \text{ Re } Z_l}{|(z_{11} + Z_s)(z_{22} + Z_l) - z_{12}z_{21}|^2}. \tag{4.69}$$

In a similar manner we may evaluate the three power gains of a two-port network in terms of the other parameter sets.

4.4 EQUIVALENT CIRCUITS FOR TWO-PORT NETWORKS

A two-port network N_1 is said to be equivalent to another two-port network N_2 if the terminal voltages and currents remain unchanged when N_1 is substituted for N_2. This implies that the same z-, y- or any of the other parameter sets may be used to characterize the performance of two equivalent two-port networks. The particular set best suited for finding the equivalent two-port network may depend on the structure of the given network and the use which is to be made of the equivalent two-port network.

In terms of the z-parameters we have

$$\begin{aligned} V_1 &= z_{11}I_1 + z_{12}I_2 \\ V_2 &= z_{21}I_1 + z_{22}I_2. \end{aligned} \tag{4.70}$$

These two relations can be represented by the equivalent circuit of Fig. 4.6(a) which involves two current-controlled voltage sources. Clearly, the network of Fig. 4.6(a) is equivalent to the general two-port network of Fig. 4.1 with respect to the available two ports. If in Fig. 4.6(a) we transform the voltage sources into current sources we get the equivalent network of Fig. 4.6(b) where, at the input end, we have an impedance z_{11} in parallel with a current-controlled current source $(z_{12}/z_{11})I_2$, and at the output end we have an impedance z_{22} in parallel with a current-controlled current source $(z_{21}/z_{22})I_1$.

To develop an equivalent circuit involving a single controlled source, consider the T-network of Fig. 4.7(a) which consists of four elements: three impedances

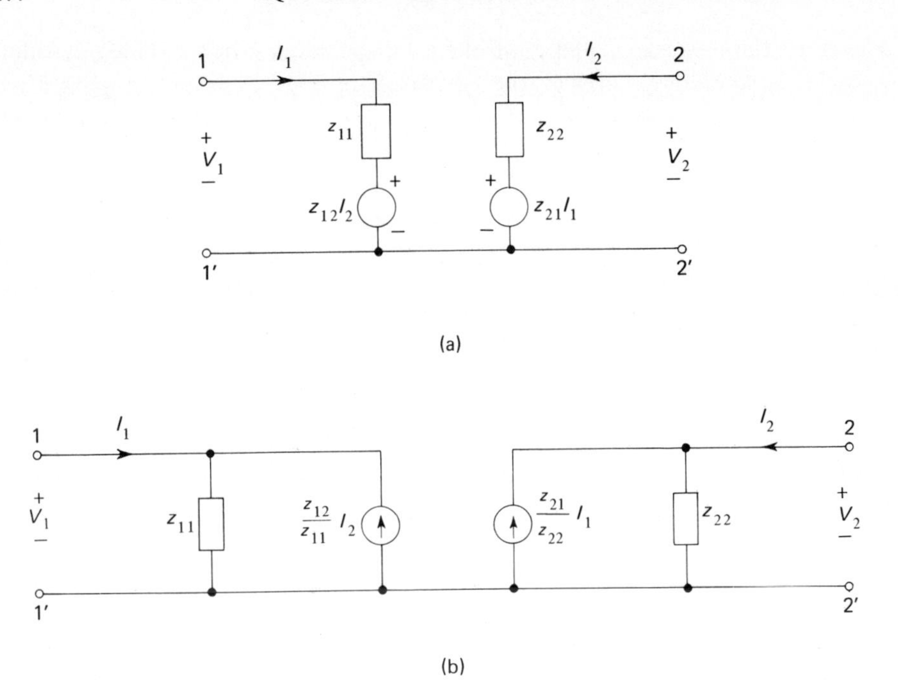

Fig. 4.6. Equivalent circuits based on the z-parameter representation.

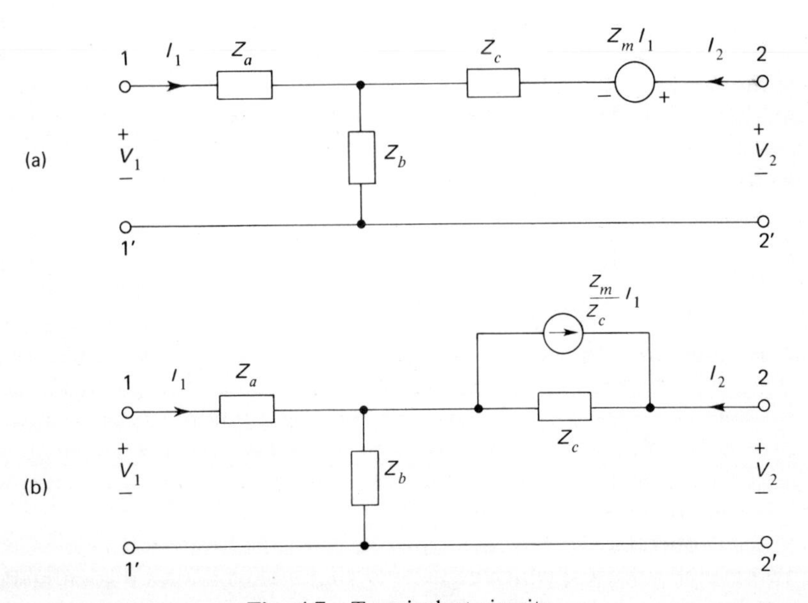

Fig. 4.7. T-equivalent circuits.

Z_a, Z_b and Z_c and a current-controlled voltage source $Z_m I_1$. This equivalent T-network, therefore, satisfies the condition of four independent parameters required to define uniquely the performance of the two-port network. Evaluating the z-parameters of the network of Fig. 4.7(a), we obtain

$$\begin{aligned} z_{11} &= Z_a + Z_b \\ z_{12} &= Z_b \\ z_{21} &= Z_m + Z_b \\ z_{22} &= Z_c + Z_b. \end{aligned} \tag{4.71}$$

Solving Eqs. (4.71) for Z_a, Z_b, Z_c and Z_m, we find that they are related to the z-parameters of the general two-port network as follows:

$$\begin{aligned} Z_a &= z_{11} - z_{12} \qquad & Z_c &= z_{22} - z_{12} \\ Z_b &= z_{12} & Z_m &= z_{21} - z_{12}. \end{aligned} \tag{4.72}$$

In Fig. 4.7(a) we may replace the series combination of Z_c and the controlled voltage source $Z_m I_1$ by the parallel combination of Z_c and a current-controlled current source $(Z_m/Z_c)I_1$; the result of this change is shown in Fig. 4.7(b). It is significant to note that for the special case of $z_{12} = z_{21}$ the controlled source vanishes from both equivalent networks of Fig. 4.7. The two equivalent T-networks of Fig. 4.7 are the basis of the circuit models shown in Fig. 4.8 which, historically, were the earliest set of models used to characterize the performance of a transistor at low frequencies.[1]

The equivalent circuits developed thus far have been derived from the voltage–current relations involving the z-parameters. Consider next the set of y-parameter equations:

$$\begin{aligned} I_1 &= y_{11}V_1 + y_{12}V_2 \\ I_2 &= y_{21}V_1 + y_{22}V_2. \end{aligned} \tag{4.73}$$

These two relations lead to the equivalent circuit of Fig. 4.9(a) which involves two voltage-controlled current sources. By transforming the current sources into voltage sources we get the equivalent circuit of Fig. 4.9(b) where, at the input end, we have an admittance y_{11} in series with a voltage-controlled voltage source $(y_{12}/y_{11})V_2$, and at the output end we have an admittance y_{22} in series with a voltage-controlled voltage source $(y_{21}/y_{22})V_1$.

For an equivalent circuit involving one controlled source only, consider the π-network of Fig. 4.10(a) which consists of four elements: three admittances Y_1, Y_2 and Y_3 and a voltage-controlled current source $Y_m V_1$. Evaluating the y-

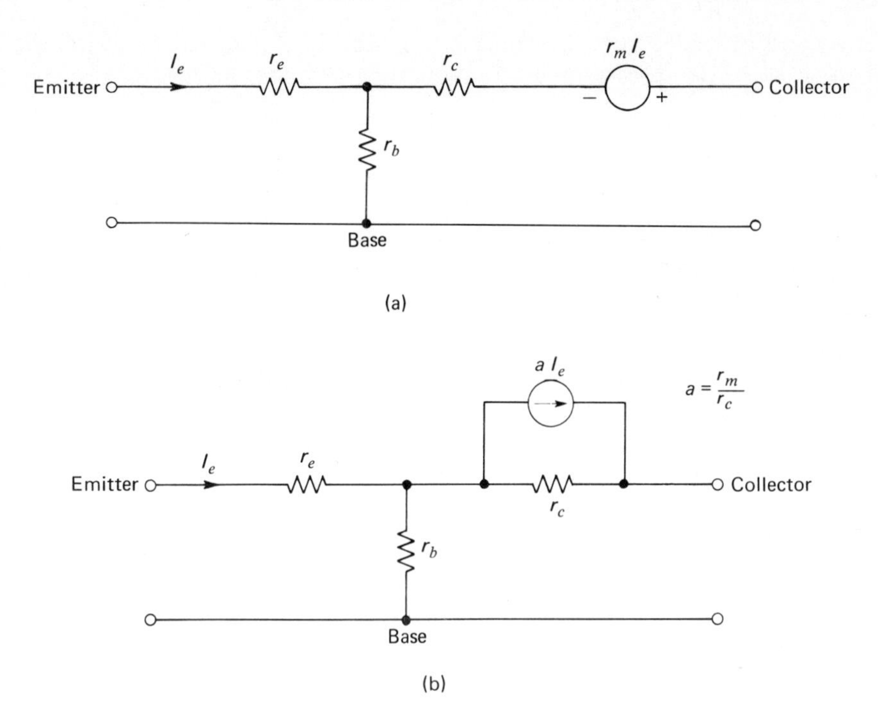

Fig. 4.8. Low-frequency *T*-circuit models of a transistor.

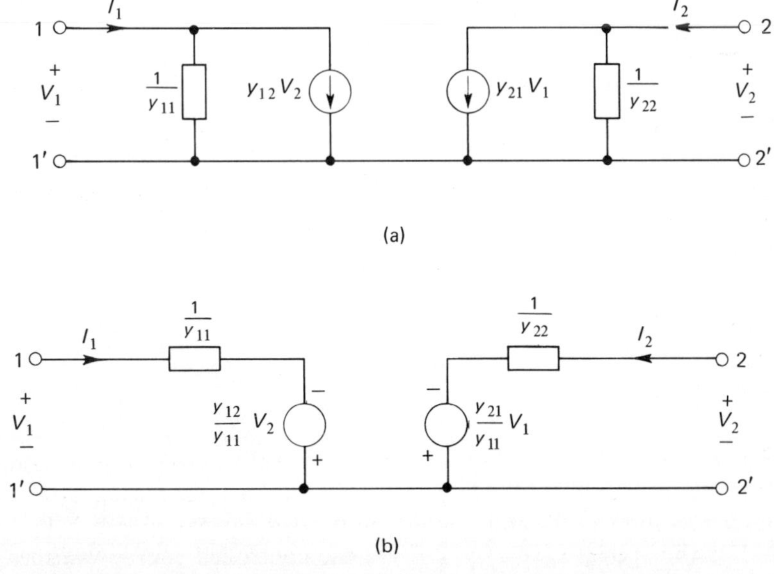

Fig. 4.9. Equivalent circuits based on the *y*-parameter representation.

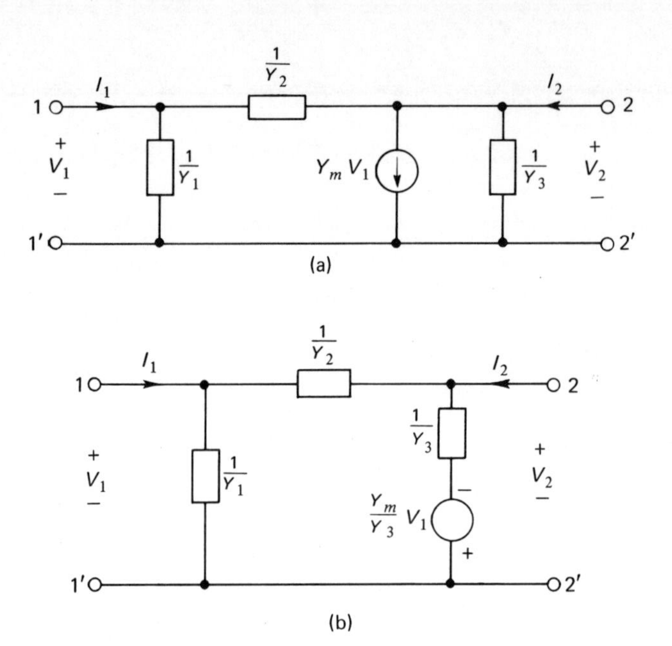

Fig. 4.10. π-equivalent circuits.

parameters of this network, we get

$$\begin{aligned} y_{11} &= Y_1 + Y_2 \\ y_{12} &= -Y_2 \\ y_{21} &= Y_m - Y_2 \\ y_{22} &= Y_2 + Y_3. \end{aligned} \tag{4.74}$$

Solving Eqs. (4.74) for Y_1, Y_2, Y_3 and Y_m, we find that they are defined in terms of the y-parameters of the general two-port network as follows:

$$\begin{aligned} Y_1 &= y_{11} + y_{12} \\ Y_2 &= -y_{12} \\ Y_3 &= y_{22} + y_{12} \\ Y_m &= y_{21} - y_{12}. \end{aligned} \tag{4.75}$$

In the equivalent π-network of Fig. 4.10(b) we have substituted the series combination of the admittance Y_3 and voltage-controlled voltage source $(Y_m/Y_3)V_1$ for the parallel combination of Y_3 and voltage–controlled current source $Y_m V_1$ of Fig. 4.10(a). For the special case of $y_{12} = y_{21}$ the controlled source vanishes from both equivalent networks of Fig. 4.10. The two equivalent π-networks of Fig. 4.10

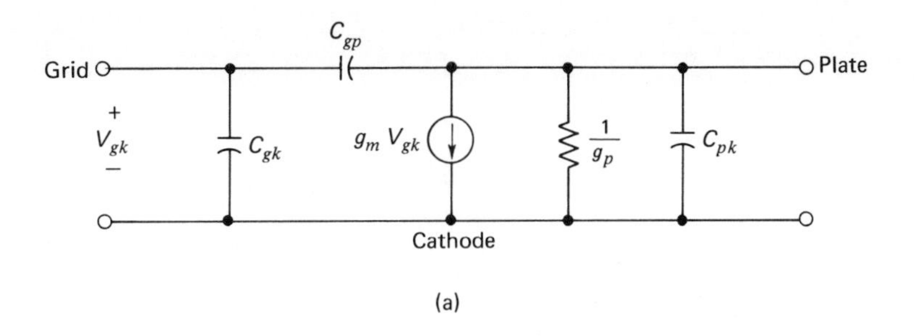

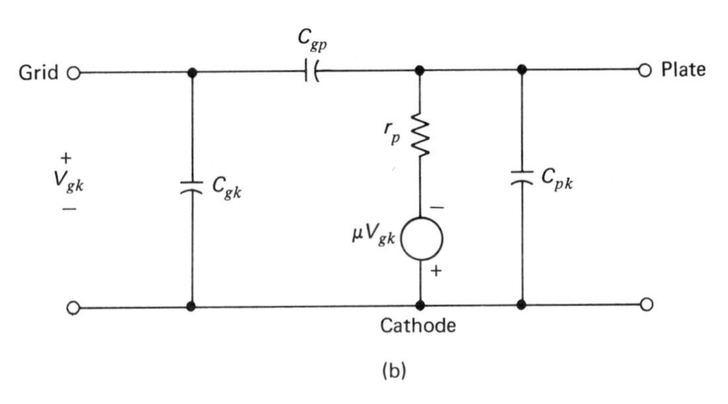

Fig. 4.11. Circuit models of a vacuum-tube triode.

are the basis of the circuit models shown in Fig. 4.11 which are used for characterizing a vacuum-tube triode, with the interelectrode capacitances of the tube included.

Another equivalent circuit for a two-port network is obtained from considering the h-parameter equations, which are repeated here for convenience:

$$\begin{aligned} V_1 &= h_{11}I_1 + h_{12}V_2 \\ I_2 &= h_{21}I_1 + h_{22}V_2. \end{aligned} \tag{4.76}$$

From these two voltage–current relations we directly deduce the so-called *hybrid equivalent circuit* shown in Fig. 4.12 involving a voltage-controlled voltage source at the input end and a current-controlled current source at the output end. This equivalent circuit is found to be particularly useful in defining the performance of a transistor, principally because of the relative ease with which the various h-parameters of a transistor can be measured directly and accurately. From the definitions given in Eqs. (4.20) we see that the measurement of h_{11} and h_{21} requires short-circuiting the output port, while the measurement of h_{12} and h_{22} requires open-circuiting the input port. Consider for example a transistor operated with its emitter terminal common to the input and output ports, as in Fig. 4.13(a);

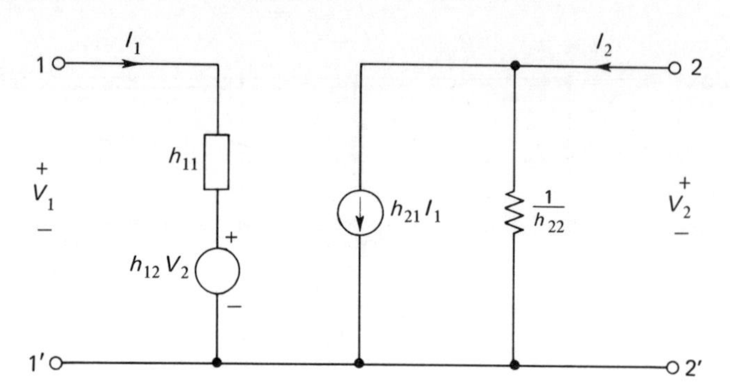

Fig. 4.12. Equivalent circuit based on h-parameters.

such a configuration is found to have a low input impedance (of the order of 1 kΩ) and a high output impedance (of the order of 50 kΩ). Therefore, it is possible to simulate fairly accurately open-circuit and short-circuit conditions across the input and output ports of the transistor, respectively, with the result that all four h-parameters are easy to measure. In the study of transistors it is normal practice to use an additional subscript to signify the particular mode in which the device is operated. For example, in the common-emitter configuration of Fig. 4.13(a) we have h_{11e}, h_{12e}, h_{21e}, and h_{22e} as the pertinent transistor h-parameters. The notation of h_{ie}, h_{re}, h_{fe} and h_{oe} is also used to symbolize the common-emitter h-parameters with

$$\begin{aligned} h_{ie} &= h_{11e} \qquad & h_{re} &= h_{12e} \\ h_{fe} &= h_{21e} \qquad & h_{oe} &= h_{22e}. \end{aligned} \tag{4.77}$$

We can thus characterize the performance of a common-emitter configuration by means of the equivalent network shown in Fig. 4.13(b).

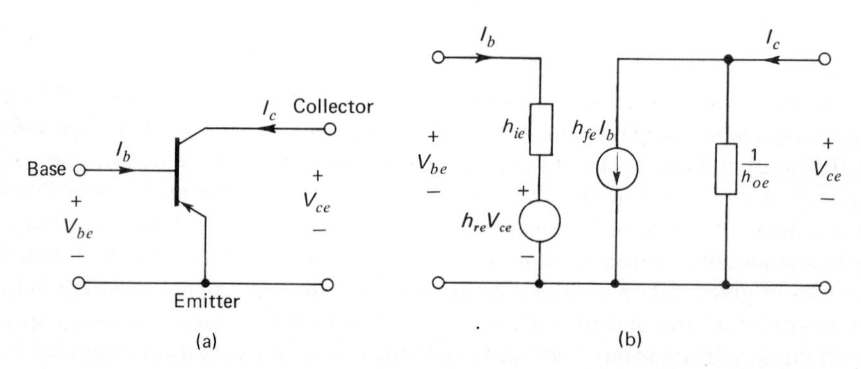

Fig. 4.13. (a) Common-emitter configuration of a transistor; (b) Hybrid model.

Finally, consider the g-parameter set of voltage–current relations

$$\begin{aligned} I_1 &= g_{11}V_1 + g_{12}I_2 \\ V_2 &= g_{21}V_1 + g_{22}I_2, \end{aligned} \tag{4.78}$$

which may be represented by the equivalent network shown in Fig. 4.14 for a general two-port network. At the input end we have a current-controlled current source and at the output end we have a voltage-controlled voltage source.

It is significant to observe that it is not possible to formulate equivalent circuits directly on the basis of the chain matrix and inverse chain matrix parameters of Eqs. (4.27) and (4.29).

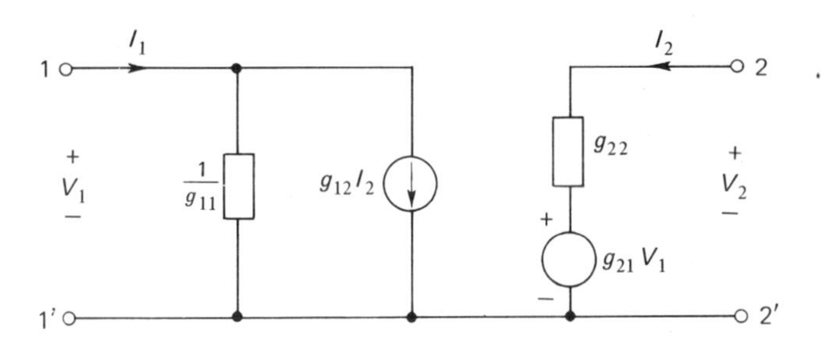

Fig. 4.14. Equivalent circuit based on g-parameters.

4.5 THE INDEFINITE ADMITTANCE MATRIX FOR A THREE-TERMINAL NETWORK

The vacuum-tube triode and transistor are basically three-terminal devices with each one having three useful modes of operation as a two-port network, depending on which of the three terminals is arranged common to the input and output ports. The vacuum-tube triode can thus be operated in the *common-cathode*, *common-plate* (that is, *cathode-follower*) and *common-grid* configurations, while the transistor can be operated in the *common-emitter*, *common-collector* (that is, *emitter-follower*) and *common-base* configurations, as illustrated in Fig. 4.15. The various configurations of the vacuum tube and transistor may be looked upon as special cases of the generalized *three-terminal network* of Fig. 4.16 where the three terminal voltages V_1, V_2 and V_3 are shown defined with respect to an arbitrary ground. The important consequence of this arrangement is that the admittance matrix of the three-terminal network is independent of orientation, and that the parameters of any specified orientation can be readily obtained as a special case.

If in Fig. 4.16 we regard the terminal voltages V_1, V_2 and V_3 as the independent variables, then each of the terminal currents I_1, I_2 and I_3 may be expressed as a

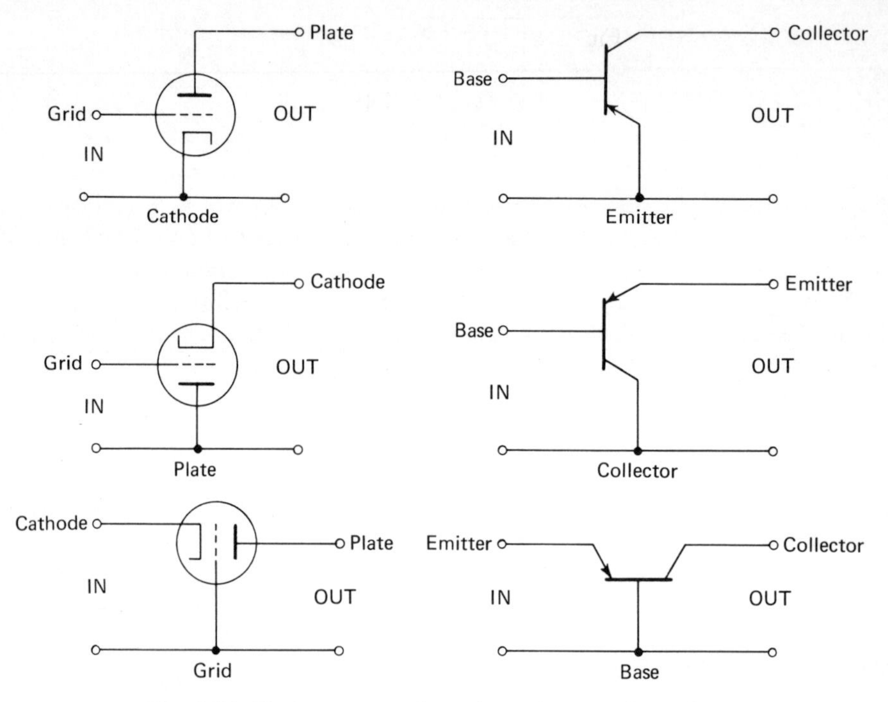

Fig. 4.15. Basic vacuum tube and transistor configurations.

linear combination of the three terminal voltages as follows:

$$\begin{aligned} I_1 &= y_{11}V_1 + y_{12}V_2 + y_{13}V_3 \\ I_2 &= y_{21}V_1 + y_{22}V_2 + y_{23}V_3 \\ I_3 &= y_{31}V_1 + y_{32}V_2 + y_{33}V_3. \end{aligned} \tag{4.79}$$

A typical admittance term y_{jk} on the right of Eqs. (4.79) is defined as the current flowing into terminal j divided by the voltage impressed between terminal k and the arbitrary ground, all other terminal voltages being reduced to zero.

In matrix form we can re-write Eqs. (4.79) as

$$\begin{bmatrix} I_1 \\ I_2 \\ I_3 \end{bmatrix} = \begin{bmatrix} y_{11} & y_{12} & y_{13} \\ y_{21} & y_{22} & y_{23} \\ y_{31} & y_{32} & y_{33} \end{bmatrix} \begin{bmatrix} V_1 \\ V_2 \\ V_3 \end{bmatrix}, \tag{4.80}$$

where the matrix of coefficients is called, after Shekel,[2] the *indefinite admittance matrix*, because the terminal which is to be used as common has been left undefined. The indefinite admittance matrix has three basic properties:

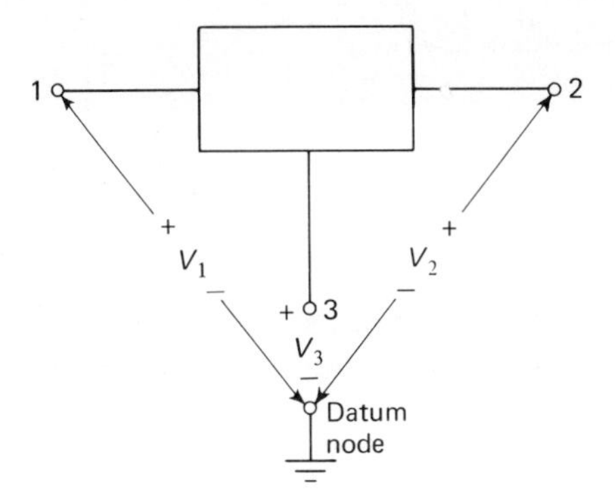

Fig. 4.16. Floating three-terminal network.

1. Each column of the indefinite admittance matrix has an algebraic sum equal to zero. This is a direct consequence of Kirchhoff's current law according to which the terminal currents I_1, I_2 and I_3 entering the network must add up to zero for any specification of the terminal voltages; so that

$$I_1 + I_2 + I_3 = 0. \tag{4.81}$$

Suppose V_2 and V_3 are specified to be zero; then from Eqs. (4.79) we have

$$\begin{aligned} I_1 &= y_{11}V_1 \\ I_2 &= y_{21}V_1 \\ I_3 &= y_{31}V_1. \end{aligned} \tag{4.82}$$

Adding these three equations and noting the condition of Eq. (4.81) we see that as V_1 does not have to be zero, then

$$y_{11} + y_{21} + y_{31} = 0, \tag{4.83}$$

which states that the sum of the admittances in the first column is zero. In a similar manner we can show that the sum of the admittances in any other column of the indefinite admittance matrix must be zero.

2. Each row of the indefinite admittance matrix has an algebraic sum equal to zero. This property follows from the fact that the point of zero potential (datum) may be chosen arbitrarily. This means that the terminal currents I_1, I_2 and I_3 remain invariant when all three terminal voltages V_1, V_2 and V_3 are changed by the same but arbitrary constant amount. Thus, if all three terminals of the network in Fig. 4.16 are grounded, clearly, no currents will flow; and if all three terminals are equally raised to some finite potential no currents will still flow. That is to say, if $V_1 = V_2 = V_3 = V$ we have $I_1 = 0$ which leads to

$$(y_{11} + y_{12} + y_{13})V = 0.$$

Therefore, for all V, we must have

$$y_{11} + y_{12} + y_{13} = 0, \tag{4.84}$$

which states that the sum of admittances in the first row of the indefinite admittance matrix is zero. Similarly, we can show that the sum of the admittances in any other row must be zero.

3. The determinant of the indefinite admittance matrix is zero. This property is a consequence of the fact that all three rows and columns of the matrix add up to zero.

If we delete the third row and third column of the indefinite admittance matrix the remaining submatrix

$$\begin{bmatrix} y_{11} & y_{12} \\ y_{21} & y_{22} \end{bmatrix} \tag{4.85}$$

is the matrix of the y-parameters obtained by arranging terminal 3 common to the input and output ports, as in Fig. 4.17(a). If the second row and second column of the indefinite admittance matrix are deleted the remaining submatrix

$$\begin{bmatrix} y_{11} & y_{13} \\ y_{31} & y_{33} \end{bmatrix} \tag{4.86}$$

is the y-matrix of the two-port network having terminal 2 common, as in Fig. 4.17(b). Similarly, if the first row and the first column of the indefinite admittance matrix are omitted the remaining submatrix

$$\begin{bmatrix} y_{22} & y_{23} \\ y_{32} & y_{33} \end{bmatrix} \tag{4.87}$$

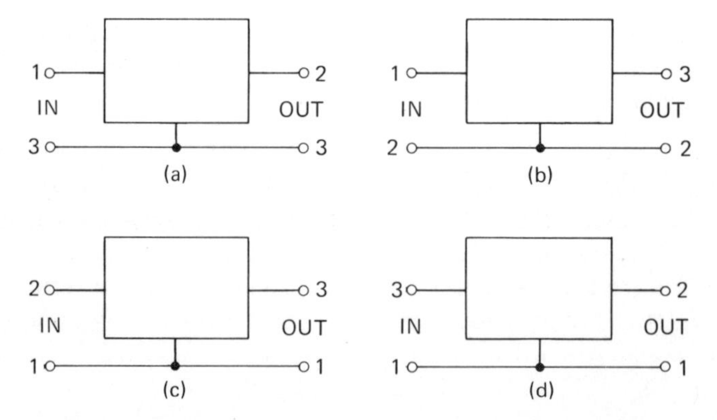

Fig. 4.17. Various orientations of a three-terminal network.

is the y-matrix of the two-port network having terminal 1 common, as in Fig. 4.17(c). If the terminal 1 is left common but terminal 3 is arranged as the input and terminal 2 is arranged as the output, as in Fig. 4.17(d), we obtain

$$\begin{bmatrix} y_{33} & y_{32} \\ y_{23} & y_{22} \end{bmatrix} \tag{4.88}$$

as the pertinent y-matrix.

It should be remarked that any $n \times n$ matrix in which the entries in each row and column add up to zero (as is the case with an indefinite admittance matrix), has the property that all of its principal minors of order $n - 1$ are equal. It follows therefore that all the y-matrices of (4.85)–(4.88) have the same determinant. That is to say, in a three-terminal network operated with a terminal common to the input and output ports the determinant of the y-matrix of the resulting two-port network is invariant to orientation of the network.

Summarizing, suppose we know the 2×2 y-matrix obtained from measurements on a three-terminal network having terminal 3, say, common; we can immediately set up the indefinite admittance matrix by adding a third row and third column so that the sum of every row and every column is zero. Then, the y-matrix of any other orientation of the three-terminal network is obtained by simply deleting the pertinent row and column from the indefinite admittance matrix.

Example 4.5. *Vacuum-tube triode.* Figure 4.18 shows the circuit model of a vacuum-tube triode, ignoring the effects of interelectrode capacitances. The tube is shown operated in its common cathode mode with the following y-matrix

$$\mathbf{Y}_k = \begin{bmatrix} 0 & 0 \\ g_m & g_p \end{bmatrix}, \tag{4.89}$$

where $g_p = 1/r_p$ is the incremental plate conductance and g_m is the mutual conductance. Adding a third row and third column to the y-matrix of Eq. (4.89) so that the elements of every row and every column add up to zero, we find that the

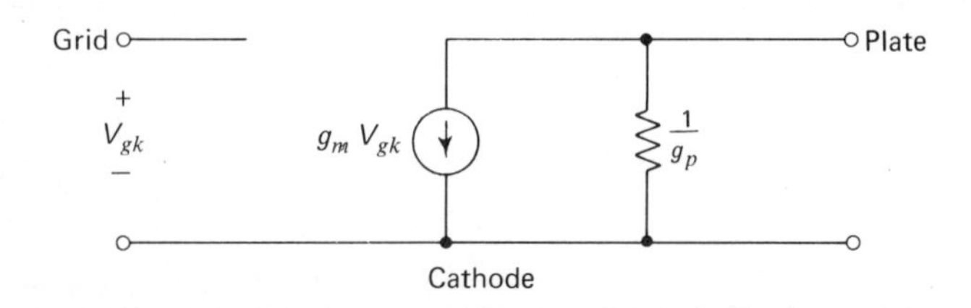

Fig. 4.18. Low-frequency circuit model of a vacuum-tube triode.

indefinite admittance matrix of a triode is

$$\begin{array}{c} \\ g \\ p \\ k \end{array} \begin{array}{c} \begin{array}{ccc} g & p & k \end{array} \\ \begin{bmatrix} 0 & 0 & 0 \\ g_m & g_p & -(g_m + g_p) \\ -g_m & -g_p & g_m + g_p \end{bmatrix} \end{array}, \tag{4.90}$$

where the symbols g, p and k have been added to identify the rows and columns corresponding to the grid, plate and cathode terminals of the triode, respectively. By deleting row p and column p from the indefinite admittance matrix of (4.90), we get

$$\mathbf{Y}_p = \begin{bmatrix} 0 & 0 \\ -g_m & g_m + g_p \end{bmatrix} \tag{4.91}$$

as the y-matrix of the cathode-follower (i.e., common-plate) configuration.

If we delete row g and column g from the indefinite admittance matrix of (4.90), the remaining submatrix is the y-matrix of a common-grid configuration operated with the plate as the input terminal and the cathode as the output terminal. In the normal operation of the common-grid configuration, however, the cathode is used as the input and the plate is used as the output; the y-matrix of this orientation is readily obtained by interchanging the p and k rows, and columns. The combined result of this change and omission of row g and column g from the indefinite admittance matrix of (4.90) is

$$\mathbf{Y}_g = \begin{bmatrix} g_m + g_p & -g_p \\ -(g_m + g_p) & g_p \end{bmatrix}. \tag{4.92}$$

Example 4.6. *Transistor.* For our next example we shall use the indefinite admittance matrix to evaluate the h-parameters of the common-collector and common-base configurations in terms of the common-emitter h-parameters. As pointed out earlier, the h-parameters are widely employed to characterize the performance of a transistor.

Let h_{ie}, h_{re}, h_{fe}, and h_{oe} denote the common emitter h-parameters. Then, changing from the h-matrix into the equivalent y-matrix with the aid of Table 4.1, we find that the common-emitter configuration has a y-matrix equal to

$$\mathbf{Y}_e = \begin{bmatrix} \dfrac{1}{h_{ie}} & -\dfrac{h_{re}}{h_{ie}} \\ \dfrac{h_{fe}}{h_{ie}} & \dfrac{\Delta_{he}}{h_{ie}} \end{bmatrix}, \tag{4.93}$$

where $\Delta_{he} = h_{ie}h_{oe} - h_{re}h_{fe}$. Adding a third row and third column to the y-matrix of Eq. (4.93) such that all the rows and columns add up to zero, we get the following indefinite admittance matrix for the transistor

$$\begin{array}{c} \\ b \\ \\ c \\ \\ e \end{array} \begin{array}{c} \begin{array}{ccc} b & c & e \end{array} \\ \begin{bmatrix} \dfrac{1}{h_{ie}} & -\dfrac{h_{re}}{h_{ie}} & \dfrac{h_{re}-1}{h_{ie}} \\[2ex] \dfrac{h_{fe}}{h_{ie}} & \dfrac{\Delta_{he}}{h_{ie}} & -\dfrac{h_{fe}+\Delta_{he}}{h_{ie}} \\[2ex] -\dfrac{1+h_{fe}}{h_{ie}} & \dfrac{h_{re}-\Delta_{he}}{h_{ie}} & \dfrac{1-h_{re}+h_{fe}+\Delta_{he}}{h_{ie}} \end{bmatrix} \end{array}, \tag{4.94}$$

where the symbols b, c and e have been included to identify the rows and columns which correspond to the base, collector and emitter terminals of the transistor, respectively. Deleting row c and column c from the indefinite admittance matrix of (4.94) yields the following y-matrix for the common-collector configuration (i.e., emitter-follower)

$$\mathbf{Y}_c = \begin{bmatrix} \dfrac{1}{h_{ie}} & \dfrac{h_{re}-1}{h_{ie}} \\[2ex] -\dfrac{1+h_{fe}}{h_{ie}} & \dfrac{1-h_{re}+h_{fe}+\Delta_{he}}{h_{ie}} \end{bmatrix}. \tag{4.95}$$

Changing from the y-matrix into the equivalent h-matrix by means of Table 4.1, and using h_{ic}, h_{rc}, h_{fc} and h_{oc} to denote the common-collector h-parameters, we get

$$\begin{bmatrix} h_{ic} & h_{rc} \\ h_{fc} & h_{oc} \end{bmatrix} = \begin{bmatrix} h_{ie} & 1-h_{re} \\ -(1+h_{fe}) & h_{oe} \end{bmatrix}. \tag{4.96}$$

In practice we usually find that $h_{re} \ll 1$, so that Eq. (4.96) may be quite justifiably simplified as follows:

$$\begin{bmatrix} h_{ic} & h_{rc} \\ h_{fc} & h_{oc} \end{bmatrix} \simeq \begin{bmatrix} h_{ie} & 1 \\ -(1+h_{fe}) & h_{oe} \end{bmatrix}. \tag{4.97}$$

Next, if we delete row b and column b from the indefinite admittance matrix of (4.94), the remaining submatrix is the y-matrix of a common-base configuration operated with the collector as input and the emitter as output. Normally, however, the common-base configuration is used with the emitter as input and collector as output. Therefore, if in the indefinite admittance matrix of (4.94) we omit row b and column b, and at the same time interchange the c and e rows and columns, we get the following y-matrix for the normal orientation of a common-base configuration

$$\mathbf{Y}_b = \begin{bmatrix} \dfrac{1 - h_{re} + h_{fe} + \Delta_{he}}{h_{ie}} & \dfrac{h_{re} - \Delta_{he}}{h_{ie}} \\ -\dfrac{h_{fe} + \Delta_{he}}{h_{ie}} & \dfrac{\Delta_{he}}{h_{ie}} \end{bmatrix}. \tag{4.98}$$

Changing from this y-matrix into an equivalent h-matrix, and using h_{ib}, h_{rb}, h_{fb} and h_{ob} to denote the common-base h-parameters, we get the result

$$\begin{bmatrix} h_{ib} & h_{rb} \\ h_{fb} & h_{ob} \end{bmatrix} = \begin{bmatrix} \dfrac{h_{ie}}{1 - h_{re} + h_{fe} + \Delta_{he}} & \dfrac{\Delta_{he} - h_{re}}{1 - h_{re} + h_{fe} + \Delta_{he}} \\ -\dfrac{h_{fe} + \Delta_{he}}{1 - h_{re} + h_{fe} + \Delta_{he}} & \dfrac{h_{oe}}{1 - h_{re} + h_{fe} + \Delta_{he}} \end{bmatrix}. \tag{4.99}$$

In addition to $h_{re} \ll 1$, in practice we also find that $\Delta_{he} \ll h_{fe}$; therefore, Eq. (4.99) may be simplified as follows:

$$\begin{bmatrix} h_{ib} & h_{rb} \\ h_{fb} & h_{ob} \end{bmatrix} \simeq \begin{bmatrix} \dfrac{h_{ie}}{1 + h_{fe}} & \dfrac{\Delta_{he} - h_{re}}{1 + h_{fe}} \\ -\dfrac{h_{fe}}{1 + h_{fe}} & \dfrac{h_{oe}}{1 + h_{fe}} \end{bmatrix}. \tag{4.100}$$

4.6 EXTERNAL CIRCUIT PROPERTIES OF BASIC VACUUM-TUBE AND TRANSISTOR CONFIGURATIONS

a) Vacuum Tube

The voltage gain K_v, input impedance Z_{in} and output impedance Z_{out} of a terminated two-port network are given by the following expressions in terms of the y-parameters (see Table 4.2):

$$K_v = -\frac{y_{21}R_l}{1 + y_{22}R_l}, \tag{4.101}$$

$$Z_{in} = \frac{1 + y_{22}R_l}{y_{11} + \Delta_y R_l}, \tag{4.102}$$

$$Z_{out} = \frac{1 + y_{11}R_s}{y_{22} + \Delta_y R_s}. \tag{4.103}$$

Using the y-matrices of Eqs. (4.89), (4.91) and (4.92) for the common-cathode, common-plate and common-grid configurations of a triode, we find that Eqs. (4.101)–(4.103) lead to the results summarized in Table 4.3. The results apply to low-frequency operation and are given in terms of the incremental plate resistance r_p and amplification factor $\mu = g_m r_p$. It is noteworthy in Table 4.3 that, because in practice $\mu \gg 1$, the common-cathode and common-grid configurations have voltage gains which are nearly equal in magnitude. The voltage gain of the common-plate (i.e., cathode-follower) configuration is close to unity for large load resistances. Furthermore, the common-plate configuration has the lowest output resistance and the common-grid configuration has the lowest input resistance.

Table 4.3. Circuit Properties of Terminated Vacuum-tube Configurations

	Common cathode	Common plate (cathode follower)	Common grid
K_v	$\dfrac{-\mu R_l}{r_p + R_l}$	$\dfrac{\mu R_l}{r_p + (1 + \mu)R_l}$	$\dfrac{(1 + \mu)R_l}{r_p + R_l}$
Z_{in}	Infinite	Infinite	$\dfrac{r_p + R_l}{1 + \mu}$
Z_{out}	r_p	$\dfrac{r_p}{1 + \mu}$	$r_p + (1 + \mu)R_s$

b) Transistor

In terms of the h-parameters the voltage gain K_v, current gain K_i, input impedance Z_{in} and output impedance Z_{out} of a terminated two-port network are given as follows (see Table 4.2).

$$K_v = \frac{-h_{21}R_l}{h_{11} + \Delta_h R_l}, \tag{4.104}$$

$$K_i = \frac{h_{21}}{1 + h_{22}R_l}, \tag{4.105}$$

$$Z_{in} = \frac{h_{11} + \Delta_h R_l}{1 + h_{22}R_l}, \tag{4.106}$$

$$Z_{out} = \frac{h_{11} + R_s}{\Delta_h + h_{22}R_s}. \tag{4.107}$$

Using these relations in conjunction with the h-parameter matrices of Eqs. (4.97) and (4.100) which pertain to the common-collector and common-base configurations, respectively, and regarding the common-emitter h-parameters as the basic transistor parameters, we obtain the results collected in Table 4.4.

Table 4.4. Circuit Properties of Terminated Transistor Configurations in Terms of Common-emitter h-parameters

	Common-emitter	Common-collector (emitter follower)	Common-base
K_v	$\frac{-h_{fe}R_l}{h_{ie}+\Delta_{he}R_l}$	$\frac{(1+h_{fe})R_l}{h_{ie}+(1+h_{fe}+h_{ie}h_{oe})R_l}$	$\frac{h_{fe}R_l}{h_{ie}+\Delta_{he}R_l}$
K_i	$\frac{h_{fe}}{1+h_{oe}R_l}$	$\frac{-(1+h_{fe})}{1+h_{oe}R_l}$	$\frac{-h_{fe}}{1+h_{fe}+h_{oe}R_l}$
Z_{in}	$\frac{h_{ie}+\Delta_{he}R_l}{1+h_{oe}R_l}$	$\frac{h_{ie}+(1+h_{fe}+h_{ie}h_{oe})R_l}{1+h_{oe}R_l}$	$\frac{h_{ie}+\Delta_{he}R_l}{1+h_{fe}+h_{oe}R_l}$
Z_{out}	$\frac{h_{ie}+R_s}{\Delta_{he}+h_{oe}R_s}$	$\frac{h_{ie}+R_s}{1+h_{fe}+h_{oe}(R_s+h_{ie})}$	$\frac{h_{ie}+(1+h_{fe})R_s}{\Delta_{he}+h_{oe}R_s}$

The results are also displayed graphically in Figs. 4.19–4.22. These four diagrams illustrate the influence of the load resistance on the voltage and current gains and the input resistance of the three basic transistor configurations, and the influence of the source resistance on the output resistance. The numerical values included in Figs. 4.19–4.22 have been calculated using the following set of typical values for the common-emitter h-parameters:

$$\begin{aligned} h_{ie} &= 1\ \text{k}\Omega \qquad h_{re} = 0{\cdot}4 \times 10^{-3} \\ h_{fe} &= 50 \qquad h_{oe} = 50\ \mu\mho \\ \Delta_{he} &= h_{ie}h_{oe} - h_{re}h_{fe} = 0{\cdot}03. \end{aligned} \tag{4.108}$$

The low-frequency properties of the three transistor configurations may thus be summarized as follows:

1. The input resistance of a common-collector (i.e., emitter-follower) stage is high, while that of a common-base stage is very low; in both cases, however, the input resistance increases with the load resistance. The input resistance of a common-emitter stage is low and decreases with increasing load resistance.
2. The output resistance of a common-collector stage is very low, while that of a common-base stage is very high; in both cases the output resistance increases with the source resistance. The output resistance of a common-emitter stage is high and decreases with increasing source resistance.
3. The common-emitter and common-base stages have large voltage gains nearly equal in magnitude, while the common-collector stage has an open-circuit voltage gain close to unity. For all three configurations the voltage gain decreases with decreasing load resistance.

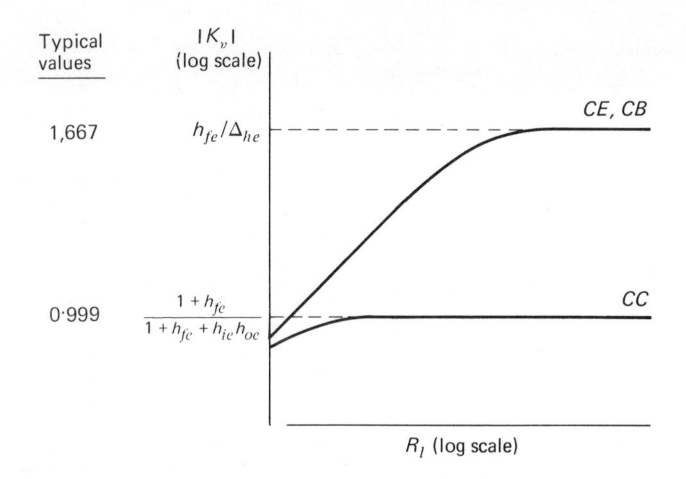

Fig. 4.19. Voltage gains of the common-emitter, common-collector, and common-base configurations for varying load resistances.

4. The common-emitter and common-collector stages have large current gains nearly equal in magnitude, while the common-base stage has a short-circuit current gain close to unity. For all three configurations the current gain decreases with increasing load resistance.
5. Only the common-emitter stage can provide a current as well as voltage gain greater than unity; it therefore provides the highest power gain for all load resistances. For low load resistances the common-collector stage provides a larger power gain than the common-base stage, while for high load resistances the common-base stage provides a larger power gain than the common-collector stage.

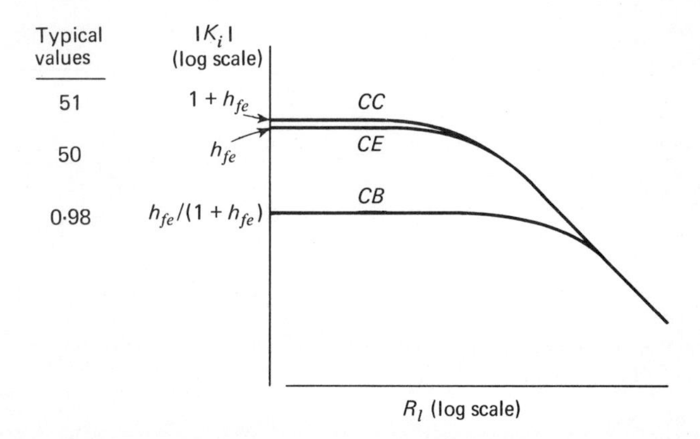

Fig. 4.20. Current gains of the common-emitter, common-collector, and common-base configurations for varying load resistances.

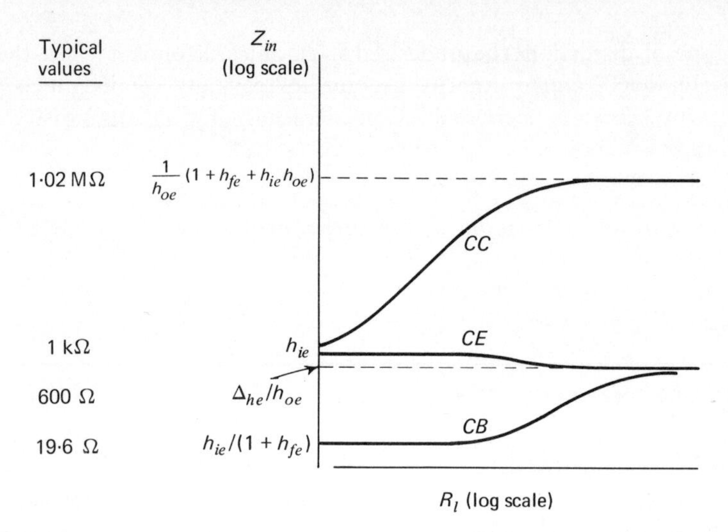

Fig. 4.21. Input resistances of the common-emitter, common-collector, and common-base configurations for varying load resistances.

4.7 NONLINEAR DISTORTION

Practical active two-port devices which are used for signal amplification inherently have nonlinear characteristics, giving rise to the generation of harmonic frequency components that are not present in the input signal. The distortion of the output signal resulting from device nonlinearity is called *nonlinear distortion.*

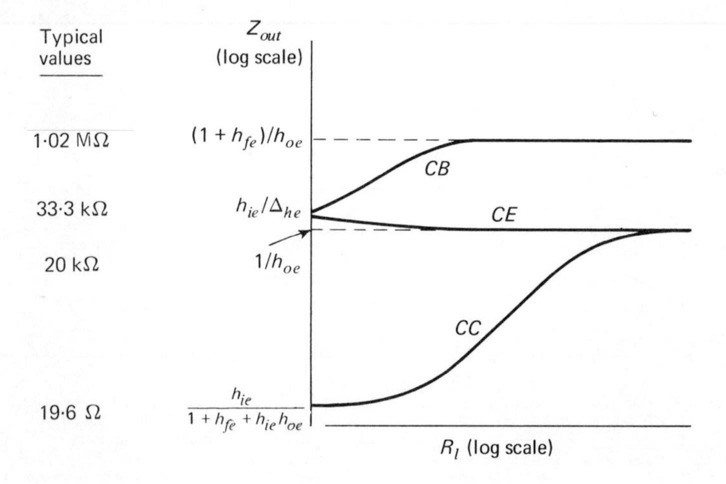

Fig. 4.22. Output resistances of the common-emitter, common-collector, and common-base configurations for varying source resistances.

In this type of distortion, the undesired harmonics disappear when the applied signal is removed, and ordinarily become increasingly significant as the level of the input signal is increased. Consideration of nonlinear distortion may proceed in two steps:

1. The development of a model involving *distortion parameters* that characterize the generation of harmonics. These parameters should be independent of the terminating source and load impedances, as is the case with the definition of the ordinary *linear* two-port parameters characterizing the transmission of desired signals.
2. Evaluation of the overall distortion performance of the network when operating between prescribed source and load impedances.

We shall assume that the applied signal level is so low that the second-harmonic distortion is less than 5%, while third- and higher-order distortion is at least one order of magnitude lower. Then, the two-port device may be regarded as *almost linear*, exhibiting only a slight deviation from linearity.[3] In order to be specific, we shall focus our attention on the transistor as an almost-linear two-port; the analysis may be, of course, readily extended to deal with other two-port devices. We shall further assume that all parameters are frequency-independent and real quantities.

The Transistor as an Almost Linear Two-port Device

The terminal voltage–current relations of a transistor, operated with emitter as the common terminal, are defined by

$$\begin{aligned} V_{BE} &= f_1(I_B, V_{CE}) \\ I_C &= f_2(I_B, V_{CE}), \end{aligned} \tag{4.109}$$

where V_{BE} and I_B denote the total voltage and current at the input port, while V_{CE} and I_C denote the total voltage and current at the output port. When the second harmonics are to be calculated, each equation in (4.109) is expanded in a Taylor series to include second-order terms; thus,

$$\begin{aligned} v_{be} &= \frac{\partial f_1}{\partial I_B} i_b + \frac{\partial f_1}{\partial V_{CE}} v_{ce} + \frac{1}{2}\frac{\partial^2 f_1}{\partial I_B^2} i_b^2 + \frac{\partial^2 f_1}{\partial I_B\, \partial V_{CE}} i_b v_{ce} + \frac{1}{2}\frac{\partial^2 f_1}{\partial V_{CE}^2} v_{ce}^2 \\ i_c &= \frac{\partial f_2}{\partial I_B} i_b + \frac{\partial f_2}{\partial V_{CE}} v_{ce} + \frac{1}{2}\frac{\partial^2 f_2}{\partial I_B^2} i_b^2 + \frac{\partial^2 f_2}{\partial I_B\, \partial V_{CE}} i_b v_{ce} + \frac{1}{2}\frac{\partial^2 f_2}{\partial V_{CE}^2} v_{ce}^2, \end{aligned} \tag{4.110}$$

where v_{be} and i_b denote the alternating voltage and current components at the input port, while v_{ce} and i_c denote the alternating voltage and current components at the output port. The choice of i_b and v_{ce} as the independent variables corresponds to characterizing the transistor by h-parameters, so that

$$\begin{aligned} h_{ie} &= \left.\frac{\partial f_1}{\partial I_B}\right|_{V_{CE}=\text{const}} \\ h_{re} &= \left.\frac{\partial f_1}{\partial V_{CE}}\right|_{I_B=\text{const}} \\ h_{fe} &= \left.\frac{\partial f_2}{\partial I_B}\right|_{V_{CE}=\text{const}} \\ h_{oe} &= \left.\frac{\partial f_2}{\partial V_{CE}}\right|_{I_B=\text{const}}. \end{aligned} \tag{4.111}$$

Accordingly, we may re-write Eqs. (4.110) as follows

$$\begin{aligned} v_{be} &= h_{ie}i_b + h_{re}v_{ce} + \frac{1}{2}\frac{\partial h_{ie}}{\partial I_B}i_b^2 + \frac{\partial h_{ie}}{\partial V_{CE}}i_b v_{ce} + \frac{1}{2}\frac{\partial h_{re}}{\partial V_{CE}}v_{ce}^2 \\ i_c &= h_{fe}i_b + h_{oe}v_{ce} + \frac{1}{2}\frac{\partial h_{fe}}{\partial I_B}i_b^2 + \frac{\partial h_{fe}}{\partial V_{CE}}i_b v_{ce} + \frac{1}{2}\frac{\partial h_{oe}}{\partial V_{CE}}v_{ce}^2. \end{aligned} \tag{4.112}$$

It is assumed that the alternating voltages and currents may be expressed as

$$\begin{aligned} v_{be} &= V_{be1}\cos\omega t + V_{be2}\cos 2\omega t \\ i_b &= I_{b1}\cos\omega t + I_{b2}\cos 2\omega t \\ v_{ce} &= V_{ce1}\cos\omega t + V_{ce2}\cos 2\omega t \\ i_c &= I_{c1}\cos\omega t + I_{c2}\cos 2\omega t. \end{aligned} \tag{4.113}$$

The approximate expressions for the dependent variables, v_{be} and i_c, are obtained by substituting the expressions including the first and second harmonics of the independent variables, i_b and v_{ce}, into the first-order terms of Eqs. (4.112), and those including the first harmonics of the independent variables into the second-order terms, thereby yielding

$$\begin{aligned} v_{be} &= h_{ie}i_b + h_{re}v_{ce} + v_{2\omega} \\ i_c &= h_{fe}i_b + h_{oe}v_{ce} + i_{2\omega} \end{aligned} \tag{4.114}$$

where $v_{2\omega}$ and $i_{2\omega}$ denote the second-harmonic distortion generators, as shown by

$$\begin{aligned} v_{2\omega} &= (a_1 I_{b1}^2 + a_2 I_{b1}V_{ce1} + a_3 V_{ce1}^2)\cos 2\omega t \\ i_{2\omega} &= (a_4 I_{b1}^2 + a_5 I_{b1}V_{ce1} + a_6 V_{ce1}^2)\cos 2\omega t. \end{aligned} \tag{4.115}$$

Equations (4.114) indicate that a transistor operating with a small degree of nonlinearity may be modelled as a linear device with a pair of distortion generators, $v_{2\omega}$ and $i_{2\omega}$, as shown in Fig. 4.23. This representation is, however, limited to the second-harmonic components.

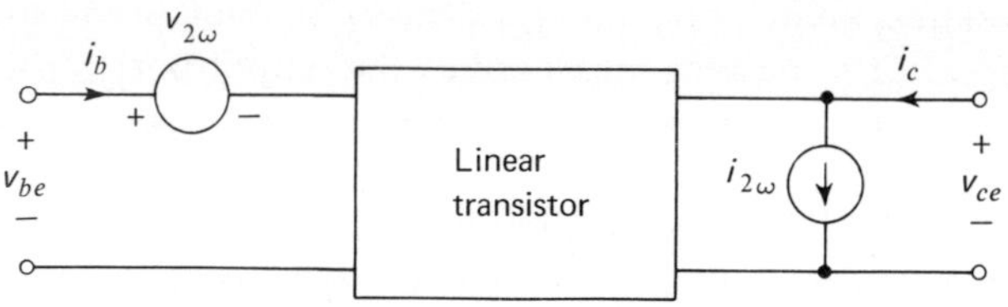

Fig. **4.23.** Model for an almost linear transistor.

The various a's constitute the desired set of *distortion parameters*, which depend solely upon variations in the transistor h-parameters produced by changes in the total values of the base current I_B and collector–emitter voltage V_{CE}, as shown by

$$a_1 = \frac{1}{4}\frac{\partial h_{ie}}{\partial I_B} \qquad a_2 = \frac{1}{2}\frac{\partial h_{ie}}{\partial V_{CE}}$$

$$a_3 = \frac{1}{4}\frac{\partial h_{re}}{\partial V_{CE}} \qquad a_4 = \frac{1}{4}\frac{\partial h_{fe}}{\partial I_B} \tag{4.116}$$

$$a_5 = \frac{1}{2}\frac{\partial h_{fe}}{\partial V_{CE}} \qquad a_6 = \frac{1}{4}\frac{\partial h_{oe}}{\partial V_{CE}}.$$

The distortion parameters may be calculated using Eqs. (4.116), or alternatively, they may be obtained by direct measurement. Thus, suppose a fundamental current of amplitude I_{b1} is applied across the input port of the transistor, with the output port shorted (i.e., $V_{ce1} = 0$), as in Fig. 4.24. Then, from Eqs. (4.115) it follows that the *input driving-point distortion parameter*, a_1, and *output transfer distortion parameter*, a_4, may be calculated from the relations,

$$a_1 = \left.\frac{V_{be2}}{I_{b1}^2}\right|_{V_{ce1}=0},$$

$$a_4 = \left.\frac{I_{c2}}{I_{b1}^2}\right|_{V_{ce1}=0}, \tag{4.117}$$

where V_{be2} denotes the amplitude of the second-harmonic component of voltage

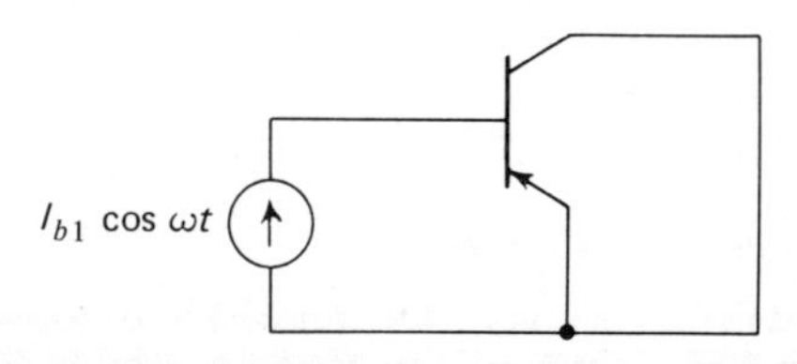

Fig. 4.24. Set-up for measuring distortion parameters a_1 and a_4.

measured across the input port, and I_{c2} denotes the amplitude of the second-harmonic component of current measured at the output port.

Suppose, next, the input port of the transistor is left open (i.e., $I_{b1} = 0$), and a fundamental voltage of amplitude V_{ce1} is applied across the output port, as in Fig. 4.25. Then, Eqs. (4.115) yield the following relations for calculating the *input transfer distortion parameter*, a_3, and the *output driving-point distortion parameter*, a_6:

$$a_3 = \left.\frac{V_{be2}}{V_{ce1}^2}\right|_{I_{b1}=0}$$
$$a_6 = \left.\frac{I_{c2}}{V_{ce1}^2}\right|_{I_{b1}=0} \tag{4.118}$$

The remaining parameters a_2 and a_5, referred to as the *input* and *output cross-product distortion parameters*, respectively, may finally be measured by simultaneously applying a fundamental input current and fundamental output voltage to the transistor.

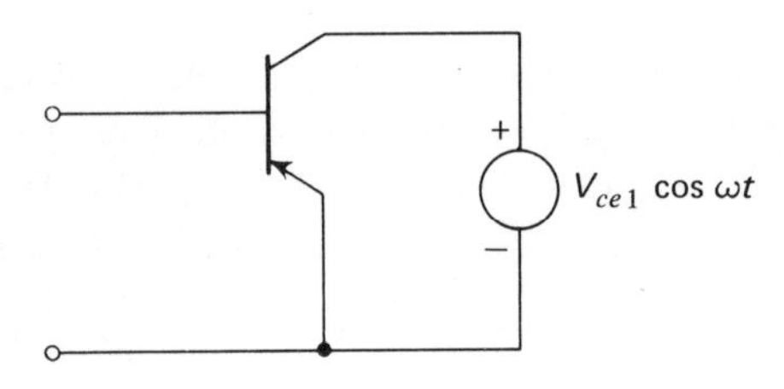

Fig. 4.25. Set-up for measuring distortion parameters a_3 and a_6.

The equivalent second-harmonic distortion generators $v_{2\omega}$ and $i_{2\omega}$ of the model of Fig. 4.23 have been defined in a manner analogous to the h-parameters. We may equally define equivalent distortion generators having a form analogous to the z-, y-, or g-parameters. Thus, in terms of $v_{2\omega}$ and $i_{2\omega}$ we can identify the equivalent distortion generators pertaining to these other forms, obtaining the results shown in Fig. 4.26 through evaluation of the port voltages or currents with the appropriate open-circuit or short-circuit termination.[4]

Distortion Factor of Transistor Amplifier

Consider Fig. 4.27 showing a transistor terminated with a voltage source $V_s \cos \omega t$ of internal resistance R_s at its input port, and a load resistance R_l at its output port. For the first harmonics of the currents and voltages in this amplifier, we

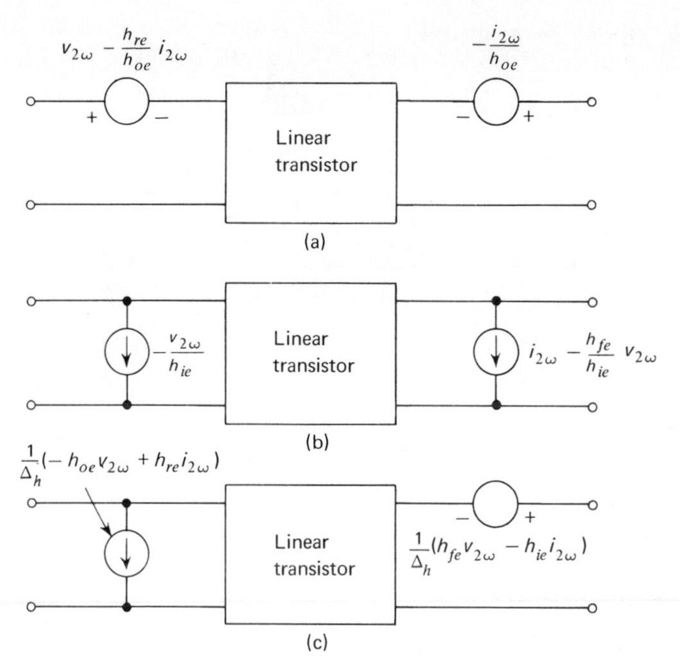

Fig. 4.26. z-, and y- and g-models for an almost linear two-port device.

may thus write

$$
\begin{aligned}
V_{ce1} &= -R_l I_{c1} \\
I_{b1} &= \frac{1 + h_{oe}R_l}{h_{fe}} I_{c1}.
\end{aligned}
\tag{4.119}
$$

Substitution of Eqs. (4.119) in (4.115) yields

$$
\begin{aligned}
v_{2\omega} &= b_1 I_{c1}^2 \cos 2\omega t \\
i_{2\omega} &= b_2 I_{c1}^2 \cos 2\omega t,
\end{aligned}
\tag{4.120}
$$

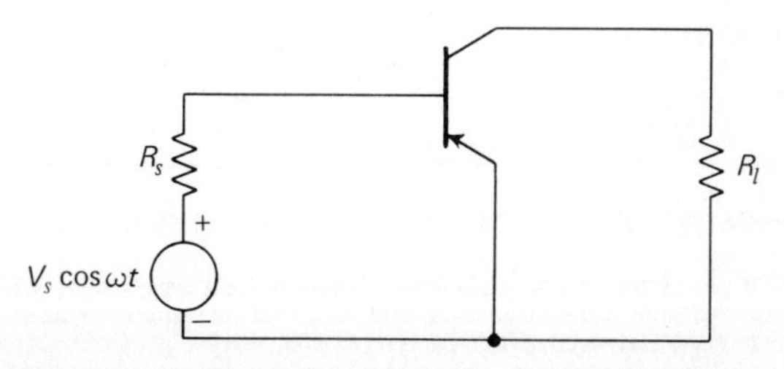

Fig. 4.27. A double-terminated transistor amplifier.

where b_1 and b_2 are defined by

$$b_1 = a_1\left(\frac{1 + h_{oe}R_l}{h_{fe}}\right)^2 - a_2\left(\frac{1 + h_{oe}R_l}{h_{fe}}\right)R_l + a_3R_l^2$$
$$b_2 = a_4\left(\frac{1 + h_{oe}R_l}{h_{fe}}\right)^2 - a_5\left(\frac{1 + h_{oe}R_l}{h_{fe}}\right)R_l + a_6R_l^2. \tag{4.121}$$

From the model of Fig. 4.23, we see that the second harmonics of the currents and voltages may be expressed, in matrix form, as follows

$$\begin{bmatrix} V_{be2} - b_1I_{c1}^2 \\ I_{c2} - b_2I_{c1}^2 \end{bmatrix} = \begin{bmatrix} h_{ie} & h_{re} \\ h_{fe} & h_{oe} \end{bmatrix} \begin{bmatrix} I_{b2} \\ V_{ce2} \end{bmatrix}, \tag{4.122}$$

with

$$V_{be2} = -R_sI_{b2}$$
$$V_{ce2} = -R_lI_{c2}. \tag{4.123}$$

Solving Eqs. (4.122) and (4.123) for I_{c2}, we therefore obtain

$$I_{c2} = \frac{-I_{c1}^2[b_1h_{fe} - b_2(R_s + h_{ie})]}{R_l\Delta_{he} + R_s(1 + h_{oe}R_l) + h_{ie}}. \tag{4.124}$$

A simple and convenient measure of distortion is the so-called *distortion factor*, denoted by k_d and defined as the ratio of the amplitude of the second harmonic component to that of the fundamental component. Hence, from Eq. (4.124) we have

$$k_d = \left|\frac{I_{c2}}{I_{c1}}\right| = \left|\frac{b_1h_{fe} - b_2(R_s + h_{ie})}{R_l\Delta_{he} + R_s(1 + h_{oe}R_l) + h_{ie}}\right| I_{c1}, \tag{4.125}$$

where b_1 and b_2 are known quantities dependent on the transistor and the load resistance as shown in Eqs. (4.121).

Equation (4.125) shows that the distortion factor is proportional to the output level, as we would expect. Also, in a common-emitter amplifier we ordinarily find that the distortion factor has a well-defined minimum which occurs for the special source resistance:

$$R_{s(min)} = \frac{b_1h_{fe}}{b_2} - h_{ie}. \tag{4.126}$$

4.8 INTERCONNECTIONS OF TWO-PORT NETWORKS

As illustrated in Fig. 4.28 a pair of two-port networks N' and N'' may be connected together in five fundamentally different ways:

a) cascade

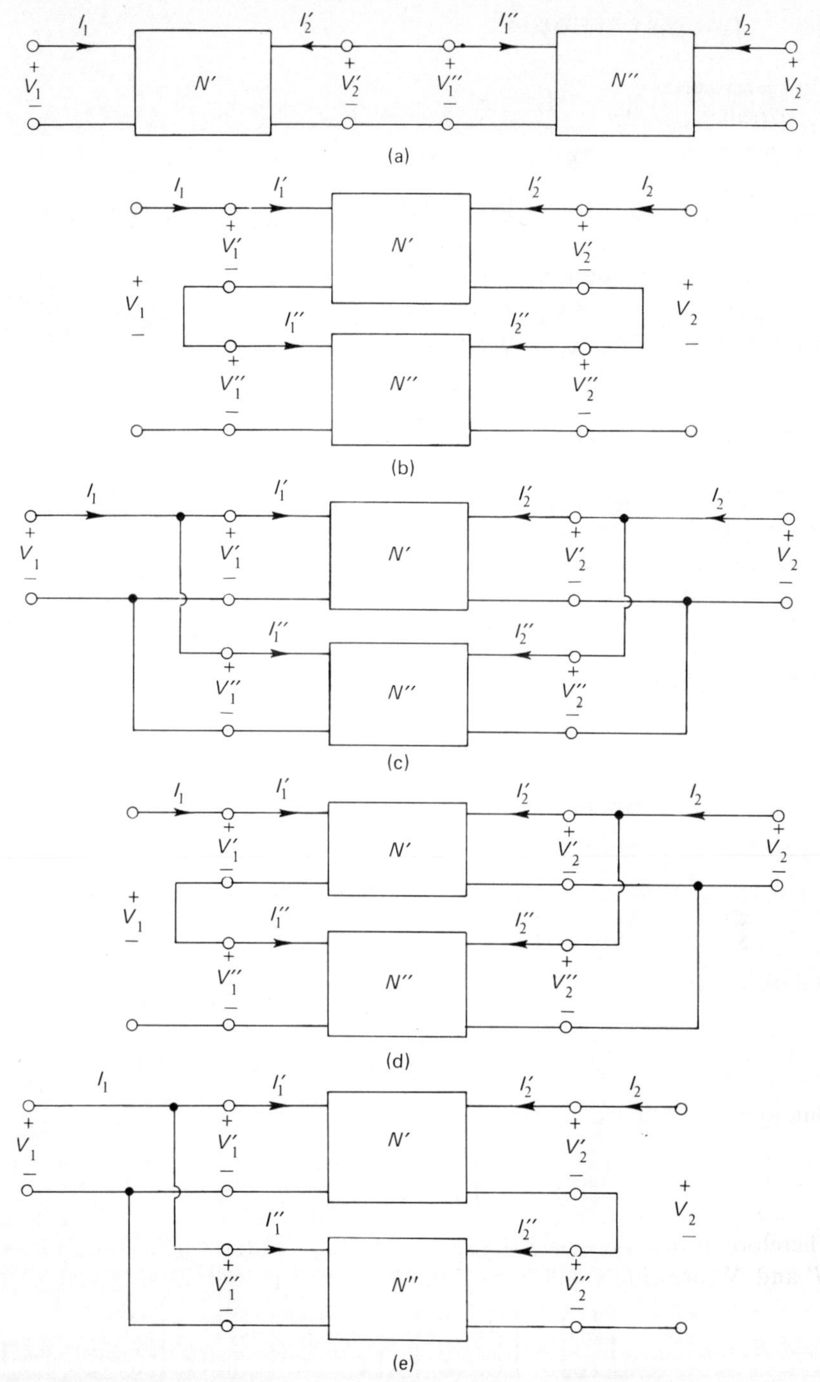

Fig. 4.28. Interconnections of a pair of two-port networks: (a) Cascade; (b) Series–series; (c) Parallel–parallel; (d) Series–parallel; (e) Parallel–series.

b) series–series
c) parallel–parallel
d) series–parallel
e) parallel–series.

The cascade connection of Fig. 4.28(a) arises, for example, when two or more amplifier stages are connected in tandem to produce increased over-all amplification. The other connections shown in parts (b) to (e) of Fig. 4.28 provide the four basic types of feedback circuits in which network N' represents an amplifier and network N'' is made up entirely of passive elements. The particularly attractive feature of these feedback connections is that the overall amplification can be made less dependent on variations in the parameters of the amplifier component due to environmental changes and aging; also, the applied feedback reduces the effects of nonlinear distortion.

In analysing the coupled network arrangements of Fig. 4.28 our task will be to determine the parameters of an equivalent two-port network for each one of them. In this evaluation we shall see that the cascade connection is most conveniently studied by means of the chain matrices, while the z-matrices are useful for dealing with the series–series connection, the y-matrices for the parallel–parallel connection, the h-matrices for the series–parallel connection, and the g-matrices for the parallel–series connection.

a) Cascade Connection

In Fig. 4.28(a) the two-port networks N' and N'' are shown connected together in tandem. Viewing this arrangement from left to right, we may use the chain matrix; thus for network N' we have

$$\begin{bmatrix} V_1 \\ I_1 \end{bmatrix} = \begin{bmatrix} A' & B' \\ C' & D' \end{bmatrix} \begin{bmatrix} V_2' \\ -I_2' \end{bmatrix} \tag{4.127}$$

and for network N''

$$\begin{bmatrix} V_1'' \\ I_1'' \end{bmatrix} = \begin{bmatrix} A'' & B'' \\ C'' & D'' \end{bmatrix} \begin{bmatrix} V_2 \\ -I_2 \end{bmatrix}. \tag{4.128}$$

But, in Fig. 4.28(a) we see that

$$\begin{bmatrix} V_2' \\ -I_2' \end{bmatrix} = \begin{bmatrix} V_1'' \\ I_1'' \end{bmatrix}. \tag{4.129}$$

Therefore, if the common voltages and currents at the junction of the networks N' and N'' are eliminated from Eqs. (4.127) and (4.128), we find that

$$\begin{bmatrix} V_1 \\ I_1 \end{bmatrix} = \begin{bmatrix} A' & B' \\ C' & D' \end{bmatrix} \begin{bmatrix} A'' & B'' \\ C'' & D'' \end{bmatrix} \begin{bmatrix} V_2 \\ -I_2 \end{bmatrix}, \tag{4.130}$$

which states that the overall chain matrix of the cascade connection of Fig. 4.28(a), viewed from the left, is equal to the product of the chain matrices of the individual two-port networks.

b) Series–Series Connection

Consider next Fig. 4.28(b) where the two-port networks N' and N'' are connected in series at both their input and output ports. In terms of the z-parameters we have for network N'

$$\begin{bmatrix} V'_1 \\ V'_2 \end{bmatrix} = \begin{bmatrix} z'_{11} & z'_{12} \\ z'_{21} & z'_{22} \end{bmatrix} \begin{bmatrix} I'_1 \\ I'_2 \end{bmatrix} \tag{4.131}$$

and for network N'' we have

$$\begin{bmatrix} V''_1 \\ V''_2 \end{bmatrix} = \begin{bmatrix} z''_{11} & z''_{12} \\ z''_{21} & z''_{22} \end{bmatrix} \begin{bmatrix} I''_1 \\ I''_2 \end{bmatrix}. \tag{4.132}$$

In Fig. 4.28(b) we see that

$$\begin{bmatrix} I_1 \\ I_2 \end{bmatrix} = \begin{bmatrix} I'_1 \\ I'_2 \end{bmatrix} = \begin{bmatrix} I''_1 \\ I''_2 \end{bmatrix}, \tag{4.133}$$

$$\begin{bmatrix} V_1 \\ V_2 \end{bmatrix} = \begin{bmatrix} V'_1 + V''_1 \\ V'_2 + V''_2 \end{bmatrix} = \begin{bmatrix} V'_1 \\ V'_2 \end{bmatrix} + \begin{bmatrix} V''_1 \\ V''_2 \end{bmatrix}. \tag{4.134}$$

Therefore, using Eqs. (4.131)–(4.134) we get

$$\begin{bmatrix} V_1 \\ V_2 \end{bmatrix} = \begin{bmatrix} z'_{11} & z'_{12} \\ z'_{21} & z'_{22} \end{bmatrix} \begin{bmatrix} I_1 \\ I_2 \end{bmatrix} + \begin{bmatrix} z''_{11} & z''_{12} \\ z''_{21} & z''_{22} \end{bmatrix} \begin{bmatrix} I_1 \\ I_2 \end{bmatrix}$$

or

$$\begin{bmatrix} V_1 \\ V_2 \end{bmatrix} = \begin{bmatrix} z'_{11} + z''_{11} & z'_{12} + z''_{12} \\ z'_{21} + z''_{21} & z'_{22} + z''_{22} \end{bmatrix} \begin{bmatrix} I_1 \\ I_2 \end{bmatrix}. \tag{4.135}$$

We thus see that the series-connected pair of two-port networks in Fig. 4.28(b) is equivalent to a single two-port network whose z-matrix is equal to the sum of the z-matrices of the individual two-port networks N' and N''. This statement is, however, valid only if the z-matrices of the individual two-port networks are not modified as a result of the interconnection. For example, in Fig. 4.29 there is a direct connection between terminals $1'$ and $2'$ of network N' but not between terminals 1 and 2 of network N''; clearly, the branch connecting terminals 1 and 2 of network N'' becomes short-circuited under interconnection and it would, therefore, be erroneous to use Eq. (4.135) for the composite network.

In order to be sure that the behaviour of each individual two-port network is not modified under interconnection, it is necessary and sufficient to establish that after the interconnection is made the current which enters the terminal of a port

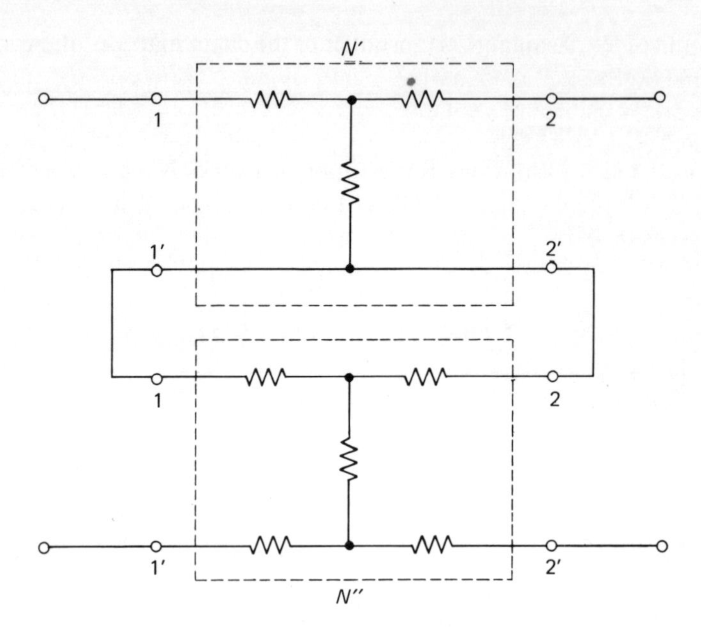

Fig. 4.29. Illustrating a case when the series–series connection is not permissible.

is identical with the current which emerges from the other terminal of the port in question.[5] In the series–series connection of Fig. 4.30 we see that the current which enters terminal 1 of the input port of network N' is I'_a, and the current which leaves terminal 1' of this network is $I'_a - I_b$. For these to be equal we have to be certain that the circulating loop current I_b, which arises from the interconnection, is zero. To test whether or not I_b is zero, an arbitrary current excitation I_1 is

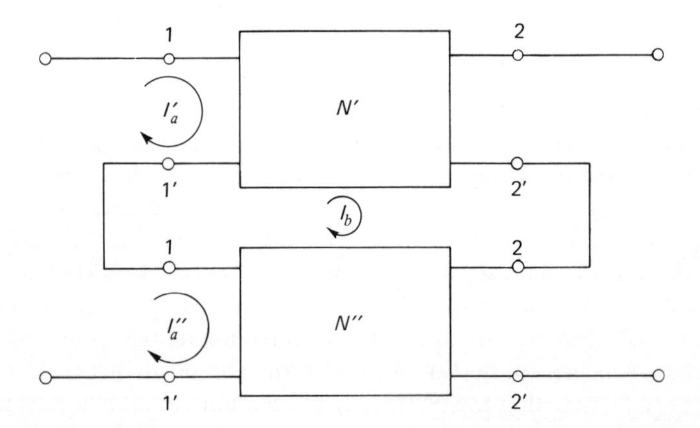

Fig. 4.30. Series–series connection with circulating loop current I_b.

connected between terminal 1′ of network N'' and terminal 1 of network N', and at the same time the connection between terminal 2 of network N'' and terminal 2′ of network N' is broken, as illustrated in Fig. 4.31(a). If the voltage difference V_a indicated in Fig. 4.31(a) is non-zero, then there will be a circulating current between the networks after the connection is made, and the interconnection is accordingly not permissible. In this way we may evaluate the effect of the interconnection on the z_{11} and z_{21} parameters of the individual networks. By carrying out a similar test at the output port, as indicated in Fig. 4.31(b), we can determine the effect of the interconnection on the z_{12} and z_{22} parameters of the individual networks. We conclude, therefore, that the series–series connection of a pair of two-port networks is permissible only if the two tests illustrated in Fig. 4.31 yield the results $V_a = 0$ and $V_b = 0$.

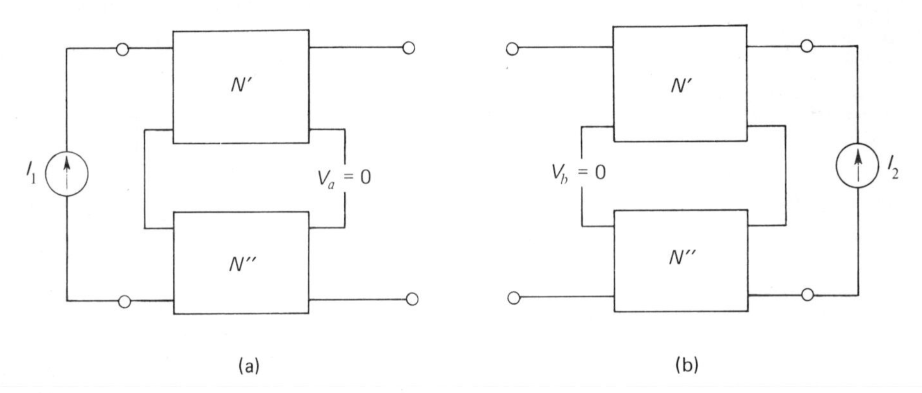

Fig. 4.31. Validity tests for the series–series connection.

c) Parallel–Parallel Connection

In Fig. 4.28(c) the two-port networks N' and N'' are connected in parallel at both their input and output ports. By describing each network in terms of its respective y-parameters we have

$$\begin{bmatrix} I'_1 \\ I'_2 \end{bmatrix} = \begin{bmatrix} y'_{11} & y'_{12} \\ y'_{21} & y'_{22} \end{bmatrix} \begin{bmatrix} V_1 \\ V_2 \end{bmatrix}, \tag{4.136}$$

$$\begin{bmatrix} I''_1 \\ I''_2 \end{bmatrix} = \begin{bmatrix} y''_{11} & y''_{12} \\ y''_{21} & y''_{22} \end{bmatrix} \begin{bmatrix} V_1 \\ V_2 \end{bmatrix}, \tag{4.137}$$

where we have noted that in Fig. 4.28(c) the networks N' and N'' have equal input and equal output voltages. Furthermore, we see that

$$\begin{bmatrix} I_1 \\ I_2 \end{bmatrix} = \begin{bmatrix} I'_1 \\ I'_2 \end{bmatrix} + \begin{bmatrix} I''_1 \\ I''_2 \end{bmatrix}. \tag{4.138}$$

Therefore, adding Eqs. (4.136) and (4.137), and then using Eq. (4.138), we get

$$\begin{bmatrix} I_1 \\ I_2 \end{bmatrix} = \begin{bmatrix} y'_{11} + y''_{11} & y'_{12} + y''_{12} \\ y'_{21} + y''_{21} & y'_{22} + y''_{22} \end{bmatrix} \begin{bmatrix} V_1 \\ V_2 \end{bmatrix}. \tag{4.139}$$

That is to say, the parallel–parallel coupled arrangement of Fig. 4.28(c) may be replaced by an equivalent two-port network whose y-matrix is equal to the sum of the y-matrices of the individual networks N' and N''. This assumes, as in the series–series connection, that the individual two-port networks are unaltered by the interconnection.

In a manner dual to the series–series connection, we may use the tests indicated in Fig. 4.32 to determine the effect of the parallel–parallel connection on the y-parameters of the individual networks. For the parallel–parallel connection of the two-port networks N' and N'' to be permissible, we must find that in Fig. 4.32 both $V_a = 0$ and $V_b = 0$.

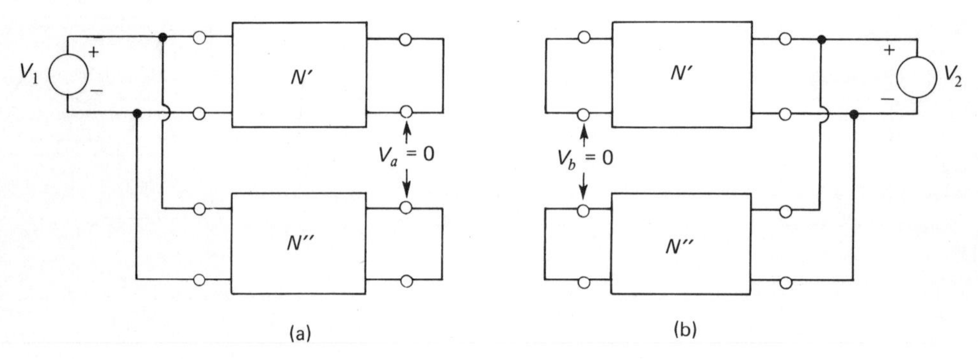

Fig. 4.32. Validity tests for the parallel–parallel connection.

d) Series–Parallel Connection

Fig. 4.28(d) shows the two-port networks N' and N'' having their input ports connected in series and their output ports connected in parallel. The overall voltage V_1 at the input port is equal to the sum of input voltages V'_1 and V''_1 of the individual networks; the overall output current I_2 is equal to the sum of their output currents I'_2 and I''_2. Therefore, if each network is described in terms of its h-parameters, we find that the overall behaviour of the coupled arrangement of Fig. 4.28(d) is expressible as follows:

$$\begin{bmatrix} V_1 \\ I_2 \end{bmatrix} = \begin{bmatrix} h'_{11} + h''_{11} & h'_{12} + h''_{12} \\ h'_{21} + h''_{21} & h'_{22} + h''_{22} \end{bmatrix} \begin{bmatrix} I_1 \\ V_2 \end{bmatrix}. \tag{4.140}$$

That is, the series–parallel connection of Fig. 4.28(d) is equivalent to a single two-port network having an h-matrix equal to the sum of the h-matrices of the individual two-port networks N' and N''. The validity tests of Fig. 4.33 may be

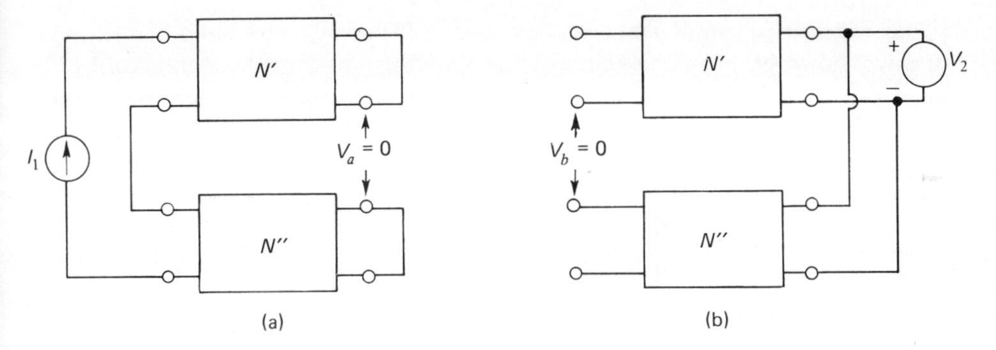

Fig. 4.33. Validity tests for the series–parallel connection.

used to establish that the h-matrices of the individual networks are not modified by the interconnection.

e) Parallel–Series Connection

Finally, consider Fig. 4.28(e) where the two-port networks N' and N'' are shown connected in parallel at their input ports and in series at their output ports. Here we see that the overall input current I_1 is equal to the sum of the input currents I_1' and I_1'' of the component networks, and the overall output voltage V_2 is equal to the sum of their output voltages V_2' and V_2''. Assuming that each of the networks N' and N'' is described in terms of its g-parameters, we find that for the composite two-port network of Fig. 4.28(e),

$$\begin{bmatrix} I_1 \\ V_2 \end{bmatrix} = \begin{bmatrix} g_{11}' + g_{11}'' & g_{12}' + g_{12}'' \\ g_{21}' + g_{21}'' & g_{22}' + g_{22}'' \end{bmatrix} \begin{bmatrix} V_1 \\ I_2 \end{bmatrix}, \tag{4.141}$$

which states that the parallel–series coupled arrangement of Fig. 4.28(e) is equivalent to a single two-port network having a g-matrix equal to the sum of the g-matrices of the individual two-port networks. The validity tests for establishing that such an interconnection is permissible are essentially the same as those of Fig. 4.33, except that the positions of the input and output ports are interchanged.

Two-port Formulation of Feedback

The parameters of active two-port devices such as vacuum tubes or transistors are subject to unavoidable variations that result from aging, changes in environmental conditions, or manufacturing tolerances. An established and extensively used technique for minimizing the effect of such parameter variations on the overall performance of an amplifier is the use of *feedback*. A feedback amplifier essentially consists of two components: an internal amplifier to provide signal amplification in the forward direction, and a passive feedback network of dependable parameters to provide a means of monitoring the actual state of the output signal. These two components can be connected together in an unlimited variety of ways; however, a group of circuits based on the non-cascade combinations of parts (b) to (e) in Fig. 4.28 is of particular interest principally because of the simplicity of these structures. In Fig. 4.34 we have presented some specific examples

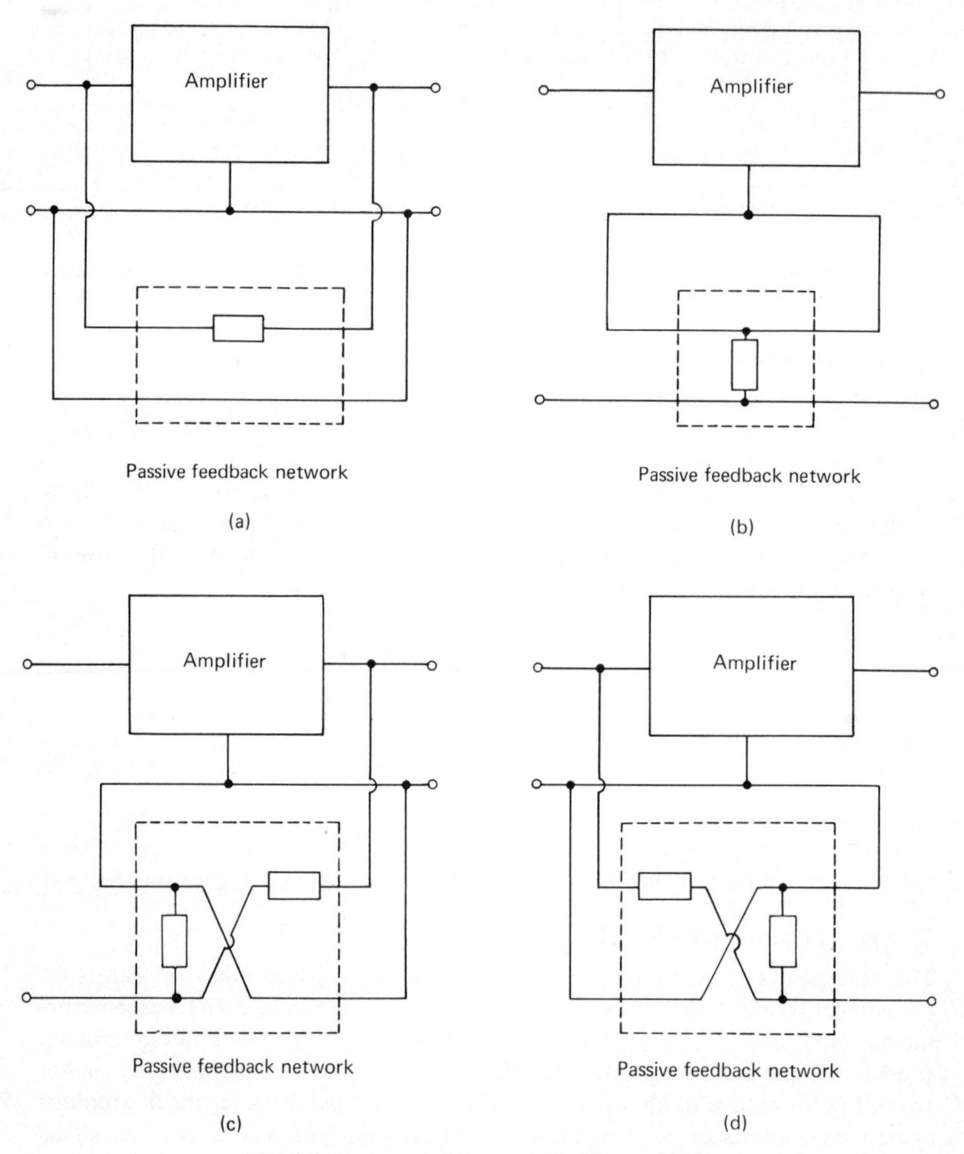

Fig. 4.34. Four basic types of feedback circuits.

of feedback circuits, with the overall parameters of each combination equal to the sum of appropriate parameters of the component networks. We thus have a simple means of modifying the parameters of a given active two-port device by the addition of parameters of a more dependable passive network.

It is significant to observe that the series and shunt feedback structures of Figs. 4.34(a) and (b) have a terminal common to the input and output ports, whereas the other two feedback structures are similar to bridge circuits with no common terminal shared by input and output. Consequently, if feedback structures such as those shown in Figs. 4.34(c) and (d) are cascaded, it would be necessary to use interstage transformers for reasons of common ground and power supply.

Example 4.7. *Series feedback amplifier.* In the transistor amplifier of Fig. 4.35(a) a resistor R_e is connected in series with the emitter terminal of a common-emitter amplifier, so that it is common to the input and output ports of the transistor and thereby enables a voltage signal proportional to the load current to be fed back to the input port. To determine the effect of the resistor R_e on the current gain, voltage gain and input impedance of the amplifier, we may view it as two subnetworks N' and N'' connected in series at their input and output ports, as in Fig. 4.35(b), with the transistor constituting network N' and the feedback resistor R_e constituting network N''. The z-matrix of network N' is, from Table 4.1, related to the common-emitter h-parameters as follows:

$$\begin{bmatrix} z'_{11} & z'_{12} \\ z'_{21} & z'_{22} \end{bmatrix} = \begin{bmatrix} \dfrac{\Delta_{he}}{h_{oe}} & \dfrac{h_{re}}{h_{oe}} \\ -\dfrac{h_{fe}}{h_{oe}} & \dfrac{1}{h_{oe}} \end{bmatrix}. \tag{4.142}$$

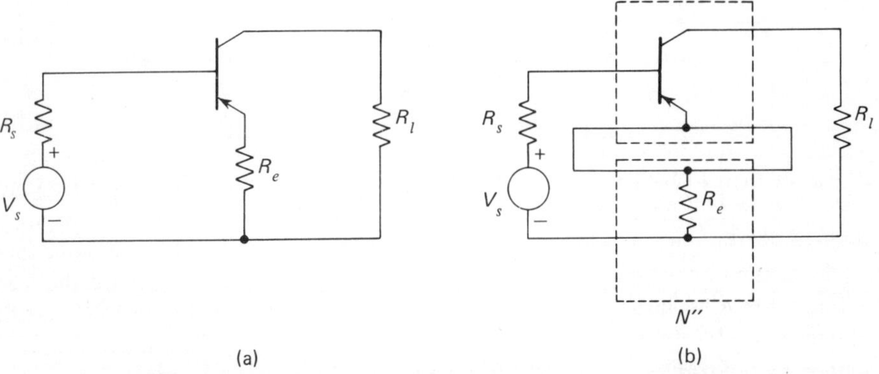

Fig. 4.35. Common-emitter amplifier with series feedback.

The z-matrix of network N'' is given by

$$\begin{bmatrix} z''_{11} & z''_{12} \\ z''_{21} & z''_{22} \end{bmatrix} = \begin{bmatrix} R_e & R_e \\ R_e & R_e \end{bmatrix}. \tag{4.143}$$

The overall z-matrix of the series–series combination of networks N' and N'' of Fig. 4.35 is the sum of the z-matrices of Eqs. (4.142) and (4.143); therefore

$$\begin{bmatrix} z_{11} & z_{12} \\ z_{21} & z_{22} \end{bmatrix} = \begin{bmatrix} \dfrac{\Delta_{he}}{h_{oe}} + R_e & \dfrac{h_{re}}{h_{oe}} + R_e \\ -\dfrac{h_{fe}}{h_{oe}} + R_e & \dfrac{1}{h_{oe}} + R_e \end{bmatrix}. \tag{4.144}$$

In practice, however, we usually find that R_e presents negligible loading at the output port of the amplifier, that is, $R_e \ll 1/h_{oe}$. This means that we would certainly find $R_e \ll h_{fe}/h_{oe}$, as h_{fe} is ordinarily large compared with unity; therefore, we may simplify Eqs. (4.144) as follows:

$$\begin{bmatrix} z_{11} & z_{12} \\ z_{21} & z_{22} \end{bmatrix} \simeq \begin{bmatrix} R_e + \dfrac{\Delta_{he}}{h_{oe}} & \dfrac{h_{re}}{h_{oe}} + R_e \\ -\dfrac{h_{fe}}{h_{oe}} & \dfrac{1}{h_{oe}} \end{bmatrix}. \tag{4.145}$$

Using the pertinent relations of Table 4.2 we find that for a load resistance R_l, the current gain, voltage gain and input impedance of the complete amplifier are as follows, respectively,

$$K_i = \frac{-z_{21}}{z_{22} + R_l} = \frac{h_{fe}}{1 + h_{oe}R_l}, \tag{4.146}$$

$$K_v = \frac{z_{21}R_l}{\Delta_z + z_{11}R_l} = \frac{-h_{fe}R_l}{(h_{ie} + R_e)(1 + h_{oe}R_l) + h_{fe}(R_e - h_{re}R_l)}, \tag{4.147}$$

$$Z_{in} = \frac{\Delta_z + z_{11}R_l}{z_{22} + R_l} = h_{ie} + R_e + \frac{h_{fe}(R_e - h_{re}R_l)}{1 + h_{oe}R_l}. \tag{4.148}$$

From Eqs. (4.146)–(4.148) we observe that the effect of the series feedback resistor R_e is to reduce the voltage gain of the amplifier and increase its input impedance; the current gain is, however, unaffected by the addition of resistor R_e. In particular, when the load R_l is small compared with $1/h_{oe}$, and the feedback resistor R_e is large compared with h_{ie}/h_{fe}, we find from Eq. (4.147) that the voltage gain closely approximates to $-R_l/R_e$ and thereby becomes practically independent of transistor parameters; the voltage gain of the amplifier is therefore said to be stabilized by the addition of resistor R_e. However, the input impedance remains sensitive to variations in transistor parameters.

Example 4.8. *Shunt feedback amplifier.* Consider next Fig. 4.36(a) where a resistor R_c is connected between the collector and base terminals of a common-emitter amplifier, so as to enable a signal proportional to the load voltage to be fed back to the input port. To evaluate the effect of resistor R_c on the current gain, voltage gain and input impedance, we may view the amplifier as the parallel–parallel connection of two sub-networks N' and N'', with the transistor constituting N' and the feedback resistor constituting N'', as in Fig. 4.36(b). The y-matrix of network N' is related to the common emitter h-parameters by

$$\begin{bmatrix} y'_{11} & y'_{12} \\ y'_{21} & y'_{22} \end{bmatrix} = \begin{bmatrix} \dfrac{1}{h_{ie}} & -\dfrac{h_{re}}{h_{ie}} \\ \dfrac{h_{fe}}{h_{ie}} & \dfrac{\Delta_{he}}{h_{ie}} \end{bmatrix}. \tag{4.149}$$

The y-matrix of network N'' is given by

$$\begin{bmatrix} y''_{11} & y''_{12} \\ y''_{21} & y''_{22} \end{bmatrix} = \begin{bmatrix} G_c & -G_c \\ -G_c & G_c \end{bmatrix}, \tag{4.150}$$

where $G_c = 1/R_c$. Therefore, adding Eqs. (4.149) and (4.150) gives the overall

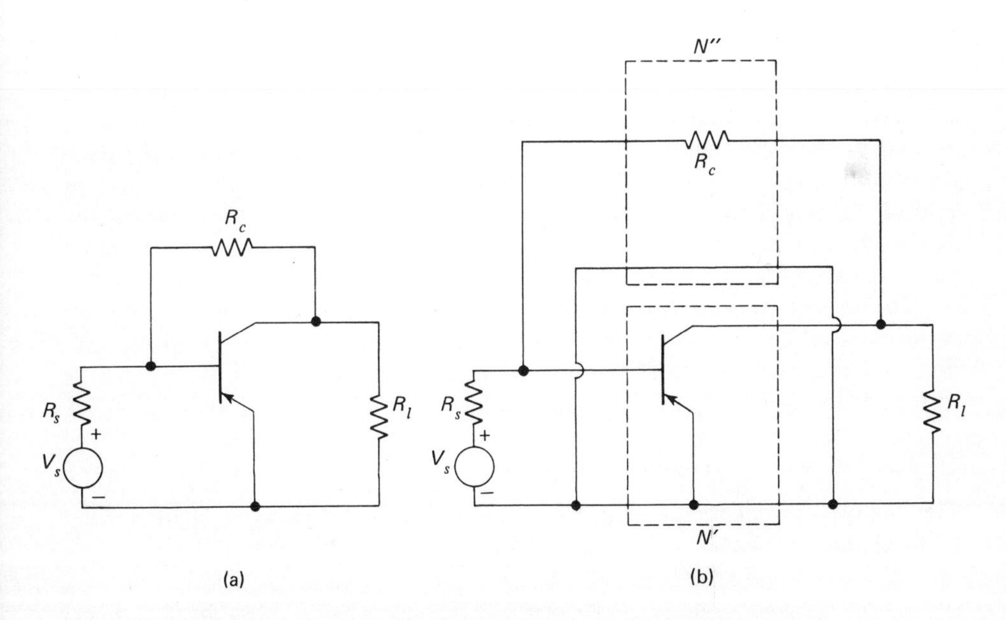

Fig. 4.36. Common-emitter amplifier with shunt feedback.

y-matrix of the parallel–parallel combination of networks N' and N'' to be

$$\begin{bmatrix} y_{11} & y_{12} \\ y_{21} & y_{22} \end{bmatrix} = \begin{bmatrix} \dfrac{1}{h_{ie}} + G_c & -\dfrac{h_{re}}{h_{ie}} - G_c \\ \dfrac{h_{fe}}{h_{ie}} - G_c & \dfrac{\Delta_{he}}{h_{ie}} + G_c \end{bmatrix}. \tag{4.151}$$

In practice we usually find that the feedback resistor R_c presents negligible loading at the input port, that is, $G_c \ll 1/h_{ie}$. This means that we may also neglect G_c compared with h_{fe}/h_{ie}, so that Eq. (4.151) simplifies into

$$\begin{bmatrix} y_{11} & y_{12} \\ y_{21} & y_{22} \end{bmatrix} \simeq \begin{bmatrix} \dfrac{1}{h_{ie}} & -\dfrac{h_{re}}{h_{ie}} - G_c \\ \dfrac{h_{fe}}{h_{ie}} & \dfrac{\Delta_{he}}{h_{ie}} + G_c \end{bmatrix}. \tag{4.152}$$

The current gain, voltage gain and input impedance of the complete amplifier are thus given as follows (see Table 4.2):

$$K_i = \frac{y_{21}G_l}{y_{11}G_l + \Delta_y} = \frac{h_{fe}G_l}{G_l + h_{oe} + G_c(1 + h_{fe})}, \tag{4.153}$$

$$K_v = \frac{-y_{21}}{y_{22} + G_l} = \frac{-h_{fe}}{h_{ie}(G_l + G_c) + \Delta_{he}}, \tag{4.154}$$

$$Z_{in} = \frac{y_{22} + G_l}{y_{11}G_l + \Delta_y} = \frac{h_{ie}(G_l + G_c) + \Delta_{he}}{G_l + h_{oe} + G_c(1 + h_{fe})}. \tag{4.155}$$

Equations (4.153)–(4.155) reveal that the effect of shunt feedback resistor R_c is to reduce the current gain and input impedance of the stage, and that the voltage gain is affected only slightly. In particular, if R_c is chosen small enough to satisfy the condition $G_c \gg (G_l + h_{oe})/h_{fe}$, then from Eq. (4.153) we find that the current gain of the amplifier closely approximates to G_l/G_c and so becomes practically independent of variations in transistor parameters; the current gain is therefore said to be stabilized by the addition of resistor R_c.

The above two examples have illustrated some of the improvements in overall performance that can result from the application of feedback. A physical explanation of the concept of feedback and a complete mathematical analysis of its various effects and design limitations will be left till Chapters 11 and 12.

4.9 FOUR SPECIAL TWO-PORT DEVICES

The voltage–current relationships for a general two-port network characterized by its chain parameters A, B, C and D are defined as follows:

$$\begin{aligned} V_1 &= AV_2 - BI_2 \\ I_1 &= CV_2 - DI_2. \end{aligned} \tag{4.156}$$

When a load impedance Z_l is connected across the output port, the conditions across the output port are constrained by the relation $V_2 = -I_2Z_l$ and, therefore, the impedance Z_{in} measured looking into the input port is

$$Z_{in} = \frac{AZ_l + B}{CZ_l + D}. \tag{4.157}$$

Four special cases of particular interest in network theory arise depending on, first, whether the input impedance is directly or inversely proportional to the load impedance and, second, whether the constant of proportionality is positive or negative.

a) Ideal Transformer

In an *ideal transformer* the input impedance is directly proportional to the load impedance, and the constant of proportionality is positive; so that

$$Z_{in} = n^2 Z_l, \tag{4.158}$$

where n is a positive real number usually referred to as the *turns ratio* of the transformer. From Eq. (4.157) we see that the condition of Eq. (4.158) is satisfied with a chain matrix of the form

$$\begin{bmatrix} A & B \\ C & D \end{bmatrix} = \begin{bmatrix} \pm n & 0 \\ 0 & \pm \dfrac{1}{n} \end{bmatrix}. \tag{4.159}$$

That is to say, in an ideal transformer we have

$$\begin{aligned} V_1 &= nV_2 \\ I_1 &= -\frac{I_2}{n}, \end{aligned} \tag{4.160}$$

according to which the input and output voltages are in phase. Alternatively, the input and output voltages of an ideal transformer may be in antiphase as indicated by

$$\begin{aligned} V_1 &= -nV_2 \\ I_1 &= \frac{I_2}{n}. \end{aligned} \tag{4.161}$$

b) Gyrator

An *ideal gyrator* is a two-port device which presents an input impedance that is proportional to the reciprocal of the terminating load impedance, with a positive constant of proportionality, as shown by[6]

$$Z_{in} = \frac{r^2}{Z_l}, \tag{4.162}$$

where r is a positive real constant called the *gyration resistance*. It is apparent that if Z_l is purely resistive, then the reflected input impedance is purely resistive too and has the same sign. If, however, Z_l is purely capacitive the input impedance is purely inductive, and vice versa. The action of the device is, therefore, to invert the impedance connected to it.

From Eq. (4.157), the condition $Z_{in} = r^2/Z_l$ is evidently satisfied with a chain matrix of the form

$$\begin{bmatrix} A & B \\ C & D \end{bmatrix} = \begin{bmatrix} 0 & r \\ \dfrac{1}{r} & 0 \end{bmatrix}. \tag{4.163}$$

With such a set of chain parameters we find that the voltage–current relations of an ideal gyrator are

$$\begin{aligned} I_1 &= gV_2 \\ I_2 &= -gV_1, \end{aligned} \tag{4.164}$$

where $g = 1/r$ is called the *gyration conductance*. Equations (4.164) show that in an ideal gyrator the current at one port is proportional to the voltage at the other port, and vice versa. The gyrator is, however, an antisymmetric device in the sense that if for signal transmission in the forward direction the input voltage is transduced into a *negative* output current (i.e., *flowing out* of the device), then for signal transmission in the reverse direction the output voltage is transduced into a *positive* input current (i.e., *flowing into* the device). The gyroscope behaves in an analogous manner in that if it is set spinning about a vertical axis such that a westward push produces a positive northward motion, then we find that a northward push produces a negative westward motion. In other words, with reference to the two mechanical co-ordinates, the gyroscope is antisymmetric. Indeed, it is from this analogy with the behaviour of a gyroscope that the electric gyrator derives its name.

The circuit symbol for a gyrator is as shown in Fig. 4.37 in which the direction of the arrow signifies the *direction of gyration*. According to this diagram, the electrical behaviour of the gyrator is as described by Eqs. (4.164). If the direction of gyration is reversed, we would have $I_1 = -gV_2$ and $I_2 = gV_1$.

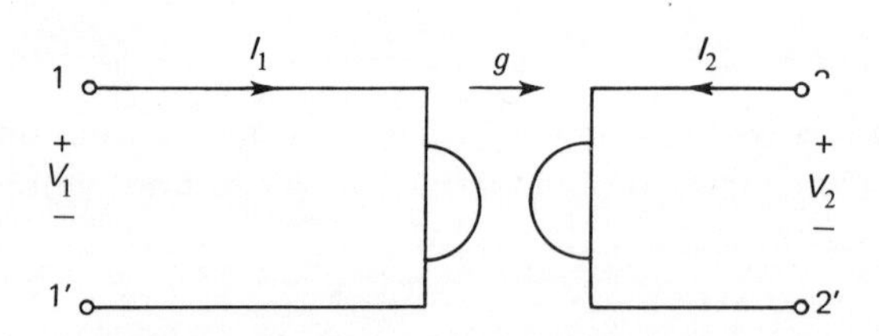

Fig. 4.37. Circuit symbol for ideal gyrator.

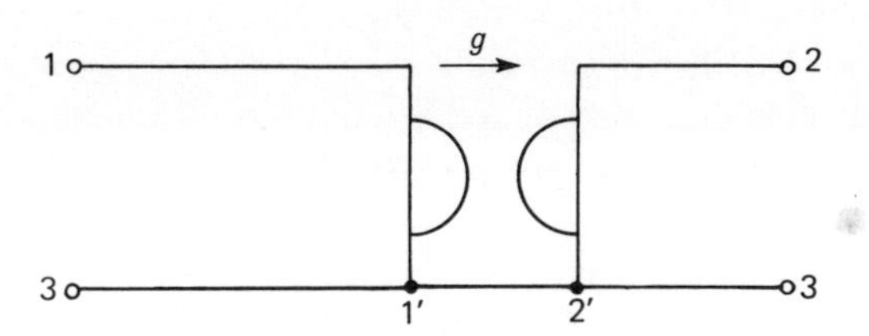

Fig. 4.38. Three-terminal gyrator with terminal 3 common.

If we short circuit terminals 1′ and 2′ of the gyrator, as in Fig. 4.38, a three-terminal device is obtained. According to Eqs. (4.164), the y-matrix of this device is

$$\begin{bmatrix} y_{11} & y_{12} \\ y_{21} & y_{22} \end{bmatrix} = \begin{bmatrix} 0 & g \\ -g & 0 \end{bmatrix}. \tag{4.165}$$

The three-terminal device may also be used with terminal 1 or 2 grounded. The indefinite admittance matrix which is obtained from the y-matrix of Eq. (4.165) by completing each row and column to zero, is therefore,

$$\begin{bmatrix} y_{11} & y_{12} & y_{13} \\ y_{21} & y_{22} & y_{23} \\ y_{31} & y_{32} & y_{33} \end{bmatrix} = \begin{bmatrix} 0 & g & -g \\ -g & 0 & g \\ g & -g & 0 \end{bmatrix}. \tag{4.166}$$

Thus, when terminal 1 is grounded, as in Fig. 4.39, we see that the y-matrix of the resulting two-port device is

$$\begin{bmatrix} y_{22} & y_{23} \\ y_{32} & y_{33} \end{bmatrix} = \begin{bmatrix} 0 & g \\ -g & 0 \end{bmatrix}. \tag{4.167}$$

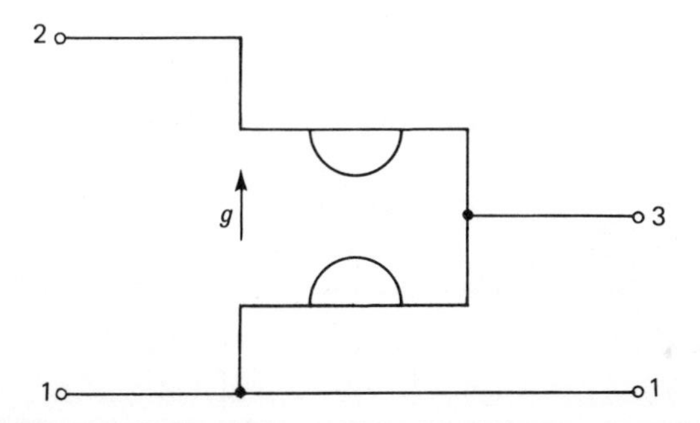

Fig. 4.39. Three-terminal gyrator with terminal 1 common.

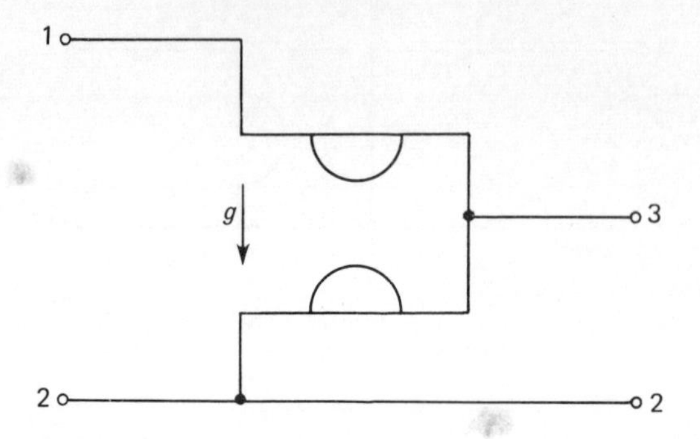

Fig. 4.40. Three-terminal gyrator with terminal 2 common.

Similarly, when terminal 2 is grounded, as in Fig. 4.40, the y-matrix is

$$\begin{bmatrix} y_{11} & y_{13} \\ y_{31} & y_{33} \end{bmatrix} = \begin{bmatrix} 0 & -g \\ g & 0 \end{bmatrix}. \tag{4.168}$$

It is therefore obvious that, no matter which terminal of the device is grounded, the resulting 2×2 matrix is always the same as that of Eq. (4.165), with a possible reversal of sign as in the case of Fig. 4.40. In other words, the three-terminal gyrator exhibits the same properties in all grounded positions. In view of this circular symmetry, a three-terminal gyrator is represented as in Fig. 4.41. The direction of gyration given in this diagram pertains to the y-matrices of Eqs. (4.165) and (4.167); in the case of Eq. (4.168) the direction of gyration is reversed.

A method of implementing the gyrator proceeds by splitting up the y-matrix as

$$\begin{bmatrix} y_{11} & y_{12} \\ y_{21} & y_{22} \end{bmatrix} = \begin{bmatrix} 0 & g \\ 0 & 0 \end{bmatrix} + \begin{bmatrix} 0 & 0 \\ -g & 0 \end{bmatrix}. \tag{4.169}$$

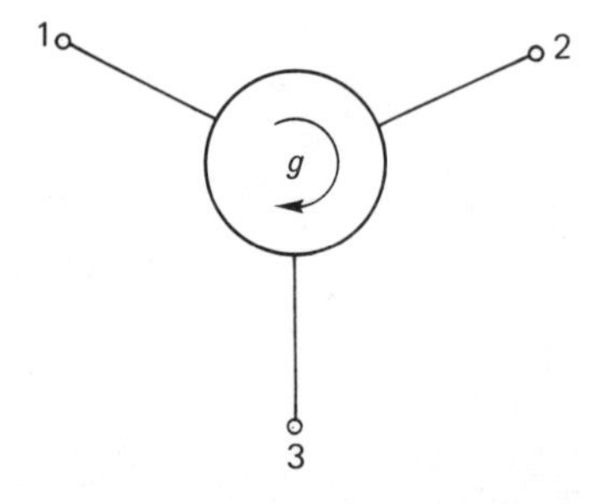

Fig. 4.41. Circuit symbol for three-terminal gyrator.

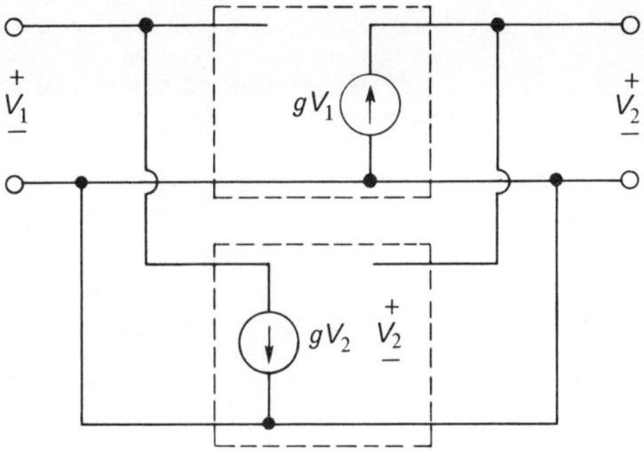

Fig. 4.42. A circuit model of the ideal gyrator.

The two component matrices of Eq. (4.169) represent voltage-controlled current sources of opposite polarity and direction of signal transmission, connected in parallel as in Fig. 4.42. This is implemented in practice by connecting two amplifiers together to form a closed loop; one amplifier has a zero phase shift from input to output whilst the other has 180° phase shift. Each amplifier has both high input impedance and high output impedance; the high impedances act to keep y_{11} and y_{22} down to insignificant magnitudes.

In order to evaluate the effect of the parasitic admittances y_{11} and y_{22} upon the performance of a practical gyrator, suppose they are non-zero and both equal to g_0 and that one of the ports of the gyrator is terminated with a perfect capacitor C. Then the impedance measured looking into the other port is the same as that of the equivalent network shown in Fig. 4.43, that is,

$$Z_{in}(j\omega) = \frac{g_0 + j\omega C}{g^2 + g_0^2 + j\omega C g_0}. \tag{4.170}$$

Assuming that $g \gg g_0$, as is ordinarily the case, and expressing $Z_{in}(j\omega)$ as

$$Z_{in}(j\omega) = R_{eq}(\omega) + j\omega L_{eq}(\omega), \tag{4.171}$$

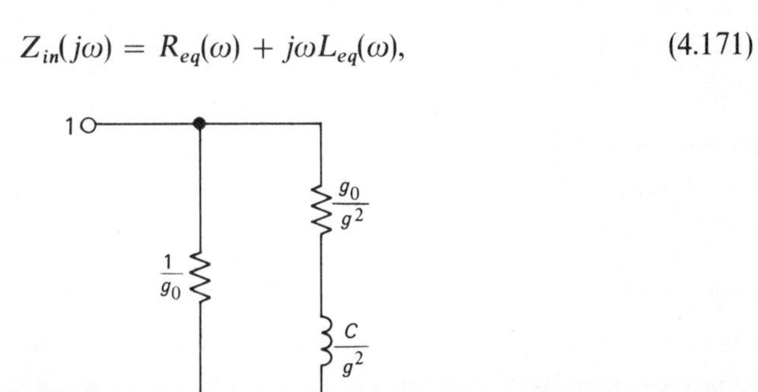

Fig. 4.43. Equivalent circuit of a non-ideal gyrator terminated with capacitor C.

we have

$$R_{eq}(\omega) = \frac{g_0(g^2 + \omega^2 C^2)}{g^4 + \omega^2 C^2 g_0^2} \tag{4.172}$$

$$L_{eq}(\omega) = \frac{Cg^2}{g^4 + \omega^2 C^2 g_0^2}. \tag{4.173}$$

The quality factor of the inductor is defined, in the usual way, by

$$Q(\omega) = \frac{\omega L_{eq}(\omega)}{R_{eq}(\omega)}. \tag{4.174}$$

Hence, Eqs. (4.172)–(4.174) give

$$Q(\omega) = \frac{\omega C g^2}{g_0(g^2 + \omega^2 C^2)}. \tag{4.175}$$

The quality factor reaches its maximum value at

$$\omega_{max} = \frac{g}{C} \tag{4.176}$$

and the maximum is given by

$$Q_{max} = \frac{g}{2g_0}. \tag{4.177}$$

Equation (4.177) indicates that a Q-factor of 500, say, demands that we have $g = 1000g_0$. Sheehan and Orchard[7] have described a high-quality gyrator circuit using a combination of field effect and bipolar transistors, which is capable of realizing Q-factors of the order of 500.

Figure 4.44 shows R_{eq}, L_{eq} and Q plotted against ω. We observe that the frequency of maximum Q is lower than the cut-off frequency of the inductance; also the Q can be made arbitrarily large, with g large, at the expense of the value of the inductance.[8]

c) Negative-impedance Converter

By definition, the input impedance of a *negative-impedance converter* (abbreviated as NIC) is proportional to the negative of the load impedance; so that

$$Z_{in} = -kZ_l, \tag{4.178}$$

where k is a positive real number called the *impedance conversion ratio.* Any negative impedance, including as special cases a negative capacitance and a negative inductance, can therefore be produced by terminating a negative-impedance converter in the positive of the required impedance. From Eq. (4.157)

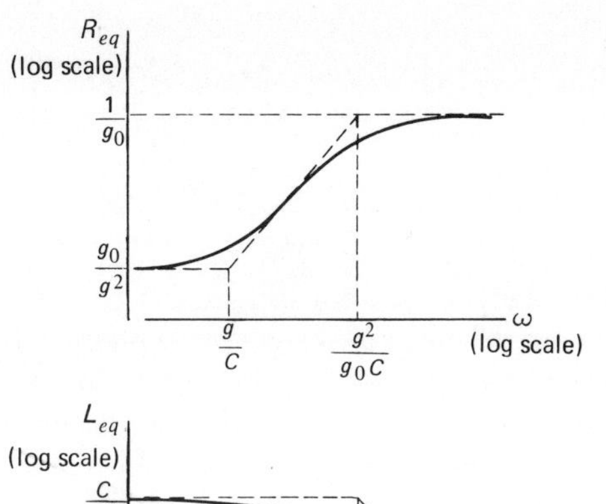

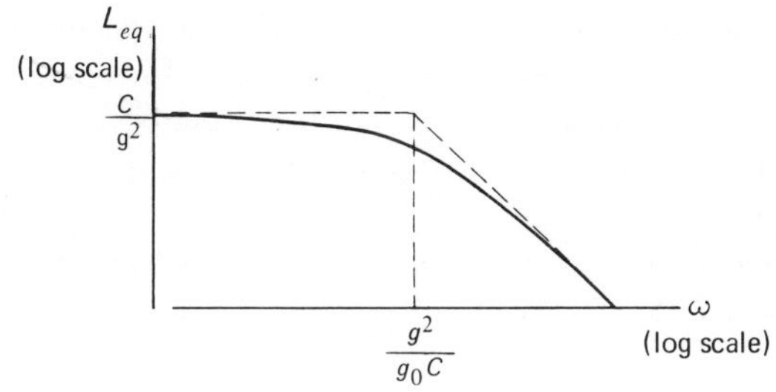

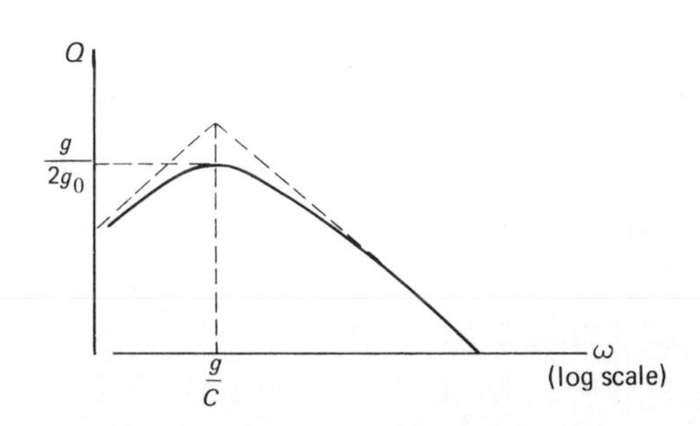

Fig. 4.44. Illustrating the frequency dependence of R_{eq}, L_{eq} and Q.

we see that the requirement of Eq. (4.178) is satisfied by the chain matrix

$$\begin{bmatrix} A & B \\ C & D \end{bmatrix} = \begin{bmatrix} \pm\sqrt{k} & 0 \\ 0 & \mp\dfrac{1}{\sqrt{k}} \end{bmatrix}. \tag{4.179}$$

which corresponds to

$$\begin{aligned} V_1 &= \sqrt{k}V_2 \\ I_1 &= \frac{I_2}{\sqrt{k}} \end{aligned} \tag{4.180}$$

or alternatively,

$$V_1 = -\sqrt{k}V_2$$
$$I_1 = -\frac{I_2}{\sqrt{k}}. \tag{4.181}$$

According to the voltage–current relations of Eqs. (4.180), the input and output voltages of the negative-impedance converter are in phase, while the direction of output current is inverted with respect to the input current. Such a negative-impedance converter is, therefore, said to be the *current-inversion type*. On the other hand, according to Eqs. (4.181) the actual direction of current flow through the device is unchanged, but the input and output voltages are inverted with respect to each other. This negative-impedance converter is, therefore, said to be of the *voltage-inversion type*.

Characterizing an ideal negative-impedance converter of the voltage inversion type by its h-parameters, we have for $k = 1$,

$$\begin{bmatrix} h_{11} & h_{12} \\ h_{21} & h_{22} \end{bmatrix} = \begin{bmatrix} 0 & -1 \\ -1 & 0 \end{bmatrix}. \tag{4.182}$$

This may be fairly closely realized using the circuit arrangement of Fig. 4.45 where, for clarity, we have omitted the biasing resistors and power supplies.[9] Assuming the two transistors to be identical and using the common-base h-parameters to characterize each transistor, we find that the overall h-matrix of the circuit in Fig. 4.45 is

$$\begin{bmatrix} h_{11} & h_{12} \\ h_{21} & h_{22} \end{bmatrix} = \begin{bmatrix} 2h_{ib} & -1 + 2h_{rb} \\ 1 + 2h_{fb} & 2h_{ob} \end{bmatrix}. \tag{4.183}$$

Recalling that ordinarily the resistance h_{ib} and conductance h_{ob} are small, h_{fb} is close to -1, and h_{rb} is small compared with unity, we see that the prescribed conditions of Eq. (4.182) are closely approached.

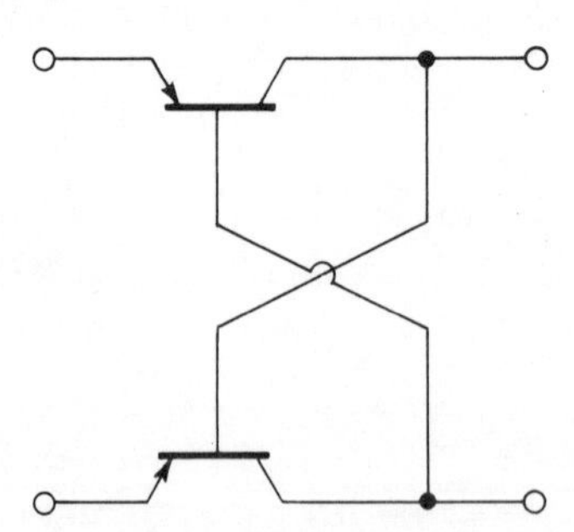

Fig. 4.45. A balanced NIC of the voltage-inversion type.

The converter of Fig. 4.45 is balanced and is therefore ideally suited for such applications as telephone repeaters. However, for other applications, it is frequently desirable to have a terminal common to the input and output ports, as in Fig. 4.46(a) which shows the idealized model of a voltage-inversion NIC, using a high-gain current amplifier.[10] The h-matrix of this circuit is obtained as

$$\begin{bmatrix} h_{11} & h_{12} \\ h_{21} & h_{22} \end{bmatrix} = \begin{bmatrix} \dfrac{R}{1+K} & \dfrac{-K}{1+K} \\ \dfrac{-K}{1+K} & \dfrac{1}{R(1+K)} \end{bmatrix}. \tag{4.184}$$

We see that as K becomes large, this h-matrix does indeed approach that of Eq. (4.182) for the ideal NIC. This model of Fig. 4.46(a) may be realized using the arrangement of Fig. 4.46(b) where the low emitter-to-base resistance h_{ib} of transistor Q_2 corresponds to the short circuit through which current I flows, and the h_{fe} of transistor Q_1 assures the realization of the desired high value for the current-transfer ratio K.

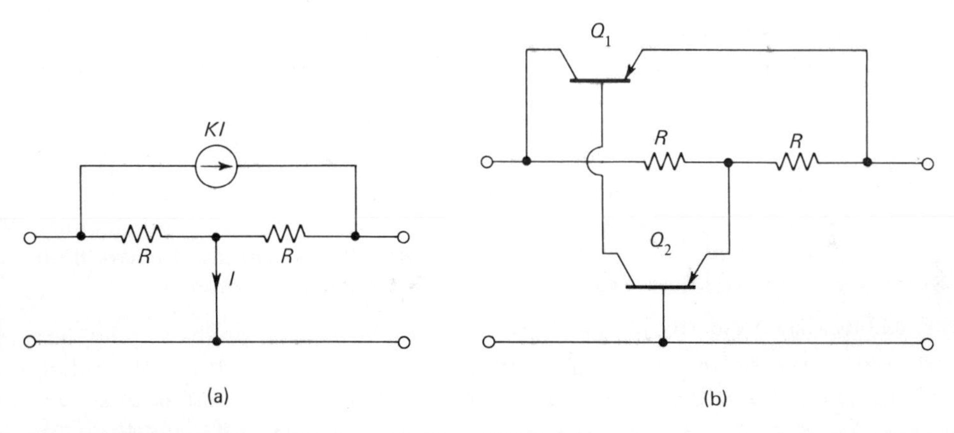

Fig. 4.46. (a) Idealized model of an unbalanced NIC of the voltage-inversion type; (b) Circuit for realizing the idealized model.

Consider next the idealized model[11] of Fig. 4.47(a) which is recognized as the dual of that shown in Fig. 4.46(a). Evaluating the h-matrix of the circuit in Fig. 4.47(a), we obtain

$$\begin{bmatrix} h_{11} & h_{12} \\ h_{21} & h_{22} \end{bmatrix} = \begin{bmatrix} 0 & 1 \\ \dfrac{K-2}{K+2} & 0 \end{bmatrix}, \tag{4.185}$$

which, for large K, approaches the h-matrix of an ideal NIC of the current-

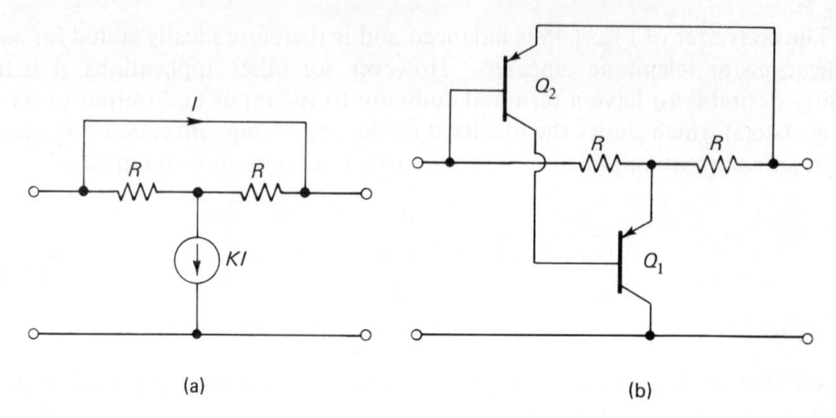

Fig. 4.47. (a) Idealized model of an unbalanced NIC of the current inversion type; (b) Circuit for realizing the idealized model.

inversion type. The idealized model of Fig. 4.47(a) may be realized using the arrangement shown in Fig. 4.47(b), where the top short-circuit is represented by the low emitter-to-base resistance h_{ib} of transistor Q_2, and the desired high current-transfer ratio K is accounted for by the h_{fe} of transistor Q_1.

From the above discussion it is apparent that in a practical NIC, be it of the voltage-inversion or current-inversion type, the parameters h_{11} and h_{22} have small, but non-zero values, thereby causing some deviation from the negative-impedance conversion properties of the ideal NIC. Hence, h_{11} and h_{22} are referred to as *parasitic parameters*. It is rather fortunate that the properties of the non-ideal NIC are such that it can serve to compensate for its own parasitics, in a manner described by Merrill.[12] In practice, we find that the parasitics are either both positive or both negative. There are, therefore, two cases to consider:

1. Suppose h_{11} and h_{22} are both positive, and that the gain product $h_{12}h_{21}$ has been adjusted to equal unity. Then the connection of two appropriate resistors, one across the input, the other in series with the output, as shown in Fig. 4.48(a), will compensate for the non-zero values of h_{11} and h_{22}.
2. If, on the other hand, h_{11} and h_{22} are both negative, we may use the arrangement shown in Fig. 4.48(b) where we have interchanged the positions of the shunt and series compensating elements.

d) Negative-impedance Inverter

In a negative-impedance inverter the input impedance is proportional to the reciprocal of the load impedance with a negative constant of proportionality, so that we may write

$$Z_{in} = -\frac{1}{G^2 Z_l}, \tag{4.186}$$

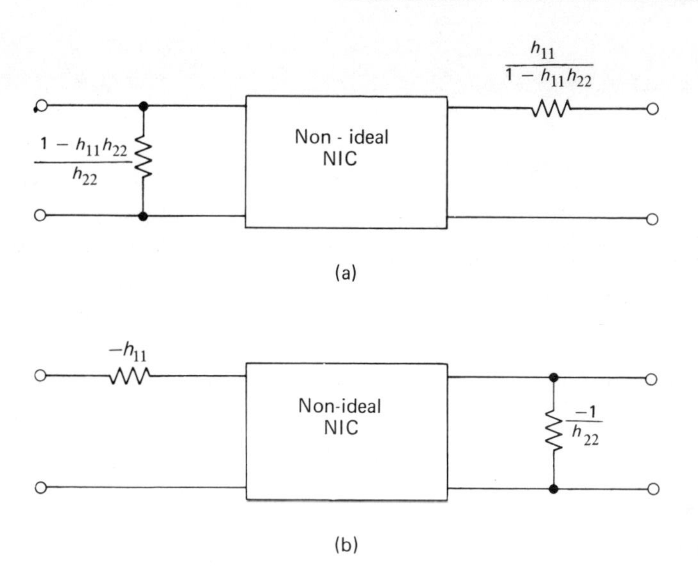

Fig. 4.48. Compensation schemes for non-ideal NIC's: (a) h_{11} and h_{22} positive; (b) h_{11} and h_{22} negative.

where G is a real constant termed the *inversion conductance*. Thus a negative impedance inverter may be used to produce a negative impedance by terminating it with the positive inverse of the required impedance. From Eq. (4.157) we see that the required condition of Eq. (4.186) is satisfied by the chain matrix

$$\begin{bmatrix} A & B \\ C & D \end{bmatrix} = \begin{bmatrix} 0 & \pm\dfrac{1}{G} \\ \mp G & 0 \end{bmatrix}. \tag{4.187}$$

The corresponding set of voltage–current relations is as follows

$$\begin{aligned} I_1 &= \mp G V_2 \\ I_2 &= \mp G V_1 . \end{aligned} \tag{4.188}$$

Therefore, unlike a gyrator, the negative-impedance inverter is symmetric in that it can be turned end for end without affecting the behaviour of a system in which the inverter is a component part.

The ideal negative-impedance inverter may be achieved using the T- or π-section of Fig. 4.49; in each case, the negative resistance element can be realized by a tunnel diode.

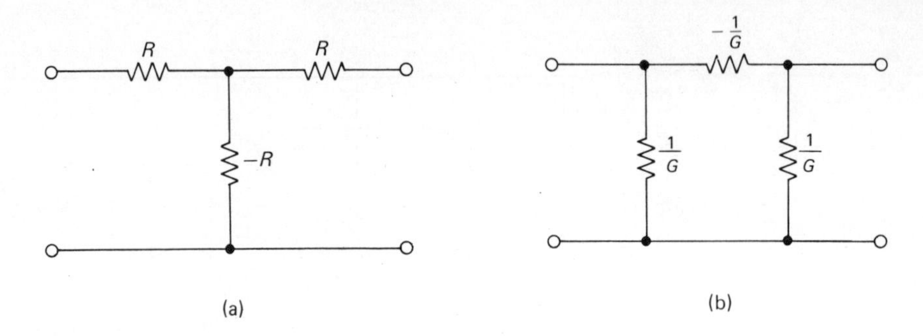

Fig. 4.49. Negative-impedance inverter circuits: (a) *T*-type; (b) π-type.

Concluding Remarks

In Table 4.5 we have summarized the properties of the ideal transformer, gyrator, negative-impedance converter and negative-impedance inverter in terms of the various parameter sets. It is noteworthy that the ideal transformer and negative-impedance converter do not possess finite z- or y-matrices, while the gyrator and negative-impedance inverter do not possess finite h- or g-matrices.

From Eqs. (4.160) and (4.164) we note that $I_1V_1 + I_2V_2 = 0$ for the ideal transformer and gyrator; both of them are therefore lossless transmission devices in that no energy is generated, dissipated or stored in either device. On the other hand, the negative-impedance converter and negative-impedance inverter are both *active* in that if either one of them is driven at the input port by a signal source and loaded at the other port with a positive resistance, the device supplies power to the load and to the driving source. If the load resistance is negative, however, then the negative-impedance converter and inverter will both absorb power from the load and the driving source.

It is also significant to observe that

1. The cascade connection of two identical gyrators is equivalent to an ideal transformer with a turns ratio $n = 1$; the cascade connection of two identical negative-impedance converters is equivalent to an ideal transformer with $n = k$; and the cascade connection of two identical negative-impedance inverters is equivalent to an ideal phase-inverting transformer with $n = 1$.
2. The cascade connection of a negative-impedance converter of conversion ratio k and a negative-impedance inverter of inversion conductance G is equivalent to a gyrator of gyration conductance $G/\sqrt{k}$.
3. The cascade connection of a gyrator of gyration conductance g and a negative-impedance inverter of inversion conductance G is equivalent to a negative-impedance converter of conversion ratio $(G/g)^2$.

Table 4.5

	$[z]$	$[y]$	$[h]$	$[g]$	A, B, C, D
Ideal transformer	Undefined	Undefined	$\begin{bmatrix} 0 & \pm n \\ \mp n & 0 \end{bmatrix}$	$\begin{bmatrix} 0 & \pm \frac{1}{n} \\ \mp \frac{1}{n} & 0 \end{bmatrix}$	$\begin{bmatrix} \pm n & 0 \\ 0 & \pm \frac{1}{n} \end{bmatrix}$
Gyrator	$\begin{bmatrix} 0 & \mp \frac{1}{g} \\ \pm \frac{1}{g} & 0 \end{bmatrix}$	$\begin{bmatrix} 0 & \mp g \\ \pm g & 0 \end{bmatrix}$	Undefined	Undefined	$\begin{bmatrix} 0 & \pm \frac{1}{g} \\ \pm g & 0 \end{bmatrix}$
Negative-impedance converter	Undefined	Undefined	$\begin{bmatrix} 0 & \pm \sqrt{k} \\ \pm \sqrt{k} & 0 \end{bmatrix}$	$\begin{bmatrix} 0 & \pm \frac{1}{\sqrt{k}} \\ \pm \frac{1}{\sqrt{k}} & 0 \end{bmatrix}$	$\begin{bmatrix} \pm \sqrt{k} & 0 \\ 0 & \mp \frac{1}{\sqrt{k}} \end{bmatrix}$
Negative-impedance inverter	$\begin{bmatrix} 0 & \mp \frac{1}{G} \\ \mp \frac{1}{G} & 0 \end{bmatrix}$	$\begin{bmatrix} 0 & \mp G \\ \mp G & 0 \end{bmatrix}$	Undefined	Undefined	$\begin{bmatrix} 0 & \pm \frac{1}{G} \\ \mp G & 0 \end{bmatrix}$

4. The cascade connection of a gyrator of gyration conductance g and a negative-impedance converter of conversion ratio k is equivalent to a negative-impedance inverter of inversion conductance $\sqrt{k}g$.

These statements may be readily verified by evaluating the product of the pertinent chain matrices.

In the next chapter and Chapter 13 of this book we shall have more to say about further properties and uses of the gyrator, negative-impedance converter and the negative-impedance inverter.

REFERENCES

1. R. F. SHEA, *Principles of Transistor Circuits*. Wiley, 1953.
2. J. SHEKEL, "Matrix Representations of Transistor Circuits," *Proc. I.R.E.*, **40,** 1493 (1952).
3. N. I. MEYER, "Nonlinear Distortion in Transistor Amplifiers at Low Signal Levels and Low Frequencies," *Proc. I.E.E.*, **104,** Part C, 108 (1957).
4. J. G. LINVILL and J. F. GIBBONS, *Transistors and Active Circuits*. McGraw-Hill, 1961.
5. E. A. GUILLEMIN, *Communication Networks*, vol. 2. Wiley, 1935.
6. B. D. H. TELLEGEN, "The Gyrator, a New Electric Network Element," *Philips Research Reports*, **3,** 81 (1948).
7. D. F. SHEEHAN and H. J. ORCHARD, "Integrable Gyrator using M.O.S. and Bipolar Transistors," *Electronics Letters*, **2,** 390 (1966).
8. T. N. RAO, P. GARY, and R. W. NEWCOMB, "Equivalent Inductance and *Q* of a Capacitor-loaded Gyrator," *I.E.E.E. Journal on Solid State Circuits*, 32 (1967).
9. J. G. LINVILL, "Transistor Negative-impedance Converters," *Proc. I.R.E.*, **41,** 725 (1953).
10. W. R. LUNDRY, "Negative-impedance Circuits—Some Basic Relations and Limitations," *Trans. I.R.E.*, **CT-4,** 132 (1957).
11. A. I. LARKY, "Negative-impedance Converters," *Trans. I.R.E.*, **CT-4,** 124 (1957).
12. J. L. MERRILL, "Theory of the Negative-impedance Converter," *B.S.T.J.*, **30,** 88 (1951).

PROBLEMS

4.1 A transistor has the following values for its common emitter h-parameters

$$h_{ie} = 1{\cdot}4\ \text{k}\Omega$$

$$h_{re} = 5 \times 10^{-4}$$

$$h_{fe} = 49$$

$$h_{oe} = 25\ \mu\mho.$$

It is operated as a common-base stage with a voltage source of internal resistance 100 Ω and a load resistance 10 kΩ. Evaluate its input and output impedances, current and voltage gains, power, available power and transducer power gains.

4.2 The two-stage transistor amplifier of Fig. P4.2 uses a cascade of two identical common-emitter stages, having the common-emitter h-parameters of Problem 4.1. Determine the input impedance and voltage gain of the complete amplifier.

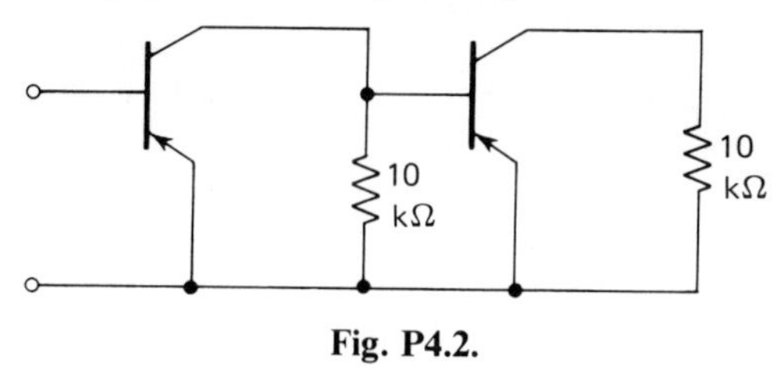

Fig. P4.2.

4.3 Figure P4.3 shows the circuit diagram of a cascode amplifier. Evaluate its overall voltage gain, given that the two tubes are identical, having the parameters $r_p = 10\,\text{k}\Omega$ and $\mu = 40$.

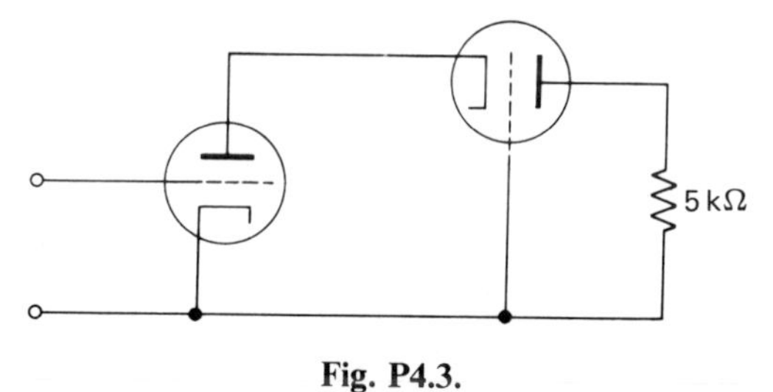

Fig. P4.3.

4.4 Evaluate the input impedance, output impedance and voltage gain of the cathode follower shown in Fig. P4.4

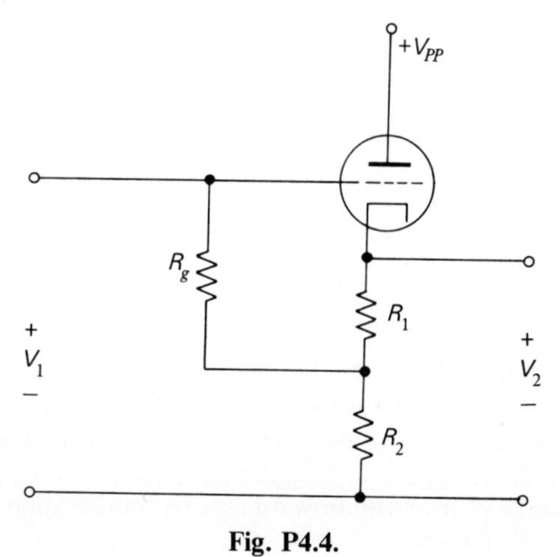

Fig. P4.4.

4.5 Determine the overall voltage gain of the cathode-coupled pair shown in Fig. P4.5. The two tubes are identical, having the parameters given in Problem 4.3.

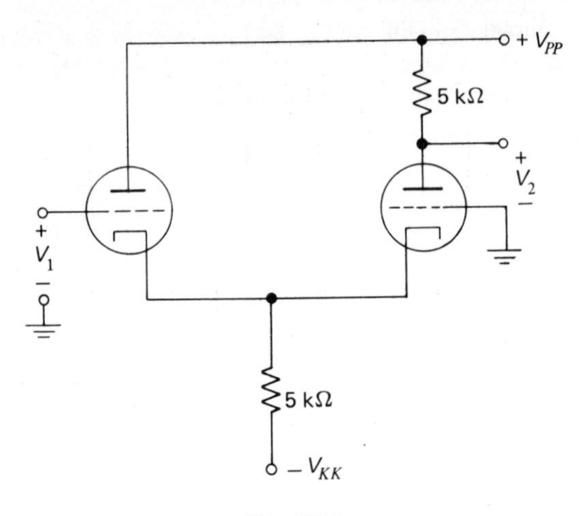

Fig. P4.5.

4.6 Evaluate the input impedance, output impedance, current gain and voltage gain of the transistor stage of Fig. P4.6. The transistor h-parameters are as given in Problem 4.1.

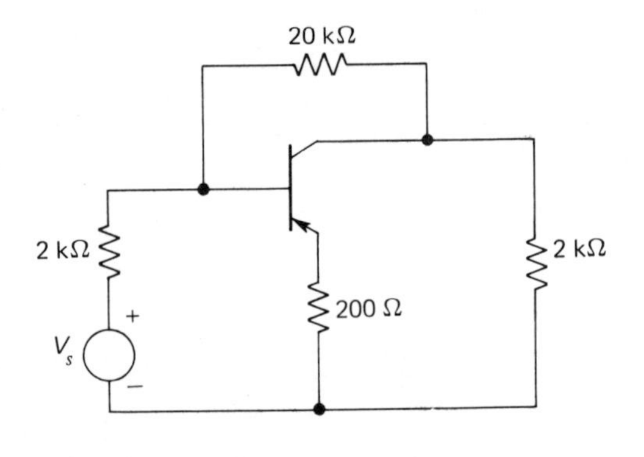

Fig. P4.6.

4.7 Fig. P4.7(a) shows the so-called *Darlington composite pair* operated in its common-collector mode; it is seen to consist basically of two common-collector stages in cascade. Determine the overall h-parameters of this configuration, and those of the common-base and common-emitter modes of operation of the pair (see parts (b) and (c) of Fig. P4.7). Discuss the desirable features of the short-circuit forward current-transfer ratio of the common-base configuration shown in Fig. P4.7(b).

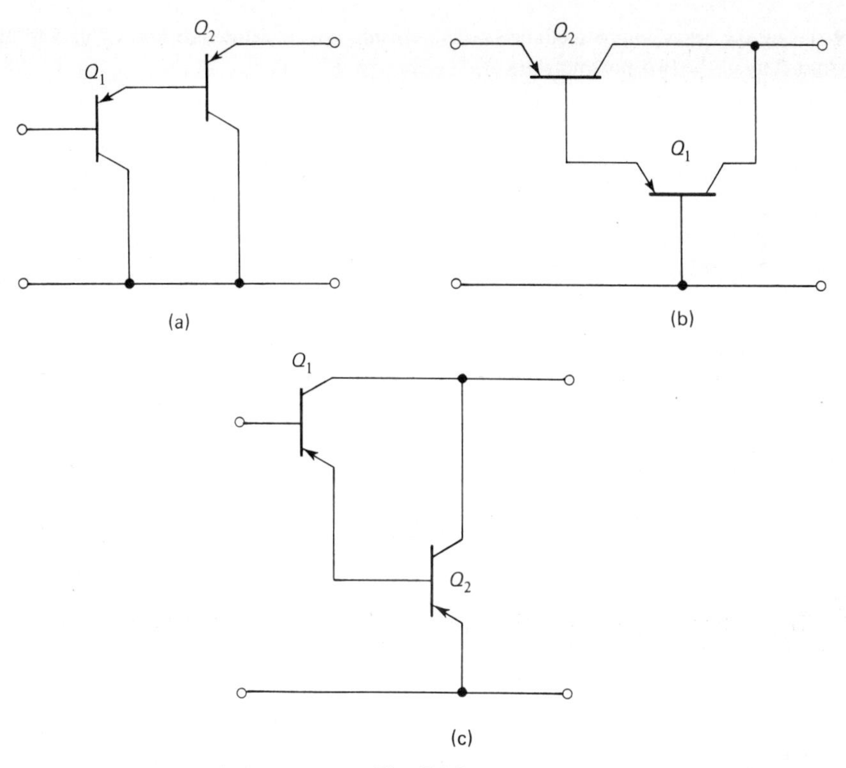

Fig. P4.7.

4.8. a) Evaluate the current gain of the two-stage amplifier of Fig. P4.8(a) and show that it is approximately equal to $-R_1/R_2$.

b) Evaluate the voltage gain of the amplifier of Fig. P4.8(b) and show that it is approximately equal to R_b/R_a.

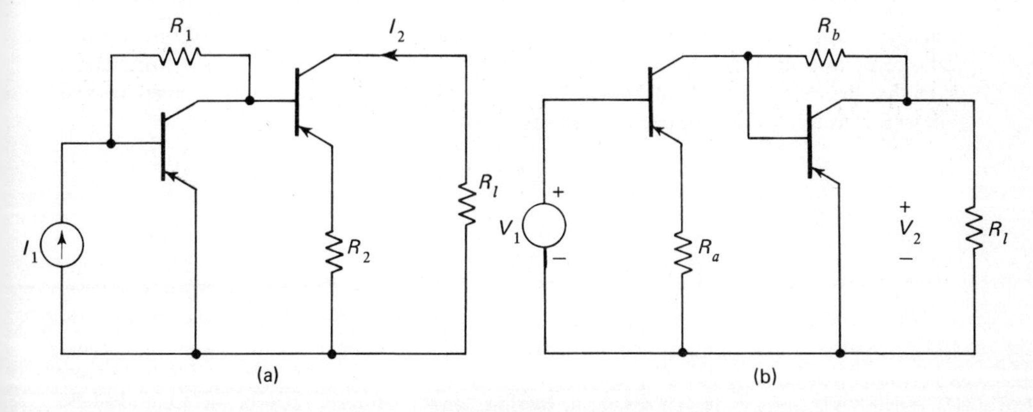

Fig. P4.8.

4.9 Using the set of four impedance measurements shown defined in Fig. P4.9, determine h-parameters of the two-port network N.

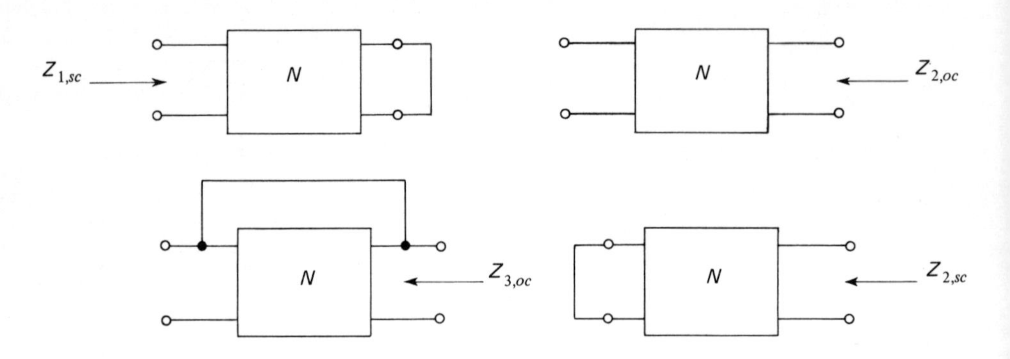

Fig. P4.9.

4.10 The fractional sensitivity of the input impedance of a two-port network with respect to variations in the load impedance is defined as follows

$$S_{Z_l}^{Z_{in}} = \frac{dZ_{in}/Z_{in}}{dZ_l/Z_l}.$$

Determine the load impedance to yield $S_{Z_l}^{Z_{in}} = 0{\cdot}1$ for a common-emitter stage having the transistor parameters of Problem 4.1. Hence, evaluate the corresponding input impedance, current and voltage gains of the stage.

4.11 The sensitivity of a gain K to variations in a given parameter W is defined by

$$S_W^K = \frac{dK/K}{dW/W}.$$

Consider a double-terminated two-port network coupling a voltage source of internal impedance Z_s to a load Z_l. Assuming that the network is characterized by its h-parameters, show that the overall gain K (defined as the ratio of load voltage to source voltage) has the following sensitivities to variations in the h-parameters:

$$S_{h_{11}}^K = -\frac{h_{11}(h_{22} + Y_l)}{\Delta}$$

$$S_{h_{12}}^K = \frac{h_{12}h_{21}}{\Delta}$$

$$S_{h_{21}}^K = \frac{(h_{11} + Z_s)(h_{22} + Y_l)}{\Delta}$$

$$S_{h_{22}}^K = -\frac{h_{22}(h_{11} + Z_s)}{\Delta},$$

where

$$\Delta = (h_{11} + Z_s)(h_{22} + Y_l) - h_{12}h_{21}.$$

4.12 Show that the NIC of Fig. P4.12(a) is of the voltage-inversion type, while that of Fig. P4.12(b) is of the current-inversion type. Evaluate the h-parameters of both configurations, and establish the conditions which must be satisfied to approximate to the ideal NIC.

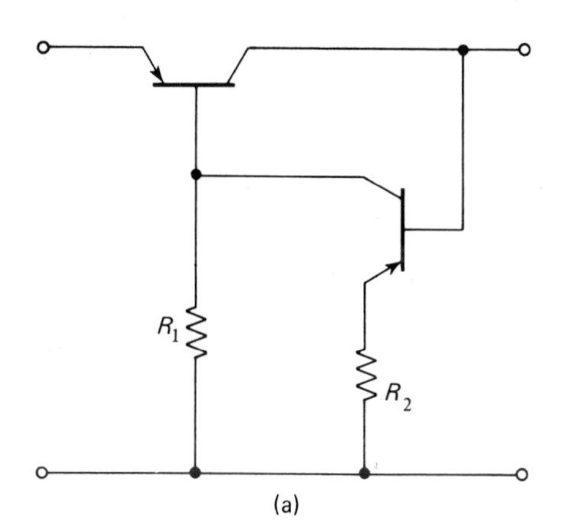

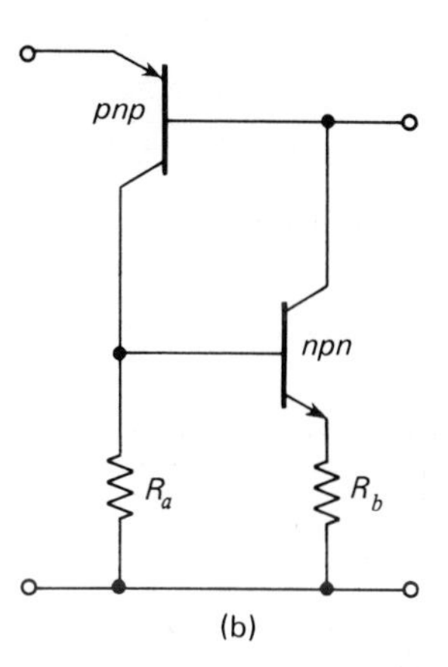

Fig. P4.12.

4.13 Determine the input impedance Z_{in} of the compensated NIC arrangement of Fig. 4.48(a) when it is terminated with a load impedance Z_l. Hence, evaluate the sensitivity of Z_{in} to variations in h_{11}, h_{22} and the gain product $h_{12}h_{21}$. Show that if

$$\frac{1}{h_{22}} \gg |Z_l| \gg h_{11},$$

then variations in $h_{12}h_{21}$ have the greatest effect on the NIC performance.[11]

4.14 The circuit of Fig. P4.14 consists of an amplifier having equal input and output resistances, and an open-circuit voltage gain μ, with the output fed back into the input using a series connection at the output port and a parallel connection at the input port. Show that if $\mu = 2$, the circuit may behave as a gyrator. What is the resultant gyration resistance?

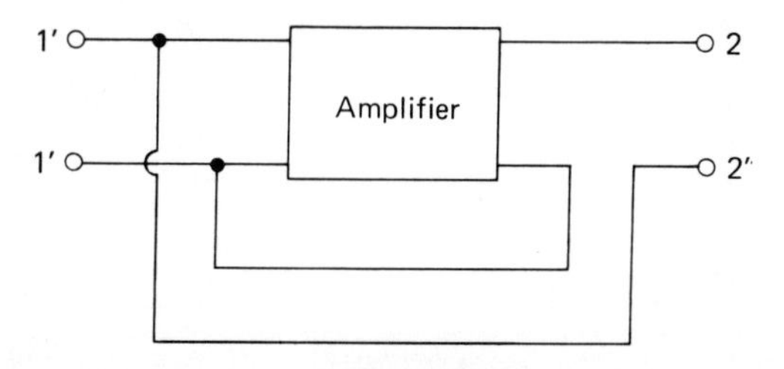

Fig. P4.14.

4.15 Show that the ladder network of Fig. P4.15 is equivalent to an ideal transformer of unity turns ratio.

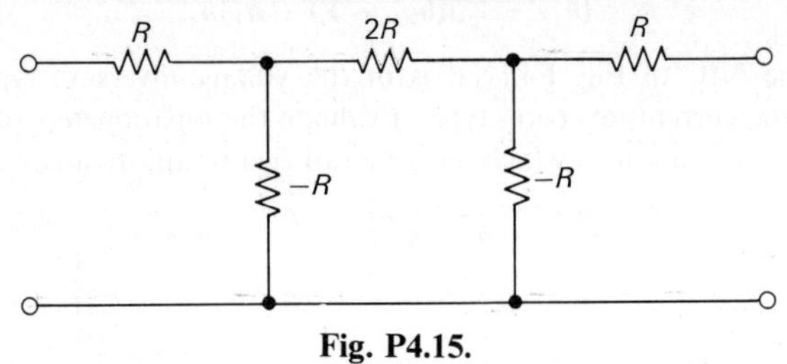

Fig. P4.15.

4.16 Show that the two networks of Fig. P4.16 are equivalent.

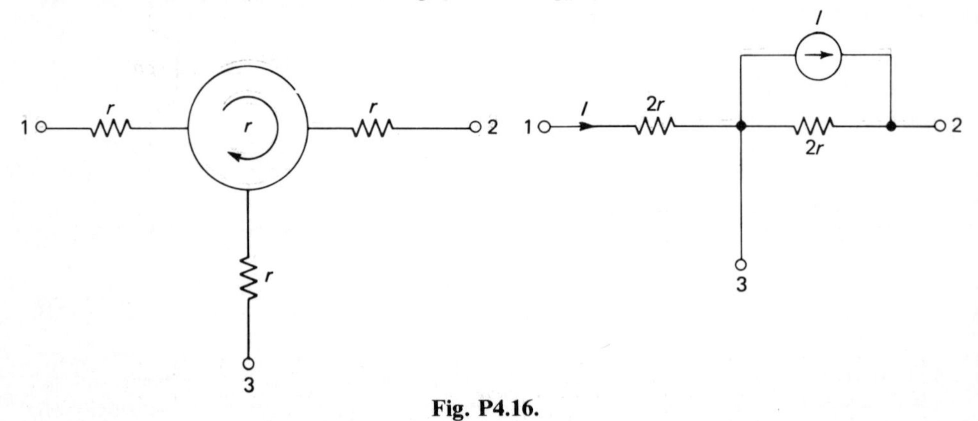

Fig. P4.16.

4.17 Determine the element values of the network shown in Fig. P4.17(b) if it is to be equivalent to that shown in part (a) of the diagram.

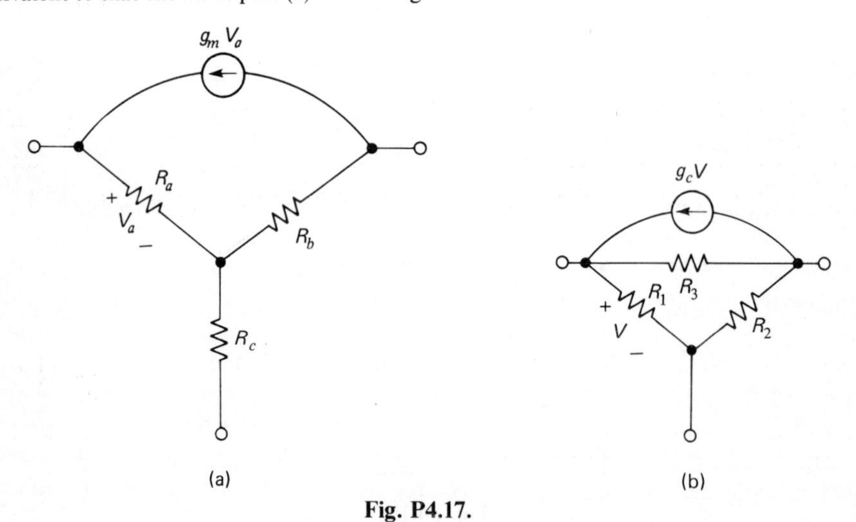

Fig. P4.17.

CHAPTER 5

FURTHER PROPERTIES OF TWO-PORT NETWORKS

5.1 INTRODUCTION

In this chapter we shall continue our study of two-port networks by considering four important functional properties: reciprocity, symmetry, absolute stability and passivity, each of which has its own particular implications; thus

1. If the two-port network is non-reciprocal we require four independent parameters to define uniquely the voltage–current relations of the network with respect to its two ports, whereas in a reciprocal two-port network only three independent parameters are needed.
2. If the two-port network is symmetrical, then only two independent parameters are needed for its complete characterization.
3. If the two-port network is not absolutely stable it becomes possible to select passive terminations for which the network is unstable.
4. The two-port network is capable of providing a maximum power gain greater than unity only if it is active.

The criteria for reciprocity, symmetry, absolute stability and passivity depend solely on the network parameters; in the following sections we will develop these criteria, and examine their possible interrelations and applications in the study of linear amplifiers and oscillators.

5.2 RECIPROCITY

Suppose we have a linear network with a single current (or voltage) excitation applied at one point and the resulting voltage (or current) response observed at another point in the network. The network is said to be *reciprocal* if the ratio of response to excitation is invariant to an interchange of the positions of the excitation and the response. This is illustrated in Fig. 5.1 for the case of a two-port network which is assumed to be initially at rest and contain no independent sources. In Fig. 5.1(a) we have an independent current source I'_1 connected across the input port, and V'_2 denotes the resulting voltage developed across the output port. In Fig. 5.1(b) we have interchanged the positions of the response and excitation. For

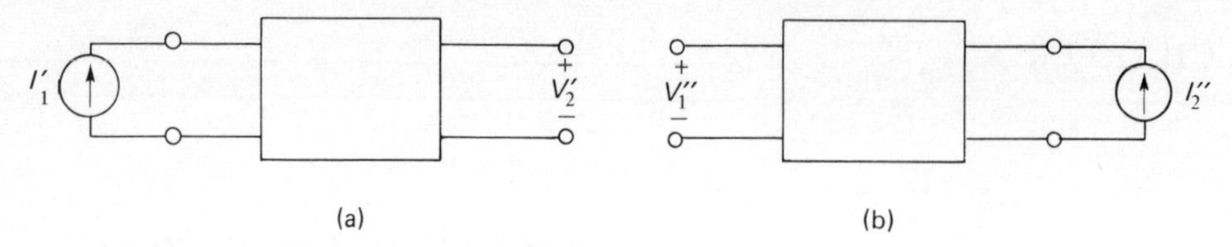

Fig. 5.1. Illustrating the reciprocity theorem in terms of z-parameters.

the two-port network to be reciprocal we must have

$$\frac{V'_2}{I'_1} = \frac{V''_1}{I''_2}. \tag{5.1}$$

However, the ratio V'_2/I'_1 is the open-circuit forward transfer impedance z_{21}, and V''_1/I''_2 is the open-circuit reverse transfer impedance z_{12}. We can, therefore, state that in terms of the z-parameters a two-port network is reciprocal if

$$z_{21} = z_{12}. \tag{5.2}$$

Alternatively, we may have an independent voltage source V'_1 connected across the input port producing a current I'_2 in a short placed between the pair of output terminals, as indicated in Fig. 5.2(a). In part (b) of this diagram we have interchanged the positions of the excitation and response. The two-port network is reciprocal if

$$\frac{I'_2}{V'_1} = \frac{I''_1}{V''_2}. \tag{5.3}$$

But, the ratio I'_2/V'_1 is the short-circuit forward transfer admittance y_{21}, and I''_1/V''_2 is the short-circuit reverse transfer admittance y_{12}. Therefore, in terms of the y-parameters we can state that a two-port network is reciprocal if

$$y_{21} = y_{12}. \tag{5.4}$$

It is significant to observe that in applying the condition for reciprocity based on Eq. (5.1) or Eq. (5.3) both voltage and current are involved. The reason for this is that when the excitation is a current and the response is a voltage, as in Fig. 5.1, then open-circuit constraints are effectively implied at both ports; and when the excitation is a voltage and the response is a current, as in Fig. 5.2, then

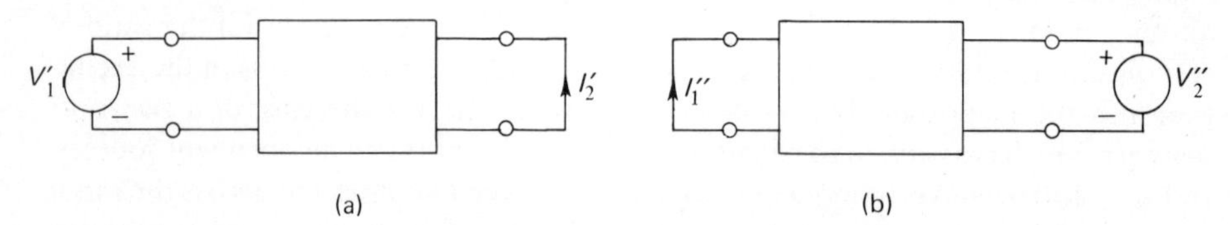

Fig. 5.2. Illustrating the reciprocity theorem in terms of y-parameters.

short-circuit constraints are implied at both ports. In both cases the terminal constraints remain unaltered when the positions of the excitation and response are interchanged. If, however, both the excitation and the response are voltages, then at the port where the excitation is applied we have a short-circuit constraint, while at the port where the response is observed we have an open-circuit constraint. Clearly, in this situation we find that an interchange of the positions of the excitation and response is accompanied by an interchange of short-circuit and open-circuit constraints at the pertinent ports, and we cannot expect the ratio of response to excitation to remain unaltered. Similarly, if both the excitation and the response are currents we have an open-circuit constraint at the port where the excitation is applied, and a short-circuit constraint at the port where the response is observed. Here again we find that an interchange of open-circuit and short-circuit constraints accompanies the interchange of the positions of the excitation and response and, accordingly, the ratio of response to excitation is changed. We conclude, therefore, that in a reciprocal network the ratio of response to excitation is invariant to an interchange of the positions of the excitation and response only if the terminal constraints remain unchanged.

Equations (5.2) and (5.4) define the criterion for reciprocity of a two-port network in terms of the z- and y-parameters, respectively. To determine the reciprocity criterion in terms of the remaining matrix parameters, we may use the matrix conversion Table 4.1 obtaining

$$h_{21} = -h_{12}, \tag{5.5}$$

$$g_{21} = -g_{12}, \tag{5.6}$$

$$AD - BC = 1. \tag{5.7}$$

For a physical interpretation of $h_{12} = -h_{21}$ as the criterion for reciprocity in terms of the h-parameters, consider the two different external connections of Fig. 5.3. In part (a) of this diagram we see that h_{21} is the short-circuit forward current ratio I'_2/I'_1 while in part (b) we see that h_{12} is the open-circuit reverse voltage ratio V''_1/V''_2. Therefore, for reciprocity we must have

$$\frac{I'_2}{I'_1} = -\frac{V''_1}{V''_2}. \tag{5.8}$$

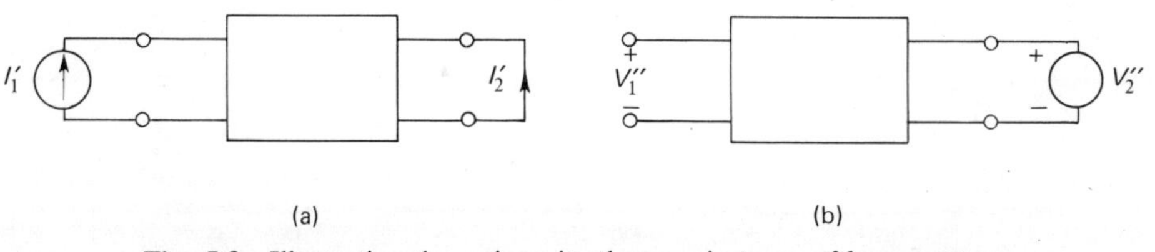

Fig. 5.3. Illustrating the reciprocity theorem in terms of h-parameters.

For a physical interpretation of the reciprocity criterion in terms of the g-parameters we see in Fig. 5.4(a) that g_{21} is the open-circuit forward voltage ratio V'_2/V'_1, while in Fig. 5.4(b) we see that g_{12} is equal to the short-circuit reverse current ratio I''_1/I''_2. Therefore the condition $g_{12} = -g_{21}$ is equivalent to

$$\frac{V'_2}{V'_1} = -\frac{I''_1}{I''_2}. \tag{5.9}$$

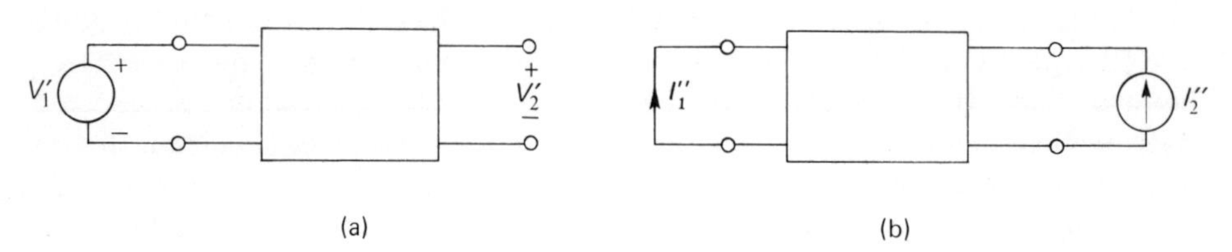

Fig. 5.4. Illustrating the reciprocity theorem in terms of g-parameters.

It is noteworthy that the relations of Eqs. (5.1), (5.3), (5.8) and (5.9) are four special cases of the following relation:

$$V''_1 I'_1 + V''_2 I'_2 = V'_1 I''_1 + V'_2 I''_2, \tag{5.10}$$

where the sets (V'_1, V'_2, I'_1, I'_2) and $(V''_1, V''_2, I''_1, I''_2)$ represent two different states of the network. In other words, if we apply the voltages V'_1 and V'_2, say, to the input and output ports, respectively, of a linear two-port network and denote the resulting input and output currents by I'_1 and I'_2, respectively, and if, next, for the same network we use a different set of input and output voltages, V''_1 and V''_2, and denote the corresponding input and output currents by I''_1 and I''_2, then these two different sets of voltages and currents will always satisfy the condition of Eq. (5.10) provided the network is reciprocal. Equation (5.10) is Tellegen's general form of the reciprocity relation for a two-port network.[1]

Reciprocity Relations in Three-terminal Networks

In Fig. 5.5 we have a three-terminal network operated with terminal 3 common to the input and output ports. We shall assume that the network is non-reciprocal,

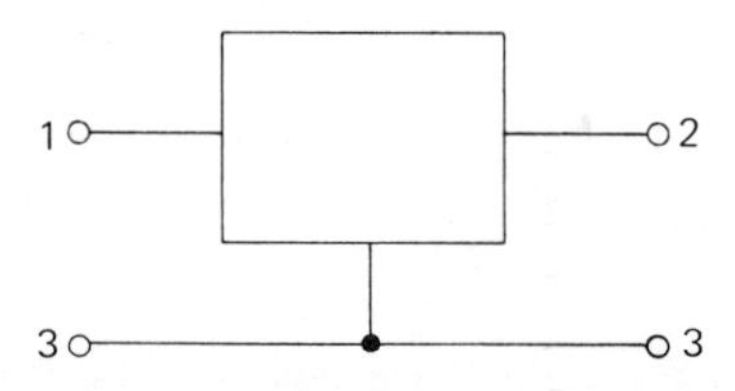

Fig. 5.5. Three-terminal network.

that is, $y_{21} \neq y_{12}$ in the y-matrix

$$\begin{bmatrix} y_{11} & y_{12} \\ y_{21} & y_{22} \end{bmatrix}$$

This admittance matrix may be expressed as the sum of a symmetric matrix and a skew-symmetric matrix as follows:

$$\begin{bmatrix} y_{11} & y_{12} \\ y_{21} & y_{22} \end{bmatrix} = \begin{bmatrix} y_{11} & \frac{1}{2}(y_{12} + y_{21}) \\ \frac{1}{2}(y_{12} + y_{21}) & y_{22} \end{bmatrix} + \begin{bmatrix} 0 & \frac{1}{2}(y_{12} - y_{21}) \\ -\frac{1}{2}(y_{12} - y_{21}) & 0 \end{bmatrix}. \tag{5.11}$$

The symmetric matrix corresponds to a reciprocal three-terminal network which is equivalent to the delta network shown in Fig. 5.6(a). The elements of the equivalent network are defined by

$$\begin{aligned} Y_a &= y_{22} + \tfrac{1}{2}(y_{12} + y_{21}) \\ Y_b &= y_{11} + \tfrac{1}{2}(y_{12} + y_{21}) \\ Y_c &= -\tfrac{1}{2}(y_{12} + y_{21}). \end{aligned} \tag{5.12}$$

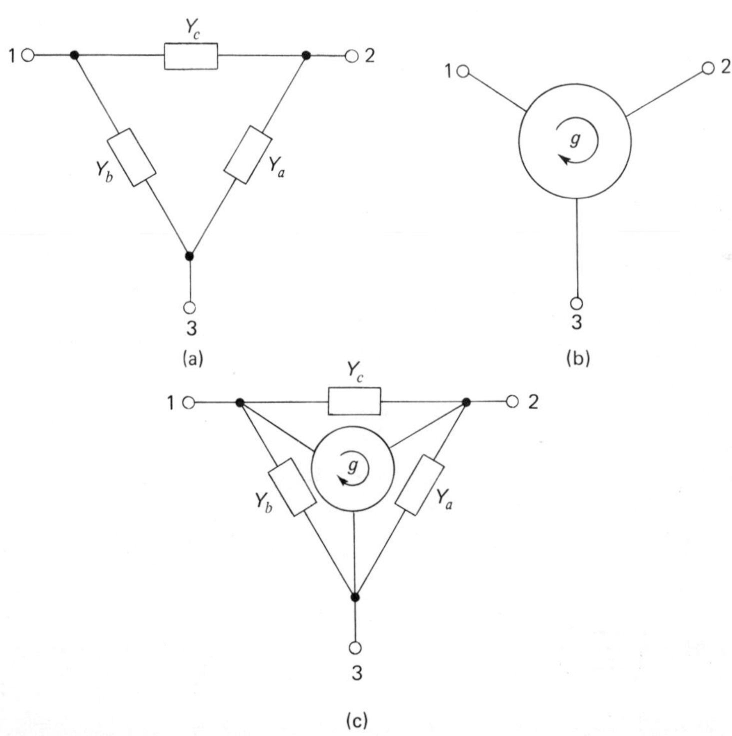

Fig. 5.6. (a) Reciprocal delta network; (b) Three-terminal gyrator; (c) Equivalent circuit of three-terminal network.

Clearly, the reciprocity of the delta network of Fig. 5.6(a) is invariant to any change of the terminal chosen common to the input and output ports.

The skew-symmetric part of Eq. (5.11) represents a three-terminal gyrator having a gyration conductance

$$g = \tfrac{1}{2}(y_{12} - y_{21}) \tag{5.13}$$

with a clockwise direction of gyration, as in Fig. 5.6(b).

A non-reciprocal three-terminal network may thus be represented as the parallel connection of a reciprocal delta network and a three-terminal gyrator, as illustrated in Fig. 5.6(c), with the gyrator representing the non-bilateral part of the device (i.e., that part which does not obey reciprocity). From the model of Fig. 5.6(c) it is easy to perceive how an ideal, unloaded gyrator may be realized.[2] It is only necessary to connect an admittance $-Y_a$ in parallel with Y_a, and in a similar manner we may compensate for the loading effects of Y_b and Y_c.

Example 5.1. *Vacuum-tube triode.* The admittance matrix of the common cathode configuration of a triode is, ignoring the interelectrode capacitances,

$$\mathbf{Y}_k = \begin{bmatrix} 0 & 0 \\ g_m & g_p \end{bmatrix}, \tag{5.14}$$

which may be expressed as follows:

$$\begin{bmatrix} 0 & 0 \\ g_m & g_p \end{bmatrix} = \begin{bmatrix} 0 & \frac{1}{2}g_m \\ \frac{1}{2}g_m & g_p \end{bmatrix} + \begin{bmatrix} 0 & -\frac{1}{2}g_m \\ \frac{1}{2}g_m & 0 \end{bmatrix}. \tag{5.15}$$

We may, therefore, represent a triode by the equivalent network shown in Fig. 5.7 in which

$$\begin{aligned} Y_a &= g_p + \tfrac{1}{2}g_m \\ Y_b &= \tfrac{1}{2}g_m \\ Y_c &= -\tfrac{1}{2}g_m \\ g &= \tfrac{1}{2}g_m. \end{aligned} \tag{5.16}$$

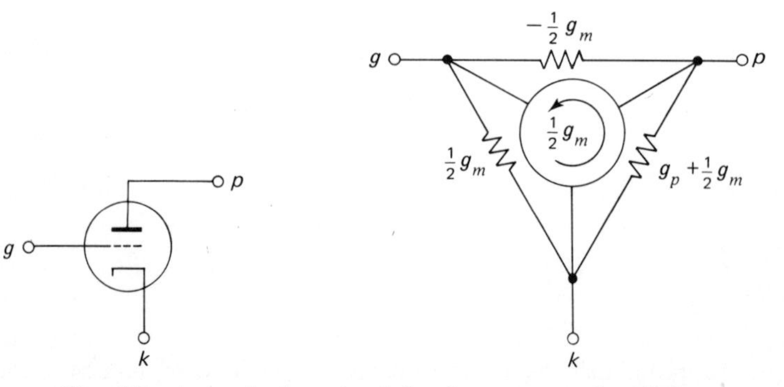

Fig. 5.7. An equivalent circuit for the vacuum-tube triode.

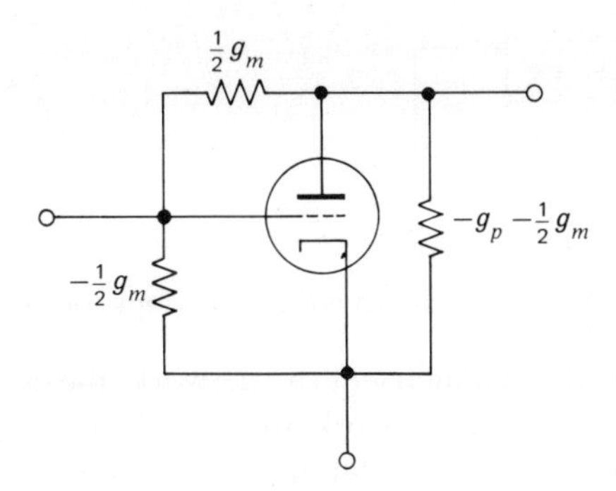

Fig. 5.8. Circuit implementation of gyrator.

It is to be noted that in Fig. 5.7 we have used a gyrator having a gyration conductance equal to $\frac{1}{2}g_m$, with a counter-clockwise direction of gyration. The external loading necessary to strip the gyrator is shown in Fig. 5.8; it corresponds to a network whose admittance matrix is the negative of the symmetric matrix in Eq. (5.15).

5.3 SYMMETRY

A two-port network is said to be *symmetrical* with respect to its input and output ports if the network can be turned end for end without influencing the electrical behaviour of a larger network in which it is embedded. In terms of the various two-port parameters, this implies that

$$z_{12} = z_{21} \quad \text{and} \quad z_{11} = z_{22} \tag{5.17}$$

$$y_{12} = y_{21} \quad \text{and} \quad y_{11} = y_{22}, \tag{5.18}$$

$$h_{12} = -h_{21} \quad \text{and} \quad h_{11}h_{22} - h_{12}h_{21} = 1, \tag{5.19}$$

$$g_{12} = -g_{21} \quad \text{and} \quad g_{11}g_{22} - g_{12}g_{21} = 1, \tag{5.20}$$

and

$$AD - BC = 1 \quad \text{and} \quad A = D. \tag{5.21}$$

We observe, therefore, that a symmetrical network requires only two independent parameters for its complete characterization. Furthermore, all symmetrical two-port networks are reciprocal; the converse is, however, not necessarily true.

For networks which are physically and electrically symmetrical there is a particularly simple procedure for calculating the parameters of interest. The procedure is based on *Bartlett's bisection theorem* which requires an examination of the behaviour of the network in question with symmetrical and antisymmetrical excitations.[3,4] Thus, consider Fig. 5.9(a) where the network is shown excited in its common mode with $V_1 = V_2 = V_c$. For this symmetrical excitation

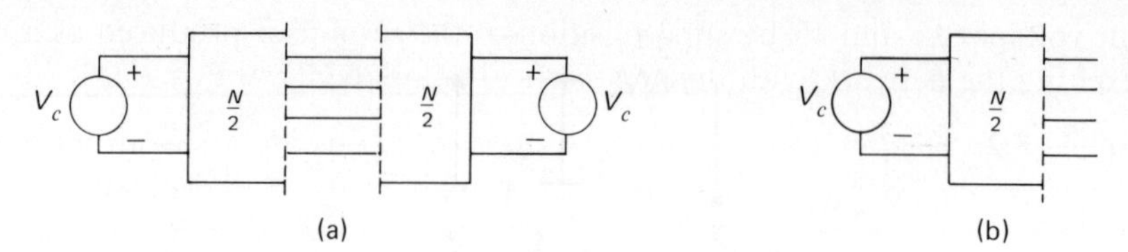

Fig. 5.9. A symmetrical network with symmetric excitations.

we find that because of the symmetry of the network, the currents in all the links connecting the two halves of the network are zero; and the currents and voltages throughout the network are unaffected if these links are cut as shown in Fig. 5.9(b). On the other hand, when the network is excited in its differential mode, we have $V_1 = V_d$ and $V_2 = -V_d$, as in Fig. 5.10(a). For this antisymmetrical excitation, the network symmetry shows that there is no potential difference between the connecting links, and the currents and voltages throughout the network are unaffected if the links are cut and joined separately in the two halves, as in Fig. 5.10(b). It thus follows that the analysis of a symmetrical network with common-mode or differential-mode excitation need only be concerned with one half of the original network, with the link currents or the link-to-link voltages in the bisected network being reduced to zero.

The common-mode and differential-mode excitations constitute a pair of *symmetrical components*. When a symmetrical network is excited by two sources V_1 and V_2 of arbitrary value the bisection theorem can still be applied, since any pair of sources V_1 and V_2 can be expressed in terms of the two symmetrical components, as shown by

$$\begin{aligned} V_1 &= V_c + V_d \\ V_2 &= V_c - V_d, \end{aligned} \tag{5.22}$$

where V_c and V_d are the common-mode and differential-mode components, respectively. When the input voltages V_1 and V_2 are known, Eqs. (5.22) can be solved for the symmetrical components to give

$$\begin{aligned} V_c &= \tfrac{1}{2}(V_1 + V_2) \\ V_d &= \tfrac{1}{2}(V_1 - V_2). \end{aligned} \tag{5.23}$$

Assuming that the network is linear, we may compute the total response to the

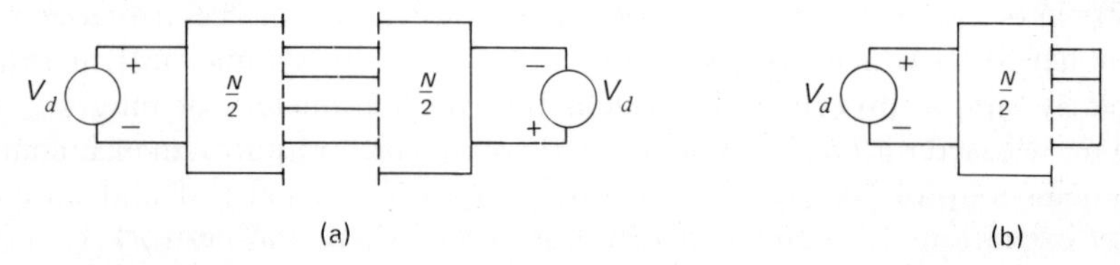

Fig. 5.10. A symmetrical network with antisymmetric excitations.

input voltages V_1 and V_2 by superposition of the responses produced as a result of exciting the network with the symmetrical components of Eqs. (5.23).

Example 5.2. *Emitter-coupled pair as a differential amplifier.* As an illustration of the use of the bisection theorem in circuit analysis, we shall evaluate the small-signal behaviour of the balanced version of the emitter-coupled pair shown in Fig. 5.11, assuming that the two transistors have identical characteristics.

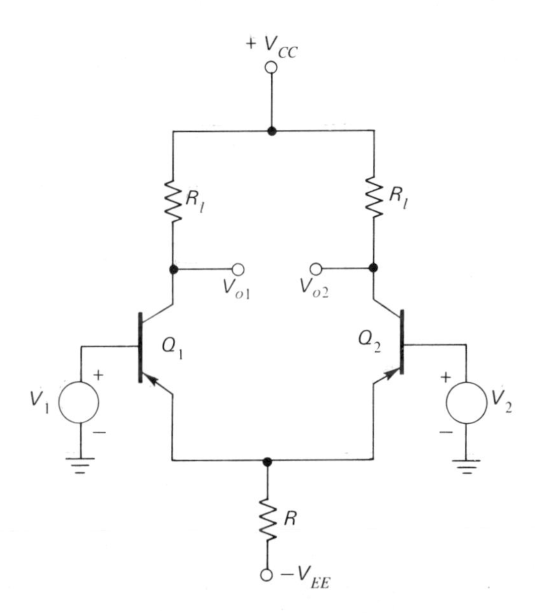

Fig. 5.11. Emitter-coupled pair.

In order to prepare the amplifier for bisection, the emitter resistor R is replaced by an equivalent pair of equal resistors $2R$ in parallel, as in Fig. 5.12(a) which pertains to a.c. operation. For common-mode excitations we bisect the amplifier as in Fig. 5.12(b) where we have retained the left-hand half only. For this excitation the currents and voltages in the right-hand half of the amplifier are obtained simply as the mirror images of those in the left-hand half. The network of Fig. 5.12(b) consists of a common-emitter stage with a feedback resistor $2R$ connected in series with its emitter terminal. The behaviour of such a stage was analysed under Example 4.7; the output voltage V'_{oc} produced by the common-mode component V_c is therefore

$$V'_{oc} = K_c V_c, \tag{5.24}$$

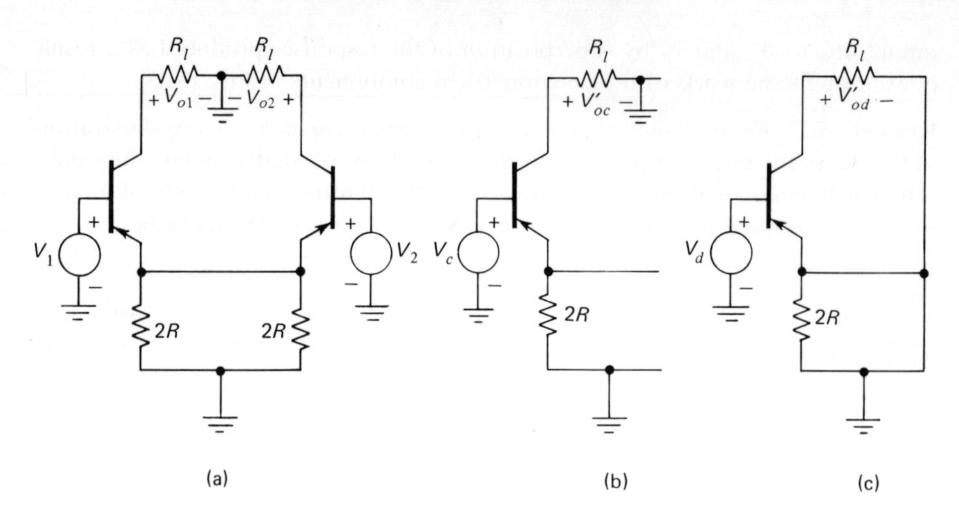

Fig. 5.12. (a) Emitter-coupled pair prepared for bisection; (b) Left-hand half of emitter-coupled pair with common-mode excitation; (c) Left-hand half of emitter-coupled pair with differential-mode excitation.

where

$$K_c = \text{voltage gain for common-mode signals}$$

$$= \frac{-h_{fe}R_l}{(h_{ie} + 2R)(1 + h_{oe}R_l) + h_{fe}(2R - h_{re}R_l)}. \tag{5.25}$$

For the differential mode excitation the amplifier is bisected as in Fig. 5.12(c). In this case the currents and voltages in the right-hand half of the amplifier are the negatives of the corresponding quantities in the left-hand half. In Fig. 5.12(c) we simply have a common-emitter stage with no emitter degeneration since the emitter resistor $2R$ is short circuited. Therefore, the output voltage V'_{od} produced by the differential mode component V_d is

$$V'_{od} = K_d V_d \tag{5.26}$$

where

$$K_d = \text{voltage gain for differential mode signals}$$

$$= \frac{-h_{fe}R_l}{h_{ie}(1 + h_{oe}R_l) - h_{fe}h_{re}R_l}. \tag{5.27}$$

When the amplifier is driven by an arbitrary set of excitations V_1 and V_2, as in Fig. 5.11, we find from Eqs. (5.23) to (5.26) that the total output voltages V_{o1} and V_{o2} developed at the collectors of transistors Q_1 and Q_2, respectively, are given as follows,

$$
\begin{aligned}
V_{o1} &= \tfrac{1}{2}(K_c + K_d)V_1 + \tfrac{1}{2}(K_c - K_d)V_2 \\
V_{o2} &= \tfrac{1}{2}(K_c - K_d)V_1 + \tfrac{1}{2}(K_c + K_d)V_2 .
\end{aligned} \tag{5.28}
$$

Hence, the voltage gain from V_1 to V_{o1}, which is the same as that from V_2 to V_{o2}, is $\frac{1}{2}(K_c + K_d)$, while the voltage gain from V_1 to V_{o2} or from V_2 to V_{o1} is $\frac{1}{2}(K_c - K_d)$.

An index of performance for a symmetric amplifier is the so-called *common-mode rejection ratio* which is defined as the ratio of the voltage gain of the amplifier with respect to common-mode signals to that with respect to differential-mode signals.[5] Thus, for the emitter-coupled pair Eqs. (5.25) and (5.27) give

$$
CMRR = \frac{K_c}{K_d} = \frac{h_{ie}(1 + h_{oe}R_l) - h_{re}h_{fe}R_l}{(h_{ie} + 2R)(1 + h_{oe}R_l) + h_{fe}(2R - h_{re}R_l)} . \tag{5.29}
$$

The smaller this ratio the less will the output voltages be affected by the common-mode variation (i.e., common change of level) of the two input voltages. This suppression of the common-mode component can be achieved, if the resistor R is chosen large enough to satisfy the condition

$$
R \gg \frac{h_{ie}(1 + h_{oe}R_l) - h_{re}h_{fe}R_l}{2(1 + h_{fe} + h_{oe}R_l)} . \tag{5.30}
$$

Then, we find that the output voltage at either collector is a function only of the difference between the input voltages and not of their average, which is a highly desirable property when the amplifier is used as a comparator or difference amplifier.

The emitter-coupled pair of Fig. 5.11 is also used with the voltage difference $V_o = V_{o1} - V_{o2}$ between the collector terminals serving as the output signal of interest. Here we find that the output V_o is directly proportional to the voltage difference $V_1 - V_2$ regardless of the common-mode component in the base-to-ground voltage. Furthermore, the gain from either input terminal to the output terminals across which the differential output voltage V_o appears is, from Eqs. (5.28), equal to K_d which is identical to the gain of a simple common-emitter amplifier and is not affected by the resistor R (except in so far as R determines the quiescent operating current and therefore affects the transistor parameters).

Effect of Small Asymmetries

Owing to variations in the parameters of vacuum tubes or transistors, which result from manufacturing tolerances and changes in environmental conditions, we find that a practical circuit will invariably depart from perfect symmetry and the application of the bisection theorem is not valid. However, if the asymmetry resulting from parameter variations is small (say, not exceeding 10 per cent), its effect on the overall performance of the circuit can be determined in a relatively simple fashion by using perturbation sources.[6] The circuit is initially assumed to be symmetrical with parameter values equal to the average values of the actual parameters in the left- and right-hand halves of the circuit. Next, a change is assumed to take place, increasing a parameter on one side of the circuit and decreasing the corresponding parameter on the other side by the same amount. The effect of this change, so long as it is relatively small, is accounted for by introducing perturbation sources in the manner described in Section 2.4.

5.4 STABILITY

Natural Frequencies and Characteristic Polynomials

A two-port network exhibits four distinct natural modes of operation corresponding to the following four possible combinations of open-circuit and short-circuit terminations:

a) both input and output ports open,
b) input port shorted and output port open,
c) input port open and output port shorted, and
d) both input and output ports shorted.

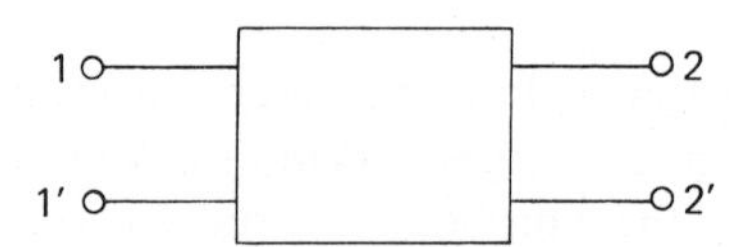

Fig. 5.13. Two-port network.

In order to establish the relations between the natural frequencies corresponding to these modes of operation and the various network parameter sets, consider the two-port network of Fig. 5.13. Let $Z_{in,oc}$ be the driving-point impedance measured at port 1–1′ with port 2–2′ open circuited, that is $Z_{in,oc} = z_{11}$. From Section 3.2 it is recalled that the open-circuit and short-circuit natural frequencies of a network with respect to a given port are, respectively, the poles and zeros of the driving-point impedance function at that port. It follows, therefore, that the poles of the impedance $Z_{in,oc}$ are the natural frequencies of the two-port network with both ports 1–1′ and 2–2′ open-circuited, as in Fig. 5.14(a); and that the zeros of $Z_{in,oc}$ are the natural frequencies with port 1–1′ short circuited and port 2–2′ open circuited, as in Fig. 5.14(b). Thus we may write

$$Z_{in,oc} = z_{11} = \frac{p_{so}(s)}{p_{oo}(s)}, \tag{5.31}$$

where $p_{so}(s)$ and $p_{oo}(s)$ are polynomials whose roots are the natural frequencies of

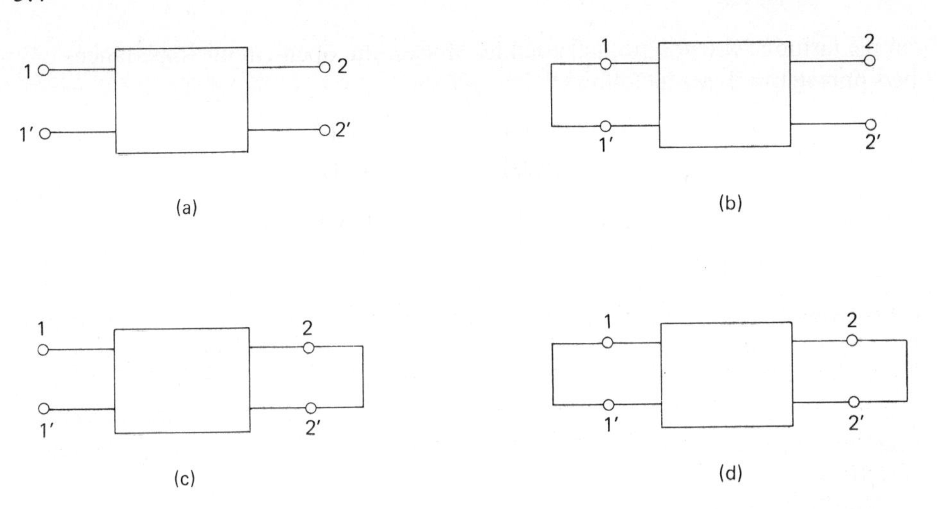

Fig. 5.14. Four natural modes of operating a two-port network: (a) Open-input–open-output; (b) Shorted-input–open-output; (c) Open-input–shorted-output; (d) Shorted-input–shorted-output.

the network for the shorted-input–open-output and open-input–open-output modes of operation, respectively.

Consider next the driving-point impedance $Z_{in,sc}$ measured at port 1–1′ with port 2–2′ short-circuited. The poles of $Z_{in,sc}$ are the natural frequencies of the network with port 1–1′ open circuited and port 2–2′ short circuited, as in Fig. 5.14(c), while the zeros of $Z_{in,sc}$ are natural frequencies with both ports short-circuited, as in Fig. 5.14(d). That is to say,

$$Z_{in,sc} = z_{11} - \frac{z_{12}z_{21}}{z_{22}} = \frac{p_{ss}(s)}{p_{os}(s)}, \tag{5.32}$$

where $p_{os}(s)$ and $p_{ss}(s)$ are polynomials whose roots are the natural frequencies for the open-input–shorted-output and shorted-input–shorted-output modes, respectively.

The behaviour of the network is completely described by the four characteristic polynomials $p_{so}(s)$, $p_{os}(s)$, $p_{oo}(s)$ and $p_{ss}(s)$, plus two other polynomials $q_{21}(s)$ and $q_{12}(s)$ whose roots correspond respectively to the frequencies of zero transmission through the network in the forward and backward directions. Thus $q_{21}(s)$ has roots at the frequencies of zero transmission at port 2–2′ when the excitation is applied at port 1–1′, and vice versa for $q_{12}(s)$. The polynomials $q_{21}(s)$ and $q_{12}(s)$ will be equal in the case of a reciprocal network, but unequal for a non-reciprocal network. For ease of reference, we shall refer to $q_{12}(s)$ and $q_{21}(s)$ as the *transmission polynomials.*

In terms of the polynomials defined above, the open-circuit impedances of a two-port network are as follows:

$$z_{11} = \frac{p_{so}(s)}{p_{oo}(s)} \qquad z_{12} = \frac{q_{12}(s)}{p_{oo}(s)}$$
$$z_{21} = \frac{q_{21}(s)}{p_{oo}(s)} \qquad z_{22} = \frac{p_{os}(s)}{p_{oo}(s)}. \tag{5.33}$$

Comparing Eqs. (5.33) with the open-circuit impedance parameters as defined by Eq. (4.3) in terms of determinants, we see that the polynomials $p_{so}(s)$, $q_{12}(s)$, $q_{21}(s)$, $p_{os}(s)$ and $p_{oo}(s)$ are closely related to the nodal-basis circuit determinant and its cofactors. Thus each polynomial as described above may differ from the related cofactor or determinant by a constant multiplier times a power of the complex variable s.

Substitution of Eqs. (5.33) in (5.32) yields

$$p_{oo}(s)p_{ss}(s) = p_{so}(s)p_{os}(s) - q_{12}(s)q_{21}(s). \tag{5.34}$$

Forming the determinant of the z-matrix from Eq. (5.33) and using Eq. (5.34) gives

$$\Delta_z = \frac{p_{so}(s)p_{os}(s) - q_{12}(s)q_{21}(s)}{p_{oo}^2(s)} = \frac{p_{oo}(s)p_{ss}(s)}{p_{oo}^2(s)} = \frac{p_{ss}(s)}{p_{oo}(s)}. \tag{5.35}$$

Analogous properties also hold for the other two-port parameter sets; thus using Eqs. (5.33) and Table 4.1 we obtain the following sets of results

$$y_{11} = \frac{p_{os}(s)}{p_{ss}(s)} \qquad y_{12} = -\frac{q_{12}(s)}{p_{ss}(s)}$$
$$y_{21} = -\frac{q_{21}(s)}{p_{ss}(s)} \qquad y_{22} = \frac{p_{so}(s)}{p_{ss}(s)} \tag{5.36}$$
$$\Delta_y = \frac{p_{oo}(s)p_{ss}(s)}{p_{ss}^2(s)} = \frac{p_{oo}(s)}{p_{ss}(s)}$$

$$h_{11} = \frac{p_{ss}(s)}{p_{os}(s)} \qquad h_{12} = \frac{q_{12}(s)}{p_{os}(s)}$$
$$h_{21} = -\frac{q_{21}(s)}{p_{os}(s)} \qquad h_{22} = \frac{p_{oo}(s)}{p_{os}(s)} \tag{5.37}$$
$$\Delta_h = \frac{p_{os}(s)p_{so}(s)}{p_{os}^2(s)} = \frac{p_{so}(s)}{p_{os}(s)}$$

$$g_{11} = \frac{p_{oo}(s)}{p_{so}(s)} \qquad g_{12} = -\frac{q_{12}(s)}{p_{so}(s)}$$

$$g_{21} = \frac{q_{21}(s)}{p_{so}(s)} \qquad g_{22} = \frac{p_{ss}(s)}{p_{so}(s)} \tag{5.38}$$

$$\Delta_g = \frac{p_{os}(s)p_{so}(s)}{p_{so}^2(s)} = \frac{p_{os}(s)}{p_{so}(s)}$$

$$A = \frac{p_{so}(s)}{q_{21}(s)} \qquad B = \frac{p_{ss}(s)}{q_{21}(s)}$$

$$C = \frac{p_{oo}(s)}{q_{21}(s)} \qquad D = \frac{p_{os}(s)}{q_{21}(s)} \tag{5.39}$$

$$AD - BC = \frac{q_{12}(s)q_{21}(s)}{q_{21}^2(s)} = \frac{q_{12}(s)}{q_{21}(s)}.$$

Examination of Eqs. (5.33)–(5.39) reveals several interesting properties:[7]

1. We note that any set of two-port network parameters such as the open-circuit impedances can be formed using only five of the polynomials $p_{so}(s)$, $p_{os}(s)$, $p_{oo}(s)$, $p_{ss}(s)$, $q_{21}(s)$ and $q_{12}(s)$; however, the sixth polynomial is formed when the determinant of the given set of parameters is formed.
2. When the determinant of a set of two-port parameters is formed in the usual way, redundant cancelling factors occur which will result in polynomials of unnecessarily high degree. Thus, division must be used to separate the redundant part out of the numerator polynomial (see Eq. 5.35).
3. If the four characteristic polynomials and the two transmission polynomials defined herein are known, we can without further calculation write down all of the two-port parameters and their determinants, each in its simplest form without redundant factors.
4. For a two-port network to be stable for all combinations of open-circuit and short-circuit terminations it is necessary that all four characteristic polynomials $p_{so}(s)$, $p_{os}(s)$, $p_{oo}(s)$ and $p_{ss}(s)$ have no roots in the right half of the s-plane; roots along the imaginary axis are permitted only if they are simple. The stability performance of the network is, however, entirely independent of where in the s-plane the roots of $q_{21}(s)$ and $q_{12}(s)$ are located, because these roots are in no way related to the natural frequencies of the network. The roots of $q_{21}(s)$ merely define the values of the complex variable s for which signal transmission in the forward direction through the network is reduced to zero, that is, the values of s for which the output port is virtually decoupled from the input port; while the roots of $q_{12}(s)$ define the values of s for which backward transmission is zero. As such, the roots of $q_{21}(s)$ and $q_{12}(s)$ may occur in any part of the s-plane.

Example 5.3. *Negative-impedance inverter.* As an illustrative example, consider the negative-impedance inverter of Fig. 5.15(a) using a tunnel diode operated in the negative-resistance part of its characteristic curve. To simplify the discussion, the tunnel diode is represented as the parallel combination of negative resistance $-r$ and junction capacitance C, as in Fig. 5.15(b). At low frequencies we have $z_{11} = z_{22} = 0$ and $z_{12} = z_{21} = -r$, which satisfy the conditions required of a negative-impedance inverter. However, as the frequency is increased the z-parameters of the circuit depart from the idealized set of conditions owing to the shunting effect of the junction capacitance, as shown by

$$\begin{aligned} z_{11} = z_{22} &= \frac{sCr^2}{sCr - 1} \\ z_{12} = z_{21} &= \frac{r}{sCr - 1}. \end{aligned} \tag{5.40}$$

Therefore,

$$\begin{aligned} p_{so}(s) &= p_{os}(s) = sCr^2 \\ p_{oo}(s) &= sCr - 1 \\ q_{12}(s) &= q_{21}(s) = r \\ p_{ss}(s) &= \frac{p_{so}(s)p_{os}(s) - q_{12}(s)q_{21}(s)}{p_{oo}(s)} = r^2(sCr + 1). \end{aligned} \tag{5.41}$$

We thus see that the characteristic polynomials $p_{so}(s)$ and $p_{os}(s)$ have each a simple root at the origin, the polynomial $p_{oo}(s)$ has a simple root at $s = 1/Cr$ in the right half plane, and the polynomial $p_{ss}(s)$ has a simple root at $s = 1/Cr$ in the left half plane. Therefore, the only natural mode of operation for which the negative-impedance inverter is unstable is when the input and output ports are both open-circuited.

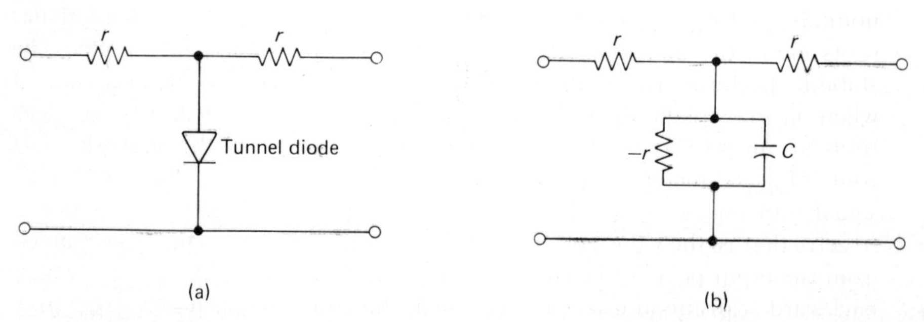

Fig. 5.15. (a) Negative-impedance inverter using tunnel diode; (b) Circuit model.

Example 5.4. *Negative-impedance converter.* Consider next the voltage-inversion type of negative-impedance converter shown in Fig. 5.16(a). To simplify the quantitative analysis of this circuit, we will assume that the two transistors are identical having the idealized circuit model of Fig. 5.16(b). On this basis, we obtain the representation of Fig. 5.16(c) for the converter. Evaluating the h-parameters of the converter, we get

$$\begin{aligned} h_{11} &= h_{22} = 0 \\ h_{12} &= -1 \\ h_{21} &= \frac{s - \omega_\alpha}{s + \omega_\alpha}, \end{aligned} \tag{5.42}$$

where ω_α is the cut-off frequency of the short-circuit forward current-transfer

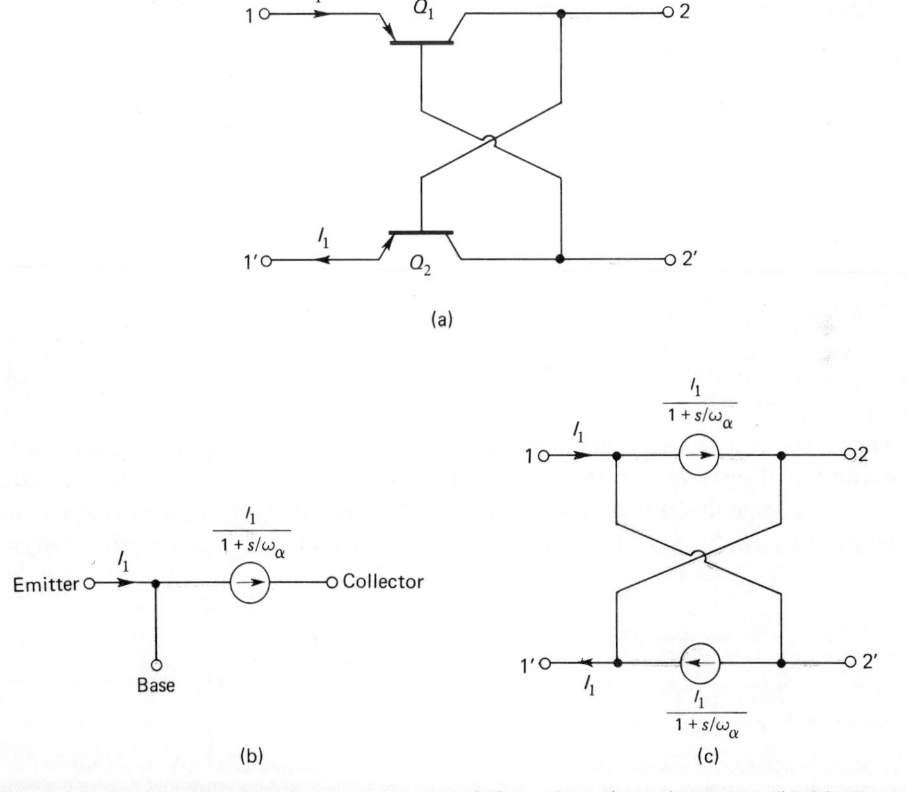

Fig. 5.16. (a) Negative-impedance converter of the voltage-inversion type; (b) Idealized model of transistor; (c) Model of the NIC.

ratio of the common-base configuration. Therefore

$$\begin{aligned} p_{ss}(s) &= p_{oo}(s) = 0, \\ p_{os}(s) &= s + \omega_\alpha, \\ q_{12}(s) &= -(s + \omega_\alpha), \\ q_{21}(s) &= -(s - \omega_\alpha), \\ p_{so}(s) &= \frac{p_{ss}(s)p_{oo}(s) + q_{12}(s)q_{21}(s)}{p_{os}(s)} = s - \omega_\alpha. \end{aligned} \tag{5.43}$$

As $p_{so}(s)$ is the only characteristic polynomial with a root in the right half plane, it follows that the only unstable natural mode of operation for the negative-impedance converter of Fig. 5.16 is when the input port is short-circuited and the output port is open-circuited. Conversely, the converter is stable if the input port is open-circuited or the output port short-circuited.

Characteristic Equation of Double-terminated Two-port Network

In Fig. 5.17 we have a two-port network driven from a voltage source V_s of internal impedance Z_s at its input port, and connected to a load Z_l at its output port. Assuming that the electrical behaviour of the two-port network is characterized by its z-parameters, we have the following equilibrium equations for the complete system

$$\begin{bmatrix} V_s \\ 0 \end{bmatrix} = \begin{bmatrix} z_{11} + Z_s & z_{12} \\ z_{21} & z_{22} + Z_l \end{bmatrix} \begin{bmatrix} I_1 \\ I_2 \end{bmatrix}. \tag{5.44}$$

Solving Eq. (5.44) for the transfer admittance I_2/V_s, we get

$$\frac{I_2}{V_s} = \frac{z_{21}}{(z_{11} + Z_s)(z_{22} + Z_l) - z_{12}z_{21}}. \tag{5.45}$$

The poles of a transfer function, defined as the ratio of response transform to excitation transform, are the natural frequencies of the system, so that for the system to be stable it is necessary that the transfer function of the system has no poles in the right half of the complex-frequency plane. The zeros of a transfer

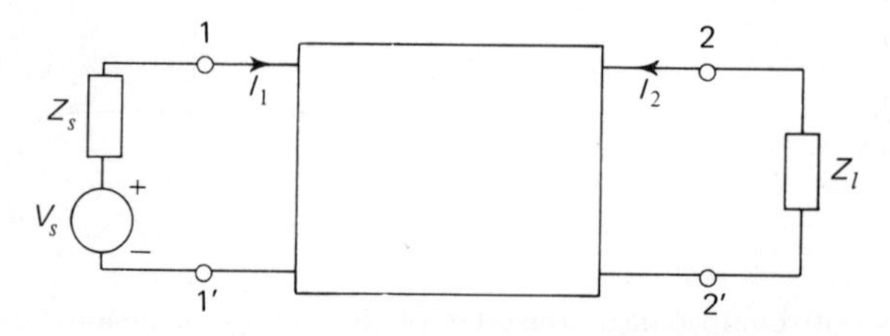

Fig. 5.17. Double-terminated two-port network.

function do not, however, have any similar physical significance; they merely represent the values of the complex frequency s for which transmission through the network is interrupted.

Therefore, the natural frequencies of the double-terminated two-port network are given by the roots of the characteristic equation

$$(z_{11} + Z_s)(z_{22} + Z_l) - z_{12}z_{21} = 0, \tag{5.46}$$

which is obtained by setting the circuit determinant, that is, the denominator of Eq. (5.45) to zero. This characteristic equation may also be expressed in the following two equivalent forms

$$Z_s + Z_{in} = 0, \tag{5.47}$$

$$Z_l + Z_{out} = 0, \tag{5.48}$$

where Z_{in} is the input impedance of the network with the output port terminated in Z_l, and Z_{out} is the output impedance with the input port terminated in Z_s, that is

$$Z_{in} = z_{11} - \frac{z_{12}z_{21}}{z_{22} + Z_l}, \tag{5.49}$$

$$Z_{out} = z_{22} - \frac{z_{12}z_{21}}{z_{11} + Z_s}. \tag{5.50}$$

From Eqs. (5.47) and (5.48) we see that at any natural frequency of a double-terminated two-port network the total impedances of the input and output loops of the network are zero. Indeed, the same thing could be said about any other closed loop of the system.

Equation (5.46) defines the characteristic equation of the system in terms of the z-parameters and the terminating source and load impedances. In a dual manner we find that the characteristic equation may be expressed in terms of the y-parameters in any of the following three equivalent forms

$$(y_{11} + Y_s)(y_{22} + Y_l) - y_{12}y_{21} = 0 \tag{5.51}$$

$$Y_s + Y_{in} = 0 \tag{5.52}$$

$$Y_l + Y_{out} = 0, \tag{5.53}$$

where $Y_s = 1/Z_s$ is the source admittance and $Y_l = 1/Z_l$ is the load admittance. The input and output admittances are given by

$$Y_{in} = y_{11} - \frac{y_{12}y_{21}}{y_{22} + Y_l} \tag{5.54}$$

$$Y_{out} = y_{22} - \frac{y_{12}y_{21}}{y_{11} + Y_s}. \tag{5.55}$$

Therefore, at any natural frequency of a double-terminated two-port network the total admittances appearing across the input and output node-pairs (and, indeed, any other node-pair of the network) are zero.

Similarly, in terms of the h- and g-parameters, respectively, the characteristic equation may be expressed as follows,

$$(h_{11} + Z_s)(h_{22} + Y_l) - h_{12}h_{21} = 0 \tag{5.56}$$

$$(g_{11} + Y_s)(g_{22} + Z_l) - g_{12}g_{21} = 0. \tag{5.57}$$

It is significant to observe that the various characteristic equations developed for a double-terminated two-port network are equivalent in the sense that they all yield the same set of natural frequencies for the network. This is readily demonstrated by substituting the expressions for the two-port parameter sets, as given in Eqs. (5.33)–(5.39), in the pertinent characteristic equation. Such a substitution reveals that the characteristic equation of a double-terminated two-port network is determined completely by specifying the four characteristic polynomials p_{os}, p_{so}, p_{oo} and p_{ss}, and the terminating source and load impedances, as shown by

$$p_{ss} + p_{os}Z_s + p_{so}Z_l + p_{oo}Z_sZ_l = 0. \tag{5.58}$$

That is to say, the natural frequencies of the network are in no way related to the transmission polynomials q_{12} and q_{21}.

Two-port Formulation of Feedback Oscillators

A feedback type of harmonic oscillator consists basically of an active network to provide amplification and a passive frequency-selective network to determine the frequency of oscillation. The two network components are interconnected as a feedback arrangement, and the parameters are chosen such that the characteristic equation of the combination has a pair of complex conjugate roots located in the right half of the s-plane, but fairly close to the imaginary axis. This requirement ensures that the oscillator is self-starting so that any transient, initiated by closing of a switch or other disturbance, will grow in amplitude until it becomes limited by the nonlinearities associated with the active device. As the oscillation builds up, the roots in the right half plane move to the left until they are located on the imaginary axis when steady state operation is achieved. For our study in this section we will consider the behaviour of a feedback oscillator only during the steady state period, and assume an ideal linear model for the circuit; such an analysis, despite its limitations, does provide us with some broad guidelines for a useful oscillator design.

As illustrated in Fig. 5.18, there are four possible ways in which the active and frequency-determining network components of a feedback oscillator may be interconnected; the cascade connection is excluded since it is not directly

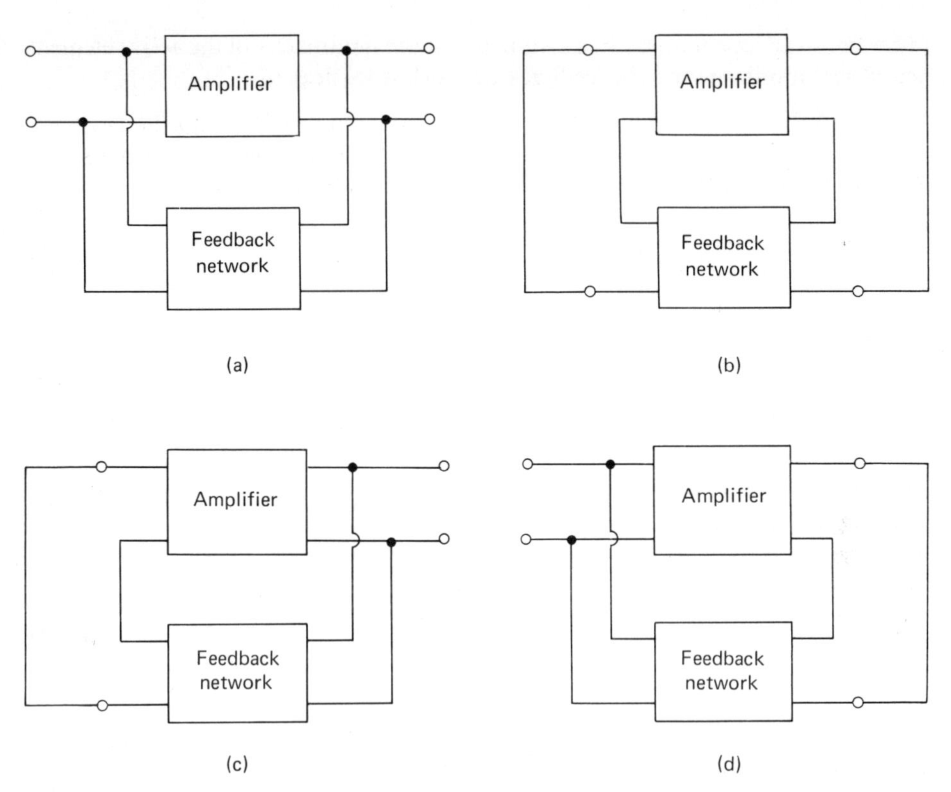

Fig. 5.18. Four basic types of feedback oscillators.

appropriate for feedback oscillators.[8] The four connections and their pertinent input and output terminations are as follows:

a) Parallel–parallel connection with the input and output ports of the combination open-circuited, as in Fig. 5.18(a).

b) Series–series connection with the input and output ports of the combination short-circuited, as in Fig. 5.18(b).

c) Series–parallel connection with the input port short-circuited and the output port open-circuited, as in Fig. 5.18(c).

d) Parallel–series connection with the input port open-circuited and the output port short-circuited, as in Fig. 5.18(d).

Therefore, using the interconnection relations of Section 4.8, we may replace the combination of active and feedback networks by an equivalent two-port network, and then use the pertinent characteristic equation to determine the condition for oscillation and frequency of oscillation. The discussion from here on will be restricted to LC oscillators. It is assumed that the parameters of the active element are all real numbers, and the feedback network is lossless.

Example 5.5. *Y-type oscillators.* In the general circuit of Fig. 5.19 the feedback network is represented by a reciprocal π-network. The y-matrices for the active element N' and feedback network N'' are, respectively, as follows

$$\mathbf{Y}_1 = \begin{bmatrix} y_i & y_r \\ y_f & y_o \end{bmatrix} \tag{5.59}$$

$$\mathbf{Y}_2 = \begin{bmatrix} \dfrac{1 + ZY}{Z} & -Y \\ -Y & \dfrac{1 + nZY}{nZ} \end{bmatrix}, \tag{5.60}$$

where it is assumed that n is a positive constant. The resultant y-matrix for the connection is therefore

$$\mathbf{Y} = \mathbf{Y}_1 + \mathbf{Y}_2 = \begin{bmatrix} y_i + \dfrac{1 + ZY}{Z} & y_r - Y \\ y_f - Y & y_o + \dfrac{1 + nZY}{nZ} \end{bmatrix}. \tag{5.61}$$

The system of Fig. 5.19 is open-circuited at both its input and output ports; hence the characteristic equation of the system is

$$\left(y_i + \frac{1 + ZY}{Z}\right)\left(y_o + \frac{1 + nZY}{nZ}\right) - (y_r - Y)(y_f - Y) = 0,$$

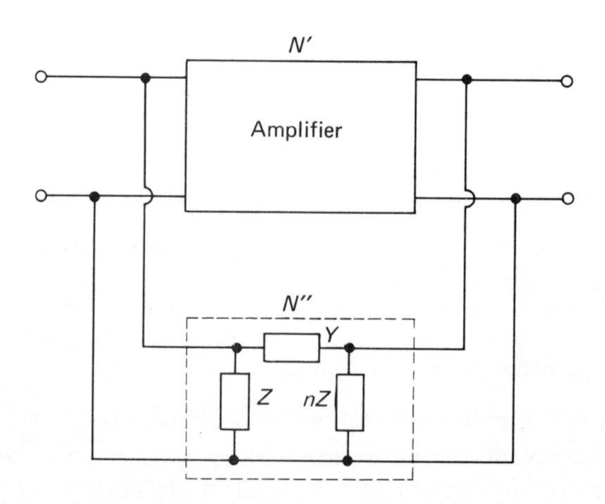

Fig. 5.19. Y-type of feedback oscillator.

or

$$\Delta_y + \frac{1 + ZY(1 + n)}{nZ^2} + \frac{y_i + ny_o + nZY\Sigma_y}{nZ} = 0, \tag{5.62}$$

where

$$\Delta_y = y_i y_o - y_r y_f, \tag{5.63}$$

$$\Sigma_y = y_i + y_r + y_f + y_o. \tag{5.64}$$

For the system to function as a sinusoidal oscillator, the characteristic equation of (5.62) is required to have a pair of conjugate roots on the imaginary axis of the s-plane. That is to say, for $s = j\omega$ the equation must have real ω-roots. Recognizing that the feedback network is purely reactive (i.e., $Z = jX$ and $Y = jB$), we may re-write Eq. (5.62) as

$$\Delta_y - \frac{1 - XB(1 + n)}{nX^2} - j\frac{y_i + ny_o - nXB\Sigma_y}{nX} = 0. \tag{5.65}$$

Clearly, for the left-hand member of Eq. (5.65) to be zero, both the real and imaginary parts must individually be zero. Thus, with all the parameters of the active element assumed real, we have

$$\Delta_y - \frac{1 - XB(1 + n)}{nX^2} = 0 \tag{5.66}$$

and

$$y_i + ny_o - nXB\Sigma_y = 0. \tag{5.67}$$

Equations (5.66) and (5.67) may be re-written as

$$\frac{1}{XB} = 1 + n + \frac{nX\Delta_y}{B} \tag{5.68}$$

and

$$y_f = \frac{y_i}{n} + ny_o + \frac{X\Delta_y}{B}(y_i + ny_o). \tag{5.69}$$

In Eq. (5.69) we have assumed that $|y_r|$ is negligibly small compared with $|y_f|$, that is, the backward transmission through the active element is considerably weaker than the forward transmission, which is ordinarily the case in practice. Equation (5.68) determines the frequency of oscillation, while Eq. (5.69) defines the necessary condition for sustained oscillations.

For the feedback network we may have $X = \omega L$ and $B = \omega C$, as in the *Hartley oscillator* of Fig. 5.20(a), or alternatively $X = -1/\omega C$, and $B = -1/\omega L$, as in the *Colpitts oscillator* of Fig. 5.20(b). In both cases a transistor operated

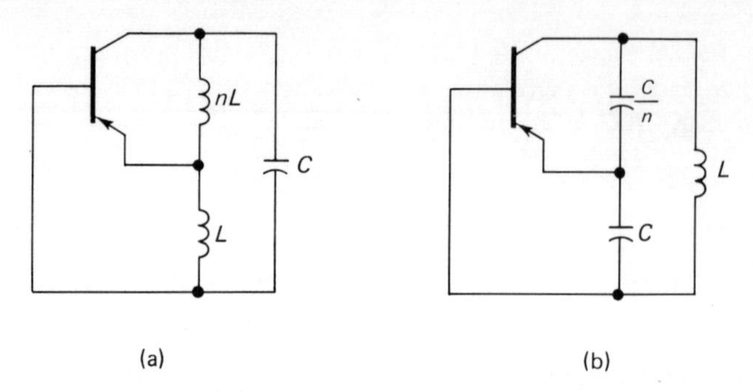

Fig. 5.20. (a) Hartley oscillator; (b) Colpitts oscillator.

in the common emitter mode is shown as the active element, so that in terms of the common-emitter h-parameters we have

$$\begin{bmatrix} y_i & y_r \\ y_f & y_o \end{bmatrix} = \begin{bmatrix} \dfrac{1}{h_{ie}} & -\dfrac{h_{re}}{h_{ie}} \\ \dfrac{h_{fe}}{h_{ie}} & \dfrac{\Delta_{he}}{h_{ie}} \end{bmatrix} \tag{5.70}$$

and

$$\Delta_y = \frac{h_{oe}}{h_{ie}}. \tag{5.71}$$

Putting $X = \omega L$ and $B = \omega C$ in Eq. (5.68), and using Eqs. (5.70) and (5.71), we find that for the Hartley oscillator the frequency of oscillation is given by

$$\frac{1}{\omega^2} = LC(1 + n) + nL^2 h_{oe}/h_{ie}. \tag{5.72}$$

The first term of the right-hand member of Eq. (5.72) represents the resonant frequency of the LC feedback network, while the second term represents the combined damping effect of the h_{ie}- and h_{oe}-parameters of the transistor. The dependence of the oscillation frequency on transistor parameters may be minimized by choosing the C/L ratio large enough to satisfy the following inequality

$$\frac{C}{L} \gg \frac{nh_{oe}}{(1 + n)h_{ie}}. \tag{5.73}$$

Assuming that this condition is satisfied, we obtain

$$\omega^2 \simeq \frac{1}{LC(1 + n)}. \tag{5.74}$$

Consider next the case of the Colpitts oscillator. Substitution of $X = -1/\omega C$ and $B = -1/\omega L$ in Eq. (5.68), and use of Eqs. (5.70) and (5.71) gives the oscillation frequency of the Colpitts oscillator of Fig. 5.20(b) to be

$$\omega^2 = \frac{1+n}{LC} + \frac{nh_{oe}}{C^2 h_{ie}}. \tag{5.75}$$

Here, again, the effect of changes in transistor parameters upon the oscillation frequency may be minimized by having C/L large enough to satisfy the inequality of (5.73). We then find that the oscillation frequency closely approximates to the resonant frequency of the LC feedback network, as shown by

$$\omega^2 \simeq \frac{1+n}{LC}. \tag{5.76}$$

The condition for sustained oscillations is the same for both the Hartley and Colpitts oscillators; thus Eq. (5.69) requires that

$$h_{fe} = \frac{1}{n} + n\Delta_{he} + \frac{Lh_{oe}}{Ch_{ie}}(1 + n\Delta_{he}). \tag{5.77}$$

If (5.73) is satisfied, the starting condition reduces to

$$h_{fe} \simeq \frac{1}{n} + n\Delta_{he}. \tag{5.78}$$

As a design procedure, we may first use the starting condition of Eq. (5.78) to select a suitable positive value for n. Then a ratio C/L is chosen so as to satisfy the inequality of (5.73) which minimizes the effect of transistor parameters on the frequency of oscillation. Next, Eq. (5.74) for the Hartley oscillator, or Eq. (5.76) for the Colpitts oscillator, is used to determine the product term LC so as to yield a specified frequency of oscillation. With C/L and LC determined, it remains to evaluate L and C individually.

Example 5.6. *Z-type oscillators.* The feedback network to be considered here is a reciprocal T-network, as shown in Fig. 5.21. Again, it is assumed that the feedback network is lossless (i.e., $Z = jX$ and $Y = jB$), and n is a positive constant. We thus obtain the following overall z-matrix for the series–series connection of Fig. 5.21

$$\mathbf{Z} = \begin{bmatrix} z_i + \dfrac{1 - XB}{jB} & z_r + jX \\ z_f + jX & z_o + \dfrac{1 - nXB}{jnB} \end{bmatrix} \tag{5.79}$$

where the quantities z_i, z_r, z_f and z_o represent the two-port parameters of the

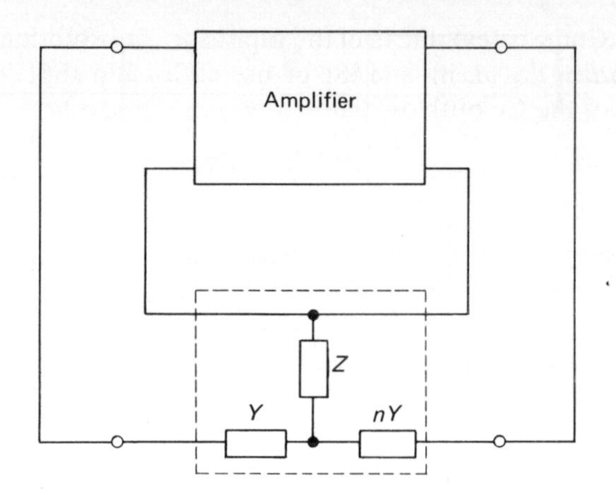

Fig. 5.21. Z-type of feedback oscillator.

active element. Therefore for sinusoidal oscillation we must have

$$\left(z_i + \frac{1 - XB}{jB}\right)\left(z_o + \frac{1 - nXB}{jnB}\right) - (z_r + jX)(z_f + jX) = 0$$

or

$$\Delta_z - \frac{1 - XB(1 + n)}{nB^2} - j\frac{z_i + nz_o - xXB\Sigma_z}{nB} = 0, \tag{5.80}$$

where

$$\Delta_z = z_i z_o - z_r z_f, \tag{5.81}$$

$$\Sigma_z = z_i - z_r - z_f + z_o. \tag{5.82}$$

Equating both the real and imaginary parts of Eq. (5.80) to zero yields the following oscillation equations for the z-type oscillators:

$$\frac{1}{XB} = 1 + n + \frac{nB\Delta_z}{X} \tag{5.83}$$

and

$$-z_f = \frac{z_i}{n} + nz_o + \frac{B\Delta_z}{X}(z_i + nz_o). \tag{5.84}$$

In Eq. (5.84) we have assumed that $|z_f| \gg |z_r|$. It is of interest to note the similarity between the oscillation equations for the Z- and Y-type oscillators.

In Fig. 5.22 we have shown two alternative oscillator circuits of the Z-type, using a common-emitter transistor amplifier as the active element. In terms of

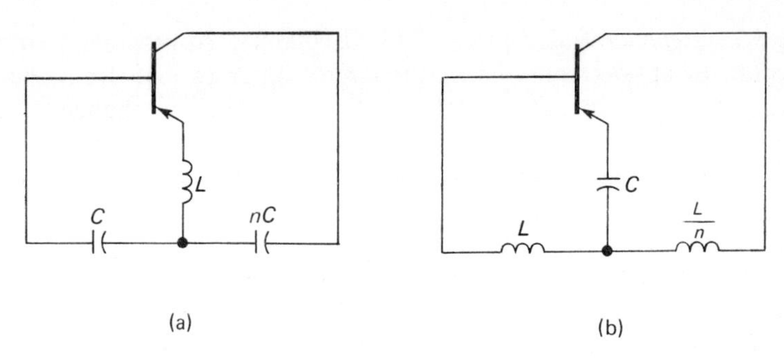

Fig. 5.22. Two transistor oscillators of the Z-type.

the common-emitter h-parameters

$$\begin{bmatrix} z_i & z_r \\ z_f & z_o \end{bmatrix} = \begin{bmatrix} \dfrac{\Delta_{he}}{h_{oe}} & \dfrac{h_{re}}{h_{oe}} \\ \dfrac{-h_{fe}}{h_{oe}} & \dfrac{1}{h_{oe}} \end{bmatrix}. \tag{5.85}$$

The oscillation frequencies for the two arrangements of Fig. 5.22 are obtained from Eq. (5.83) as

$$\omega^2 = \frac{1}{LC(1+n) + nC^2 h_{ie}/h_{oe}} \quad \text{(for Fig. 5.22a)}, \tag{5.86}$$

$$\omega^2 = \frac{1}{LC}(1+n) + \frac{nh_{ie}}{L^2 h_{oe}} \quad \text{(for Fig. 5.22b)}. \tag{5.87}$$

Here we find that in order to minimize the effect of changes in transistor parameters on the oscillation frequency, the L/C ratio must be chosen large enough to satisfy the inequality

$$\frac{L}{C} \gg \frac{nh_{ie}}{(1+n)h_{oe}}. \tag{5.88}$$

The starting condition for the two z-type oscillators of Fig. 5.22 is the same; thus from Eq. (5.84) we obtain

$$h_{fe} = \frac{\Delta_{he}}{n} + n + \frac{Ch_{ie}}{Lh_{oe}}(\Delta_{he} + n). \tag{5.89}$$

Assuming that the inequality of (5.88) is satisfied, the starting condition reduces to

$$h_{fe} \simeq \frac{\Delta_{he}}{n} + n. \tag{5.90}$$

Example 5.7. *Reactive stabilization of LC oscillators.* In practice the frequency of oscillation of an electronic oscillator circuit tends to drift from the desired value owing to changes in the parameters of the active devices produced by changes in environmental conditions, aging, etc. The effects of the active devices upon the oscillation frequency may be reduced by the proper choice of element values for the frequency selective feedback network. This can be achieved, for example, by choosing a large C/L ratio in the Y-type oscillators, but a large L/C ratio in the Z-type oscillators. Also, in the Hartley oscillator the frequency stability may be improved further by introducing mutual inductance between the two inductive elements and having them tightly coupled magnetically, so that the coupling factor is very close to unity.

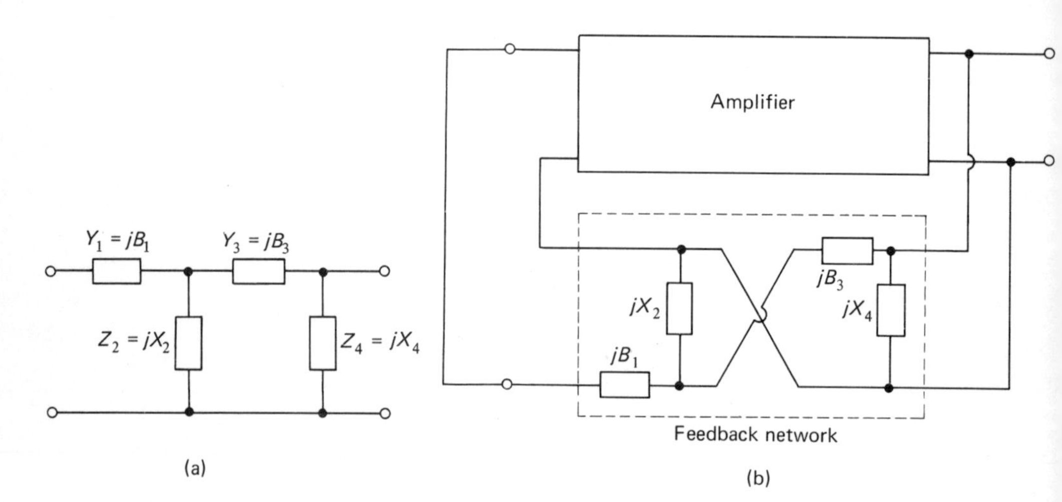

Fig. 5.23. Reactive stabilization of LC feedback oscillators.

Another procedure for minimizing frequency drift in an LC oscillator is to use a lossless four-element ladder network[9] for the frequency-determining feedback component of the oscillator, as in Fig. 5.23(a). Such a network may be considered as either the feedback π-section of a Y-type oscillator with a frequency stabilizing element in the form of a series branch of admittance Y_1, or the feedback T-section of a Z-type oscillator with a frequency stabilizing element in the form of a shunt branch of impedance Z_4. In both cases the additional element, which is redundant to the oscillation function, has been incorporated to make the frequency of oscillation independent of parameters of the active element.

An oscillator using the network of Fig. 5.23(a) in its feedback path is most conveniently analysed by considering the arrangement as the series–parallel connection of the active and feedback network components, with the input port short-circuited and output port open-circuited, as in Fig. 5.23(b). The overall h-matrix of the combination is therefore

$$\mathbf{H} = \begin{bmatrix} h_i + \dfrac{1 - X_2(B_1 + B_3)}{jB_1(1 - X_2B_3)} & h_r + \dfrac{X_2B_3}{1 - X_2B_3} \\ h_f - \dfrac{X_2B_3}{1 - X_2B_3} & h_o + \dfrac{1 - B_3(X_2 + X_4)}{jX_4(1 - X_2B_3)} \end{bmatrix}, \tag{5.91}$$

where the quantities h_i, h_r, h_f and h_o represent the h-parameters of the amplifier component. For sinusoidal oscillation, the determinant of the h-matrix in Eq. (5.91) must be zero. Consider first the imaginary part of this determinant; setting it to zero gives the frequency-determining equation as

$$\frac{h_i}{X_4}[1 - B_3(X_2 + X_4)] + \frac{h_o}{B_1}[1 - X_2(B_1 + B_3)] = 0. \tag{5.92}$$

For perfect frequency stabilization, the frequency of oscillation must be independent of the parameters h_i and h_o contributed by the active device. This can be achieved only if at the desired frequency of oscillation we simultaneously have

$$1 - B_3(X_2 + X_4) = 0 \tag{5.93}$$

and

$$1 - X_2(B_1 + B_3) = 0. \tag{5.94}$$

Equations (5.93) and (5.94) together determine the frequency of oscillation and the necessary condition for perfect frequency stabilization, which assumes a perfectly lossless passive feedback network.

Consider next the real part of the determinant of the h-matrix in Eq. (5.91); equating it to zero gives the necessary minimum value of h_{fe} for oscillation as

$$h_f = \frac{B_3}{B_1} + \frac{B_1}{B_3}\Delta_h. \tag{5.95}$$

In Eq. (5.95) we have assumed that the amplifier has $h_f \gg h_r$, which ordinarily is the case.

For the feedback ladder network we may use a low-pass or high-pass filter as in Fig. 5.24, where a common-emitter amplifier is shown as the active element. Application of Eqs. (5.93)–(5.95) to these two arrangements yields

	Fig. 5.24(a)	Fig. 5.24(b)
Frequency of oscillation	$\omega^2 = \dfrac{1}{C_2}\left(\dfrac{1}{L_1} + \dfrac{1}{L_3}\right)$	$\omega^2 = \dfrac{1}{L_2(C_1 + C_3)}$
Condition for frequency stabilization	$L_1C_2 = L_3C_4$	$C_1L_2 = C_3L_4$
Condition for oscillation	$h_{fe} = \dfrac{L_1}{L_3} + \dfrac{L_3}{L_1}\Delta_{he}$	$h_{fe} = \dfrac{C_3}{C_1} + \dfrac{C_1}{C_3}\Delta_{he}$

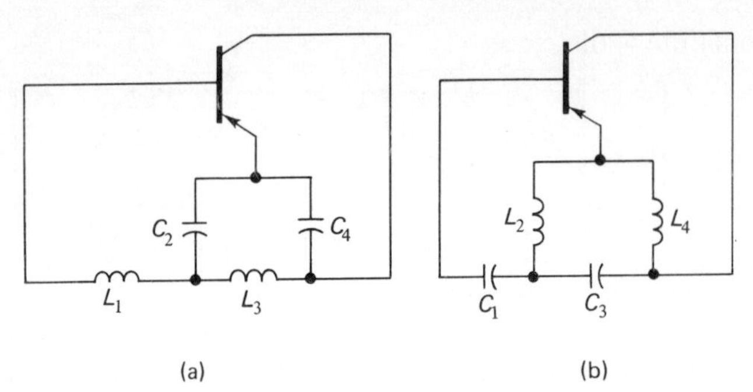

Fig. 5.24. Two stabilized *LC* feedback oscillators.

5.5 POTENTIAL INSTABILITY AND ABSOLUTE STABILITY

A two-port network is said to be *potentially unstable* if it is possible to select uncoupled passive source and load impedances which, when connected across the input and output ports of the network, respectively, produce an unstable system. The transistor is an example of a potentially unstable device in that, if properly terminated at its input and output ports, it may become unstable (i.e., may oscillate) within certain frequency ranges, even in the absence of external feedback. In a transistor the problem of potential instability arises because it is a non-unilateral device, in that a signal applied to the output port of a transistor amplifier results in a response at the input port. The existence of this internal feedback within the transistor is expressed by the fact that in the z-, y-, h- or g-matrix representation of the device the parameter with subscript 12 (that is, z_{12}, y_{12}, h_{12}, or g_{12}) is different from zero.

In some instances the property of potential instability is a desirable one to have as, for example, in the so-called *tuned plate–tuned grid oscillator* which oscillates by virtue of the grid-to-plate interelectrode capacitance of the vacuum-tube triode. In the design of linear amplifiers, however, potential instability can pose a serious problem. For example, in a high-frequency narrow-band-pass amplifier, where the transistor parameters are ordinarily complex and the susceptances of the terminating source and load admittances assume a wide range of values, it is often found that the amplifier is not only potentially unstable, but is very likely to become actually unstable. In such a situation, actual instability is overcome by a proper choice of the loading conditions at the input and output ports, or by diminishing by some external means the effect of the internal feedback parameter which is responsible for the potential instability of the device.

When a two-port network remains stable under all possible uncoupled passive terminations it is said to be *absolutely* or *unconditionally stable*, the criterion for which depends on the parameters of the two-port network alone. A necessary and sufficient condition for the absolute stability of a two-port network is that the one-port networks of Fig. 5.25, resulting from any passive output and input termina-

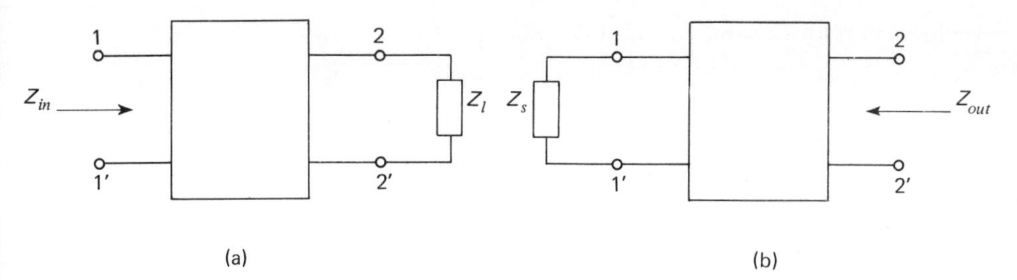

Fig. 5.25. One-port networks resulting from terminating a two-port network with: (a) load impedance Z_l at the output; (b) source impedance Z_s at the input.

tions, are themselves passive. That is to say, a two-port network is absolutely stable if its input (or output) impedance for any passive load (or source) impedance is a positive real function. The reason for this requirement is that if both the input impedance Z_{in} and the source impedance Z_s are positive real, then the total impedance $Z_s + Z_{in}$ of the input loop is positive real too, and it would not be possible for $Z_{in} + Z_s$ to have zeros in the right half plane. Since the zeros of $Z_{in} + Z_s$ are natural frequencies of the terminated two-port network, it follows that for all possible passive terminations a two-port network will not have any natural frequency in the right half plane; that is, the two-port network will be absolutely stable provided that the input impedance Z_{in} is a positive real function. Equivalently, we can state that a two-port network is absolutely stable if its output impedance, for any passive source impedance, is a positive real function.

From Chapter 3 we recall that a rational function $Z(s)$ is positive real if, in addition to being real for real s, it fulfils the following requirements:

1. $Z(s)$ has no poles in the right half of the complex-frequency plane.
2. Any poles of $Z(s)$ on the imaginary axis are simple with real and positive residues. (5.96)
3. $\text{Re}\,[Z(j\omega)] \geq 0$ for all ω, where Re denotes the real part of the expression which follows it.

The input and output impedances of a two-port network must both satisfy the conditions of (5.96) for all possible passive terminations if the two-port network is to be absolutely stable. When Z_{in} satisfies conditions 1 and 2 it is in effect implied that the one-port network of Fig. 5.25(a) is open-circuit stable with respect to the port 1–1′ for any passive Z_l. In Eq. (5.49) we see that the zeros of $z_{22} + Z_l$ are poles of Z_{in}; therefore, z_{22} must be a positive real function if Z_{in} is to satisfy conditions 1 and 2 for any passive Z_l. Similarly, since the zeros of $z_{11} + Z_s$ are poles of Z_{out} (see Eq. 5.50) it follows that Z_{out} satisfies conditions 1 and 2, and the one-port network of Fig. 5.25(b) is open-circuit stable with respect to the port 2–2′ for any passive Z_s, provided that z_{11} is a positive real function. The positive realness of both z_{11} and z_{22}, in turn, requires that

a) $z_{11}(s)$ and $z_{22}(s)$ have no poles in the right half plane.
b) Any poles of $z_{11}(s)$ and $z_{22}(s)$ on the imaginary axis are simple with real and positive residues.
c) $\text{Re}\,[z_{11}(j\omega)] \geq 0$ and $\text{Re}\,[z_{22}(j\omega)] \geq 0$ for all ω. (5.97)

The conditions of (5.97) only assure the absolute stability of the two-port network under open-circuit conditions across the input and output ports; for absolute stability under all possible passive terminations it is necessary that $Z_{in}(j\omega)$ and $Z_{out}(j\omega)$ satisfy condition 3 of (5.96). In general, the z-parameters, the load impedance and $Z_{in}(j\omega)$ have complex values; thus let

$$z_{11} = r_{11} + jx_{11} \tag{5.98}$$

$$z_{22} = r_{22} + jx_{22} \tag{5.99}$$

$$z_{12}z_{21} = M + jN \tag{5.100}$$

$$Z_l = R_l + jX_l \tag{5.101}$$

$$Z_{in} = R_{in} + jX_{in}. \tag{5.102}$$

Substituting into Eq. (5.49), we get

$$R_{in} + jX_{in} = r_{11} + jx_{11} - \frac{M + jN}{(r_{22} + R_l) + j(x_{22} + X_l)}, \tag{5.103}$$

from which we find that the real part R_{in} of the input impedance is given by

$$R_{in} = r_{11} - \frac{M(r_{22} + R_l) + N(x_{22} + X_l)}{(r_{22} + R_l)^2 + (x_{22} + X_l)^2}. \tag{5.104}$$

Let

$$\Gamma = \frac{R_{in}}{r_{11}}[(r_{22} + R_l)^2 + (x_{22} + X_l)^2]. \tag{5.105}$$

Assuming that the condition $r_{11} \geq 0$ is satisfied (see condition (c) of (5.97)) we see that the quantity Γ has the same sign as R_{in}. Eliminating R_{in} from Eqs. (5.104) and (5.105) gives

$$\Gamma = \left[R_l + \left(r_{22} - \frac{M}{2r_{11}}\right)\right]^2 + \left[X_l + \left(x_{22} - \frac{N}{2r_{11}}\right)\right]^2 - \frac{M^2 + N^2}{4r_{11}^2}, \tag{5.106}$$

which represents the equation of a paraboloid of revolution,[10] as illustrated in Fig. 5.26(a). When Γ is maintained constant, Eq. (5.105) represents a circle in a plane above the (R_l, X_l)-plane by a distance equal to the constant Γ-value, as

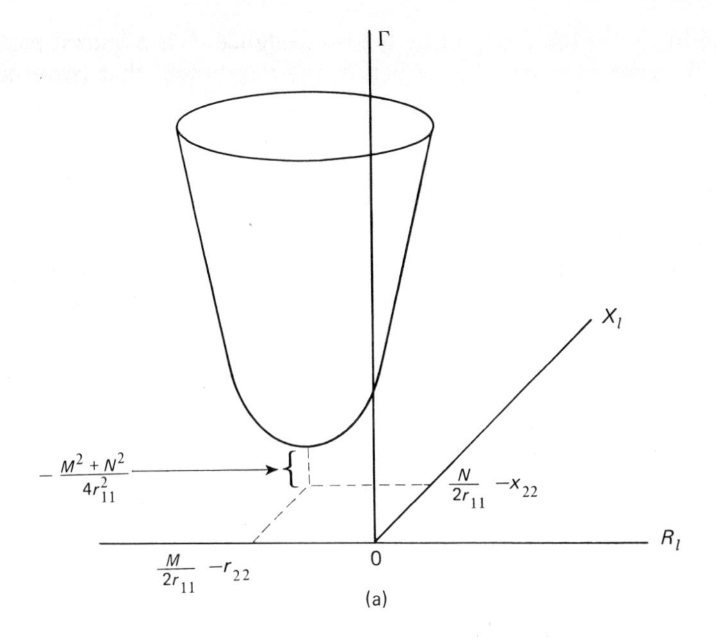

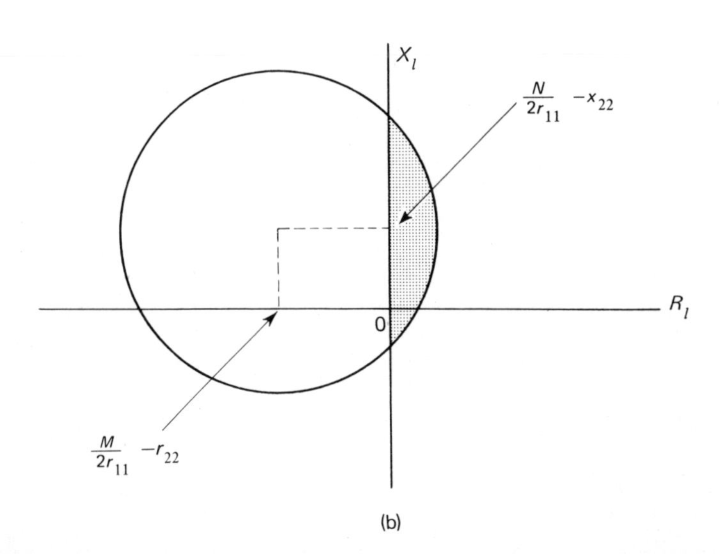

Fig. 5.26. (a) Paraboloid of revolution describing Eq. (5.106); (b) Circle in the (R_l, X_l)-plane for Γ = constant.

in Fig. 5.26(b). Since the terminating load impedance Z_l is assumed passive, we have $R_l \geq 0$. Therefore, only the region of the paraboloid that corresponds to the shaded area in Fig. 5.26(b) can be used. For $R_l = 0$, Eq. (5.106) reduces to

$$\Gamma = \left(r_{22} - \frac{M}{2r_{11}}\right)^2 + \left[X_l + \left(x_{22} - \frac{N}{2r_{11}}\right)\right]^2 - \frac{M^2 + N^2}{4r_{11}^2}. \tag{5.107}$$

This represents the equation of the parabola cut out from the paraboloid of revolution by the (Γ, X_l)-plane (see Fig. 5.27). As Γ has the same sign as the input resistance R_{in}, the vertex of the parabola in Fig. 5.27 must stay above the X_l-axis if R_{in} is to remain non-negative. In other words, we must have

$$\left(r_{22} - \frac{M}{2r_{11}}\right)^2 - \frac{M^2 + N^2}{4r_{11}^2} \geq 0, \tag{5.108}$$

which may be re-written in the form

$$2r_{11}r_{22} - (M + \sqrt{M^2 + N^2}) \geq 0. \tag{5.109}$$

From the definitions of Eqs. (5.98)–(5.100) we see that the condition of Eq. (5.109) means the following

$$2\text{Re}\, z_{11}\, \text{Re}\, z_{22} - \text{Re}(z_{12}z_{21}) - |z_{12}z_{21}| \geq 0. \tag{5.110}$$

This condition is symmetrical with respect to the subscripts 1 and 2; it will therefore apply equally to the output impedance Z_{out}.

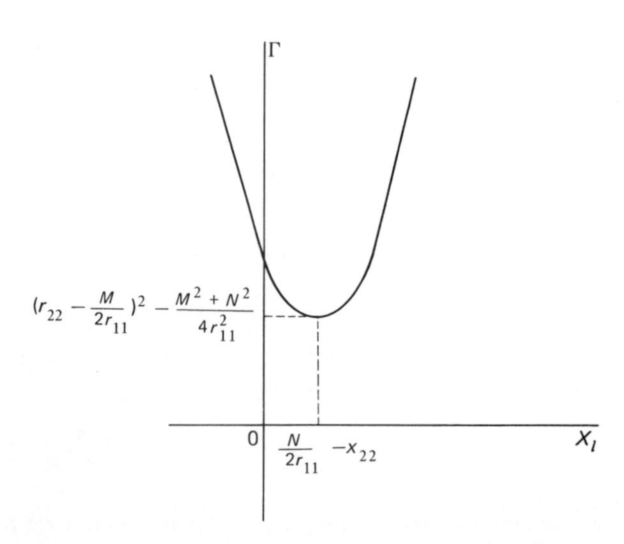

Fig. 5.27. Parabola cut out from the paraboloid of revolution by the (Γ, R_l)-plane.

Combining the conditions of (5.97) and the condition of (5.110), we may now state that a two-port network is absolutely stable if and only if

a) $z_{11}(s)$ and $z_{22}(s)$ have no poles in the right half plane;
b) Any poles of $z_{11}(s)$ and $z_{22}(s)$ on the imaginary axis are simple with real and positive residues;
c) For all real values of ω we have

$$\begin{aligned} \text{Re } z_{11} &\geq 0, \\ \text{Re } z_{22} &\geq 0, \\ 2\text{Re } z_{11} \text{ Re } z_{22} - \text{Re}(z_{12}z_{21}) - |z_{12}z_{21}| &\geq 0. \end{aligned} \tag{5.111}$$

Conditions (a) and (b), together with the first two of conditions (c) assure the absolute stability of the two-port network when the input or output port is open-circuited; the last of conditions (c) assures its absolute stability for all other passive terminations. These conditions constitute *Llewellyn's criterion for absolute stability.*[11] If any of the conditions is not satisfied, the network is potentially unstable.

The above analysis has been entirely in terms of the z-parameters. In a dual manner we may formulate the necessary and sufficient conditions for the absolute stability of a general two-port network in terms of the y-parameters as follows:

a) $y_{11}(s)$ and $y_{22}(s)$ have no poles in the right half plane.
b) Any poles of $y_{11}(s)$ and $y_{22}(s)$ on the imaginary axis are simple with positive and real residues.
c) For all real values of ω,

$$\begin{aligned} \text{Re } y_{11} &\geq 0, \\ \text{Re } y_{22} &\geq 0, \\ 2\text{Re } y_{11} \text{ Re } y_{22} - \text{Re}(y_{12}y_{21}) - |y_{12}y_{21}| &\geq 0. \end{aligned} \tag{5.112}$$

Conditions (a) and (b), together with the first two of conditions (c) assure the absolute stability of the two-port network when the input or output port is short-circuited; the last of conditions (c) assures its absolute stability for all other passive terminations. The network is potentially unstable if any of the conditions is violated.

In the same way we may also postulate the criterion for absolute stability of a two-port network in terms of the remaining h- and g-matrix parameters.

Special Cases

a) *Reciprocal networks.* If the two-port network is reciprocal, we have $z_{12} = z_{21}$, for which the third of conditions (c) in (5.111) reduces to

$$\text{Re } z_{11} \text{ Re } z_{22} - [\text{Re } z_{12}]^2 \geq 0. \tag{5.113}$$

b) *Networks without z- and y-matrices.* It is sometimes found that a two-port network does not possess a finite z- or y-matrix. This arises when the parameters h_{11} and h_{22} of the h-matrix of the network are both zero, or equivalently the parameters B and C in the chain matrix are zero, as in an ideal transformer and ideal negative-impedance converter. In order to develop a criterion for the absolute stability of such networks, we note from condition (c) in (5.111) that absolute stability in terms of the h-parameters requires

$$2\text{Re}\, h_{11}\, \text{Re}\, h_{22} - \text{Re}(h_{12}h_{21}) - |h_{12}h_{21}| \geq 0. \tag{5.114}$$

When $h_{11} = 0$ and $h_{22} = 0$, the condition (5.114) simplifies as

$$-\text{Re}(h_{12}h_{21}) - |h_{12}h_{21}| \geq 0, \tag{5.115}$$

which, for finite values of h_{12} and h_{21}, can only be satisfied if the phase angles of h_{12} and h_{21} add up to 180 degrees. Equivalently, in terms of the chain parameters the requirement for absolute stability is that A and D must have equal phase angles.

Unilateralization

As a result of the internal feedback existing within a non-unilateral active two-port device (as represented, for example, by the parameter y_{12} in the y-matrix characterization) we find that

1. The input and output driving-point impedances of the network are functions of the load and source impedances, respectively.
2. The network may oscillate even in the absence of any external feedback. Thus, from the third line of condition (c) in (5.112) we see that if at any frequency the internal feedback parameter y_{12} is large enough for the sum $\text{Re}(y_{12}y_{21}) + |y_{12}y_{21}|$ to be greater than $2\text{Re}\, y_{11}\, \text{Re}\, y_{22}$, then the device is potentially unstable at that frequency.

These two phenomena are particularly serious in the case of high-frequency multistage tuned amplifiers using transistors. In such an amplifier we find that changes in the impedance of any one of the tuned interstage coupling networks are reflected and transmitted through the preceding and following transistor stages, respectively, thereby affecting the tuning of other coupling networks in the amplifier and making the alignment process laboriously difficult. Also, in some cases the interaction between stages is great enough for the amplifier to become actually unstable.

The undesirable effects of internal feedback may be eliminated by connecting a passive two-port network to the non-unilateral active device in question such that the resultant backward transmission through the combination is zero. The combination network is then unilateralized in that it transmits electric signals in the forward direction only. As shown in Table 5.1 there are four basic types of

Table 5.1

Type of unilateral-ization	Connection	Condition for unilateral-ization	Unilateralized parameters
Y	y_i y_r y_f y_o Y $1:n$	$y_r = -\frac{Y}{n}$	$\mathbf{Y} = \begin{bmatrix} y_i - ny_r & 0 \\ y_f - y_r & y_o - \frac{y_r}{n} \end{bmatrix}$
Z	z_i z_r z_f z_o Z $1:n$	$z_r = nZ$	$\mathbf{Z} = \begin{bmatrix} z_i + \frac{z_r}{n} & 0 \\ z_f - z_r & z_o + nz_r \end{bmatrix}$
H	h_i h_r h_f h_o Y Z	$h_r = \frac{ZY}{1 + ZY}$	$\mathbf{H} = \begin{bmatrix} h_i + \frac{h_r}{Y} & 0 \\ h_f + h_r & h_o + \frac{h_r}{Z} \end{bmatrix}$
G	g_i g_r g_f g_o Y Z	$g_r = \frac{-ZY}{1 + ZY}$	$\mathbf{G} = \begin{bmatrix} g_i - \frac{g_r}{Z} & 0 \\ g_f + g_r & g_o - \frac{g_r}{Y} \end{bmatrix}$

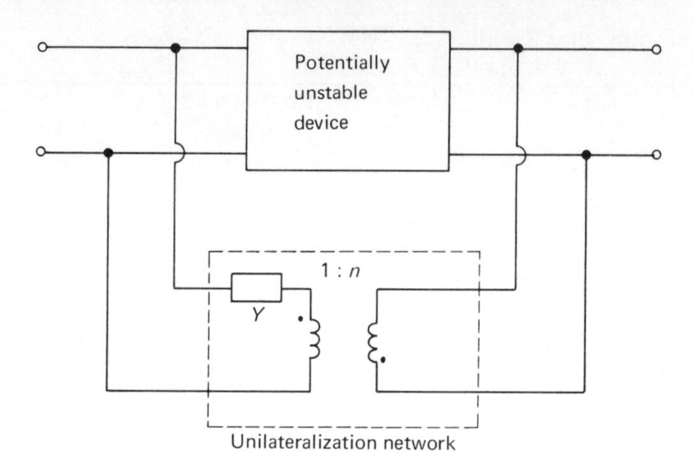

Fig. 5.28. Y-type of unilateralization.

unilateralizations to consider.[12] In the Y-type of unilateralized circuit, reproduced in Fig. 5.28 for convenience, the phase-inverting transformer is used in order that only passive elements are needed to construct the Y-network. It is assumed that the transformer is ideal, so that the overall y-matrix of the parallel–parallel connection of Fig. 5.28 is

$$\mathbf{Y} = \begin{bmatrix} y_i + Y & y_r + \dfrac{Y}{n} \\ y_f + \dfrac{Y}{n} & y_o + \dfrac{Y}{n^2} \end{bmatrix}. \tag{5.116}$$

The condition for unilateralization is therefore

$$y_{12} = y_r + \frac{Y}{n} = 0$$

or

$$Y = -ny_r. \tag{5.117}$$

Substitution of Eq. (5.117) in Eq. (5.116) yields the following y-matrix for the unilateralized network

$$\mathbf{Y} = \begin{bmatrix} y_i - ny_r & 0 \\ y_f - y_r & y_o - \dfrac{y_r}{n} \end{bmatrix}. \tag{5.118}$$

For the unilateralized two-port network, the requirements for absolute stability are that the original device be stable in the shorted-input–shorted-output mode, and that for all ω

$$\begin{aligned} \operatorname{Re}(y_i - ny_r) &\geq 0 \\ \operatorname{Re}\left(y_o - \frac{y_r}{n}\right) &\geq 0. \end{aligned} \tag{5.119}$$

For most transistors, $\operatorname{Re} y_i$ and $\operatorname{Re} y_o$ in the common-emitter and common-base configurations are positive over the entire useful frequency range, and $\operatorname{Re} y_r$ is negative so that the unilateralized network would be stable for all positive values of transformer turns ratio n. Negative values of n need not be considered since this would require Y to contain active elements in order to satisfy Eq. (5.117).

The remaining types of unilateralizations can be evaluated in a similar manner; their general properties are given in Table 5.1. In the H- and G-types, one of the elements Z or Y is usually made resistive while the other element contains both resistive and reactive elements. Furthermore, we usually find that h_r or g_r has a magnitude small compared with unity, so that $|ZY| \ll 1$.

Example 5.8. *Common-emitter amplifier.* A transistor operated in the common-emitter configuration has the following set of y-parameters measured at 1 MHz.

$$\begin{aligned} y_{ie} &= (2{\cdot}5 + j2)\ \mathrm{m}\mho \\ y_{re} &= (-2 - j25)\ \mu\mho \\ y_{fe} &= (30 - j4)\ \mathrm{m}\mho \\ y_{oe} &= (30 + j80)\ \mu\Omega. \end{aligned}$$

Therefore,

$$\begin{aligned} \operatorname{Re} y_{ie} &= 2{\cdot}5\ \mathrm{m}\mho \\ \operatorname{Re} y_{oe} &= 30\ \mu\mho \\ 2\operatorname{Re} y_{ie} \operatorname{Re} y_{oe} - \operatorname{Re}(y_{re}y_{fe}) - |y_{re}y_{fe}| &= -440. \end{aligned}$$

Since the third one of conditions (c) in (5.112) is violated, the device is potentially unstable at 1 MHz.

To overcome the potential instability of the common-emitter amplifier we may use the Y-type of unilateralization which is well-suited for transistor applications. The turns ratio n of the phase-inverting transformer is under the designer's control; therefore, choosing $n = 10$, Eq. (5.117) gives the required element Y to be

$$Y = -ny_{re} = 20 + j250\ \mu\mho$$

which, at the operating frequency of 1 MHz, represents the admittance of a 50 kΩ resistor in parallel with a 40 pF capacitor, as in Fig. 5.29. The y-parameters

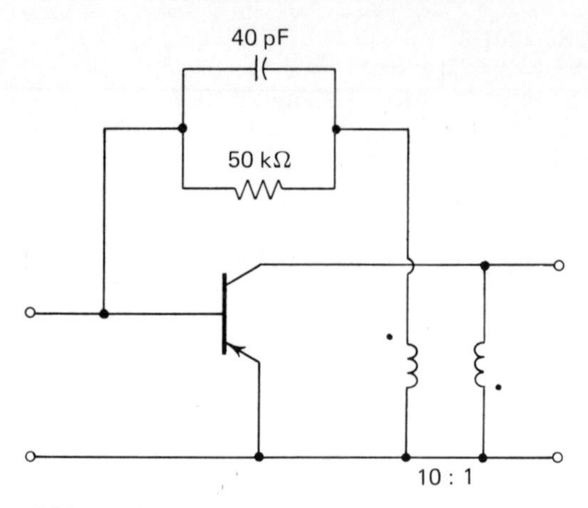

Fig. 5.29. Unilateralized common-emitter configuration.

of this network are, from Eq. (5.118), as follows

$$y_{11} = (2{\cdot}52 + j2{\cdot}25)\ \mathrm{m℧}$$
$$y_{12} = 0$$
$$y_{21} = (30 - j3{\cdot}97)\ \mathrm{m℧}$$
$$y_{22} = (30{\cdot}2 + j82{\cdot}5)\ \mu℧.$$

Some Design Considerations of Tuned Transistor Amplifiers

Some of the important factors which determine the performance of a tuned amplifier can be derived with reference to Fig. 5.30 which shows a common-emitter tuned amplifier (the transistor is designated as the black box). It is assumed that the transistor has been rendered unilateral and, as such, has the following overall y-matrix (see Eq. 5.118)

$$\begin{bmatrix} y_{11} & y_{12} \\ y_{21} & y_{22} \end{bmatrix} = \begin{bmatrix} y_{ie} - ny_{re} & 0 \\ y_{fe} - y_{re} & y_{oe} - \dfrac{y_{re}}{n} \end{bmatrix}, \tag{5.120}$$

where y_{ie}, y_{re}, y_{fe} and y_{oe} are the common-emitter y-parameters, and n is the turns ratio of the phase-inverting transformer (see Fig. 5.28). Over a narrow band of frequencies, the variations in the common-emitter y-parameters are so small in practice that the real and imaginary parts of each parameter can be justifiably assumed constant. The forward transfer admittance y_{fe} has a somewhat lagging phase angle with a second-order effect that can be neglected provided the operating frequency is much lower than the ω_T of the transistor. Furthermore, since

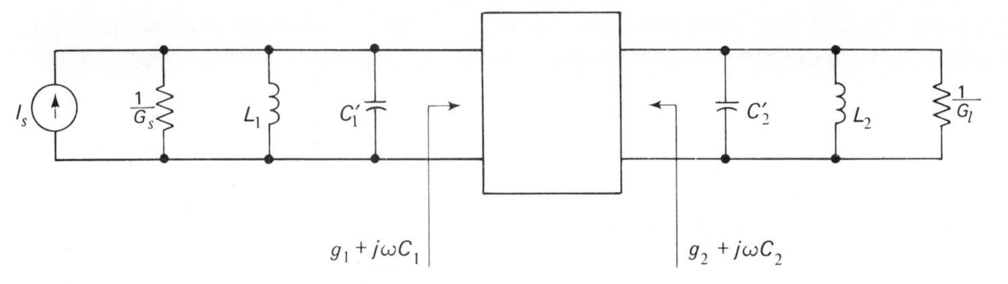

Fig. 5.30. Tuned transistor amplifier.

$|y_{fe}| \gg |y_{re}|$, it follows that Eq. (5.120) may be expressed in the form

$$\begin{bmatrix} y_{11} & y_{12} \\ y_{21} & y_{22} \end{bmatrix} = \begin{bmatrix} g_1 + j\omega C_1 & 0 \\ g_{fe} & g_2 + j\omega C_2 \end{bmatrix}, \tag{5.121}$$

where

$$g_1 = \mathrm{Re}(y_{ie} - ny_{re}), \tag{5.122}$$

$$g_2 = \mathrm{Re}\left(y_{oe} - \frac{y_{re}}{n}\right), \tag{5.123}$$

$$\omega C_1 = \mathrm{Im}(y_{ie} - ny_{re}), \tag{5.124}$$

$$\omega C_2 = \mathrm{Im}\left(y_{oe} - \frac{y_{re}}{n}\right), \tag{5.125}$$

with Im signifying the imaginary part of the expression that follows it. However, owing to component tolerances and the variations that arise in transistor parameters as a result of production spread and changes in environmental conditions, we find that perfect unilateralization cannot be maintained in every one of a number of similarly designed tuned-amplifier stages. The departure from perfect unilateralization is caused largely by variations in the capacitive components of the transistor parameter y_{re} and the unilateralizing element, and may therefore be represented by a capacitance δC connected between the input and output terminals,[13] as in Fig. 5.31. The result is that the amplifier circuit as a whole is now no longer unilateral.

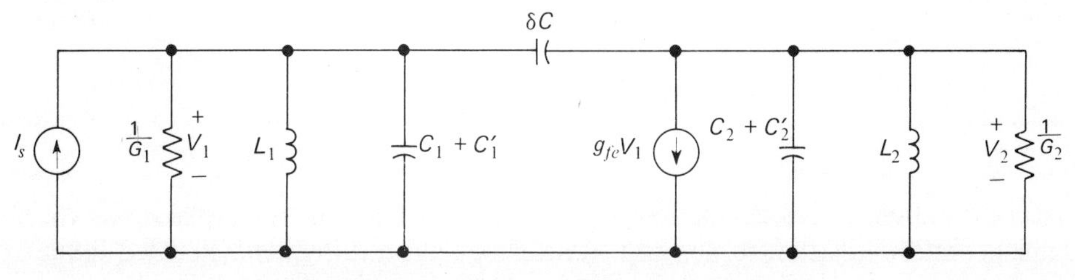

Fig. 5.31. Circuit model of tuned amplifier.

In Fig. 5.31 we have also assumed that the resonant circuits at the amplifier input and output are synchronously tuned, their total tuned parallel conductances being given as

$$G_1 = g_1 + G_s, \tag{5.126}$$

$$G_2 = g_2 + G_l. \tag{5.127}$$

Therefore, neglecting the shunting effects of the capacitance δC at the input and output ports, we find that the natural frequencies of the complete circuit are given as the roots of the characteristic equation

$$\left[1 + Q\left(\frac{s}{\omega_0} + \frac{\omega_0}{s}\right)\right]^2 + \frac{s\delta C g_{fe}}{G_1 G_2} = 0, \tag{5.128}$$

where

$$\omega_0^2 = \frac{1}{L_1(C_1' + C_1)} = \frac{1}{L_2(C_2' + C_2)}, \tag{5.129}$$

$$Q = \frac{1}{\omega_0 G_1 L_1} = \frac{1}{\omega_0 G_2 L_2}. \tag{5.130}$$

If the bandwidth is a small fraction of the centre frequency ω_0, which is the case when the loaded Q-factor is high, then we may restrict our attention to the immediate neighbourhood of $s = j\omega_0$. Let a new variable λ be defined as follows

$$\lambda = s - j\omega_0,$$

or

$$s = \lambda + j\omega_0. \tag{5.131}$$

The origin of the λ-plane corresponds to $s = j\omega_0$ in the s-plane; therefore, in the neighbourhood of $s = j\omega_0$, we have $|\lambda| \ll 1$ and

$$\begin{aligned}\frac{s}{\omega_0} + \frac{\omega_0}{s} &= \frac{\lambda}{\omega_0} + j1 + \frac{1}{\dfrac{\lambda}{\omega_0} + j1} \\ &= \frac{\lambda}{\omega_0} + j1 - j\left(1 - \frac{\lambda}{j\omega_0} + \cdots\right) \\ &\simeq 2\frac{\lambda}{\omega_0}.\end{aligned} \tag{5.132}$$

This result is referred to as the *narrow-band approximation.* It indicates that a natural frequency located in the neighbourhood of $j\omega_0$ in the s-plane has the same geometric relation to $j\omega_0$ as a transformed natural frequency located in the vicinity of the origin of the λ-plane, except that it is shrunk by a factor of 2.

For tuned amplifiers whose bandwidth is small compared with the centre frequency ω_0, a good approximation to the characteristic equation of (5.128), in the vicinity of $s = j\omega_0$, is therefore

$$\left(1 + \frac{2Q}{\omega_0}\lambda\right)^2 + \frac{j\omega_0 \delta C g_{fe}}{G_1 G_2} = 0$$

or

$$\left(\frac{\lambda}{\omega_0}\right)^2 + \frac{1}{Q}\left(\frac{\lambda}{\omega_0}\right) + \frac{1}{4Q^2}\left(1 + \frac{j\omega_0 \delta C g_{fe}}{G_1 G_2}\right) = 0. \tag{5.133}$$

Equation (5.133) is a quadratic in λ, with roots at

$$\lambda_1, \lambda_2 = \frac{\omega_0}{2Q}\left(-1 \pm \sqrt{\frac{-j\omega_0 \delta C g_{fe}}{G_1 G_2}}\right). \tag{5.134}$$

However, the square root of $-j$ is equal to $(1 - j)/\sqrt{2}$; hence, we have

$$\lambda_1, \lambda_2 = \frac{\omega_0}{2Q}[-1 \pm \sqrt{k}(1 - j)] \tag{5.135}$$

where k is directly proportional to the feedback capacitance δC, as defined by

$$k = \frac{\omega_0 \delta C g_{fe}}{2G_1 G_2}. \tag{5.136}$$

Equation (5.135) indicates that increasing the magnitude of k, starting from zero, causes the roots λ_1 and λ_2 to separate on a line making an angle of $-45°$ with the real axis of the λ-plane, as in Fig. 5.32. When $k < 1$, the roots λ_1 and

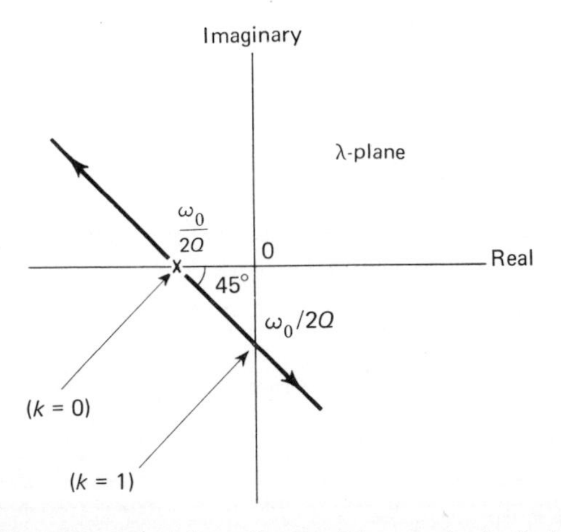

Fig. 5.32. Migration of natural frequencies in λ-plane.

λ_2 both lie in the left half of the λ-plane and the amplifier is stable. When $k > 1$, one of the roots moves into the right half plane and the amplifier becomes actually unstable. When $k = 1$, it is critically stable. The factor k is thus a measure of the stability status of the amplifier; it is appropriate to refer to it as the *stability factor*.

Since the amplifier is on the threshold of instability for $k = 1$, it follows from Eq. (5.136) that the minimum value of feedback capacitance δC required for oscillation is

$$\delta C_{crit} = \frac{2G_1G_2}{\omega_0 g_{fe}}. \tag{5.137}$$

Hence, the stability factor k is equal to the actual value of δC divided by δC_{crit}, that is

$$k = \frac{\delta C}{\delta C_{crit}}. \tag{5.138}$$

Equation (5.137) applies for a single amplifier stage. If several similar amplifier stages are connected in cascade, the value of δC_{crit} per stage diminishes by the factors 2, 2·61 and 3 for two, three and four amplifier stages, respectively.[14]

Another undesirable effect of the feedback capacitance δC is that it causes a skew in the selectivity characteristic of the amplifier. For a quantitative evaluation of this effect, we find from Fig. 5.31 that the output voltage V_2 is given by

$$V_2 = \frac{-g_{fe}I_s/G_1G_2}{\left[1 + Q\left(\dfrac{s}{\omega_0} + \dfrac{\omega_0}{s}\right)\right]^2 + \dfrac{s\delta C g_{fe}}{G_1G_2}}. \tag{5.139}$$

Using the narrow band approximation of Eq. (5.132) and setting $s = j\omega$, we get

$$V_2 \simeq \frac{-g_{fe}I_s/G_1G_2}{1 - x^2 + j2(k + x)}, \tag{5.140}$$

where

$$x = \frac{2Q(\omega - \omega_0)}{\omega_0}. \tag{5.141}$$

For $k = 0$ and $x = 0$ (i.e., $\omega = \omega_0$), the output voltage V_2 is equal to $-g_{fe}I_s/G_1G_2$. The response of the amplifier, normalized with respect to $-g_{fe}I_s/G_1G_2$, is therefore

$$y = \frac{V_2}{-g_{fe}I_s/G_1G_2} = \frac{1}{1 - x^2 + j2(k + x)}. \tag{5.142}$$

Evaluating the magnitude of y, we get

$$|y| = \frac{1}{[(1 + x^2)^2 + 4k(k + 2x)]^{1/2}}. \tag{5.143}$$

In Eq. (5.143) the term $4k(k + 2x)$ is responsible for the skew in the frequency characteristic, the skew becoming more pronounced with increasing k. This is illustrated in Fig. 5.33 where we have shown $|y|$ plotted against x for $k = 0, 0{\cdot}25$ and $0{\cdot}5$.

Summarizing, the special value δC_{crit} of feedback capacitance, as given by Eq. (5.137), is the minimum value required to produce oscillation. A smaller feedback capacitance, while not sufficiently large to produce oscillation, may introduce objectionable skew in the selectivity characteristic and may make the alignment process difficult. It is generally desirable, then, to design the amplifier in such a manner that the maximum net feedback capacitance δC which can be expected from normal parameter limits is considerably smaller than δC_{crit}.

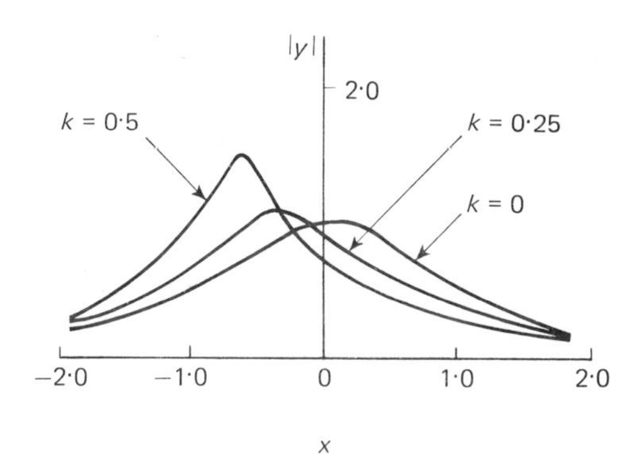

Fig. 5.33. Effect of feedback capacitance δC on the selectivity characteristic.

5.6 PASSIVITY

In previous sections of this chapter and the preceding one, we have on many occasions referred to a specific two-port network as passive or active. In order to justify such a distinction we must have a criterion by means of which we can determine whether a given two-port network is indeed passive or active. A linear two-port network is defined to be passive if for all excitations the total energy delivered to the network at its input and output ports is non-negative. Let $i_1(t)$ and $v_1(t)$ denote the instantaneous values of the current and voltage at the input port, and $i_2(t)$ and $v_2(t)$ denote the instantaneous current and voltage at the output port. Then, the total energy $\mathscr{E}(t)$ delivered to the network is

$$\mathscr{E}(t) = \int_{-\infty}^{t} [v_1(\tau)i_1(\tau) + v_2(\tau)i_2(\tau)]\, d\tau. \tag{5.144}$$

The two-port network is passive if $\mathscr{E}(t)$ is non-negative; otherwise, the network is active. In order to translate this passivity criterion in the time domain into an equivalent criterion in terms of the two-port parameters in the complex-frequency domain, we shall find it instructive to recall the principal result of Section 3.4 dealing with the passivity of a one-port network. There we showed that the energy requirement of

$$\mathscr{E}(t) = \int_{-\infty}^{t} v(\tau)i(\tau)\, d\tau \geq 0 \tag{5.145}$$

is equivalent to $\text{Re}\,[Z(s)] \geq 0$ for $\text{Re}\, s \geq 0$. The driving-point impedance $Z(s)$ of the one-port network may be expressed as

$$Z(s) = \frac{V(s)}{I(s)} = \frac{V(s)}{I(s)} \frac{I^*(s)}{I^*(s)} = \frac{V(s)I^*(s)}{|I(s)|^2}, \tag{5.146}$$

where $I^*(s)$ is the complex conjugate of $I(s)$, and $|I(s)|$ is its magnitude. Since $|I(s)|^2$ is positive for all values of s, it follows that the one-port network is passive if

$$\text{Re}[V(s)I^*(s)] \geq 0 \qquad \text{for} \quad \text{Re}\, s \geq 0. \tag{5.147}$$

Therefore, by analogy with the case of a one-port network, we see from Eq. (5.144) that a two-port network is passive if

$$\text{Re}[V_1(s)I_1^*(s) + V_2(s)I_2^*(s)] \geq 0 \qquad \text{for} \quad \text{Re}\, s \geq 0. \tag{5.148}$$

The transformed voltage–current relations of a two-port network in terms of the z-parameters are as follows:

$$\begin{aligned} V_1(s) &= z_{11}(s)I_1(s) + z_{12}(s)I_2(s) \\ V_2(s) &= z_{21}(s)I_1(s) + z_{22}(s)I_2(s). \end{aligned} \tag{5.149}$$

Hence, eliminating the voltage transforms from (5.148) and (5.149) we find that a two-port network is passive if

$$\text{Re}\,[F(s)] \geq 0 \qquad \text{for} \quad \text{Re}\, s \geq 0, \tag{5.150}$$

where the function $F(s)$ is defined by

$$\begin{aligned} F(s) = {} & z_{11}(s)I_1(s)I_1^*(s) + z_{12}(s)I_1^*(s)I_2(s) + z_{21}(s)I_1(s)I_2^*(s) \\ & + z_{22}(s)I_2(s)I_2^*(s). \end{aligned} \tag{5.151}$$

Normally, $F(s)$ is real for real s. We conclude, therefore, that a linear two-port network is passive if and only if the function $F(s)$ of Eq. (5.151) is positive real. As such, $F(s)$ cannot have poles in the right half of the complex-frequency plane. Therefore, none of the z-parameters of a passive two-port network is permitted to have poles in the right half plane. Moreover, if $F(s)$ has any simple poles on the imaginary axis, then for $F(s)$ to be positive real it is necessary that the residues of the z-parameters satisfy certain *residue conditions.*[15] Suppose $F(s)$ has a simple pole at $s = j\omega_0$ with a residue equal to k_0, and let k_{11}, k_{12}, k_{21} and k_{22} denote the residues of $z_{11}(s)$, $z_{12}(s)$, $z_{21}(s)$ and $z_{22}(s)$, respectively, at this pole. Then we may expand the $F(s)$ and the z's in Laurent series about the point $s = j\omega_0$; in the immediate neighbourhood of $s = j\omega_0$ we can approximate each series with its dominant term, so that

$$\frac{k_0}{s - j\omega_0} = \frac{k_{11}}{s - j\omega_0} I_1(j\omega_0) I_1^*(j\omega_0) + \frac{k_{12}}{s - j\omega_0} I_1^*(j\omega_0) I_2(j\omega_0)$$
$$+ \frac{k_{21}}{s - j\omega_0} I_1(j\omega_0) I_2^*(j\omega_0) + \frac{k_{22}}{s - j\omega_0} I_2(j\omega_0) I_2^*(j\omega_0);$$

that is,

$$k_0 = k_{11} I_1(j\omega_0) I_1^*(j\omega_0) + k_{12} I_1^*(j\omega_0) I_2(j\omega_0) + k_{21} I_1(j\omega_0) I_2^*(j\omega_0)$$
$$+ k_{22} I_2(j\omega_0) I_2^*(j\omega_0). \tag{5.152}$$

From the properties of a positive real function we know that the residue k_0 must be a real and positive number. Likewise, k_{11} and k_{22} must be real and positive numbers since the driving-point impedances $z_{11}(s)$ and $z_{22}(s)$ are positive real functions. Furthermore, since $I_1(j\omega_0) I_1^*(j\omega_0) = |I_1(j\omega_0)|^2$ and $I_2(j\omega_0) I_2^*(j\omega_0) = |I_2(j\omega_0)|^2$ are also real and positive, it follows from Eq. (5.152) that

$$k_{12} I_1^*(j\omega_0) I_2(j\omega_0) + k_{21} I_1(j\omega_0) I_2^*(j\omega_0)$$

must be real and positive too. However, $I_1^*(j\omega_0) I_2(j\omega_0)$ and $I_1(j\omega_0) I_2^*(j\omega_0)$ are complex conjugates; hence the residues k_{12} and k_{21} must likewise be complex conjugates:

$$k_{21} = k_{12}^*. \tag{5.153}$$

Accordingly, Eq. (5.152) is a *Hermitian form*;* as such, it may be reduced to its diagonal form as follows:

$$k_0 = \left[I_1(j\omega_0) + \frac{k_{12}}{k_{11}} I_2(j\omega_0) \right] \left[I_1^*(j\omega_0) + \frac{k_{21}}{k_{11}} I_2^*(j\omega_0) \right] k_{11}$$
$$+ I_2(j\omega_0) I_2^*(j\omega_0) \frac{k_{11} k_{22} - k_{12} k_{21}}{k_{11}}$$

or

$$k_0 = |a_1|^2 k_{11} + |a_2|^2 \frac{k_{11} k_{22} - k_{12} k_{21}}{k_{11}}, \tag{5.154}$$

where

$$a_1 = I_1(j\omega_0) + \frac{k_{12}}{k_{11}} I_2(j\omega_0)$$
$$a_2 = I_2(j\omega_0). \tag{5.155}$$

---

* A Hermitian form is an expression of the form $\Sigma\, x_i^* x_j h_{ij}$ in which the matrix of coefficients, h_{ij} known as a Hermitian matrix, has the property that $h_{ij} = h_{ji}^*$. For further details see S. Perlis, *Theory of matrices*, p. 98, Addison-Wesley, 1952.

The form of Eq. (5.154) is positive definite (that is, $k_0 \geq 0$ for all values of a_1 and a_2) only if

$$\begin{aligned} k_{11} &\geq 0 \\ k_{11}k_{22} - k_{12}k_{21} &\geq 0 \end{aligned} \tag{5.156}$$

(which also imply that $k_{22} \geq 0$). These are the residue conditions which the simple poles of the z-parameters on the imaginary axis, if any, must satisfy in a passive two-port network.

Lastly, the function $F(s)$ must also satisfy the condition $\text{Re}\,[F(j\omega)] \geq 0$ for all ω, if it is to be a positive real function. The real part of $F(j\omega)$ may be evaluated from the relation

$$\text{Re}\,[F(j\omega)] = \tfrac{1}{2}[F(j\omega) + F^*(j\omega)]. \tag{5.157}$$

Putting $s = j\omega$ in Eq. (5.151) gives

$$\begin{aligned} F(j\omega) = {} & z_{11}(j\omega)I_1(j\omega)I_1^*(j\omega) + z_{12}(j\omega)I_1^*(j\omega)I_2(j\omega) \\ & + z_{21}(j\omega)I_1(j\omega)I_2^*(j\omega) + z_{22}(j\omega)I_2(j\omega)I_2^*(j\omega). \end{aligned} \tag{5.158}$$

The complex conjugate of this $F(j\omega)$ is

$$\begin{aligned} F^*(j\omega) = {} & z_{11}^*(j\omega)I_1^*(j\omega)I_1(j\omega) + z_{12}^*(j\omega)I_1(j\omega)I_2^*(j\omega) \\ & + z_{21}^*(j\omega)I_1^*(j\omega)I_2(j\omega) + z_{22}^*(j\omega)I_2^*(j\omega)I_2(j\omega). \end{aligned} \tag{5.159}$$

Substitution of Eqs. (5.158) and (5.159) in Eq. (5.157) yields the following Hermitian form

$$\begin{aligned} \text{Re}\,[F(j\omega)] = {} & z'_{11}(j\omega)I_1(j\omega)I_1^*(j\omega) + z'_{12}(j\omega)I_1^*(j\omega)I_2(j\omega) \\ & + z'_{21}(j\omega)I_1(j\omega)I_2^*(j\omega) + z'_{22}(j\omega)I_2(j\omega)I_2^*(j\omega), \end{aligned} \tag{5.160}$$

where

$$\begin{bmatrix} z'_{11}(j\omega) & z'_{12}(j\omega) \\ z'_{21}(j\omega) & z'_{22}(j\omega) \end{bmatrix} = \frac{1}{2}\begin{bmatrix} z_{11}(j\omega) + z_{11}^*(j\omega) & z_{12}(j\omega) + z_{21}^*(j\omega) \\ z_{21}(j\omega) + z_{12}^*(j\omega) & z_{22}(j\omega) + z_{22}^*(j\omega) \end{bmatrix}. \tag{5.161}$$

In Eq. (5.161) we see that the primed z-matrix is a Hermitian matrix equal to half the sum of the z-matrix and the transpose of its complex conjugate. In a general two-port network all the z-parameters have complex values at $s = j\omega$; thus, let

$$\begin{bmatrix} z_{11}(j\omega) & z_{12}(j\omega) \\ z_{21}(j\omega) & z_{22}(j\omega) \end{bmatrix} = \begin{bmatrix} r_{11} + jx_{11} & r_{12} + jx_{12} \\ r_{21} + jx_{21} & r_{22} + jx_{22} \end{bmatrix}. \tag{5.162}$$

The transpose of the complex conjugate of this z-matrix is

$$\begin{bmatrix} z_{11}^*(j\omega) & z_{21}^*(j\omega) \\ z_{12}^*(j\omega) & z_{22}^*(j\omega) \end{bmatrix} = \begin{bmatrix} r_{11} - jx_{11} & r_{21} - jx_{21} \\ r_{12} - jx_{12} & r_{22} - jx_{22} \end{bmatrix}. \tag{5.163}$$

Adding Eqs. (5.162) and (5.163), and then using the definition of Eq. (5.161) we find that in terms of the z-parameters the Hermitian z-matrix of a two-port network is

$$\begin{bmatrix} z'_{11}(j\omega) & z'_{12}(j\omega) \\ z'_{21}(j\omega) & z'_{22}(j\omega) \end{bmatrix} = \begin{bmatrix} r_{11} & \begin{pmatrix} \frac{1}{2}(r_{12}+r_{21}) \\ +\frac{j}{2}(x_{12}-x_{21}) \end{pmatrix} \\ \begin{pmatrix} \frac{1}{2}(r_{12}+r_{21}) \\ -\frac{j}{2}(x_{12}-x_{21}) \end{pmatrix} & r_{22} \end{bmatrix}. \quad (5.164)$$

The Hermitian form of Eq. (5.160) may be reduced to its diagonal form as follows:

$$\operatorname{Re}[F(j\omega)] = |b_1|^2 z'_{11}(j\omega) + |b_2|^2 \frac{z'_{11}(j\omega)z'_{22}(j\omega) - z'_{12}(j\omega)z'_{21}(j\omega)}{z'_{11}(j\omega)}, \quad (5.165)$$

where

$$b_1 = I_1(j\omega) + \frac{z'_{12}(j\omega)}{z'_{11}(j\omega)} I_2(j\omega)$$
$$b_2 = I_2(j\omega). \quad (5.166)$$

The form of Eq. (5.165) is positive definite (that is, $\operatorname{Re}[F(j\omega)] \geq 0$ for all values of b_1 and b_2) only if,

$$z'_{11}(j\omega) \geq 0 \quad \text{for all } \omega$$
$$z'_{11}(j\omega)z'_{22}(j\omega) - z'_{12}(j\omega)z'_{21}(j\omega) \geq 0 \quad \text{for all } \omega \quad (5.167)$$

(which also imply that $z'_{22}(j\omega) \geq 0$ for all ω). Substitution of the primed z's of Eq. (5.164) in (5.167) gives

$$r_{11} \geq 0$$
$$r_{22} \geq 0 \quad (5.168)$$
$$4r_{11}r_{22} - (r_{12} + r_{21})^2 - (x_{12} - x_{21})^2 \geq 0$$

for all ω.

We conclude, therefore, that a linear two-port network is passive if and only if

a) The z-parameters have no poles in the right half plane.

b) Any poles of the z-parameters on the imaginary axis are simple, and the residues of the z-parameters at these poles satisfy the following conditions:

$$
\begin{aligned}
&k_{11} \geq 0, \\
&k_{22} \geq 0 \\
k_{11}k_{22} - k_{12}k_{21} &\geq 0, \quad \text{with } k_{21} = k_{12}^*.
\end{aligned}
\tag{5.169}
$$

c) The real and imaginary parts of the z-parameters satisfy the following conditions for all ω

$$
\begin{aligned}
&\text{Re}\, z_{11} \geq 0 \\
&\text{Re}\, z_{22} \geq 0 \\
4\,\text{Re}\, z_{11}\, \text{Re}\, z_{22} - [\text{Re}\, z_{12} &+ \text{Re}\, z_{21}]^2 - [\text{Im}\, z_{12} - \text{Im}\, z_{21}]^2 \geq 0.
\end{aligned}
$$

The network is active if it is not passive. Conditions (a), (b) and (c) of (5.169) constitute *Raisbeck's passivity criterion.*[16]

The conditions of (5.169) define the necessary and sufficient conditions for the passivity of a two-port network in terms of the z-parameters. In a similar manner we can express the passivity criterion in terms of the y-, h-, or g-parameters.

A special case occurs when the network does not possess a finite z- or y-matrix; this arises when the parameters h_{11} and h_{22} in the h-matrix are both zero, as in the ideal transformer and ideal negative-impedance converter. In order to develop a criterion for the passivity of such networks, we see from condition (c) in (5.169) that passivity in terms of the h-parameters requires

$$4\,\text{Re}\, h_{11}\, \text{Re}\, h_{22} - [\text{Re}\, h_{12} + \text{Re}\, h_{21}]^2 - [\text{Im}\, h_{12} - \text{Im}\, h_{21}]^2 \geq 0. \tag{5.170}$$

When $h_{11} = 0$ and $h_{22} = 0$, the condition (5.170) simplifies as

$$-[\text{Re}\, h_{12} + \text{Re}\, h_{21}]^2 - [\text{Im}\, h_{12} - \text{Im}\, h_{21}]^2 \geq 0, \tag{5.171}$$

which can only be satisfied if both terms in the left-hand member are simultaneously zero, that is,

$$
\begin{aligned}
\text{Re}\, h_{12} &= -\text{Re}\, h_{21} \\
\text{Im}\, h_{12} &= \text{Im}\, h_{21}.
\end{aligned}
\tag{5.172}
$$

In other words, a two-port network for which both h_{11} and h_{22} are zero is passive if h_{12} is the negative of the complex conjugate of h_{21}. Equivalently, in terms of the chain parameters passivity requires A to be the complex conjugate of $1/D$.

The Stability–Activity Diagram

Comparison of the passivity conditions of (5.169) with the absolute stability conditions of (5.111) indicates that the passivity criterion implies conditions (a) and (b) and the first two of conditions (c) of (5.111). Furthermore, the last of conditions (c) in (5.111) for absolute stability can be manipulated into the form

$$\frac{r_{121}}{\sqrt{r_{11}r_{22}}} \leq 1, \tag{5.173}$$

where r_{121} is the real part of $\sqrt{z_{12}z_{21}}$, while the last of conditions (c) in (5.169) for passivity can be expressed in the form

$$\frac{r_{121}^2}{r_{11}r_{22}} + \frac{(|z_{12}| - |z_{21}|)^2}{4r_{11}r_{22}} \leq 1. \tag{5.174}$$

In the so-called *stability–activity diagram*[17] of Fig. 5.34 we have represented (5.173) and (5.174) in a two-dimensional diagram by choosing the dimensionless parameters $r_{121}/\sqrt{r_{11}r_{22}}$ and $||z_{12}| - |z_{21}||/2\sqrt{r_{11}r_{22}}$ as the two co-ordinates. The boundary between the regions of absolute stability and potential instability is represented by the vertical line at $(r_{121}/\sqrt{r_{11}r_{22}}) = 1$, while the boundary between the regions of passivity and activity is represented by the quadrant of a circle of unit radius. We thus observe that the conditions for passivity imply the conditions for absolute stability; the conditions for absolute stability do not however, imply all the conditions for passivity. In other words, all passive two-port networks are absolutely stable, but not all absolutely stable two-port networks are passive. Furthermore, all potentially unstable two-port networks are necessarily active but not all active two-port networks are potentially unstable.

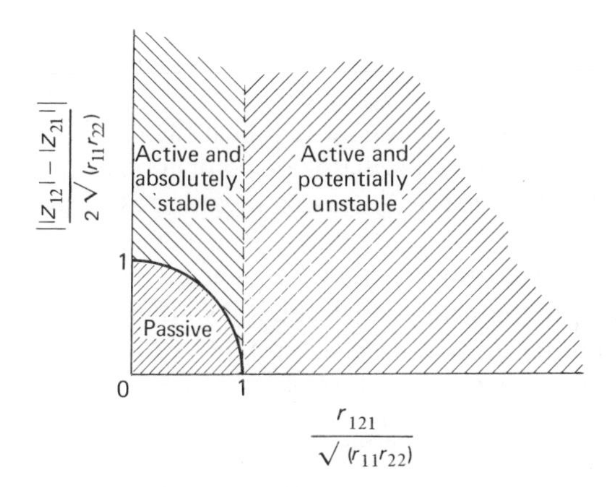

Fig. 5.34. Stability–activity diagram.

In general, as the frequency is varied, a non-unilateral active two-port device having frequency-dependent parameters (e.g., a transistor) may move between the various classes, including the class of passive networks. This is illustrated in Fig. 5.35 showing the locus of $||z_{12}| - |z_{21}||/(2\sqrt{r_{11}r_{22}})$ against $r_{121}/\sqrt{r_{11}r_{22}}$

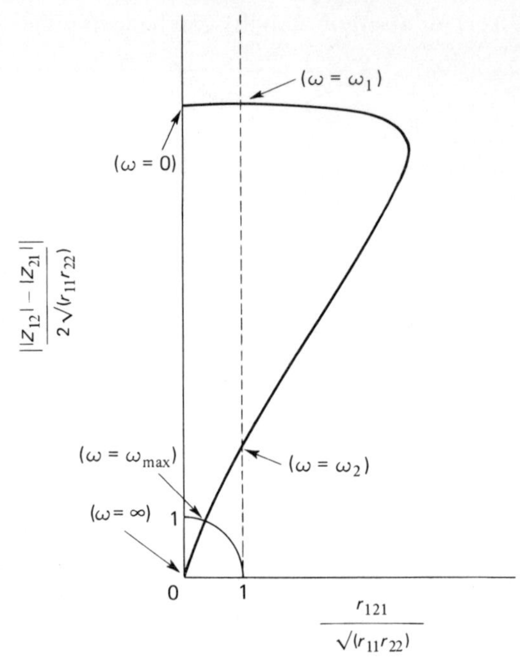

Fig. 5.35. Stability–activity diagram of a transistor.

with frequency as a running parameter. Three critical frequencies ω_1, ω_2 and ω_{max} may be identified. The frequency range from ω_1 to ω_2 defines the frequencies of potential instability, while at ω_{max} the device ceases to be active. Clearly, ω_{max} defines the maximum frequency up to which the device may be used as the active component of a feedback oscillator or be operated as an amplifier with a gain greater than unity; so that ω_{max} may be regarded as a figure of merit for the device.

For a transistor operating at high frequencies with the emitter terminal common, the maximum frequency of potentiál instability, ω_2, and the maximum frequency of oscillation, ω_{max}, can be determined from the simplified hybrid-π model of Fig. 5.36, obtaining

$$\omega_2 \simeq \frac{1}{2r_{b'}C_{b'e}} \tag{5.175}$$

and

$$\omega_{max} \simeq \sqrt{\frac{g_m}{4r_{b'}C_{b'c}C_{b'e}}}\,. \tag{5.176}$$

The expression for the minimum frequency of potential instability, ω_1, is more involved, requiring use of the complete hybrid-π model. (See Problem 5.14).

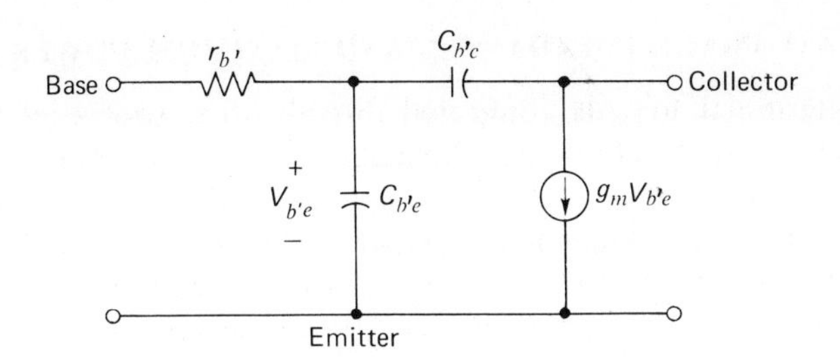

Fig. 5.36. Simplified hybrid-π model.

Other Implications of the Passivity Criterion

Consider next the case of a reciprocal two-port network for which we have $z_{12} = z_{21}$ and $y_{12} = y_{21}$. In such a network we find that the conditions for passivity and absolute stability become identical. In other words, a reciprocal two-port network that is absolutely stable is necessarily passive. However, all passive networks are not necessarily reciprocal. The ideal gyrator is an example of a passive but non-reciprocal two-port network; this is readily demonstrated by considering the z-matrix of an ideal gyrator:

$$\mathbf{Z} = \begin{bmatrix} 0 & r \\ -r & 0 \end{bmatrix}, \tag{5.177}$$

where r is the gyration resistance. Equation (5.177) satisfies the passivity criterion, but not the reciprocity criterion.

A gyrator can be activated by choosing the z-matrix

$$\mathbf{Z} = \begin{bmatrix} 0 & r_1 \\ -r_2 & 0 \end{bmatrix}, \tag{5.178}$$

in which $r_1 \neq r_2$. The *active gyrator* is non-reciprocal and has similar impedance-inverting properties as the ideal gyrator; in addition, it can provide amplification, depending upon the direction of signal transmission through the device.

Although the majority of active two-port networks are non-reciprocal, activity does not necessarily imply non-reciprocity. Thus, the negative-impedance inverter is, for example, active but reciprocal. An ideal negative-impedance inverter has the following z-matrix:

$$\mathbf{Z} = \begin{bmatrix} 0 & R \\ R & 0 \end{bmatrix}, \tag{5.179}$$

which clearly satisfies the reciprocity criterion, but violates the passivity criterion.

5.7 CONJUGATE-IMAGE IMPEDANCES AND MAXIMUM POWER GAIN

When a two-terminal load is connected directly to a source of signal, as in Fig. 5.37, and it is required to deliver the maximum amount of power to the load, we find it necessary to make the load impedance equal to the complex conjugate of the source impedance. This form of impedance matching, to secure maximum power transfer from the source to the load, is called matching on the conjugate-image basis. In order to extend this idea to the case of a double-terminated

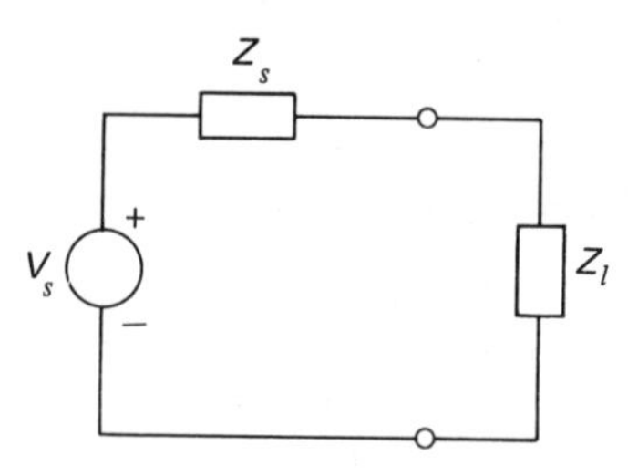

Fig. 5.37. Two-terminal load connected directly to source of signal.

two-port network consider an infinite cascade of two-port networks, a portion of which is shown in Fig. 5.38. In this diagram, network N_a is the given network, and network N_b is identical to network N_a except that the open-circuit impedances (i.e., the z-parameters) of network N_b are conjugates of those of network N_a. Also, each network N_b is reversed so that similar terminals are connected to network N_a at each junction. In an infinite cascade of this form, provided there is some dissipation, we find that at any junction of two networks the impedance looking to the right is the complex conjugate of the impedance looking to the left. The cascade connection of Fig. 5.38 may be compared with the corresponding

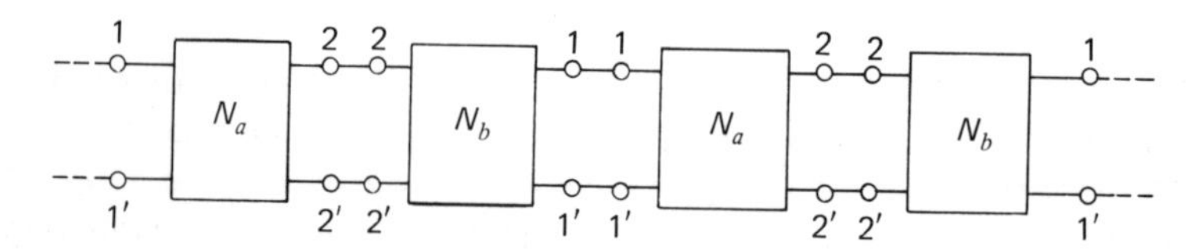

Fig. 5.38. A chain of two-port networks connected on the conjugate-image basis.

cascade of an infinite number of identical networks on the normal-image basis. In the special case of purely resistive two-port networks, the two cascade connections become equivalent to one another.

From Fig. 5.38 we deduce the following definitions for the *conjugate-image impedances*[18] of a two-port network: if a two-port network is so connected between such a source and load that the input impedance is the conjugate of the source impedance, while the output impedance is the conjugate of the load impedance, then the terminating source and load impedances are the conjugate-image impedances of the network. This is illustrated diagrammatically in Fig. 5.39 where Z_{c1} and Z_{c2} are the conjugate-image impedances.

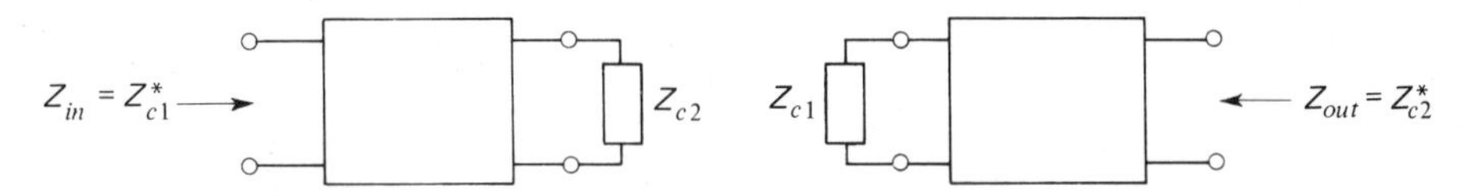

Fig. 5.39. Illustrating the definitions of the conjugate-image impedances of a two-port network.

The conjugate-image impedances define the optimum passive terminations for which an absolutely stable active two-port network attains its maximum available power gain. In fact, for such a set of terminations the power gain, available power gain and transducer power gain assume a common maximum value. We can, therefore, determine the maximum available power gain and the optimum terminations for a given two-port network using any one of three alternative procedures:[19]

1) We can find the load impedance which results in the maximum power gain G_{max}, and then determine the input impedance of the two-port network terminated with the optimum load impedance; the complex conjugate of this input impedance gives the required value of optimum source impedance.
2) We can find the source impedance which results in the maximum available power gain and then determine the output impedance of the two-port network terminated with the optimum source impedance; the complex conjugate of this output impedance gives the required value of optimum load impedance.
3) We can find the terminating source and load impedances for which the two-port network is conjugately matched at its input and output ports.

The third procedure, which is the only one requiring simultaneous considerations of the source and load impedances, is quite involved when the two-port parameters are complex; in such a case it is simpler to use procedure (1) or (2). In the following evaluation we shall use procedure (1).

In terms of the z-parameters the power gain G is given by (see Eq. 4.62)

$$G = \frac{|z_{21}|^2 \operatorname{Re} Z_l}{|z_{22} + Z_l|^2 \operatorname{Re}\left(z_{11} - \dfrac{z_{12}z_{21}}{z_{22} + Z_l}\right)}. \tag{5.180}$$

When the two-port parameters and the load impedance Z_l are complex, Eq. (5.180) yields

$$G = \frac{R_l(r_{21}^2 + x_{21}^2)/r_{11}}{(r_{22} + R_l)^2 + (x_{22} + X_l)^2 - \dfrac{M}{r_{11}}(r_{22} + R_l) - \dfrac{N}{r_{11}}(x_{22} + X_l)}, \tag{5.181}$$

where

$$M + jN = z_{12}z_{21};$$

that is,

$$\begin{aligned} M &= r_{12}r_{21} - x_{12}x_{21} \\ N &= x_{12}r_{21} + x_{21}r_{12}. \end{aligned} \tag{5.182}$$

To evaluate the maximum value of the power gain, we set the partial derivatives $\partial G/\partial R_l$ and $\partial G/\partial X_l$ to zero. This results in the following optimum value for the load impedance

$$Z_{c2} = \sqrt{r_{22}^2 - \frac{M^2}{r_{11}^2} - \frac{N^2}{4r_{11}^2}} + j\left(\frac{M}{2r_{11}} - x_{22}\right). \tag{5.183}$$

The optimum source impedance Z_{c1} is equal to the complex conjugate of the input impedance in Eq. (5.184) which results when the output port is terminated in Z_{c2}:

$$Z_{in} = z_{11} - \frac{z_{12}z_{21}}{z_{22} + Z_{c2}}. \tag{5.184}$$

Therefore, using Eqs. (5.183) and (5.184) we get

$$Z_{c1} = \sqrt{r_{11}^2 - \frac{M^2}{r_{22}^2} - \frac{N^2}{4r_{22}^2}} + j\left(\frac{M}{2r_{22}} - x_{11}\right). \tag{5.185}$$

The maximum value G_{max} of the power gain is obtained by substituting Eq. (5.183) in (5.181), which yields

$$G_{max} = \frac{|z_{21}|^2}{2r_{11}r_{22} - M + \sqrt{(2r_{11}r_{22} - M)^2 - (M^2 + N^2)}}. \tag{5.186}$$

Clearly, the maximum power gain G_{max} is a meaningful quantity only when the following condition is satisfied

$$2r_{11}r_{22} - M - \sqrt{M^2 + N^2} \geq 0$$

or

$$2 \operatorname{Re} z_{11} \operatorname{Re} z_{22} - \operatorname{Re}(z_{12}z_{21}) - |z_{12}z_{21}| \geq 0, \tag{5.187}$$

which is recognized to be the third of conditions (c) in the criterion for absolute stability (see 5.111). In addition, the real parts of both z_{11} and z_{22} must be non-negative. In other words, the method of conjugate-image impedance matching and the associated concept of maximum power gain are meaningful only if, at the frequency of interest, the device is absolutely stable.

From Eq. (5.186) we see that the maximum power gain G_{max} is greater than unity provided we have

$$|z_{21}|^2 > 2r_{11}r_{22} - M + \sqrt{(2r_{11}r_{22} - M)^2 - (M^2 + N^2)}. \tag{5.188}$$

After some manipulations we find that the condition of (5.188) is equivalent to the requirement

$$4r_{11}r_{22} - (r_{12} + r_{21})^2 - (x_{12} - x_{21})^2 < 0, \tag{5.189}$$

which violates the third of conditions (c) in (5.169) for passivity. This means that only an active two-port network is capable of providing a maximum power gain greater than unity.

It is of interest to note that when the two-port network is unilateral (i.e., the reverse transfer impedance z_{12} is zero), Eq. (5.186) reduces to:

$$G_{max} = \frac{|z_{21}|^2}{4 \operatorname{Re} z_{11} \operatorname{Re} z_{22}}. \tag{5.190}$$

For this special case, the optimum source impedance $Z_s = z_{11}^*$ and the optimum load impedance $Z_l = z_{22}^*$.

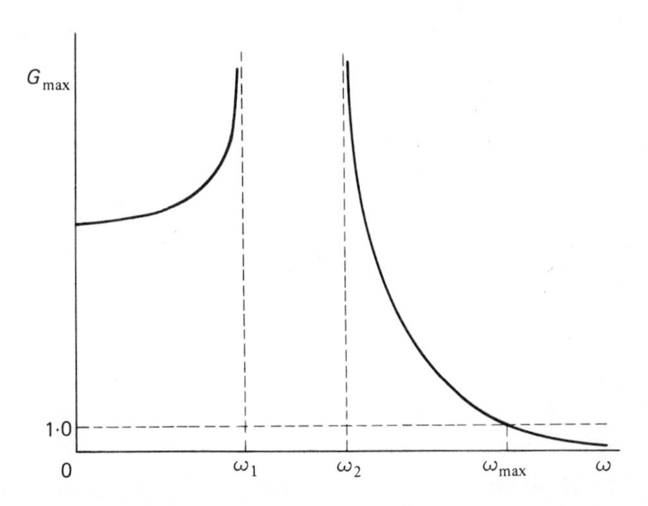

Fig. 5.40. Illustrating frequency dependence of maximum power gain for a non-unilateral active device.

Concluding Remarks

In a physical active two-port device (e.g., a transistor) with frequency-dependent parameters and some degree of internal feedback, we ordinarily find that the maximum power gain G_{max} varies with frequency in the manner illustrated in Fig. 5.40. The frequencies ω_1 and ω_2 define the range within which the device is potentially unstable; hence, G_{max} is non-definable for $\omega_1 \leq \omega \leq \omega_2$. Outside this frequency range, the device is absolutely stable, and a definable G_{max} becomes realizable with a unique set of optimum source and load impedances. At frequencies higher than ω_{max} the device ceases to be active and the maximum power gain takes on a value less than unity; ω_{max} thus defines the maximum frequency up to which it is possible to use the device as an amplifier or the active component of an oscillator.

REFERENCES

1. B. D. H. Tellegen, "A General Network Theorem with Applications," *Philips Research Reports*, **7,** 259 (1952).
2. J. Shekel, "The Gyrator as a Three-terminal Element," *Proc. I.R.E.*, **41,** 1014 (1953).
3. A. C. Bartlett, "An Extension of a Property of Artificial Lines," *Phil. Mag.* **4,** 902 (1927).
4. O. Brune, "Note on Bartlett's Bisection Theorem for Four-terminal Electrical Networks," *Phil. Mag.* **14,** 806 (1932).
5. G. E. Valley and H. Wallman, *Vacuum-tube Amplifiers*. McGraw-Hill, 1948
6. R. D. Thornton *et al.*, *Multistage Transistor Circuits*. Wiley, 1965.
7. G. L. Matthaei, "Some Simplifications for Analysis of Linear Circuits," *Trans. I.R.E.* **CT-4,** 120 (1957).
8. A. J. Cote, "Matrix Analysis of Oscillators and Transistor Applications," *Trans. I.R.E.* **CT-5,** 181 (1958).
9. H. Jefferson, "Stabilization of Feedback Oscillators," *Wireless Engineer* **22,** 384 (1945).
10. A. P. Stern, "Considerations of the Stability of Active Elements and Applications to Transistors," *I.R.E. Convention Record*, part 2, 46 (1956).
11. F. B. Llewellyn, "Some Fundamental Properties of Transmission Systems," *Proc. I.R.E.* **40,** 271 (1952).
12. A. J. Cote, "Evaluation of Transistor Neutralization Networks," *Trans. I.R.E.* **CT-5,** 95 (1958).
13. D. D. Holmes and T. D. Stanley, "Stability Considerations in Transistor Intermediate-frequency Amplifiers," *Transistors*, I, R.C.A. (1956).
14. J. Thompson, "Oscillation in Tuned R.F. Amplifiers," *Proc. I.R.E.* **19,** 421 (1931).
15. L. de Pian, *Linear Active Network Theory*. Prentice-Hall, 1962.
16. G. Raisbeck, "A Definition of Passive Linear Networks in Terms of Time and Energy," *J. Appl. Phys.* **25,** 1510 (1954).

17. T. Fjallbrant, "Activity and Stability of Linear Networks," *Trans. I.E.E.E.* **CT-12,** 12 (1965).
18. S. Roberts, "Conjugate-image Impedances," *Proc. I.R.E.* **34,** 198P (1946).
19. J. G. Linvill and J. F. Gibbons, *Transistors and Active Circuits*. (McGraw-Hill, 1961).
20. D. F. Page and A. R. Boothroyd, "Instability in Two-port Active Networks," *Trans. I.R.E.* **CT-5,** 133 (1958).
21. A. J. Cote, "Matrix Analysis of *RL* and *RC* Oscillators," *Trans. I.R.E.* **CT-6,** 232 (1959).

PROBLEMS

5.1 Determine the z-parameters of the two-port network shown in Fig. P5.1. Hence, find the value of control parameter g for which the network is reciprocal.

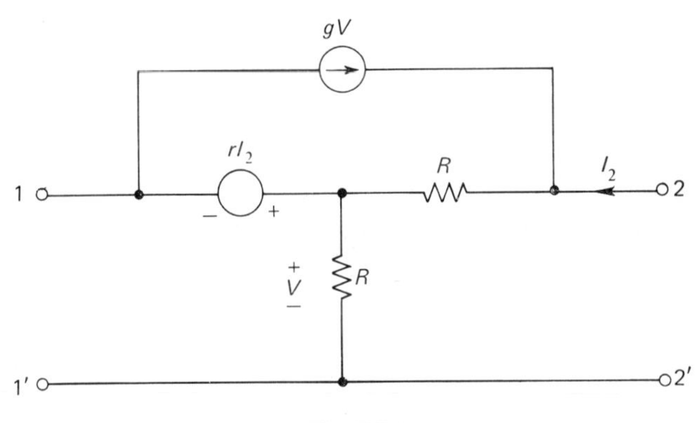

Fig. P5.1.

5.2 It is proposed to represent a non-reciprocal three-terminal network by an equivalent circuit consisting of a gyrator and three impedances, as in Fig. P5.2. Determine the elements of this circuit.

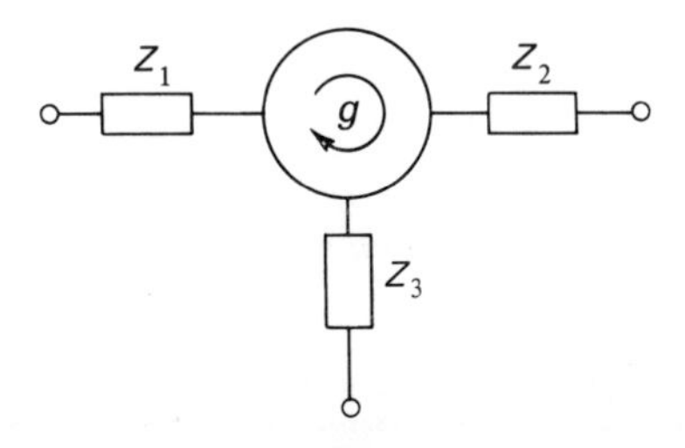

Fig. P5.2.

5.3 For the symmetrical cathode-coupled pair circuit of Fig. P5.3, evaluate the current which flows through the ammeter M of internal resistance 100 Ω. It is assumed that the two tubes are identical having $r_p = 8$ kΩ and $\mu = 40$.

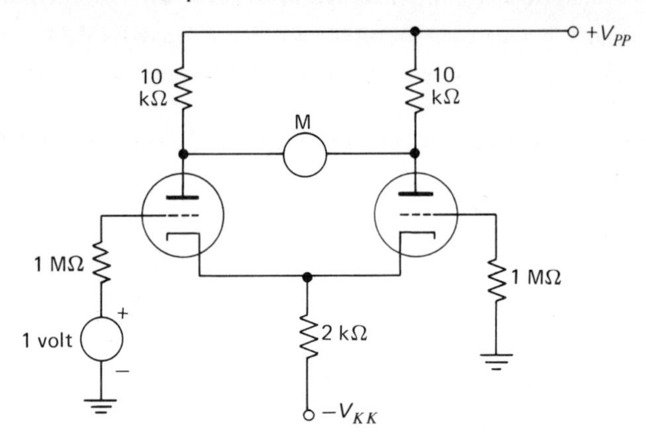

Fig. P5.3.

5.4 In part (a) of Fig. P5.4 a coil of inductance 10 mH and resistance 10 Ω is connected between the base and collector terminals of a common-emitter stage. Evaluate the range of frequencies for which the circuit is potentially unstable. If the stage is terminated with parallel RC networks, as in part (b) of Fig. P5.4 determine the ratio C_2/C_1 for which the complete circuit is on the threshold of instability. If $C_1 = 0{\cdot}01$ μF, what is the frequency of oscillation corresponding to this condition? The transistor parameters may be assumed to be real, having the following set of values

$$h_{ie} = 500\ \Omega \qquad h_{re} = 10^{-4}$$

$$h_{fe} = 100 \qquad h_{oe} = 20\ \mu\mho.$$

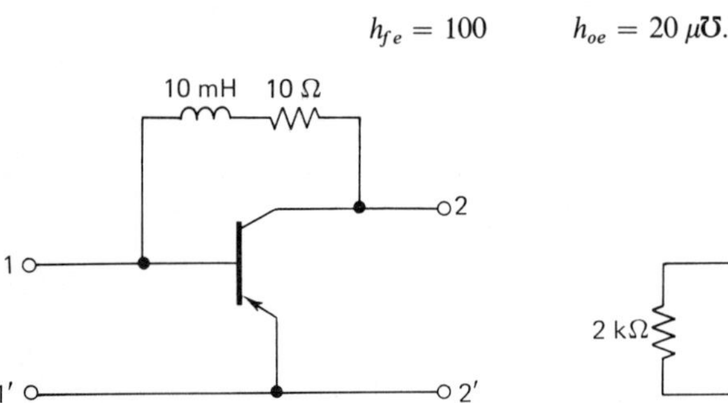

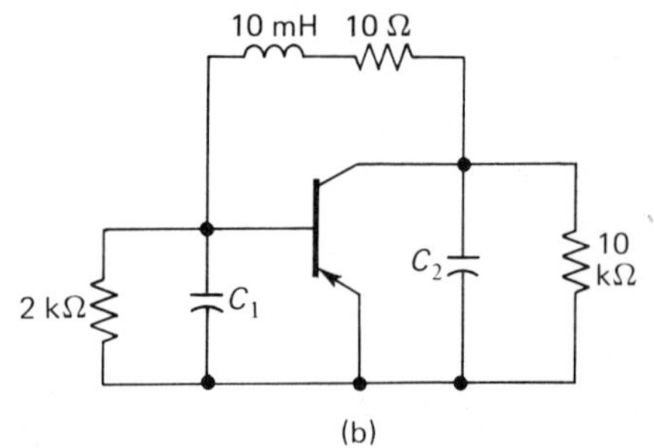

Fig. P5.4.

5.5 A certain two-port network has the following z-parameters

$$z_{11} = 625\ \Omega \qquad z_{12} = 601\ \Omega$$

$$z_{21} = 0{\cdot}98\ M\Omega \qquad z_{22} = 1\ M\Omega.$$

Is the network active? Evaluate the optimum source and load resistances and the corresponding maximum power gain of the network.

5.6 Evaluate the characteristic polynomials of the negative-impedance inverter circuits shown in Fig. P5.6. Hence, determine the terminations for which they are unstable.

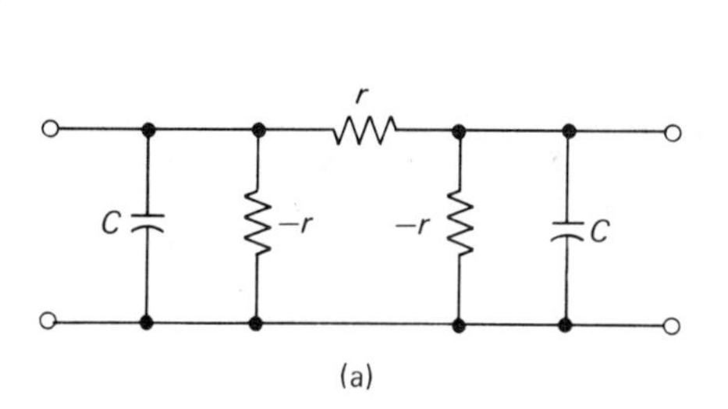

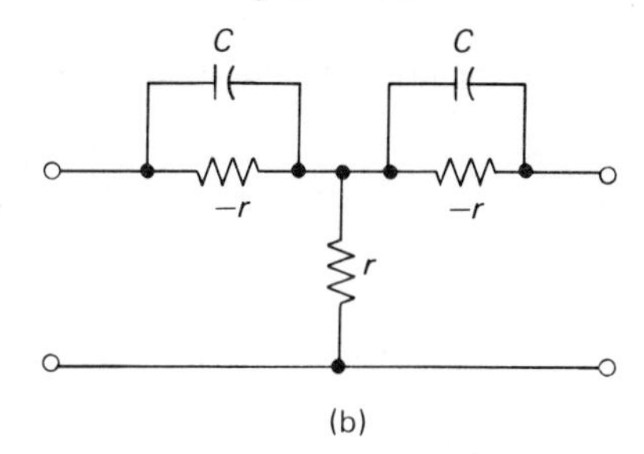

Fig. P5.6.

5.7 The maximum potential instability of an active two-port network is defined as that condition in which the network is unstable with the maximum possible load conductance.[20] Denoting this special value of load conductance by G_l^0, show that in terms of the y-parameters and the source admittance Y_s,

$$G_l^0 = \frac{|y_{21} - y_{12}|^2}{4G_{11}} + \frac{\operatorname{Re} y_{12} \operatorname{Re} y_{21}}{G_{11}} - \operatorname{Re} y_{22}$$

where

$$G_{11} = \operatorname{Re}(y_{11} + Y_s).$$

5.8 Design a frequency-stabilized Colpitts oscillator having a frequency of oscillation 10 KHz and using a transistor with parameters as given in Problem 5.4.

5.9 In Fig. P5.9 we have shown the H- and G-types of feedback oscillators.[21] For each configuration, determine the frequency of oscillation and the condition for sustained oscillations. Suggest specific transistor amplifiers which would be appropriate for the active elements of these two configurations.

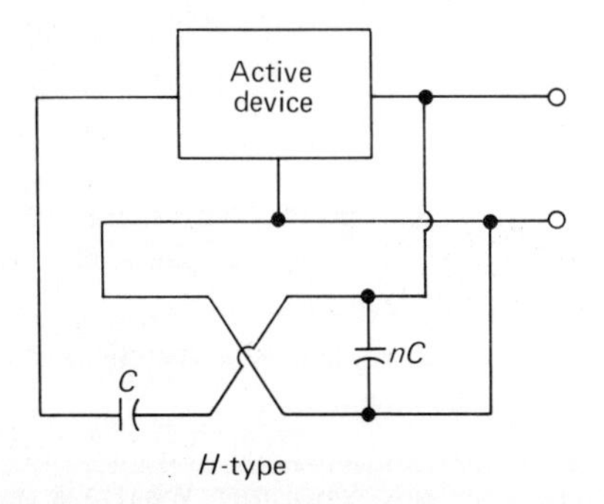

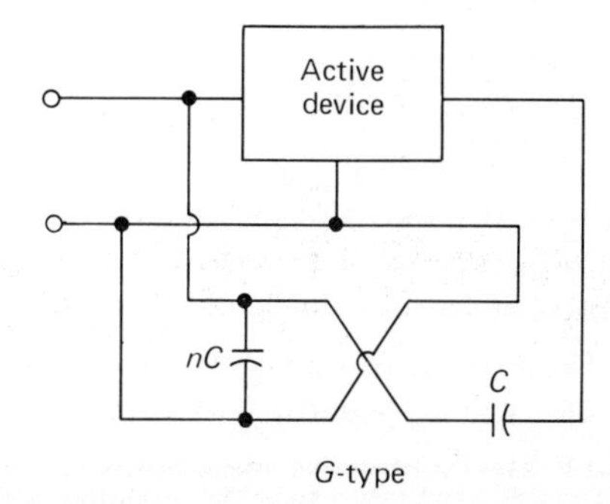

Fig. P5.9.

5.10 In Fig. P5.10 ports 1–1′ and 2–2′ of a two-port network are shorted together. Show that the structure is on the threshold of instability if in terms of the y-parameters

$$\Sigma_y = 0$$

or, equivalently, if in terms of the chain parameters

$$(A - 1)(D - 1) - BC = 0.$$

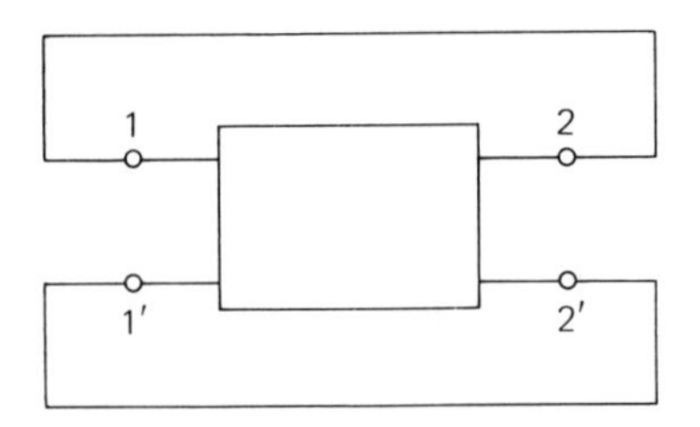

Fig. P5.10.

5.11 Show that in the unilateralized circuit of Fig. 5.28 the transducer gain is a maximum when the turns ratio of the phase-inverting transformer has the optimum value

$$n_{op} = \sqrt{\frac{\text{Re } y_i}{\text{Re } y_o}}$$

5.12 Design a tuned transistor amplifier with a centre frequency of 10 MHz and bandwidth of 500 KHz, given that the measured transistor parameters are as follows

$$\begin{aligned} y_{ie} &= 1{\cdot}5\underline{/43°} \text{ m℧} \\ y_{re} &= -0{\cdot}16\underline{/78°} \text{ m℧} \\ y_{fe} &= 25\underline{/-40°} \text{ m℧} \\ y_{oe} &= 1\underline{/54°} \text{ m℧}. \end{aligned}$$

Assume that the transistor parameters may vary by ± 30 per cent, and the components available for unilateralization have a tolerance of ± 10 per cent. The skew in the selectivity characteristic must not exceed that corresponding to a stability factor $k = 0{\cdot}25$.

5.13 If the y_{fe} of a transistor is assumed to have a phase shift ϕ, determine the effect of ϕ on the stability and frequency response of a single-stage tuned amplifier.

5.14 Fig. P5.14 shows the complete hybrid-π model of a transistor operated with the emitter terminal common. Plot the stability–activity diagram for this transistor. What are the frequencies of potential instability and maximum frequency of oscillation?

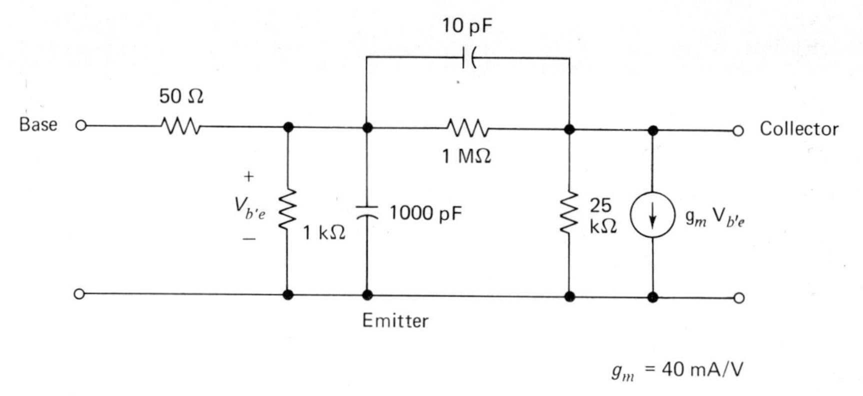
10 pF
50 Ω
Base
Collector
1 MΩ
+
$V_{b'e}$
−
1 kΩ
1000 pF
25 kΩ
$g_m V_{b'e}$
Emitter
g_m = 40 mA/V

Fig. P5.14.

CHAPTER 6

THE SCATTERING MATRIX

6.1 INTRODUCTION

The behaviour of a linear two-port network may be described in terms of the *z*-, *y*-, *h*-, *g*-, or chain-matrix parameters, as was done in the previous two chapters; the choice of a particular parameter set is usually determined by the nature of the problem being considered. However, it becomes increasingly difficult to measure these parameter sets at microwave frequencies where lead inductance and capacitance make short- and open-circuits rather difficult to attain. Moreover, in the case of a transistor, the simulation of a short-circuit frequently causes the device to oscillate under test. These measurement difficulties may be overcome by using the *scattering parameters* which are measured with the device embedded between resistive terminations.

The scattering parameters form a matrix of transformation between incident-wave and reflected-wave amplitudes at the network ports for a prescribed set of port terminations. These wave amplitudes are defined as linear combinations of the port voltages and currents in such a manner as to represent directly the power flow in the network. The scattering parameters are thus found to be particularly useful for handling problems of power transfer in networks designed to be terminated with prescribed loads. Also, scattering parameters exist for all passive, and most active, networks even when the given network does not possess an open-circuit impedance or short-circuit admittance matrix.

We shall begin our study of the scattering matrix formalism of network behaviour by first considering the simple case of a one-port network, and then extend the results obtained to multi-port networks. For the sake of simplicity, however, the examples will deal with networks of frequency-invariant properties or with network calculations at a single frequency.

6.2 SCATTERING RELATIONS FOR A ONE-PORT NETWORK

Consider a linear one-port network of driving-point impedance Z connected to a voltage source V_s of internal resistance R_0, as shown in Fig. 6.1. The resultant port voltage V and current I are generally chosen as the variables of interest. However, we may equally choose any linear transformation of V and I as long

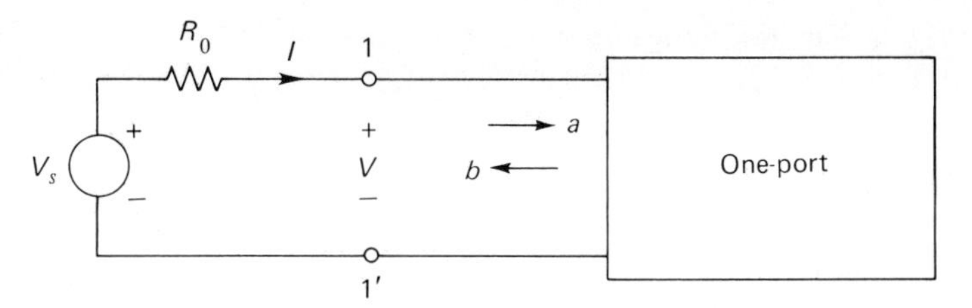

Fig. 6.1. Illustrating the definition of scattering variables of a one-port network.

as the transformation is not singular, i.e., as long as the inverse transformation exists. We may thus define *incident* and *reflected power waves* a and b, with reference to the resistance R_0, which is assumed to be positive, as follows,[1]

$$\begin{aligned} a &= \frac{1}{2\sqrt{R_0}}(V + R_0 I) \\ b &= \frac{1}{2\sqrt{R_0}}(V - R_0 I). \end{aligned} \tag{6.1}$$

It is apparent that the units of the variables a and b are (volt-amperes)$^{1/2}$. With a fixed R_0, if V and I are given, a and b are readily calculated from Eqs. (6.1). On the other hand, if a and b are given, V and I are obtained from the inverse transformation

$$\begin{aligned} V &= \sqrt{R_0}(a + b) \\ I &= \frac{1}{\sqrt{R_0}}(a - b). \end{aligned} \tag{6.2}$$

Thus, any result in terms of one set of variables can readily be converted to that in terms of the other set of variables.

The ratio of reflected to incident wave for a one-port network is defined as the *reflection coefficient* or *scattering parameter* of the network; thus,

$$\rho = \frac{b}{a}. \tag{6.3}$$

Using Eqs. (6.1) and (6.3) we obtain

$$\rho = \frac{Z - R_0}{Z + R_0}, \tag{6.4}$$

where $Z = V/I$ is the driving-point impedance of the network measured at port 1–1′. When the reference resistance R_0 and the real part of the load Z have the same sign, $|\rho| < 1$ and when they have opposite signs, $|\rho| > 1$. When the load is purely reactive, $|\rho| = 1$.

Referring to Fig. 6.1, we see that

$$V = V_s - R_0 I. \tag{6.5}$$

Substituting this into the first relation of Eq. (6.1), and evaluating the square of the magnitude, we get

$$|a|^2 = \frac{|V_s|^2}{4R_0}, \tag{6.6}$$

which is recognized as the average value of the maximum available power of the source, resulting when the load is matched to the source (i.e., $Z = R_0$). When V_s is set equal to zero, the incident wave a becomes zero also.

The average value of the actual power entering the one-port network, P_l, is equal to $\text{Re}\,(VI^*)$; hence, from Eqs. (6.2) we obtain

$$\begin{aligned} P_l &= |a|^2 - |b|^2 \\ &= |a|^2(1 - |\rho|^2). \end{aligned} \tag{6.7}$$

Equation (6.7) shows that when the load is not matched, i.e., the condition $Z = R_0$ is not satisfied, a part of the incident power $|a|^2$ is reflected back to the source. This reflected power is given by $|b|^2$ so that the net power absorbed by the network is equal to $|a|^2 - |b|^2$. In other words, $|\rho|^2$ provides a quantitative measure of the deviation of the actual network behaviour from the matched condition.

Moreover, from Eq. (6.7) it is apparent that if the one-port network is passive, then at any real frequency ω, the magnitude of the reflection coefficient is bounded by unity, as shown by

$$|\rho(j\omega)| \leq 1. \tag{6.8}$$

Otherwise, the network is active.

6.3 THE SCATTERING REPRESENTATION OF AN n-PORT NETWORK

The basic definitions of Eqs. (6.1) pertaining to the incident and reflected waves in a one-port may readily be extended to an n-port network by employing matrix elements rather than scalars. Thus, consider a linear n-port network loaded as in Fig. 6.2 and let $\mathbf{a}$, $\mathbf{b}$, $\mathbf{V}$ and $\mathbf{I}$ be $n \times 1$ column vectors whose jth elements are respectively, a_j, b_j, V_j, I_j at the jth port of the network. Then, $\mathbf{a}$ and $\mathbf{b}$ may be expressed in terms of $\mathbf{V}$ and $\mathbf{I}$ as follows:

$$\begin{aligned} \mathbf{a} &= \tfrac{1}{2}\mathbf{R}_0^{-1/2}(\mathbf{V} + \mathbf{R}_0\mathbf{I}) \\ \mathbf{b} &= \tfrac{1}{2}\mathbf{R}_0^{-1/2}(\mathbf{V} - \mathbf{R}_0\mathbf{I}), \end{aligned} \tag{6.9}$$

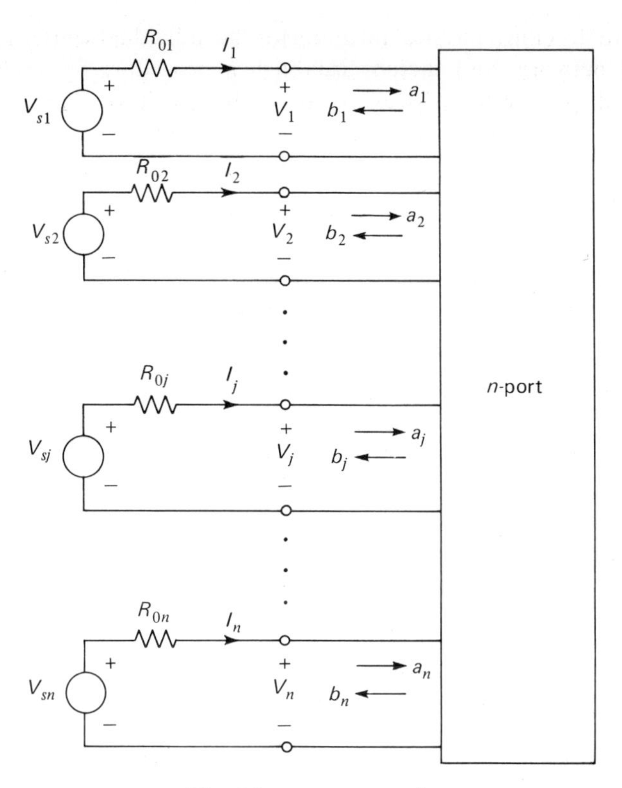

Fig. 6.2. n-port network.

where $\mathbf{R}_0$ is an $n \times n$ diagonal matrix whose jth diagonal element is given by R_{0j}. It is assumed that $R_{0j} > 0$ for $j = 1, 2, \ldots, n$, so that $\mathbf{R}_0$ is positive definite.

Since there is a linear relation between $\mathbf{V}$ and $\mathbf{I}$ given by

$$\mathbf{V} = \mathbf{ZI}, \tag{6.10}$$

where $\mathbf{Z}$ is the *open-circuit impedance matrix* of the n-port network, and since $\mathbf{a}$ and $\mathbf{b}$ are the results of a linear transformation of $\mathbf{V}$ and $\mathbf{I}$, it follows that there must be a linear relation between $\mathbf{a}$ and $\mathbf{b}$. This relation may be expressed in the matrix form

$$\mathbf{b} = \mathbf{Sa}, \tag{6.11}$$

where $\mathbf{S}$ is called the *scattering matrix* of the n-port network. The various components which make up $\mathbf{S}$ are called the *scattering parameters* of the n-port network. It should be noted that the scattering matrix is defined with respect to the reference resistance matrix $\mathbf{R}_0$ which is quite arbitrary; however, once $\mathbf{R}_0$ is chosen, it is fixed.

In order to develop a physical meaning for the individual scattering parameters of an n-port network, and thereby establish a procedure for their evaluation, suppose the jth port of the network is driven by a voltage source and the other ports are terminated with their respective reference resistances, as shown in Fig. 6.3. Then, we have

$$V_j = V_{sj} - R_{0j}I_j \tag{6.12}$$

and

$$V_k = -R_{0k}I_k, \quad \text{for} \quad k \neq j. \tag{6.13}$$

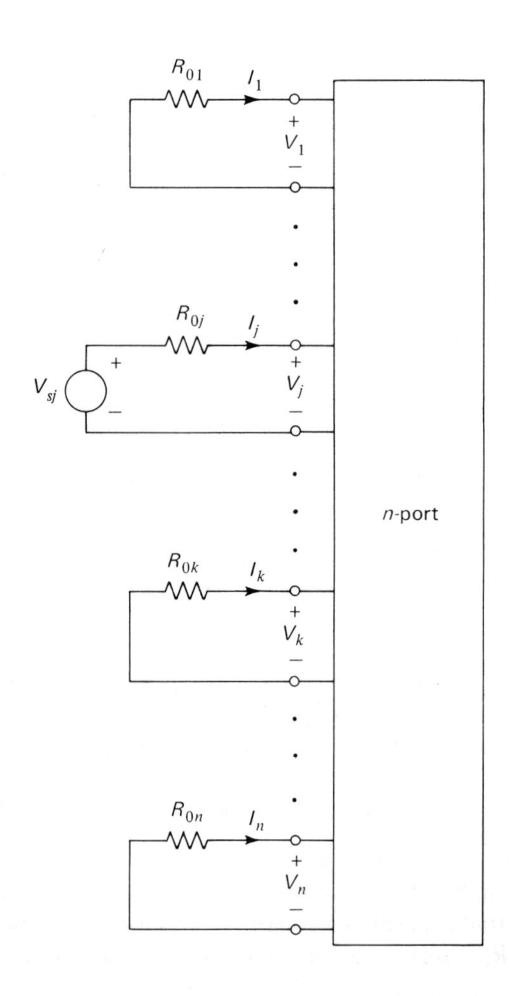

Fig. 6.3. Illustrating the meaning of scattering parameters of an n-port network.

According to the definition of an incident wave, the constraints of Eqs. (6.12) and (6.13), respectively, are equivalent to

$$a_j = \frac{V_{sj}}{2\sqrt{R_{0j}}} \tag{6.14}$$

and

$$a_k = 0, \quad \text{for} \quad k \neq j. \tag{6.15}$$

Hence, the scattering equations for Fig. 6.3 become

$$\begin{aligned} b_1 &= S_{11}a_1 + S_{12}a_2 + \cdots + S_{1n}a_n = S_{1j}a_j \\ b_2 &= S_{21}a_1 + S_{22}a_2 + \cdots + S_{2n}a_n = S_{2j}a_j \\ &\vdots \qquad\qquad\qquad\qquad\qquad\qquad\qquad \vdots \\ b_n &= S_{n1}a_1 + S_{n2}a_2 + \cdots + S_{nn}a_n = S_{nj}a_j \end{aligned} \tag{6.16}$$

from which the following definitions for the scattering parameters of an n-port are obtained

$$S_{jj} = \frac{b_j}{a_j}, \quad \text{for} \quad a_k = 0 \quad \text{and} \quad j \neq k \tag{6.17}$$

and

$$S_{kj} = \frac{b_k}{a_j}, \quad \text{for} \quad a_k = 0 \quad \text{and} \quad j \neq k. \tag{6.18}$$

In particular, Eq. (6.17) shows that each diagonal element S_{jj} of the scattering matrix is equal to the reflection coefficient at the jth port when the n-port network is terminated in its reference resistances; that is,

$$S_{jj} = \frac{Z_{jj} - R_{0j}}{Z_{jj} + R_{0j}}, \tag{6.19}$$

where $Z_{jj} = V_j/I_j$ is the driving-point impedance measured at the jth port when all other ports of the network are terminated in their reference resistances.

The parameter S_{kj}, for $k \neq j$, is called the *transmission coefficient* from port j to port k. To evaluate it, we first note that according to the definition of a reflected wave, the constraint of Eq. (6.13) is equivalent to

$$b_k = -\sqrt{R_{0k}}\, I_k, \qquad \text{for} \quad k \neq j. \tag{6.20}$$

Hence, using Eqs. (6.14), (6.18) and (6.20), we obtain

$$S_{kj} = -2\sqrt{R_{0j}R_{0k}}\,\frac{I_k}{V_{sj}}. \tag{6.21}$$

The squared magnitude of this expression, i.e., $|S_{kj}|^2$ is recognized to be the transducer power gain of the terminated network from the source at port j to the load R_{0k} at port k.

If the n-port network is reciprocal, $S_{kj} = S_{jk}$, that is, the scattering matrix is symmetric, and this condition is written as

$$\mathbf{S} = \mathbf{S}^t, \tag{6.22}$$

where $\mathbf{S}^t$ denotes the transpose of the scattering matrix.

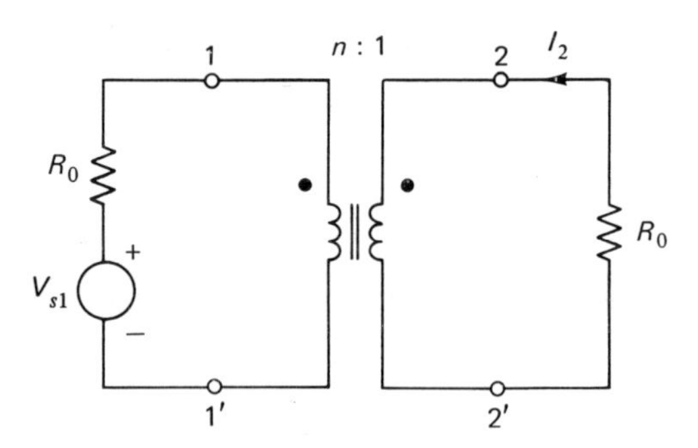

Fig. 6.4. Terminated ideal transformer.

Example 6.1. *Ideal transformer.* Consider an ideal transformer of turns ratio $n{:}1$, operating between equal terminating resistances, as shown in Fig. 6.4. To evaluate S_{11}, we note that with port 2 terminated in R_0, the input impedance is equal to $n^2 R_0$; hence, from Eq. (6.19)

$$S_{11} = \frac{n^2 - 1}{n^2 + 1}. \tag{6.23}$$

Next, to evaluate S_{21}, we note that with a voltage source V_{s1} driving port 1, the resultant current at port 2 is

$$I_2 = -\frac{nV_{s1}}{R_0(n^2 + 1)}. \tag{6.24}$$

Hence, from Eq. (6.21),

$$S_{21} = \frac{2n}{n^2 + 1}. \tag{6.25}$$

In a similar manner, we may evaluate the remaining scattering parameters S_{12} and S_{22}. The scattering matrix for the ideal transformer with respect to the reference resistances shown in Fig. 6.4 is therefore

$$\begin{bmatrix} S_{11} & S_{12} \\ S_{21} & S_{22} \end{bmatrix} = \frac{1}{n^2+1}\begin{bmatrix} n^2-1 & 2n \\ 2n & 1-n^2 \end{bmatrix}. \tag{6.26}$$

In the matched case (i.e., when $n = 1$), Eq. (6.26) simplifies as

$$\begin{bmatrix} S_{11} & S_{12} \\ S_{21} & S_{22} \end{bmatrix} = \begin{bmatrix} 0 & 1 \\ 1 & 0 \end{bmatrix}. \tag{6.27}$$

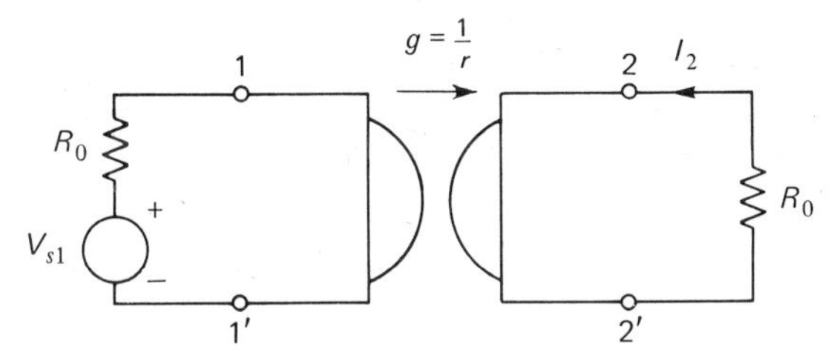

Fig. 6.5. Terminated ideal gyrator.

Example 6.2. *Ideal gyrator.* Consider, next, an ideal gyrator of gyration resistance r operating between equal resistances, as shown in Fig. 6.5. With port 2 terminated in R_0, the input impedance is equal to r^2/R_0, so that

$$S_{11} = \frac{r^2 - R_0^2}{r^2 + R_0^2}. \tag{6.28}$$

With a voltage source V_{s1} driving port 1, the resultant output current is equal to $-2rV_{s1}/(r^2 + R_0^2)$, so that

$$S_{21} = \frac{2rR_0}{r^2 + R_0^2}. \tag{6.29}$$

We may similarly evaluate S_{12} and S_{22}. The scattering matrix of the ideal gyrator is therefore

$$\begin{bmatrix} S_{11} & S_{12} \\ S_{21} & S_{22} \end{bmatrix} = \frac{1}{r^2 + R_0^2}\begin{bmatrix} r^2 - R_0^2 & -2rR_0 \\ 2rR_0 & r^2 - R_0^2 \end{bmatrix}. \tag{6.30}$$

When $r = R_0$, the network is matched and Eq. (6.30) simplifies as

$$\begin{bmatrix} S_{11} & S_{12} \\ S_{21} & S_{22} \end{bmatrix} = \begin{bmatrix} 0 & -1 \\ 1 & 0 \end{bmatrix}. \tag{6.31}$$

Example 6.3. *Hybrid coil.* The hybrid coil consists of an ideal three-winding transformer network with four ports, as shown in Fig. 6.6. It is assumed that the

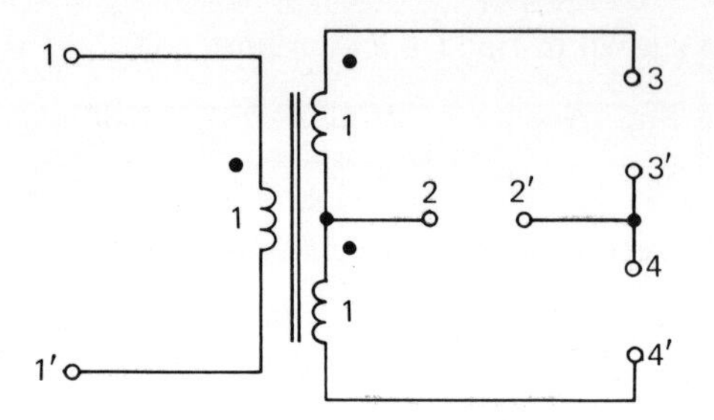

Fig. 6.6. Hybrid coil.

reference resistance matrix is specified as

$$\mathbf{R}_0 = \begin{bmatrix} kR_0 & 0 & 0 & 0 \\ 0 & kR_0 & 0 & 0 \\ 0 & 0 & R_0 & 0 \\ 0 & 0 & 0 & R_0 \end{bmatrix}. \tag{6.32}$$

Using the definitions of Eqs. (6.19) and (6.21), the scattering matrix of the hybrid coil with respect to this resistance matrix is obtained as[2]

$$\mathbf{S} = \begin{bmatrix} \alpha & 0 & \beta & \beta \\ 0 & \alpha & \beta & -\beta \\ \beta & \beta & -\alpha & 0 \\ \beta & -\beta & 0 & -\alpha \end{bmatrix}, \tag{6.33}$$

with

$$\alpha = \frac{1 - 2k}{1 + 2k}, \tag{6.34}$$

$$\beta = \frac{2\sqrt{k}}{1 + 2k}. \tag{6.35}$$

When the network is matched, there are no reflections at the four ports, so that $S_{11} = S_{22} = S_{33} = S_{44} = 0$. In Eq. (6.33) this occurs when $\alpha = 0$ or $k = \frac{1}{2}$. Thus, if in Fig. 6.6 the ports 1 and 2 are terminated in resistors of $R_0/2$, while ports 3 and 4 are terminated in resistors of R_0, the network is perfectly matched at all ports. With $\alpha = 0$ and $\beta = 1/\sqrt{2}$, the resultant scattering matrix of a matched hybrid coil is evidently

$$\mathbf{S} = \frac{1}{\sqrt{2}} \begin{bmatrix} 0 & 0 & 1 & 1 \\ 0 & 0 & 1 & -1 \\ 1 & 1 & 0 & 0 \\ 1 & -1 & 0 & 0 \end{bmatrix}. \tag{6.36}$$

6.4 CRITERION FOR PASSIVITY OF AN n-PORT NETWORK

The average power P_j entering an n-port network through port j is equal to $|a_j|^2 - |b_j|^2$. Thus, the total average power absorbed by the network is given by

$$\sum_{j=1}^{n} P_j = \sum_{j=1}^{n} |a_j|^2 - |b_j|^2$$
$$= \mathbf{a}^+\mathbf{a} - \mathbf{b}^+\mathbf{b}. \tag{6.37}$$

Substitution of Eq. (6.11) in (6.37) yields

$$\sum_{j=1}^{n} P_j = \mathbf{a}^+(\mathbf{1} - \mathbf{S}^+\mathbf{S})\mathbf{a}, \tag{6.38}$$

where $\mathbf{1}$ is an $n \times n$ unit matrix, and the superscript $+$ indicates the complex-conjugate transposed matrix. Since $\mathbf{a}$ is arbitrary, Eq. (6.38) reveals that if the n-port network is passive, then at all frequencies,

$$\mathbf{1} - \mathbf{S}^+\mathbf{S} \geq \mathbf{0}, \tag{6.39}$$

where $\mathbf{0}$ is an $n \times n$ null matrix.

If, in addition, the n-port network is lossless, then whether the network is reciprocal or not,

$$\mathbf{S}^+\mathbf{S} = \mathbf{1}. \tag{6.40}$$

A matrix which satisfies Eq. (6.40) is called a *unitary matrix*. An important property of a unitary matrix $\mathbf{S}$ is that if $\mathbf{t}_i$ is the ith column vector of $\mathbf{S}$, then

$$\mathbf{t}_i^+\mathbf{t}_j = \delta_{ij}, \tag{6.41}$$

where δ_{ij} is the *Kronecker delta* defined by

$$\delta_{ij} = \begin{cases} 0, & \text{for} \quad i \neq j \\ 1, & \text{for} \quad i = j. \end{cases}$$

Scattering Properties of a Lossless Two-port Network

Consider a lossless two-port network with the scattering matrix,

$$\mathbf{S} = \begin{bmatrix} S_{11} & S_{12} \\ S_{21} & S_{22} \end{bmatrix}.$$

For the network to be lossless, $\mathbf{S}$ must be a unitary matrix. If, therefore, we apply Eq. (6.40), or equivalently Eq. (6.41), to the two-port network, we obtain three independent equations:

$$\begin{aligned} |S_{11}|^2 + |S_{21}|^2 &= 1, \\ |S_{12}|^2 + |S_{22}|^2 &= 1, \\ S_{11}^*S_{12} + S_{21}^*S_{22} &= 0. \end{aligned} \tag{6.42}$$

From the last condition, we have

$$S_{12} = -\frac{S_{22}}{S_{11}^*} S_{21}^* . \tag{6.43}$$

If, next, we subtract the first condition from the second, and then use Eq. (6.43) and simplify, we get

$$|S_{11}| = |S_{22}| . \tag{6.44}$$

Equation (6.44) shows that, for a resistively terminated lossless two-port network, the magnitude of the reflection coefficient at one port is equal to that at the other port.

Also, from Eqs. (6.43) and (6.44),

$$|S_{12}| = |S_{21}| . \tag{6.45}$$

Thus, in a lossless network the magnitude of the transmission coefficient is the same in both directions, so that in such a network non-reciprocity can only arise by virtue of the fact that the phase angles of S_{12} and S_{21} need not be equal.

It is obvious that an ideal transformer, which is reciprocal and characterized by the scattering matrix of Eq. (6.26), and the ideal gyrator, which is non-reciprocal and characterized by the scattering matrix of Eq. (6.30), satisfy the conditions and consequences of Eqs. (6.42), thereby confirming the lossless nature of both devices.

6.5 SCATTERING–IMPEDANCE RELATIONS

Elimination of the vectors **a**, **b** and **V** from Eqs. (6.9)–(6.11) yields

$$\mathbf{R}_0^{-1/2}(\mathbf{Z} - \mathbf{R}_0)\mathbf{I} = \mathbf{S}\mathbf{R}_0^{-1/2}(\mathbf{Z} + \mathbf{R}_0)\mathbf{I},$$

from which the following expression for the scattering matrix **S** in terms of the open-circuit impedance matrix of the n-port network, **Z**, and the reference resistance matrix $\mathbf{R}_0$ is obtained,

$$\mathbf{S} = \mathbf{R}_0^{-1/2}(\mathbf{Z} - \mathbf{R}_0)(\mathbf{Z} + \mathbf{R}_0)^{-1}\mathbf{R}_0^{1/2} \tag{6.46}$$

or equivalently,

$$\mathbf{S} = \mathbf{1} - 2\,\mathbf{R}_0^{1/2}(\mathbf{Z} + \mathbf{R}_0)^{-1}\mathbf{R}_0^{1/2} \tag{6.47}$$

where **1** is an $n \times n$ unit matrix. Conversely, we may express **Z** in terms of **S**, as shown by

$$\mathbf{Z} = \mathbf{R}_0^{1/2}(\mathbf{1} - \mathbf{S})^{-1}(\mathbf{1} + \mathbf{S})\mathbf{R}_0^{1/2}. \tag{6.48}$$

In a similar manner, we can obtain relations between the scattering and short-circuit admittance matrices of the n-port network by noting that the short-circuit admittance and open-circuit impedance matrices are the inverses of each other.

Scattering Parameter Relationships for a Two-port Network

To illustrate the application of Eqs. (6.46) and (6.48), consider a two-port network terminated at its input and output ports in the resistances R_{01} and R_{02}, respectively. Let

$$\mathbf{Z} = \begin{bmatrix} z_{11} & z_{12} \\ z_{21} & z_{22} \end{bmatrix}$$

denote the open-circuit impedance (i.e., z-) matrix of the two-port network normalized with respect to R_{01} and R_{02}. To achieve this normalization, the input and output driving-point impedance parameters are divided by R_{01} and R_{02}, respectively, while the forward and reverse transfer impedance parameters are both divided by $\sqrt{R_{01}R_{02}}$. Then, applying Eqs. (6.46) and (6.48), we obtain the results listed in Table 6.1.

This table also includes the interrelations between the scattering and normalized short-circuit admittance (i.e., y-) parameters of the two-port network. To normalize the y-parameters with respect to the reference resistances R_{01} and R_{02}, the input and output driving-point admittance parameters are multiplied

Table 6.1. Conversion Relations For Scattering Parameters

S-parameters in terms of z- and y-parameters	z- and y-parameters in terms of scattering parameters
$S_{11} = \dfrac{(z_{11} - 1)(z_{22} + 1) - z_{12}z_{21}}{(z_{11} + 1)(z_{22} + 1) - z_{12}z_{21}}$	$z_{11} = \dfrac{(1 + S_{11})(1 - S_{22}) + S_{12}S_{21}}{(1 - S_{11})(1 - S_{22}) - S_{12}S_{21}}$
$S_{12} = \dfrac{2z_{12}}{(z_{11} + 1)(z_{22} + 1) - z_{12}z_{21}}$	$z_{12} = \dfrac{2S_{12}}{(1 - S_{11})(1 - S_{22}) - S_{12}S_{21}}$
$S_{21} = \dfrac{2z_{21}}{(z_{11} + 1)(z_{22} + 1) - z_{12}z_{21}}$	$z_{21} = \dfrac{2S_{21}}{(1 - S_{11})(1 - S_{22}) - S_{12}S_{21}}$
$S_{22} = \dfrac{(z_{11} + 1)(z_{22} - 1) - z_{12}z_{21}}{(z_{11} + 1)(z_{22} + 1) - z_{12}z_{21}}$	$z_{22} = \dfrac{(1 + S_{22})(1 - S_{11}) + S_{12}S_{21}}{(1 - S_{11})(1 - S_{22}) - S_{12}S_{21}}$
$S_{11} = \dfrac{(1 - y_{11})(1 + y_{22}) - y_{12}y_{21}}{(1 + y_{11})(1 + y_{22}) - y_{12}y_{21}}$	$y_{11} = \dfrac{(1 + S_{22})(1 - S_{11}) + S_{12}S_{21}}{(1 + S_{11})(1 + S_{22}) - S_{12}S_{21}}$
$S_{12} = \dfrac{-2y_{12}}{(1 + y_{11})(1 + y_{22}) - y_{12}y_{21}}$	$y_{12} = \dfrac{-2S_{12}}{(1 + S_{11})(1 + S_{22}) - S_{12}S_{21}}$
$S_{21} = \dfrac{-2y_{21}}{(1 + y_{11})(1 + y_{22}) - y_{12}y_{21}}$	$y_{21} = \dfrac{-2S_{21}}{(1 + S_{11})(1 + S_{22}) - S_{12}S_{21}}$
$S_{22} = \dfrac{(1 + y_{11})(1 - y_{22}) - y_{12}y_{21}}{(1 + y_{11})(1 + y_{22}) - y_{12}y_{21}}$	$y_{22} = \dfrac{(1 + S_{11})(1 - S_{22}) + S_{12}S_{21}}{(1 + S_{11})(1 + S_{22}) - S_{12}S_{21}}$

by R_{01} and R_{02}, respectively, while the forward and reverse transfer admittance parameters are both multiplied by $\sqrt{R_{01}R_{02}}$.

6.6 NETWORK CALCULATIONS WITH SCATTERING PARAMETERS

The performance of an n-port network terminated in arbitrary impedances not equal to the reference impedances may readily be determined in terms of scattering parameters by supposing that the terminating impedances act as generators of incident and reflected waves at the pertinent ports. Thus, if Z_k denotes the actual terminating impedance at port k, the reflection coefficient of this load with respect to the reference resistance R_{0k} is given by

$$\rho_k = \frac{Z_k - R_{0k}}{Z_k + R_{0k}}. \tag{6.49}$$

Since, for the terminations, the roles of the network input and output waves are interchanged, we may write

$$\rho_k = \frac{a_k}{b_k}. \tag{6.50}$$

These are the boundary conditions to be inserted into Eq. 6.11 in order to determine the resultant network performance.

Properties of an Arbitrarily Terminated Two-port Network

Consider a two-port network terminated in a load Z_l at port 2, as shown in Fig. 6.7. The scattering relations describing the two-port network, written in full, are

$$\begin{bmatrix} b_1 \\ b_2 \end{bmatrix} = \begin{bmatrix} S_{11} & S_{12} \\ S_{21} & S_{22} \end{bmatrix} \begin{bmatrix} a_1 \\ a_2 \end{bmatrix}, \tag{6.51}$$

in which it is assumed that the scattering parameters have been computed with respect to the reference resistance matrix

$$\mathbf{R}_0 = \begin{bmatrix} R_{01} & 0 \\ 0 & R_{02} \end{bmatrix}.$$

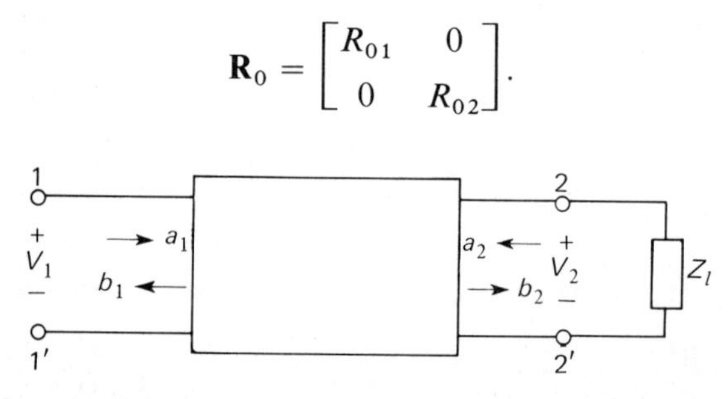

Fig. 6.7. Two-port network terminated at port 2.

The load may thus be described by the reflection coefficient ρ_l defined as

$$\rho_l = \frac{Z_l - R_{02}}{Z_l + R_{02}}. \tag{6.52}$$

For the structure of Fig. 6.7, we may determine the input reflection coefficient $\rho_{in} = b_1/a_1$ with respect to R_{01} by inserting the constraint $a_2 = \rho_l b_2$ into Eq. (6.51) and then solving for the ratio b_1/a_1:

$$\rho_{in} = S_{11} + \frac{\rho_l S_{12} S_{21}}{1 - \rho_l S_{22}}. \tag{6.53}$$

Also, the forward transmission coefficient $\tau_{21} = b_2/a_1$ of the terminated network from port 1 to port 2 is given by

$$\tau_{21} = \frac{S_{21}}{1 - \rho_l S_{22}}. \tag{6.54}$$

If the termination is a perfect match (i.e., $Z_l = R_{02}$ or $\rho_l = 0$) we obtain the expected results $\rho_{in} = S_{11}$ and $\tau_{21} = S_{21}$.

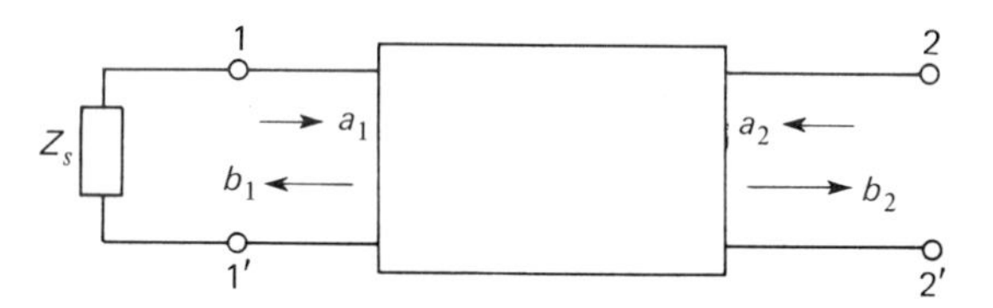

Fig. 6.8. Two-port network terminated at port 1.

Similarly, if the network is terminated with an arbitrary impedance Z_s at port 1, as in Fig. 6.8, the resultant output reflection coefficient $\rho_{out} = b_2/a_2$ is

$$\rho_{out} = S_{22} + \frac{\rho_s S_{12} S_{21}}{1 - \rho_s S_{11}} \tag{6.55}$$

and the reverse transmission coefficient $\tau_{12} = b_1/a_2$ from port 2 to port 1 is

$$\tau_{12} = \frac{S_{12}}{1 - \rho_s S_{11}}, \tag{6.56}$$

where ρ_s is the source reflection coefficient defined by

$$\rho_s = \frac{Z_s - R_{01}}{Z_s + R_{01}}. \tag{6.57}$$

Here again, if the termination is a perfect match (i.e., $Z_s = R_{01}$ or $\rho_s = 0$) we obtain $\rho_{out} = S_{22}$ and $\tau_{12} = S_{12}$, as expected.

Returning to the situation depicted in Fig. 6.7, the voltage gain $K_v = V_2/V_1$ of the network with an arbitrary load Z_l is determined by observing that

$$V_1 = \sqrt{R_{01}}(a_1 + b_1) = a_1\sqrt{R_{01}}(1 + \rho_{in}) \tag{6.58}$$

and

$$V_2 = \sqrt{R_{02}}(a_2 + b_2) = b_2\sqrt{R_{02}}(1 + \rho_l). \tag{6.59}$$

Therefore, using Eqs. (6.58) and (6.59), together with (6.53) and (6.54), we obtain

$$K_v = \frac{V_2}{V_1} = \sqrt{\frac{R_{02}}{R_{01}}}\frac{S_{21}(1 + \rho_l)}{1 + S_{11} - \rho_l(S_{22} + \Delta_S)}, \tag{6.60}$$

where

$$\Delta_S = S_{11}S_{22} - S_{12}S_{21}. \tag{6.61}$$

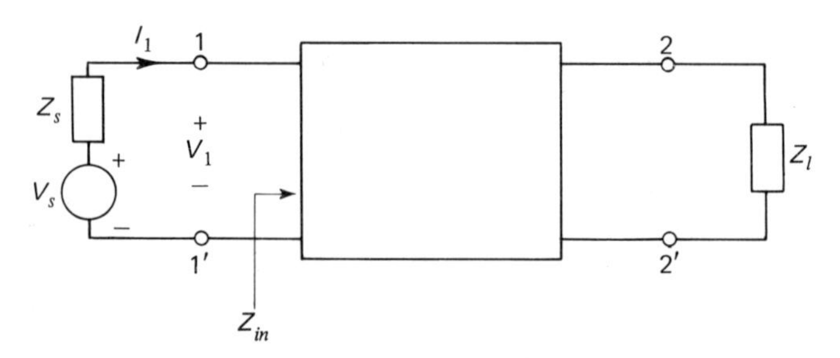

Fig. 6.9. Double-terminated two-port network.

In the double-terminated system of Fig. 6.9, a parameter of paı ticular interest is the transducer power gain G_T defined as the ratio of power delivered to the load, P_l, to power available from the source, P_{avs}. Since the wave incident on the load is equal to b_2 it follows from Eq. (6.7) that

$$P_l = |b_2|^2(1 - |\rho_l|^2). \tag{6.62}$$

Referring to Fig. 6.9, the power available from the source is

$$P_{avs} = \frac{|V_s|^2}{4 \operatorname{Re} Z_s}. \tag{6.63}$$

Thus, Eqs. (6.62) and (6.63), together with (6.53) and (6.54), give

$$G_T = \frac{P_l}{P_{avs}} = 4(1 - |\rho_l|^2) \operatorname{Re} Z_s \left|\frac{a_1 S_{21}}{V_s(1 - \rho_l S_{22})}\right|^2. \tag{6.64}$$

To determine a_1 in terms of V_s, consider the situation depicted in Fig. 6.10 where Z_{in} denotes the input impedance of the network terminated with Z_l. The input

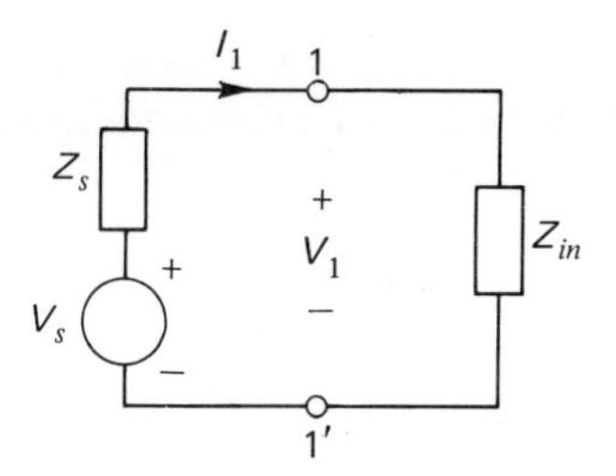

Fig. 6.10. Illustrating the input-port conditions.

voltage and current are evidently

$$V_1 = \frac{Z_{in}V_s}{Z_{in} + Z_s} \tag{6.65}$$

$$I_1 = \frac{V_s}{Z_{in} + Z_s}. \tag{6.66}$$

The resultant incident wave at the input port is therefore

$$a_1 = \frac{1}{2\sqrt{R_{01}}}(V_1 + R_{01}I_1) = \frac{V_s(Z_{in} + R_{01})}{2\sqrt{R_{01}}(Z_{in} + Z_s)}. \tag{6.67}$$

Substituting Eq. (6.67) in (6.64), and noting that $Z_{in} = R_{01}(1 + \rho_{in})/(1 - \rho_{in})$ and $Z_s = R_{01}(1 + \rho_s)/(1 - \rho_s)$, the following expression for the transducer power gain is obtained

$$G_T = \frac{|S_{21}|^2(1 - |\rho_s|^2)(1 - |\rho_l|^2)}{|(1 - \rho_l S_{22})(1 - \rho_s \rho_{in})|^2}$$

or

$$G_T = \frac{|S_{21}|^2(1 - |\rho_s|^2)(1 - |\rho_l|^2)}{|(1 - \rho_s S_{11})(1 - \rho_l S_{22}) - \rho_s \rho_l S_{12} S_{21}|^2}. \tag{6.68}$$

Equation (6.68) indicates that if, at some frequency, the condition $\rho_s \rho_{in} = 1$ is satisfied, then the transducer power gain becomes infinite, signifying actual instability of the system with resultant oscillation at that frequency. If, on the other hand, $|\rho_{in}\rho_s| < 1$, then the system is stable, regardless of the phase angles of ρ_{in} and ρ_s.

A two-port device is absolutely stable (i.e., stable for all possible uncoupled passive terminations) if the real parts of its input and output impedances remain positive when the passive load and source impedances, respectively, are changed arbitrarily. In terms of the scattering-matrix formalism, this requires $|\rho_{in}|$ to be less than unity when ρ_l is changed arbitrarily, but keeping $|\rho_l| < 1$. Similarly,

$|\rho_{out}| < 1$ is required for all $|\rho_s| < 1$. Using Eqs. (6.53) and (6.55), a little manipulation shows that the necessary and sufficient conditions for the absolute stability of a two-port device, at a specified frequency, are given by[3]

$$\begin{aligned} 1 - |S_{11}|^2 - |S_{12}S_{21}| &\geq 0 \\ 1 - |S_{22}|^2 - |S_{12}S_{21}| &\geq 0 \\ 1 - |S_{11}|^2 - |S_{22}|^2 - 2|S_{12}S_{21}| + |\Delta_S|^2 &\geq 0. \end{aligned} \tag{6.69}$$

In the double-terminated system of Fig. 6.9, the transducer power gain attains its maximum value when the two-port device is conjugate-image matched at both ports simultaneously, that is, $\rho_s = \rho_{in}^*$ and $\rho_l = \rho_{out}^*$. After a rather straightforward but lengthy calculation, we obtain

$$G_{max} = \left|\frac{S_{21}}{S_{12}}\right|(\eta - \sqrt{\eta^2 - 1}), \tag{6.70}$$

where

$$\eta = \frac{1 - |S_{11}|^2 - |S_{22}|^2 + |\Delta_S|^2}{2|S_{12}S_{21}|}. \tag{6.71}$$

If the two-port device is absolutely stable, it is guaranteed that $\eta > 1$ and G_{max} is thus a meaningful quantity.

For the special case $|S_{12}S_{21}| = 0$, the device is absolutely stable if $|S_{11}| < 1$ and $|S_{22}| < 1$; and the expression for G_{max} simplifies as

$$G_{max} = \frac{|S_{21}|^2}{(1 - |S_{11}|^2)(1 - |S_{22}|^2)}. \tag{6.72}$$

This maximum gain is attained when $\rho_s = S_{11}^*$ and $\rho_l = S_{22}^*$.

Reflection Type of Negative-resistance Amplifier

When a network is terminated with a one-port device that has a driving-point impedance with negative real part (e.g., as exhibited by a tunnel diode), the wave incident upon the termination is reflected at increased amplitude. This rather suggests that signal amplification may be obtained if some means is provided to separate the resultant incident and reflected waves. A common way of implementing this separation process is to use a *three-port circulator*, which, in its ideal form, is a lossless and non-reciprocal device described by the scattering matrix (see Fig. 6.11)

$$\mathbf{S} = \begin{bmatrix} 0 & 0 & 1 \\ 1 & 0 & 0 \\ 0 & 1 & 0 \end{bmatrix}. \tag{6.73}$$

Thus, with matched terminations, power entering port 1 of the circulator all

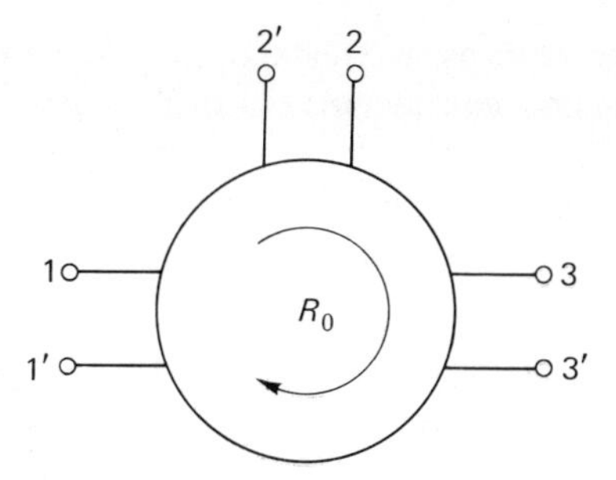

Fig. 6.11. Three-port circulator.

emerges at port 2; power entering port 2 all emerges at port 3; power entering port 3 all emerges at port 1.

Suppose port 2 of the circulator is terminated in a negative resistance $-r_d$ and the network is driven and loaded as shown in Fig. 6.12. The mismatch at port 2 gives rise to a reflection coefficient ρ_d which is greater than unity in magnitude, as shown by

$$\rho_d = \frac{-r_d - R_0}{-r_d + R_0} = \frac{r_d + R_0}{r_d - R_0}, \tag{6.74}$$

where R_0 is the characteristic resistance of the circulator.

With **S** defined as in Eq. (6.73), we may write the scattering relations for the arrangement of Fig. 6.12 as follows

$$\begin{aligned} b_1 &= a_3 \\ b_2 &= a_1 \\ b_3 &= a_2. \end{aligned} \tag{6.75}$$

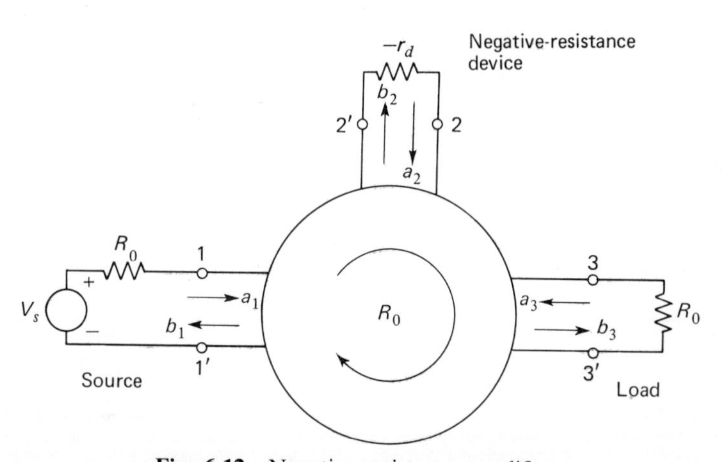

Fig. 6.12. Negative-resistance amplifier.

As for the boundary conditions, we have $a_2 = \rho_d b_2$ and $a_3 = 0$, the latter condition assuming a matched load at port 3. Inserting these boundary conditions into Eq. (6.75), we get $b_1 = 0$ and $b_3 = \rho_d a_1$. Thus, under matched conditions at the input and output ports, the amplifier is unilateral, with a transducer power gain equal to the square of the magnitude of the reflection coefficient ρ_d at port 2.

Concluding Remarks

To conclude, the scattering matrix representation of an n-port network leads to simple relations, namely $\mathbf{b} = \mathbf{S}\mathbf{a}$, between an incident wave vector $\mathbf{a}$ and reflected wave vector $\mathbf{b}$, when the various ports of the network are terminated in their reference resistances. The scattering matrix $\mathbf{S}$ may be interpreted in terms of waves entering or leaving the network under matched conditions. We thus find that the scattering parameters have the nature of a reflection or transmission coefficient. As the conditions for the existence of the scattering matrix are not very restrictive, we often find it possible to use the scattering matrix in situations where other network descriptions would fail. Also, since the scattering matrix is an exclusive property of the n-port network, it is always possible to analyse the system even when the various ports of the network are terminated in arbitrary impedances.

REFERENCES

1. H. J. Carlin, "The Scattering Matrix in Network Theory," *Trans. I.R.E.* **CT-3,** 88 (1956).
2. H. J. Carlin and A. B. Giordano, *Network Theory*. Prentice-Hall, 1964.
3. K. Kurokawa, "Power Waves and the Scattering Matrix," *Trans. I.E.E.E.* **MTT-13,** 194 (1965).
4. H. J. Carlin, "Singular Network Elements," *Trans. I.E.E.E.* **CT-11,** 67 (1964).
5. P. C. J. Hill, "Some Applications of the Scattering Matrix," *Proc. I.E.E.* **112,** 15 (1965).

PROBLEMS

6.1 Evaluate the scattering parameters for the networks shown in Fig. P6.1, assuming equal port terminations.

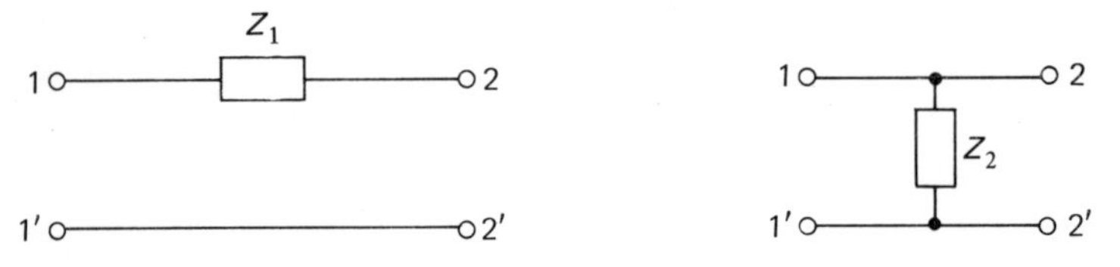

Fig. P6.1

6.2 (a) Show that for the voltage-controlled voltage source of Fig. P6.2(a) the scattering matrix is

$$\mathbf{S} = \begin{bmatrix} 1 & 0 \\ 2\mu & 1 \end{bmatrix}.$$

(b) Show that for the current-controlled current source of Fig. P6.2(b) the scattering matrix is

$$\mathbf{S} = \begin{bmatrix} -1 & 0 \\ 2\alpha & 1 \end{bmatrix}.$$

In both cases, assume the ports are terminated in equal resistances.

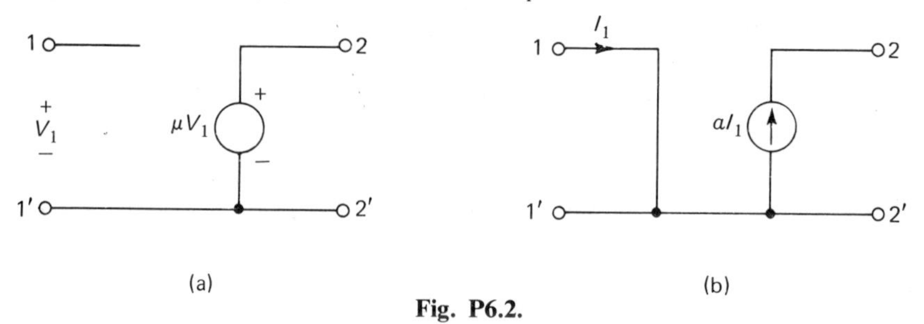

Fig. P6.2.

6.3 Show that the scattering parameters and normalized h-parameters of a two-port network are related by

$$\begin{bmatrix} S_{11} & S_{12} \\ S_{21} & S_{22} \end{bmatrix} = \begin{bmatrix} \dfrac{(h_{11}-1)(h_{22}+1)-h_{12}h_{21}}{(h_{11}+1)(h_{22}+1)-h_{12}h_{21}} & \dfrac{2h_{12}}{(h_{11}+1)(h_{22}+1)-h_{12}h_{21}} \\ \dfrac{-2h_{21}}{(h_{11}+1)(h_{22}+1)-h_{12}h_{21}} & \dfrac{(1+h_{11})(1-h_{22})+h_{12}h_{21}}{(h_{11}+1)(h_{22}+1)-h_{12}h_{21}} \end{bmatrix}$$

$$\begin{bmatrix} h_{11} & h_{12} \\ h_{21} & h_{22} \end{bmatrix} = \begin{bmatrix} \dfrac{(1+S_{11})(1+S_{22})-S_{12}S_{21}}{(1-S_{11})(1+S_{22})+S_{12}S_{21}} & \dfrac{2S_{12}}{(1-S_{11})(1+S_{22})+S_{12}S_{21}} \\ \dfrac{-2S_{21}}{(1-S_{11})(1+S_{22})+S_{12}S_{21}} & \dfrac{(1-S_{22})(1-S_{11})-S_{12}S_{21}}{(1-S_{11})(1+S_{22})+S_{12}S_{21}} \end{bmatrix}.$$

How is the normalization of the h-parameters obtained?

6.4 Evaluate the power gain of a two-port device in terms of the scattering parameters.

6.5 Measurements on a transistor operating at 159 MHz and between 50 Ω terminating resistances yield the following set of scattering parameters:

$$S_{11} = 0{\cdot}24\underline{/-120°}$$
$$S_{12} = 0{\cdot}05\underline{/+110°}$$
$$S_{21} = 2{\cdot}2\underline{/+70°}$$
$$S_{22} = 0{\cdot}8\underline{/-30°}$$

a) Is this device absolute stable?
b) What is the maximum transducer power gain attainable from the device?
c) It is proposed to use this transistor as part of a tuned amplifier operating between 50 Ω terminations, with matching sections as shown in Fig. P6.5. By choosing $\rho_s = S_{11}^*$ and $\rho_l = S_{22}^*$, determine the required element values of both matching sections. What is the transducer power gain of the designed amplifier?

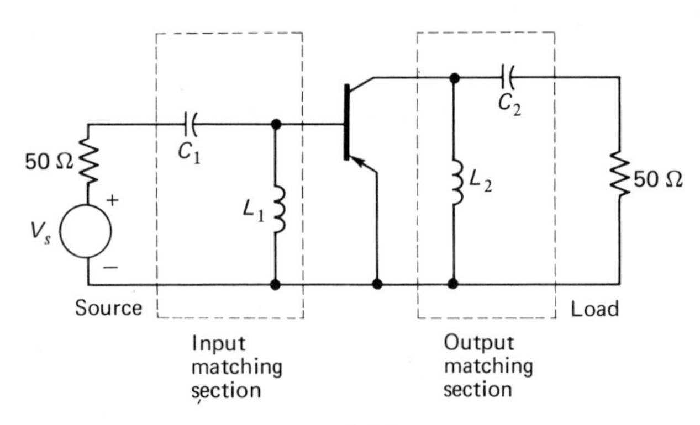

Fig. P6.5.

6.6 a) A *nullator* is a one-port device characterized by $V = I = 0$. Show that the terminated circulator arrangement of Fig. P6.6(a) behaves at port 1 as a nullator.[4]

b) A *norator* is a one-port device in which the voltage and current are completely independent. Show that if the terminations at ports 2 and 3 of the circulator are interchanged, as in Fig. P6.6(b), the resultant behaviour at port 1 is the same as that in a norator.

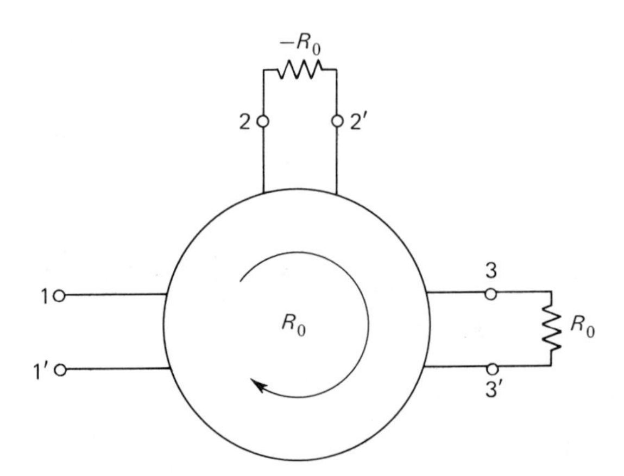

Fig. P6.6(a).

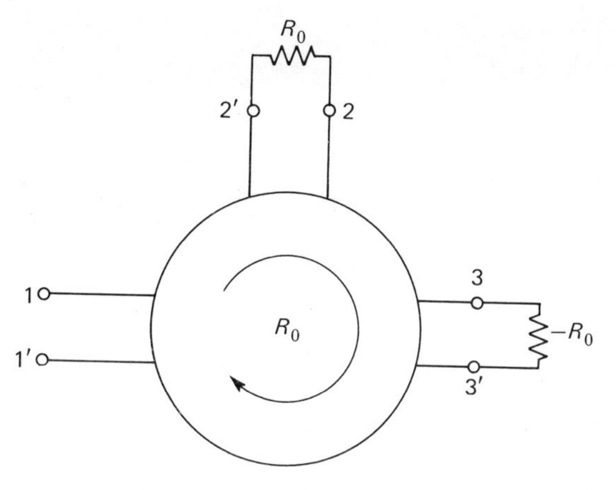

Fig. P6.6(b).

6.7 In a reflection type of negative-resistance amplifier, the load presents a mismatch, that is, $\rho_l \neq 0$. Evaluate the input reflection coefficient and transducer power gain of the resultant amplifier.[5]

6.8 In Fig. P6.8 the input and output ports of a two-port network are directly coupled together. Show that for sustained oscillations,

$$(1 - S_{12})(1 - S_{21}) - S_{11}S_{22} = 0.$$

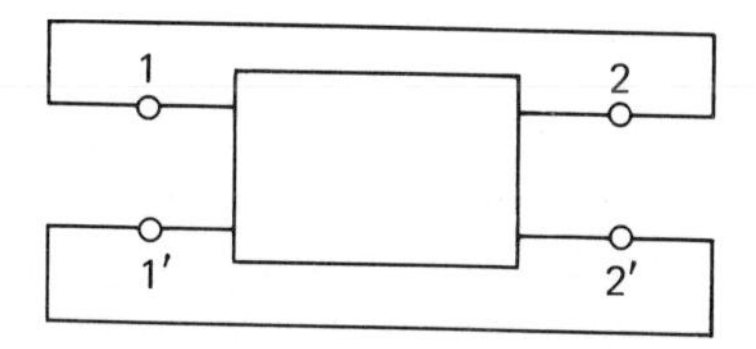

Fig. P6.8.

CHAPTER 7

GAIN–PHASE ANALYSIS

7.1 INTRODUCTION

It is customary practice to specify the dynamic performance of a linear network in terms of its transient response to a generalized excitation, or alternatively in terms of its steady-state frequency response obtained by subjecting the network to a sinusoidal excitation of constant amplitude. It is the purpose of this chapter to examine certain aspects of frequency-response analysis, deferring consideration of the time response to Chapters 8 and 9. When the pole and zero locations of a given system function are known, the associated frequency response is best studied by constructing the well-known *Bode diagram* which is obtained by adding the componental responses arising from the individual poles and zeros. For a more detailed study of the frequency response, however, we shall need to establish a number of integral relationships that are satisfied by the real and imaginary components of certain classes of physical system functions. These relationships, which are derived by utilizing Cauchy's theory of integration in the complex-frequency plane, are found to be useful in estimating the fundamental limitations imposed upon the behaviour of physical networks by parasitic elements, and provide guidelines for their compensation, as encountered in the practical design of wideband and feedback amplifiers. We may also use the relations to compute the imaginary component of the frequency response when the real component is specified in the entire frequency spectrum; or we may choose to specify the real component in one portion of the frequency spectrum and the imaginary component in the remainder of the frequency spectrum, and then use the relations to complete the characteristic by determining the imaginary component in the first range of frequencies and the real component in the second range of frequencies.

7.2 BODE DIAGRAM

The transfer function $H(s)$ of a linear time-invariant network is defined as the ratio of the Laplace transform of the response function $r(t)$ to that of the excitation function $e(t)$, under zero initial conditions. In general, each of $e(t)$ and $r(t)$ can be a voltage or a current, so that $H(s)$ may represent a transfer-voltage ratio, transfer-current ratio, transfer impedance or transfer admittance. The transfer function

$H(s)$ is a real, rational function of the complex-frequency variable s. In terms of its poles and zeros, $H(s)$ can be expressed in the following factored form

$$H(s) = K_0 \frac{\prod_{i=1}^{m} \left(1 - \frac{s}{z_i}\right)}{\prod_{i=1}^{n} \left(1 - \frac{s}{p_i}\right)}, \tag{7.1}$$

where K_0 is a scale factor; $z_1, z_2, \ldots, z_m$ are the zeros of $H(s)$; and $p_1, p_2, \ldots, p_n$ are the poles of $H(s)$. For values of s restricted to the imaginary axis (i.e., $s = j\omega$), we have

$$H(j\omega) = K_0 \frac{\prod_{i=1}^{m} \left(1 - \frac{j\omega}{z_i}\right)}{\prod_{i=1}^{n} \left(1 - \frac{j\omega}{p_i}\right)}. \tag{7.2}$$

The multiplication and division of the various numerator and denominator factors in Eq. (7.2), representing the zeros and poles, are transformed into addition and subtraction by considering the logarithm of $H(j\omega)$ rather than $H(j\omega)$ itself as the function of interest, as shown by

$$\log_\varepsilon H(j\omega) = \log_\varepsilon K_0 + \sum_{i=1}^{m} \log_\varepsilon\left(1 - \frac{j\omega}{z_i}\right) - \sum_{i=1}^{n} \log_\varepsilon\left(1 - \frac{j\omega}{p_i}\right). \tag{7.3}$$

Let $H(j\omega)$ be expressed in its polar form,

$$H(j\omega) = |H(j\omega)|\varepsilon^{j \operatorname{ang} H(j\omega)}. \tag{7.4}$$

It is then apparent that we may write

$$\log_\varepsilon H(j\omega) = A(\omega) + jB(\omega), \tag{7.5}$$

with

$$A(\omega) = \log_\varepsilon |H(j\omega)| \tag{7.6}$$

and

$$B(\omega) = \operatorname{ang} H(j\omega). \tag{7.7}$$

The function $A(\omega)$ is called the *logarithmic gain* or simply *gain function* with the *neper* as its unit, while $B(\omega)$ is called the *phase function* with the *radian* as its unit. Equation (7.5) thus reveals that the gain and phase functions are the real and imaginary parts of the logarithm of the transfer function, respectively. It is noteworthy that the gain function may also be expressed in *decibels* as follows:

$$A'(\omega) = 20 \log_{10} |H(j\omega)|. \tag{7.8}$$

The two gain functions $A(\omega)$ and $A'(\omega)$ are related by

$$A'(\omega) = 8{\cdot}69\ A(\omega). \tag{7.9}$$

In other words, one neper is equal to 8·69 decibels.

Returning to Eq. (7.3), and separating the real and imaginary parts, we get

$$A'(\omega) = 20 \log_{10} K_0 + \sum_{i=1}^{m} 20 \log_{10} \left| 1 - \frac{j\omega}{z_i} \right| - \sum_{i=1}^{n} 20 \log_{10} \left| 1 - \frac{j\omega}{p_i} \right|, \tag{7.10}$$

$$B(\omega) = \sum_{i=1}^{m} \text{ang} \left(1 - \frac{j\omega}{z_i} \right) - \sum_{i=1}^{n} \text{ang} \left(1 - \frac{j\omega}{p_i} \right), \tag{7.11}$$

where it is assumed that K_0 is a positive quantity. We thus observe that by plotting separate response curves for the components of $20 \log_{10} H(j\omega)$, the gain as well as phase contributions of the various zero and pole factors can be combined by simple addition, so that the effect of changes in individual parameters becomes readily apparent.

Consider the typical zero factor $(1 - j\omega/z_i)$. There are two possible cases to consider: z_i real and z_i complex. Let us begin with the case of a real zero, assumed in the left half plane; so that $z_i = -\sigma_i$ and the factor in question is $(1 + j\omega/\sigma_i)$. We find that the associated gain function in decibel units is $10 \log_{10}[1 + (\omega/\sigma_i)^2]$, and the phase function is $\tan^{-1}(\omega/\sigma_i)$. In Fig. 7.1 we have shown these functions plotted against ω/σ_i on a logarithmic scale. We see that the gain response is characterized by two asymptotes, a low-frequency asymptote of zero slope and a high-frequency asymptote having a slope of 6 decibels per octave.* The point of intersection of the two asymptotes is called the *corner frequency* or *break frequency*; it occurs when $\omega = \sigma_i$. We also see that the two asymptotes together provide a broken-line approximation to the actual gain curve. The maximum error occurs at $\omega = \sigma_i$ and equals 3 decibels. At both $\omega = \sigma_i/2$ and $\omega = 2\sigma_i$, that is, at an octave below and above σ_i, the error is equal to 1 decibel. The actual gain curve can thus be sketched with reasonable accuracy from the asymptotic approximations by taking account of the errors involved.

A real pole factor will contribute gain and phase characteristics similar to those of Fig. 7.1 except that they are the negative of those due to a zero factor. Since real poles and zeros characterize networks combining resistance and one kind of reactance (inductance or capacitance), we shall refer to the gain responses contributed by such factors as *RX slopes.*[1] When a transfer function has real poles and zeros only, and each pole and zero factor is represented by its asymptotic curve, we obtain an approximate representation for the overall gain characteristic in the form of a series of straight line segments.

In the case of a pair of complex conjugate zeros z_i and z_i^* located at $-\omega_i(\cos\theta_i \pm j \sin\theta_i)$, as in Fig. 7.2, the significant scaled zero factor to consider

* An octave is an interval in which the frequency changes in the ratio of two-to-one.

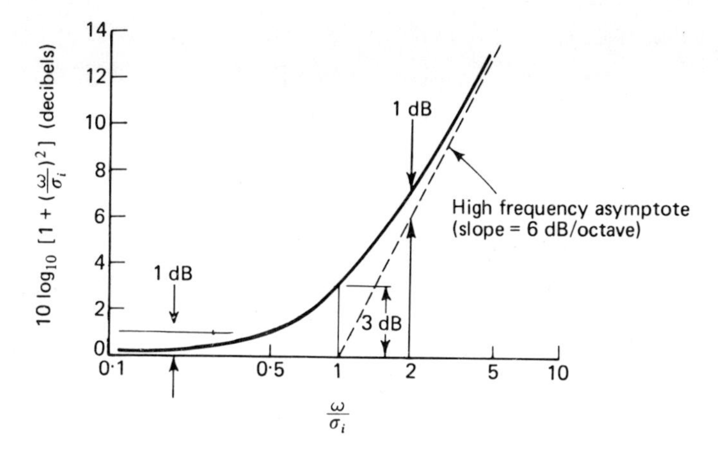

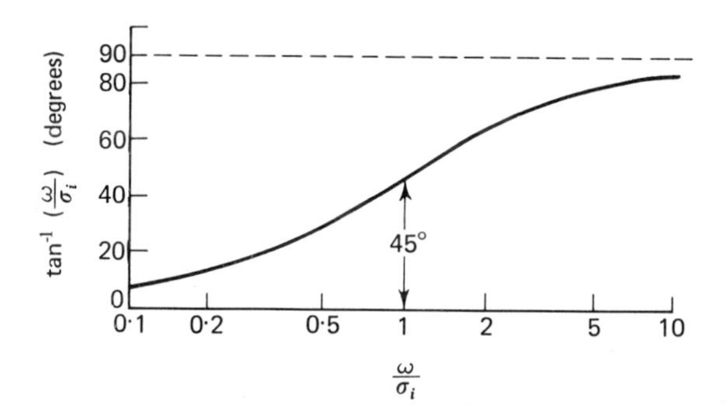

Fig. 7.1. Gain–phase plots for simple zero factor $(1 + s/\sigma_i)$.

is:

$$\left(1 + \frac{2\zeta_i s}{\omega_i} + \frac{s^2}{\omega_i^2}\right),$$

where $\zeta_i = \cos\theta_i$. For $s = j\omega$, the contribution of such a factor to the gain in decibels is:

$$10\log_{10}\left[1 + \frac{2\omega^2}{\omega_i^2}(2\zeta_i^2 - 1) + \frac{\omega^4}{\omega_i^4}\right]$$

and its contribution to the phase is:

$$\tan^{-1}\left[2\zeta_i\Big/\left(\frac{\omega_i}{\omega} - \frac{\omega}{\omega_i}\right)\right].$$

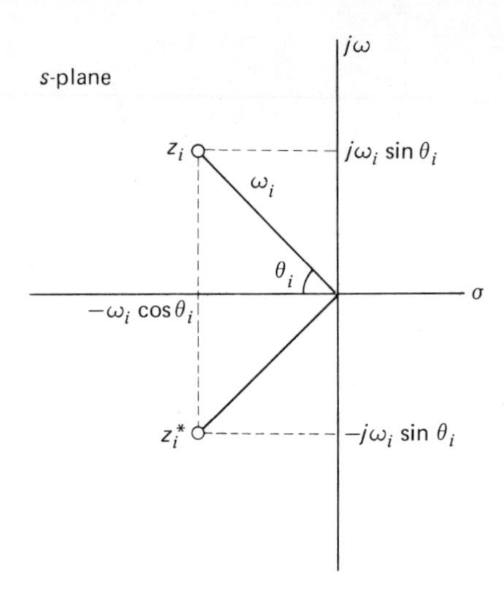

Fig. 7.2. A pair of complex conjugate zeros.

These functions are shown plotted in Fig. 7.3 against ω/ω_i on a logarithmic scale, and for different values of ζ_i.

7.3 MINIMUM-PHASE AND NON-MINIMUM-PHASE NETWORKS

The design of equalizers, wideband and feedback amplifiers in the frequency domain frequently involves the shaping of gain and phase responses; we need a method of constructing one response if the other is given. The problem of finding, let us say, the phase response from a given gain response is easily solved when the gain response pertains to a known pole–zero pattern. However, when the given gain response is arbitrarily postulated or is possibly a measured response, it may be impossible to ascertain the actual phase response because any number of phase characteristics can correspond to a given gain response, except in a restricted class of two-port networks known as *minimum-phase networks.* In these networks, the properties of which are well-suited to equalizer and amplifier applications, there does exist a definite relation between the gain and phase response.

A stable linear two-port network is said to be of the *minimum-phase* type if its transfer function has no zeros in the right half plane; otherwise, it is of the *non-minimum-phase type.* In general, any non-minimum-phase function with zeros in the right half plane can be expressed as the product of an *all-pass* function which contains the zeros in the right half plane and a minimum-phase function which contains no such zeros. This identity is illustrated by the pole–zero patterns of Fig. 7.4. An all-pass function is a special kind of non-minimum-phase function whose zeros and poles are the negatives of one another. Clearly, along the

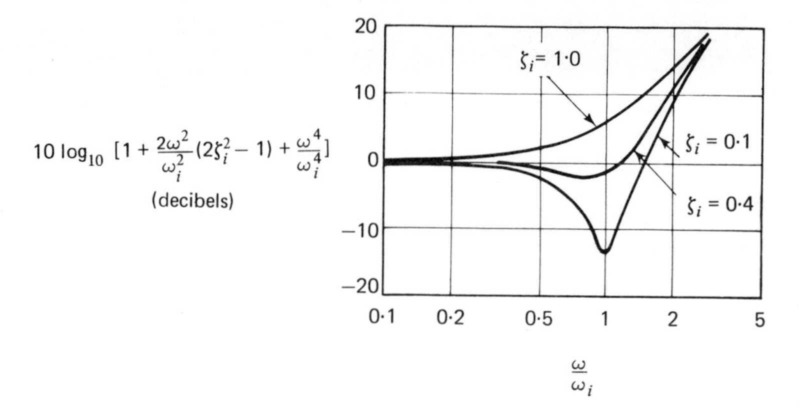

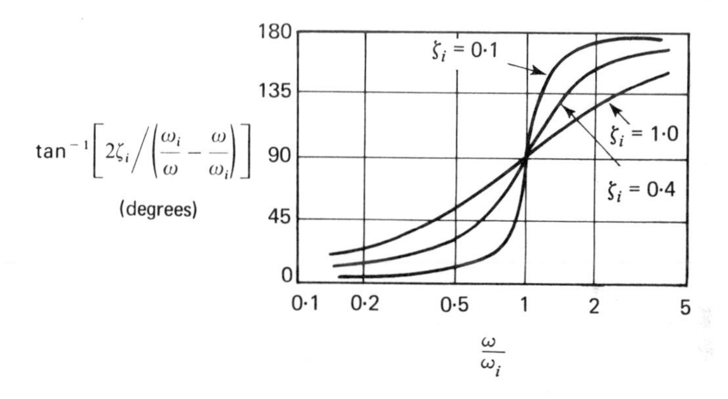

Fig. 7.3. Gain–phase plots for zero factor $(1 + 2\zeta_i(s/\omega_i) + s^2/\omega_i^2)$.

imaginary axis of the s-plane an all-pass function has a magnitude equal to unity for all values of ω; its phase shift is, however, frequency dependent. For example, the all-pass function $(s - a)/(s + a)$ has a phase shift equal to $-2\tan^{-1}(\omega/a)$, which continuously decreases from 0° at zero frequency to −180° at infinite frequency. It follows, therefore, that in Fig. 7.4 the minimum-phase function $H(s)$ and non-minimum-phase function $H'(s)$ have exactly the same magnitudes for $s = j\omega$, but they differ in phase by an amount equal to the phase shift of the all-pass function $H_0(s)$. In other words, for a given gain characteristic the minimum-phase transfer function has the minimum phase shift possible, and any change in the gain characteristic produces a corresponding change in the phase characteristic, and vice versa. This implies that in the case of a minimum-phase

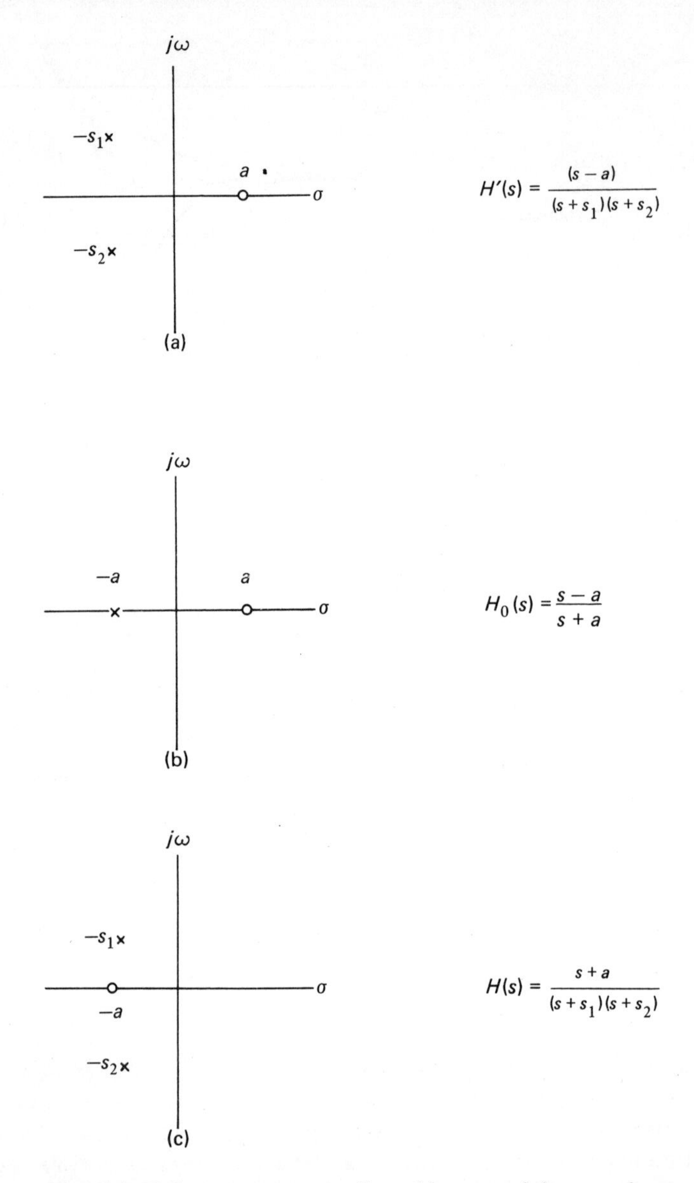

Fig. 7.4. Pole–zero patterns for: (a) non-minimum phase function; (b) all-pass function; (c) minimum phase function.

transfer function there exists a unique relationship between the associated gain and phase characteristics, such that we can obtain some useful information about one characteristic when the other is specified over the complete frequency spectrum or at extreme frequencies.

When a transfer function has a zero in the right half plane, the implication is that at some complex frequency $\sigma_k + j\omega_k$ with positive σ_k, the transmission through the network in the pertinent direction is reduced to zero. In a passive reciprocal ladder such as that of Fig. 7.5, transmission through the network can only be reduced to zero when either the impedance of a series branch becomes infinite or when the impedance of a shunt branch becomes zero. Since the poles and zeros of the branch impedances must occur in the left half plane if the branch impedances are passive, it follows that any passive ladder network is a minimum-phase network.

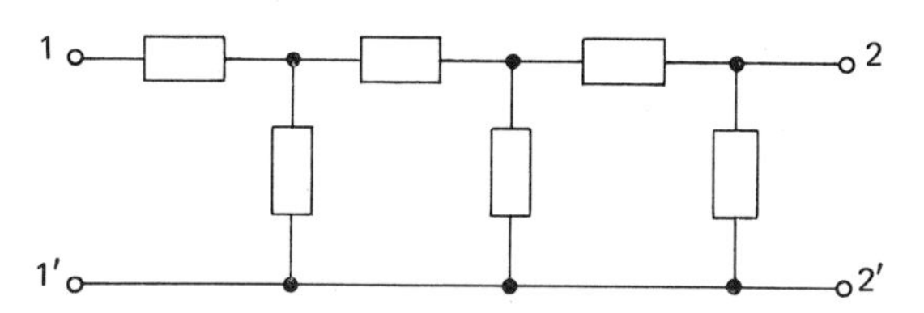

Fig. 7.5. Ladder network.

For a physical network, passive or active, to exhibit a transmission zero in the right half plane, there must be at least two separate paths connecting the input and output ports of the network, so that at some value of s the currents delivered to the load by these paths cancel out. The lattice, bridged-T, and twin-T networks shown in Fig. 7.6 evidently fulfil this requirement and may, therefore, have a non-minimum-phase characteristic, depending upon the particular values of the elements.

It is also of interest to note that if a transfer function $H(s)$ has poles of multiplicity m_1 and m_2 at zero and infinite frequency, respectively, then for $H(s)$ to be of the minimum-phase type it is necessary that the net phase change between zero and infinite frequency be equal to $(m_1 + m_2)(\pi/2)$ rad. In particular, if $H(s)$ has a finite magnitude at both zero and infinite frequency, the net phase change must be zero. On the other hand, if $H(s)$ is non-minimum phase, the net phase change is always positive.[2]

7.4 SOME CONSEQUENCES OF CAUCHY'S THEOREM

The key concept in the mathematical development of the relationship between gain and phase, the various forms of which were first used in network theory by Bode, is *Cauchy's Theorem* of complex variable theory.[3] This theorem states that if a function $f(s)$ of the complex variable s is analytic within a closed contour in the s-plane and also on the contour itself, then the integral of the function $f(s)$ taken around that contour is equal to zero. In our study we shall be concerned with a contour which consists of the entire imaginary axis of the s-plane and an infinite semicircular arc to the right. Along the imaginary axis infinitesimal

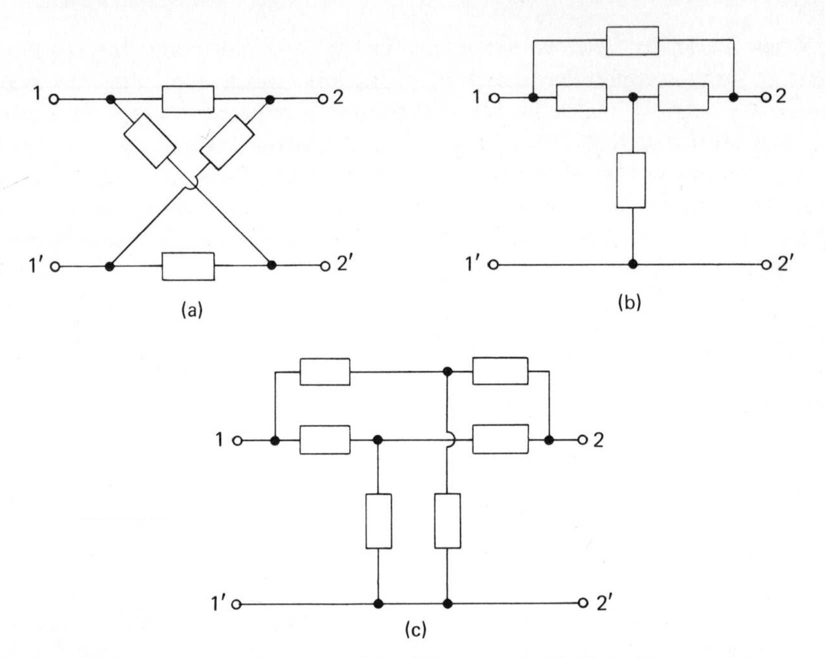

Fig. 7.6. (a) Lattice; (b) Bridged-T network; (c) Twin-T network.

semicircular indentations are introduced so as to avoid any singularities that may be found there. The reason for choosing such a contour is that since a minimum-phase transfer function $H(s)$ has no poles or zeros in the right half plane, then its logarithm will be analytic in this region. We may thus use Cauchy's theorem to study the integral of $\log_\varepsilon H(s)$ along the complete contour. The integrals around the small indentations and the large semicircular part of the contour may readily be evaluated; in terms of these results the integral of $\log_\varepsilon H(j\omega)$ along the imaginary axis may then be expressed.

Let $f(s)$ be a rational function that is analytic everywhere in the right half plane, and has a finite value equal to $f(\infty)$ when s is infinitely large. It is also assumed that any singularities of $f(s)$ at a point $j\omega_0$ on the imaginary axis are of such a nature that $(s - j\omega_0)f(s)$ vanishes as s approaches $j\omega_0$. Thus, $f(s)$ may represent the natural logarithm of a minimum-phase transfer function. Since it is permissible for a minimum phase function to have poles and zeros along the imaginary axis, its logarithm will have logarithmic singularities there. Such singularities may, however, be dismissed as they contribute nothing to the contour integral.*

* Suppose $H(s)$ has a pole or zero at $s = j\omega_0$; then $f(s) = \log_\varepsilon H(s)$ will have a logarithmic singularity at $s = j\omega_0$. As s approaches $j\omega_0$, $f(s)$ approaches $-\infty$. However, in the neighbourhood of $s = j\omega_0$ the function $f(s)$ increases so slowly that the limit of $(s - j\omega_0)f(s)$ is zero as s approaches $j\omega_0$.

It should be mentioned that the function $f(s)$ may also represent an open-circuit stable driving-point impedance function, a short-circuit stable driving-point admittance function, or a stable transfer function, provided that none of these functions has poles on the imaginary axis of the s-plane.

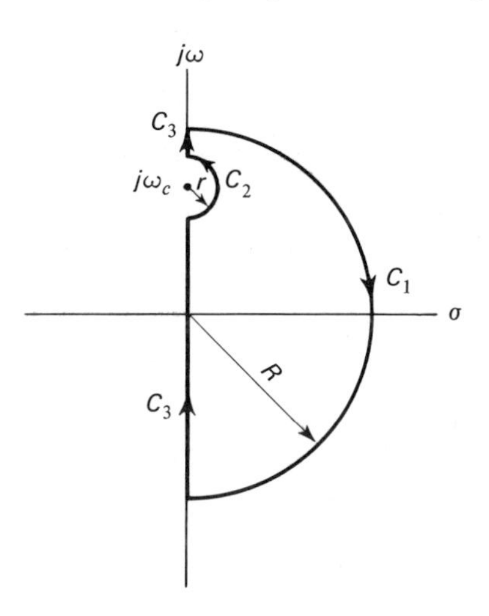

Fig. 7.7. Contour for deriving the integral relationships between the gain and phase components of a minimum phase transfer function.

In order to develop the relations between the real and imaginary components of $f(j\omega)$ we shall integrate the related function $f(s)/(s - j\omega_c)$ around the closed contour C of Fig. 7.7, where ω_c is any value of ω. The contour C consists of three parts: the large semicircle C_1 of radius R, the small semicircular indentation C_2 (of radius r) around the point $s = j\omega_c$, and the imaginary axis C_3. The indentation C_2 has been introduced to avoid the pole of the function $f(s)/(s - j\omega_c)$ at the point $s = j\omega_c$ and so ensure that this function is analytic at all points on and within the contour C. From Cauchy's theorem it follows that:

$$\oint \frac{f(s)}{s - j\omega_c}\, ds = 0, \tag{7.12}$$

where the path of integration is the contour C of Fig. 7.7. The contour integral on the left of Eq. (7.12) may be expressed as the sum of three integrals evaluated along the paths C_1, C_2, and C_3. Along the semicircle C_1 we have $s = R\varepsilon^{j\theta}$ and $ds = jR\varepsilon^{j\theta}\, d\theta$; hence,

$$\int_{C_1} \frac{f(s)}{s - j\omega_c}\, ds = \int_{\pi/2}^{-\pi/2} \frac{f(R\varepsilon^{j\theta})}{R\varepsilon^{j\theta} - j\omega_c} jR\varepsilon^{j\theta}\, d\theta. \tag{7.13}$$

Now let R become infinitely large, so that

$$\int_{C_1} \frac{f(s)}{s - j\omega_c}\, ds = jf(\infty) \int_{\pi/2}^{-\pi/2} d\theta. \tag{7.14}$$
$$= -j\pi f(\infty).$$

When $s = j\omega$, we have

$$f(j\omega) = A(\omega) + jB(\omega), \tag{7.15}$$

where the real part $A(\omega)$ is an even function of ω, and the imaginary part $B(\omega)$ is an odd function of ω. If $f(\infty)$ is to be finite, the imaginary part must be zero at infinite frequency, so that $f(\infty) = A(\infty)$ and Eq. (7.14) becomes

$$\int_{C_1} \frac{f(s)}{s - j\omega_c}\, ds = -j\pi A(\infty). \tag{7.16}$$

Next, along the small semicircle C_2 we have $s = j\omega_c + r\varepsilon^{j\theta}$, and $ds = jr\varepsilon^{j\theta}\, d\theta$; therefore, the integral round C_2 is:

$$\int_{C_2} \frac{f(s)}{s - j\omega_c}\, ds = \int_{-\pi/2}^{\pi/2} \frac{f(j\omega_c + r\varepsilon^{j\theta})}{r\varepsilon^{j\theta}}\, jr\varepsilon^{j\theta}\, d\theta. \tag{7.17}$$

Now let r approach zero, so that

$$\int_{C_2} \frac{f(s)}{s - j\omega_c}\, ds = jf(j\omega_c) \int_{-\pi/2}^{\pi/2} d\theta$$
$$= j\pi f(j\omega_c), \tag{7.18}$$

where $f(j\omega_c)$ is the value of $f(s)$ at the point $s = j\omega_c$.

Along the imaginary axis we have $s = j\omega$ and $ds = j\, d\omega$; therefore, the integral along the portion C_3 of the imaginary axis is:

$$\int_{C_3} \frac{f(s)}{s - j\omega_c}\, ds = \int_{-R}^{\omega_c - r} \frac{f(j\omega)}{\omega - \omega_c}\, d\omega + \int_{\omega_c + r}^{R} \frac{f(j\omega)}{\omega - \omega_c}\, d\omega. \tag{7.19}$$

In the limit as $r \to 0$ and $R \to \infty$, Eq. (7.19) reduces to:

$$\int_{C_3} \frac{f(s)}{s - j\omega_c}\, ds = \int_{-\infty}^{\infty} \frac{f(j\omega)}{\omega - \omega_c}\, d\omega. \tag{7.20}$$

The integral along the contour C is equal to the sum of the three integrals of Eqs. (7.16), (7.18) and (7.20). Hence, adding these three contributions and using Eq. (7.12), we get:

$$\int_{-\infty}^{\infty} \frac{f(j\omega)}{\omega - \omega_c}\, d\omega = j\pi[A(\infty) - f(j\omega_c)]. \tag{7.21}$$

If we equate the reals and imaginaries in Eq. (7.21), there results

$$A(\omega_c) = A(\infty) - \frac{1}{\pi}\int_{-\infty}^{\infty} \frac{B(\omega)}{\omega - \omega_c}\, d\omega, \tag{7.22}$$

$$B(\omega_c) = \frac{1}{\pi}\int_{-\infty}^{\infty} \frac{A(\omega)}{\omega - \omega_c}\, d\omega. \tag{7.23}$$

Equation (7.23) states that if the real part of a network function $f(j\omega)$ is specified over all frequencies, then its imaginary part is completely determined; while Eq. (7.22) states that if the imaginary part is specified over all frequencies, the real part is completely determined except for the additive constant $A(\infty)$. Both equations assume that the function $f(s)$ is analytic in the right half plane, and that if it has a singularity at $s = j\omega_0$ on the imaginary axis, the limit $(s - j\omega_0)f(s)$ approaches zero as s approaches $j\omega_0$.

It should be noted that in Eqs. (7.22) and (7.23) the integration along the imaginary axis of the s-plane must avoid the pole at $s = j\omega_c$ in a symmetrical manner. This will yield the *principal values* of the integrals on the right. Thus in Eq. (7.22) we have

$$\int_{-\infty}^{\infty} \frac{B(\omega)}{\omega - \omega_c}\, d\omega = \lim_{r\to 0}\left[\int_{-\infty}^{\omega_c - r} \frac{B(\omega)}{\omega - \omega_c}\, d\omega + \int_{\omega_c + r}^{\infty} \frac{B(\omega)}{\omega - \omega_c}\, d\omega\right]. \tag{7.24}$$

In a similar manner, we may evaluate the integral in Eq. (7.23).

The functions which are of interest to us have real and imaginary parts that are even and odd functions of the frequency ω, respectively. By making use of this property we can develop alternative forms of the relations, in which the integration extends over only the upper half of the imaginary axis. From Eq. (7.22) we have

$$A(\omega_c) = A(\infty) - \frac{1}{\pi}\int_{-\infty}^{0} \frac{B(\omega)}{\omega - \omega_c}\, d\omega - \frac{1}{\pi}\int_{0}^{\infty} \frac{B(\omega)}{\omega - \omega_c}\, d\omega. \tag{7.25}$$

If in the first of the two integrals we replace ω by $-\omega$ and change the limits of integration accordingly, and use the odd function property of $B(\omega)$, that is, $B(-\omega) = -B(\omega)$, there results

$$\int_{-\infty}^{0} \frac{B(\omega)}{\omega - \omega_c}\, d\omega = -\int_{\infty}^{0} \frac{B(\omega)}{\omega + \omega_c}\, d\omega = \int_{0}^{\infty} \frac{B(\omega)}{\omega + \omega_c}\, d\omega. \tag{7.26}$$

Substitution of Eq. (7.26) in Eq. (7.25) yields:

$$\begin{aligned} A(\omega_c) &= A(\infty) - \frac{1}{\pi}\int_{0}^{\infty} B(\omega)\left(\frac{1}{\omega - \omega_c} + \frac{1}{\omega + \omega_c}\right) d\omega \\ &= A(\infty) - \frac{2}{\pi}\int_{0}^{\infty} \frac{\omega B(\omega)}{\omega^2 - \omega_c^2}\, d\omega. \end{aligned} \tag{7.27}$$

In a similar manner, starting with Eq. (7.23) and making use of the even-function property of $A(\omega)$, that is, $A(-\omega) = A(\omega)$, we get

$$B(\omega_c) = \frac{2\omega_c}{\pi}\int_0^\infty \frac{A(\omega)}{\omega^2 - \omega_c^2}\,d\omega. \tag{7.28}$$

However, by direct integration we may verify that

$$\int_0^\infty \frac{d\omega}{\omega^2 - \omega_c^2} = 0. \tag{7.29}$$

Therefore, we may add any convenient constants to the numerators of the integrands in Eqs. (7.27) and (7.28) without affecting the values of the integrals. Thus, adding the constant $-\omega_c B(\omega_c)$ to the numerator of the integrand in Eq. (7.27) and $-A(\omega_c)$ to the numerator of the integrand in Eq. (7.28) so as to remove the singularities of the integrands at $\omega = \omega_c$, we obtain

$$A(\omega_c) = A(\infty) - \frac{2}{\pi}\int_0^\infty \frac{\omega B(\omega) - \omega_c B(\omega_c)}{\omega^2 - \omega_c^2}\,d\omega, \tag{7.30}$$

$$B(\omega_c) = \frac{2\omega_c}{\pi}\int_0^\infty \frac{A(\omega) - A(\omega_c)}{\omega^2 - \omega_c^2}\,d\omega. \tag{7.31}$$

We will next consider the limiting forms which these two relations assume when the frequency ω_c approaches zero or infinity. The relation of Eq. (7.31) is later on modified into a form particularly suitable for the graphical computation of the phase shift associated with the gain characteristic of a minimum-phase transfer function.

The Phase-integral Theorem

When the frequency ω_c approaches zero we find that Eq. (7.30) leads to the result

$$A(\infty) - A(0) = \frac{2}{\pi}\int_0^\infty \frac{B(\omega)}{\omega}\,d\omega. \tag{7.32}$$

In practice, it is found more convenient to use the logarithm of frequency as the variable of interest; let us therefore define

$$u = \log_\varepsilon \omega. \tag{7.33}$$

Then we find that $du = d\omega/\omega$. Also $u = -\infty$ when $\omega = 0$, and $u = \infty$ when $\omega = \infty$. Hence, Eq. (7.32) gives

$$\int_{-\infty}^{\infty} B(\omega)\,du = \frac{\pi}{2}[A(\infty) - A(0)]. \tag{7.34}$$

Equation (7.34) states that the total area under the imaginary component (i.e.

phase), plotted on a logarithmic frequency scale, is equal to $\pi/2$ times the difference between the limiting values assumed by the real component (i.e., gain) at zero and infinite frequency, and that it is entirely independent of the manner in which the real part varies between these two limits. This result is known as the *phase-integral theorem*; it is illustrated by Fig. 7.8. If the change in the gain characteristic is concentrated in a narrow portion of the frequency spectrum the associated phase characteristic rises to a sharp peak, while if the change in the gain characteristic is more gradual the phase characteristic is broad and flat. For a given total change in the gain characteristic, however, the area under the phase characteristic is always the same.

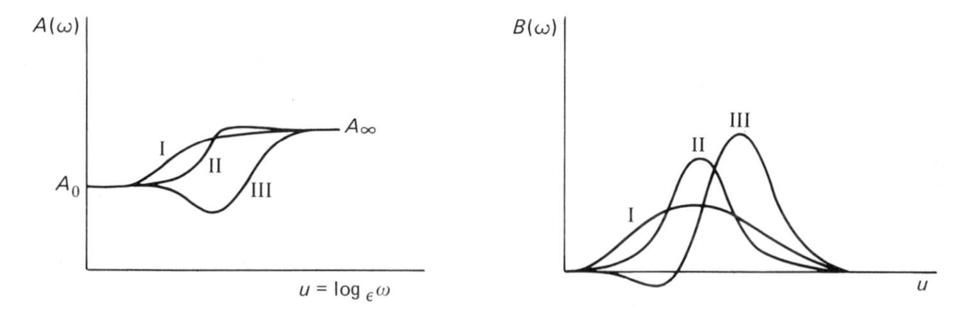

Fig. 7.8. Illustrating the phase-integral theorem.

The basic theorem also applies to a driving-point impedance (or admittance) function, in which case the real component is resistance (or conductance) and the imaginary component is reactance (or susceptance); in this form, Eq. (7.34) is referred to as the *reactance-integral theorem.*

Gain-integral Theorem

If in Eq. (7.31) we multiply both sides by ω_c, and then let ω_c become infinitely large, there results

$$\lim_{\omega_c \to \infty} [\omega_c B(\omega_c)] = -\frac{2}{\pi} \int_0^\infty [A(\omega) - A(\infty)]\, d\omega, \tag{7.35}$$

that is,

$$\int_0^\infty [A(\omega) - A(\infty)]\, d\omega = -\frac{\pi}{2} \lim_{\omega \to \infty} [\omega B(\omega)]. \tag{7.36}$$

Equation (7.36) states that the total area under the curve of the real component versus frequency, shifted vertically by an amount $A(\infty)$, is determined entirely by the high-frequency behaviour of the imaginary component. This result is known as the *gain-integral theorem* when the function of interest is the logarithm of a minimum-phase transfer function, and as the *resistance-integral theorem* in the case of a driving-point impedance (or admittance) function. The theorem, however, finds greater application in its latter form.

Example 7.1. *Gain–bandwidth product.* To illustrate the usefulness of the resistance-integral theorem, consider the unilateral amplifier shown in Fig. 7.9 where the output port is shunted with a capacitance C. It is assumed the impedance

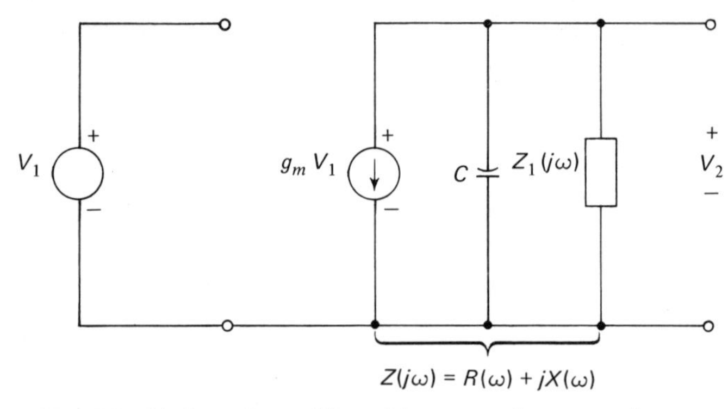

Fig. 7.9. Unilateral amplifier with output-shunt capacitance.

$Z_1(j\omega)$ to the right of the capacitor does not vanish at infinite frequency, so that C represents the total shunt capacitance across the output port. We then have $R(\infty) = 0$ and

$$\lim_{\omega \to \infty} [\omega X(\omega)] = -\frac{1}{C}. \tag{7.37}$$

Hence, applying the resistance-integral theorem to the total impedance $Z(j\omega)$ yields

$$\int_0^\infty R(\omega)\, d\omega = \frac{\pi}{2C}. \tag{7.38}$$

In Fig. 7.9 we see, however, that the amplifier has a voltage gain $K_v(j\omega)$ equal to $-g_m Z(j\omega)$, where the minus sign represents the 180° phase shift which the amplifier produces at low frequencies. Thus,

$$R(\omega) = -\frac{1}{g_m} \operatorname{Re}[K_v(j\omega)]. \tag{7.39}$$

Substitution of Eq. (7.39) in Eq. (7.38) gives

$$\frac{g_m}{C} = -\frac{2}{\pi} \int_0^\infty \operatorname{Re}[K_v(j\omega)]\, d\omega. \tag{7.40}$$

Therefore, the ratio g_m/C completely determines the area enclosed by the real part of the voltage gain plotted versus frequency. That is to say, the *gain–bandwidth product* of the amplifier is fixed by the value of g_m/C; the larger this ratio, the better the amplifier. This means that for a specified value of g_m/C we can have a large voltage gain over a narrow band of frequencies or a small voltage gain over a wide band of frequencies. The ratio g_m/C may therefore be regarded as a *figure of merit* for the amplifier.

Gain-slope Theorem

The phase-integral and gain-integral theorems specify two specialized forms of the relations between the gain and phase characteristics of a minimum-phase transfer function. The first theorem defines the integral of the phase characteristic in terms of the behaviour of the associated gain characteristic at zero and infinite frequencies, while the second theorem defines the integral of the gain characteristic in terms of the phase characteristic at infinite frequency. From the design standpoint a desirable formula would be one in which either the gain or phase at one frequency is determined from the other specified at all frequencies. The simultaneous control of gain and phase would then become unnecessary, since the desired characteristics would be interrelated and only one need be synthesized in detail. In general, however, gain measurements can be made with more convenience and greater accuracy than phase measurements; therefore, the type of formula required is one which enables computation of the phase characteristic corresponding to a gain characteristic that is prescribed over the complete frequency spectrum. One form of such a desired formula is given by Eq. (7.31). It is, however, more convenient to change to a logarithmic frequency scale by using the relation $u = \log_\varepsilon(\omega/\omega_c)$; then we find that Eq. (7.31) modifies to

$$B(\omega_c) = \frac{1}{\pi}\int_{-\infty}^{\infty} \frac{A(\omega) - A(\omega_c)}{\sinh u}\, du. \tag{7.41}$$

Integrating the right-hand side of Eq. (7.41) by parts, we obtain

$$\begin{aligned} B(\omega_c) = &-\frac{1}{\pi}\left[(A(\omega) - A(\omega_c))\log_\varepsilon\left(\coth\frac{u}{2}\right)\right]_{-\infty}^{\infty} \\ &+\frac{1}{\pi}\int_{-\infty}^{\infty} \frac{dA(\omega)}{du}\log_\varepsilon\left(\coth\frac{u}{2}\right) du. \end{aligned} \tag{7.42}$$

Now $\coth(u/2) = (\varepsilon^u + 1)/(\varepsilon^u - 1)$; hence, when $u = \infty$, $\coth(u/2) = 1$ and $\log_\varepsilon(\coth u/2) = 0$. Further, when $u = -\infty$, $\coth(u/2) = -1 = \varepsilon^{j\pi}$ and so $\log_\varepsilon(\coth u/2) = j\pi$. Also $A(\omega) = A(0)$ when $u = -\infty$ because this corresponds to $\omega = 0$. Therefore, the integrated part of Eq. (7.42) reduces to $j(A(0) - A(\omega_c))$.

Next, for negative values of u, we note that $\coth(u/2)$ is negative and its logarithm is complex; so that when $u < 0$,

$$\log_\varepsilon\left(\coth\frac{u}{2}\right) = \log_\varepsilon\left(\coth\frac{|u|}{2}\right) + j\pi. \tag{7.43}$$

Therefore,

$$\frac{1}{\pi}\int_{-\infty}^{\infty}\frac{dA(\omega)}{du}\log_\varepsilon\left(\coth\frac{u}{2}\right)du = \frac{1}{\pi}\int_{-\infty}^{\infty}\frac{dA(\omega)}{du}\log_\varepsilon\left(\coth\frac{|u|}{2}\right)du + j\int_{-\infty}^{0}\frac{dA(\omega)}{du}du$$

$$= \frac{1}{\pi}\int_{-\infty}^{\infty}\frac{dA(\omega)}{du}\log_\varepsilon\left(\coth\frac{|u|}{2}\right)du + j(A(\omega_c) - A(0)).$$

(7.44)

Combining the results of Eqs. (7.42) to (7.44) we finally get:

$$B(\omega_c) = \frac{1}{\pi}\int_{-\infty}^{\infty}\frac{dA(\omega)}{du}\log_\varepsilon\left(\coth\frac{|u|}{2}\right)du. \tag{7.45}$$

This relation is known as the *gain-slope theorem*; in effect, it implies that the phase shift of a minimum-phase transfer function is proportional to the slope of the associated gain characteristic plotted on a logarithmic frequency scale. Since the integration includes the complete frequency spectrum, it follows that the phase shift at any point depends upon the slopes of the gain characteristic at all parts of the spectrum, the relative importance of any gain-slope being determined by the *weighting factor*

$$\log_\varepsilon\left(\coth\frac{|u|}{2}\right) = \log_\varepsilon\left|\frac{\omega + \omega_c}{\omega - \omega_c}\right|. \tag{7.46}$$

This term is shown plotted in Fig. 7.10 where we see that it becomes logarithmically infinite at $\omega = \omega_c$. At frequencies high compared with ω_c the weighting factor approximates to $2\omega_c/\omega$, and at frequencies low compared with ω_c it approximates to $2\omega/\omega_c$. Therefore, the slope of the gain characteristic at a frequency ω close to ω_c, where the phase shift is being computed, has a much larger effect than the gain slope at a more remote frequency. More specifically, the phase shift at a frequency ω_c is effectively determined by the slope of the gain characteristic in a frequency range extending from $\omega_c/10$ to $10\omega_c$.

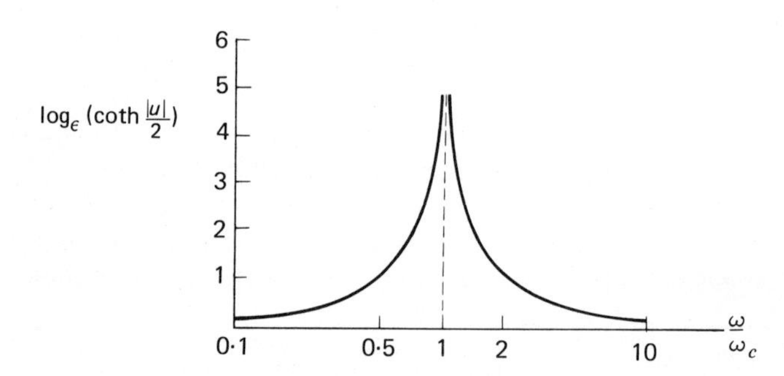

Fig. 7.10. Weighting factor $\log_\varepsilon(\coth|\mu|/2)$.

Example 7.2. *Semi-infinite slope.* To illustrate the usefulness of Eq. (7.45) we shall evaluate the phase characteristic associated with the so-called *semi-infinite slope* shown in part (a) of Fig. 7.11 where we see that the gain is zero up to $\omega_c = \omega_0$; thereafter, it has a *unit slope*. Unit slope is a change of one neper per unit change of u, which is equal to 6 decibels per octave. The semi-infinite slope is observed to be the same as the asymptote of the RX slope, except that the inclined portion can be assigned any steepness whatsoever.

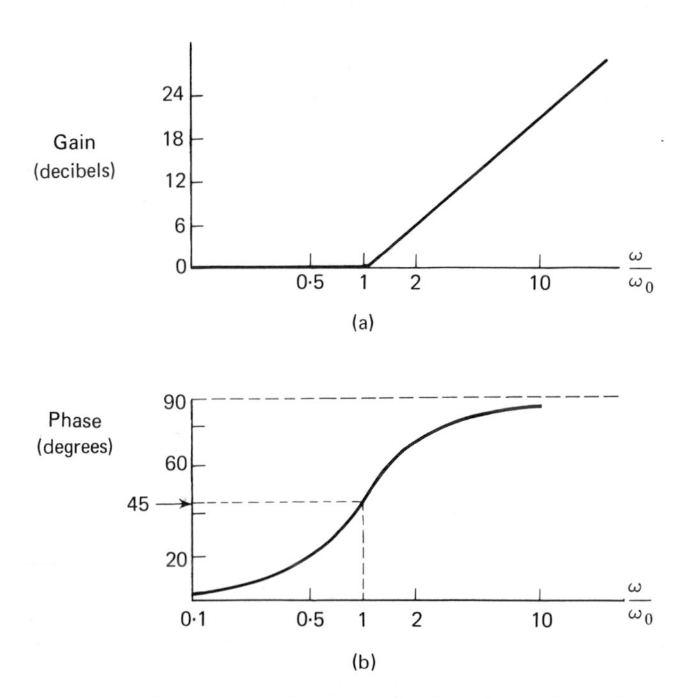

Fig. 7.11. (a) Semi-infinite slope; (b) Associated phase characteristic.

Let $u_c = \log_\varepsilon(\omega_c/\omega_0)$; then in the range below u_c the slope is zero and the integrand in Eq. (7.45) vanishes, and so we find that the phase shift at any frequency ω_c is

$$B(\omega_c) = \frac{1}{\pi}\int_{u_c}^{\infty} \log_\varepsilon\left(\coth\frac{|u|}{2}\right) du. \tag{7.47}$$

This definite integral has been computed and tabulated by Thomas.[4] The results of the integration are shown plotted in part (b) of Fig. 7.11. It is significant to observe that if ω_c is low compared with the corner frequency ω_0, then we have:

$$B(\omega_c) \simeq \frac{2\omega_c}{\pi\omega_0} \qquad \text{for} \quad \omega_c \ll \omega_0. \tag{7.48}$$

That is to say, at low frequencies the phase associated with a semi-infinite slope varies practically in a linear manner with frequency. In the case of an RX slope, on the other hand, the phase shift is given by

$$B(\omega_c) \simeq \frac{\omega_c}{\omega_0} \qquad \text{for} \quad \omega_c \ll \omega_0. \tag{7.49}$$

We thus see that the difference between the two expressions of Eqs. (7.48) and (7.49) is the scale factor $2/\pi$; this is due to the abrupt change in the gain slope at ω_0 in the case of the semi-infinite slope.

A Graphical Procedure for Computing the Phase Characteristic from the Gain Characteristic

A particularly useful application of the semi-infinite slope is in the graphical evaluation of the phase shift $B(\omega)$ when the gain $A(\omega)$ is known in analytic or graphical form. The procedure is based on the assumption that the gain–frequency characteristic will be approximated by a series of straight line segments, which is always possible provided enough straight lines are used. This is illustrated in Fig. 7.12(a) where we see that the gain characteristic, plotted on a logarithmic frequency scale, is fairly well approximated by four straight line segments. In choosing the set of straight lines it is ordinarily sufficient to represent correctly the major trends in the gain characteristic. The gain-slope theorem, as expressed by Eq. (7.45), indicates that the relationship between gain and phase involves a smoothing or averaging out of the gain characteristic. Therefore, provided the major trends are correctly represented, the phase characteristics corresponding to the actual gain characteristic and to its straight-line approximation will be much more nearly equal than are the actual and approximate gain characteristics.

The straight-line approximation to the gain characteristic is next resolved into the sum of semi-infinite slopes in the manner indicated in Fig. 7.12(b), which applies to the approximation given in part (a) of the diagram. Using Fig. 7.11(b), or Thomas's tables for greater accuracy, the minimum phase shift is found for each component semi-infinite slope, and these are summed to obtain the composite phase response. It should, however, be noted that for a semi-infinite slope of a steepness of, say, k units of 6 decibels per octave, the phase response of the unit slope is simply multiplied by k.

Gain and Phase Characteristics Prescribed in Different Frequency Ranges

Another important formula relating the gain and phase components is concerned with the determination of the complete characteristic when the gain is specified in one portion of the frequency spectrum and the phase in the remainder. Suppose, for example, the gain is known at frequencies below ω_0 and the phase in the range above ω_0. The problem to be solved, then, is that of completing the characteristic by calculating the phase in the range below ω_0 and the gain above this point. For this evaluation, we may consider as the function of interest,

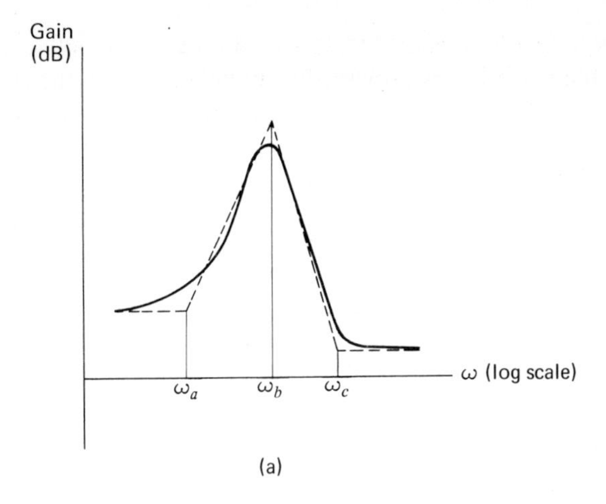

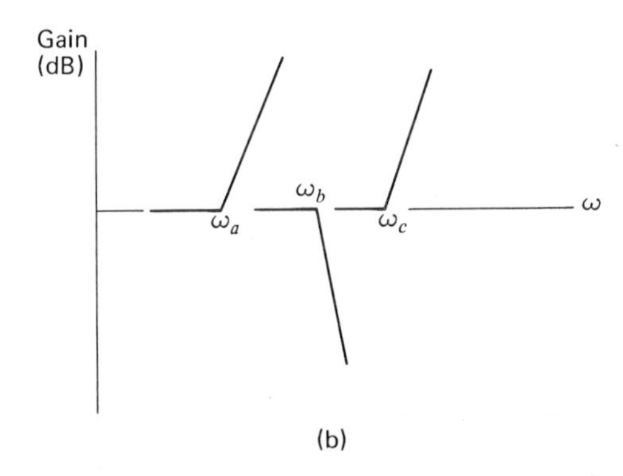

Fig. 7.12. Illustrating the graphical computation of phase characteristic from prescribed gain characteristic.

$f(s)/\sqrt{1 + s^2/\omega_0^2}$, which has branch points at $s = \pm j\omega_0$. The branch cuts may be located entirely in the left half plane, so that they are of no concern. It is assumed that for $s = j\omega$ the positive square root of $\sqrt{1 - \omega^2/\omega_0^2}$ is taken, that is, $\sqrt{1 - \omega^2/\omega_0^2}$ is a positive real quantity for $-\omega_0 < \omega < \omega_0$, it is a positive imaginary for $\omega > \omega_0$ and a negative imaginary for $\omega < -\omega_0$. Since $f(j\omega) = A(\omega) + jB(\omega)$ we may, therefore, write:

$$\frac{f(j\omega)}{\sqrt{1 - \omega^2/\omega_0^2}} = \begin{cases} \dfrac{A(\omega)}{\sqrt{1 - \omega^2/\omega_0^2}} + j\dfrac{B(\omega)}{\sqrt{1 - \omega^2/\omega_0^2}}, & \text{for} \quad \omega < \omega_0 \\ \\ \dfrac{B(\omega)}{\sqrt{\omega^2/\omega_0^2 - 1}} - j\dfrac{A(\omega)}{\sqrt{\omega^2/\omega_0^2 - 1}}, & \text{for} \quad \omega > \omega_0, \end{cases} \tag{7.50}$$

where all the square roots on the right-hand side are positive real quantities. Any of the formulae developed previously can be applied to the present situation if we replace the original $A(\omega)$ and $B(\omega)$, respectively, by the real and imaginary parts of the expression given by Eq. (7.50). If we make these substitutions in Eq. (7.28), for example, we get,

$$\int_0^{\omega_0} \frac{A(\omega)}{\sqrt{1-\omega^2/\omega_0^2}} \cdot \frac{d\omega}{(\omega^2-\omega_c^2)} + \int_{\omega_0}^{\infty} \frac{B(\omega)}{\sqrt{\omega^2/\omega_0^2-1}} \frac{d\omega}{(\omega^2-\omega_c^2)}$$

$$= \frac{\pi B(\omega_c)}{2\omega_c\sqrt{1-\omega_c^2/\omega_0^2}}, \quad \text{for} \quad \omega_c < \omega_0$$

$$= \frac{-\pi A(\omega_c)}{2\omega_c\sqrt{\omega_c^2/\omega_0^2-1}}, \quad \text{for} \quad \omega_c > \omega_0. \tag{7.51}$$

Originally we assumed that $A(\omega)$ was known below ω_0 and $B(\omega)$ above ω_0; hence, the two integrations in Eq. (7.51) can be carried out and the remaining portions of both characteristics evaluated.

Example 7.3. *Constant-phase cut-off.* Suppose the gain $A(\omega)$ is specified to be constant at a level equal to A_0 nepers up to a frequency ω_0, and from ω_0 upwards the phase shift $B(\omega)$ is specified to be constant at a level equal to $-k\pi/2$ radians. Substitution in Eq. (7.51) gives the gain and phase-shift characteristics in the rest of the frequency spectrum as

$$B(\omega_c) = -k\sin^{-1}\left(\frac{\omega_c}{\omega_0}\right), \quad \text{for} \quad \omega_c < \omega_0$$

$$A(\omega_c) = A_0 - k\log_\varepsilon\left[\frac{\omega_c}{\omega_0} + \sqrt{\left(\frac{\omega_c}{\omega_0}\right)^2 - 1}\right], \quad \text{for} \quad \omega_c > \omega_0, \tag{7.52}$$

which are shown plotted in Fig. 7.13 for the case of $A_0 = 3{\cdot}46$ nepers ($=30$ decibels) and $k = \frac{5}{3}$. We shall refer to this characteristic as a *constant-phase cut-off.*

When the frequency ω_c becomes large compared with ω_0 we find from Eq. (7.52) that the gain approximates to:

$$A(\omega_c) \simeq A_0 - k\log_\varepsilon\left(\frac{2\omega_c}{\omega_0}\right), \quad \text{for} \quad \omega_c \gg \omega_0. \tag{7.53}$$

Therefore, at high frequencies, the gain decreases asymptotically at the rate of $-6k$ decibels per octave. If this asymptote is extended to the left, it intersects the flat portion of the gain characteristic at $\omega_0/2$, as can be deduced by setting $A(\omega_c) = A_0$ in Eq. (7.53). We thus see that the gain characteristic of Fig. 7.13 bears a resemblance to a semi-infinite slope, except that the flat portion extends one octave higher in frequency. This device may be specified in terms of either the gain slope or the constant phase value, (a slope of $6k$ decibels per octave corresponding to a constant phase shift of $k\pi/2$ radians).

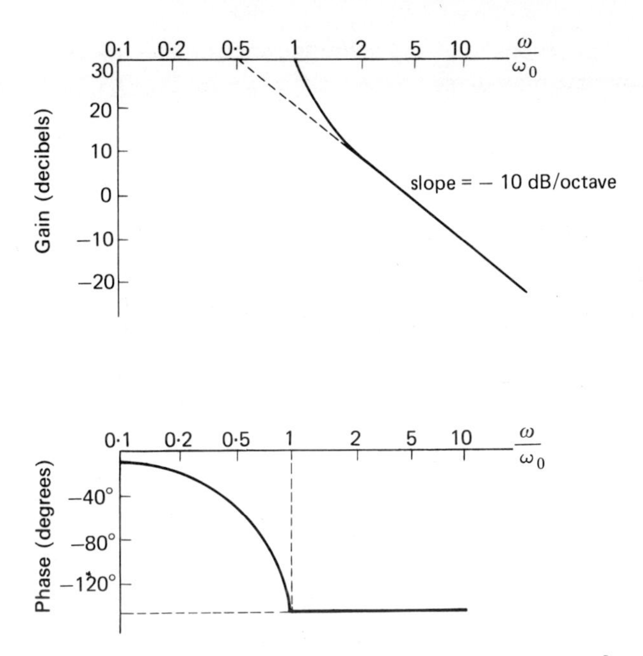

Fig. 7.13. Constant-phase cut-off characteristic for $k = \frac{5}{3}$.

Example 7.4. *Synthesis of optimum inter-stage one-port network.* Consider the situation depicted in Fig. 7.14, representing the model of a two-stage pentode amplifier. The total parasitic capacitance, that is, the sum of the output capacitance of the first stage and the input capacitance of the second stage is represented by C; and the elements introduced into the inter-stage by design are represented by $Z_1(j\omega)$. The total effective inter-stage impedance $Z(j\omega)$ is thus constrained to have a leading shunt capacitance of value C, and the problem is to synthesize $Z_1(j\omega)$ so as to realize the largest possible constant gain over the frequency range between zero and ω_0. These requirements are evidently satisfied if $Z(j\omega)$ has a magnitude equal to some constant R_0 inside the useful frequency band,

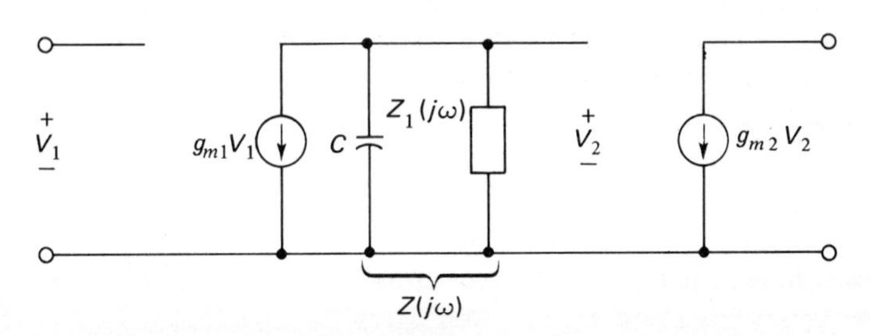

Fig. 7.14. Two-stage amplifier with inter-stage network.

and a phase angle equal to $-\pi/2$ at frequencies outside the band. To find the rest of the complete inter-stage characteristic, we use Eq. (7.52), obtaining

$$\begin{aligned} \text{ang } Z(j\omega) &= -\sin^{-1}\left(\frac{\omega}{\omega_0}\right), \qquad \text{for} \quad 0 < \omega < \omega_0 \\ |Z(j\omega)| &= \frac{R_0}{(\omega/\omega_0) + \sqrt{(\omega/\omega_0)^2 - 1}}, \qquad \text{for} \quad \omega > \omega_0. \end{aligned} \tag{7.54}$$

These results can be combined to yield the single formula

$$Z(j\omega) = \frac{R_0}{(j\omega/\omega_0) + \sqrt{1 - \omega^2/\omega_0^2}} \tag{7.55}$$

or in terms of the admittance function,

$$Y(j\omega) = \frac{1}{R_0}\left(\frac{j\omega}{\omega_0} + \sqrt{1 - \frac{\omega^2}{\omega_0^2}}\right). \tag{7.56}$$

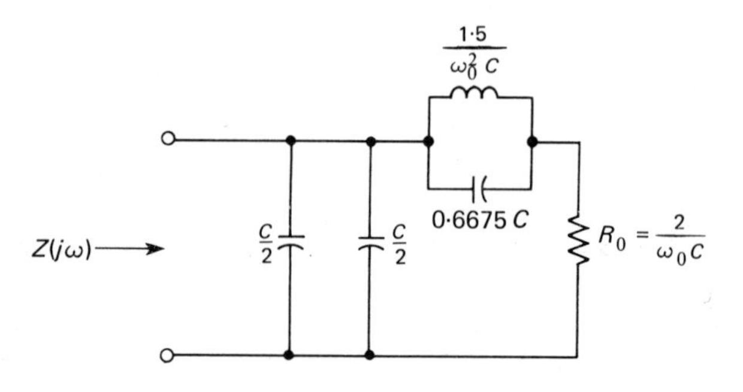

Fig. 7.15. A realization of the optimum inter-stage network.

Equation (7.56) defines the admittance of the optimum inter-stage network. It is identified to be the admittance of a capacitance $1/\omega_0 R_0$ in parallel with the mid-shunt image admittance of a low-pass filter of the constant-k type, with a cut-off frequency ω_0. For practical purposes, such a structure may be approximated by a finite network; the *Wheeler network* of Fig. 7.15 shows one such approximation.[2] The relation

$$R_0 = \frac{2}{\omega_0 C}$$

fixes the maximum constant magnitude of inter-stage impedance achievable over a given frequency band and with a given parasitic capacitance.

Concluding Remarks

To conclude, a transfer function $H(s)$ is of the minimum-phase type if and only if:

1. $H(s)$ has no poles or zeros in the right half of the s-plane.
2. Any poles of $H(s)$ which occur on the imaginary axis are simple. There is, however, no necessity that zeros of $H(s)$ on the imaginary axis be simple.

For $s = j\omega$, we may write $\log_\varepsilon H(j\omega) = A(\omega) + jB(\omega)$. In a minimum-phase network, the real part $A(\omega)$ (i.e., the gain function) completely determines the imaginary part $B(\omega)$ (i.e., the phase function), while $B(\omega)$ determines $A(\omega)$ within an arbitrary constant equal to $A(\infty)$. Consequently, in a minimum-phase network no change in the phase characteristic can be made without, at the same time, affecting the associated gain characteristic.

In Table 7.1 we have tabulated, for convenient reference, the various forms of the integral relations developed between $A(\omega)$ and $B(\omega)$. These relations are of great practical importance in the synthesis of equalizer networks, and shaping the loop gain of single-loop feedback amplifiers.

Table 7.1. Integral Relations between $A(\omega)$ and $B(\omega)$

$$
\begin{aligned}
A(\omega_c) - A(\infty) &= -\frac{1}{\pi}\int_{-\infty}^{\infty} \frac{B(\omega)}{\omega - \omega_c}\,d\omega \\
&= -\frac{2}{\pi}\int_{0}^{\infty} \frac{\omega B(\omega) - \omega_c B(\omega_c)}{\omega^2 - \omega_c^2}\,d\omega \\
&= \frac{1}{\pi\omega_c}\int_{0}^{\infty} \frac{d(\omega B)}{d\omega}\log_\varepsilon\left|\frac{\omega - \omega_c}{\omega + \omega_c}\right| d\omega \\
&= -\frac{1}{\pi\omega_c}\int_{-\infty}^{\infty} \frac{d(\omega B)}{du}\log_\varepsilon\left(\coth\frac{|u|}{2}\right) du
\end{aligned}
$$

$$
\begin{aligned}
B(\omega_c) &= \frac{1}{\pi}\int_{-\infty}^{\infty} \frac{A(\omega)}{\omega - \omega_c}\,d\omega \\
&= \frac{2\omega_c}{\pi}\int_{0}^{\infty} \frac{A(\omega) - A(\omega_c)}{\omega^2 - \omega_c^2}\,d\omega \\
&= -\frac{1}{\pi}\int_{0}^{\infty} \frac{dA}{d\omega}\log_\varepsilon\left|\frac{\omega - \omega_c}{\omega + \omega_c}\right| d\omega \\
&= \frac{1}{\pi}\int_{-\infty}^{\infty} \frac{dA}{du}\log_\varepsilon\left(\coth\frac{|u|}{2}\right) du
\end{aligned}
$$

REFERENCES

1. W. A. LYNCH, "The Stability Problem in Feedback Amplifiers," *Proc. I.R.E.* **39,** 1000 (1951).
2. H. W. BODE, *Network Analysis and Feedback Amplifier Design*. Van Nostrand, 1945.
3. E. A. GUILLEMIN, *The Mathematics of Circuit Analysis*. Wiley, 1949.
4. D. E. THOMAS, "Tables of Phase Associated with a Semi-infinite Unit Slope of Attenuation," *B.S.T.J.* 870 (1947).
5. J. G. THOMASON, *Linear Feedback Analysis*. Pergamon, 1956.
6. G. G. GOURIET, "Two Theorems Concerning Group Delay with Practical Applications to Delay Correction," *Proc. I.E.E.* **104,** 240 (1957).

PROBLEMS

7.1 Figure P7.1 shows a lossless two-port network coupling a load R_l to a current source of shunt capacitance C. Using the resistance-integral theorem, show that for a flat transmission characteristic over a frequency band between zero and ω_0, we must have[2]

$$\left|\frac{V_2}{I_s}\right|^2 = \frac{\pi R_l}{2\omega_0 C}.$$

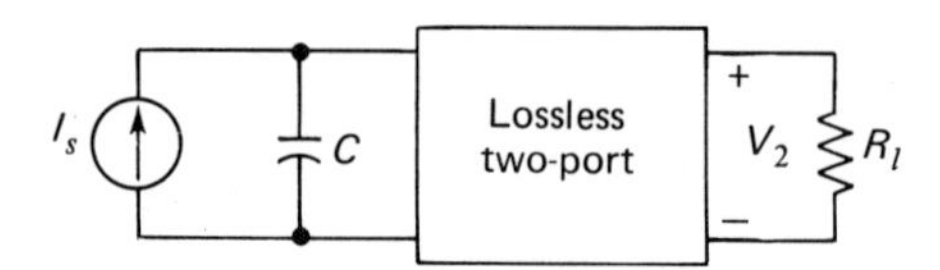

Fig. P7.1.

7.2 Using the phase-integral theorem, show that the insertion of the inductor L in the network of Fig. P7.2 introduces a phase advance into the frequency response.

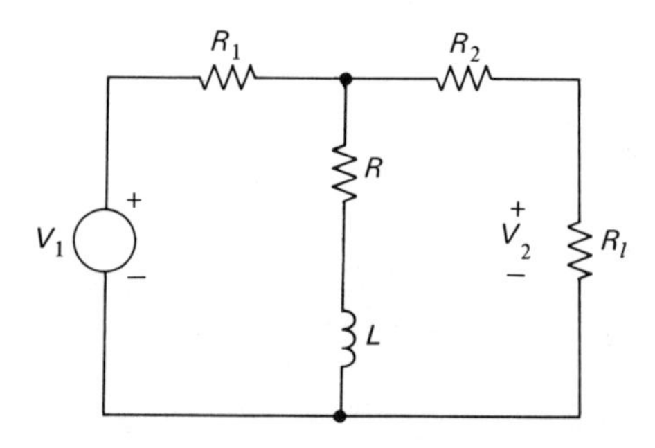

Fig. P7.2.

7.3 Evaluate the overall transfer function of the two-path network of Fig. P7.3; hence, show the network is of the non-minimum-phase type.[5] Construct a Bode diagram for the network.

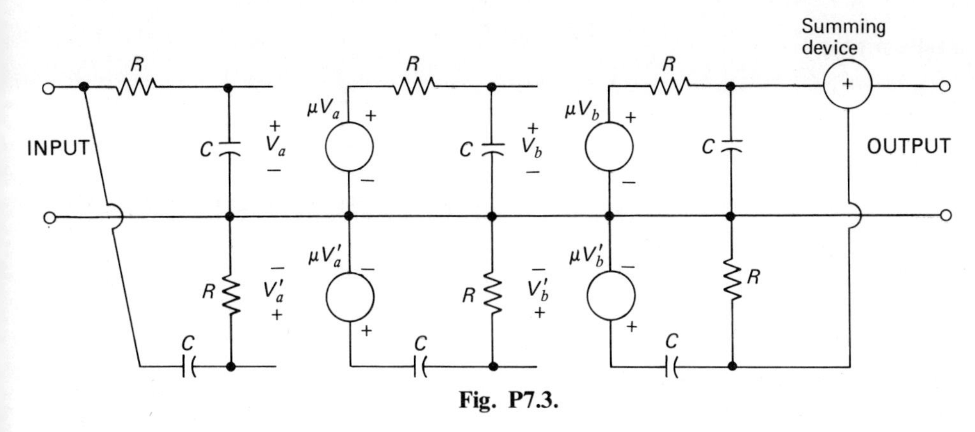

Fig. P7.3.

7.4 Measurements on a minimum-phase network yield the gain response shown plotted in Fig. P7.4. Evaluate the associated phase response.

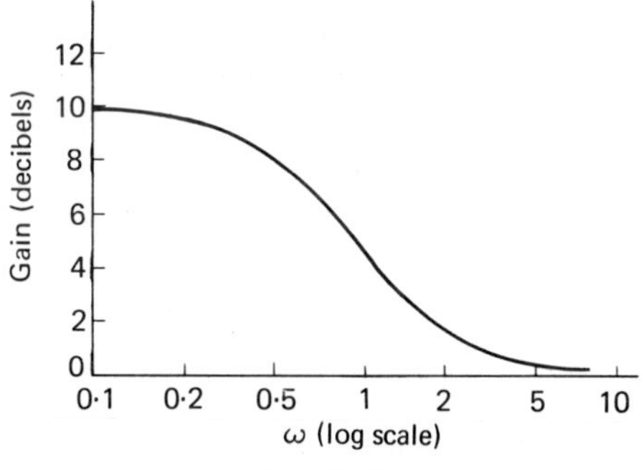

Fig. P7.4.

7.5 Consider $\dot{f}(s)/f(s)$, in which $f(s)$ is a transfer function and $\dot{f}(s)$ is its first derivative with respect to s. Using Cauchy's theorem, show that[6]

$$\dot{A}(\omega_c) = \frac{1}{\pi}\int_{-\infty}^{\infty} \frac{\tau(\omega)}{\omega - \omega_c}\, d\omega$$

$$\tau(\omega_c) = \frac{1}{\pi}\int_{-\infty}^{\infty} \frac{\dot{A}(\omega_c)}{\omega - \omega_c}\, d\omega$$

where $\dot{A}(\omega) = dA/d\omega$ is the gain slope, and $\tau(\omega) = -dB/d\omega$ is the group delay. It is assumed that $f(j\omega) = A(\omega) + jB(\omega)$.

CHAPTER 8

TRANSIENT RESPONSE

8.1 INTRODUCTION

The transient behaviour of any linear network is contained implicitly in the transfer function of the network in the sense that if the transfer function is known, then the response of the network to a specified excitation can be determined by computing the inverse Laplace transform of the product of the transfer function and excitation transform. In this chapter, we shall be mainly concerned with the transient behaviour of low-pass amplifiers. It will be shown that if the response of the amplifier to a unit-step function consists of a *monotonic* rise to a final constant value, then simple expressions for the *delay* and *rise time* of the amplifier can be computed directly from the transfer function. Moreover, such a formulation enables us to assess analytically the overall transient behaviour of a multistage amplifier in terms of the individual stage responses in a rather simple fashion, and also establish useful relations between the steady-state frequency response and step response of the amplifier. However, before proceeding to develop the mechanics of the method, it is rather instructive to examine the procedures of conventional transient-response analysis.

8.2 STEP RESPONSE OF LOW-PASS AMPLIFIERS

The transfer function $H(s)$ of a linear, time-invariant network comprising lumped elements is, by definition, equal to the ratio of the Laplace transform of the response (i.e., output) to that of the excitation (i.e., input), under zero initial conditions. Thus, we may write

$$H(s) = \frac{R(s)}{E(s)} \tag{8.1}$$

or

$$R(s) = E(s)H(s), \tag{8.2}$$

where $R(s)$ is the response transform defined by

$$R(s) = \mathscr{L}[r(t)] = \int_0^\infty r(t)\, \varepsilon^{-st}\, dt \tag{8.3}$$

and $E(s)$ is the excitation transform similarly defined. The signals $e(t)$ and $r(t)$ may each represent a voltage or a current.

In a low-pass amplifier the useful frequency band extends down to zero frequency, as shown by the transfer function:

$$H(s) = K_0 \frac{1 + a_1 s + a_2 s^2 + \cdots + a_n s^n}{1 + b_1 s + b_2 s^2 + \cdots + b_m s^m}, \tag{8.4}$$

where the a's and b's are real constants; K_0 is a scale factor equal to the gain of the amplifier at zero frequency; and $m > n$. For convenience in analysis, we shall use the *normalized* transfer function

$$H_n(s) = \frac{H(s)}{K_0} = \frac{1 + a_1 s + a_2 s^2 + \cdots + a_n s^n}{1 + b_1 s + b_2 s^2 + \cdots + b_m s^m}. \tag{8.5}$$

When the amplifier is initially at rest and if the excitation is a unit-step function $u(t)$ (with a Laplace transform equal to $1/s$), the resultant response function, denoted by $h_n^{-1}(t)$, may be determined by means of the inverse Laplace transformation,

$$h_n^{-1}(t) = \frac{1}{2\pi j} \int_{c-j\infty}^{c+j\infty} \frac{1}{s} H_n(s)\, \varepsilon^{st}\, ds, \tag{8.6}$$

where c is a constant greater than σ_a, the abscissa of absolute convergence for $H_n(s)$. In a stable amplifier, the poles of $H_n(s)$ all lie in the left half of the s-plane; we may therefore choose $c > 0$.

Evaluation of the inversion integral in Equation (8.6) requires the use of the calculus of residues. A simpler method of obtaining $h_n^{-1}(t)$ which is adequate for a great majority of cases of interest, is to express the response transform

$$\mathscr{L}[h_n^{-1}(t)] = \frac{1}{s} H_n(s) \tag{8.7}$$

as a sum of partial fractions, and then identify each term with a corresponding time function with the aid of a table of transform pairs.* Thus, if the poles of $H_n(s)$ are located at $s = s_1, s_2, \ldots, s_p$, and the order of the pole at $s = s_i$ is q_i, the partial fraction expansion of $(1/s)H_n(s)$, in general, takes the following form[1]

$$\frac{1}{s} H_n(s) = \frac{1}{s} + \sum_{i=1}^{p} \sum_{j=1}^{q_i} \frac{k_{ij}}{(s - s_i)^{q_i - j + 1}}, \tag{8.8}$$

where the inner sum with the index j accounts for each of the q_i terms associated with a particular pole s_i, while the outer sum with the index i accounts for each of the p poles. In Eq. (8.8) we have $q_1 + q_2 + \cdots + q_p = m$, where m is the order of the denominator polynomial of $H_n(s)$. The residue k_{ij} may be evaluated as

* A short table of Laplace transform pairs is given in Chapter 1.

follows: for $q_i = 1$, that is, a simple pole at $s = s_i$,

$$k_{ii} = \left[\frac{s - s_i}{s} H_n(s)\right]_{s=s_i}. \tag{8.9}$$

For a pole of order q_i at $s = s_i$,

$$k_{ij} = \frac{1}{(j-1)!}\left[\frac{d^{j-1}}{ds^{j-1}} \frac{(s - s_i)^{q_i}}{s} H_n(s)\right]_{s=s_i}. \tag{8.10}$$

Using the expansion of Eq. (8.8), the step response $h_n^{-1}(t)$ for $t \geq 0$ can now be written as

$$h_n^{-1}(t) = 1 + \sum_{i=1}^{p} \sum_{j=1}^{q_i} \frac{k_{ij}}{(q_i - j)!} t^{q_i - j} \varepsilon^{s_i t}. \tag{8.11}$$

It is noteworthy that, as a result of normalization, the step response has a final value of unity, which may readily be verified by using the final value theorem of the Laplace transformation.*

Example 8.1. *Shunt-compensated inter-stage network.* Consider a single-stage pentode amplifier having the model shown in Fig. 8.1 where an inductor L has been added in series with the load resistor R to compensate for the high-frequency effects of the total parasitic capacitance C. The amplifier has a voltage gain

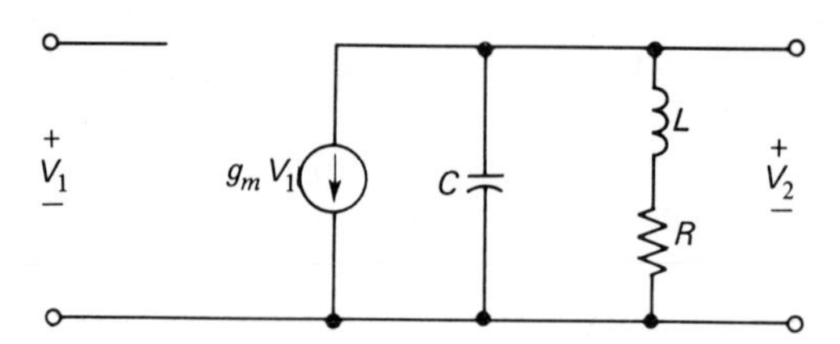

Fig. 8.1. Shunt-compensated inter-stage network.

equal to $-g_m R$ at zero frequency. Normalizing the transfer-voltage ratio V_2/V_1 of the amplifier with respect to $-g_m R$, we obtain,

$$H_n(s) = \frac{1 + m\tau_0 s}{1 + \tau_0 s + m\tau_0^2 s^2}, \tag{8.12}$$

where

$$\tau_0 = RC, \tag{8.13}$$

$$m = \frac{L}{CR^2}. \tag{8.14}$$

* The final value theorem states that if $F(s)$ is the Laplace transform of $f(t)$, then $f(\infty) = \lim_{s \to 0} sF(s)$.

Thus, $H_n(s)$ has a zero at $s = -1/m\tau_0$, and two poles located at

$$s_1, s_2 = \frac{1}{2m\tau_0}[-1 \pm \sqrt{1 - 4m}]. \tag{8.15}$$

We may therefore identify three possible cases, depending upon the value of parameter m:

1. When $m < \frac{1}{4}$, the poles of $H_n(s)$ are real and different, and the step response of the amplifier is *overdamped*.
2. When $m = \frac{1}{4}$, a second-order pole occurs at $s = -2/\tau_0$, and the step response is *critically damped.*
3. When $m > \frac{1}{4}$, the two poles are complex conjugates, and the step response is *underdamped* or *oscillatory*.

The pole–zero patterns and corresponding step-response curves pertaining to these three cases are plotted in Fig. 8.2. We see that the critically damped response exhibits the fastest possible monotonic rise. Also, as m is increased the speed of rise is increased, with oscillations appearing for values of m greater than $\frac{1}{4}$.

8.3 FORCED RESPONSE TO PERIODIC NONSINUSOIDAL EXCITATIONS

In the study of linear networks, a periodic nonsinusoidal function (e.g., a square wave) is often used as the excitation. Let $e(t)$ be a periodic function of period T,

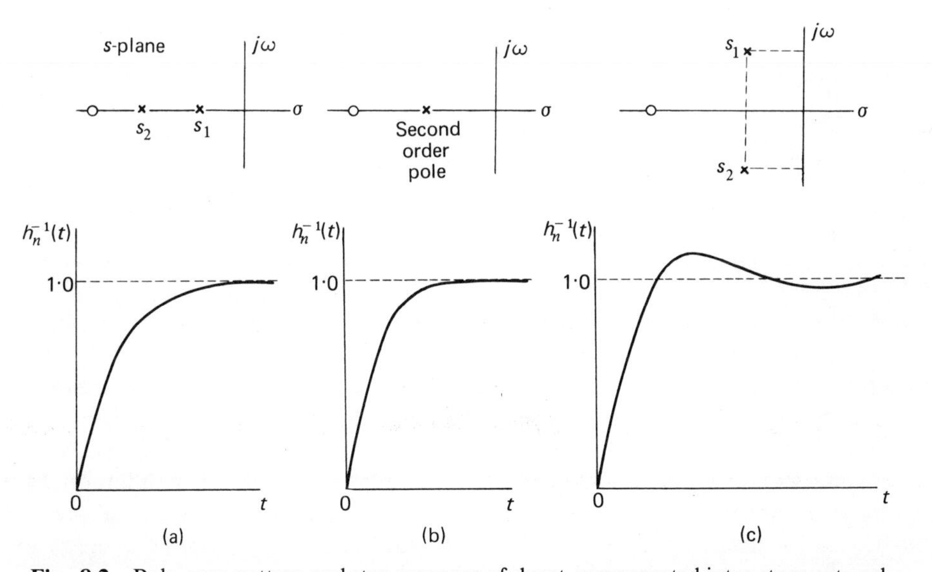

Fig. 8.2. Pole–zero pattern and step response of shunt-compensated inter-stage network: (a) Overdamped, i.e., $m < \frac{1}{4}$; (b) Critically damped, i.e., $m = \frac{1}{4}$; (c) Underdamped, i.e. $m > \frac{1}{4}$.

and $e_1(t)$ be the function describing $e(t)$ during the first period. Then, we may write

$$e(t) = e_1(t) + e_1(t - T) + e_1(t - 2T) + \cdots. \tag{8.16}$$

It is assumed that $e(t)$ is zero for $t < 0$ and that $e_1(t)$ is transformable. To evaluate the Laplace transform of $e(t)$, we may use the *shifting theorem* which states that delay in the time domain by an amount a corresponds to multiplication by the factor ε^{-as} in the complex-frequency domain. Thus, taking the Laplace transform of Eq. (8.16), we obtain

$$\begin{aligned} E(s) = \mathscr{L}[e(t)] &= (1 + \varepsilon^{-Ts} + \varepsilon^{-2Ts} + \cdots)E_1(s) \\ &= \frac{E_1(s)}{1 - \varepsilon^{-Ts}}, \end{aligned} \tag{8.17}$$

where $E_1(s)$ is the Laplace transform of the function corresponding to the first period. The Laplace transform of a transformable periodic function is therefore fully determined by the first period of the function.

Now, suppose the function $e(t)$ is applied to a network whose transfer function is $H(s)$; the product of the transfer function and the excitation transform from Eq. (8.17) gives the response transform as

$$\begin{aligned} R(s) &= E(s)H(s) \\ &= \frac{E_1(s)}{1 - \varepsilon^{-Ts}} H(s). \end{aligned} \tag{8.18}$$

The complete response function for all t is therefore

$$r(t) = \mathscr{L}^{-1}\left[\frac{E_1(s)}{1 - \varepsilon^{-Ts}} H(s)\right]. \tag{8.19}$$

Examination of Eq. (8.19) indicates that the response $r(t)$ consists of two components: natural and forced. The natural-response component is characterized by the poles of the transfer function $H(s)$ (which are natural frequencies of the network), while the forced-response component is characterized by the poles of the excitation transform. We shall assume that the transfer function $H(s)$ has no poles on the imaginary axis of the s-plane (including the point at infinity), so that the natural response component eventually decays to zero and only the forced response component remains as the steady-state response of the network.

The function $E_1(s)$, pertaining to the first period, is ordinarily differentiable everywhere in the s-plane, so that the zeros of the denominator factor $1 - \varepsilon^{-Ts}$ in Eq. (8.17) make up all the poles of the excitation transform. The zeros of $1 - \varepsilon^{-Ts}$ occur when

$$\varepsilon^{-Ts} = 1$$

or

$$s = j2k\pi/T,$$

where k is an integer. The excitation transform $E(s)$ therefore has an infinite number of simple poles on the imaginary axis.

If we were to proceed to evaluate the response function $r(t)$ in the usual manner (i.e., expand $R(s)$ in partial fractions and then determine the inverse Laplace transform term by term), we would find, in general, that $r(t)$ contains terms of the form $k_\nu \varepsilon^{s_\nu t}$ (contributed by the poles of $H(s)$, with s_ν denoting a typical pole and k_ν the residue of $E(s)H(s)$ at that pole) plus an infinite number of sinusoidal terms. In other words, the partial-fraction expansion procedure yields an expression for the response function $r(t)$ that is in the form of a transient term plus a Fourier series, representing the natural-response and the periodic forced-response components, respectively. However, a simpler procedure is to recognize that since the forced response is periodic and of the same period as the excitation function, then clearly the forced response in the first period is precisely of the same form as the forced response in any other period. It is therefore sufficient to determine the forced response in the first period only. Such an evaluation may proceed as follows:[2,3]

1. The natural-response component, denoted as $r_N(t)$, is determined for all t. Since the natural response is characterized by the poles of the transfer function $H(s)$, we may write for all t

$$r_N(t) = \sum_\nu k_\nu \varepsilon^{s_\nu t}, \tag{8.20}$$

where s_ν denotes a pole of $H(s)$, assumed simple, and k_ν denotes the residue of $[E_1(s)/(1 - \varepsilon^{-Ts})]H(s)$ at this pole.

2. The function $r_1(t)$ that describes the total response during the first period is evaluated from the relation

$$r_1(t) = \mathscr{L}^{-1}[E_1(s)H(s)] \tag{8.21}$$

3. The forced-response component $r_F(t)$ during the first period is determined in closed form by subtracting $r_N(t)$ from $r_1(t)$; thus,

$$r_F(t) = r_1(t) - r_N(t) \tag{8.22}$$

for $0 \le t \le T$.

Example 8.2. *Square-wave response of transistor amplifier.* To illustrate this procedure, consider first the periodic square voltage wave $v_1(t)$ of Fig. 8.3(a). As illustrated in part (b) of the diagram, the function $v_{11}(t)$ describing the first period of this waveform may be represented as the sum of three step functions, alternately positive and negative; so that we may write

$$v_{11}(t) = u(t) - 2u\left(t - \frac{T}{2}\right) + u(t - T), \tag{8.23}$$

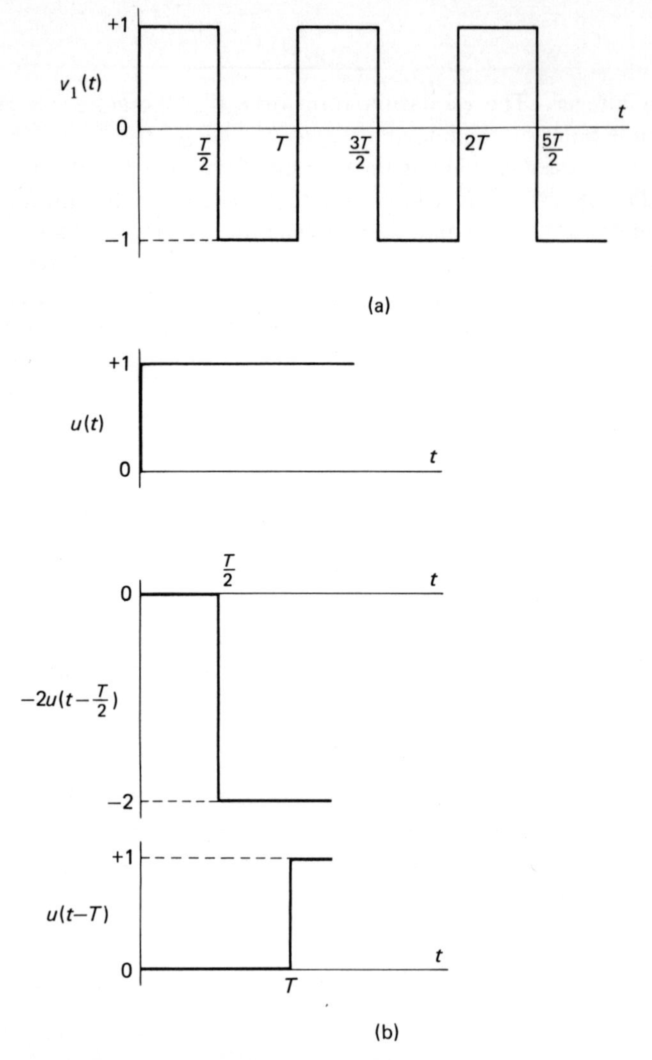

Fig. 8.3. (a) Periodic square wave; (b) Representation of the first period of the square wave as the sum of three step functions.

the Laplace transform of which is

$$
\begin{aligned}
V_{11}(s) &= \frac{1}{s}(1 - 2\varepsilon^{-(T/2)s} + \varepsilon^{-Ts}) \\
&= \frac{1}{s}(1 - \varepsilon^{-(T/2)s})^2.
\end{aligned}
\tag{8.24}
$$

The excitation transform is therefore

$$\begin{aligned} V_1(s) &= \frac{V_{11}(s)}{1 - \varepsilon^{-Ts}} \\ &= \frac{1 - \varepsilon^{-(T/2)s}}{s(1 + \varepsilon^{-(T/2)s})}. \end{aligned} \tag{8.25}$$

Now, consider Fig. 8.4(a) depicting a common-emitter amplifier coupling a load resistance R_l to a voltage source that supplies $v_1(t)$. To determine the resultant response, the transistor is first replaced by a simplified hybrid-π model, as in Fig. 8.4(b). The internal feedback existing around the controlled source, due to the collector capacitance $C_{b'c}$, gives rise to the well-known *Miller effect* as a result of which the collector capacitance has the effect of a larger capacitance

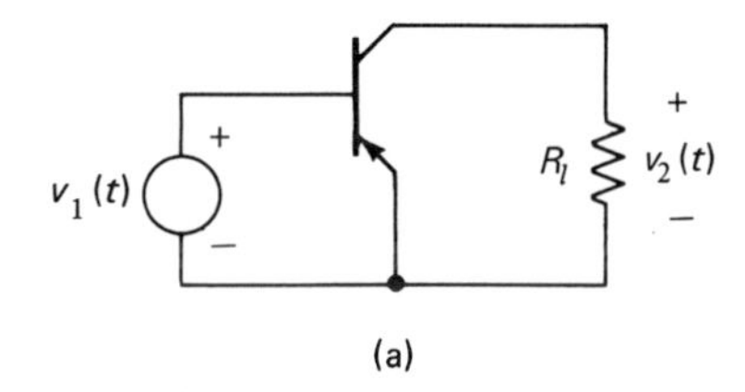

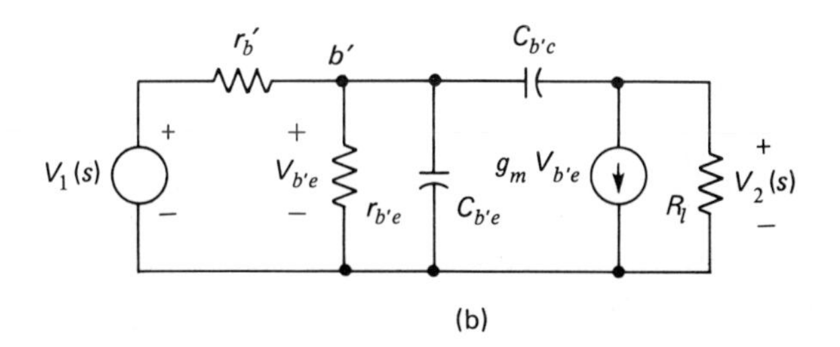

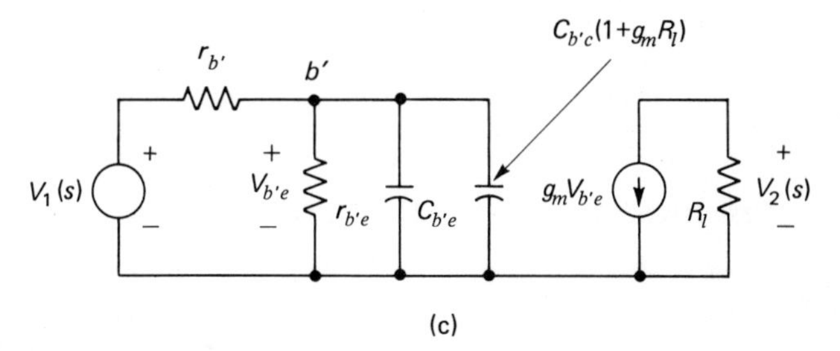

Fig. 8.4. (a) Common-emitter amplifier; (b) High frequency model of the amplifier; (c) Simplified model.

at the input port. More precisely, in so far as the conditions at the input port are concerned, we may replace $C_{b'c}$ by an equivalent capacitance equal to $C_{b'c}(1 + g_m R_l)$ connected between node b' and ground, as shown in Fig. 8.4(c). We thus have

$$H(s) = \frac{V_2(s)}{V_1(s)} = \frac{K_0}{1 + \tau_1 s}, \tag{8.26}$$

where

$$K_0 = \frac{-g_m r_{b'e} R_l}{r_{b'} + r_{b'e}} \tag{8.27}$$

$$\tau_1 = \frac{r_{b'} r_{b'e}}{r_{b'} + r_{b'e}} [C_{b'e} + C_{b'c}(1 + g_m R_l)]. \tag{8.28}$$

From Eqs. (8.25) and (8.26), the response transform is

$$V_2(s) = \frac{K_0(1 - \varepsilon^{-(T/2)s})}{s(1 + \tau_1 s)(1 + \varepsilon^{-(T/2)s})}, \tag{8.29}$$

which may be expressed in the form of a series as follows

$$\begin{aligned} V_2(s) &= \frac{K_0}{s(1 + \tau_1 s)}(1 - \varepsilon^{-(T/2)s})(1 - \varepsilon^{-(T/2)s} + \varepsilon^{-Ts} - \varepsilon^{-(3T/2)s} + \cdots) \\ &= \frac{K_0}{s(1 + \tau_1 s)}(1 - 2\varepsilon^{-(T/2)s} + 2\varepsilon^{-Ts} - 2\varepsilon^{-(3T/2)s} + \cdots). \end{aligned} \tag{8.30}$$

Recognizing that a term of the form ε^{-as} corresponds to a delay of a seconds in the time domain, it follows that the transform of the response function during the first period can be evaluated from Eq. (8.30) by disregarding all terms containing ε^{-as} with $a \geq T$. That is, the transform of the total response function during the first period is given by

$$V_{21}(s) = \frac{K_0}{s(1 + \tau_1 s)}(1 - 2\varepsilon^{-(T/2)s}). \tag{8.31}$$

Hence, the inverse transform of Eq. (8.31) yields

$$v_{21}(t) = \begin{cases} K_0(1 - \varepsilon^{-t/\tau_1}) & \text{for } 0 < t \leq \dfrac{T}{2} \\ K_0[-1 + \varepsilon^{-t/\tau_1}(2\varepsilon^{T/2\tau_1} - 1)] & \text{for } \dfrac{T}{2} \leq t \leq T. \end{cases} \tag{8.32}$$

The natural response component results from the pole at $s_1 = -1/\tau_1$. The residue k_1 of $V_2(s)$ at this pole is obtained from Eq. (8.29) as

$$k_1 = -\frac{K_0(1 - \varepsilon^{T/2\tau_1})}{1 + \varepsilon^{T/2\tau_1}}. \tag{8.33}$$

Hence,

$$v_{2N}(t) = k_1 \varepsilon^{s_1 t} = -\frac{K_0(1 - \varepsilon^{T/2\tau_1})}{1 + \varepsilon^{T/2\tau_1}} \varepsilon^{-t/\tau_1}. \tag{8.34}$$

Subtraction of $v_{2N}(t)$ from $v_{21}(t)$ gives the forced-response component during the first period as

$$v_{2F}(t) = \begin{cases} K_0\left[1 - \left(\dfrac{2\varepsilon^{T/2\tau_1}}{\varepsilon^{T/2\tau_1} + 1}\right)\varepsilon^{-t/\tau_1}\right] & \text{for} \quad 0 \le t \le \dfrac{T}{2} \\ K_0\left[-1 + \left(\dfrac{2\varepsilon^{T/\tau_1}}{\varepsilon^{T/2\tau_1} + 1}\right)\varepsilon^{-t/\tau_1}\right] & \text{for} \quad \dfrac{T}{2} \le t \le T. \end{cases} \tag{8.35}$$

For all other periods, we have

$$v_{2F}(t + kT) = v_{2F}(t), \tag{8.36}$$

where k is an integer. In Fig. 8.5 we have shown $v_{2F}(t)$ plotted against t/τ_1 for three different values of $T/2\tau_1$. It is apparent that for a true reproduction of the input waveform, the time constant τ_1 of the amplifier must be very small compared with the half period of the input square wave.

8.4 THE DEFINITION OF RISE TIME AND DELAY

It is evident that the three types of step-response curves shown in Fig. 8.2 possess two common features: a delay which occurs before the response is well under way, and a finite time of rise. Thus, for many purposes the step response of a low-pass amplifier may be sufficiently well described in terms of its *delay* and *rise time.* In experimental work, the delay τ_D is usually defined as the time required for the response to reach 50 per cent of its final value measured from the application of the input step; and the rise time τ_R is usually defined as the time required for the response to increase from 10 to 90 per cent of its final value. These definitions are illustrated in Fig. 8.6 together with the commonly used definition of *overshoot.*

Although these definitions of delay and rise time lend themselves readily to measurement in the laboratory, they are extremely awkward for making computations, or for evaluating the relative merits of various methods of compensating an amplifier to improve its speed of rise. The difficulty, of course, lies in the necessity for computing the step-response curve for each case under consideration.

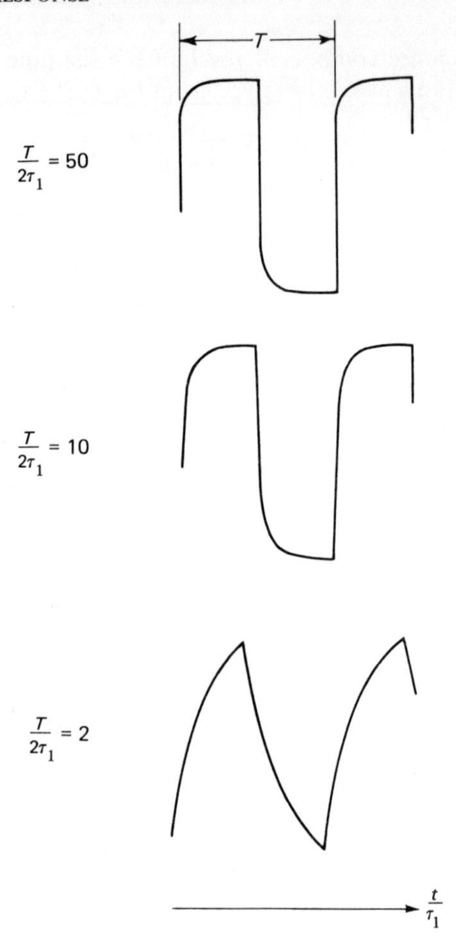

Fig. 8.5. Illustrating the forced response of common-emitter amplifier to a square wave for different $T/(2\tau_1)$.

Indeed, unless a digital computer is available, such an undertaking can often be quite formidable.

However, in the important class of low-pass amplifiers with a step response that is a *monotonic* function of time (i.e., non-decreasing), this computational difficulty is overcome by following Elmore's analysis[4] which yields essentially the same results for delay and rise time as obtained from the previous definitions. Since the time derivative of the response of a low-pass amplifier to a unit-step function is equal to the response $h_n(t)$ to a unit-impulse function, the step response $h_n^{-1}(t)$ will be monotonic over the range of t from 0 to ∞, provided $h_n(t)$ is non-negative throughout this range. Equivalently, the step response is monotonic if it exhibits no overshoot.

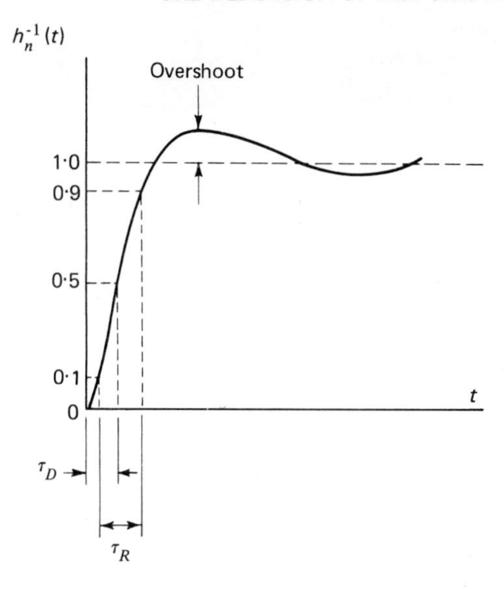

Fig. 8.6. Illustrating the definitions of rise time, delay and overshoot.

According to Elmore, the delay τ_D is defined as the *centre of gravity* of the area bounded by the impulse response of the amplifier, that is,

$$\tau_D = \frac{\int_0^\infty th_n(t)\,dt}{\int_0^\infty h_n(t)\,dt}. \tag{8.37}$$

Since, for a normalized step response that has a final value of unity, as in Fig. 8.7,

$$\int_0^\infty h_n(t)\,dt = 1, \tag{8.38}$$

the formula for the delay takes the simple form,

$$\tau_D = \int_0^\infty th_n(t)\,dt. \tag{8.39}$$

Except for the case of a very asymmetrical response curve, the result of using Eq. (8.39) differs but little from that obtained from the previous definition of delay τ_D.

Next, we observe that the shorter the rise time τ_R, the narrower (and higher) the impulse response. It is reasonable, therefore, to define τ_R as proportional to

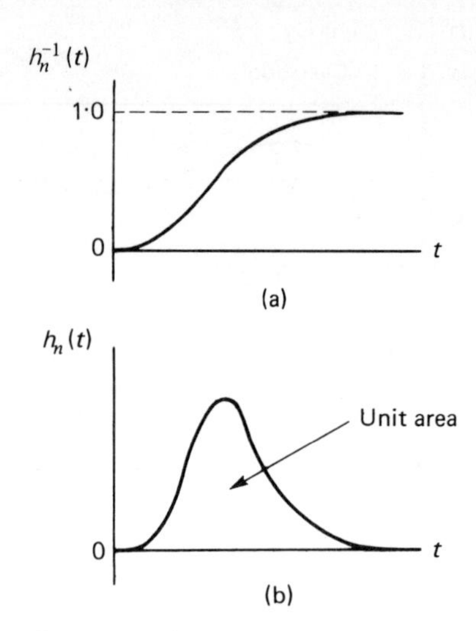

Fig. 8.7. Illustrating the monotonic transient response: (a) Step response; (b) Impulse response.

the *radius of gyration* of the area bounded by $h_n(t)$, that is,

$$\tau_R^2 = \frac{\text{const.} \int_0^\infty (t - \tau_D)^2 h_n(t)\, dt}{\int_0^\infty h_n(t)\, dt} \tag{8.40}$$

which, in view of Eq. (8.38), simplifies as

$$\tau_R^2 = \text{const.} \int_0^\infty (t - \tau_D)^2 h_n(t)\, dt. \tag{8.41}$$

In other words, the rise time is proportional to the *standard deviation* of the impulse response $h_n(t)$. Elmore chooses the constant of proportionality as 2π for the following reason: the impulse response of an infinite number of resistance–capacitance-coupled amplifier stages approaches a *Gaussian error function*, so that in order to make the result of applying the new definition of rise time to such an amplifier agree with that obtained for the rise time based on the maximum slope of the step-response curve, we should have

$$\tau_R = \frac{1}{h_n(t)|_{max}} = \sqrt{2\pi} \quad [\text{radius of gyration of } h_n(t)]. \tag{8.42}$$

This expresses the relation between the height and the radius of gyration of a Gaussian error curve of unit area.

We may now express Eq. (8.41) in the alternative form

$$\tau_R^2 = 2\pi\left[\int_0^\infty t^2 h_n(t)\,dt - \tau_D^2\right]. \tag{8.43}$$

It is important to note that in Eqs. (8.39) and (8.43) it is assumed that $h_n(t)$ is non-negative throughout the range of t from 0 to ∞; otherwise, they both become meaningless.

The usefulness of Elmore's definitions of delay and rise time lies in the fact that it is a simple matter to evaluate them directly from the transfer function $H_n(s)$. To demonstrate this, we first note that $H_n(s)$ is related to the impulse response $h_n(t)$ by the direct Laplace transformation

$$H_n(s) = \int_0^\infty h_n(t)\,\varepsilon^{-st}\,dt. \tag{8.44}$$

If, next, we expand ε^{-st} in a power series in st, and then integrate term by term, we obtain

$$\begin{aligned} H_n(s) &= \int_0^\infty h_n(t)\left[1 - st + \frac{(st)^2}{2!} - \cdots\right] dt \\ &= 1 - s\int_0^\infty th_n(t)\,dt + \frac{s^2}{2!}\int_0^\infty t^2 h_n(t)\,dt - \cdots. \end{aligned} \tag{8.45}$$

Substitution of Eqs. (8.39) and (8.43) in Eq. (8.45) yields

$$H_n(s) = 1 - s\tau_D + \frac{s^2}{2!}\left(\frac{\tau_R^2}{2\pi} + \tau_D^2\right) - \cdots. \tag{8.46}$$

If also we expand Eq. (8.5) in ascending powers of s by simple division, we get

$$H_n(s) = 1 - (b_1 - a_1)s + (b_1^2 - a_1 b_1 + a_2 - b_2)s^2 - \cdots. \tag{8.47}$$

Comparing Eqs. (8.45) and (8.47), and equating the coefficients of the s and s^2 terms, we obtain the desired results:

$$\tau_D = b_1 - a_1 \tag{8.48}$$

$$\tau_R = \sqrt{2\pi[b_1^2 - a_1^2 + 2(a_2 - b_2)]}. \tag{8.49}$$

Although the applicability of these two formulae is limited to linear, time-invariant networks with a monotonic step response, there are many practical situations where this kind of response is specifically sought as a design objective.

Example 8.3. *Shunt-compensated inter-stage network* (*continued*). The step response of the amplifier in Fig. 8.1 is clearly monotonic if it is not underdamped, i.e., $m \leq \frac{1}{4}$ (this keeps the poles of $H_n(s)$ on the negative real axis of the s-plane).

From Eq. (8.12), defining the system function of the amplifier, we note that

$$\begin{aligned} a_1 &= m\tau_0 \\ b_1 &= \tau_0 \\ b_2 &= m\tau_0^2. \end{aligned} \tag{8.50}$$

Therefore, insertion of these parameters in Eqs. (8.48) and (8.49) yields

$$\begin{aligned} \tau_D &= \tau_0(1 - m) \\ \tau_R &= \tau_0\sqrt{2\pi(1 - 2m - m^2)}. \end{aligned} \tag{8.51}$$

When $m = 0$, corresponding to an uncompensated amplifier stage, $\tau_D = \tau_0$ and $\tau_R = 2{\cdot}51\tau_0$. When $m = \frac{1}{4}$, corresponding to critical shunt compensation, $\tau_D = 3\tau_0/4$ and $\tau_R = 1{\cdot}66\tau_0$. To express the improvement realized by compensating the amplifier it is convenient to define the *rise-time figure of merit* η as the ratio of the rise time of the uncompensated amplifier to that of the compensated amplifier. Clearly, $\eta = \frac{2{\cdot}51}{1{\cdot}66} = 1{\cdot}51$ for a shunt-compensated amplifier adjusted for critical damping.

Example 8.4. *Relation between rise time and bandwidth.* Consider a linear amplifier the normalized transfer function of which is defined by

$$H_n(j\omega) = \varepsilon^{-c\omega^2}\varepsilon^{-j\omega\tau_D}, \tag{8.52}$$

where τ_D is a constant delay for all frequencies; and c is a constant related to the *half-power bandwidth B* as follows

$$c = \frac{\log_\varepsilon 2}{8\pi^2 B^2}, \tag{8.53}$$

with B defined (in cycles per second) from zero to the frequency at which $H_n(j\omega)$ drops to $1/\sqrt{2}$ of its zero-frequency value. As illustrated in Fig. 8.8, the magnitude response is in the form of a Gaussian error function, while the phase response is linear. An amplifier having such a characteristic is known as a *Gaussian amplifier.*

Our interest in the Gaussian amplifier is merely theoretical: to establish the fundamental relationships that exist between the transient and frequency response. The impulse response of this amplifier is a Gaussian error function also, as shown by[5]

$$h_n(t) = \frac{1}{2\sqrt{\pi c}}\varepsilon^{-(t-\tau_D)^2/4c}. \tag{8.54}$$

As indicated in Fig. 8.9, the impulse response $h_n(t)$ is non-negative for any t. Hence, the step response is monotonic, so that we may legitimately apply Elmore's definition to evaluate the rise time of the idealized Gaussian amplifier. Thus,

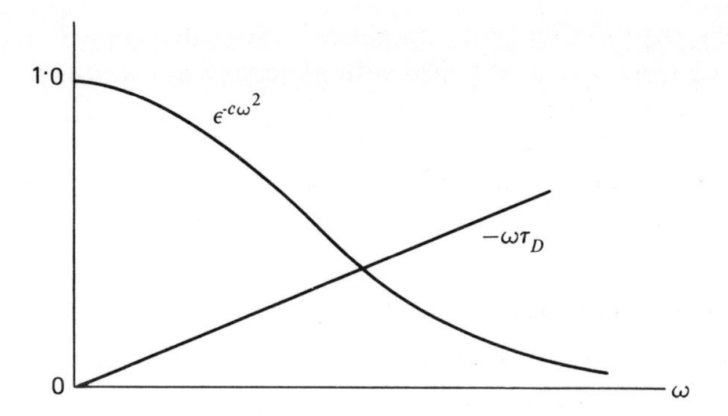

Fig. 8.8. Frequency response of idealized Gaussian amplifier.

putting $j\omega = s$ in Eq. (8.52), and expanding $H_n(s)$ in ascending powers of s, we obtain

$$H_n(s) = 1 - s\tau_D + \frac{s^2}{2}(2c + \tau_D^2) - \cdots. \tag{8.55}$$

Therefore, comparing Eqs. (8.46) and (8.55) we obtain

$$\frac{\tau_R^2}{2\pi} = 2c$$

or

$$\tau_R B = \left(\frac{\log_\varepsilon 2}{2\pi}\right)^{1/2} = 0{\cdot}332. \tag{8.56}$$

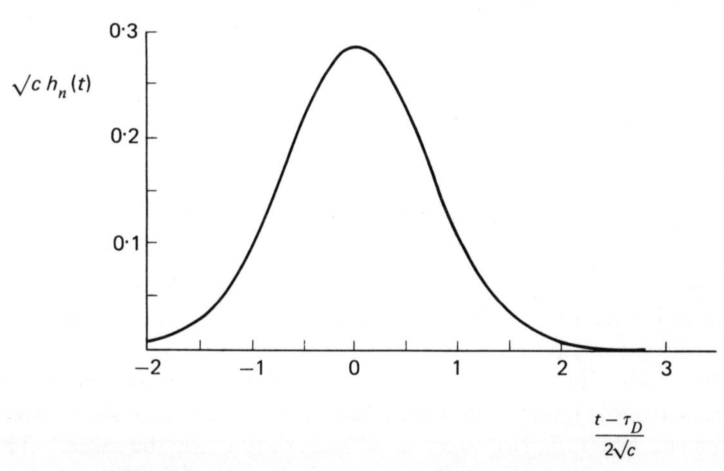

Fig. 8.9. Impulse response of Gaussian amplifier.

In other words, the rise time of the amplifier is inversely proportional to its bandwidth, so that a faster rise is obtained with a greater bandwidth.

It should be noted that the idealized Gaussian amplifier as defined by the transfer function of Eq. (8.52) is, strictly speaking, physically non-realizable. However, an approximate realization can be achieved using a sequence of cascaded low-pass *RC* networks with isolation amplifiers (see Problem 8.8).

8.5 MULTI-STAGE AMPLIFIERS

Consider a multi-stage amplifier consisting of N stages in cascade, with each stage having a monotonic transient response. Assuming unilateral stages, so that there is no interaction between adjacent ones, the overall transfer function of the entire amplifier is equal to the product of the transfer functions of the individual stages. Let the normalized transfer function of the ith stage be denoted by $H_{ni}(s)$, and its delay and rise time by τ_{Di} and τ_{Ri}, respectively. Then, according to Eq. (8.46) we may expand $H_{ni}(s)$ in the series

$$H_{ni}(s) = 1 - s\tau_{Di} + \frac{s^2}{2!}\left(\frac{\tau_{Ri}^2}{2\pi} + \tau_{Di}^2\right) - \cdots. \tag{8.57}$$

The overall transfer function of the complete amplifier therefore takes the form

$$\begin{aligned} H_n(s) &= \prod_{i=1}^{N} H_{ni}(s) \\ &= 1 - s\sum_i \tau_{Di} + \frac{s^2}{2!}\left[\sum_i \frac{\tau_{Ri}^2}{2\pi} + \sum_i \tau_{Di}^2 + \sum_{i\neq j} \tau_{Di}\tau_{Dj}\right] - \cdots. \end{aligned} \tag{8.58}$$

By using Eq. (8.46) again, we find that the overall delay and rise times of the complete amplifier are given by

$$\tau_D = \sum_{i=1}^{N} \tau_{Di} \tag{8.59}$$

and

$$\tau_R = \left[\sum_{i=1}^{N} \tau_{Ri}^2\right]^{1/2}. \tag{8.60}$$

Thus, Eq. (8.59) shows that in a multi-stage amplifier, the overall delay is the sum of the individual stage delays, while Eq. (8.60) shows that the overall rise time is the root mean square of the individual stage rise times.

Another result of interest is that in a cascade of similar and unilateral amplifier stages (e.g., pentode stages) the shortest possible overall rise time for a given overall gain is achieved when the rise times of all stages are made the same.[4] The proof requires minimizing τ_R in Eq. (8.60) (through the use of Lagrange's method of undetermined multipliers) subject to the condition that for a specified overall gain,

$$\prod_{i=1}^{N} \tau_{Ri} = \text{const.} \tag{8.61}$$

which results from the fact that the rise time of any stage in the amplifier is proportional to the gain of the stage, since both quantities are proportional to the load resistor. Thus, if τ_{R1} is the rise time of each stage, the overall rise time of an N-stage amplifier takes the form

$$\tau_R = \sqrt{N}\tau_{R1} \tag{8.62}$$

8.6 CONVOLUTION IN THE TIME DOMAIN

The method of *convolution* provides another approach for evaluating the transient response of a linear network by the integration of real-time functions. The basis of the method is to resolve the excitation function into a continuum of impulses, compute the response of the network to each impulse independently, and then obtain the total response as the superposition of the individual impulse responses. Consider a continuous function $e(t)$, which is zero for $t < 0$, applied as the input to a linear, time-invariant network of impulse response $h(t)$. Approximating the function $e(t)$ by a staircase function that is the sum of rectangular pulses of equal width Δt, as indicated by the dashed lines in Fig. 8.10, we may write

$$e(t) \simeq \sum_{k=0}^{\infty} e(k\Delta t)\left[\frac{u(t - k\Delta t) - u(t - k\Delta t - \Delta t)}{\Delta t}\right]\Delta t. \tag{8.63}$$

Clearly, as the incremental width Δt is made smaller, the approximation becomes progressively better, the product $k\Delta t$ assumes more closely the role of a continuous

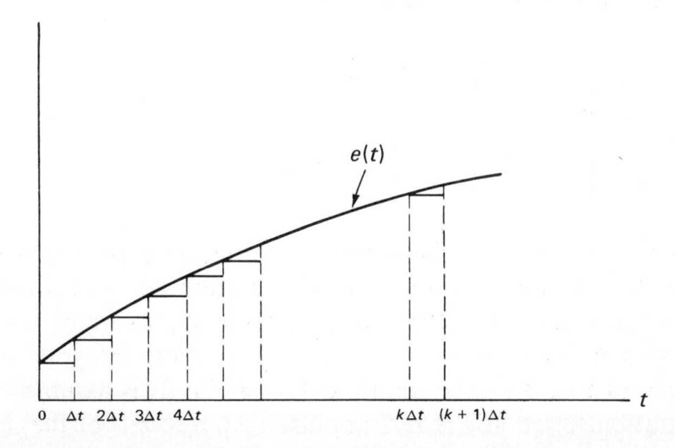

Fig. 8.10. Approximation of $e(t)$ by a sum of rectangular pulses.

variable τ (say), and the pulse defined by

$$\frac{u(t - k\Delta t) - u(t - k\Delta t - \Delta t)}{\Delta t}$$

approaches more closely a unit-impulse located at time $k\Delta t$. Thus, in the limit, as $\Delta t \to 0$, we find that $\Delta t \to d\tau$ and the summation of Eq. (8.63) defines the integral

$$e(t) = \int_0^\infty e(\tau)\delta(t - \tau)\, d\tau. \tag{8.64}$$

Equation (8.64) indicates that the function $e(t)$ is resolvable into a continuum of impulses, each of which is of the form $e(\tau)\delta(t - \tau)\, d\tau$. If $h(t)$ defines the response of the network to a unit impulse applied at $t = 0$, then invoking linearity and time-invariance, the response to an impulse of strength $e(\tau)\, d\tau$ applied at time τ is equal to $e(\tau)h(t - \tau)\, d\tau$. Summing the individual impulse responses gives the total response of the network as

$$r(t) = \int_0^\infty e(\tau)h(t - \tau)\, d\tau. \tag{8.65}$$

The relationship of Eq. (8.65) is called the *convolution integral* expressing the convolution of functions $e(t)$ and $h(t)$. We thus speak of $e(t)$ and $h(t)$ being convolved to yield $r(t)$. The convolution integral may also be expressed in the form

$$r(t) = \int_0^\infty h(\tau)e(t - \tau)\, d\tau. \tag{8.66}$$

The equivalence of Eqs. (8.65) and (8.66) is readily demonstrated by a simple change of variables. It is therefore immaterial whether we convolve $e(t)$ and $h(t)$ or vice versa.

For physically realizable networks, $h(t - \tau) = 0$ for $\tau > t$, so that we may terminate the integration at $\tau = t$ as shown by

$$r(t) = \int_0^t e(\tau)h(t - \tau)\, d\tau, \tag{8.67}$$

$$r(t) = \int_0^t h(\tau)e(t - \tau)\, d\tau. \tag{8.68}$$

Using Eq. (8.67) or (8.68) it can be shown[6] that the transform of $r(t)$ is equal to the product of the transforms of $e(t)$ and $h(t)$. In other words, convolution in the time domain is equivalent to multiplication in the complex-frequency domain.

For a physical interpretation of the convolution integral, consider the specific example of Fig. 8.11 pertaining to Eq. (8.67). It is assumed that we are given the excitation function $e(t)$ and impulse response $h(t)$ of the network, and the problem is to determine the response $r(t)$ at any time t. As indicated in Fig. 8.11,

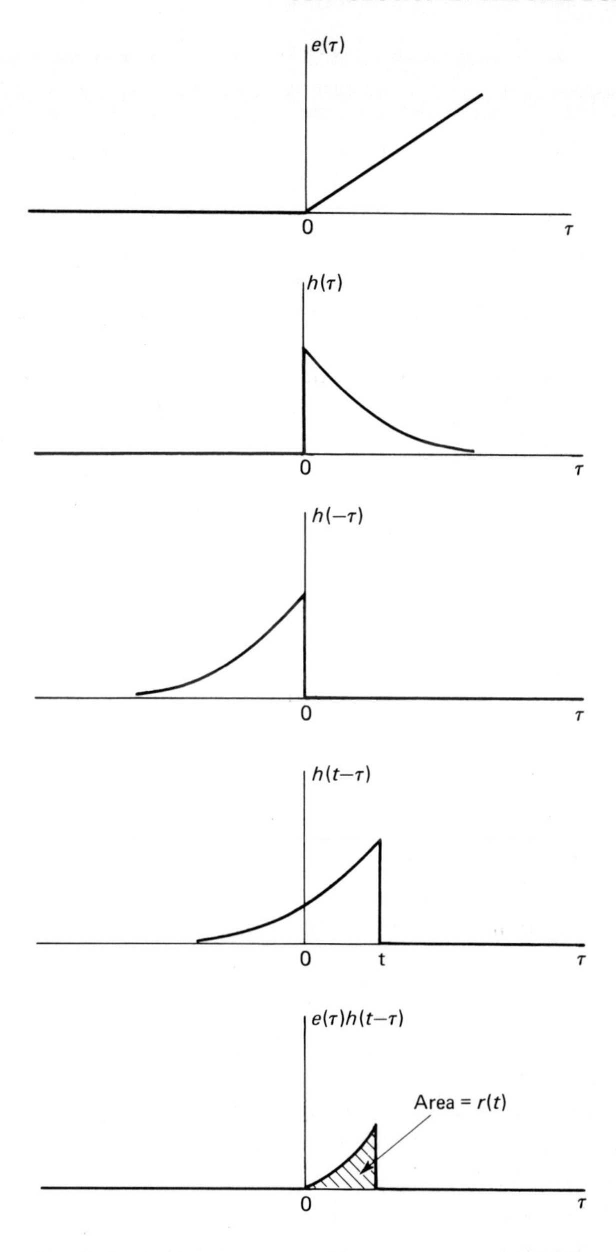

Fig. 8.11. Illustrating the physical interpretation of the convolution integral.

the function $e(\tau)$ is identical with $e(t)$ except for a change of variable from t to τ. As for $h(t - \tau)$ as a function of τ, we can obtain it from $h(t)$ by a combined process of reflection (about the vertical axis) and translation (to the right). The response $r(t)$, that is, the convolution of $e(t)$ and $h(t)$ is finally obtained as the area under the product of $e(\tau)$ and $h(t - \tau)$. Clearly, this area is a function of t, since the value of t determines the relative positions of $e(\tau)$ and $h(t - \tau)$ along the τ-axis. We thus see that the impulse response $h(t)$ plays the role of a *weighting function* in the sense that the value of the response at any given time is equal to a weighted integral over the entire past history of the excitation, weighted according to the impulse response.

The convolution integral provides the basis for a precise definition of stability of a linear network.[7] The network is defined to be stable if the response function is bounded for all bounded excitation functions. If therefore we specify

$$|e(t)| \leq M, \tag{8.69}$$

where M is some positive real finite number, then substitution of Eq. (8.69) in Eq. (8.66) yields

$$|r(t)| \leq M \int_0^\infty |h(\tau)|\, d\tau. \tag{8.70}$$

The necessary and sufficient condition for $r(t)$ to be bounded for a bounded $e(t)$ is therefore

$$\int_0^\infty |h(t)|\, dt < \infty. \tag{8.71}$$

In other words, the network is stable if and only if the impulse response is absolutely integrable from zero to infinity.

The impulse response of a linear, time-invariant network can be written in the form

$$h(t) = \sum_\nu k_\nu\, \varepsilon^{s_\nu t} \tag{8.72}$$

which assumes that the natural frequencies of the network, s_ν, are all distinct. It is thus apparent that for the network to be stable, the condition that $h(t)$ be absolutely integrable is equivalent to the requirement that all the s_ν's have negative real parts (that is, all natural frequencies of the network be located in the left half of the s-plane).

Example 8.5. *Response of common-emitter amplifier to a ramp.* Consider the excitation function

$$e(t) = t, \tag{8.73}$$

representing a ramp of unit slope, applied to the common-emitter amplifier of

Fig. 8.4(a). From Eq. (8.26), the impulse response of the amplifier is obtained as

$$h(t) = \frac{K_0}{\tau_1}\varepsilon^{-t/\tau_1}. \tag{8.74}$$

Hence, the use of Eq. (8.67) yields

$$\begin{aligned} r(t) &= \int_0^t K_0 \frac{\tau}{\tau_1}\varepsilon^{-(t-\tau)/\tau_1}\,d\tau \\ &= K_0\varepsilon^{-t/\tau_1}\int_0^t \frac{\tau}{\tau_1}\varepsilon^{\tau/\tau_1}\,d\tau. \end{aligned} \tag{8.75}$$

Integrating by parts,

$$\begin{aligned} r(t) &= K_0\varepsilon^{-t/\tau_1}\left(\tau\varepsilon^{\tau/\tau_1}\Big|_0^t - \int_0^t \varepsilon^{\tau/\tau_1}\,d\tau\right) \\ &= K_0\tau_1\left(\frac{t}{\tau_1} - 1 + \varepsilon^{-t/\tau_1}\right). \end{aligned} \tag{8.76}$$

Figure 8.12 shows the response $r(t)$ plotted against time.

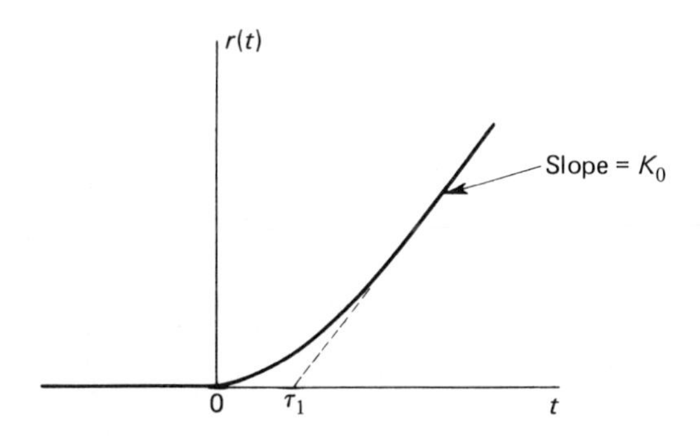

Fig. 8.12. Response of common-emitter amplifier to a ramp input.

Numerical Convolution

A rather useful technique for convolving two given functions numerically is the so-called *multiplication method* which lends itself readily to programming on a digital computer.[8] The two functions to be convolved are sampled at equal time increments and the values of the functions at the sampling instants are tabulated in two rows one above the other. Then the elements in the first row are multiplied one at a time by each of the elements in the second row. The leading element of each row of multiplication products thus obtained is located immediately below the particular element of the second row that is multiplying the first row. The result of convolution at each sampling instant is finally obtained by numerically integrating the multiplication products of each of the columns. This procedure is best illustrated by means of an example.

Example 8.6. Consider the problem of numerically convolving the two time functions of Eqs. (8.73) and (8.74). Assuming a sampling period of $0{\cdot}2\tau_1$ seconds, we may construct Table 8.1, as described above. At each sampling instant, the approximate value of the convolution product $r(t)$ is obtained from the pertinent column using the trapezoidal rule for numerical integration. Figure 8.13

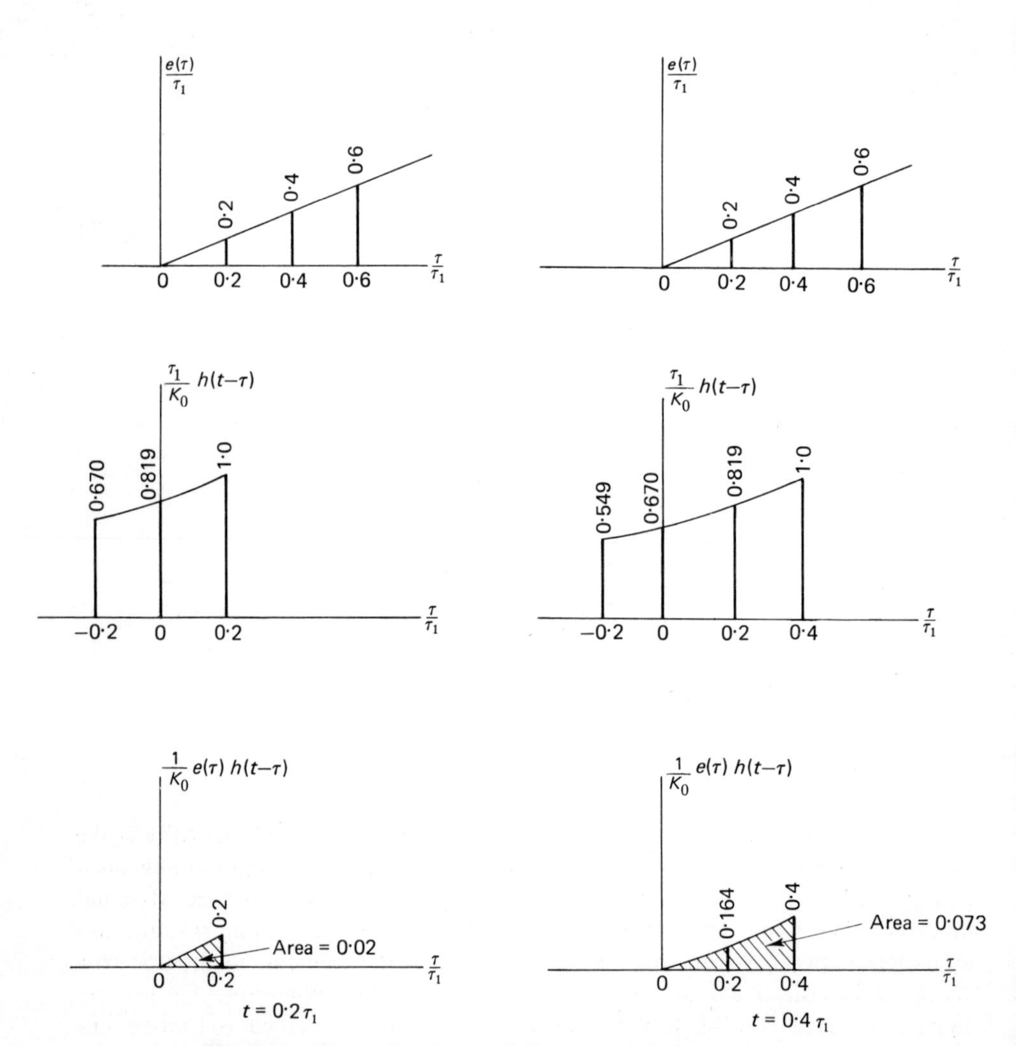

Fig. 8.13. Illustrating the multiplication method of convolution.

Table 8.1. Numerical Convolution of the $e(t)$ and $h(t)$ of Eqs. (8.73) and (8.74) for a Sampling Period of $0{\cdot}2\tau_1$ seconds

$\frac{t}{\tau_1}$	0	0·2	0·4	0·6	0·8	1·0	1·2	1·4	1·6	1·8	2·0
$\frac{1}{\tau_1}e(t)$	0	0·2	0·4	0·6	0·8	1·0	1·2	1·4	1·6	1·8	2·0
$\frac{\tau_1}{K_0}h(t)$	1	0·819	0·670	0·549	0·449	0·368	0·301	0·247	0·202	0·165	0·135
	0	0·2	0·4	0·6	0·8	1·0	1·2	1·4	1·6	1·8	2·0
		0	0·164	0·328	0·491	0·654	0·819	0·983	1·147	1·310	1·472
			0	0·134	0·268	0·402	0·536	0·670	0·804	0·938	1·072
				0	0·110	0·220	0·329	0·439	0·549	0·659	0·769
					0	0·090	0·180	0·269	0·359	0·449	0·539
						0	0·074	0·147	0·221	0·294	0·368
							0	0·060	0·120	0·181	0·241
								0	0·049	0·099	0·148
									0	0·040	0·081
										0	0·033
											0
$\frac{r(t)}{K_0\tau_1}$	0	0·02	0·073	0·152	0·254	0·373	0·508	0·654	0·810	0·974	1·144
$\frac{r(t)}{K_0\tau_1}$ (actual value)*	0	0·019	0·070	0·149	0·249	0·368	0·501	0·647	0·802	0·965	1·135

* Exact expression: $r(t) = K_0\tau_1\left(\frac{t}{\tau_1} - 1 + \varepsilon^{-t/\tau_1}\right)$.

illustrates graphically the operations involved in computing $r(t)$ at two selected sampling instants.

In Table 8.1 it is apparent that the values of $r(t)$ resulting from the numerical convolution are fairly close to the actual values as obtained from Eq. (8.76). Clearly at the cost of increased computational effort, the accuracy of the numerical convolution can be further improved, if so required, by reducing the sampling period.

Concluding Remarks

The convolution integral demonstrates that once we know the impulse response $h(t)$ of a linear, time-invariant network, formulation of the response function $r(t)$ to an arbitrary excitation function $e(t)$ follows directly. The integral is particularly useful when evaluation of the response function using the usual Laplace transform theory breaks down. This occurs when

1. the excitation $e(t)$ does not possess a rational algebraic Laplace transform,
2. the excitation is known only graphically, or
3. the excitation is of such a complex waveform that evaluation of the response as the inverse Laplace transform of the product $E(s)H(s)$ is practically impossible.

In such cases, provided we know the impulse response $h(t)$, by calculation or measurement, then we can determine the response $r(t)$ by working directly with $e(t)$ and $h(t)$ in the time domain, using the process of numerical convolution as previously described, or some other variations of it.[2]

REFERENCES

1. M. F. Gardner and J. L. Barnes, *Transients in Linear Systems*. Wiley, 1942.
2. J. G. Truxal, *Automatic Feedback Control System Synthesis*. McGraw-Hill, 1955.
3. S. Seshu and N. Balabanian, *Linear Network Analysis*. Wiley, 1959.
4. W. C. Elmore, "The Transient Response of Damped Linear Networks with Particular Regard to Wideband Amplifiers," *J. Appl. Phys.* **19,** 55 (1948).
5. A. Papoulis, *The Fourier Integral and its Applications*. McGraw-Hill, 1962.
6. D. K. Cheng, *Analysis of Linear Systems*. Addison-Wesley, 1959.
7. S. J. Mason and H. J. Zimmerman, *Electronic Circuits, Signals and Systems*. Wiley, 1960.
8. G. R. Cooper and C. D. McGillem, *Methods of Signal and System Analysis*. Holt, Rinehart and Winston, 1967.
9. J. M. Pettit and M. M. McWhorter, *Electronic Amplifier Circuits*. McGraw-Hill, 1961.
10. J. H. Mulligan, "The Effect of Pole and Zero Locations on the Transient Response of Linear Dynamic Systems," *Proc. I.R.E.* **37,** 516 (1949).
11. G. E. Valley and H. Wallman, *Vacuum-tube Amplifiers*. McGraw-Hill, 1948.

PROBLEMS

8.1 Determine the forced (steady-state) response of the amplifier of Fig. 8.4(a) when the excitation function is in the form of a sawtooth, as in Fig. P8.1.

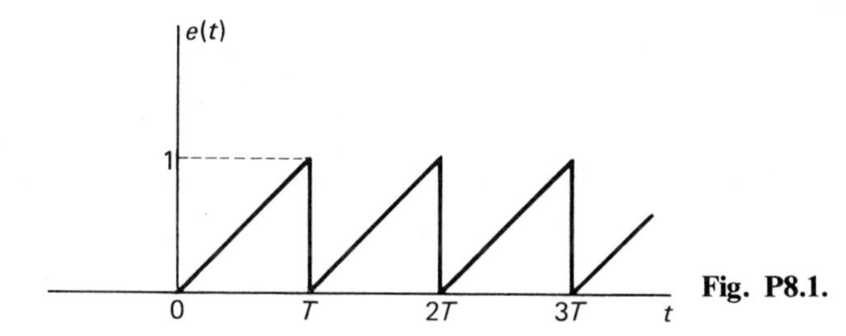

Fig. P8.1.

8.2 The so-called Doba network[9] of Fig. P8.2 has a linear phase characteristic. Develop an expression for the step response; hence, evaluate the 10 to 90 per cent rise time and percentage overshoot.

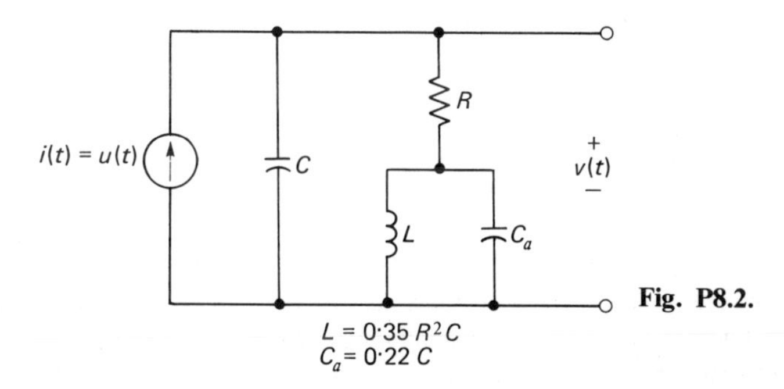

Fig. P8.2.

8.3 Evaluate the impulse response of a low-pass amplifier having the pole pattern shown in Fig. P8.3. Hence, show that the step response is monotonic if the real pole does not lie to the left of the line joining the two complex poles.[10]

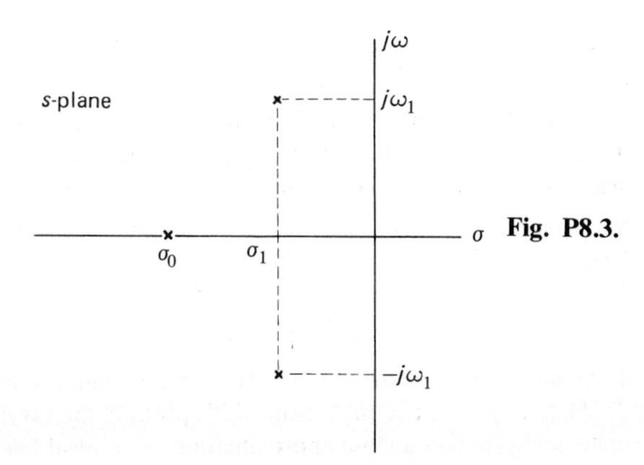

Fig. P8.3.

8.4 Figure P8.4 shows a shunt-compensated transistor amplifier.[9] Determine its transfer function, and hence evaluate the inductance L required for critical damping. Using Elmore's definition, what is the corresponding rise time of the amplifier?

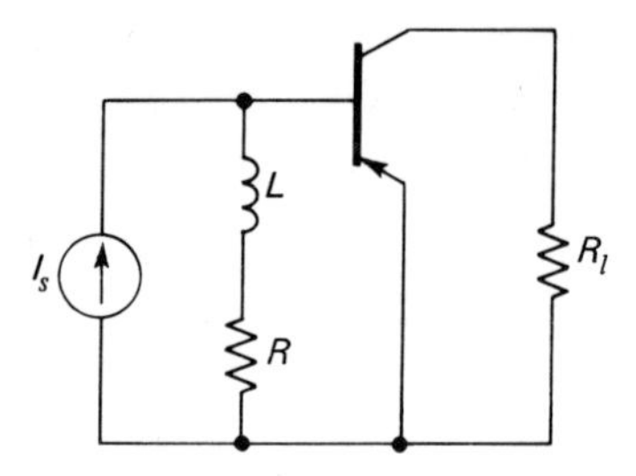

Fig. P8.4.

8.5 The intrinsic common-base short-circuit current gain of a transistor is given by

$$\alpha(s) = \operatorname{sech} \frac{W}{L_p}(1 + s\tau_p)^{1/2},$$

where W = effective width of base region,
L_p = diffusion length of holes,
τ_p = lifetime of holes.
By expanding $\alpha(s)$ in powers of s, determine the rise time τ_R.

8.6 Show that in any linear stable low-pass network with transfer function $H_n(s)$ the delay and rise time are related to the magnitude function $M(\omega) = |H_n(j\omega)|$ and phase function $B(\omega) = \text{ang}\, H_n(j\omega)$ of the network by

$$\tau_D = -\left.\frac{dB(\omega)}{d\omega}\right|_{\omega=0}$$

$$\tau_R^2 = -2\pi \left.\frac{d^2M(\omega)}{d\omega^2}\right|_{\omega=0}.$$

Use Elmore's definitions for τ_D and τ_R.

8.7 Consider an amplifier consisting of N identical non-interacting stages. Assuming the inter-stage coupling to be of the form shown in Fig. 8.1, find the optimum gain per stage to achieve a specified overall gain with the minimum overall rise time.[4]

8.8 In the arrangement of Fig. P8.8, each isolation amplifier has unity gain, infinite input impedance and zero output impedance. By choosing

$$\tau_0 = \frac{1}{2\pi B}\sqrt{2^{1/N} - 1},$$

where $\tau_0 = CR$, the half-power bandwidth of the entire amplifier, for any N, is held constant at some value B. Determine the rise time–bandwidth product for varying N. Show that for $N = 6$, the arrangement provides a close approximation to an ideal Gaussian amplifier.[11]

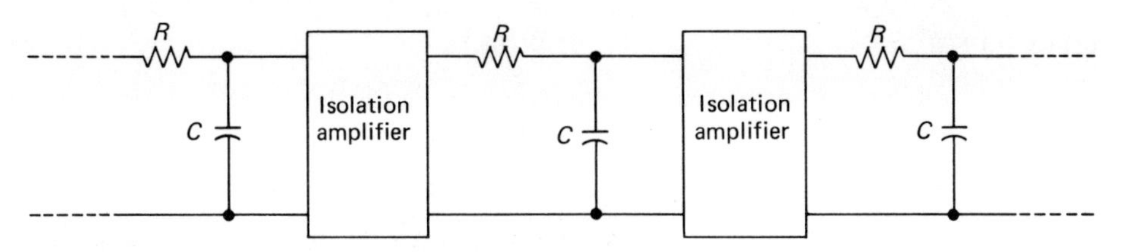

Fig. P8.8.

8.9 A linear network has the impulse response

$$h(t) = \delta(t) - \varepsilon^{-t},$$

where $\delta(t)$ is a unit impulse. Using the convolution integral, determine the response to an excitation in the form of a rectangular pulse, as in Fig. P8.9.

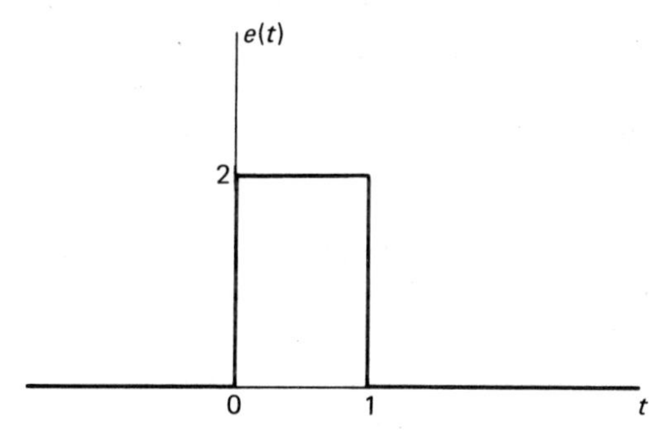

Fig. P8.9.

8.10 Using the process of numerical convolution, compute the response of a three-stage amplifier having the impulse response

$$h(t) = \frac{t^2}{2}\varepsilon^{-t}$$

when the excitation is

$$e(t) = \varepsilon^{-t} \sin t.$$

CHAPTER 9

STATE-VARIABLE APPROACH

9.1 INTRODUCTION

In Section 2.5 we developed a systematic procedure for formulating the state equations of a linear, time-invariant network. It was shown that with inductor currents and capacitor voltages chosen as the state variables, the dynamic behaviour of the network may be fully described in terms of a minimal set of first-order differential equations. The basic difference between the state-variable approach and the approach based on the transfer function as pursued in the previous two chapters is that in the transfer function approach the emphasis is placed upon the input–output relationship of the network which is assumed to be initially relaxed, with no attention paid to the behaviour of any of the dynamic variables that may exist in the interior of the network. On the other hand, in the state-variable approach the time dependences of all the state variables are determined by solving the pertinent state equations, and complete information on the behaviour of the network under consideration is thereby obtained, including the effect of initial conditions. The central problem in determining the time response by state-variable techniques is that of computing the so-called *state-transition matrix* which describes the transition of the network from one state to another. As will be shown in the following sections, the nature of all solutions to the state equations, whether driven or source free, depend upon this matrix.

9.2 TIME RESPONSE

The most general form of the state equations of a linear, time-invariant network is as follows[1]

$$\frac{d\mathbf{x}(t)}{dt} = \mathbf{A}\mathbf{x}(t) + \mathbf{B}\mathbf{e}(t), \tag{9.1}$$

$$\mathbf{r}(t) = \mathbf{C}\mathbf{x}(t) + \mathbf{D}\mathbf{e}(t), \tag{9.2}$$

where $\mathbf{x}(t)$ is the *state vector*; $\mathbf{e}(t)$ is the *excitation* or *input vector*; $\mathbf{r}(t)$ is the *response* or *output vector.* $\mathbf{A}$ is the *characteristic matrix* of the network, while $\mathbf{B}$ is a matrix coupling the input to the various state variables. The matrix $\mathbf{C}$ represents the coupling of the various state variables to the output, while the matrix $\mathbf{D}$ represents any direct coupling that may exist between the input and output ports of the network. In some physical networks $\mathbf{D}$ is zero.

Consider first the source-free form of the state equations, that is,

$$\frac{d\mathbf{x}(t)}{dt} = \mathbf{A}\mathbf{x}(t). \tag{9.3}$$

The solution to this homogeneous system of first-order differential equations is

$$\mathbf{x}(t) = \varepsilon^{\mathbf{A}t}\mathbf{x}(0), \tag{9.4}$$

where $\mathbf{x}(0)$ is a column vector whose elements are the initial conditions of the components of the state vector $\mathbf{x}(t)$, and $\varepsilon^{\mathbf{A}t}$ is a matrix defined by a power series in $\mathbf{A}$ as

$$\begin{aligned}\varepsilon^{\mathbf{A}t} &= \mathbf{1} + \mathbf{A}t + \mathbf{A}^2\frac{t^2}{2!} + \mathbf{A}^3\frac{t^3}{3!} + \cdots \\ &= \sum_{n=0}^{\infty} \mathbf{A}^n\frac{t^n}{n!}.\end{aligned} \tag{9.5}$$

This series can be shown[1] to converge absolutely for all finite t and $\mathbf{A}$. Equation (9.5) can be substituted into Eq. (9.3) to verify that it is a solution.

To obtain the complete solution to Eq. (9.1), assume[2]

$$\mathbf{x}(t) = \varepsilon^{\mathbf{A}t}\mathbf{f}(t), \tag{9.6}$$

where $\mathbf{f}(t)$ is to be determined. Substituting in Eq. (9.1) and premultiplying by $\varepsilon^{-\mathbf{A}t}$, we obtain

$$\frac{d\mathbf{f}(t)}{dt} = \varepsilon^{-\mathbf{A}t}\mathbf{B}\mathbf{e}(t). \tag{9.7}$$

Hence, integrating from $-\infty$ to t yields

$$\mathbf{f}(t) = \int_{-\infty}^{t} \varepsilon^{-\mathbf{A}\tau}\mathbf{B}\mathbf{e}(\tau)\,d\tau. \tag{9.8}$$

From Eqs. (9.6) and (9.8),

$$\begin{aligned}\mathbf{x}(t) &= \varepsilon^{\mathbf{A}t}\int_{-\infty}^{t} \varepsilon^{-\mathbf{A}\tau}\mathbf{B}\mathbf{e}(\tau)\,d\tau \\ &= \varepsilon^{\mathbf{A}t}\int_{-\infty}^{0} \varepsilon^{-\mathbf{A}\tau}\mathbf{B}\mathbf{e}(\tau)\,d\tau + \int_{0}^{t} \varepsilon^{\mathbf{A}(t-\tau)}\mathbf{B}\mathbf{e}(\tau)\,d\tau.\end{aligned} \tag{9.9}$$

Putting $t = 0$ in Eq. (9.9), the initial state established at time $t = 0$ by some previous driving function is obtained as

$$\mathbf{x}(0) = \int_{-\infty}^{0} \varepsilon^{-\mathbf{A}\tau}\mathbf{B}\mathbf{e}(\tau)\,d\tau. \tag{9.10}$$

Equation (9.9) may thus be rewritten as

$$\mathbf{x}(t) = \varepsilon^{\mathbf{A}t}\mathbf{x}(0) + \int_0^t \varepsilon^{\mathbf{A}(t-\tau)}\mathbf{B}\mathbf{e}(\tau)\,d\tau. \tag{9.11}$$

The term $\varepsilon^{\mathbf{A}t}\mathbf{x}(0)$, which is contributed by the initial state, is called the *zero-input-state response*, while the term

$$\int_0^t \varepsilon^{\mathbf{A}(t-\tau)}\mathbf{B}\mathbf{e}(\tau)\,d\tau,$$

which is contributed by the excitation, is called the *forced-state response* of the network.

9.3 STATE-TRANSITION MATRIX

In Eq. (9.4) pertaining to a source-free network we see that the matrix $\varepsilon^{\mathbf{A}t}$ plays the role of an operator which takes the initial state $\mathbf{x}(0)$ into the state $\mathbf{x}(t)$ at time t. For this reason, $\varepsilon^{\mathbf{A}t}$ is called the *state-transition matrix*, which, for convenience, is usually written as $\boldsymbol{\phi}(t)$. Thus, with

$$\boldsymbol{\phi}(t) = \varepsilon^{\mathbf{A}t} \tag{9.12}$$

we may express the complete time response of the network, from Eq. (9.11), in the equivalent form

$$\mathbf{x}(t) = \boldsymbol{\phi}(t)\mathbf{x}(0) + \int_0^t \boldsymbol{\phi}(t-\tau)\mathbf{B}\mathbf{e}(\tau)\,d\tau. \tag{9.13}$$

For a physical interpretation of the state-transition matrix, let

$$\boldsymbol{\phi}(t) = \begin{bmatrix} \phi_{11}(t) & \phi_{12}(t) & \cdots & \phi_{1q}(t) \\ \phi_{21}(t) & \phi_{22}(t) & \cdots & \phi_{2q}(t) \\ \cdot & \cdot & & \cdot \\ \cdot & \cdot & & \cdot \\ \cdot & \cdot & & \cdot \\ \phi_{q1}(t) & \phi_{q2}(t) & \cdots & \phi_{qq}(t) \end{bmatrix}. \tag{9.14}$$

Then, the ith term $x_i(t)$ of the zero-input state response $\boldsymbol{\phi}(t)\mathbf{x}(0)$ may be expressed as

$$x_i(t) = \sum_{j=1}^{q} \phi_{ij}(t)x_j(0), \tag{9.15}$$

which indicates that the element $\phi_{ij}(t)$ of the state-transition matrix may be determined by placing a unit initial condition on state variable x_j, and zero initial conditions on all other state variables. Then, with $\mathbf{e}(t) = \mathbf{0}$, the $x_i(t)$ is equal to $\phi_{ij}(t)$.

We now consider some properties of the state-transition matrix $\boldsymbol{\phi}(t)$:

1. Putting $t = 0$ in the series expansion of Eq. (9.5) for $\boldsymbol{\phi}(t) = \varepsilon^{\mathbf{A}t}$, we obtain
$$\boldsymbol{\phi}(0) = \mathbf{1}. \tag{9.16}$$

2. Since
$$\mathbf{x}(t) = \boldsymbol{\phi}(t)\mathbf{x}(0)$$
is a unique solution of the homogeneous state equations (i.e., with $\mathbf{e}(t) = \mathbf{0}$), we may write
$$\mathbf{x}(t_1 + t_2) = \boldsymbol{\phi}(t_1 + t_2)\mathbf{x}(0)$$
and
$$\begin{aligned}\mathbf{x}(t_1 + t_2) &= \boldsymbol{\phi}(t_1)\mathbf{x}(t_2)\\ &= \boldsymbol{\phi}(t_1)\boldsymbol{\phi}(t_2)\mathbf{x}(0).\end{aligned}$$
Therefore,
$$\boldsymbol{\phi}(t_1 + t_2)\mathbf{x}(0) = \boldsymbol{\phi}(t_1)\boldsymbol{\phi}(t_2)\mathbf{x}(0)$$
or
$$\boldsymbol{\phi}(t_1 + t_2) = \boldsymbol{\phi}(t_1)\boldsymbol{\phi}(t_2). \tag{9.17}$$
Equation (9.17) states that $\boldsymbol{\phi}(t)$ has the multiplicative property.

3. If in Eq. (9.17) we set $t_1 = -t_2 = t$, we obtain
$$\boldsymbol{\phi}(0) = \boldsymbol{\phi}(t)\boldsymbol{\phi}(-t).$$
Since $\boldsymbol{\phi}(0) = \mathbf{1}$, it therefore follows that the inverse of the state-transition matrix satisfies the relation
$$\boldsymbol{\phi}^{-1}(t) = \boldsymbol{\phi}(-t). \tag{9.18}$$

Sylvester's Theorem

The computation of the state-transition matrix $\boldsymbol{\phi}(t)$ using the series expansion of $\varepsilon^{\mathbf{A}t}$ as in Eq. (9.5) is not practical except in very special cases or for small values of t. For computational purposes a closed form of solution for $\boldsymbol{\phi}(t)$ is more desirable. Sylvester's theorem[3] provides a useful method for computing the state-transition matrix using strictly numerical processes in the time domain. The theorem may be stated as follows: If $f(\mathbf{A})$ is a matrix polynomial in $\mathbf{A}$, and if the square matrix $\mathbf{A}$ has q distinct eigenvalues, then $f(\mathbf{A})$ may be expressed as
$$f(\mathbf{A}) = \sum_{i=1}^{q} f(s_i)\mathbf{N}(s_i), \tag{9.19}$$

where the s_i's are the *eigenvalues* of **A** and

$$\mathbf{N}(s_i) = \prod_{\substack{j=1 \\ j \neq i}}^{q} \frac{(s_j\mathbf{1} - \mathbf{A})}{(s_j - s_i)}. \tag{9.20}$$

The eigenvalues of the matrix **A** are roots of the characteristic equation obtained by setting the determinant of the matrix $(s\mathbf{1} - \mathbf{A})$ equal to zero, that is

$$\det(s\mathbf{1} - \mathbf{A}) = \prod_{i=1}^{q} (s - s_i). \tag{9.21}$$

Clearly, either the eigenvalues are real numbers or if they are complex they occur in conjugate pairs.

Since $\boldsymbol{\phi}(t) = \varepsilon^{\mathbf{A}t}$ is expressible in terms of a convergent series as in Eq. (9.5), we may apply Sylvester's theorem to compute $\boldsymbol{\phi}(t)$, obtaining

$$\boldsymbol{\phi}(t) = \sum_{i=1}^{q} \varepsilon^{s_i t}\mathbf{N}(s_i). \tag{9.22}$$

It should be noted that the eigenvalues of **A** are identical to the natural frequencies of the network. In a stable network, therefore, all the eigenvalues of matrix **A** are located in the left half of the s-plane; eigenvalues along the imaginary axis are permitted only if they are simple.

Example 9.1. *RC–NIC network.* Consider the network shown in Fig. 9.1 involving a current-inversion negative-impedance converter of unity conversion ratio. For the specific element values given, we obtain the state equations

$$\frac{d}{dt}\begin{bmatrix} v_1(t) \\ v_2(t) \end{bmatrix} = \begin{bmatrix} -2 & 1 \\ -1 & \frac{1}{6} \end{bmatrix} \begin{bmatrix} v_1(t) \\ v_2(t) \end{bmatrix} + \begin{bmatrix} 1 \\ 0 \end{bmatrix} v_s(t), \tag{9.23}$$

where $v_1(t)$ and $v_2(t)$ represent the voltages across the input and output capacitors,

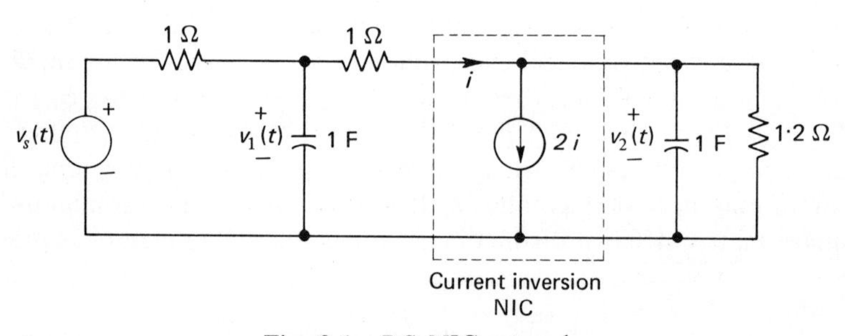

Fig. 9.1. *RC*–NIC network.

respectively. Hence,

$$\mathbf{A} = \begin{bmatrix} -2 & 1 \\ -1 & \frac{1}{6} \end{bmatrix}, \qquad \mathbf{B} = \begin{bmatrix} 1 \\ 0 \end{bmatrix}. \tag{9.24}$$

The eigenvalues of matrix $\mathbf{A}$ are roots of the characteristic equation:

$$s^2 + \tfrac{11}{6}s + \tfrac{2}{3} = 0,$$

that is, $s_1 = -\frac{1}{2}$ and $s_2 = -\frac{4}{3}$. Thus, the systematic application of Sylvester's theorem yields:

$$\mathbf{N}(s_1) = \frac{s_2\mathbf{1} - \mathbf{A}}{s_2 - s_1} = \begin{bmatrix} -\frac{4}{5} & \frac{6}{5} \\ -\frac{6}{5} & \frac{9}{5} \end{bmatrix}$$

$$\mathbf{N}(s_2) = \frac{s_1\mathbf{1} - \mathbf{A}}{s_1 - s_2} = \begin{bmatrix} \frac{9}{5} & -\frac{6}{5} \\ \frac{6}{5} & -\frac{4}{5} \end{bmatrix}$$

$$\begin{aligned}
\boldsymbol{\phi}(t) &= \varepsilon^{s_1 t}\mathbf{N}(s_1) + \varepsilon^{s_2 t}\mathbf{N}(s_2) \\
&= \varepsilon^{-(1/2)t}\begin{bmatrix} -\frac{4}{5} & \frac{6}{5} \\ -\frac{6}{5} & \frac{9}{5} \end{bmatrix} + \varepsilon^{-(4/3)t}\begin{bmatrix} \frac{9}{5} & -\frac{6}{5} \\ \frac{6}{5} & -\frac{4}{5} \end{bmatrix} \\
&= \begin{bmatrix} -\frac{4}{5}\varepsilon^{-(1/2)t} + \frac{9}{5}\varepsilon^{-(4/3)t} & \frac{6}{5}\varepsilon^{-(1/2)t} - \frac{6}{5}\varepsilon^{-(4/3)t} \\ -\frac{6}{5}\varepsilon^{-(1/2)t} + \frac{6}{5}\varepsilon^{-(4/3)t} & \frac{9}{5}\varepsilon^{-(1/2)t} - \frac{4}{5}\varepsilon^{-(4/3)t} \end{bmatrix}.
\end{aligned} \tag{9.25}$$

The natural or source-free state response of the network is therefore

$$\begin{aligned}
\begin{bmatrix} v_1(t) \\ v_2(t) \end{bmatrix} &= \boldsymbol{\phi}(t)\begin{bmatrix} v_1(0) \\ v_2(0) \end{bmatrix} \\
&= \begin{bmatrix} -\frac{4}{5}v_1(0) + \frac{6}{5}v_2(0) \\ -\frac{6}{5}v_1(0) + \frac{9}{5}v_2(0) \end{bmatrix}\varepsilon^{-(1/2)t} + \begin{bmatrix} \frac{9}{5}v_1(0) - \frac{6}{5}v_2(0) \\ \frac{6}{5}v_1(0) - \frac{4}{5}v_2(0) \end{bmatrix}\varepsilon^{-(4/3)t}.
\end{aligned} \tag{9.26}$$

From Eq. (9.26) we see that if the initial conditions are chosen so that $v_1(0) = \frac{2}{3}v_2(0)$, then only the natural mode $\varepsilon^{-(1/2)t}$ is excited in both $v_1(t)$ and $v_2(t)$. If, on the other hand, the initial conditions are chosen so that $v_1(0) = \frac{3}{2}v_2(0)$, then only the mode $\varepsilon^{-(4/3)t}$ is excited.

The zero-state response of the network is obtained from Eq. (9.13) with the initial state $\mathbf{x}(0)$ taken to be zero; thus

$$\begin{aligned}
\begin{bmatrix} v_1(t) \\ v_2(t) \end{bmatrix} &= \int_0^t \boldsymbol{\phi}(t-\tau)\mathbf{B}\mathbf{e}(\tau)\,d\tau \\
&= \int_0^t \begin{bmatrix} -\frac{4}{5}\varepsilon^{-(t-\tau)/2} + \frac{9}{5}\varepsilon^{-4(t-\tau)/3} \\ -\frac{6}{5}\varepsilon^{-(t-\tau)/2} + \frac{6}{5}\varepsilon^{-4(t-\tau)/3} \end{bmatrix} v_s(\tau)\,d\tau.
\end{aligned} \tag{9.27}$$

Order of Complexity

The *order of complexity* of a network is defined to be the number of independent state variables that characterize the dynamic behaviour of the network.[4] Equivalently, we may define the order of complexity to be the number of finite natural frequencies the network possesses, the dimension of the **A**-matrix, or the dimension of the state-transition matrix $\boldsymbol{\phi}(t)$. Thus, from the analysis pursued in Section 2.5 it is apparent that the order of complexity of a general *LCR* network is equal to the number of reactive elements less the number of independent capacitor-only loops, less the number of independent inductor-only cut-sets. The presence of controlled sources may also cause a reduction in the order of complexity of the network; in this case, however, the order of complexity becomes critically dependent upon the control parameters of the controlled sources responsible for such a reduction.[5]

9.4 A METHOD FOR COMPUTING THE APPROXIMATE TIME RESPONSE[6]

The so-called *discrete-time approximation* provides a step-by-step method to compute an approximate solution to the state equation (9.1) without determining the state transition matrix. The method is based on the assumption that during an increment of time Δt the derivative $d\mathbf{x}/dt$ may be considered to be approximately constant provided Δt is sufficiently small. By definition, we have

$$\frac{d\mathbf{x}(t)}{dt} = \lim_{\Delta t \to 0} \frac{\mathbf{x}(t + \Delta t) - \mathbf{x}(t)}{\Delta t}.$$

Hence, if t is divided into small increments $\Delta t = \tau$, we may write approximately

$$\frac{d\mathbf{x}(t)}{dt} \simeq \frac{\mathbf{x}(t + \tau) - \mathbf{x}(t)}{\tau}. \tag{9.28}$$

Substitution of Eq. (9.28) in (9.1) thus yields

$$\frac{\mathbf{x}(t + \tau) - \mathbf{x}(t)}{\tau} \simeq \mathbf{Ax}(t) + \mathbf{Be}(t). \tag{9.29}$$

Solving Eq. (9.29) for $\mathbf{x}(t + \tau)$, we obtain

$$\mathbf{x}(t + \tau) \simeq (\tau\mathbf{A} + \mathbf{1})\mathbf{x}(t) + \tau\mathbf{Be}(t). \tag{9.30}$$

With $t = 0, \tau, 2\tau, 3\tau, \ldots$ defining the successive time intervals of interest, we may rewrite Eq. (9.30) in the form

$$\mathbf{x}[(1 + k)\tau] \simeq (\tau\mathbf{A} + \mathbf{1})\mathbf{x}(k\tau) + \tau\mathbf{Be}(k\tau), \tag{9.31}$$

where $k = 0, 1, 2, 3, \ldots$. The recurrence formula of Eq. (9.31) reveals that an approximate value for the state vector $\mathbf{x}(t)$ at $t = (1 + k)\tau$ can be obtained in terms of the values of $\mathbf{x}(t)$ and the input vector $\mathbf{e}(t)$ at $t = k\tau$. Such a calculation is ideally suited for programming on a digital computer. The computation obviously becomes increasingly more accurate with reduced τ.

Example 9.2. *RC–NIC network* (continued). As an illustrative example, we shall use the discrete time approximation to evaluate the zero-input state response of the RC–NIC network of Fig. 9.1, with

$$A = \begin{bmatrix} -2 & 1 \\ -1 & \frac{1}{6} \end{bmatrix}.$$

We shall assume that the initial conditions are $v_1(0) = 0$ and $v_2(0) = 1$ volt. For reasonable accuracy, τ should be chosen less than one-fifth of the smallest time constant of the network. With the eigenvalues of the matrix **A** previously determined as $-\frac{1}{2}$ and $-\frac{4}{3}$, the time constants of the network are 2 and $\frac{3}{4}$ sec, respectively. We may thus choose $\tau = 0{\cdot}1$ sec. Then, with $\mathbf{e}(t) = \mathbf{0}$, Eq. (9.31) gives the state response for $t = \tau = 0{\cdot}1$ sec, or $k = 0$, to be

$$\begin{bmatrix} v_1(0{\cdot}1) \\ v_2(0{\cdot}1) \end{bmatrix} = \begin{bmatrix} 0{\cdot}8 & 0{\cdot}1 \\ -0{\cdot}1 & 1{\cdot}017 \end{bmatrix} \begin{bmatrix} v_1(0) \\ v_2(0) \end{bmatrix} = \begin{bmatrix} 0{\cdot}1 \\ 1{\cdot}017 \end{bmatrix}.$$

Next, the state response for $t = 2\tau = 0{\cdot}2$ sec, or $k = 1$, is

$$\begin{bmatrix} v_1(0{\cdot}2) \\ v_2(0{\cdot}2) \end{bmatrix} = \begin{bmatrix} 0{\cdot}8 & 0{\cdot}1 \\ -0{\cdot}1 & 1{\cdot}017 \end{bmatrix} \begin{bmatrix} 0{\cdot}1 \\ 1{\cdot}017 \end{bmatrix} = \begin{bmatrix} 0{\cdot}182 \\ 1{\cdot}025 \end{bmatrix},$$

and so on for $k = 2, 3, 4, \ldots$. Table 9.1 lists the approximate values of $v_1(t)$ and

Table 9.1

Time t(sec)	$v_1(t)$		$v_2(t)$	
	Approximate	Actual	Approximate	Actual
0	0	0	1·000	1·000
0·1	0·100	0·091	1·017	1·010
0·2	0·182	0·167	1·025	1·016
0·3	0·248	0·229	1·024	1·014
0·4	0·300	0·28	1·016	1·005
0·5	0·342	0·319	1·003	0·990
0·6	0·374	0·35	0·986	0·975

$v_2(t)$ as calculated in this manner, and, for comparison, their actual values as calculated using Eq. (9.26), up to $k = 5$, or $t = 0{\cdot}6$ sec. Close agreement with the actual values is observed.

Figure 9.2 shows $v_2(t)$ plotted against $v_1(t)$, using the set of actual values, with t as a running parameter. The curve describing motion of the tip of the state vector is called a *trajectory*, and the v_1v_2-plane is called a *phase plane*. Obviously, for each set of initial conditions there is a unique trajectory.

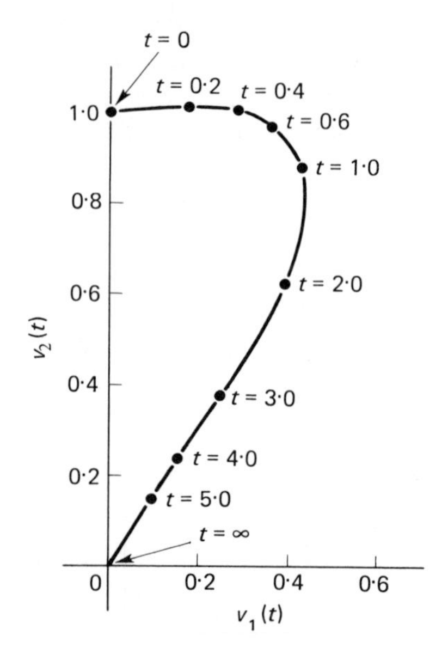

Fig. 9.2. A trajectory for the *RC*–NIC network of Fig. 9.1.

Nonlinear Networks

The method of discrete-time approximation for computing the state vector $\mathbf{x}(t)$ is also applicable to nonlinear networks. Thus, consider a source-free network characterized by the nonlinear state equation

$$\frac{d\mathbf{x}(t)}{dt} = f(\mathbf{x}(t)). \tag{9.32}$$

The derivative $d\mathbf{x}/dt$ is defined approximately by Eq. (9.28). Hence, substitution of Eq. (9.28) in (9.32) yields

$$\mathbf{x}(t + \tau) \simeq \tau f(\mathbf{x}(t)) + \mathbf{x}(t)$$

or equivalently, with $t = k\tau$,

$$\mathbf{x}[(1 + k)\tau] \simeq \tau f(\mathbf{x}(k\tau)) + \mathbf{x}(k\tau), \tag{9.33}$$

where $k = 0, 1, 2, 3, \ldots$. Thus, starting from an initial state $\mathbf{x}(0)$, we may compute the state response of the network at the successive time intervals $\tau, 2\tau, 3\tau, \ldots$ in a step-by-step manner.

Example 9.3. *Negative-resistance oscillator.* To illustrate the use of Eq. (9.33), consider the parallel tuned *LCR* circuit of Fig. 9.3, with the *I–V* characteristic of the negative-resistance device defined by the cubic equation

$$i_R(t) = -av(t) + bv^3(t) \tag{9.34}$$

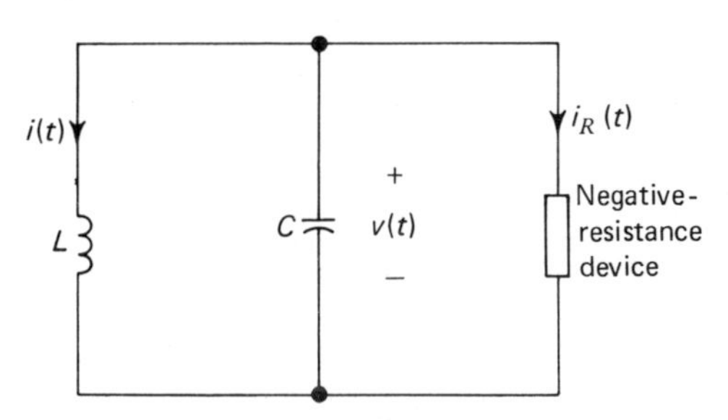

Fig. 9.3. Negative-resistance oscillator.

where a and b are positive constants. Choosing the current through the inductor, $i(t)$, and voltage across the capacitor, $v(t)$, as the state variables, we obtain the state equations*

$$\frac{d}{dt}\begin{bmatrix} i(t) \\ v(t) \end{bmatrix} = \begin{bmatrix} \dfrac{1}{L}v(t) \\ \dfrac{1}{C}\{-i(t) + av(t) - bv^3(t)\} \end{bmatrix}. \tag{9.35}$$

To be specific, suppose we have the following set of scaled element values:

$$L = 1\text{H},$$
$$C = 1\text{F},$$
$$a = 1\mho,$$
$$b = \tfrac{1}{3}\,\text{amp/volt}^3$$

---

* If we eliminate $i(t)$ in Eq. (9.35) we get the nonlinear second-order differential equation

$$C\frac{d^2v}{dt^2} - (a - 3bv^2)\frac{dv}{dt} + \frac{v}{L} = 0$$

which is recognized as the well-known Van der Pol equation.

for which Eq. (9.35) takes on the form

$$\frac{d}{dt}\begin{bmatrix} i(t) \\ v(t) \end{bmatrix} = \begin{bmatrix} v(t) \\ -i(t) + v(t) - \frac{1}{3}v^3(t) \end{bmatrix}. \tag{9.36}$$

Then, assuming $\Delta t = \tau = 0{\cdot}1$ second we may use Eq. (9.33) to compute the two trajectories shown plotted in Fig. 9.4 corresponding to the two different sets of initial conditions indicated in the diagram. In both cases we see that after a sufficiently long time the trajectories assume the form of a closed curve known as the *limit cycle.* In other words, a limit cycle represents the stationary oscillatory state of the oscillator. A common feature of self-starting oscillators is that their stationary oscillatory state is completely independent of the initial conditions but depends uniquely on the parameters of the oscillator circuit.

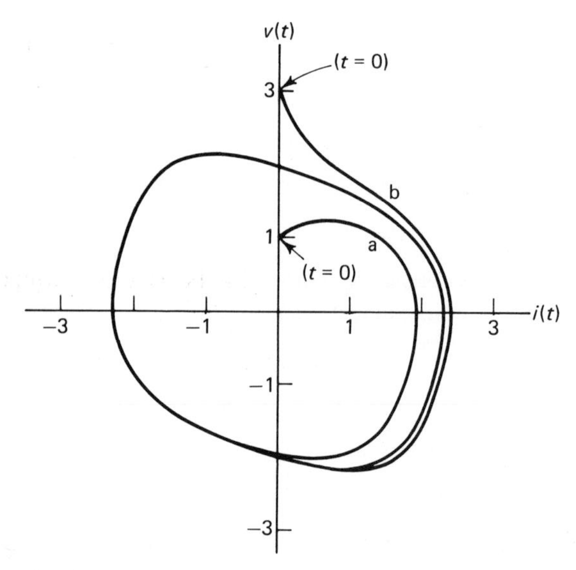

Fig. 9.4. Trajectories for Eq. (9.36) for the initial conditions: (a) $\dot{v}(0) = 1$ V, $i(0) = 0$; (b) $v(0) = 3$ V, $i(0) = 0$.

9.5 FREQUENCY RESPONSE

Taking the Laplace transforms of both sides of Eqs. (9.1) and (9.2), we obtain

$$s\mathbf{X}(s) - \mathbf{x}(0) = \mathbf{AX}(s) + \mathbf{BE}(s) \tag{9.37}$$

$$\mathbf{R}(s) = \mathbf{CX}(s) + \mathbf{DE}(s), \tag{9.38}$$

where $\mathbf{X}(s)$, $\mathbf{R}(s)$ and $\mathbf{E}(s)$ are column vectors whose components are the Laplace transforms of the corresponding components of the vectors $\mathbf{x}(t)$, $\mathbf{r}(t)$ and $\mathbf{e}(t)$, respectively; $\mathbf{x}(0)$ is the initial state vector. Hence,

$$\mathbf{X}(s) = (s\mathbf{1} - \mathbf{A})^{-1}\mathbf{x}(0) + (s\mathbf{1} - \mathbf{A})^{-1}\mathbf{BE}(s). \tag{9.39}$$

The state vector $\mathbf{x}(t)$ is then found by determining the inverse Laplace transform of $\mathbf{X}(s)$ in the usual way, which provides another procedure for determining $\mathbf{x}(t)$.

For purposes of frequency-response analysis, we may assume the network to be initially at rest, so that $\mathbf{x}(0) = \mathbf{0}$. Then, Eq. (9.39) simplifies as

$$\mathbf{X}(s) = (s\mathbf{1} - \mathbf{A})^{-1}\mathbf{BE}(s). \tag{9.40}$$

Substitution of Eq. (9.40) in Eq. (9.38) yields

$$\begin{aligned}\mathbf{R}(s) &= [\mathbf{C}(s\mathbf{1} - \mathbf{A})^{-1}\mathbf{B} + \mathbf{D}]\mathbf{E}(s)\\ &= \mathbf{H}(s)\mathbf{E}(s),\end{aligned} \tag{9.41}$$

where

$$\mathbf{H}(s) = \mathbf{C}(s\mathbf{1} - \mathbf{A})^{-1}\mathbf{B} + \mathbf{D}. \tag{9.42}$$

The matrix $\mathbf{H}(s)$ is called the *transfer matrix* of the network. The ijth element $H_{ij}(s)$ of this matrix is the Laplace transform of the response that results at the ith output port of the network due to a unit impulse applied at the jth input port.

Comparison of Eq. (9.11) with (9.39) reveals that the state transition matrix is the inverse Laplace transform of the matrix $(s\mathbf{1} - \mathbf{A})^{-1}$, that is,

$$\varepsilon^{\mathbf{A}t} = \boldsymbol{\phi}(t) = \mathscr{L}^{-1}[(s\mathbf{1} - \mathbf{A})^{-1}] \tag{9.43}$$

or

$$\boldsymbol{\Phi}(s) = \mathscr{L}[\boldsymbol{\phi}(t)] = (s\mathbf{1} - \mathbf{A})^{-1} = \frac{\operatorname{adj}(s\mathbf{1} - \mathbf{A})}{\det(s\mathbf{1} - \mathbf{A})}, \tag{9.44}$$

where $\operatorname{adj}(s\mathbf{1} - \mathbf{A})$ is the adjoint matrix of $(s\mathbf{1} - \mathbf{A})$, which is obtained by replacing each element of $(s\mathbf{1} - \mathbf{A})$ by its cofactor and transposing. If we assume distinct eigenvalues and take the inverse Laplace transform of $\boldsymbol{\Phi}(s)$, we obtain

$$\boldsymbol{\phi}(t) = \sum_{i=1}^{q} \varepsilon^{s_i t} \frac{\operatorname{adj}(s_i\mathbf{1} - \mathbf{A})}{\prod\limits_{\substack{j=1\\ j\neq i}}^{q} (s_i - s_j)}. \tag{9.45}$$

A case of particular interest is that of a single-input–single-output network that possesses no direct coupling between the input and output ports. For such a case, Eq. (9.42) simplifies as

$$H(s) = \mathbf{c}^t(s\mathbf{1} - \mathbf{A})^{-1}\mathbf{b} = \mathbf{c}^t\boldsymbol{\Phi}(s)\mathbf{b}, \tag{9.46}$$

where $\mathbf{c}$ and $\mathbf{b}$ are both column matrices; $\mathbf{c}^t$ is a row matrix, being the transpose of $\mathbf{c}$.

It should be noted that the transfer function of a network, $H(s)$, is unique, whether it is obtained by using Eq. (9.46) or by any other procedure. On the other hand, a given transfer function $H(s)$ can, in general, be associated with several sets of state equations, so that there is no unique set of state equations to describe the network.[1]

Example 9.4. *The RC–NIC network* (continued). Assuming that in Fig. 9.1 the voltage $v_2(t)$ provides the desired output, we obtain

$$\mathbf{b} = \begin{bmatrix} 1 \\ 0 \end{bmatrix}$$

$$\mathbf{c} = \begin{bmatrix} 0 \\ 1 \end{bmatrix}.$$

The **A**-matrix was previously determined as

$$\mathbf{A} = \begin{bmatrix} -2 & 1 \\ -1 & \frac{1}{6} \end{bmatrix}.$$

The Laplace transform of the state-transition matrix is therefore

$$\mathbf{\Phi}(s) = (s\mathbf{1} - \mathbf{A})^{-1} = \begin{bmatrix} s+2 & -1 \\ 1 & s-\frac{1}{6} \end{bmatrix}^{-1}$$

$$= \frac{\begin{bmatrix} s-\frac{1}{6} & 1 \\ -1 & s+2 \end{bmatrix}}{s^2 + \frac{11}{6}s + \frac{2}{3}}.$$

Thus, the use of Eq. (9.46) gives

$$H(s) = \frac{V_2(s)}{V_s(s)} = \frac{-1}{s^2 + \frac{11}{6}s + \frac{2}{3}}.$$

9.6 CONTROLLABILITY AND OBSERVABILITY OF NATURAL MODES[2,7]

The natural modes of a network are said to be completely *controllable* if they are all coupled to the input of the network. There are, however, situations where a mode or modes of the network may be independent of the driving function; such modes are said to be *non-controllable.* If a network possesses non-controllable modes, it is quite possible to find the zero-input state response of the network unstable and yet find the forced-state response stable.

A related phenomenon is that of *observability* of modes. The natural modes of a network are said to be completely *observable* if they are all coupled to the specified output of the network. Any modes of the network that are uncoupled from the output, independently of the initial conditions, are said to be *non-*

observable at that output. Here again if the network has non-observable modes in a given output and input, it is clearly possible for the network to be unstable and yet select an output that appears stable.

Controllability

To determine the necessary and sufficient condition for the complete controllability of the natural modes, consider a single-input network represented by the state equations:

$$\frac{d\mathbf{x}(t)}{dt} = \mathbf{A}\mathbf{x}(t) + \mathbf{b}e(t). \tag{9.47}$$

Assuming that the eigenvalues of **A** are distinct, there exists a nonsingular matrix **M**, known as the *modal matrix*, which transforms **A** into diagonal form as shown by[3]

$$\mathbf{M}\mathbf{A}\mathbf{M}^{-1} = \mathbf{\Lambda} \tag{9.48}$$

where $\mathbf{\Lambda}$ is a diagonal matrix composed of the eigenvalues $s_1, s_2, \ldots, s_q$ of matrix **A**, or the natural frequencies. Such a transformation is called a *similarity transformation.* Let us define a new set of variables $\boldsymbol{\xi}(t)$ related to $\mathbf{x}(t)$ by

$$\boldsymbol{\xi}(t) = \mathbf{M}\mathbf{x}(t) \tag{9.49}$$

or

$$\mathbf{x}(t) = \mathbf{M}^{-1}\boldsymbol{\xi}(t). \tag{9.50}$$

Then, substituting Eq. (9.50) in (9.47) and premultiplying by **M**, we get

$$\frac{d\boldsymbol{\xi}(t)}{dt} = \mathbf{M}\mathbf{A}\mathbf{M}^{-1}\boldsymbol{\xi}(t) + \mathbf{M}\mathbf{b}e(t)$$

$$= \mathbf{\Lambda}\boldsymbol{\xi}(t) + \boldsymbol{\beta}e(t), \tag{9.51}$$

where

$$\boldsymbol{\beta} = \mathbf{M}\mathbf{b} \tag{9.52}$$

is a column matrix with a number of elements equal to the number of natural modes of the network. The ith term $d\xi_i(t)/dt$ of Eq. (9.51) may be written as

$$\frac{d\xi_i(t)}{dt} = s_i\xi_i(t) + \beta_i e(t). \tag{9.53}$$

It is thus apparent that each natural mode occurs in $\boldsymbol{\xi}(t)$, and hence in the complete state response $\mathbf{x}(t)$, if and only if all the elements of $\boldsymbol{\beta}$ are nonzero.

Suppose we now define the square matrix

$$\mathbf{P} = [\boldsymbol{\beta} \quad \boldsymbol{\Lambda\beta} \quad \boldsymbol{\Lambda}^2\boldsymbol{\beta} \quad \dots \quad \boldsymbol{\Lambda}^{q-1}\boldsymbol{\beta}]$$

$$= \begin{bmatrix} \beta_1 & s_1\beta_1 & \cdots & s_1^{q-1}\beta_1 \\ \beta_2 & s_2\beta_2 & \cdots & s_2^{q-1}\beta_2 \\ \cdot & \cdot & & \cdot \\ \cdot & \cdot & & \cdot \\ \cdot & \cdot & & \cdot \\ \beta_q & s_q\beta_q & \cdots & s_q^{q-1}\beta_q \end{bmatrix}. \tag{9.54}$$

Clearly, with the eigenvalues $s_1, s_2, \dots, s_q$ assumed to be distinct, the matrix $\mathbf{P}$ is nonsingular if and only if $\boldsymbol{\beta}$ has a nonzero element, that is, if and only if all the natural modes are controllable. But,

$$\boldsymbol{\beta} = \mathbf{Mb}$$

$$\boldsymbol{\Lambda\beta} = \mathbf{MAM}^{-1}\mathbf{Mb} = \mathbf{MAb}$$

$$\cdots\cdots\cdots\cdots\cdots\cdots\cdots\cdots$$

$$\boldsymbol{\Lambda}^{q-1}\boldsymbol{\beta} = \mathbf{MA}^{q-1}\mathbf{b};$$

we may therefore write

$$\mathbf{P} = \mathbf{M}[\mathbf{b} \ \mathbf{Ab} \ \mathbf{A}^2\mathbf{b} \cdots \mathbf{A}^{q-1}\mathbf{b}]. \tag{9.55}$$

Since $\mathbf{M}$ is nonsingular, it follows that the necessary and sufficient condition for the natural modes of a linear, time-invariant network to be completely controllable is that the matrix

$$[\mathbf{b} \ \mathbf{Ab} \ \mathbf{A}^2\mathbf{b} \dots \mathbf{A}^{q-1}\mathbf{b}]$$

be nonsingular, that is, of rank q, where q is the order of complexity of the network. If the network contains non-controllable modes, then the nullity of this matrix is precisely equal to the number of such modes.*

This criterion for controllability is equally valid when the matrix $\mathbf{A}$ has repeated eigenvalues. In the completely general case, we may follow a proof similar to that outlined above, with the modification that the matrix $\mathbf{A}$ is transformed by means of a similarity transformation to a *Jordan Canonical matrix.*[2] In this case, however, it should be noted that if an eigenvalue s_i is of order m_i, then the associated natural modes are $\varepsilon^{s_it}, t\varepsilon^{s_it}, \dots, t^{m_i-1}\varepsilon^{s_it}$.

When the network has a total of m inputs, we have

$$\frac{d\mathbf{x}(t)}{dt} = \mathbf{Ax}(t) + \sum_{i=1}^{m} \mathbf{b}_i e_i(t)$$

$$= \mathbf{Ax}(t) + \mathbf{Be}(t), \tag{9.56}$$

* The rank of a matrix is the maximum number of linearly independent rows (or columns) of the matrix. For a $q \times q$ matrix the rank plus nullity equals q.

where $\mathbf{B}$ is an $m \times q$ matrix. Thus, the necessary and sufficient condition for all the natural frequencies of the network to be controllable by any input $e_i(t)$ is that the matrix

$$[\mathbf{b}_i \;\; \mathbf{A}\mathbf{b}_i \;\; \mathbf{A}^2\mathbf{b}_i \ldots \mathbf{A}^{q-1}\mathbf{b}_i]$$

be nonsingular for $i = 1, 2, \ldots, m$. That is to say, the necessary and sufficient condition for all the natural frequencies of a multi-input network to be controllable by the input vector $\mathbf{e}(t)$ is that the matrix

$$[\mathbf{B} \;\; \mathbf{AB} \;\; \mathbf{A}^2\mathbf{B} \ldots \mathbf{A}^{q-1}\mathbf{B}]$$

be of a rank equal to the order of complexity of the network, q.

Observability

Suppose the network has a single output $r(t)$ which is related to the state vector $\mathbf{x}(t)$ by

$$r(t) = \mathbf{c}^t\mathbf{x}(t), \tag{9.57}$$

where $\mathbf{c}^t$ is a row matrix. If, as before, we assume that the eigenvalues of matrix $\mathbf{A}$ are all distinct, we may eliminate $\mathbf{x}(t)$ between Eqs. (9.50) and (9.57), obtaining

$$r(t) = \boldsymbol{\gamma}^t\boldsymbol{\xi}(t). \tag{9.58}$$

where

$$\boldsymbol{\gamma}^t = \mathbf{c}^t\mathbf{M}^{-1}. \tag{9.59}$$

Since the transpose of the product of two matrices is equal to the product of the transposes of the matrices in reverse order, we may also write

$$\boldsymbol{\gamma} = (\mathbf{M}^t)^{-1}\mathbf{c}. \tag{9.60}$$

Equation (9.58) indicates that all natural modes of the network will be observed in the output $r(t)$ provided all the elements of the column matrix $\boldsymbol{\gamma}$ are non-zero. Suppose we define a new square matrix $\mathbf{Q}$ of dimension q as follows:

$$\mathbf{Q} = [\boldsymbol{\gamma} \;\; \boldsymbol{\Lambda\gamma} \;\; \boldsymbol{\Lambda}^2\boldsymbol{\gamma} \ldots \boldsymbol{\Lambda}^{q-1}\boldsymbol{\gamma}]$$
$$= \begin{bmatrix} \gamma_1 & s_1\gamma_1 & \cdots & s_1^{q-1}\gamma_1 \\ \gamma_2 & s_2\gamma_2 & & s_2^{q-1}\gamma_2 \\ \cdot & \cdot & & \cdot \\ \cdot & \cdot & & \cdot \\ \cdot & \cdot & & \cdot \\ \gamma_q & s_q\gamma_q & \cdots & s_q^{q-1}\gamma_q \end{bmatrix}. \tag{9.61}$$

Clearly, the matrix $\mathbf{Q}$ as defined above is nonsingular if and only if γ_i is non-zero for $i = 1, 2, \ldots, q$. Using Eqs. (9.48) and (9.60), and noting that $\boldsymbol{\Lambda} = \boldsymbol{\Lambda}^t$ (since $\boldsymbol{\Lambda}$

is a diagonal matrix), we find that

$$\gamma = (\mathbf{M}^t)^{-1}\mathbf{c}$$

$$\Lambda\gamma = \Lambda^t\gamma = (\mathbf{M}^t)^{-1}\mathbf{A}^t\mathbf{M}^t(\mathbf{M}^t)^{-1}\mathbf{c} = (\mathbf{M}^t)^{-1}\mathbf{A}^t\mathbf{c}$$

$$\cdots\cdots\cdots\cdots\cdots\cdots\cdots\cdots\cdots\cdots\cdots\cdots$$

$$\Lambda^{q-1}\gamma = (\mathbf{M}^t)^{-1}(\mathbf{A}^t)^{q-1}\mathbf{c}.$$

Since the matrix $\mathbf{M}$ is nonsingular, it thus follows that the necessary and sufficient condition for the natural modes of a linear, time-invariant network to be completely observable in a specified output and input is that the $q \times q$ matrix

$$[\mathbf{c} \quad \mathbf{A}^t\mathbf{c} \quad (\mathbf{A}^t)^2 \quad \mathbf{c} \ldots (\mathbf{A}^t)^{q-1}\mathbf{c}]$$

be of rank q, where $\mathbf{A}^t$ is the transpose of $\mathbf{A}$.

In the case of a network with n outputs, we may write

$$\mathbf{r}(t) = \begin{bmatrix} r_1(t) \\ r_2(t) \\ \cdot \\ \cdot \\ \cdot \\ r_n(t) \end{bmatrix} = \begin{bmatrix} \mathbf{c}_1^t \\ \mathbf{c}_2^t \\ \cdot \\ \cdot \\ \cdot \\ \mathbf{c}_n^t \end{bmatrix} \mathbf{x}(t) = \mathbf{C}\mathbf{x}(t) \tag{9.62}$$

where $\mathbf{C}$ is an $n \times q$ matrix. Thus, the necessary and sufficient condition for all the natural frequencies of the network to be observable in any output $r_i(t)$ is that the matrix

$$[\mathbf{c}_i \quad \mathbf{A}^t\mathbf{c}_i \quad (\mathbf{A}^t)^2\mathbf{c}_i \ldots (\mathbf{A}^t)^{q-1}\mathbf{c}_i]$$

be nonsingular for $i = 1, 2, \ldots, n$. In other words, the necessary and sufficient condition for all the natural frequencies of a multi-output network to be observable in the output vector $\mathbf{r}(t)$ is that the matrix

$$[\mathbf{C}^t \quad \mathbf{A}^t\mathbf{C}^t \quad (\mathbf{A}^t)^2\mathbf{C}^t \ldots (\mathbf{A}^t)^{q-1}\mathbf{C}^t]$$

be of a rank equal to the order of complexity of the network, q.

Example 9.5. *An RC network with negative elements.* Consider the network shown in Fig. 9.5 containing two negative-resistance elements. Such a network may, for example, represent the circuit model of two tunnel diodes connected in series with a positive-resistance element. The state equations for the network are obtained as

$$\frac{d}{dt}\begin{bmatrix} v_1(t) \\ v_2(t) \end{bmatrix} = \begin{bmatrix} 0 & -1 \\ -1 & 0 \end{bmatrix} \begin{bmatrix} v_1(t) \\ v_2(t) \end{bmatrix} + \begin{bmatrix} 1 \\ 1 \end{bmatrix} v_s(t). \tag{9.63}$$

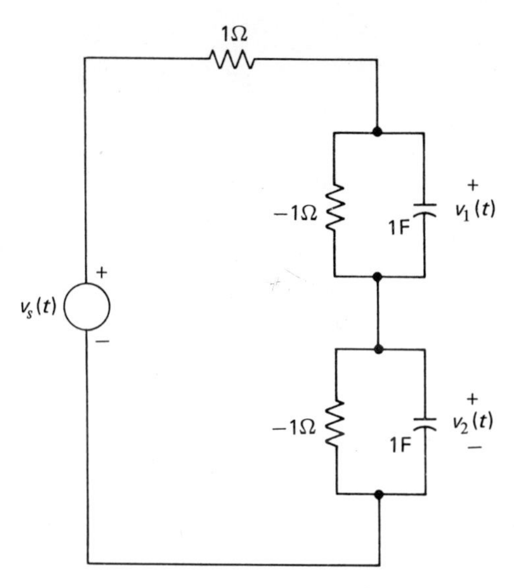

Fig. 9.5. $\pm RC$ network.

Hence,

$$\mathbf{A} = \begin{bmatrix} 0 & -1 \\ -1 & 0 \end{bmatrix}$$

$$\mathbf{b} = \begin{bmatrix} 1 \\ 1 \end{bmatrix}.$$

The order of complexity of the network is obviously two, so that there are two natural modes to consider. However,

$$[\mathbf{b} \quad \mathbf{Ab}] = \begin{bmatrix} 1 & -1 \\ 1 & -1 \end{bmatrix}$$

which is evidently singular, with a nullity of one. Hence, the network has one non-controllable mode. This can be confirmed by evaluating the forced-state response of the network. We thus find that the network has two natural modes: ε^{-t} and ε^{t}, with the mode ε^{-t} coupled to the source and therefore controllable, and the mode ε^{t} uncoupled from the source and therefore non-controllable.

To investigate the observability of these two modes, assume that the output $r(t) = v_1(t)$, so that

$$\mathbf{c} = \begin{bmatrix} 1 \\ 0 \end{bmatrix}.$$

Hence,

$$[\mathbf{c} \quad \mathbf{A}^t\mathbf{c}] = \begin{bmatrix} 1 & 0 \\ 0 & -1 \end{bmatrix},$$

which is nonsingular, indicating that the modes ε^{-t} and ε^{t} are both observable in $v_1(t)$. Similarly, we can show that both modes are also observable in $v_2(t)$. Consider, next, the voltage developed across the $+1\,\Omega$-resistor as another possible output, so that

$$r(t) = -v_1(t) - v_2(t) + v_s(t), \tag{9.64}$$

that is,

$$\mathbf{c} = \begin{bmatrix} -1 \\ -1 \end{bmatrix}.$$

In this case, we have

$$[\mathbf{c} \quad \mathbf{A}^t\mathbf{c}] = \begin{bmatrix} -1 & 1 \\ -1 & 1 \end{bmatrix}$$

which is singular, with a nullity of one. This means that, independently of the initial conditions, one natural mode of the network is non-observable in the voltage across the $+1\,\Omega$-resistor. More specifically, by examining the zero-input state response of the network we find that the mode that is non-observable in the voltage across the $+1\,\Omega$-resistor is ε^{t}. In other words, the network of Fig. 9.5 has an unstable zero-input response in $v_1(t)$ or $v_2(t)$; and yet, independently of the initial conditions, it appears to have a stable response across the $+1\,\Omega$-resistor.

Concluding Remarks

From the viewpoint of controllability and observability of natural modes, we may classify electric networks into four types:

a) completely controllable and completely observable
b) completely controllable but unobservable
c) completely observable but uncontrollable
d) uncontrollable and unobservable.

If the network is completely controllable, all the natural modes are coupled to the input and it is possible to transfer the network from any initial state $\mathbf{x}(0)$ to any desired state in a finite length of time by applying an appropriate input. If the network is completely observable, all the natural modes are coupled to the output and it is possible to identify the initial state $\mathbf{x}(0)$ of the network by observing the output $\mathbf{r}(t)$ for a finite length of time. In this sense, the concepts of controllability and observability are the dual of each other.

REFERENCES

1. L. A. ZADEH and C. A. DESOER, *Linear System Theory*. McGraw-Hill, 1963.
2. R. J. SCHWARZ and B. FRIEDLAND, *Linear Systems*. McGraw-Hill, 1965.
3. P. M. DERUSSO, R. J. ROY, C. M. CLOSE, *State Variables for Engineers*. Wiley, 1965.
4. P. R. BRYANT, "The Order of Complexity of Electrical Networks," *Proc. I.E.E.* **106C**, 174 (1959).
5. C. POTTLE, "State Space Techniques for General Active Network Analysis," in *System Analysis by Digital Computer*. Wiley, 1966.
6. R. C. DORF, *Modern Control Systems*. Addison-Wesley, 1967.
7. R. E. KALMAN, Y. C. HO and K. S. NARENDRA, "Controllability of Linear Dynamical Systems," in *Contributions to Differential Equations*. Interscience Publishers, 1962.

PROBLEMS

9.1 Develop the state equations for the network shown in Fig. P9.1. Hence, evaluate:
a) The zero-input state response
b) The forced state response produced by a unit-step function
c) The transfer impedance $V_2(s)/I_1(s)$.

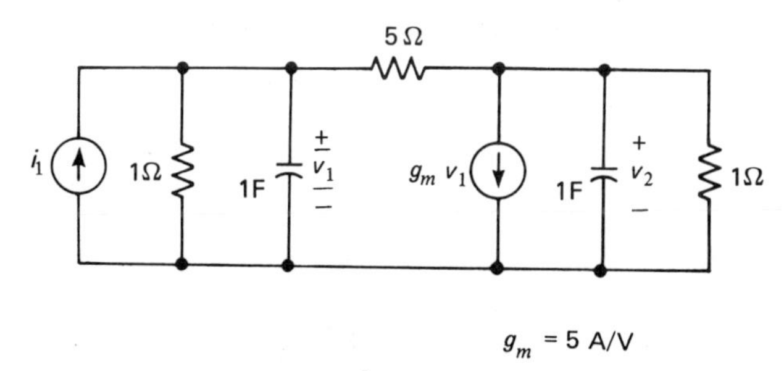

Fig. P9.1.

9.2 Evaluate the state-transition matrix for the network shown in Fig. P9.2 using
a) Sylvester's theorem
b) The Laplace transform method.

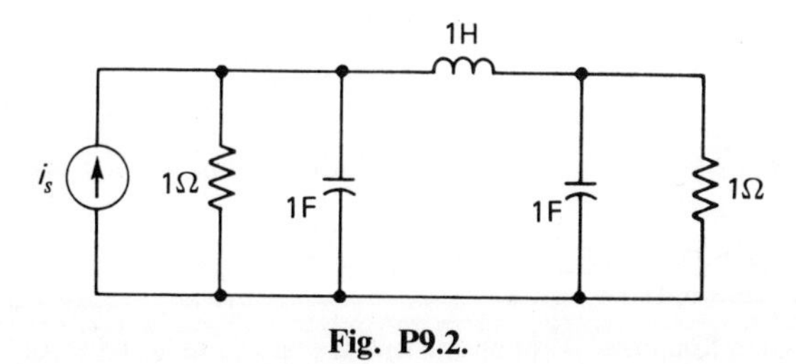

Fig. P9.2.

9.3 Using the discrete-time approximation, determine the zero-input state response of the network shown in Fig. P9.1 given that $v_1(0) = v_2(0) = 1$ volt.

9.4 A linear, time-invariant network has the state-transition matrix

$$\boldsymbol{\phi}(t) = \begin{bmatrix} \varepsilon^{-t} & t\varepsilon^{-t} \\ 0 & \varepsilon^{-t} \end{bmatrix}.$$

Find the **A**-matrix of the network.

9.5 State whether or not the following matrices are realizable as state-transition matrices by linear, time-invariant networks, giving reasons for your answer:

$$\begin{bmatrix} \varepsilon^{-t} & 2t\varepsilon^{-t} \\ 4t\varepsilon^{-t} & 3\varepsilon^{-t} \end{bmatrix}$$

$$\begin{bmatrix} \varepsilon^{-t} & 2\varepsilon^{-t} \\ 4\varepsilon^{-t} & 3\varepsilon^{-t} \end{bmatrix}$$

$$\begin{bmatrix} \varepsilon^{-t} & 0 \\ 0 & 3\varepsilon^{-t} \end{bmatrix}$$

$$\begin{bmatrix} t\varepsilon^{-t} & 2\varepsilon^{-t} \\ 4\varepsilon^{-t} & 3t\varepsilon^{-t} \end{bmatrix}.$$

9.6 What is the order of complexity of the network shown in Fig. P9.6? Determine the condition necessary for complete controllability of all modes of the network.

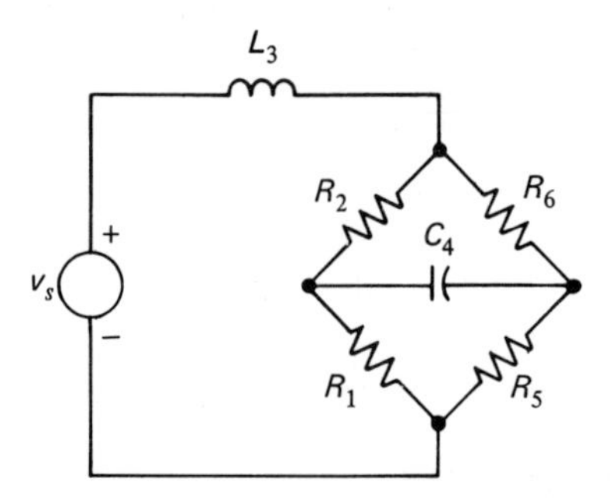

Fig. P9.6.

9.7 In the network of Fig. P9.7 determine the inductor current $i(t)$ and capacitor voltage $v(t)$ for:

a) $v_{s1}(t) = v_{s2}(t) = u(t)$
b) $v_{s1}(t) = u(t)$ and $v_{s2}(t) = -u(t)$,

where $u(t)$ is a unit-step function. Assume the initial conditions to be zero.

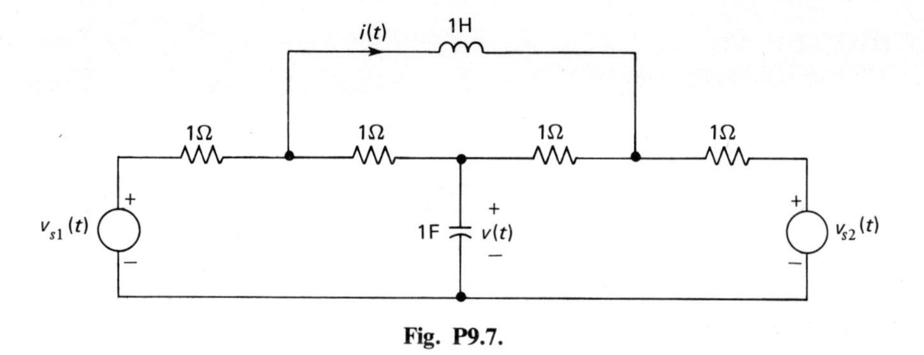

Fig. P9.7.

CHAPTER 10

SIGNAL-FLOW GRAPHS

10.1 INTRODUCTION

The analysis of a linear network reduces ultimately to the solution of a set of linear algebraic equations describing the equilibrium state of the network. The solution may be obtained by conventional algebraic methods or, alternatively, we can use the method of *signal-flow graphs* which was originated by Mason.[1,2] The signal-flow graph is particularly attractive in the study of linear physical networks for three reasons:

1. It makes possible a graphical visualization of the relationships amongst the variables of the network.
2. It provides a systematic procedure for manipulation of the variables of interest.
3. It offers a means of solving the pertinent equations directly by inspection of the graph.

In the following sections we shall consider the properties and algebra of signal-flow graphs. Although we shall mainly use electronic circuits as illustrative examples, the theory of signal-flow graphs is equally applicable to the solution of problems in automatic control, mechanics, conditional probability, etc.

10.2 THE SIGNAL-FLOW GRAPH

A signal-flow graph is a network of *directed branches* which are interconnected at certain points called *nodes*. A typical node j has an associated *node signal* x_j. A typical directed branch jk originates at node j and terminates upon node k and has an associated *branch transmittance* t_{jk} specifying the manner in which the signal x_k at node k depends upon the signal x_j at node j. The flow of signals in the various parts of the graph is dictated by the following three basic rules which are portrayed graphically in Fig. 10.1:

a) Signal flows along a branch only in the direction defined by the arrow and is multiplied by the transmittance of that branch (see Fig. 10.1a).
b) A node signal is equal to the algebraic sum of all signals entering the pertinent node via the incoming branches (see Fig. 10.1b).

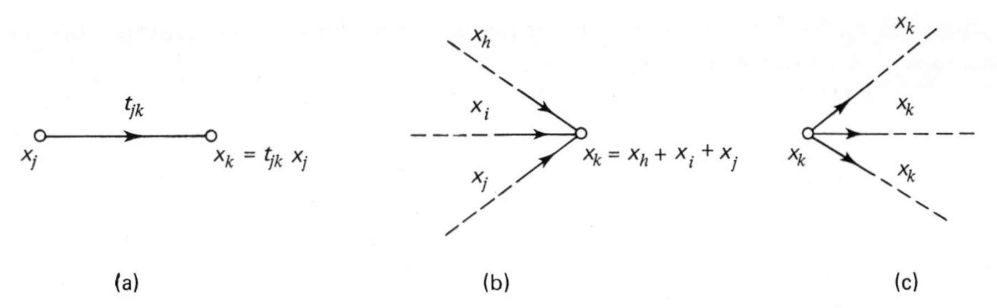

Fig. 10.1. Illustrating three basic properties of signal-flow graphs.

c) The signal at a node is applied to each outgoing branch which originates from that node, the transmission being entirely independent of the transmittances of the outgoing branches (see Fig. 10.1c).

To illustrate how a set of simultaneous equations can be represented by a signal-flow graph, consider the case of a linear system having only one independent variable x_0, and n dependent variables $x_1, x_2, \ldots, x_n$ which are related as follows:

$$\begin{aligned} -t_{01}x_0 &= a_{11}x_1 + a_{12}x_2 + \cdots + a_{1n}x_n \\ 0 &= a_{21}x_1 + a_{22}x_2 + \cdots + a_{2n}x_n \\ &\cdots\cdots\cdots\cdots\cdots\cdots \\ 0 &= a_{n1}x_1 + a_{n2}x_2 + \cdots + a_{nn}x_n. \end{aligned} \tag{10.1}$$

These equations may be re-written in the form

$$\begin{aligned} x_1 &= t_{01}x_0 + t_{11}x_1 + t_{21}x_2 + \cdots + t_{n1}x_n \\ x_2 &= \phantom{t_{01}x_0 +{}} t_{12}x_1 + t_{22}x_2 + \cdots + t_{n2}x_n \\ &\cdots\cdots\cdots\cdots\cdots\cdots \\ x_n &= \phantom{t_{01}x_0 +{}} t_{1n}x_1 + t_{2n}x_2 + \cdots + t_{nn}x_n, \end{aligned} \tag{10.2}$$

where the various t's are related to the given a's by

$$\begin{bmatrix} t_{11} & t_{21} & \cdots & t_{n1} \\ t_{12} & t_{22} & \cdots & t_{n2} \\ \cdot & \cdot & & \cdot \\ \cdot & \cdot & & \cdot \\ \cdot & \cdot & & \cdot \\ t_{1n} & t_{2n} & \cdots & t_{nn} \end{bmatrix} = \begin{bmatrix} 1 + a_{11} & a_{12} & \cdots & a_{1n} \\ a_{21} & 1 + a_{22} & \cdots & a_{2n} \\ \cdot & \cdot & & \cdot \\ \cdot & \cdot & & \cdot \\ \cdot & \cdot & & \cdot \\ a_{n1} & a_{n2} & \cdots & 1 + a_{nn} \end{bmatrix}. \tag{10.3}$$

Equations (10.2) can be represented by a signal-flow graph having nodes $0, 1, 2, \ldots, n$; the variable x_k in the equations corresponds to the node k in the graph, and the coefficient t_{jk} corresponds to the transmittance of the branch jk

which originates from node j and terminates upon node k. An example for the case $n = 3$ is shown in Fig. 10.2.

In Eqs. (10.2) we see that each dependent variable is expressed explicitly as the *effect* due to all other variables of the system acting as *causes.* Clearly, therefore, a signal-flow graph is essentially a *cause-and-effect* form of representing a set of linear algebraic equations. In this formulation, however, it is important to note that a given dependent node signal acts as an effect in one equation only; in all other equations that particular node signal assumes the role of a cause.

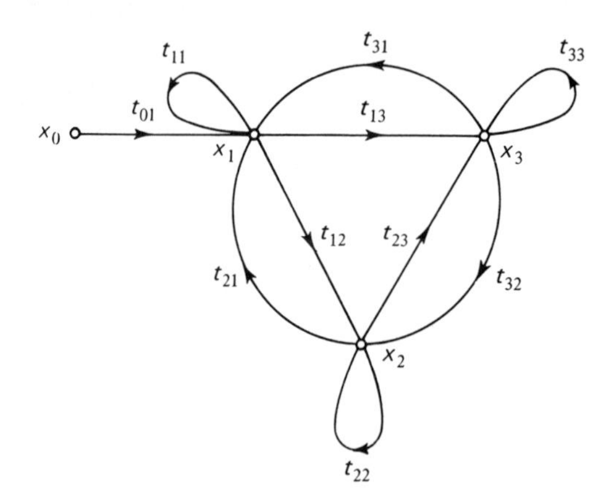

Fig. 10.2. Signal-flow graph for a set of three simultaneous equations.

In constructing the signal-flow graph for a circuit we are guided by the desire to maintain a correspondence between the graph and the flow of signals in the various parts of the actual circuit. Furthermore, we must define as many variables as are needed to construct the graph completely. As an illustration we will consider two examples involving a ladder network and an emitter follower.

Example 10.1. *Ladder network.* The equilibrium equations of the ladder network shown in Fig. 10.3(a) are, in terms of the loop currents I_1 and I_2, defined as follows:

$$\begin{aligned} V_1 &= (Z_1 + Z_a)I_1 - Z_aI_2 \\ 0 &= -Z_aI_1 + (Z_a + Z_2 + Z_b)I_2. \end{aligned} \tag{10.4}$$

Re-writing Eqs. (10.4) in a *cause-and-effect* form, we have

$$\begin{aligned} I_1 &= \frac{1}{Z_1 + Z_a}V_1 + \frac{Z_a}{Z_1 + Z_a}I_2 \\ I_2 &= \frac{Z_a}{Z_a + Z_2 + Z_b}I_1, \end{aligned} \tag{10.5}$$

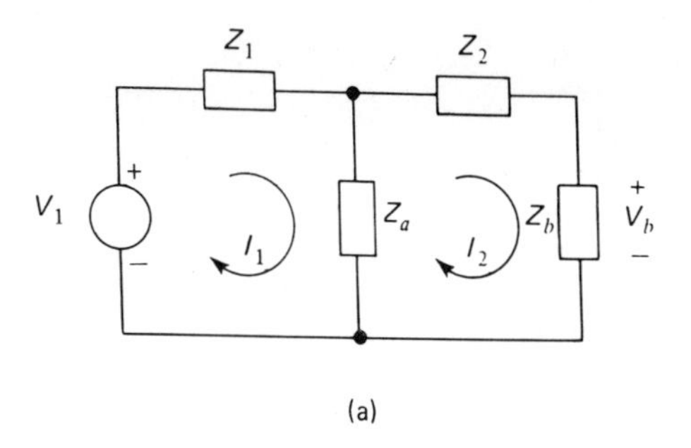

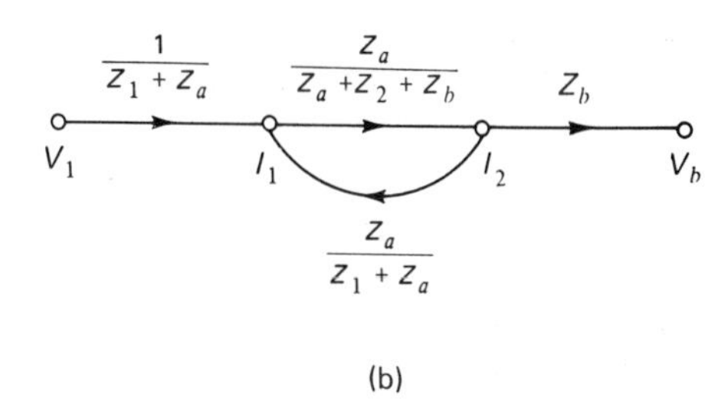

Fig. 10.3. (a) Ladder network analysed on the loop basis; (b) Signal-flow graph representation.

which, together with the relation $V_b = Z_b I_2$, may be represented by the signal-flow graph shown in part (b) of Fig. 10.3.

If, however, we set up the equilibrium equations of the network in terms of the branch currents flowing through Z_1 and Z_2, and the node voltages V_1, V_a and V_b, as in Fig. 10.4(a), we get

$$\begin{aligned} I_1 &= Y_1(V_1 - V_a) \\ V_a &= Z_a(I_1 - I_2) \\ I_2 &= Y_2(V_a - V_b) \\ V_b &= Z_b I_2, \end{aligned} \tag{10.6}$$

where $Y_1 = 1/Z_1$ and $Y_2 = 1/Z_2$. Equations (10.6) lead directly to the signal-flow graph representation shown in Fig. 10.4(b).

We see, therefore, that it is possible to construct different signal-flow graphs for the same network depending upon the variables chosen and the method of writing the equilibrium equations which describe the behaviour of the network.

Example 10.2. *Emitter follower.* The schematic diagram of an emitter follower is shown in part (a) of Fig. 10.5. In part (b) we have used a hybrid model

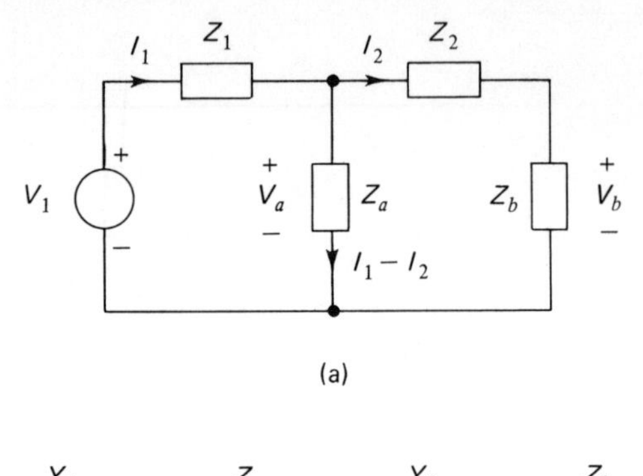

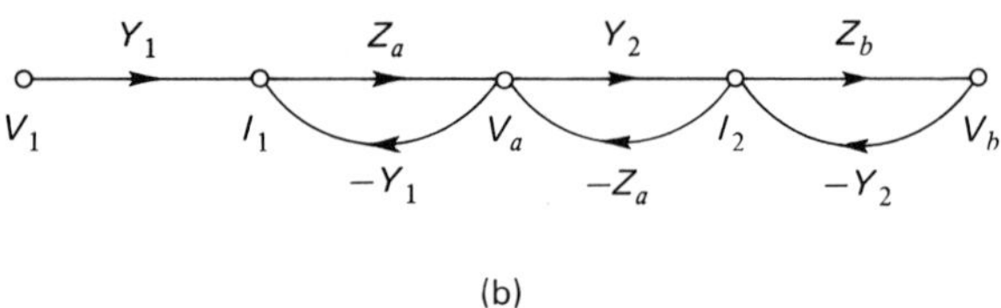

Fig. 10.4. (a) Ladder network analysed in terms of branch currents; (b) Signal-flow graph representation.

to represent the transistor for small-signal operation. Assuming that V_s, I_b, I_e and V_o are chosen as the variables of interest, we have

$$I_b = \frac{1}{R_s + h_{ie}}[V_s - (1 - h_{re})V_o]$$

$$I_e = (1 + h_{fe})I_b - h_{oe}V_o \tag{10.7}$$

$$V_o = I_e R_l.$$

The first equation specifies the manner in which the source voltage V_s causes the input base current I_b; the second equation specifies the output emitter current I_e resulting from I_b; and the third equation defines the output voltage resulting from I_e. The complete signal-flow graph for the emitter follower is therefore as shown in part (c) of Fig. 10.5.

Further Definitions and Terminology

In a signal-flow graph a node with only outgoing branches is referred to as a *source node* of the graph; the signal associated with such a node is necessarily an independent variable. A node having one or more incoming branches is a *dependent node* whether or not it has outgoing branches as well. When a dependent node has no outgoing branches, however, it is referred to as a *sink node.* Clearly, the total number of nodes in a graph is equal to the total number of variables in the set of simultaneous equations which the graph represents.

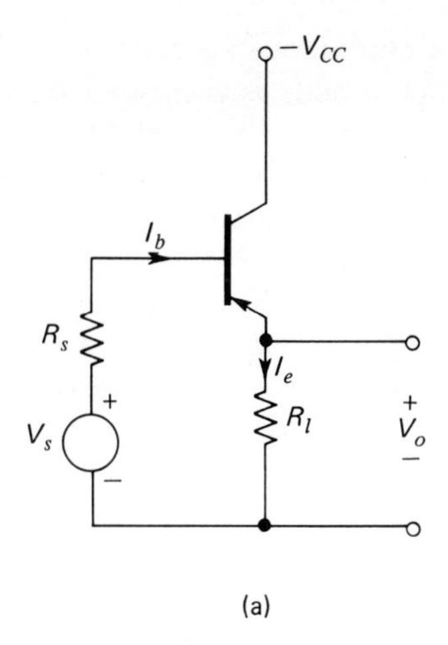

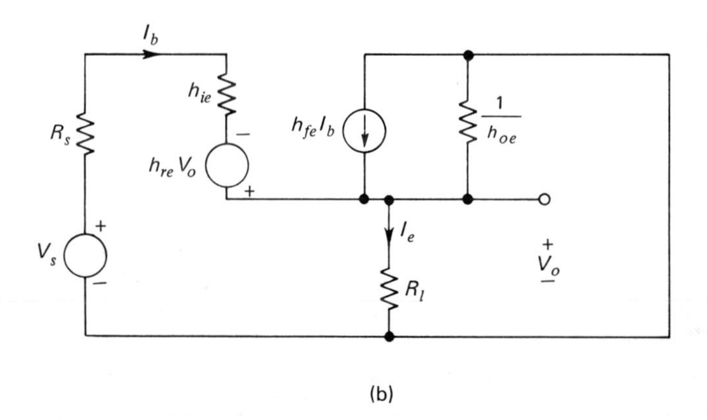

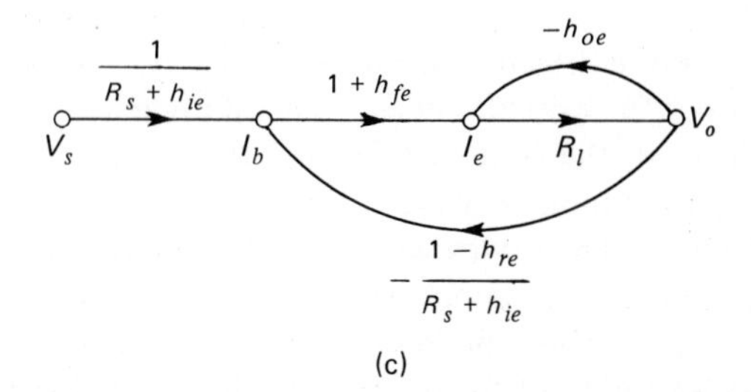

Fig. 10.5. (a) Emitter follower; (b) Circuit model; (c) Signal-flow graph.

A *path* is defined as a continuous succession of branches, traversed in the arrow direction, along which a node is encountered only once. The *path transmittance*, denoted by P, is equal to the product of the branch transmittances which constitute the associated path. In a signal-flow graph we may find that there are many possible different paths originating at a specified node j and terminating upon another node k, or only one possible path or none at all.

A simple closed path along which no node is encountered more than once per traversal is called a *feedback loop*. The *loop transmittance*, denoted by L, is equal to the product of the branch transmittances constituting that loop. In Fig. 10.2, for example, the signal-flow graph has three closed branches t_{11}, t_{22} and t_{33} (sometimes called *self loops*) and three loops $t_{12}t_{21}$, $t_{23}t_{32}$ and $t_{13}t_{31}$.

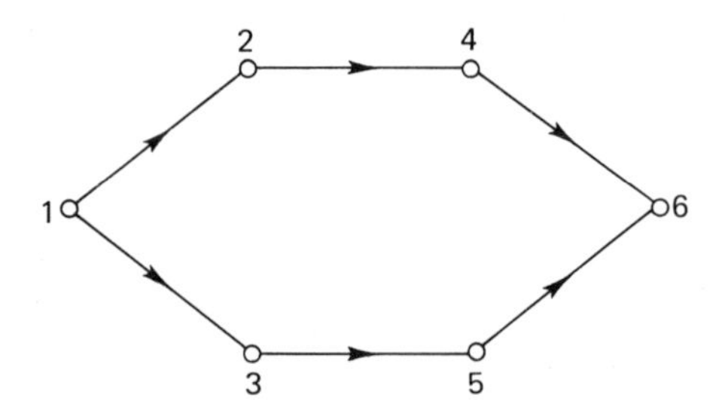

Fig. 10.6. Cascade graph.

Signal-flow graphs may, in general, be classified into *cascade graphs* and *feedback graphs*. In a cascade graph there are no feedback loops at all; Fig. 10.6 shows a simple example of a cascade graph. A feedback graph, on the other hand, is a signal-flow graph containing one or more feedback loops; Fig. 10.7 shows two examples of such graphs. An important characteristic of a feedback graph is a number called the *index* which is a measure of its relative complexity. It is defined as the minimum number of nodes that must be split so as to eliminate all feedback loops in the graph. In other words, the index of a feedback graph is a measure of the number of independent feedback loops. Splitting of a node implies dividing the node in question into two half nodes, one half node assuming the function of a new source to which all the branches leaving that node are attached, the other half node assuming the function of a new sink to which all the branches entering that node are attached. The nodes which must be split to interrupt all feedback loops in a graph are called *essential nodes*. The two feedback graphs shown in parts (a) and (b) of Fig. 10.7 both possess three feedback loops. However, the graph in Fig. 10.7(a) has an index of unity, since all feedback loops pass through the essential node 2; the result of splitting this node is given in Fig. 10.7(c). The graph in Fig. 10.7(b), on the other hand, has an index of two since all three feedback loops can only be removed by simultaneously splitting nodes 1 and 2, as illustrated in Fig. 10.7(d). Clearly, therefore, the index of a feedback graph depends not only

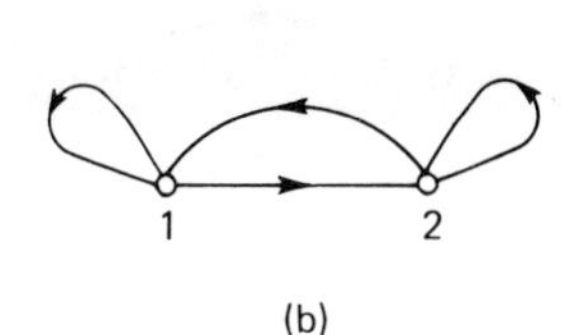

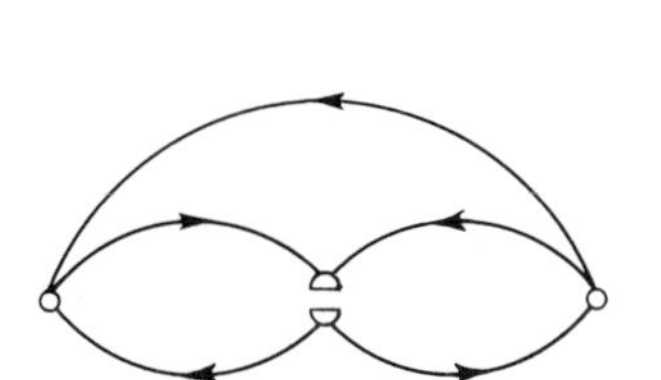

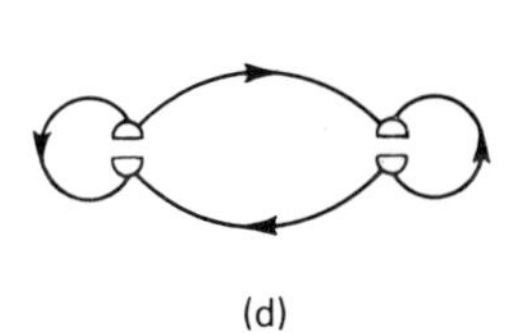

Fig. 10.7. (a) Feedback graph having one essential node; (b) Feedback graph having two essential nodes; (c) Graph of part (a) with essential node 2 split; (d) Graph of part (b) with essential nodes 1 and 2 split.

upon the number of feedback loops contained in a graph but also upon the degree of coupling between the loops. This characteristic of a signal-flow graph of the feedback type is particularly important in the study of feedback systems.

10.3 THE ALGEBRA OF SIGNAL-FLOW GRAPHS

In manipulating a signal-flow graph with the aim of reducing it to a single branch connecting the source to the sink, we can make use of certain topological properties of the graph. Figure 10.8 illustrates four *elementary equivalences* which can be readily verified by writing the corresponding algebraic equations. These equivalences are sufficient for the complete reduction of a cascade graph in which there are no closed chains of dependency. As an illustrative example, consider the cascade graph shown in part (a) of Fig. 10.9; by successively absorbing nodes x_2, x_3, x_4 and x_5, we ultimately obtain the result given in part (e) of the diagram.

When, however, the graph contains one or more feedback loops, application of the elementary equivalences will ultimately lead to the appearance of one or more self loops. This is illustrated in Fig. 10.10 where the reduced graph with a self loop at node 2 results from using equivalence (d).

To determine the effect of a self loop at some node, consider Fig. 10.11 where we have

$$\begin{aligned} x_1 &= x_0 + tx_1 \\ x_2 &= x_1. \end{aligned} \tag{10.8}$$

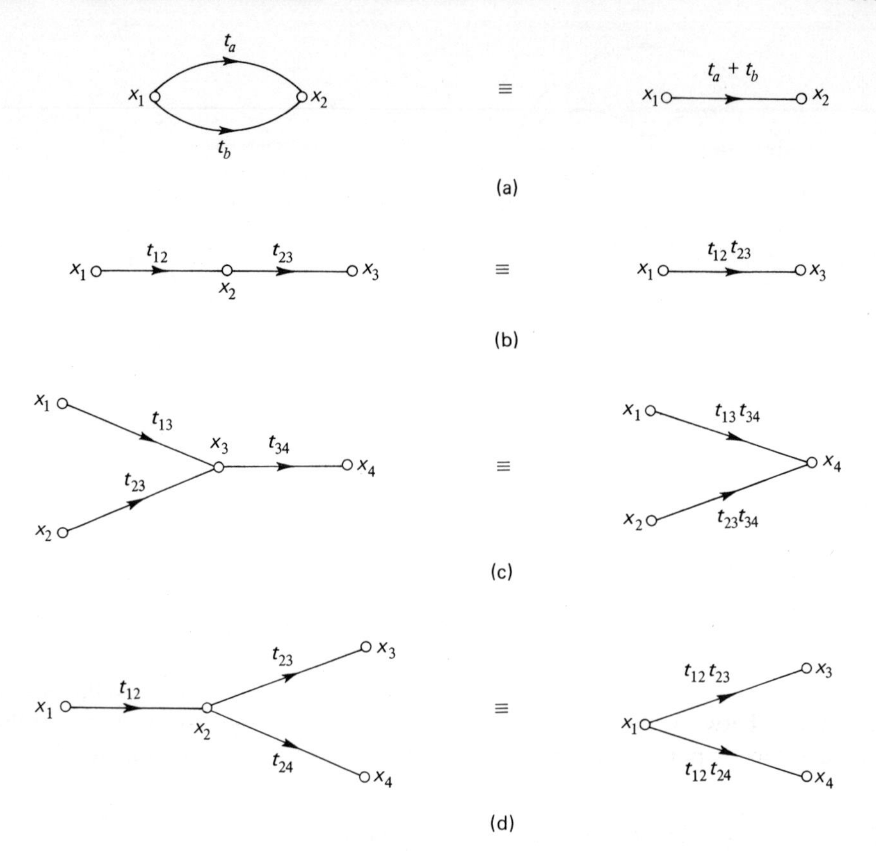

Fig. 10.8. Four elementary equivalences.

Eliminating x_1 and expressing x_2 in terms of x_0, we get

$$x_2 = \frac{1}{1 - t} x_0. \tag{10.9}$$

If the self loop were not present the signal path from x_0 to x_2 would have a transmittance of unity. When the self-loop is added, the transmittance of the path from x_0 to x_2 assumes the value $1/(1 - t)$. We see, therefore, that the effect of a self loop of transmittance t is to divide the signal by the factor $(1 - t)$ as it is transmitted through the node.

Example 10.3. *Emitter follower* (continued). A signal-flow graph for the emitter follower was constructed earlier, and for convenience it is reproduced in part (a) of Fig. 10.12. To determine the overall gain using a step-by-step reduction process,

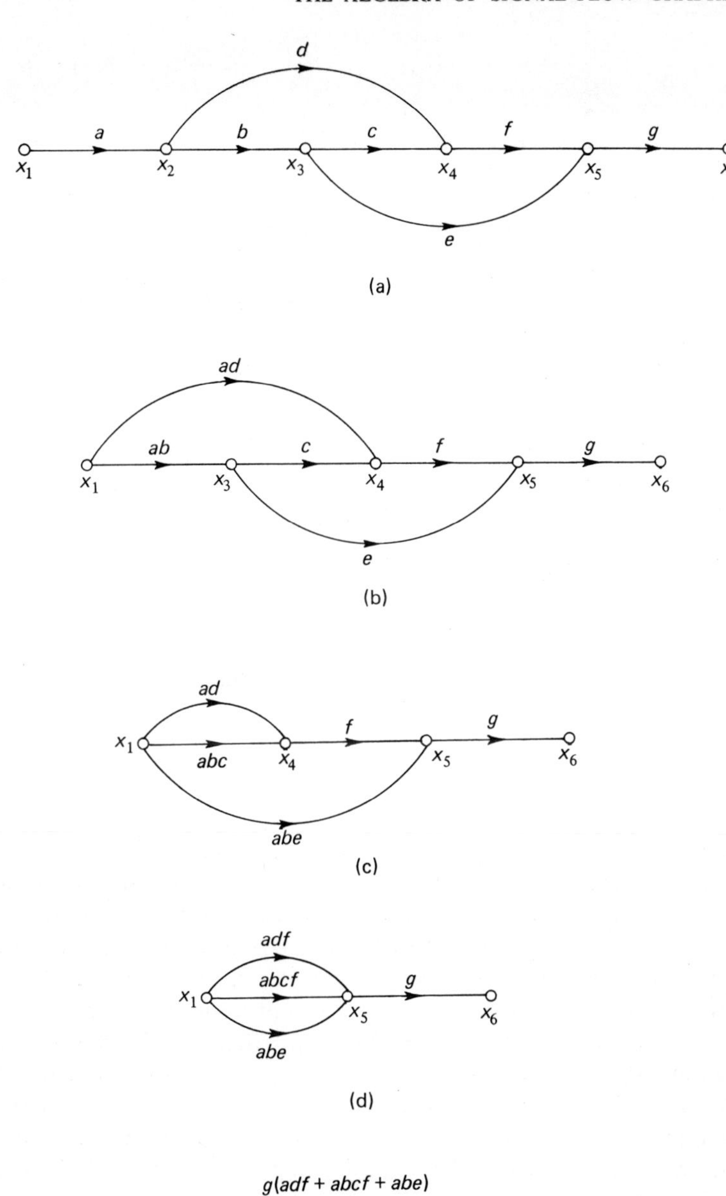

Fig. 10.9. Illustrating the step-by-step reduction of a cascade graph.

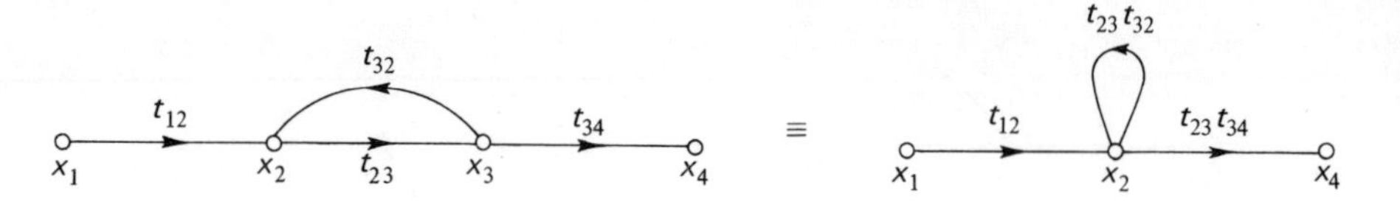

Fig. 10.10. A feedback graph.

we may follow the series of steps illustrated in Fig. 10.12 and listed below:

1. Absorbing the node I_e, by the application of equivalence (c), yields the graph shown in Fig. 10.12(b) which has a self loop at node V_o.
2. The node I_b is next absorbed in the same manner as I_e, obtaining the graph of Fig. 10.12(c) where we now have a second self loop at V_o.
3. The two self loops at the node V_o in Fig. 10.12(c) appear in parallel; we may therefore replace them with an equivalent single self loop, as in Fig. 10.12(d).
4. Finally, eliminating the self loop in Fig. 10.12(d), we find that the overall gain of the emitter follower of Fig. 10.5(a) is given by

$$K = \frac{V_o}{V_s} = \frac{R_l(1 + h_{fe})}{(R_s + h_{ie})(1 + h_{oe}R_l) + R_l(1 + h_{fe})}, \tag{10.10}$$

where we have ignored the effect of h_{re} as it is, in practice, small compared with unity.

Mason's Direct Rule

The *graph transmittance* or *graph gain* K, defined as the signal appearing at the sink node per unit signal applied at the source node, can be determined in either one of two ways: With the aid of the five basic equivalences given in Figs. 10.8 and 10.11 we may reduce the signal-flow graph in a step-by-step manner to a single branch connecting the source to sink; the transmittance of this branch gives the required graph transmittance K. The use of this procedure was demonstrated in Example 10.3 dealing with the emitter follower. Alternatively, the solution can be obtained in a direct manner by inspection of the graph and applying *Mason's direct rule.* In order to develop ideas leading to the formulation of this rule, let us examine the formation patterns for the graph transmittance of the example shown in Fig. 10.13(a). This graph exhibits three feedback loops whose

Fig. 10.11. Evaluating the effect of self loop.

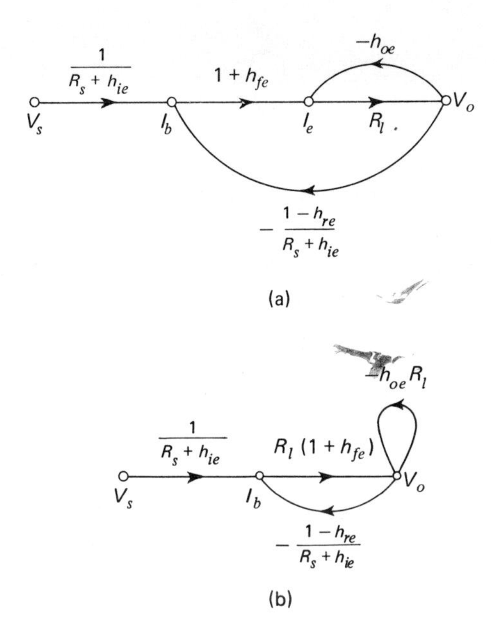

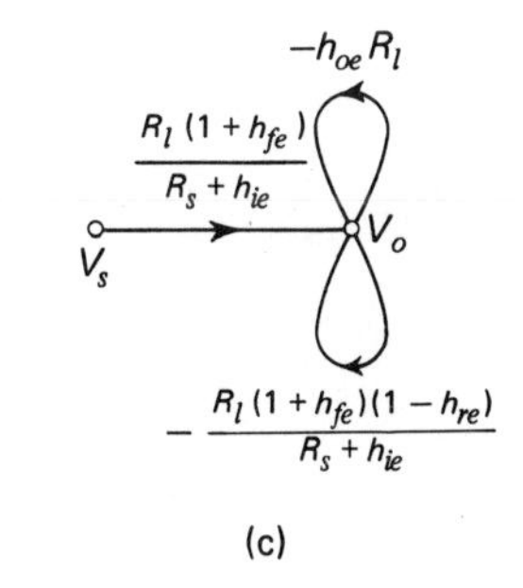

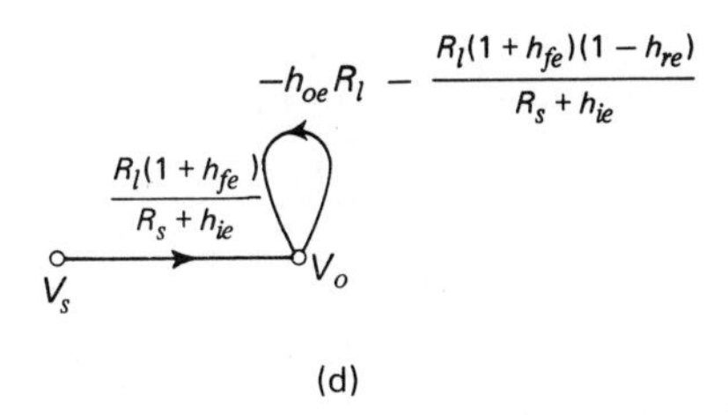

Fig. 10.12. Step-by-step reduction of signal-flow graph of emitter follower.

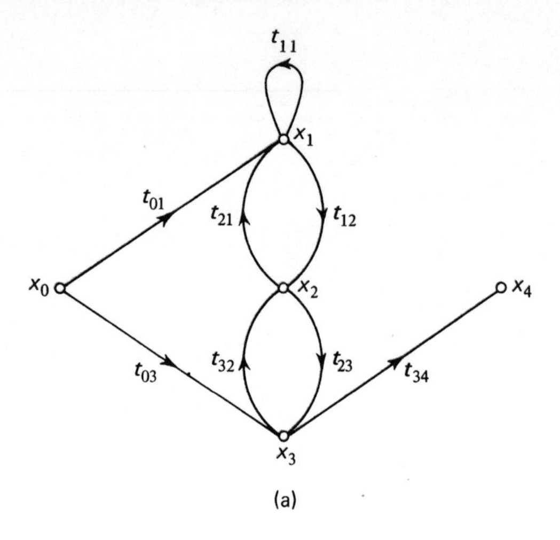

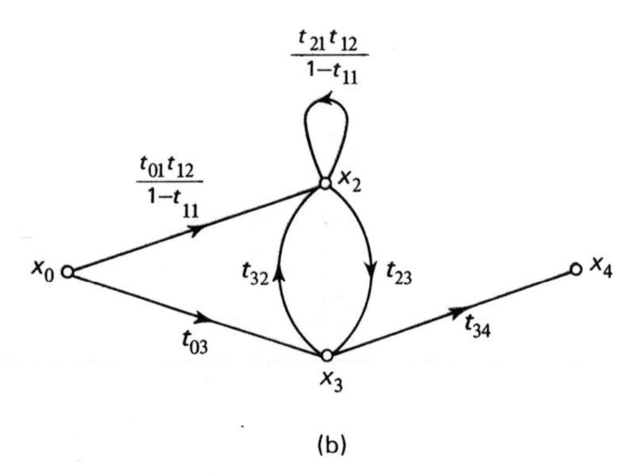

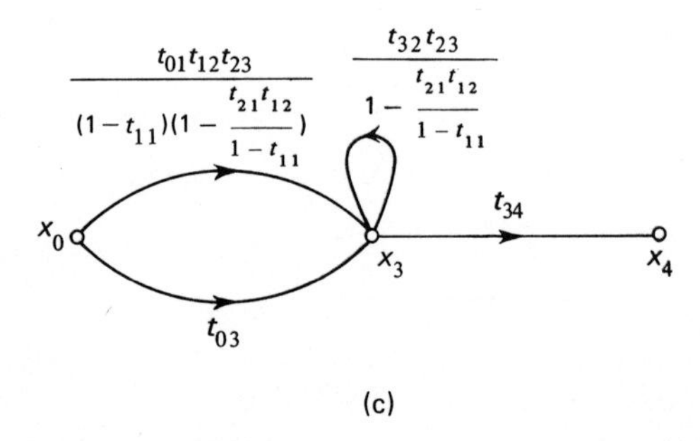

Fig. 10.13. Example used to illustrate the formation patterns for the graph transmittance.

transmittances are

$$\begin{aligned} L_1 &= t_{11} \\ L_2 &= t_{21}t_{12} \\ L_3 &= t_{32}t_{23}. \end{aligned} \tag{10.11}$$

There are two forward paths connecting the source x_0 to sink x_4 and whose transmittances are

$$\begin{aligned} P_1 &= t_{01}t_{12}t_{23}t_{34} \\ P_2 &= t_{03}t_{34}. \end{aligned} \tag{10.12}$$

By absorbing node x_1 we get the reduced graph shown in part (b) of Fig. 10.13. Subsequent absorption of node x_2 leads to the simpler form shown in part (c) from which we see, by inspection, that the graph transmittance $K = (x_4/x_0)$ is

$$K = \frac{t_{34}\left\{t_{03} + \dfrac{t_{01}t_{12}t_{23}}{(1 - t_{11})[1 - t_{21}t_{12}/(1 - t_{11})]}\right\}}{1 - \dfrac{t_{32}t_{23}}{1 - t_{21}t_{12}/(1 - t_{11})}}. \tag{10.13}$$

After clearing the numerator and denominator of fractions, and then substituting the definitions of Eqs. (10.11) and (10.12), we get

$$K = \frac{P_1 + P_2[1 - (L_1 + L_2)]}{1 - (L_1 + L_2 + L_3) + L_1L_3}. \tag{10.14}$$

From this expression we observe that, in general, the graph transmittance is in the form of a sum of all forward path transmittances, P_k, each of which is multiplied by a certain weighting factor, Δ_k, with the total sum divided by a quantity Δ which is wholly concerned with the transmittances of the different feedback loops and their mutual interactions. Furthermore, Δ does not contain the transmittance products of loops which touch each other in the graph. On the basis of these observations we may formulate the general expression for the graph transmittance K as follows:

$$K = \frac{1}{\Delta}\sum_{k=1}^{n} P_k\Delta_k, \tag{10.15}$$

where

P_k = transmittance of the kth forward path from source to sink

$\Delta = 1 -$ (sum of all individual loop transmittances)

$+$ (sum of transmittance products of all possible sets of nontouching loops taken two at a time)

$-$ (sum of transmittance products of all possible sets of nontouching loops taken three at a time)

$+ \cdots$

and

Δ_k = the value of Δ for that portion of the graph not touching the kth forward path.

The form of Eq. (10.15) suggests that we call Δ the *determinant* of the graph, and call Δ_k the *cofactor* of forward path k. Equation (10.15) is called, after its originator, *Mason's direct rule.* An algebraic proof of the rule is given elsewhere.[3]

The graph determinant Δ involves only the feedback loops in the graph, being determined by their individual transmittances and their interaction. Thus, in the case of Fig. 10.13(a) we have

$$\Delta = 1 - (L_1 + L_2 + L_3) + L_1 L_3. \tag{10.16}$$

This graph has three feedback loops, and the sum of their individual transmittances gives the term $(L_1 + L_2 + L_3)$. When the loops are taken two at a time we find only one possible set of non-touching loops which lead to the term $L_1 L_3$. This is the final term in the graph determinant of Fig. 10.13(a) because there are no possible sets of three or more non-touching loops.

The cofactor Δ_k is concerned only with those loops which do not touch the path k; so that if this path, together with all touching branches, is removed from the graph, the determinant of the remaining subgraph gives the path cofactor Δ_k. If the path in question touches all loops of the graph, or if the graph contains no loops at all, then the path cofactor is unity. In the case of Fig. 10.13(a) we have two possible forward paths, one of which ($P_1 = t_{01}t_{12}t_{23}t_{34}$) touches all three loops of the graph; therefore $\Delta_1 = 1$. The other forward path ($P_2 = t_{03}t_{34}$) touches the third loop, and L_3 is therefore absent from the associated path cofactor, so that $\Delta_2 = 1 - (L_1 + L_2)$.

If the graph contains several independent sources, their combined effect may be determined by considering the effect of each source individually, and then applying the principle of superposition.

Example 10.4. *Miller integrator.* The transistor version of the so-called Miller integrator is shown in Fig. 10.14(a). This circuit produces an output signal which approximates to the integral of the input time function. The incremental circuit model is given in part (b), assuming that the electrical behaviour of the transistor is characterized by the common-emitter h-parameters. In terms of the variables shown defined in Fig. 10.14(b), we may express the equilibrium equations in a cause-and-effect form as follows

$$\begin{aligned} I_s &= \frac{1}{R}(V_s - V_{be}) \qquad & V_{be} &= h_{ie}I_b + h_{re}V_{ce} \\ I_b &= I_s - I_f \qquad & I_f &= sC(V_{be} - V_{ce}) \\ V_{ce} &= \frac{-h_{fe}I_b + I_f}{h_{oe} + G_l}. \end{aligned} \tag{10.17}$$

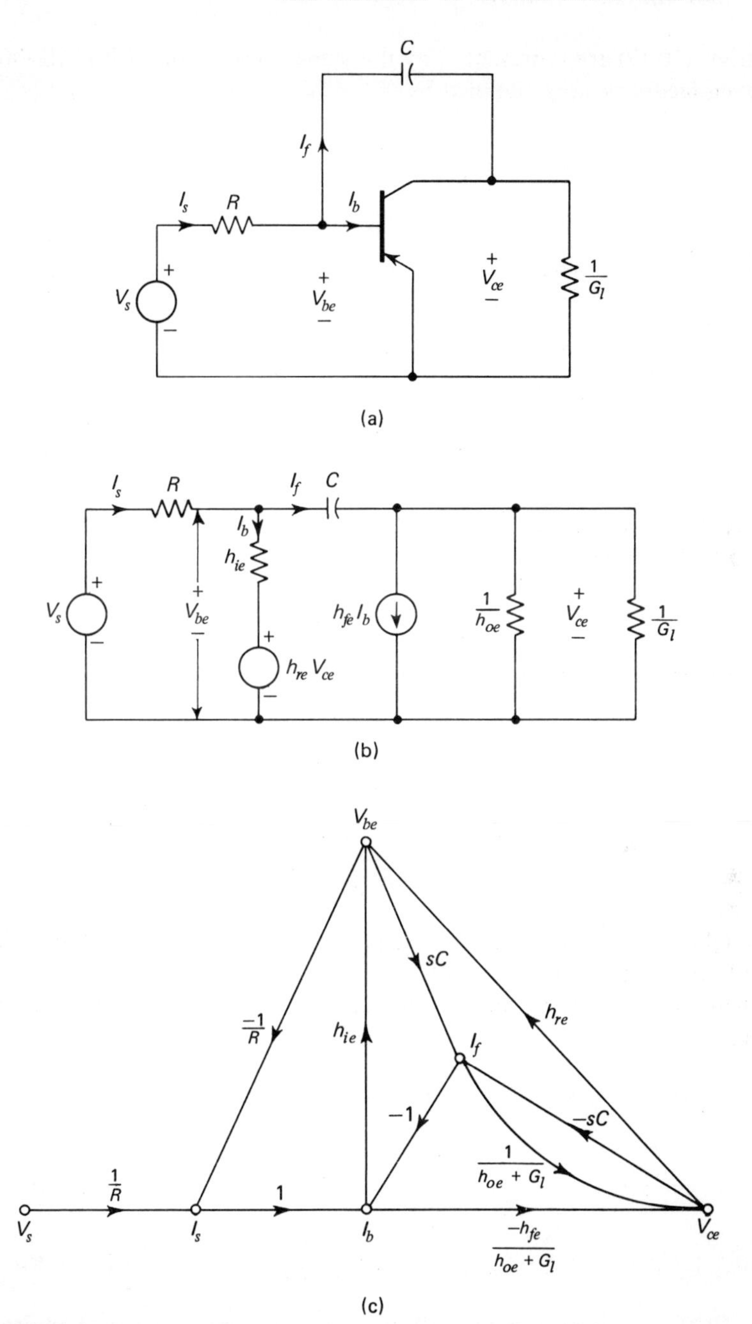

Fig. 10.14. (a) Miller integrator; (b) Circuit model; (c) Signal-flow graph.

Equations (10.17) are represented by the signal-flow graph of Fig. 10.14(c) which has seven feedback loops defined by

$$\begin{aligned} L_1 &= -\frac{h_{ie}}{R} & L_4 &= -\frac{sC}{h_{oe} + G_l} \\ L_2 &= -sCh_{ie} & L_5 &= -\frac{sCh_{fe}}{h_{oe} + G_l} \\ L_3 &= \frac{sCh_{re}}{h_{oe} + G_l} & L_6 &= \frac{h_{fe}h_{re}}{R(h_{oe} + G_l)} \\ L_7 &= \frac{sCh_{fe}h_{re}}{h_{oe} + G_l}. \end{aligned} \tag{10.18}$$

In Fig. 10.14(c) we see that L_1 and L_4 constitute the only pair of non-touching loops; the graph determinant is therefore

$$\Delta = 1 - (L_1 + L_2 + L_3 + L_4 + L_5 + L_6 + L_7) + L_1L_4. \tag{10.19}$$

Furthermore, there are two possible paths P_1 and P_2 from the source V_s to output V_{ce}, as defined by

$$\begin{aligned} P_1 &= \frac{-h_{fe}}{R(h_{oe} + G_l)} \\ P_2 &= \frac{sCh_{ie}}{R(h_{oe} + G_l)}. \end{aligned} \tag{10.20}$$

The weighting factor for each of these paths is unity, because both of them touch all the feedback loops in the graph.

Therefore, applying Mason's direct rule gives the overall transmittance $K = V_{ce}/V_s$ as

$$K = \frac{P_1 + P_2}{\Delta}$$

or

$$K = \frac{-h_{fe} + sCh_{ie}}{(h_{oe} + G_l)(R + h_{ie}) - h_{re}h_{fe} + sCR[1 + h_{fe} + h_{ie}(h_{oe} + G_l + 1/R)]}. \tag{10.21}$$

In Eq. (10.21) we have ignored the effect of loops L_3 and L_7, as the magnitude of h_{re} is negligibly small compared with unity. From Eq. (10.21) we see that for the network of Fig. 10.14(a) to operate as an integrator, the feedback capacitor C must be chosen such that, at all frequencies of interest,

$$\frac{(h_{oe} + G_l)(R + h_{ie}) - h_{re}h_{fe}}{R[1 + h_{fe} + h_{ie}(h_{oe} + G_l + 1/R)]} \ll \omega C \ll \frac{h_{fe}}{h_{ie}}.$$

Example 10.5. *Voltage-feedback pair.* As another example illustrating the application of Mason's direct rule, consider the so-called *voltage-feedback pair* of Fig. 10.15(a). It consists of two common-emitter stages connected in cascade in which a signal proportional to the output voltage is fed back to the input end of the amplifier by means of resistors R_1 and R_2. The amplifier has a high input impedance and a low output impedance, which make it most suited for voltage amplification. That is to say, in formulating a graph for the amplifier, the input voltage V_1 should appear as the source node.

In order to simplify the analysis, we shall neglect the effects of parameters h_{re} and h_{oe} for both transistors. Then in terms of the variables shown defined, we

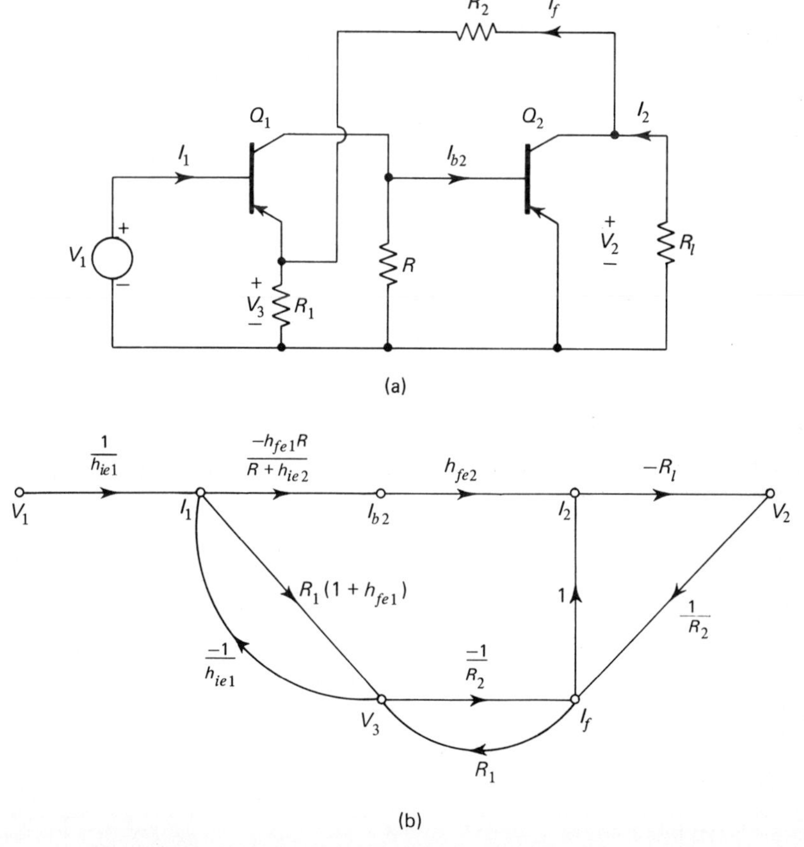

Fig. 10.15. (a) Voltage-feedback pair; (b) Signal-flow graph.

may write, by inspection,

$$
\begin{aligned}
I_1 &= \frac{V_1}{h_{ie1}} - \frac{V_3}{h_{ie1}} \\
I_{b2} &= \frac{-h_{fe1}R}{R + h_{ie2}} I_1 \\
I_2 &= h_{fe2} I_{b2} + I_f \\
V_2 &= -I_2 R_l \\
V_3 &= R_1 I_f + R_1(1 + h_{fe1}) I_1 \\
I_f &= \frac{V_2}{R_2} - \frac{V_3}{R_2},
\end{aligned}
\tag{10.22}
$$

where the parameters h_{ie1} and h_{fe1} refer to transistor $Q1$, while h_{ie2} and h_{fe2} refer to transistor $Q2$.

Representing Eqs. (10.22) with a signal-flow graph, we obtain the diagram shown in Fig. 10.15(b) which possesses four feedback loops defined by

$$
\begin{aligned}
L_1 &= -\frac{R_1(1 + h_{fe1})}{h_{ie1}} \\
L_2 &= -\frac{R_1}{R_2} \\
L_3 &= -\frac{R_l}{R_2} \\
L_4 &= -\frac{h_{fe1} h_{fe2} R R_1 R_l}{R_2 h_{ie1}(R + h_{ie2})}.
\end{aligned}
\tag{10.23}
$$

Since there is only one pair of non-interacting loops contributed by L_1 and L_3, it follows that

$$\Delta = 1 - (L_1 + L_2 + L_3 + L_4) + L_1 L_3. \tag{10.24}$$

In Fig. 10.15(b) we also see that there are two paths connecting V_1 to V_2:

$$
\begin{aligned}
P_1 &= \frac{h_{fe1} h_{fe2} R R_l}{h_{ie1}(R + h_{ie2})} \\
P_2 &= \frac{R_1 R_l (1 + h_{fe1})}{h_{ie1} R_2}.
\end{aligned}
\tag{10.25}
$$

The weighting factor for P_1 is $\Delta_1 = 1 - L_2$ while for P_2 the weighting factor is unity.

Therefore, applying Mason's direct rule yields

$$K_v = \frac{V_2}{V_1} = \frac{P_1(1 - L_2) + P_2}{\Delta}$$

or

$$K_v = \frac{h_{fe1}h_{fe2}RR_l(R_1 + R_2) + R_1R_l(R + h_{ie2})(1 + h_{fe1})}{\{(R + h_{ie2})[h_{ie1}R_1 + (R_2 + R_l)(h_{ie1} + R_1 + h_{fe1}R_1)] + h_{fe1}h_{fe2}RR_1R_l\}}. \tag{10.26}$$

Equation (10.26) indicates that if the h_{fe}'s of the transistors are large, the voltage gain of the circuit approximates to the limiting value of $(R_1 + R_2)/R_1$ and thereby becomes practically independent of transistor properties.

Example 10.6. *Current-feedback pair.* In the so-called *current-feedback pair* shown in Fig. 10.16(a) the resistors R_1 and R_2 enable a signal proportional to the

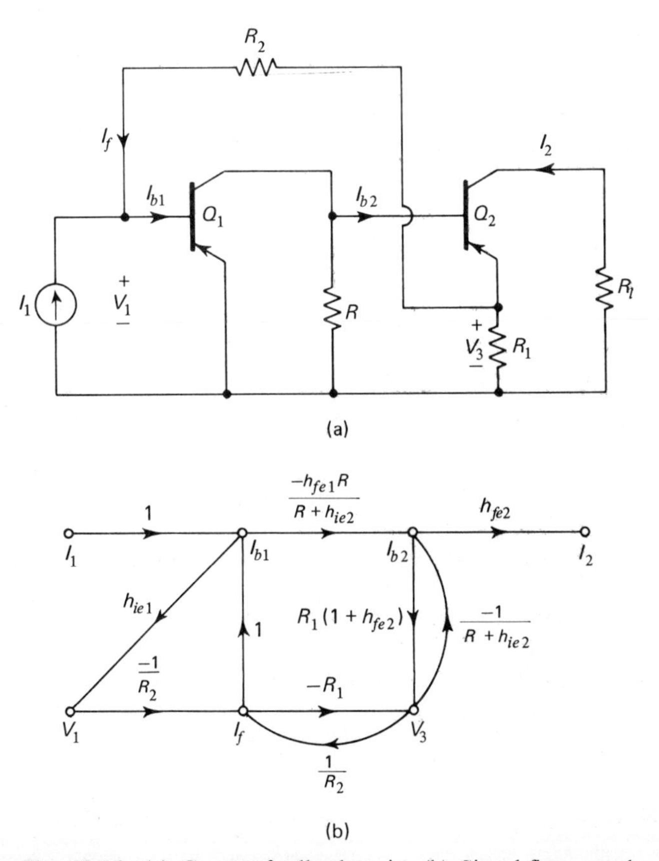

Fig. 10.16. (a) Current-feedback pair; (b) Signal-flow graph.

output current to be fed back to the input port of the amplifier. This amplifier has a low input impedance and a high output impedance. It is therefore most suited for current amplification, so that in the graph formulation the input current I_1 should appear as the source node.

Here again, to simplify the analysis we shall neglect the effects of parameters h_{re} and h_{oe} for both transistors. We thus obtain the equations

$$\begin{aligned} I_{b1} &= I_1 + I_f \\ I_{b2} &= -\frac{h_{fe1} R I_{b1}}{R + h_{ie2}} - \frac{V_3}{R + h_{ie2}} \\ I_2 &= h_{fe2} I_{b2} \\ V_1 &= h_{ie1} I_{b1} \\ I_f &= \frac{V_3}{R_2} - \frac{V_1}{R_2} \\ V_3 &= R_1(1 + h_{fe2}) I_{b2} - R_1 I_f, \end{aligned} \tag{10.27}$$

which yield the signal-flow graph of Fig. 10.16(b). In this diagram, there are four feedback loops defined as follows:

$$\begin{aligned} L_1 &= -\frac{h_{ie1}}{R_2} \\ L_2 &= -\frac{R_1}{R_2} \\ L_3 &= -\frac{R_1(1 + h_{fe2})}{R + h_{ie2}} \\ L_4 &= -\frac{h_{fe1}(1 + h_{fe2}) R R_1}{R_2(R + h_{ie2})}. \end{aligned} \tag{10.28}$$

Loops L_1 and L_3 constitute the only pair of non-touching loops; therefore,

$$\Delta = 1 - (L_1 + L_2 + L_3 + L_4) + L_1 L_3. \tag{10.29}$$

Also, there are two paths connecting I_1 to I_2:

$$P_1 = -\frac{h_{fe1} h_{fe2} R}{R + h_{ie2}}, \tag{10.30}$$

for which the weighting factor is

$$\Delta_1 = 1 - L_2. \tag{10.31}$$

The transmittance of the second path is

$$P_2 = \frac{-h_{ie1}h_{fe2}R_1}{R_2(R + h_{ie2})}, \tag{10.32}$$

for which the weighting factor Δ_2 is equal to unity.

Hence, applying Mason's direct rule, we obtain

$$K_i = \frac{I_2}{I_1} = \frac{P_1(1 - L_2) + P_2}{\Delta}$$

or

$$K_i = \frac{-h_{fe1}h_{fe2}R(R_1 + R_2) - h_{ie1}h_{fe2}R_1}{(R + h_{ie2})(R_1 + h_{ie1} + R_2) + R_1(1 + h_{fe2})(R_2 + h_{ie1} + h_{fe1}R)}. \tag{10.33}$$

For large values of h_{fe}'s, the current gain of the amplifier approximates to the limiting value of $-(R_1 + R_2)/R_1$, thereby becoming practically independent of transistor properties.

Concluding Remarks

To conclude, a signal-flow graph provides a graphical portrayal of the structure of a physical network in terms of nodes and directed branches, emphasizing the causal relationships between the variables chosen for the analysis of the network. In general, a signal-flow graph is not unique in the sense that a given network can usually be represented by many different signal-flow graphs, depending upon the set of variables chosen and the way the network equations are written. However, the various signal-flow graphs representing a given network have precisely the same overall gain.

Mason's rule provides a direct means of evaluating the overall gain of a signal-flow graph without requiring any reduction or manipulation of the graph. The rule merely requires a knowledge of the loop and path transmittances of the graph, and the way the various loops interact with one another.

REFERENCES

1. S. J. Mason, "Feedback Theory—Some Properties of Signal-flow Graphs," *Proc. I.R.E.* **41,** 1144 (1953).
2. S. J. Mason, "Feedback Theory—Further Properties of Signal-flow Graphs," *Proc. I.R.E.* **44,** 920 (1956).
3. C. S. Lorens, *Flowgraphs*. McGraw-Hill, 1964.

PROBLEMS

10.1 Determine the transmittance x_3/x_1 for the graph shown in Fig. P10.1, using
a) A step-by-step reduction process,
b) Mason's direct rule.

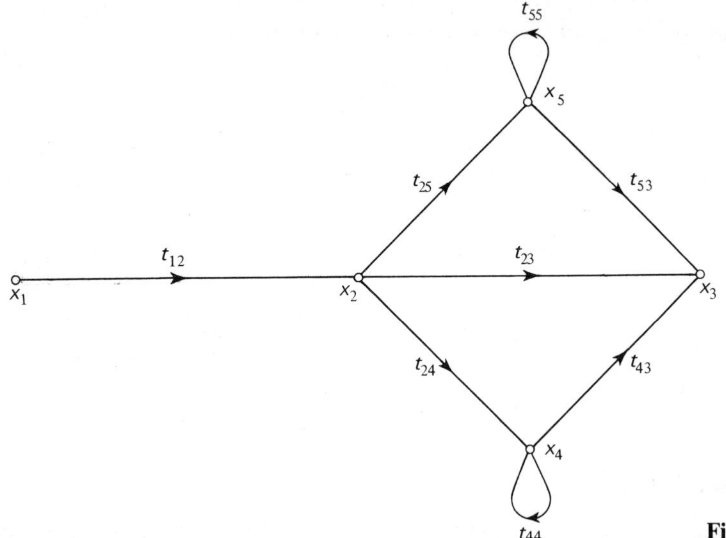

Fig. P10.1.

10.2 Construct a signal-flow graph for the cascode pair shown in Fig. P10.2; then evaluate V_2/V_1.

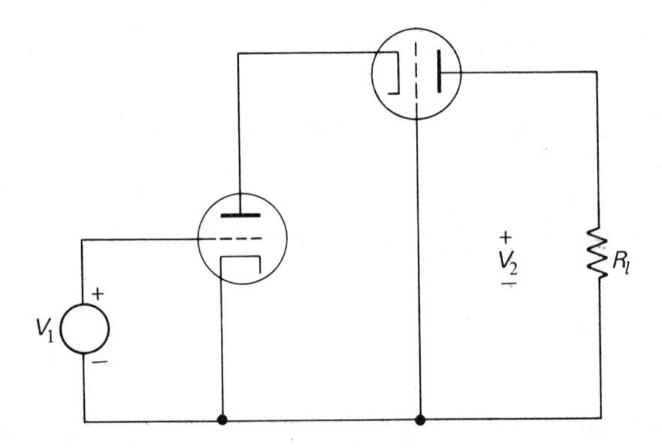

Fig. P10.2.

10.3 Construct signal-flow graphs for the transistor amplifier configurations shown in parts (a) to (d) of Fig. P10.3, using the simplified hybrid model given in part (e) of the diagram. Applying Mason's direct rule, determine the overall gain for each configuration.

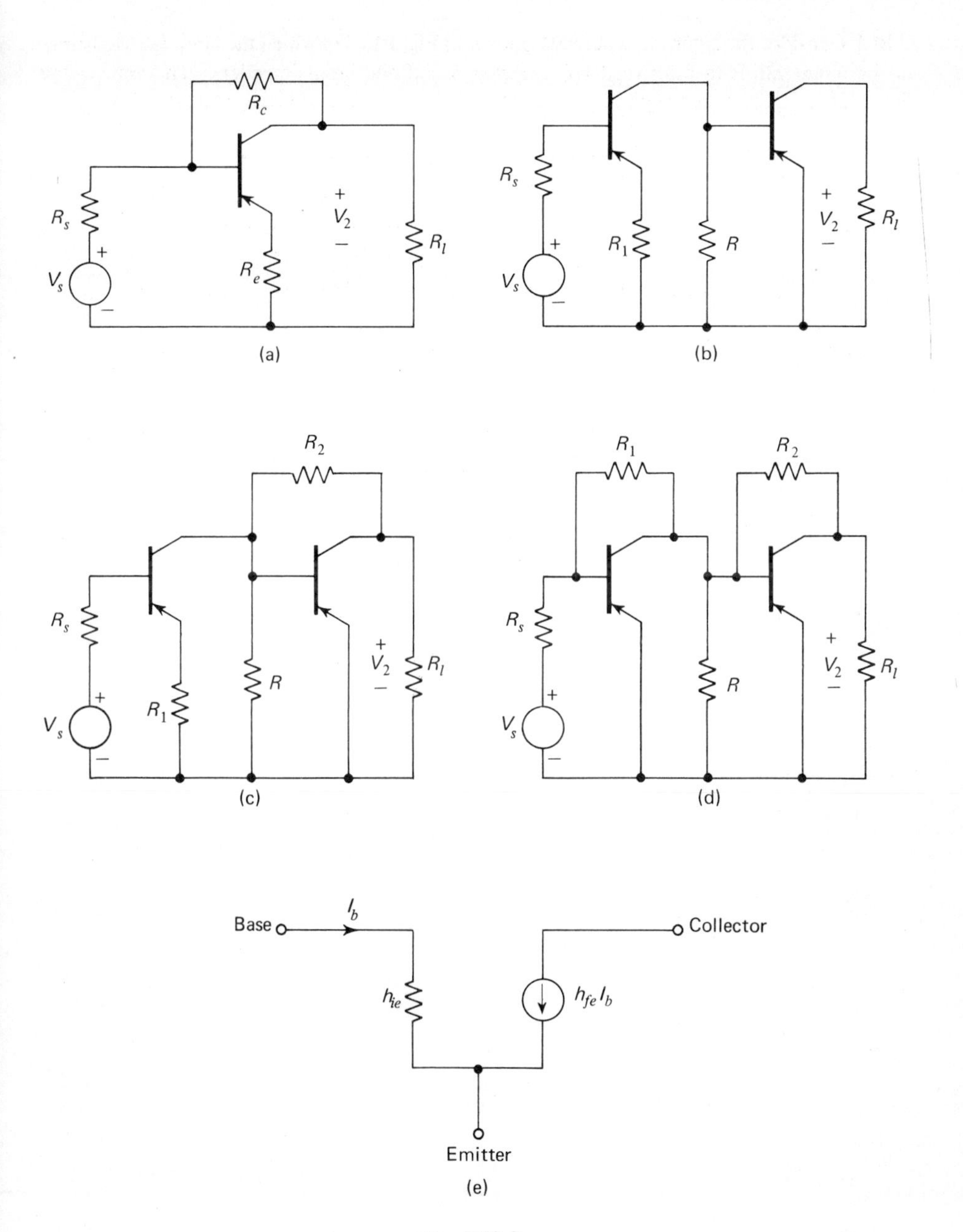

R_c
R_s
V_s
R_e
V_2
R_l
(a)
R_s
V_s
R_1
R
V_2
R_l
(b)
R_2
R_s
V_s
R_1
R
V_2
R_l
(c)
R_1
R_2
R_s
V_s
R
V_2
R_l
(d)
Base
I_b
h_{ie}
$h_{fe} I_b$
Collector
Emitter
(e)

Fig. P10.3.

10.4 Consider the computing network shown in Fig. P10.4 in which the amplifier is assumed to be unilateral, with a forward voltage gain K, infinite input impedance and zero output impedance. Construct a signal-flow graph for this network, and hence evaluate V_2/V_1.

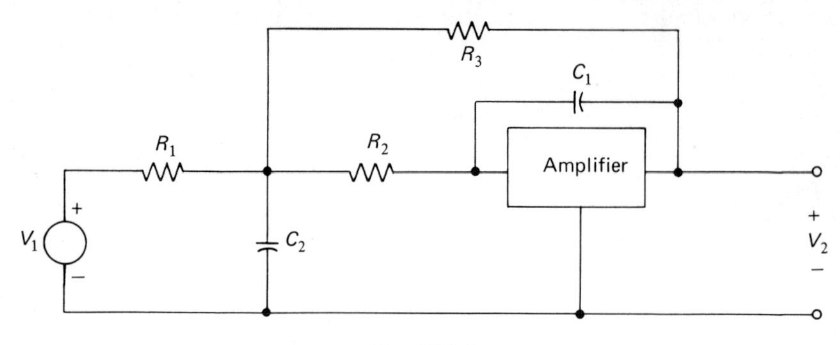

Fig. P10.4.

CHAPTER 11

FEEDBACK THEORY

11.1 INTRODUCTION

Feedback is said to exist in a system if there is a closed sequence of cause-and-effect inter-relationships between certain variables of the system. By adopting this broad definition of feedback it may be argued that such matters as absence or presence of feedback in a system and the number of feedback loops are decided by the manner in which the behaviour of the system is formulated. To illustrate this vagueness in the definition of feedback, consider Fig. 11.1 showing the common-cathode and common-plate configurations of a vacuum-tube triode. If we use the common-cathode model of Fig. 11.2 for the analysis we get the signal-flow graph representations of Fig. 11.3 on the basis of which we may classify the common-cathode configuration as a non-feedback system, and the common-plate configuration as a single-loop feedback system with a negative-loop transmission. It is, however, equally possible to analyse the two configurations using the common-plate model of Fig. 11.4. Then we find that the common-cathode configuration, as represented by the flow graph of Fig. 11.5(a), has a single-loop with a positive-loop transmission, whereas the flow-graph representation of Fig. 11.5(b) which pertains to a common-plate configuration has no feedback loop at all. We see, therefore, that whether or not feedback is present in a system is a matter of viewpoint, depending on how the problem is formulated.

The fact that feedback is a matter of viewpoint may also be illustrated by considering a reciprocal passive ladder network. Such a network is ordinarily classified as a non-feedback system; yet according to the signal-flow graph representation of Fig. 10.4, the ladder network behaves as a multiple-loop feedback system.

In our study of feedback we shall, however, be largely concerned with active networks in which physical feedback loops are purposely added to produce a specific change in the performance of the system. It is in this sense that we ordinarily speak of a feedback system. The simplest form of a feedback system may, as in the idealized block diagram of Fig. 11.6, be regarded as a combination of a linear amplifier and a passive feedback network by means of which a controllable fraction of the output signal is fed back to the input. Feedback may thus be used to make the gain of the amplifier less sensitive to variations in the parameters of the active components, to reduce the effects of noise and nonlinear

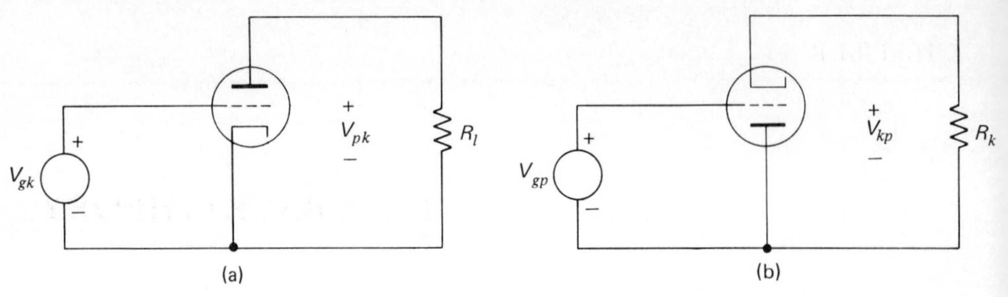

Fig. 11.1. (a) Common-cathode amplifier; (b) Common-plate amplifier.

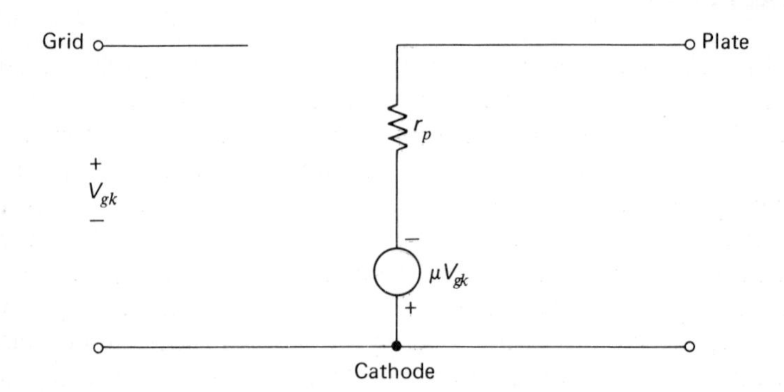

Fig. 11.2. Common-cathode model.

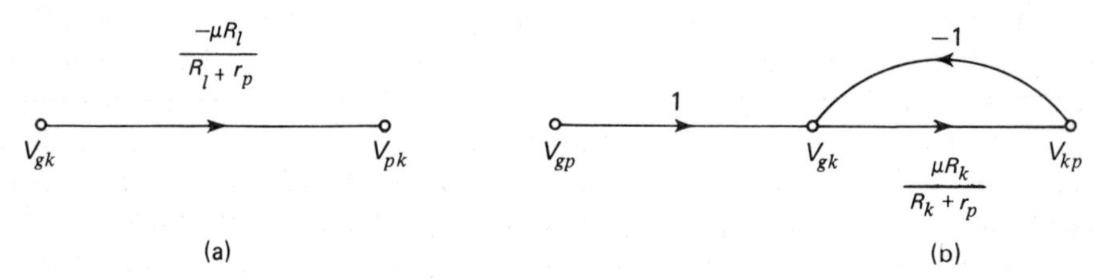

Fig. 11.3. Signal-flow graphs for (a) Common-cathode amplifier, and (b) Common-plate amplifier, based on the model of Fig. 11.2.

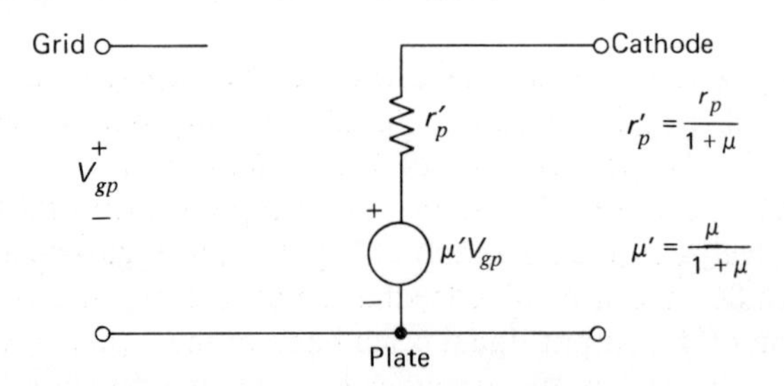

Fig. 11.4. Common-plate model.

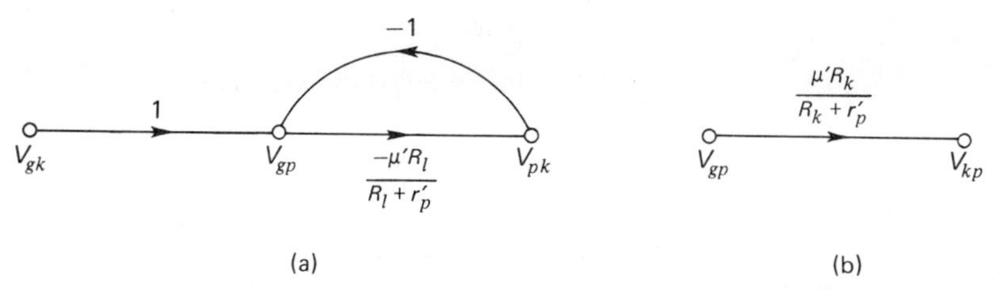

Fig. 11.5. Signal-flow graphs for (a) Common-cathode amplifier, and (b) Common-plate amplifier, based on the model of Fig. 11.4.

distortion, to control the level of the driving-point impedance at a certain point of the system, or to control the stability (or instability) of the system.

The behaviour of a feedback system, like any other electrical system, may be determined by applying the conventional loop or nodal method of network analysis, the basic principles of which are in no way affected by the presence of feedback in the system. Alternatively, the feedback system may be analysed using matrix methods or signal-flow graphs, as demonstrated in previous chapters. Yet another method of analysis is based on a mathematical theory of feedback which not only places the feedback in evidence, but also provides a quantitative measure of the various effects of feedback on system performance and enables their evaluation in a unified manner. This theory of feedback may be developed algebraically in terms of the circuit determinant, as originally carried out by Bode,[1] or in terms of signal-flow graphs as first proposed by Mason[2] and later extended by Truxal.[3] The use of signal-flow graphs has the clear advantage of providing an invaluable aid in interpreting and understanding the various results of the feedback theory. In this chapter we shall study both approaches, starting with the Mason–Truxal approach.

11.2 BASIC FEEDBACK CONCEPTS

Consider the network of Fig. 11.7 where x_s represents an independent source connected across the port 1–1′, and x_l denotes the resulting signal developed

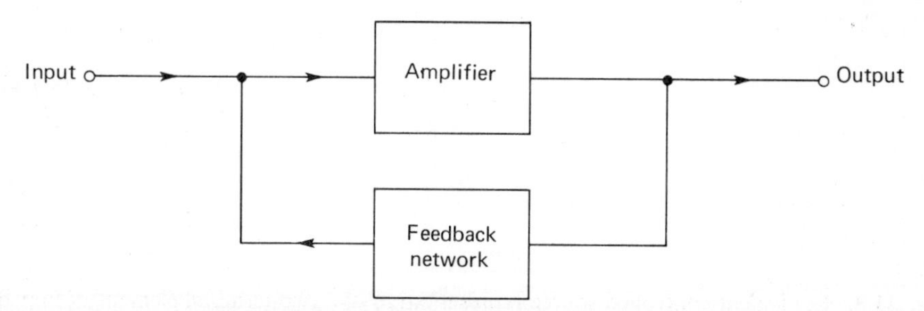

Fig. 11.6. Block diagram of single-loop feedback system.

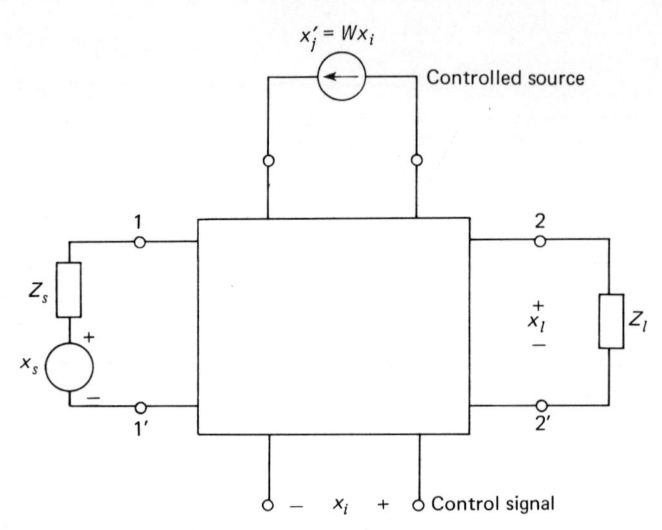

Fig. 11.7. Linear active network with controlled source $x'_j = Wx_i$ singled out for special consideration.

across the load. We have singled out a controlled source x'_j of strength Wx_i for special consideration, W being the control parameter and x_i the control signal. The various sources and signals can be currents or voltages; in Fig. 11.7 we have shown x_s as a voltage source, x'_j as a current source, and both x_i and x_l as voltage signals for the purpose of illustration only. By introducing appropriate definitions for the parameter W and the control signal x_i the generalized controlled source x'_j may be used to represent any one of the four controlled-source types which appear in the circuit models of such active devices as vacuum tubes and transistors. Furthermore, it may also be used to represent a bilateral two-terminal element of impedance Z by considering such an element as a current-controlled voltage source, as in Fig. 11.8(b), or as a voltage-controlled current source, as in Fig. 11.8(c);

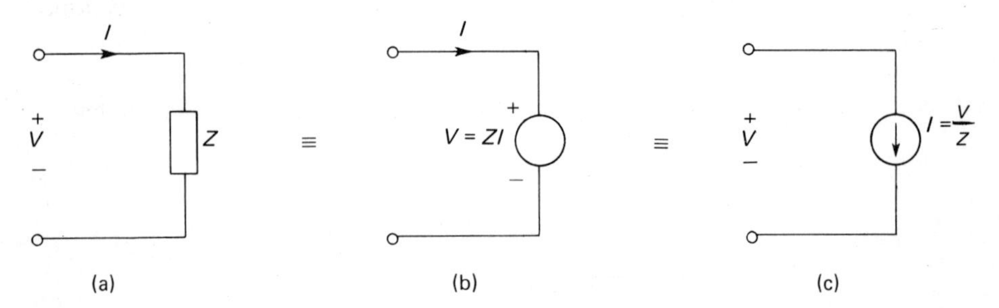

Fig. 11.8. Equivalent controlled source representation of a two-terminal element of impedance Z.

in both cases the control signal and controlled source appear at the same terminal-pair.

Assuming that the network of Fig. 11.7 is linear, we may apply the principle of superposition to add the separate effects of the independent source x_s and controlled source x'_j upon both the load signal x_l and the control signal x_i; so that

$$\begin{aligned} x_l &= t_{sl}x_s + t_{jl}x'_j \\ x_i &= t_{si}x_s + t_{ji}x'_j, \end{aligned} \tag{11.1}$$

where $x'_j = Wx_i$ and the transmittances t_{sl}, t_{jl}, t_{si}, and t_{ji} are all independent of the control parameter W. Equations (11.1) may be represented by the signal-flow graph of Fig. 11.9. In this diagram we see that the transmittance t_{sl} accounts for all paths from the independent source to the load that by-pass the controlled source x'_j; it is therefore called the *direct* or *leakage transmittance.* We also see that the transmittance t_{ji} enables a certain portion of the controlled source x'_j to be fed back to its control signal x_i; it is therefore responsible for the feedback acting around the controlled source. If t_{ji} is zero, then we have no feedback acting around the controlled source. The diagram of Fig. 11.9 is called the *fundamental feedback-flow graph* as it is the basis of the mathematical theory of feedback.

From Eqs. (11.1) we obtain the following definitions for the various transmittances

$$\begin{aligned} t_{sl} &= \left.\frac{x_l}{x_s}\right|_{x'_j=0} & t_{jl} &= \left.\frac{x_l}{x'_j}\right|_{x_s=0} \\ t_{si} &= \left.\frac{x_i}{x_s}\right|_{x'_j=0} & t_{ji} &= \left.\frac{x_i}{x'_j}\right|_{x_s=0}. \end{aligned} \tag{11.2}$$

Therefore, when $x_s = 1$ and the control parameter W is set equal to zero (that is, $x'_j = 0$), we find that $t_{sl} = x_l$ and $t_{si} = x_i$. On the other hand, when the independent source x_s is zero and the controlled source x'_j is replaced with an independent source of unit strength, we have $t_{jl} = x_l$ and $t_{ji} = x_i$.

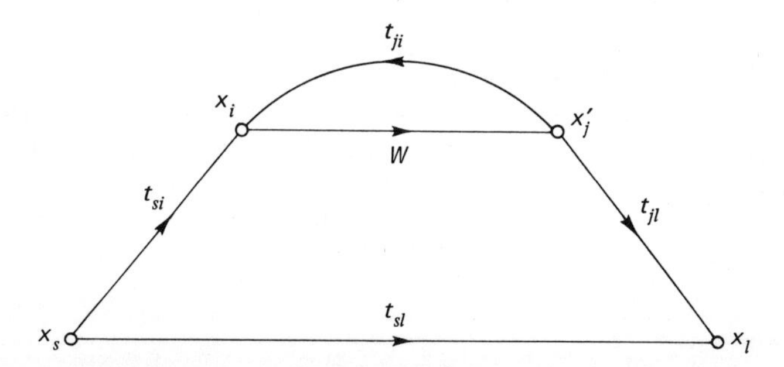

Fig. 11.9. Fundamental feedback-flow graph.

In Fig. 11.9 we see, by inspection, that the *closed-loop transmittance* or overall gain of the system is given by

$$K = \frac{x_l}{x_s} = t_{sl} + \frac{t_{si}Wt_{jl}}{1 - Wt_{ji}}. \tag{11.3}$$

This relation is called the *fundamental feedback equation* from which most of the important properties of linear feedback circuits may be derived.

For a specified element W, the problem of evaluating the closed-loop transmittance is thus reduced to one of evaluating the four transmittances t_{sl}, t_{si}, t_{jl} and t_{ji}, the first two of which are found for the simplifying condition of W equal to zero, and the latter two are found for x_s equal to zero.

It is noteworthy that if $|Wt_{ji}| \gg 1$, the closed-loop transmittance K approaches the limiting value of $(t_{sl} - t_{si}t_{jl}/t_{ji})$ and thereby becomes practically independent of variations in the parameter W.

Example 11.1. *An active bridged-T network.* As an example illustrating the above analysis by superposition, consider the active bridged-T network of Fig. 11.10 due to Lampard.[4,5] The interesting property of this network is that as the control parameter g_m approaches infinity, the network approaches (in so far as its input impedance and transfer-voltage ratio are concerned) the passive constant-resistance bridged-T network of Fig. 11.11 in which $Z_aZ_b = R^2$.

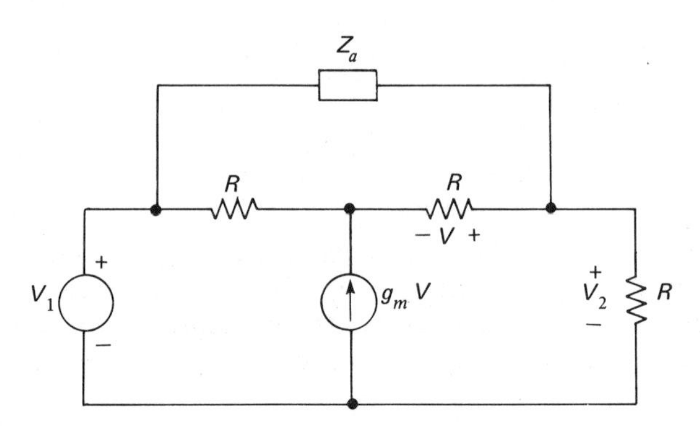

Fig. 11.10. Active bridged-T network.

Designating the control parameter g_m in Fig. 11.10 as the element W, we may make the identifications $x_i = V$ and $x'_j = g_mV$. Also, we note that $x_s = V_1$ and $x_l = V_2$. Therefore, to determine the transmittances t_{sl} and t_{si}, we put $g_m = 0$ and $V_1 = 1$, as in Fig. 11.12, and so we readily find that:

$$t_{sl} = \frac{2R + Z_a}{2R + 3Z_a}$$
$$t_{si} = \frac{-Z_a}{2R + 3Z_a}. \tag{11.4}$$

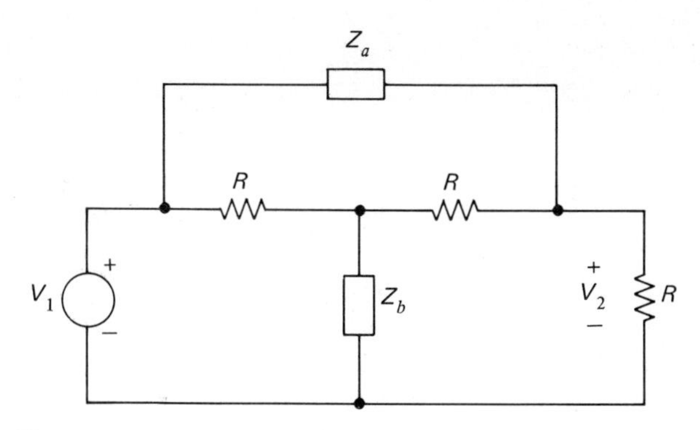

Fig. 11.11. Passive constant-resistance bridged-T network.

Next, for the determination of transmittances t_{jl} and t_{ji}, we put $V_1 = 0$ and replace the controlled source $g_m V$ with a unit current source, as in Fig. 11.13. We thus obtain:

$$t_{jl} = \frac{RZ_a}{2R + 3Z_a}$$
$$t_{ji} = -\frac{R(R + Z_a)}{2R + 3Z_a}. \tag{11.5}$$

Substitution of Eqs. (11.4) and (11.5) in (11.3) yields the transfer-voltage ratio for the active network of Fig. 11.10 as:

$$\frac{V_2}{V_1} = \frac{(2R + Z_a) + g_m R^2}{2R + 3Z_a + g_m R(R + Z_a)}. \tag{11.6}$$

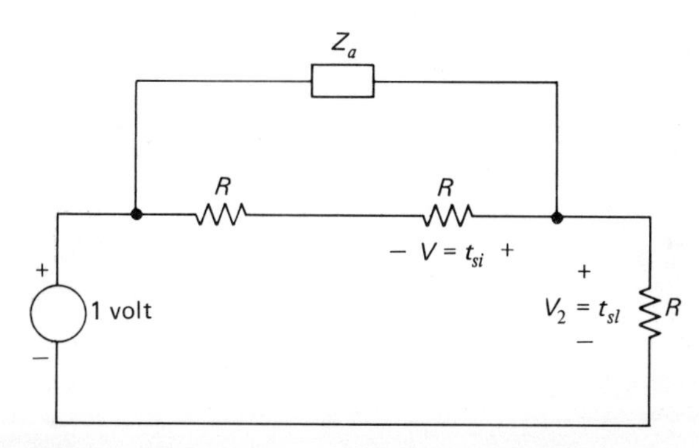

Fig. 11.12. Network for evaluating transmittances t_{sl} and t_{si}.

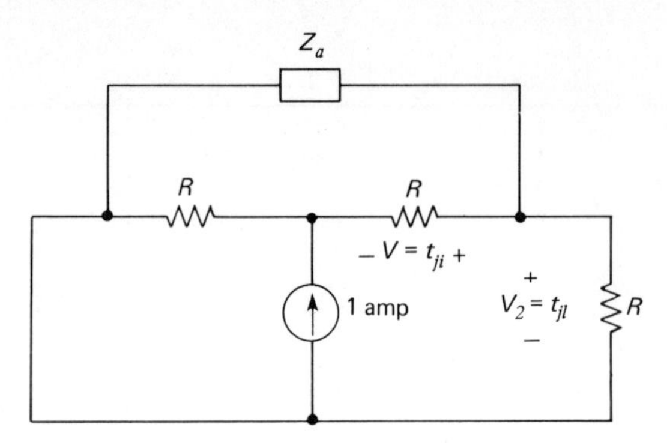

Fig. 11.13. Network for evaluating transmittances t_{jl} and t_{ji}.

Therefore, as $g_m \to \infty$, we find that

$$\frac{V_2}{V_1} \to \frac{R}{R + Z_a}, \tag{11.7}$$

which is the transfer-voltage ratio for the passive bridged-T network of Fig. 11.11 having $Z_a Z_b = R^2$.

Return Ratio and Return Difference

Suppose in Fig. 11.9 the feedback loop is opened by breaking it at the branch with transmittance W. This is illustrated in Fig. 11.14 where we have included two points a and a' on either side of the break. Also, let the independent source x_s be reduced to zero, and a unit signal be transmitted from point a'. The *return ratio* T for the element W is defined as the negative of the signal returned to

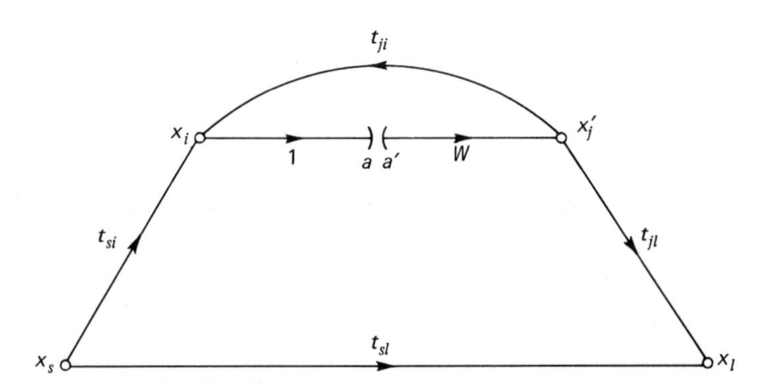

Fig. 11.14. The fundamental feedback-flow graph with feedback loop opened.

point a on the other side of the break; so that

$$T = -Wt_{ji}. \tag{11.8}$$

The negative sign is introduced to conform with established conventions in feedback amplifier theory. The product term Wt_{ji} is the loop transmittance for the element W; hence, the return ratio is the negative of the loop transmittance.

If in the second line of Eqs. (11.1) we put $x_s = 0$ and $x'_j = W$ we find that $x_i = Wt_{ji}$. This, therefore, suggests another useful way of evaluating the return ratio: if in a linear circuit we have a controlled source Wx_i, W being the control parameter and x_i the control signal, then the return ratio for W is equal to the negative of the variable x_i developed in that system which results from replacing the controlled source in question by an independent source of strength W. This procedure is equivalent to breaking the loop and applying a unit signal at the input port of the controlled source, and measuring the signal developed across the break.

In Fig. 11.14 the difference between the transmitted unit signal and the returned signal under the condition $x_s = 0$ is called the *return difference* F for the element W. It is related to the loop transmittance and return ratio as follows:

$$F = 1 - Wt_{ji} = 1 + T. \tag{11.9}$$

The return difference for an element W is a quantitative measure of the amount of feedback acting around W. When the loop transmittance Wt_{ji} is a negative quantity, so that the return difference F is greater than unity, the feedback is said to be *negative* or *degenerative*. On the other hand, when the loop transmittance is positive the feedback is said to be *positive* or *regenerative*. From these definitions and from Eq. (11.3) it therefore follows that negative feedback reduces the magnitude of the closed-loop transmittance of the amplifier, while positive feedback increases it.

The amount of feedback, as represented by the return difference, is often expressed in decibels. Thus, by definition, the number of decibels is equal to $-20 \log_{10} F$. The sign of the feedback, positive or negative, is accordingly the same as the sign of the logarithm of the return difference.

Null Return Difference

Another useful feedback concept is the *null return difference* F' defined as the return difference which results under the condition that the independent source x_s is adjusted to reduce the load signal x_l to zero. Thus, with a unit signal transmitted from point a' in Fig. 11.14, so that $x'_j = W$, and with $x_l = 0$, we find that the required x_s is $-Wt_{jl}/t_{sl}$ and the corresponding value of the signal returned to point a is

$$x_i = Wt_{ji} - \frac{t_{si}Wt_{jl}}{t_{sl}}. \tag{11.10}$$

The null return difference is equal to the unit transmitted signal minus the returned signal of Eq. (11.10), so that

$$F' = 1 - Wt_{ji} + \frac{t_{si}Wt_{jl}}{t_{sl}}$$

or

$$F' = F + \frac{t_{si}Wt_{jl}}{t_{sl}}. \tag{11.11}$$

Using Eqs. (11.3), (11.9) and (11.11), and recognizing that the direct transmittance t_{sl} is equal to the special value K^0 of the closed-loop transmittance which results when W is zero, we obtain the following compact formula for the closed-loop transmittance K in terms of the two return differences

$$K = t_{sl}\frac{F'}{F} = K^0\frac{F'}{F}. \tag{11.12}$$

Example 11.2. *Return differences of active bridged-T network.* Consider evaluation of the return difference and null return difference with reference to g_m in the active bridged-T network of Fig. 11.10 previously analysed. Replacing the controlled source g_mV with an independent current source of g_m amp, and short-circuiting the input port (i.e., $V_1 = 0$), as in Fig. 11.15, we find that the returned signal V is equal to $-g_mR(R + Z_a)/(2R + 3Z_a)$; so that the return difference for g_m is given by

$$F = 1 + \frac{g_mR(R + Z_a)}{2R + 3Z_a}. \tag{11.13}$$

To determine the null return difference, we may consider the network of Fig. 11.16 in which the controlled source g_mV has again been replaced with an

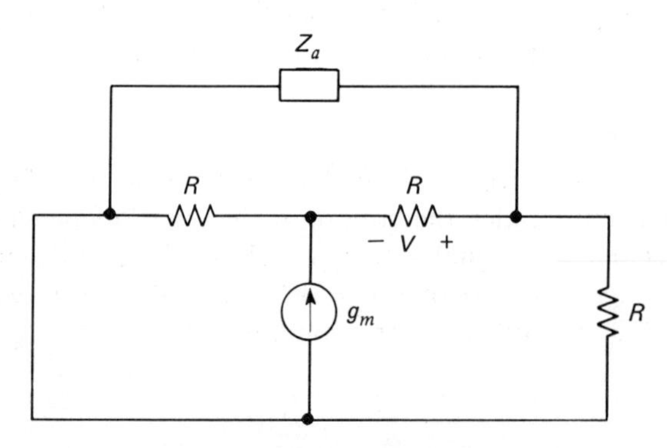

Fig. 11.15. Network for evaluating return difference for g_m.

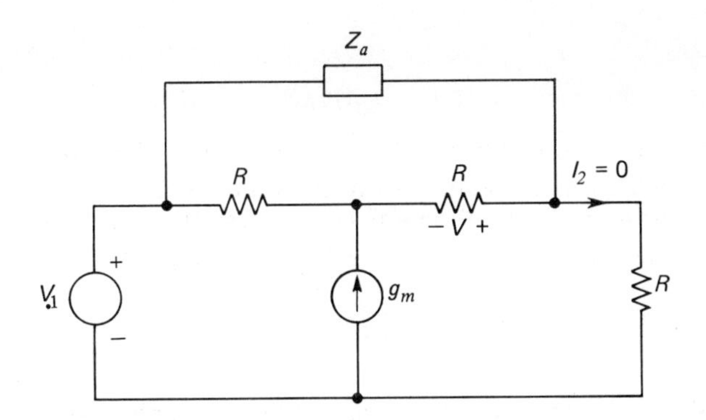

Fig. 11.16. Network for evaluating null return difference for g_m.

independent current source of g_m amp. In this case, however, the input voltage V_1 is not zero, but rather is adjusted to make the output voltage V_2 equal to zero. But if $V_2 = 0$, then we must also have $I_2 = 0$ so that from Kirchhoff's voltage law we obtain

$$V + R\left(g_m + \frac{V}{R}\right) + Z_a\frac{V}{R} = 0$$

or

$$V = -\frac{g_m R^2}{2R + Z_a}.$$

The null return difference for g_m is therefore

$$F' = 1 + \frac{g_m R^2}{2R + Z_a}. \tag{11.14}$$

This result may also be obtained by substituting the transmittances of Eqs. (11.4) and (11.5) in Eq. (11.11).

Complementary Return Difference[6]

The *complementary return difference*, denoted by $\bar{F}$, is defined by the requirement that when a unit signal is applied to a' in Fig. 11.14, $\bar{F}$ is the difference between the unit transmitted signal and the returned signal under the condition that the source signal x_s is adjusted so that $x_s + \bar{x}_l = 0$, in which $\bar{x}_l$ denotes the value of load signal when $x_s = 0$. From Fig. 11.14 we see that under these conditions $\bar{x}_l = Wt_{jl}$ and $x_i = Wt_{ji} - t_{si}Wt_{jl}$. The complementary return difference for W is therefore

$$\bar{F} = 1 - Wt_{ji} + t_{si}Wt_{jl}$$

or

$$\bar{F} = F + t_{si}Wt_{jl}. \tag{11.15}$$

The complementary return difference is of utility when the direct transmittance t_{sl} is zero. In such cases the null return difference is not defined. Thus, when $t_{sl} = 0$ we find from Eqs. (11.3), (11.9) and (11.15) that the expression for the closed-loop transmittance takes on the following special form

$$K = \frac{\bar{F} - F}{F}. \tag{11.16}$$

Feedback Analysis of a Two-port Network and the Non-uniqueness of Return Difference[7]

From Chapter 4 it is recalled that the external behaviour of a two-port network can be defined in terms of the z-, y-, h- or g-matrix representations. In each of these four cases the parameter with subscript 21, that is, the parameter z_{21}, y_{21}, h_{21} or g_{21} accounts for the forward transmission of the input signal. On the other hand, the parameter with subscript 12, that is, z_{12}, y_{12}, h_{12} or g_{12} accounts for signal transmission in the reverse direction, thereby giving rise to a certain amount of internal feedback within the two-port network. This is clearly illustrated in part (b) of Fig. 11.17 which shows a signal-flow graph representation for a double terminated two-port network in terms of the y-parameters. From Fig. 11.17(b) we see, by inspection, that the overall transfer impedance of the network is given by

$$\frac{V_2}{I_s} = \frac{-y_{21}}{(y_{11} + Y_s)(y_{22} + Y_l) - y_{12}y_{21}} \tag{11.17}$$

and that the return difference for the element y_{21} is

$$F_{y_{21}} = 1 - \frac{y_{12}y_{21}}{(y_{11} + Y_s)(y_{22} + Y_l)}. \tag{11.18}$$

This is a measure of the amount of internal feedback acting around the element y_{21}. When the source admittance Y_s or the load admittance Y_l is very large (that is, the input or output port is short-circuited) the return difference $F_{y_{21}}$ approaches unity and the effect of internal feedback within the two-port network is diminished.

The equivalent circuit representation shown in part (a) of Fig. 11.17 involves two controlled sources; it is, however, equally possible to represent the two-port network by means of the π-equivalent circuit of Fig. 11.18(a) which involves a single controlled source $Y_m V_1$ responsible for the major part of forward transmission through the network. This second representation leads to the signal-flow graph of Fig. 11.18(b) for the double-terminated two-port network. In this diagram we observe that when the controlled source $Y_m V_1$ is inactivated by setting Y_m equal to zero, we are left with a finite leakage transmission from the source to the load through the passive element $-y_{12}$ which is common to the input and

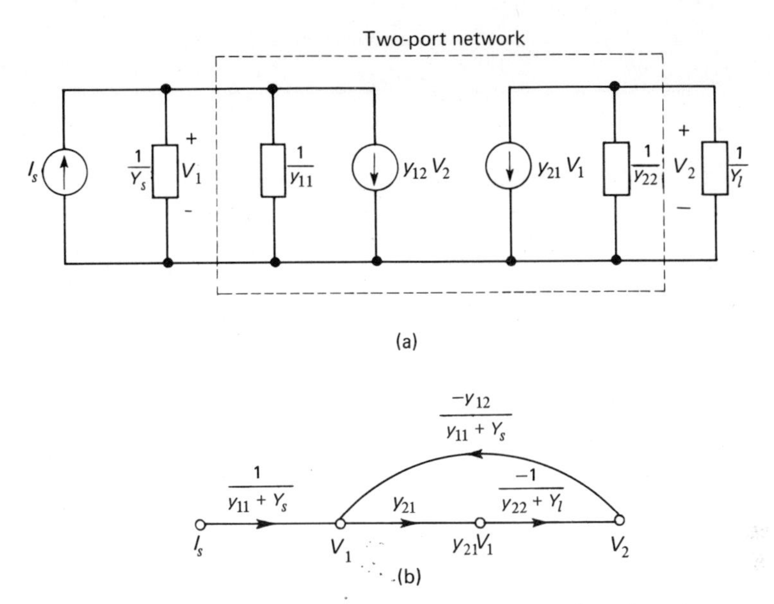

Fig. 11.17. (a) Model of a double-terminated two-port network based on the y-parameters; (b) Signal-flow graph.

output nodes of the network (see Fig. 11.18a). This is in direct contrast to the signal-flow graph of Fig. 11.17(b) where all transmission from the source to the load is interrupted when we put $y_{21} = 0$.

From Fig. 11.18(b) we find that the overall transfer impedance of the network is

$$\frac{V_2}{I_s} = -\frac{y_{12}}{D} - \frac{Y_m[(y_{22} + Y_l)/D][(y_{11} + Y_s)/D]}{1 - Y_m(y_{12}/D)}, \tag{11.19}$$

where

$$D = (y_{11} + Y_s)(y_{22} + Y_l) - y_{12}^2. \tag{11.20}$$

Since $Y_m = y_{21} - y_{12}$, we find that after some manipulations Eq. (11.19) simplifies to the same form as that of Eq. (11.17). This is, of course, to be expected since the overall transfer function of a two-port network is invariant to the choice of the circuit model adopted to represent its external behaviour. But the return difference F_{Y_m} for the element Y_m, which is given by

$$F_{Ym} = 1 - \frac{y_{12}Y_m}{D} \tag{11.21}$$

has in general a value quite different from the return difference $F_{y_{21}}$ for y_{21}. This implies, therefore, that as there is no unique equivalent circuit representation of a two-port device, we cannot speak meaningfully of the return difference of the

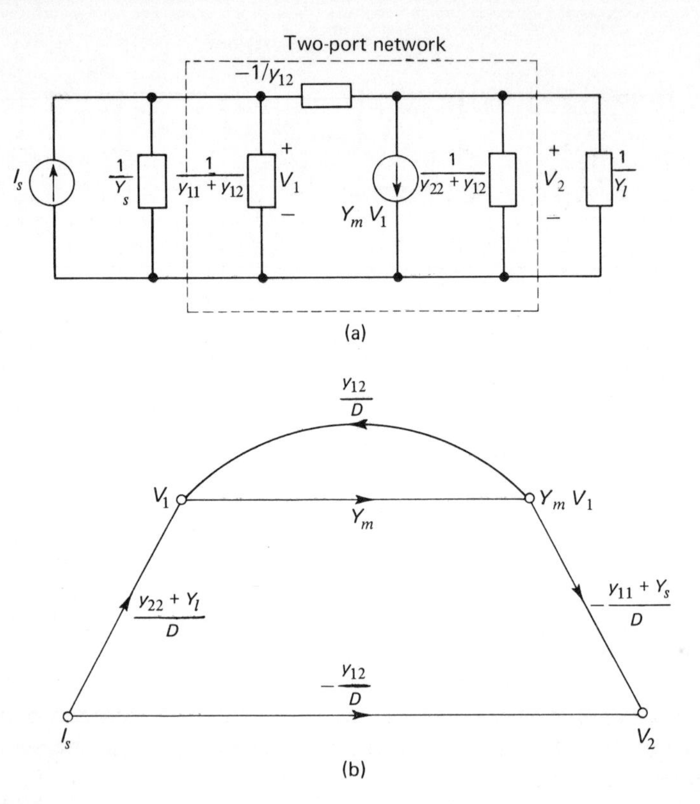

Fig. 11.18. (a) π-model of a double-terminated two-port network; (b) Signal-flow graph.

device itself. The term return difference has significance only with respect to some specific parameter of the device. In other words, to speak of a return difference presupposes that some definite controlled source has been fixed upon as the basis for a feedback analysis, and this, in turn, implies that some definite circuit model has been chosen for the device.

A Practical Method for Measuring the Return Ratio[7,8]

In order to measure the return ratio for an element W associated with a specified controlled source Wx_i we have to derive from the original feedback amplifier a new circuit in which the feedback acting around the controlled source Wx_i has been removed, so that it no longer influences its own control signal. In the new circuit, however, we have to ensure that the controlled source in question continues to drive the same system of network elements which it sees during normal operation. The practical implementation of this requirement rather restricts the choice of the element W for which the return ratio can be measured, because in

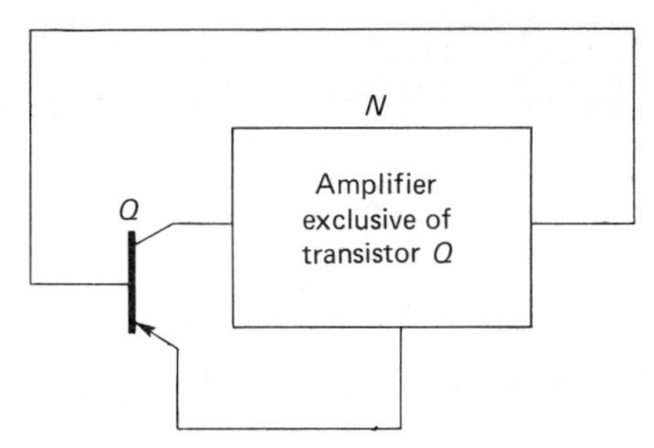

Fig. 11.19. Feedback amplifier with transistor Q singled out for special consideration.

a physical system the controlled source and control both form part of a single device which cannot be physically decomposed. Indeed, the return ratio is directly measurable in practice only if the controlled source in question is of such a nature that all forward transmission through the device is interrupted when that source is put equal to zero. Thus it is, at least in principle, an easy matter to construct a circuit for directly measuring the return ratio for the z_{21}-, y_{21}-, h_{21}- or g_{21}-parameter of a two-port device. As an illustration, suppose in Fig. 11.19 it is required to measure the return ratio with reference to the forward short-circuit transfer admittance y_{fe} of transistor Q operated in the common-emitter connection. The network N in Fig. 11.19 represents the complete feedback amplifier exclusive of the transistor Q in question. Let the feedback loop be broken at the input to the common-emitter stage Q. As illustrated in Fig. 11.20, this is achieved by connecting an independent voltage source V_1 across the input port of transistor Q and terminating the other end of the feedback loop with an admittance y_{ie} shunted by a controlled current source $y_{re}V_2$. We thus ensure that the controlled current source $y_{fe}V_1$ in the y-parameters model of transistor Q in Fig. 11.20 appears as the only driving function in the circuit and is independent of the voltage V_1' developed at the (original) controlling node. Furthermore, the controlled source $y_{fe}V_1$ drives a replica of what it sees during normal operation. When the applied V_1 is a unit voltage, the negative of the returned voltage V_1' gives the required return ratio with reference to y_{fe} of the common-emitter stage Q. In Fig. 11.20 the controlled source $y_{re}V_2$ accounts for the internal feedback within transistor Q in terms of the y-parameters. It can be neglected only if, at all frequencies of interest, the externally applied feedback is large compared with the internal feedback. If we find that we are not justified in ignoring the effect of the controlled source $y_{re}V_2$, we can simulate it by using an additional transistor Q', as in Fig. 11.21. This additional transistor should match as closely as possible the common-emitter stage in question. Furthermore, the admittance Y_N in Fig. 11.21 should approximate the admittance which the remaining portion of the feedback amplifier presents to the output port of transistor Q under consideration (with the external

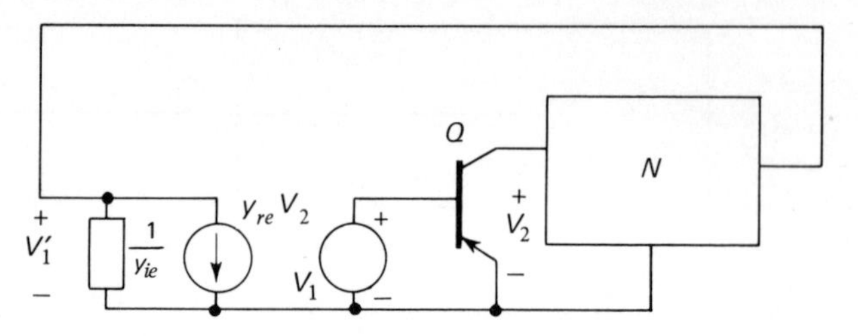

Fig. 11.20. Network for evaluating the return difference for y_{fe} of transistor Q.

feedback loop opened). In Fig. 11.21 we have assumed that $|Y_N| \gg |y_{re}|$ and $|y_{ie}| \gg |y_{re}|$, both of which are assumptions perfectly justified in a common-emitter stage. Also, in Fig. 11.21 it is noteworthy that the element $-y_{re}$ is positive real since the admittance y_{re} of a transistor is ordinarily negative.

Ideally, a voltage source has zero internal impedance; therefore, in Fig. 11.21 we can join together the two base terminals of transistors Q and Q', and feed them both from a single voltage source of zero (or very low) internal impedance. In this way, we may avoid the need for two separate voltage sources.

11.3 AN ALGEBRAIC METHOD OF EVALUATING THE RETURN DIFFERENCE AND NULL RETURN DIFFERENCE

The return difference and null return difference for a specified element W can be evaluated using Eqs. (11.9) and (11.11) which require a knowledge of t_{sl}, t_{si}, t_{ji} and t_{jl}. These four transmittances can, in turn, be determined using the definitions of Eqs. (11.2); the transmittances t_{sl} and t_{si} being evaluated under the simplifying condition that the specified element W be set equal to zero, and the transmittances t_{jl} and t_{ji} under the condition of the independent source x_s set equal to zero. Alternatively, we may evaluate the return difference and null return difference from the set of equilibrium equations describing the behaviour of the circuit.

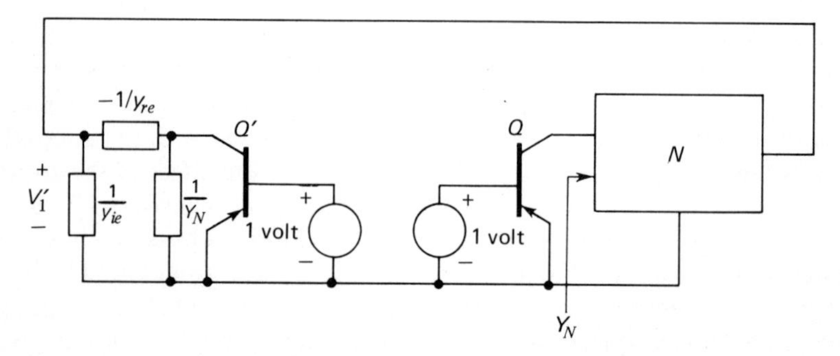

Fig. 11.21. Experimental set-up for measuring the return difference for y_{fe} of transistor Q.

Thus, consider a set of simultaneous linear equations of the form

$$\begin{bmatrix} x_s \\ 0 \\ \cdot \\ \cdot \\ \cdot \\ 0 \\ 0 \\ \cdot \\ \cdot \\ \cdot \\ 0 \end{bmatrix} = \begin{bmatrix} a_{11} & a_{12} & \cdots & a_{1i} & a_{1j} & \cdots & a_{1l} \\ a_{21} & a_{22} & \cdots & a_{2i} & a_{2j} & \cdots & a_{2l} \\ \cdot & \cdot & & \cdot & \cdot & & \cdot \\ \cdot & \cdot & & \cdot & \cdot & & \cdot \\ \cdot & \cdot & & \cdot & \cdot & & \cdot \\ a_{i1} & a_{i2} & \cdots & a_{ii} & a_{ij} & \cdots & a_{il} \\ a_{j1} & a_{j2} & \cdots & W + a_{ji} & a_{jj} & \cdots & a_{jl} \\ \cdot & \cdot & & \cdot & \cdot & & \cdot \\ \cdot & \cdot & & \cdot & \cdot & & \cdot \\ \cdot & \cdot & & \cdot & \cdot & & \cdot \\ a_{l1} & a_{l2} & \cdots & a_{li} & a_{lj} & \cdots & a_{ll} \end{bmatrix} \begin{bmatrix} x_1 \\ x_2 \\ \cdot \\ \cdot \\ \cdot \\ x_i \\ x_j \\ \cdot \\ \cdot \\ \cdot \\ x_l \end{bmatrix}, \tag{11.22}$$

in which x_s is an independent variable; and $x_1, x_2, \ldots, x_l$ are dependent variables. Equations (11.22) may represent the loop (or nodal) equations of a linear circuit with x_s denoting the only independent voltage (or current) source appearing in loop 1 (or at node 1); $x_1, x_2, \ldots, x_l$ denoting the various loop currents (or node voltages); and the matrix of coefficients on the right representing the loop impedance (or nodal-admittance) matrix of the network. In Eq. (11.22) we have assumed that a specified element W occurs only in the jth row and ith column of the matrix of coefficients. It is always possible to write the network equations such that this assumption is satisfied, except when the element W is a mutual inductance. We can isolate the contribution due to the element W by re-writing Eq. (11.22) in the following equivalent form:

$$\begin{bmatrix} x_s \\ 0 \\ \cdot \\ \cdot \\ \cdot \\ 0 \\ -x_j' \\ \cdot \\ \cdot \\ \cdot \\ 0 \end{bmatrix} = \begin{bmatrix} a_{11} & a_{12} & \cdots & a_{1i} & a_{1j} & \cdots & a_{1l} \\ a_{21} & a_{22} & \cdots & a_{2i} & a_{2j} & \cdots & a_{2l} \\ \cdot & \cdot & & \cdot & \cdot & & \cdot \\ \cdot & \cdot & & \cdot & \cdot & & \cdot \\ \cdot & \cdot & & \cdot & \cdot & & \cdot \\ a_{i1} & a_{i2} & \cdots & a_{ii} & a_{ij} & \cdots & a_{il} \\ a_{j1} & a_{j2} & \cdots & a_{ji} & a_{jj} & \cdots & a_{jl} \\ \cdot & \cdot & & \cdot & \cdot & & \cdot \\ \cdot & \cdot & & \cdot & \cdot & & \cdot \\ \cdot & \cdot & & \cdot & \cdot & & \cdot \\ a_{l1} & a_{l2} & \cdots & a_{li} & a_{lj} & \cdots & a_{ll} \end{bmatrix} \begin{bmatrix} x_1 \\ x_2 \\ \cdot \\ \cdot \\ \cdot \\ x_i \\ x_j \\ \cdot \\ \cdot \\ \cdot \\ x_l \end{bmatrix}, \tag{11.23}$$

where $x'_j = Wx_i$. As such, x'_j may be looked upon as a controlled source located in the jth loop (or at the jth node) and dependent upon the signal x_i in the ith loop (or at the ith node). In effect, Eq. (11.23) represents the equilibrium equations of the network with the controlled source x'_j treated as if it were independent. Assuming that x_l represents the output signal of the system, and solving Eq. (11.23) for x_l and the control signal x_i by Cramer's rule, we get

$$\begin{aligned} x_l &= \frac{\Delta_{1l}^0}{\Delta^0} x_s - \frac{\Delta_{jl}}{\Delta^0} x'_j \\ x_i &= \frac{\Delta_{1i}}{\Delta^0} x_s - \frac{\Delta_{ji}}{\Delta^0} x'_j, \end{aligned} \tag{11.24}$$

where Δ^0 is the determinant of the matrix of coefficients on the right of Eq. (11.23); it represents the special value of the circuit determinant Δ of the original system described by Eq. (11.22) when $W = 0$. In Eqs. (11.24) we have made use of the fact that $\Delta_{jl}^0 = \Delta_{jl}$, $\Delta_{1i}^0 = \Delta_{1i}$ and $\Delta_{ji}^0 = \Delta_{ji}$. On comparing Eqs. (11.1) and (11.24) we see that the transmittances t_{sl}, t_{jl}, t_{si} and t_{ji} of the fundamental feedback-flow graph of Fig. 11.9 are given by:

$$\begin{aligned} t_{sl} &= \frac{\Delta_{1l}^0}{\Delta^0} & t_{jl} &= -\frac{\Delta_{jl}}{\Delta^0} \\ t_{si} &= \frac{\Delta_{1i}}{\Delta^0} & t_{ji} &= -\frac{\Delta_{ji}}{\Delta^0}. \end{aligned} \tag{11.25}$$

Therefore, the return difference for the element W is

$$F = 1 - Wt_{ji} = 1 + W\frac{\Delta_{ji}}{\Delta^0}. \tag{11.26}$$

But $\Delta^0 + W\Delta_{ji}$ is the value of the circuit determinant of the original system, that is, when the element W has its normal value (see Eq. 11.22); therefore, we may also write Eq. (11.26) as

$$F = \frac{\Delta}{\Delta^0}, \tag{11.27}$$

which states that the return difference for any element W in a linear circuit is equal to the ratio of the values assumed by the circuit determinant when the specified element W has its normal operating value and when W vanishes.

To determine the null return difference we may use Eq. (11.11) which, in conjunction with Eqs. (11.25), yields

$$\begin{aligned} F' &= 1 + W\frac{\Delta_{ji}}{\Delta^0} - W\frac{\Delta_{1i}\Delta_{jl}}{\Delta^0\Delta_{1l}^0} \\ &= 1 + \frac{W}{\Delta^0\Delta_{1l}^0}(\Delta_{1l}^0\Delta_{ji} - \Delta_{1i}\Delta_{jl}). \end{aligned} \tag{11.28}$$

This can be simplified by noting that*

$$\Delta^0 \Delta_{1l,ji} = \Delta^0_{1l}\Delta_{ji} - \Delta_{1i}\Delta_{jl} \tag{11.29}$$

Therefore, using Eqs. (11.28) and (11.29) we get the result

$$F' = 1 + W\frac{\Delta_{1l,ji}}{\Delta^0_{1l}}. \tag{11.30}$$

However, $\Delta^0_{1l} + W\Delta_{1l,ji}$ is the value of the cofactor Δ_{1l}; it follows therefore that

$$F' = \frac{\Delta_{1l}}{\Delta^0_{1l}}, \tag{11.31}$$

which states that the null return difference for any element W is equal to the ratio of the values assumed by the cofactor Δ_{1l} when the specified element W has its normal value and when W vanishes. This assumes that the independent source x_s is located in the first loop (or node), and that the lth loop current (or node voltage) constitutes the load signal x_l. The result of Eq. (11.31) may also be verified in another way. The closed-loop transmittance $K = x_l/x_s$ of the complete circuit is, from Eq. (11.22), given by

$$K = \frac{\Delta_{1l}}{\Delta}. \tag{11.32}$$

The direct transmittance t_{sl} is therefore

$$t_{sl} = K|_{W=0} = \frac{\Delta^0_{1l}}{\Delta^0}. \tag{11.33}$$

But, from Eq. (11.12) we have that the null return difference is

$$F' = F\frac{K}{t_{sl}}. \tag{11.34}$$

Hence, use of Eq. (11.27) and Eqs. (11.32)–(11.34) shows that, as before, the null return difference F' is equal to the ratio $\Delta_{1l}/\Delta^0_{1l}$.

Relation Between Return Differences for Two Elements[1]

A formula that is found useful in feedback calculations is the relationship between the return differences for two elements in a circuit under normal operating

* Equation (11.29) is a special case of the following general identity from the theory of determinants:

$$\Delta\Delta_{ab,cd} = \Delta_{ab}\Delta_{cd} - \Delta_{ad}\Delta_{cb},$$

where Δ is the determinant of any $l \times l$ matrix of coefficients, a and c are any two rows of the matrix, and b and d are any two columns of the matrix.

conditions and the return difference which would be obtained for each element if the other were to vanish. Suppose W_1 and W_2 are the two elements in question. To express the fact that the circuit determinant is a function of both W_1 and W_2, we may write it as $\Delta(W_1, W_2)$; so that $\Delta(0, W_2)$ denotes the determinant which results when W_1 is zero, $\Delta(W_1, 0)$ the determinant when W_2 is zero, and $\Delta(0, 0)$ the determinant when W_1 and W_2 are both zero.

The return differences for W_1 and W_2 are as follows, respectively:

$$F_1 = \frac{\Delta(W_1, W_2)}{\Delta(0, W_2)}$$
$$F_2 = \frac{\Delta(W_1, W_2)}{\Delta(W_1, 0)}. \tag{11.35}$$

The ratio F_1/F_2 is therefore

$$\frac{F_1}{F_2} = \frac{\Delta(W_1, 0)}{\Delta(0, W_2)} = \frac{\Delta(W_1, 0)/\Delta(0, 0)}{\Delta(0, W_2)/\Delta(0, 0)}$$

or

$$\frac{F_1}{F_2} = \frac{(F_1)_{W_2=0}}{(F_2)_{W_1=0}}. \tag{11.36}$$

Equation (11.36) is the desired relationship; it is particularly useful in the evaluation of the return difference with reference to one element if the return difference with reference to another element is known.

11.4 PROPERTIES OF FEEDBACK

a) Sensitivity

One of the most important properties of negative feedback is its ability to make the closed-loop transmittance of an amplifier less sensitive to variations in the parameters of the active devices, which result from manufacturing tolerances, aging, or environmental changes. The sensitivity S, as defined in Eq. (11.37), provides a useful quantitative measure for the degree of dependence of the closed-loop transmittance K upon a specified parameter W:

$$S = \frac{d(\log K)}{d(\log W)}$$

or

$$S = \frac{dK/K}{dW/W}. \tag{11.37}$$

In other words, the sensitivity of the transmittance K with respect to a specified element W is equal to the ratio of the fractional change in K over the fractional change in W which produces the initial change in W. It is assumed that all changes concerned are differentially small.

The expression for the sensitivity may also be written in the form

$$S = \frac{W}{K}\frac{dK}{dW}. \tag{11.38}$$

But,

$$K = t_{sl} + \frac{t_{si}Wt_{jl}}{1 - Wt_{ji}}, \tag{11.39}$$

where the transmittances t_{sl}, t_{si}, t_{jl} and t_{ji} are all independent of W; therefore, straightforward differentiation with respect to W gives the derivative dK/dW as

$$\frac{dK}{dW} = \frac{t_{si}t_{jl}}{(1 - Wt_{ji})^2}. \tag{11.40}$$

Substitution of this result in Eq. (11.38) yields:

$$S = \frac{Wt_{si}t_{jl}}{K(1 - Wt_{ji})^2}. \tag{11.41}$$

If next we substitute Eq. (11.39) in (11.41), and note that $t_{sl} = K^0$ and $F = 1 - Wt_{ji}$, we obtain:

$$S = \frac{1}{F}\left(1 - \frac{K^0}{K}\right). \tag{11.42}$$

Another useful expression for the sensitivity is obtained by noting that K^0/FK is equal to the reciprocal of the null return difference F'; so that

$$S = \frac{1}{F} - \frac{1}{F'}. \tag{11.43}$$

If the direct transmittance K^0 is zero or, equivalently, the null return difference F' is infinite, then we find that $S = 1/F$. That is to say, the sensitivity and reciprocal of the return difference are equal for any element whose vanishing reduces transmission through the system as a whole to zero. This condition is, for example, closely satisfied in a single-loop feedback amplifier consisting of a cascade of common-emitter or common-cathode stages, with a passive network providing the backward transmission path for the feedback signal. If in such a structure any one of the tubes or transistors fails, the forward transmission is effectively reduced to zero, except for the current leaking through the feedback network into the load. However, this leakage is ordinarily so much smaller than the normal output that it may be justifiably neglected.

In general, for a specified element, the reciprocal of return difference will give a conservative estimate of sensitivity provided the absolute value of K^0 is not greater than that of K, whilst it is a pessimistic estimate if K^0 is nearly equal to K in magnitude as well as phase.

Example 11.3. *Sensitivity of active bridged-T network.* Equation (11.43) provides a direct basis for evaluating the sensitivity of the transfer-voltage ratio of the active bridged-T network with respect to changes in the control parameter g_m. The required return difference and null return difference are given by Eqs. (11.13) and (11.14), respectively. The resulting sensitivity of the network is therefore:

$$S = \frac{1}{1 + g_m R(R + Z_a)/(2R + 3Z_a)} - \frac{1}{1 + g_m R^2/(2R + Z_a)}$$

or

$$S = \frac{-g_m Z_a^2 R}{(2R + Z_a + g_m R^2)(2R + 3Z_a + g_m R^2 + g_m R Z_a)}. \tag{11.44}$$

For large values of g_m, we have:

$$S \to \frac{-Z_a^2}{g_m R^2(R + Z_a)}. \tag{11.45}$$

Example 11.4. *Combined positive and negative feedback.* For an amplifier consisting of two or more stages connected in cascade it is sometimes beneficial to employ positive local feedback around each stage and negative overall feedback around the complete forward path, as illustrated by the signal-flow graph in Fig. 11.22. The overall transmittance of this configuration is:

$$K = \frac{\mu_1 \mu_2}{(1 - \mu_1 \beta_1)(1 - \mu_2 \beta_2) - \mu_1 \mu_2 \beta_3}. \tag{11.46}$$

As there is no transmission from input to output when μ_1 is set to zero, it follows that the sensitivity of K with respect to μ_1 is equal to the reciprocal of the return

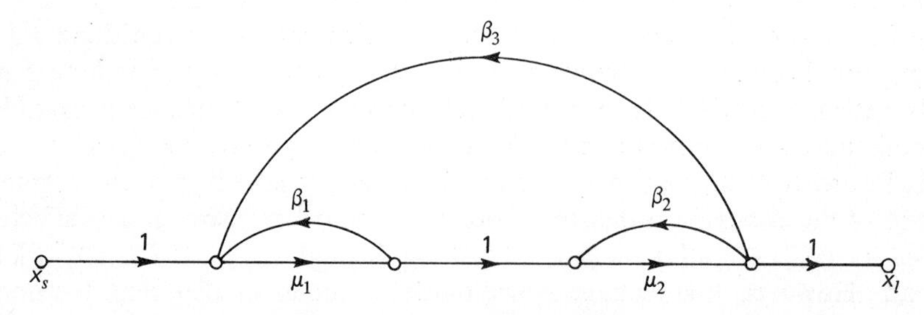

Fig. 11.22. Multiple-loop feedback amplifier with combined positive and negative feedback.

difference for μ_1; therefore,

$$S_1 = \frac{1}{1 - \mu_1\left(\beta_1 + \dfrac{\mu_2\beta_3}{1 - \mu_2\beta_2}\right)}, \tag{11.47}$$

from which we find that:

$$\begin{aligned} S_1 &= -\frac{1 - \mu_2\beta_2}{\mu_1\mu_2\beta_3} \quad &\text{for} \quad \mu_1\beta_1 = 1, \\ S_1 &= 0 \quad &\text{for} \quad \mu_2\beta_2 = 1. \end{aligned} \tag{11.48}$$

When $\mu_2\beta_2 = 1$ the amount of feedback acting around the element μ_1 is infinite, because of the positive local feedback acting around μ_2. Similarly, the sensitivity of K with respect to μ_2 is zero when $\mu_1\beta_1 = 1$. We see, therefore, that if $\mu_1\beta_1 = \mu_2\beta_2 = 1$, the overall transmittance K is $-1/\beta_3$ and both sensitivities are zero. For this reason, the configuration of Fig. 11.22 is sometimes referred to as a *zero-sensitivity system.*

b) Reduction of Noise and Nonlinear Distortion

Another property of negative feedback of fundamental engineering importance is that it can, under certain circumstances, reduce the effect of noise and nonlinear distortion generated within an amplifier. To demonstrate this effect, consider first the case of a linear amplifier without feedback represented by the signal-flow graph of Fig. 11.23(a) in which the extraneous signal x_n represents noise or non-

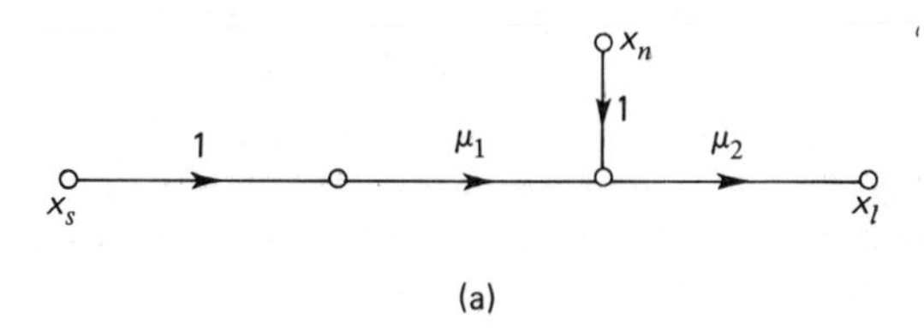

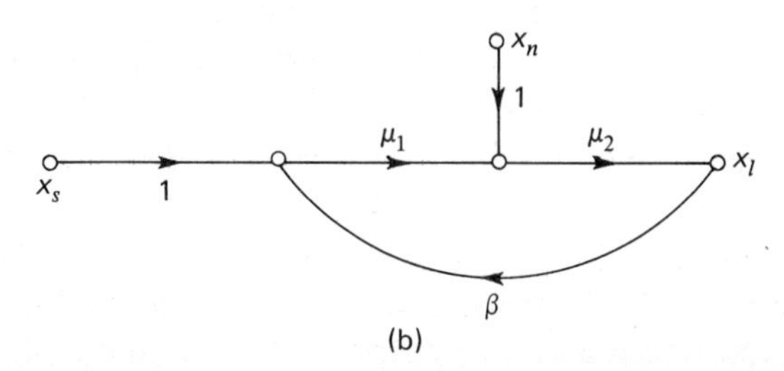

Fig. 11.23. Signal-flow graph of linear amplifier with extraneous signal x_n: (a) without feedback, (b) with feedback.

linear distortion introduced at some arbitrary point within the amplifier. The input signal is x_s, and the output containing both signal and noise is

$$x_l = \mu_1 \mu_2 x_s + \mu_2 x_n. \tag{11.49}$$

The output *signal-to-noise ratio* in the absence of feedback is therefore equal to $\mu_1 x_s / x_n$.

Consider next the flow graph in part (b) of Fig. 11.23 where feedback is shown applied to the amplifier. In this case the output signal due to the combined action of x_s and x_n is

$$x_l = \frac{\mu_1 \mu_2 x_s}{1 - \mu_1 \mu_2 \beta} + \frac{\mu_2 x_n}{1 - \mu_1 \mu_2 \beta}, \tag{11.50}$$

where the first term represents the contribution of the input signal x_s to the output, and the second term represents the contribution of the noise x_n. The output signal-to-noise ratio with feedback is therefore equal to $\mu_1 x_s / x_n$ which is identical with the result obtained in the absence of feedback.

The application of negative feedback does, however, help to improve the output signal-to-noise ratio in an indirect manner.[9] Clearly, this ratio, which is equal to $\mu_1 x_s / x_n$ with or without feedback, may be increased through increasing the magnitude of the input signal x_s relative to that of the noise x_n, or through increasing the gain μ_1 which precedes the point where the noise is injected. In the amplifier without feedback, such an increase is often precluded as the resulting increased output signal may cause excessive distortion, or overloading, in the output stage of the amplifier. If, on the other hand, negative feedback is applied and at the same time one or both of the input signal x_s and the gain μ_1 are increased so that the output signal remains at the same level as it had before the application of feedback, then from Eqs. (11.49) and (11.50) we see that the net effect at the output will be to reduce the contribution of the extraneous signal x_n by the factor $1/(1 - \mu_1 \mu_2 \beta)$. We conclude, therefore, that provided the output signal power of the amplifier is maintained constant with and without feedback, the application of negative feedback increases the output signal-to-noise ratio by the factor $(1 - \mu_1 \mu_2 \beta)$. This improvement in performance is possible, however, only if the extraneous signal x_n is generated within the amplifier. Thus, if x_n originates at the input port of the amplifier we have a case equivalent to $\mu_1 = 1$; then the signal and noise components add up directly and are amplified by the same amount, in which case the application of negative feedback would prove to be of no help at all.

Example 11.5. *Combined positive and negative feedback* (continued). Another interesting property of the feedback amplifier described by Fig. 11.22 is that if $\mu_1 \beta_1$ is equal to unity, then all of the output distortion due to the μ_2-amplifier is theoretically eliminated. To demonstrate this property, consider the signal-flow graph of Fig. 11.24 where we have included a distortion generator, x_n, at the

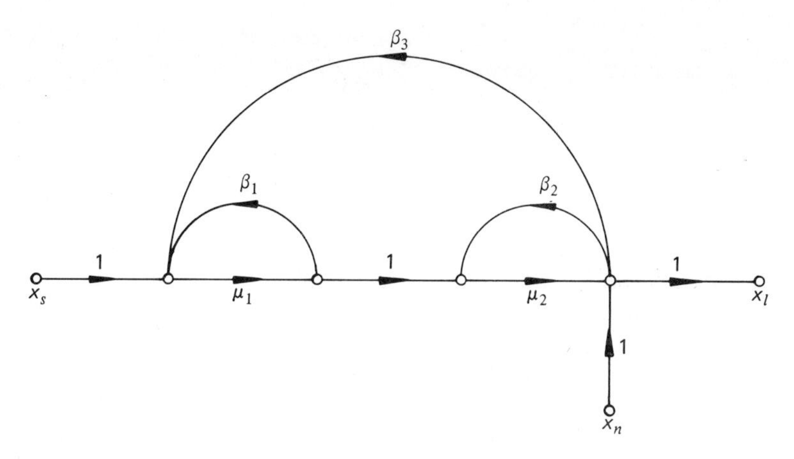

Fig. 11.24. Illustrating the effect of output distortion on amplifier with combined positive and negative feedback.

output end of the μ_2-amplifier. The output signal due to the combined action of the input signal, x_s, and x_n is given by

$$x_l = Kx_s + \frac{(1 - \mu_1\beta_1)x_n}{(1 - \mu_1\beta_1)(1 - \mu_2\beta_2) - \mu_1\mu_2\beta_3}, \tag{11.51}$$

where K is defined by Eq. (11.46). We see that the second term on the right of Eq. (11.51) contains the factor $(1 - \mu_1\beta_1)$ in its numerator. Clearly, therefore, if $\mu_1\beta_1 = 1$, the distortion component x_n is completely eliminated from the output x_l.

It should, however, be noted that a feedback amplifier combining positive local feedbacks (with $\mu_1\beta_1 = \mu_2\beta_2 = 1$) and negative overall feedback, as in Fig. 11.22, can only be conditionally stable, and if either μ_1 or μ_2 should vanish, the amplifier will become unstable.[10] We shall have more to say about conditional stability in the next chapter.

c) Control of Driving-point Impedances

Feedback is commonly used to control the driving-point impedances at various points of an amplifier. For example, we may use negative feedback to influence the input and output impedances which an amplifier presents to its source and load, respectively, the reason being that the levels of these impedances directly determine the efficiency of power transfer from the source to the amplifier input and from the amplifier output to the load.

To determine the influence of feedback on the driving-point impedance Z at an arbitrarily selected port k–k' of a linear circuit, consider Fig. 11.25 where the network N contains no independent sources, and a current source I_k is shown connected across the port k–k'. If V_k is the resulting voltage developed across

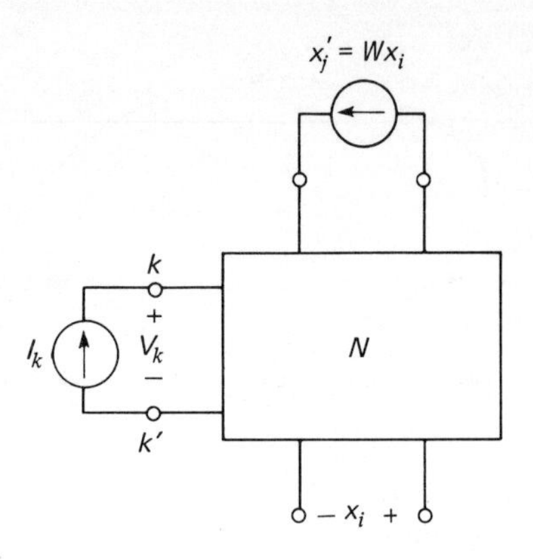

Fig. 11.25. Illustrating evaluation of the driving-point impedance at port k–k'.

this port, we find that $Z = V_k/I_k$. Let us suppose that a generalized controlled source $x_j' = Wx_i$ has been picked from the interior of network N for special consideration. Then, adding the separate effects of the sources I_k and x_j', we obtain:

$$\begin{aligned} V_k &= Z^0 I_k + t_{jk} x_j' \\ x_i &= t_{ki} I_k + t_{ji} x_j', \end{aligned} \tag{11.52}$$

where $x_j' = Wx_i$. The various transmittances are defined as follows:

$$\begin{aligned} Z^0 &= \left.\frac{V_k}{I_k}\right|_{x_j'=0} & t_{jk} &= \left.\frac{V_k}{x_j'}\right|_{I_k=0} \\ t_{ki} &= \left.\frac{x_i}{I_k}\right|_{x_j'=0} & t_{ji} &= \left.\frac{x_i}{x_j'}\right|_{I_k=0}. \end{aligned} \tag{11.53}$$

In particular, we see that Z^0 is the value assumed by the driving-point impedance Z when signal flow through the controlled source is reduced to zero, which may be achieved by setting the control parameter W equal to zero.

Equations (11.52) are represented by the signal-flow graph shown in Fig. 11.26 where, by inspection, we find that:

$$Z = \frac{V_k}{I_k} = Z^0 + \frac{t_{ki} W t_{jk}}{1 - W t_{ji}} \tag{11.54}$$

or

$$Z = Z^0 \frac{1 - W t_{ji} + W \dfrac{t_{ki} t_{jk}}{Z^0}}{1 - W t_{ji}}. \tag{11.55}$$

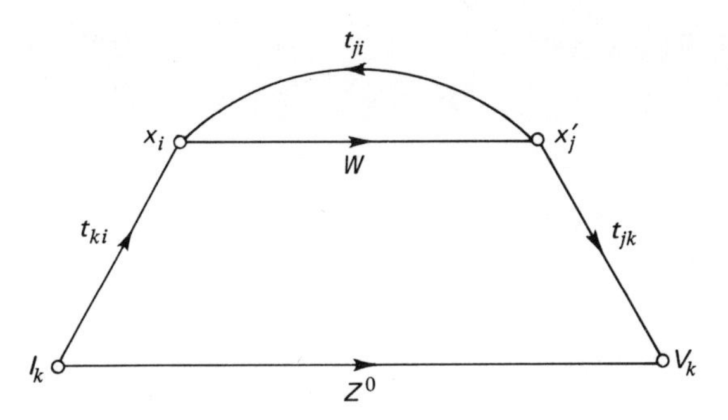

Fig. 11.26. Signal-flow graph representing Eqs. (11.52).

However, from the signal-flow graph of Fig. 11.26 we see that $1 - Wt_{ji}$ is the return difference for the element W with $I_k = 0$, that is, with the port k–k' open circuited; so that

$$F_{oc} = 1 - Wt_{ji}. \tag{11.56}$$

In other words, if we open-circuit the port k–k' and replace the controlled source Wx_i with an independent source W, as in Fig. 11.27, and if x_{oc} denotes the special value of the variable x_i developed in the resulting system, then we have

$$F_{oc} = 1 - x_{oc}. \tag{11.57}$$

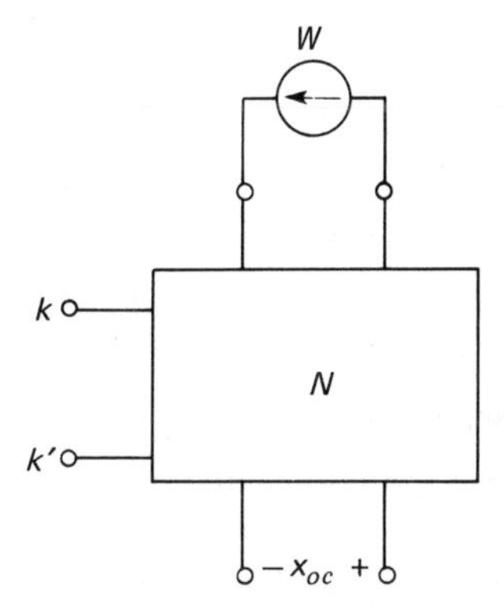

Fig. 11.27. Network for evaluating the returned signal x_{oc} obtained with port k–k' open-circuited.

The term $1 - Wt_{ji} + Wt_{ki}t_{jk}/Z^0$ is the return difference for the element W with $V_k = 0$, that is, with the port k–k' short circuited; so that

$$F_{sc} = 1 - Wt_{ji} + W\frac{t_{ki}t_{jk}}{Z^0}. \tag{11.58}$$

Equivalently, we may write

$$F_{sc} = 1 - x_{sc}, \tag{11.59}$$

where x_{sc} is the special value of the variable x_i developed in the system of Fig. 11.28 in which we have short circuited the port k–k' and replaced the controlled source Wx_i with an independent source W.

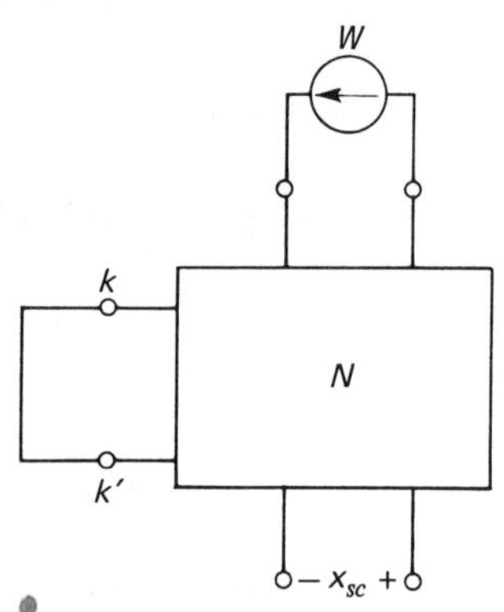

Fig. 11.28. Network for evaluating returned signal x_{sc} with port k–k' short circuited.

Therefore, substitution of Eqs. (11.56) and (11.58) in Eq. (11.55) yields:

$$Z = Z^0 \frac{F_{sc}}{F_{oc}}. \tag{11.60}$$

This relation is known as *Blackman's formula.*[11] Comparison of the signal-flow graphs shown in Figs. 11.9 and 11.26 reveals that the relation of Eq. (11.60) is simply a restatement of Eq. (11.12) in terms of the graph of Fig. 11.26 for the calculation of driving-point impedance Z at a given port k–k'.

One of the uses to which Blackman's formula may be put is in the determination of return difference by impedance measurements. However, since this formula involves the two return differences F_{sc} and F_{oc}, only one of which may be identified with the return difference to be determined, one of these return differences must be known.

In the case of single-loop feedback amplifiers it is possible to choose terminals k–k' so that either F_{sc} or F_{oc} is unity. If $F_{oc} = 1$ and $F_{sc} = F$, where F is the return difference under normal operating conditions, then

$$Z = Z^0 F$$

or

$$F = \frac{Z}{Z^0}. \tag{11.61}$$

On the other hand, if

$$F_{sc} = 1 \quad \text{and} \quad F_{oc} = F,$$

we have

$$Z = \frac{Z^0}{F}$$

or

$$F = \frac{Z^0}{Z}. \tag{11.62}$$

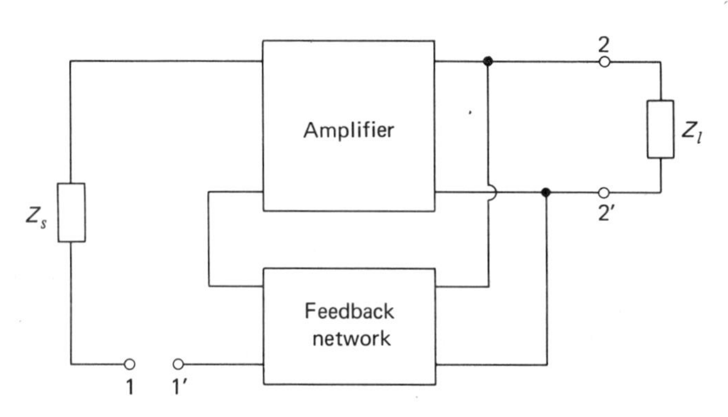

Fig. 11.29. Series–shunt feedback amplifier.

Figure 11.29 shows a feedback structure in which the internal amplifier and feedback network components are connected in series at the input end, and in parallel at the output end. At the port 1–1′ in this diagram we see that the conditions for Eq. (11.61) are obviously fulfilled. Hence, if the impedance measurements are made at these terminals the required return difference is given by Eq. (11.61). On the other hand, at the port 2–2′ in Fig. 11.29 the conditions for Eq. (11.62) are obviously fulfilled. Hence, if the impedance measurements are made at these terminals, the required return difference is given by Eq. (11.62). Furthermore, Eqs. (11.61) and (11.62) state that the impedance measured in any series line is multiplied by the normal value of the return difference, while the impedance measured across the path of the feedback loop is divided by the same return difference.

Example 11.6. *Bridge feedback*. The effects of series and shunt feedback may be combined in a single amplifier in a variety of ways; Fig. 11.30 shows a particular example in which R_c is the series feedback resistance, while the inductive potential

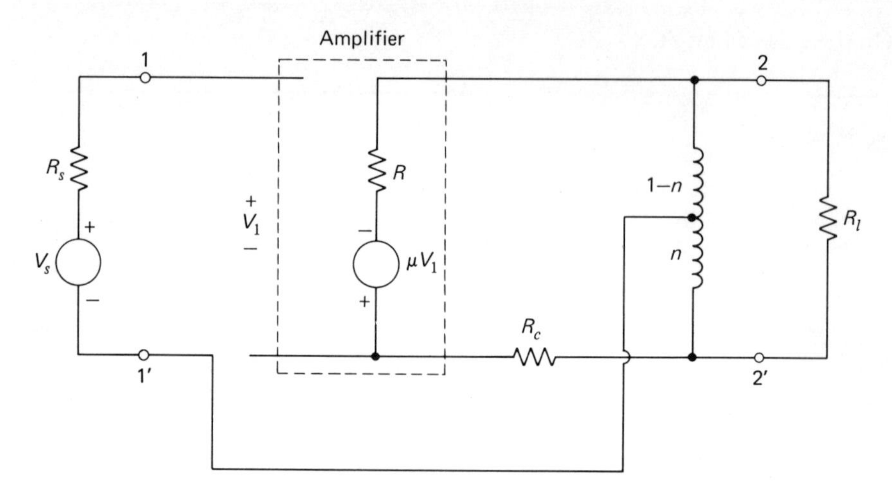

Fig. 11.30. Bridge feedback amplifier.

divider across the load R_l is used for the shunt feedback. The series combination of the resistance R and the controlled voltage source μV_1 is assumed to represent the Thévenin equivalent circuit for the internal amplifier with respect to its output port. The potential divider, and the resistances R_c and R are seen to constitute the four arms of a Wheatstone bridge; for this reason the feedback structure of Fig. 11.30 is usually referred to as a *bridge feedback amplifier.*

Assuming that the inductive potential divider is ideal, we find from Fig. 11.30, by inspection, that with reference to μ and port 2–2',

$$Z_{out}^0 = Z_{out}|_{\mu=0} = R_c + R \tag{11.63}$$

$$F_{oc} = 1 + \mu n \tag{11.64}$$

$$F_{sc} = 1 + \frac{\mu R_c}{R_c + R}. \tag{11.65}$$

Therefore, use of Eqs. (11.60) and Eqs. (11.63)–(11.65) gives the output impedance Z_{out} of the feedback amplifier as

$$Z_{out} = \frac{R + (1 + \mu)R_c}{1 + \mu n}. \tag{11.66}$$

This output impedance is clearly independent of μ if F_{oc} equals F_{sc}. Equating these values from Eqs. (11.64) and (11.65) shows that the bridge must be balanced, that is, the condition

$$\frac{R}{R_c} = \frac{1-n}{n} \tag{11.67}$$

must be satisfied if Z_{out} is to be independent of μ. Furthermore, this adjustment makes the return difference for μ independent of the load impedance, a condition which is not true for either the series or shunt feedback alone.

Assuming that the bridge is balanced, and therefore, the condition of Eq. (11.67) is satisfied, the output impedance Z_{out} assumes the special value of R_c/n. Then, by choosing the load resistance R_l equal to R_c/n, we have a perfect impedance match between the load and the output port of the feedback amplifier. In practice, however, we find that the output resistance R of the internal amplifier is quite variable and, therefore, very likely to depart from the desired value of $R_c(1 - n)/n$, which results in a certain degree of impedance mismatch at the output port. A quantitative measure of this impedance mismatch is the output *reflection coefficient*,

$$\rho_{out} = \frac{Z_{out} - R_l}{Z_{out} + R_l}. \tag{11.68}$$

Suppose $R_l = R_c/n$, and R has a value different from $R_c(1 - n)/n$ so that the bridge is unbalanced. Substitution of Eq. (11.66) in Eq. (11.68) yields:

$$\rho_{out} = \frac{(n-1)R_c + nR}{(n+1)R_c + nR + 2\mu n R_c}. \tag{11.69}$$

When μ is zero the reflection coefficient ρ_{out} takes on the special value

$$\rho_{out}^0 = \frac{(n-1)R_c + nR}{(n+1)R_c + nR}. \tag{11.70}$$

The ratio of ρ_{out}^0 to ρ_{out} is thus equal to:

$$\frac{\rho_{out}^0}{\rho_{out}} = 1 + \frac{2\mu n R_c}{(n+1)R_c + nR}. \tag{11.71}$$

The right-hand member of Eq. 11.71 is recognized to be simply the normal return difference for the element μ; so that for an unbalanced bridge we have:

$$\rho_{out} = \frac{\rho_{out}^0}{F}. \tag{11.72}$$

Now the reflection coefficient ρ_{out}^0 is always less than unity in absolute value. Hence, from Eq. (11.72) it follows that ρ_{out} can be made negligibly small by applying a large amount of negative feedback. In other words, in an unbalanced bridge if the feedback is large the complete amplifier presents to the load an impedance close to R_c/n even though the actual output resistance of the internal amplifier may depart considerably from the value required to satisfy the balance condition of Eq. (11.67). However, once the balance of the bridge has been disturbed the applied feedback is no longer independent of the load impedance.

Example 11.7. *Input impedance of active bridged-T network.* As another example to exemplify the relationship of Eq. (11.60), we will evaluate the input impedance of the active bridged-T network in Fig. 11.10 considered previously. With reference to the control parameter g_m, we find from Fig. 11.10 that

$$Z_{in}^0 = \frac{R(3Z_a + 2R)}{2R + Z_a} \tag{11.73}$$

$$F_{sc} = 1 + \frac{g_m R(R + Z_a)}{2R + 3Z_a} \tag{11.74}$$

$$F_{oc} = 1 + \frac{g_m R(R + Z_a)}{2R + Z_a}. \tag{11.75}$$

Equation (11.60) consequently gives the impedance measured at the input port of the active network in Fig. 11.10 as

$$Z_{in} = \frac{R[2R + 3Z_a + g_m R(R + Z_a)]}{2R + Z_a + g_m R(R + Z_a)}. \tag{11.76}$$

As g_m approaches infinity, we see that Z_{in} approaches R, which is the same as that of a passive constant resistance bridged-T network with Z_a as the bridging impedance.

11.5 MULTIPLE-LOOP FEEDBACK CIRCUITS

In the preceding sections of this chapter we have seen that the concept of return difference plays a central role in the theory of linear feedback circuits in that it is directly related to the sensitivity of a circuit's response to imperfections of one of its elements, the determination of its transmission and driving-point properties, and, as we shall see in the next chapter, its stability status. In this section we shall generalize the concepts of return difference and null return difference for a single element, by considering the presence of a multiplicity of controlled sources. The results so obtained are applicable to linear circuits which possess a multiplicity of physical feedback loops so that signals can be returned to some point inside the circuit by more than one path. The results are, however, by no means restricted to situations of this type.

Consider the network of Fig. 11.31 having a single independent source x_s connected across the input port, and a total of n controlled sources constrained by the matrix relation

$$\mathbf{x}' = \mathbf{W}\mathbf{x}, \tag{11.77}$$

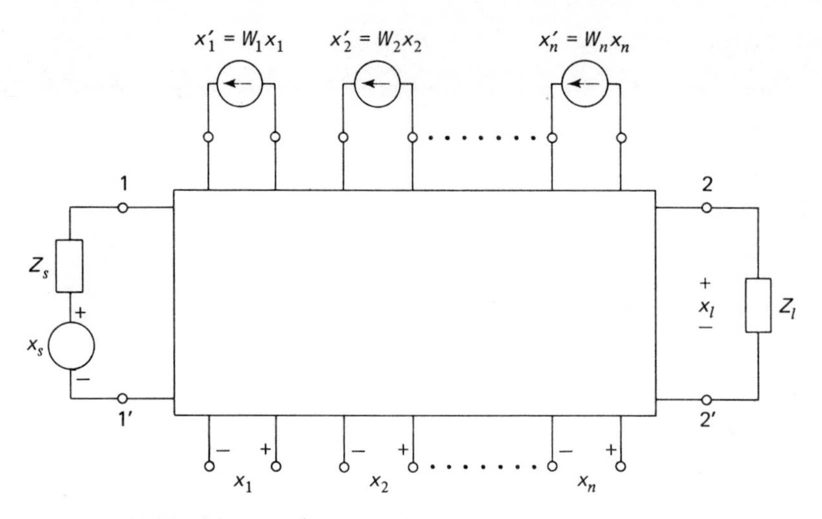

Fig. 11.31. Linear active network with multiple controlled sources.

where

$$\mathbf{x} = \text{control signal vector} = \begin{bmatrix} x_1 \\ x_2 \\ \cdot \\ \cdot \\ \cdot \\ x_n \end{bmatrix}.$$

$$\mathbf{W} = \text{diagonal matrix composed of control parameters} = \begin{bmatrix} W_1 & 0 & \cdots & 0 \\ 0 & W_2 & \cdots & 0 \\ \cdot & \cdot & & \cdot \\ \cdot & \cdot & & \cdot \\ \cdot & \cdot & & \cdot \\ 0 & 0 & \cdots & W_n \end{bmatrix}.$$

The load signal developed across the port 2–2′ is represented by x_l. Assuming that the network is linear, we may apply the superposition theorem to obtain:

$$\begin{aligned} x_l &= K^0 x_s + \mathbf{Q}\mathbf{x}' \\ \mathbf{x} &= \mathbf{P} x_s + \boldsymbol{\beta}\mathbf{x}', \end{aligned} \qquad (11.78)$$

where

P = column matrix expressing dependence of control signals upon the independent source x_s

$$= \begin{bmatrix} t_{s1} \\ t_{s2} \\ \cdot \\ \cdot \\ \cdot \\ t_{sn} \end{bmatrix},$$

Q = row matrix expressing dependence of load signal x_l upon the controlled sources

$$= [t_{1l} \quad t_{2l} \quad \cdots \quad t_{nl}],$$

β = square matrix defining influence of the outputs of controlled sources upon the values of the control signals

$$= \begin{bmatrix} t_{11} & t_{21} & \cdots & t_{n1} \\ t_{12} & t_{22} & \cdots & t_{n2} \\ \cdot & \cdot & & \cdot \\ \cdot & \cdot & & \cdot \\ \cdot & \cdot & & \cdot \\ t_{1n} & t_{2n} & \cdots & t_{nn} \end{bmatrix}.$$

Equations (11.78), together with the constraint in Eq. (11.77), may be represented by the signal-flow graph of Fig. 11.32 in which **P**, **Q**, **β** and K^0 are referred to as *flow matrices.* K^0 is a constant and will be defined later.

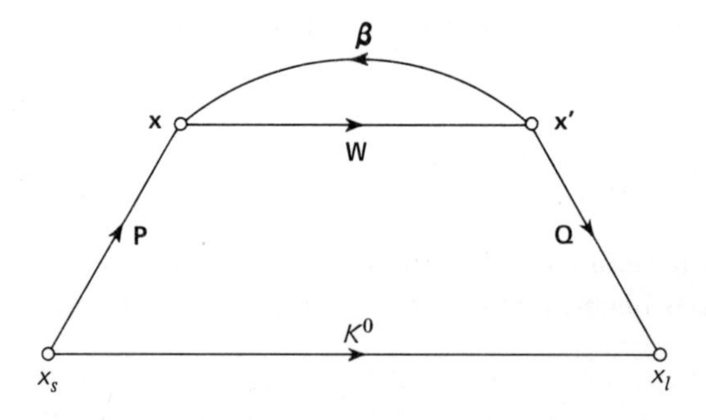

Fig. 11.32. Feedback-flow graph representing Eq. (11.78).

Return Ratio and Return Difference Matrices

The negative of the product of the two square matrices $\boldsymbol{\beta}$ and $\mathbf{W}$ is termed the *return ratio matrix*, $\mathbf{T}$, that is

$$\mathbf{T} = -\boldsymbol{\beta}\mathbf{W} \tag{11.79}$$

or in expanded form:

$$\mathbf{T} = -\begin{bmatrix} t_{11}W_1 & t_{21}W_2 & \cdots & t_{n1}W_n \\ t_{12}W_1 & t_{22}W_2 & \cdots & t_{n2}W_n \\ \cdot & \cdot & & \cdot \\ \cdot & \cdot & & \cdot \\ \cdot & \cdot & & \cdot \\ t_{1n}W_1 & t_{2n}W_2 & \cdots & t_{nn}W_n \end{bmatrix}. \tag{11.80}$$

In order to develop a physical circuit interpretation for the return ratio matrix, we shall partition the square matrix on the right of Eq. (11.80) into n column matrices as follows:

$$\mathbf{T} = [\mathbf{T}_1 \quad \mathbf{T}_2 \quad \cdots \quad \mathbf{T}_n], \tag{11.81}$$

where

$$\mathbf{T}_k = -\begin{bmatrix} t_{k1}W_k \\ t_{k2}W_k \\ \cdot \\ \cdot \\ \cdot \\ t_{kn}W_k \end{bmatrix}, \qquad k = 1, 2, \ldots, n. \tag{11.82}$$

If the independent source x_s is reduced to zero, and all the controlled sources except $x'_k = W_k x_k$ are inactivated, and if the controlled source x'_k so left is itself replaced by an independent source of strength W_k, then from the second line of Eq. (11.78) we see that the kth column matrix $\mathbf{T}_k$ of the return ratio matrix is equal to the negative of the column matrix $\mathbf{x}$ of variables developed in the resulting system. This definition of the return ratio matrix is a generalization of the definition of the return ratio for a single element.

The *return difference matrix* $\mathbf{F}$ for $\mathbf{W}$, which was first used in network theory by Tasny-Tschiassny[12], is defined by

$$\mathbf{F} = \mathbf{1} + \mathbf{T}, \tag{11.83}$$

where $\mathbf{1}$ is an $n \times n$ unit matrix. The return difference F_k for a given element W_k, with all other W's having their normal operating values, is given by

$$F_k = \frac{\Delta^F}{\Delta^F_{kk}}, \tag{11.84}$$

where Δ^F is the determinant of the return difference matrix $\mathbf{F}$, and Δ^F_{kk} is the cofactor of F_{kk} in $\mathbf{F}$. The relation of Eq. (11.84) is a direct consequence of Eq. (11.27).

Null Return Difference Matrix

The *null return-difference matrix* $\mathbf{F}'$ is evaluated under the condition that the independent source x_s is adjusted so that the load signal x_l is zero. Then from the first line of Eq. (11.78) we find that the required x_s is

$$x_s = -\frac{1}{K^0}\mathbf{Q}\mathbf{x}'. \tag{11.85}$$

Substitution of Eq. (11.85) in the second line of Eq. (11.78) yields

$$\mathbf{x} = \boldsymbol{\beta}\mathbf{x}' - \frac{1}{K^0}\mathbf{PQ}\mathbf{x}'. \tag{11.86}$$

Suppose all the controlled sources except $x_1' = W_1 x_1$ are inactivated, and x_1' is itself replaced by an independent source W_1. Then, if the column matrix $\mathbf{T}_1'$ denotes the negative of the column matrix $\mathbf{x}$ of variables developed in the resulting system adjusted for $x_l = 0$, we find from Eq. (11.86) that:

$$\mathbf{T}_1' = -\boldsymbol{\beta}\begin{bmatrix} W_1 \\ 0 \\ 0 \\ \cdot \\ \cdot \\ \cdot \\ 0 \end{bmatrix} + \frac{1}{K^0}\mathbf{PQ}\begin{bmatrix} W_1 \\ 0 \\ 0 \\ \cdot \\ \cdot \\ \cdot \\ 0 \end{bmatrix}. \tag{11.87}$$

Similarly, if the column matrix $\mathbf{T}_2'$ denotes the negative of the column matrix $\mathbf{x}$ of variables developed in the system that results from inactivating all controlled sources except x_2' which is itself replaced by an independent source W_2, then

$$\mathbf{T}_2' = -\boldsymbol{\beta}\begin{bmatrix} 0 \\ W_2 \\ 0 \\ \cdot \\ \cdot \\ \cdot \\ 0 \end{bmatrix} + \frac{1}{K^0}\mathbf{PQ}\begin{bmatrix} 0 \\ W_2 \\ 0 \\ \cdot \\ \cdot \\ \cdot \\ 0 \end{bmatrix}. \tag{11.88}$$

If in a similar manner we treat the remaining $n - 2$ controlled sources and so obtain the column matrices $\mathbf{T}'_3, \ldots, \mathbf{T}'_n$, we may then define the *null return difference matrix* $\mathbf{F}'$ for $\mathbf{W}$ as follows:

$$\mathbf{F}' = \mathbf{1} + [\mathbf{T}'_1 \quad \mathbf{T}'_2 \quad \cdots \quad \mathbf{T}'_n]$$

$$= \mathbf{1} - \boldsymbol{\beta}\begin{bmatrix} W_1 & 0 & \cdots & 0 \\ 0 & W_2 & \cdots & 0 \\ \cdot & \cdot & & \cdot \\ \cdot & \cdot & & \cdot \\ \cdot & \cdot & & \cdot \\ 0 & 0 & \cdots & W_n \end{bmatrix} + \frac{1}{K^0}\mathbf{PQ}\begin{bmatrix} W_1 & 0 & \cdots & 0 \\ 0 & W_2 & \cdots & 0 \\ \cdot & \cdot & & \cdot \\ \cdot & \cdot & & \cdot \\ \cdot & \cdot & & \cdot \\ 0 & 0 & \cdots & W_n \end{bmatrix}$$

$$= \mathbf{1} - \boldsymbol{\beta}\mathbf{W} + \frac{1}{K^0}\mathbf{PQW}$$

that is,

$$\mathbf{F}' = \mathbf{F} + \frac{1}{K^0}\mathbf{PQW}. \tag{11.89}$$

The null return difference F'_k for the element W_k, with all other W's having their normal operating values, is given by

$$F'_k = \frac{\Delta^{F'}}{\Delta^{F'}_{kk}} \tag{11.90}$$

where $\Delta^{F'}$ is the determinant of the null return difference matrix, and $\Delta^{F'}_{kk}$ is the cofactor of F'_{kk} in $\mathbf{F}'$.

Complementary Return Difference Matrix[6]

The *complementary return difference matrix*, denoted by $\overline{\mathbf{F}}$, is defined as the return difference matrix which results when the input signal x_s is adjusted so that $x_s + \bar{x}_l = 0$, in which $\bar{x}_l$ denotes the value of the output signal x_l when $x_s = 0$. This matrix is readily obtained from the expression of Eq. (11.89) for the null return difference matrix by setting the direct transmittance K^0 equal to unity; that is,

$$\overline{\mathbf{F}} = \mathbf{F} + \mathbf{PQW}. \tag{11.91}$$

The complementary return difference matrix is found to be particularly useful when K^0 is zero, in which case the null return difference matrix does not exist.

11.6 GENERALIZED FEEDBACK FORMULAE

a) Closed-loop Transmittance

The closed-loop transmittance K is equal to the ratio of the load signal x_l to the independent source signal x_s with all controlled sources operating as normal.

Substitution of $\mathbf{x}' = \mathbf{W}\mathbf{x}$ in the second line of Eq. (11.78), and use of Eqs. (11.79) and (11.83) yields:

$$\mathbf{F}\mathbf{x} = \mathbf{P}x_s. \tag{11.92}$$

Pre-multiplying both sides of Eq. (11.92) with the inverse matrix $\mathbf{F}^{-1}$, we get

$$\mathbf{x} = \mathbf{F}^{-1}\mathbf{P}x_s. \tag{11.93}$$

Thus, the first line of Eq. (11.78) gives the load signal x_l to be:

$$\begin{aligned} x_l &= K^0 x_s + \mathbf{QWx} \\ &= K^0 x_s + \mathbf{QWF}^{-1}\mathbf{P}x_s. \end{aligned} \tag{11.94}$$

Therefore, the closed-loop transmittance $K = x_l/x_s$ is

$$K = K^0 + \mathbf{QWF}^{-1}\mathbf{P}, \tag{11.95}$$

where K^0 is the special value of the closed-loop transmittance which results when all the controlled sources are inactivated, that is, $\mathbf{W} = \mathbf{0}$.

Another compact expression for the closed-loop transmittance in terms of the determinants of the return difference and null return difference matrices may be obtained as follows: The inverse of the return difference matrix $\mathbf{F}$ and its adjoint are related by:

$$\mathbf{F}^{-1} = \frac{\text{adj}\,(\mathbf{F})}{\Delta^F}, \tag{11.96}$$

which assumes that the determinant Δ^F of the return difference matrix is non-zero. Substitution of Eq. (11.96) in (11.95) therefore gives

$$K = K^0 + \frac{1}{\Delta^F}\mathbf{QW}\,\text{adj}(\mathbf{F})\mathbf{P}$$

or,

$$K = \frac{K^0}{\Delta^F}\left[\Delta^F + \frac{1}{K^0}\mathbf{QW}\,\text{adj}(\mathbf{F})\mathbf{P}\right]. \tag{11.97}$$

But the determinant of the null return difference matrix is given by*

$$\begin{aligned} \Delta^{F'} &= \det\left[\mathbf{F} + \frac{1}{K^0}\mathbf{PQW}\right] \\ &= \Delta^F + \frac{1}{K^0}\mathbf{QW}\,\text{adj}(\mathbf{F})\mathbf{P}. \end{aligned} \tag{11.98}$$

---

* Equation (11.98) is a special case of the following general identity

$$\det[\mathbf{G} + \mathbf{IJ}] = \det \mathbf{G} + \mathbf{J}\,\text{adj}\,(\mathbf{G})\mathbf{I}$$

where $\mathbf{G}$, $\mathbf{I}$ and $\mathbf{J}$ respectively denote matrices of orders $n \times n$, $n \times 1$ and $1 \times n$. For a proof, see Ref. 6.

Therefore, using Eqs. (11.97) and (11.98) we obtain

$$K = K^0 \frac{\Delta^{F'}}{\Delta^F}. \tag{11.99}$$

The result of Eq. (11.99) may also be obtained by the repeated application of the formula of Eq. (11.12) to include the combined effect of all the controlled sources within the network.[13] Suppose we focus our attention on W_1 as a variable element and assume that all the other W's are held fixed at their normal operating values; then from Eq. (11.12) we have

$$K(W_1, W_2, W_3, \ldots, W_n) = K(0, W_2, W_3, \ldots, W_n) \frac{F'_1(W_1, W_2, W_3, \ldots, W_n)}{F_1(W_1, W_2, W_3, \ldots, W_n)}, \tag{11.100}$$

where $K(0, W_2, W_3, \ldots, W_n)$ is the special value of the closed-loop transmittance which results when only the element W_1 is set equal to zero; and $F_1(W_1, W_2, W_3, \ldots, W_n)$ and $F'_1(W_1, W_2, W_3, \ldots, W_n)$ are the return difference and null return difference, respectively, for W_1 with all the other W's having their normal operating values. Equation (11.12) may next be applied to $K(0, W_2, W_3, \ldots, W_n)$ where the dependence on the element W_2 is studied to yield:

$$K(0, W_2, W_3, \ldots, W_n) = K(0, 0, W_3, \ldots, W_n) \frac{F'_2(0, W_2, W_3, \ldots, W_n)}{F_2(0, W_2, W_3, \ldots, W_n)}. \tag{11.101}$$

The process is repeated until all the n controlled sources have been included. In this manner we find that the closed-loop transmittance of a linear circuit containing n controlled sources may be expressed in the form

$$K(W_1, W_2, \ldots, W_n) = K^0 \left(\frac{F'_1(W_1, W_2, \ldots, W_n)}{F_1(W_1, W_2, \ldots, W_n)} \right) \left(\frac{F'_2(0, W_2, \ldots, W_n)}{F_2(0, W_2, \ldots, W_n)} \right) \cdots \left(\frac{F'_n(0, 0, \ldots, 0, W_n)}{F_n(0, 0, \ldots, 0, W_n)} \right), \tag{11.102}$$

where, as remarked earlier, K^0 is the value to which the closed-loop transmittance reduces when all the W's are zero. The use of Eqs. (11.84), (11.90) and (11.102) consequently gives

$$K(W_1, W_2, \ldots, W_n) = K^0 \left(\frac{\Delta^{F'}}{\Delta^{F'}_{11}} \cdot \frac{\Delta^F_{11}}{\Delta^F} \right) \left(\frac{\Delta^{F'}_{11}}{\Delta^{F'}_{1122}} \cdot \frac{\Delta^F_{1122}}{\Delta^F_{11}} \right) \cdots \left(\frac{\Delta'_{1122\cdots(n-1)(n-1)}}{\Delta_{1122\cdots(n-1)(n-1)}} \right). \tag{11.103}$$

After cancellation of common factors between the various numerators and denominators on the right of Eq. (11.103), we find that ultimately it reduces to the same form as Eq. (11.99).

In general, the use of Eq. (11.99) requires less computational effort than Eq. (11.95) when evaluating the closed-loop transmittance K, so long as the direct transmittance K^0 does not vanish. This is particularly true if the return difference matrix is of order 3 or higher.

If, however, K^0 is zero, then Eq. (11.95) reduces to

$$K = \mathbf{QWF}^{-1}\mathbf{P}, \tag{11.104}$$

which may readily be manipulated into the following form, using Eq. (11.96) and the identity in the footnote on page 420,

$$K = \frac{\Delta^{\bar{F}} - \Delta^F}{\Delta^F}, \tag{11.105}$$

where $\Delta^{\bar{F}}$ is the determinant of the complementary return difference matrix.

b) Sensitivity

Suppose that the elements $W_1, W_2, \ldots, W_n$ change by small incremental amounts thereby causing the matrix $\mathbf{W}$ to change by $d\mathbf{W}$. This, in turn, will cause the return difference matrix $\mathbf{F}$ to change by $d\mathbf{F}$. Since the direct transmittance K^0 is independent of the elements $W_1, W_2, \ldots, W_n$, it follows from Eq. (11.95) that the corresponding change dK in the closed-loop transmittance is

$$dK = \mathbf{Q}(d\mathbf{W}\,.\,\mathbf{F}^{-1} + \mathbf{W}\,.\,d\mathbf{F}^{-1})\mathbf{P}. \tag{11.106}$$

But the differential of the inverse matrix $\mathbf{F}^{-1}$ is[14]

$$d\mathbf{F}^{-1} = -\mathbf{F}^{-1}\,.\,d\mathbf{F}\,.\,\mathbf{F}^{-1}. \tag{11.107}$$

Therefore,

$$dK = \mathbf{Q}(d\mathbf{W}\,.\,\mathbf{F}^{-1} - \mathbf{WF}^{-1}\,.\,d\mathbf{F}\,.\,\mathbf{F}^{-1})\mathbf{P}. \tag{11.108}$$

Next, using Eqs. (11.79), (11.83) and (11.108) we get

$$dK = \mathbf{QWF}^{-1}\left[\frac{dW}{W}\right]\mathbf{F}^{-1}\mathbf{P} \tag{11.109}$$

where we have

$$\left[\frac{dW}{W}\right] = \begin{bmatrix} \frac{dW_1}{W_1} & 0 & \cdots & 0 \\ 0 & \frac{dW_2}{W_2} & \cdots & 0 \\ \cdot & \cdot & & \cdot \\ \cdot & \cdot & & \cdot \\ \cdot & \cdot & & \cdot \\ 0 & 0 & \cdots & \frac{dW_n}{W_n} \end{bmatrix}. \tag{11.110}$$

Having evaluated K from Eq. (11.95) and dK from Eq. (11.109), we can then evaluate the required fractional change dK/K in the closed-loop transmittance.

Alternatively, we may generalize the formula of Eq. (11.43) as follows: if the element W_k is changed by the incremental amount dW_k and the remaining W's are held constant, then the corresponding fractional change in the closed-loop transmittance is

$$\frac{dK}{K} = \left(\frac{1}{F_k} - \frac{1}{F'_k}\right)\frac{dW_k}{W_k}, \tag{11.111}$$

where F_k is the return difference and F'_k is the null return difference for W_k with all the other W's having their normal operating values. Substitution of Eqs. (11.84) and (11.90) in Eq. (11.111) yields:

$$\frac{dK}{K} = \left(\frac{\Delta_{kk}^F}{\Delta^F} - \frac{\Delta_{kk}^{F'}}{\Delta^{F'}}\right)\frac{dW_k}{W_k}. \tag{11.112}$$

If, therefore, all the elements $W_1, W_2, \ldots, W_n$ were allowed to change simultaneously, we find that the total fractional change in the closed-loop transmittance is given by

$$\frac{dK}{K} = \sum_{k=1}^{n}\left(\frac{\Delta_{kk}^F}{\Delta^F} - \frac{\Delta_{kk}^{F'}}{\Delta^{F'}}\right)\frac{dW_k}{W_k}. \tag{11.113}$$

c) Driving-point Impedance

By analogy with Eq. (11.99) we may state that the driving-point impedance Z measured looking into any port of a linear feedback circuit is given by:

$$Z = Z^0\frac{\Delta^{F_{sc}}}{\Delta^{F_{oc}}}, \tag{11.114}$$

where Z^0 is the special value of the driving-point impedance which results when all the specified elements $W_1, W_2, \ldots, W_n$ are zero. $\Delta^{F_{sc}}$ is the determinant of the *short-circuit return difference matrix* $\mathbf{F}_{sc}$ for $\mathbf{W}$ and subject to the condition that the terminals where the impedance Z is measured are short-circuited. $\Delta^{F_{oc}}$ is the determinant of the *open-circuit return difference matrix* $\mathbf{F}_{oc}$ for $\mathbf{W}$ and subject to the condition that the terminals where Z is measured are open-circuited. Comparison with Eq. (11.60) shows that the relationship of Eq. (11.114) is a generalization of Blackman's formula.

Equation (11.114) provides the basis of a useful procedure for the indirect measurement of return difference. Suppose that in Fig. 11.33(a) it is required to measure the return difference $F_{y_{fe}}$ with reference to the y_{fe} of transistor Q. The network N represents the complete feedback amplifier exclusive of the transistor Q. Representing this transistor by its common-emitter y-parameters model, and regarding y_{fe} and y_{re} as the control parameters of interest, we see that when the

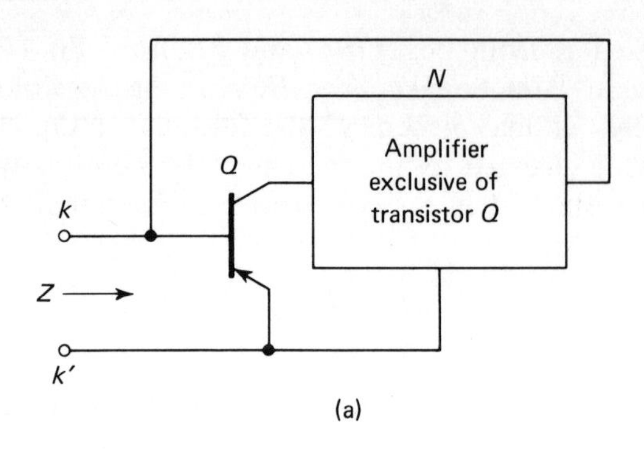

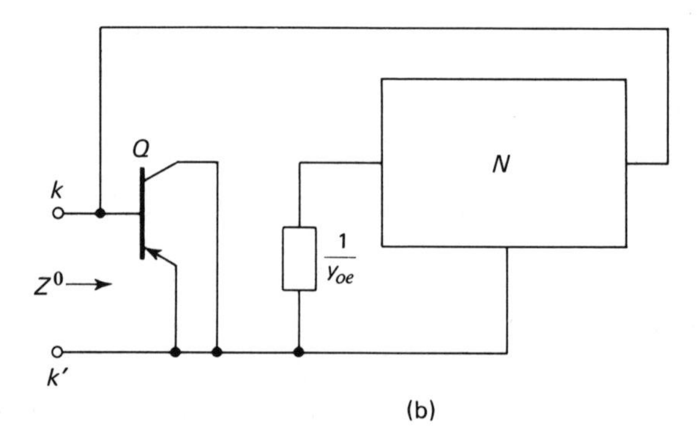

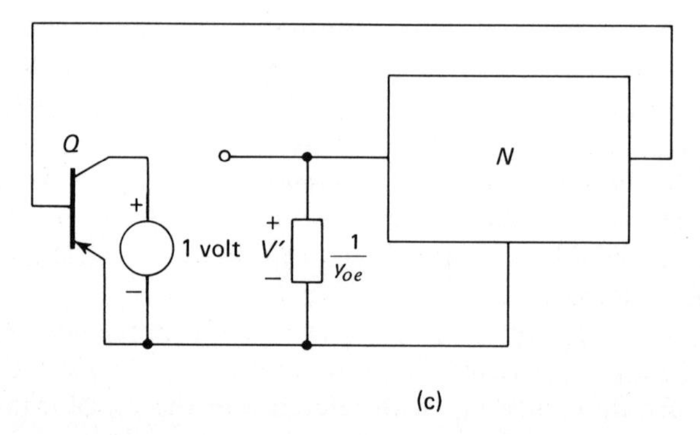

Fig. 11.33. Illustrating the use of impedance measurements for evaluating the return difference for y_{fe} of transistor Q; (a) Set-up for measuring impedance Z; (b) Set-up for measuring impedance Z^0; (c) Set-up for measuring return difference F_{22} for y_{re} with $y_{fe} = 0$.

port k–k' is open-circuited, the resulting return difference matrix is exactly the same as that found under normal operating conditions, so that

$$\Delta^{F_{oc}} = \Delta^F = \begin{vmatrix} F_{11} & F_{21} \\ F_{12} & F_{22} \end{vmatrix}. \tag{11.115}$$

The elements F_{11} and F_{12} are evaluated with $y_{re} = 0$, while F_{21} and F_{22} are evaluated with $y_{fe} = 0$.

When, however, the port k–k' is short-circuited, we see from Fig. 11.33(a) that the flow of signal around the feedback loop is interrupted; accordingly, we have $\Delta^{F_{sc}} = 1$. Therefore, Eq. (11.114) yields

$$Z = \frac{Z^0}{\Delta^F}, \tag{11.116}$$

where Z^0 is the value of the driving-point impedance Z at port k–k' which results when y_{fe} and y_{re} are both set equal to zero. The required return difference for y_{fe} of transister Q is, however, given by

$$F_{y_{fe}} = \frac{\Delta^F}{F_{22}}. \tag{11.117}$$

Hence, from Eqs. (11.116) and (11.117),

$$F_{y_{fe}} = \frac{Z^0}{ZF_{22}}. \tag{11.118}$$

To measure Z^0 the feedback loop is opened at the collector terminal of transistor Q, and the broken loop is terminated simply with the y_{oe} of transistor Q, as in Fig. 11.33(b). We thus see that for the measurement of impedance Z, the feedback loop is closed, while, for the measurement of Z^0, the feedback loop is open. To measure F_{22}, which is the return difference for y_{re} when y_{fe} is zero, we can use the arrangement shown in Fig. 11.33(c), where the feedback loop has again been broken at the collector of transistor Q and the broken loop is terminated with y_{oe}. When a unit voltage is applied across the output port of transistor Q, we find that F_{22} is equal to unity minus the voltage V' developed across y_{oe}.

In practice, we ordinarily find that F_{22} is indistinguishable from unity. This is so because the unit signal applied to the output port of transistor Q in Fig. 11.33(c) would undergo such heavy attenuation on being transmitted in the reverse direction through the stage Q and network N, that the returned voltage V' would be practically zero. We are therefore perfectly justified in reducing Eq. (11.118) to the simpler form,

$$F_{y_{fe}} = \frac{Z^0}{Z}. \tag{11.119}$$

Although the impedance Z^0 is independent of the internal feedback associated with transistor Q, the effect of this internal feedback is largely accounted for in the evaluation of the impedance Z.

Example 11.8. *Emitter-to-emitter feedback.* To exemplify the generalized relationships of Eqs. (11.99), (11.113) and (11.114), we will evaluate the closed-loop transmittance of the *emitter-to-emitter feedback amplifier* shown in Fig. 11.34, its

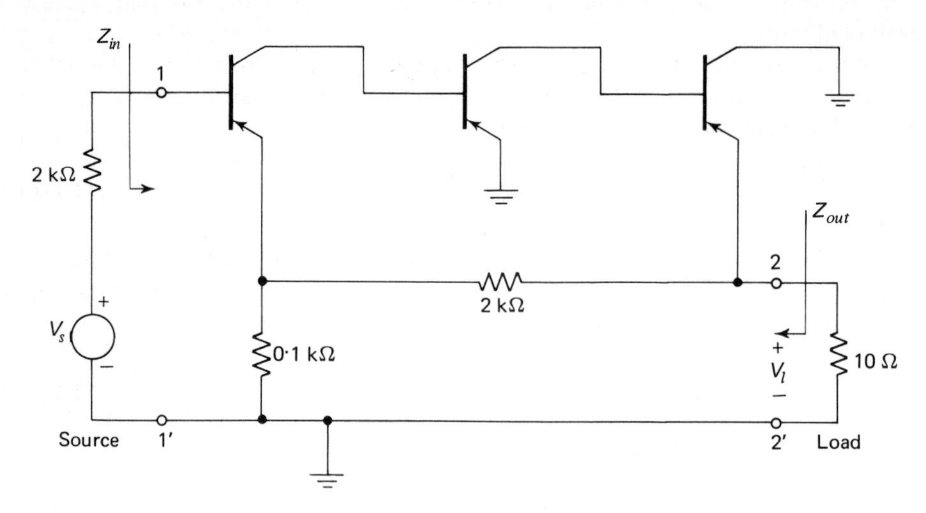

Fig. 11.34. Emitter-to-emitter feedback amplifier.

sensitivity to variations in h_{fe}'s of the transistors, and its input and output impedances. It is assumed that the behaviour of each transistor may be represented by the simplified hybrid model of Fig. 11.35, ignoring the effects of h_{re} and h_{oe}.

Choosing h_{fe}'s of the transistors as the control parameters, we have the base currents as the control signals; so that

$$\mathbf{W} = \begin{bmatrix} h_{fe1} & 0 & 0 \\ 0 & h_{fe2} & 0 \\ 0 & 0 & h_{fe3} \end{bmatrix}$$

$$\mathbf{x} = \begin{bmatrix} I_{b1} \\ I_{b2} \\ I_{b3} \end{bmatrix}$$

$$\mathbf{x}' = \begin{bmatrix} h_{fe1}I_{b1} \\ h_{fe2}I_{b2} \\ h_{fe3}I_{b3} \end{bmatrix}.$$

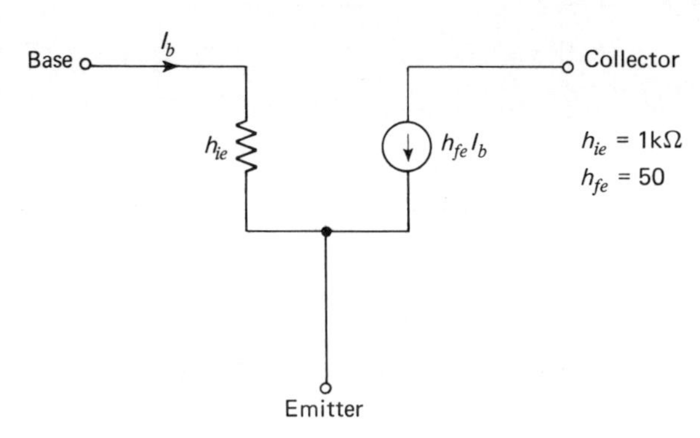

Fig. 11.35. Simplified hybrid model of transistor.

To apply Eq. (11.99) for evaluating the overall voltage gain $K = V_l/V_s$, we require the direct transmittance K^0 and the determinants of the return difference and null return difference matrices. To calculate the direct transmittance K^0, we inactivate all three transistors by reducing their h_{fe}'s to zero, as in Fig. 11.36. Then, with $V_s = 1$ volt, we directly find from this diagram that:

$$K^0 = 1{\cdot}53 \times 10^{-4}.$$

Next, the elements of the return difference matrix may be determined under the condition $V_s = 0$ in three steps, as depicted in Fig. 11.37. In each part of the diagram, two of the controlled sources $h_{fe1}I_{b1}$, $h_{fe2}I_{b2}$ and $h_{fe3}I_{b3}$ are inactivated, and the remaining one is replaced with an independent current source of a strength equal to the associated h_{fe}. The calculated values of the returned base currents pertaining to the three parts of Fig. 11.37 are contained in Table 11.1.

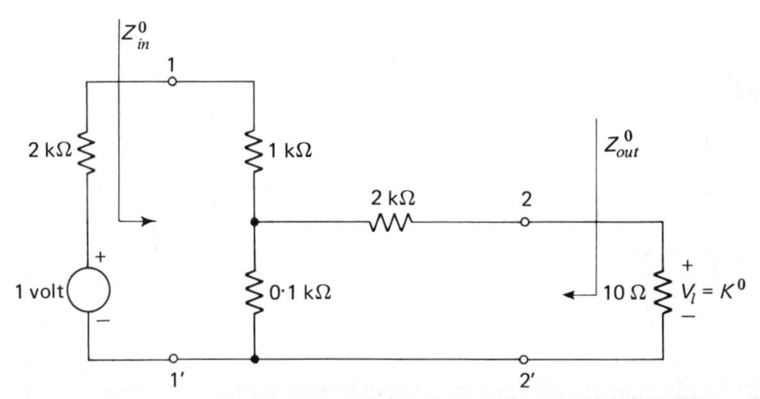

Fig. 11.36. Network for evaluating the direct transmittance K^0.

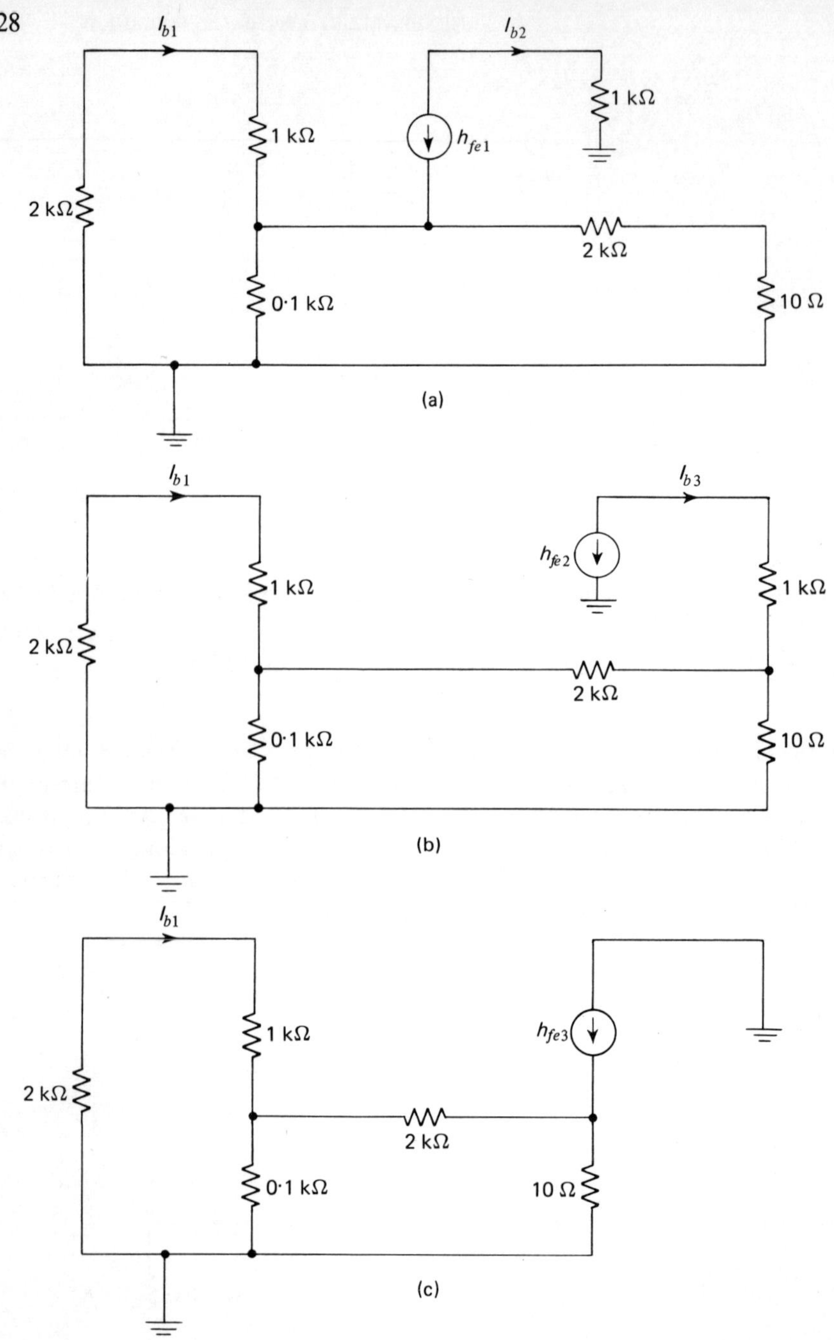

Fig. 11.37. Evaluating the elements of return difference matrix: (a) $h_{fe1} \neq 0$, $h_{fe2} = h_{fe3} = 0$; (b) $h_{fe2} \neq 0$, $h_{fe1} = h_{fe3} = 0$; (c) $h_{fe3} \neq 0$, $h_{fe1} = h_{fe2} = 0$.

Table 11.1. Calculation of Return Difference Matrix, With $V_s = 0$

	$h_{fe1} \neq 0$, $h_{fe2} = h_{fe3} = 0$	$h_{fe2} \neq 0$, $h_{fe1} = h_{fe3} = 0$	$h_{fe3} \neq 0$, $h_{fe1} = h_{fe2} = 0$
I_{b1}	$-1{\cdot}54$	$0{\cdot}0077$	$-0{\cdot}0077$
I_{b2}	-50	0	0
I_{b3}	0	-50	0

From Table 11.1 we immediately obtain the return difference matrix:

$$\mathbf{F} = \begin{bmatrix} 2{\cdot}54 & -0{\cdot}0077 & 0{\cdot}0077 \\ 50 & 1 & 0 \\ 0 & 50 & 1 \end{bmatrix},$$

the determinant of which is equal to

$$\Delta^F = 22{\cdot}1.$$

Similarly, the evaluation of the null return difference matrix may be carried out in three steps, as illustrated in Fig. 11.38. In this case, however, the source voltage V_s is adjusted to make the output voltage V_l equal to zero. Under this condition we obtain the returned base currents contained in Table 11.2.

Table 11.2. Calculation of Null Return Difference Matrix, With $V_l = 0$

	$h_{fe1} \neq 0$, $h_{fe2} = h_{fe3} = 0$	$h_{fe2} \neq 0$, $h_{fe1} = h_{fe3} = 0$	$h_{fe3} \neq 0$, $h_{fe1} = h_{fe2} = 0$
I_{b1}	-50	1050	-1050
I_{b2}	-50	0	0
I_{b3}	0	-50	0

From Table 11.2 we then find that the null return difference matrix is:

$$\mathbf{F}' = \begin{bmatrix} 51 & -1050 & 1050 \\ 50 & 1 & 0 \\ 0 & 50 & 1 \end{bmatrix},$$

the determinant of which is equal to

$$\Delta^{F'} = 2{\cdot}68 \times 10^6.$$

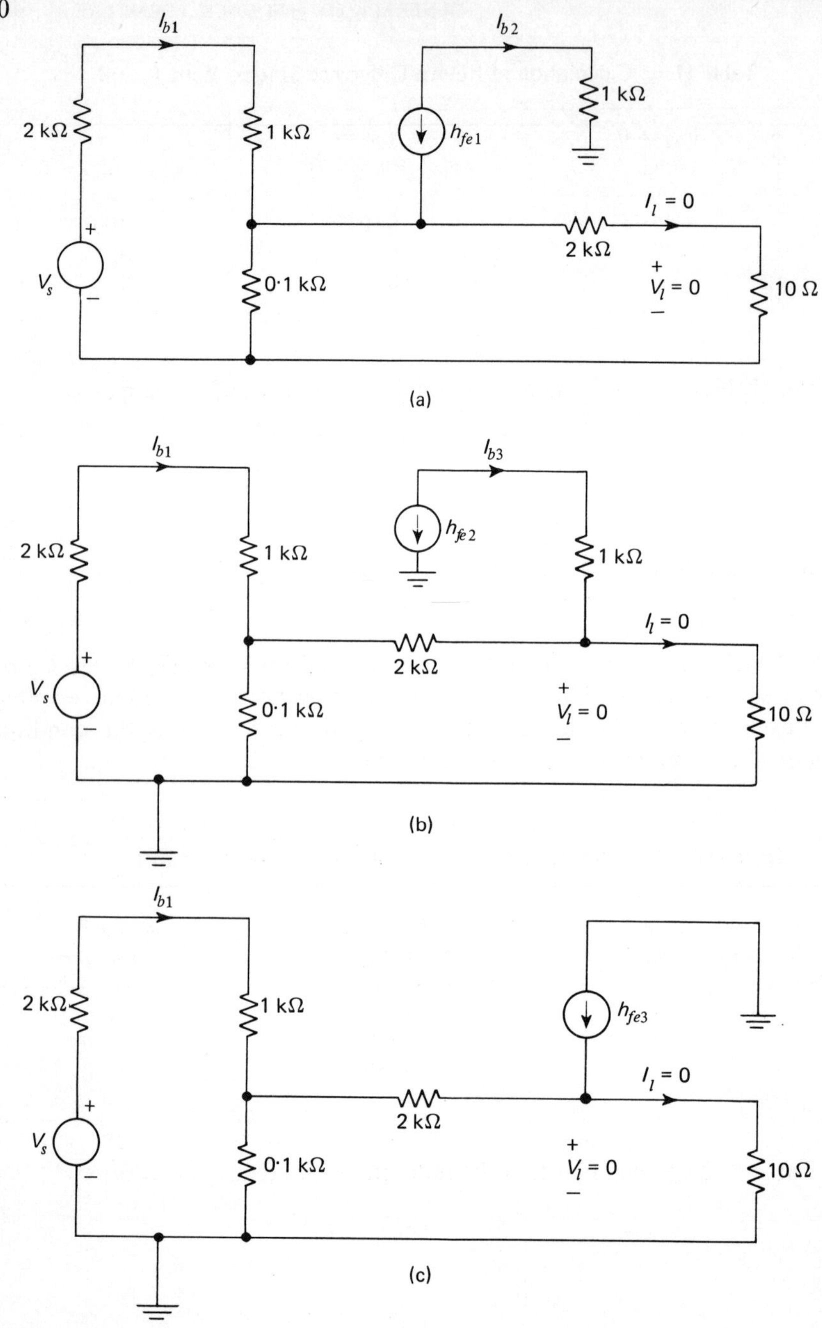

Fig. 11.38. Evaluating the elements of null return difference matrix: (a) $h_{fe1} \neq 0$, $h_{fe2} = h_{fe3} = 0$; (b) $h_{fe2} \neq 0$, $h_{fe1} = h_{fe3} = 0$; (c) $h_{fe3} \neq 0$, $h_{fe1} = h_{fe2} = 0$.

Equation (11.99) consequently gives the overall voltage gain of the amplifier as:

$$K = 18{\cdot}5.$$

It should be noted that the null return difference matrix could have also been evaluated using Eq. (11.89); in general, however, the procedure described above is simpler so long as the direct transmittance K^0 is non-zero.

The effect of variations in the h_{fe}'s of the transistors may be determined using Eq. (11.113); we have

$$\Delta_{11}^F = \begin{vmatrix} 1 & 0 \\ 50 & 1 \end{vmatrix} = 1$$

$$\Delta_{22}^F = \begin{vmatrix} 2{\cdot}54 & 0{\cdot}0077 \\ 0 & 1 \end{vmatrix} = 2{\cdot}54$$

$$\Delta_{33}^F = \begin{vmatrix} 2{\cdot}54 & -0{\cdot}0077 \\ 50 & 1 \end{vmatrix} = 2{\cdot}92$$

and

$$\Delta_{11}^{F'} = \begin{vmatrix} 1 & 0 \\ 50 & 1 \end{vmatrix} = 1$$

$$\Delta_{22}^{F'} = \begin{vmatrix} 51 & 1050 \\ 0 & 1 \end{vmatrix} = 51$$

$$\Delta_{33}^{F'} = \begin{vmatrix} 51 & -1050 \\ 50 & 1 \end{vmatrix} = 5{\cdot}25 \times 10^4.$$

Therefore, from Eq. (11.113) it follows that the total fractional change in K due to simultaneous variations in h_{fe1}, h_{fe2} and h_{fe3} is given by

$$\frac{dK}{K} = 0{\cdot}045\frac{dh_{fe1}}{h_{fe1}} + 0{\cdot}115\frac{dh_{fe2}}{h_{fe2}} + 0{\cdot}112\frac{dh_{fe3}}{h_{fe3}}.$$

To determine the input impedance Z_{in} using Eq. (11.114), we see from Fig. 11.36 that when all three transistors are dead,

$$Z_{in}^0 = 1{\cdot}095\ \mathrm{k}\Omega.$$

Also, from Fig. 11.37 we readily find that the short-circuit and open-circuit return difference matrices with respect to the input port 1–1′ are given as follows, respectively:

$$\mathbf{F}_{1,sc} = \begin{bmatrix} 5{\cdot}35 & -0{\cdot}0216 & 0{\cdot}0216 \\ 50 & 1 & 0 \\ 0 & 50 & 1 \end{bmatrix}$$

$$\mathbf{F}_{1,oc} = \begin{bmatrix} 1 & 0 & 0 \\ 50 & 1 & 0 \\ 0 & 50 & 1 \end{bmatrix}.$$

The required determinants of these two matrices are therefore,

$$\Delta^{F_{1,sc}} = 60{\cdot}4$$

$$\Delta^{F_{1,oc}} = 1.$$

Equation (11.114) thus gives the input impedance of the amplifier under normal operating conditions to be:

$$Z_{in} = 66\ \text{k}\Omega.$$

Evaluating the output impedance measured at port 2–2′ in a similar manner, we obtain

$$Z_{out}^{\mathbf{0}} = 2{\cdot}1\ \text{k}\Omega$$

$$\mathbf{F}_{2,sc} = \begin{bmatrix} 2{\cdot}54 & 0 & 0 \\ 50 & 1 & 0 \\ 0 & 50 & 1 \end{bmatrix}$$

$$\mathbf{F}_{2,oc} = \begin{bmatrix} 2{\cdot}61 & -1{\cdot}61 & 1{\cdot}61 \\ 50 & 1 & 0 \\ 0 & 50 & 1 \end{bmatrix}$$

$$\Delta^{F_{2,sc}} = 2{\cdot}54$$

$$\Delta^{F_{2,oc}} = 4{\cdot}1 \times 10^3$$

$$Z_{out} = 1{\cdot}29\ \Omega.$$

Concluding Remarks

From the above example we see that the analysis of a multiple-loop feedback circuit is greatly simplified by using the concepts of return difference and null return difference matrices. The particular virtue of this method of analysis is that it enables a complicated circuit to be subdivided into a number of much simpler circuits, in which the externally applied excitation and the controlled sources within the network are considered one at a time, with the controlled sources being treated as if they were independent. Consequently, the required evaluation can, in many cases, be carried out almost by inspection. Also, throughout the analysis a physical view of the problem under study is maintained.

Furthermore, it should be emphasized that although Eqs. (11.78), upon which the method of analysis is based, exhibit feedback in the sense that the outputs of the controlled sources influence the value of the control signals, as illustrated by the signal-flow graph of Fig. 11.32, it is not necessary that feedback exist in the network in the usual physical sense. For example, we may have a network consisting entirely of one-port elements, with the controlled sources representing a certain subset of these elements. In such a situation, feedback arises in the topological characterization of the network merely as a result of the way in which the equilibrium equations are formulated.

REFERENCES

1. H. W. BODE, *Network Analysis and Feedback Amplifier Design*. Van Nostrand, 1945.
2. S. J. MASON, "Feedback Theory—Some Properties of Signal-flow Graphs," *Proc. I.R.E.* **41,** 1114 (1953).
3. J. G. TRUXAL, *Automatic Feedback Control System Synthesis*. McGraw-Hill, 1955.
4. A. G. STUART and D. G. LAMPARD, "Bridge Networks Incorporating Active Elements and Applications to Network Synthesis," *Trans. I.E.E.E.* **CT-10,** 357 (1963).
5. L. A. ZADEH, "Multipole Analysis of Active Networks," *Trans. I.R.E.* **CT-4,** 97 (1957).
6. I. W. SANDBERG, "On the Theory of Linear Multi-loop Feedback Systems," *B.S.T.J.* **42,** 355 (1963).
7. R. F. HOSKINS, "Definition of Loop Gain and Return Difference in Transistor Feedback Amplifiers," *Proc. I.E.E.* **112,** 1995 (1965).
8. F. H. BLECHER, "Design Principles for Single-loop Transistor Feedback Amplifiers," *Trans. I.R.E.* **CT-4,** 145 (1957).
9. T. S. GRAY, *Applied Electronics*. Wiley, 1954.
10. F. H. BLECHER, "Transistor Multiple-loop Feedback Amplifiers," *Proc. National Electronics Conference* **13,** 19 (1957).
11. R. B. BLACKMAN, "Effect of Feedback on Impedance," *B.S.T.J.* **22,** 269 (1943).
12. L. TASNY-TSCHIASSNY, "The Return Difference Matrix in Linear Networks," *Proc. I.E.E.* **100,** part IV, 39 (1953).
13. J. H. MULLIGAN, "Signal Transmission in Non-reciprocal Systems," *Symposium Proceedings*, Polytechnic Institute of Brooklyn, **10,** 125 (1960).
14. R. A. FRAZER, W. J. DUNCAN and A. R. COLLAR, *Elementary Matrices*. C.U.P., 1938.
15. I. W. SANDBERG, "Synthesis of Driving-point Impedance with Active Networks," *B.S.T.J.* **39,** 947 (1960).
16. D. M. TAUB, "Differential Amplifier Based on Feedback Pair," *Proc. I.E.E.* **112,** 709 (1965).

PROBLEMS

11.1 Consider the emitter-coupled pair shown in Fig. P11.1. By replacing the common resistor R with an equivalent current-controlled voltage source, determine:

a) The return difference for the resistor R

b) The voltage gain of the amplifier

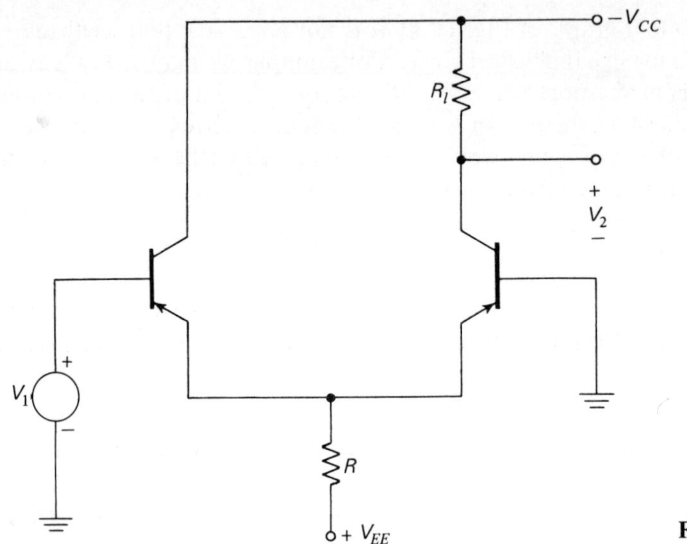

Fig. P11.1.

11.2 For the cathode follower of Fig. P11.2, evaluate:

a) The return difference and null return difference with reference to the amplification factor μ,

b) The voltage gain V_2/V_1,

c) The input and output impedances.

The tube parameters are as follows: $r_p = 8\text{ k}\Omega$ and $\mu = 30$.

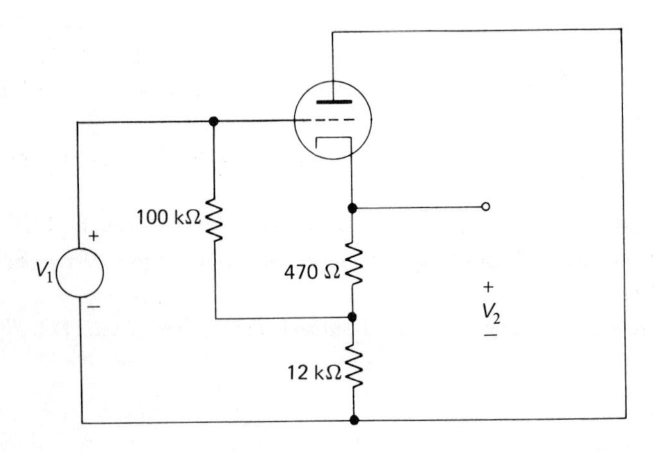

Fig. P11.2.

11.3 Figure P11.3 shows the signal-flow graph for a two-parallel path feedback amplifier. Determine the condition which the cross-coupling transmittance β_{12} must satisfy if the system is to have zero sensitivity with respect to both μ_1 and μ_2.

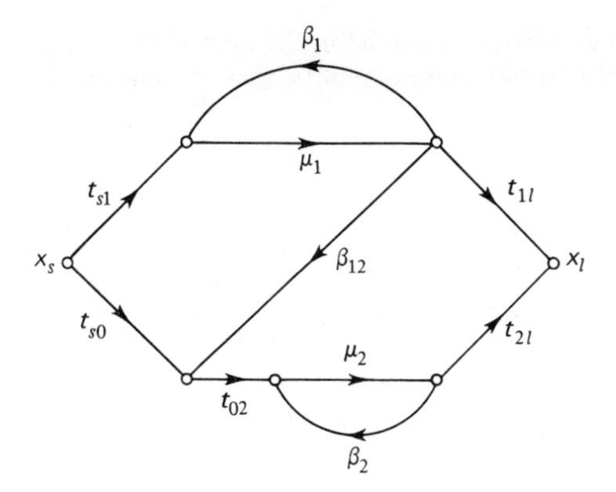

Fig. P11.3.

11.4 In the compound-feedback stage of Fig. P11.4 the series and shunt forms of local feedback are combined together. Assuming that the effects of transistor parameters h_{re} and h_{oe} are negligible, determine the necessary condition for the input impedance to be independent of h_{fe}. What are the corresponding output impedance and closed-loop gain of the amplifier?

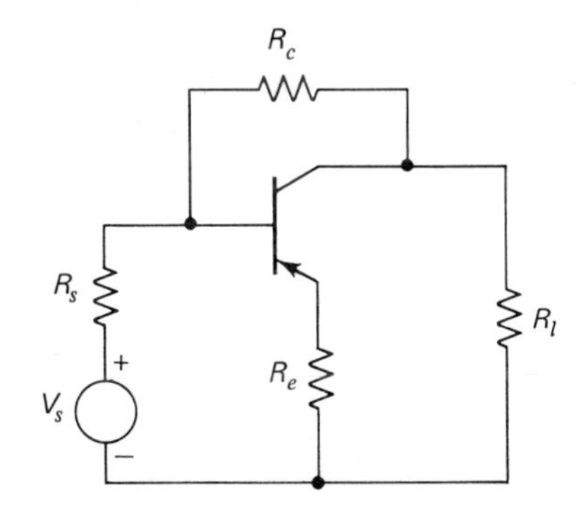

Fig. P11.4.

11.5 Using Blackman's formula, determine the impedance measured at the port 1–1′ of the network in Fig. P11.5.

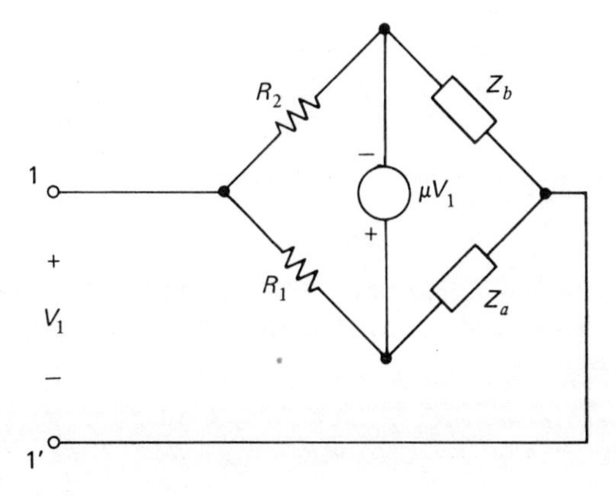

Fig. P11.5.

11.6 Determine the impedance measured at the port 1–1′ of the active network shown in Fig. P11.6 which is used for the synthesis of driving-point impedances.[15]

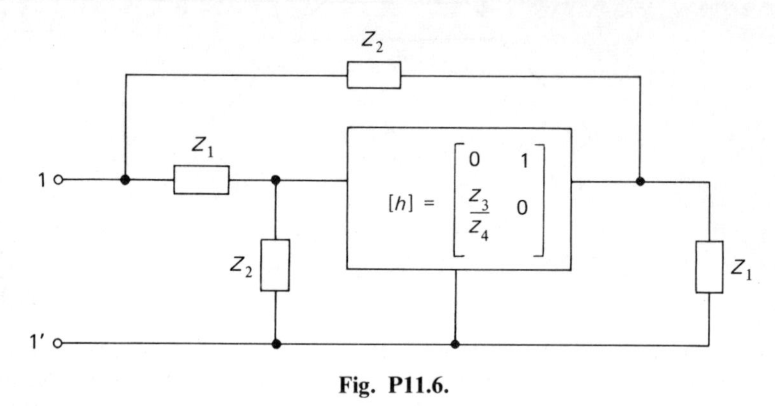

Fig. P11.6.

11.7 The two signal-flow graphs shown in Fig. P11.7 represent alternatives for a feedback amplifier involving three identical stages, each of gain μ. Both configurations employ negative feedback to make the overall transmittance of the amplifier less dependent on changes in μ. Determine which of the two configurations has a lower sensitivity with respect to μ, assuming that they provide equal overall transmittances.

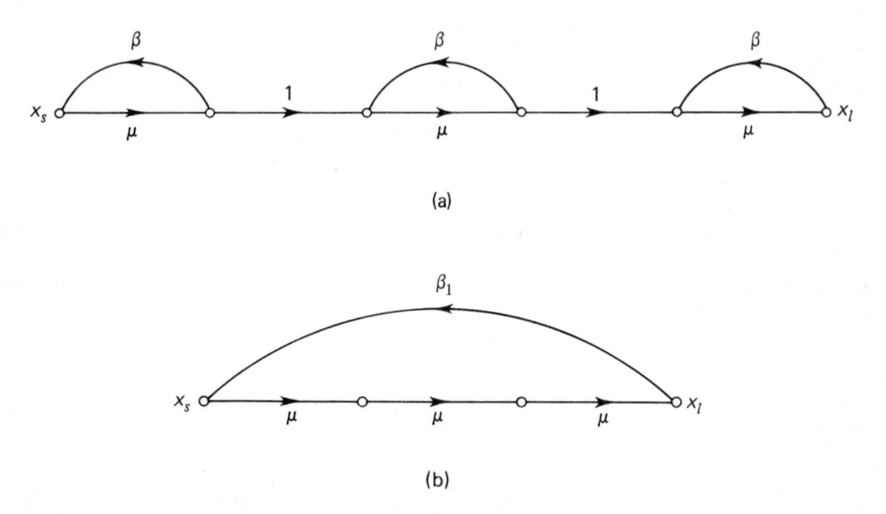

Fig. P11.7.

11.8 Assuming that the transformer of the split-load stage of Fig. P11.8 is ideal, and that the transistor parameters h_{re} and h_{oe} have negligible effects, show that the return difference with reference to h_{fe} is:

$$F = 1 + \frac{n_1 h_{fe} R_l(n_1 + n_2)}{n_1^2 R_l + h_{ie} + R_s}.$$

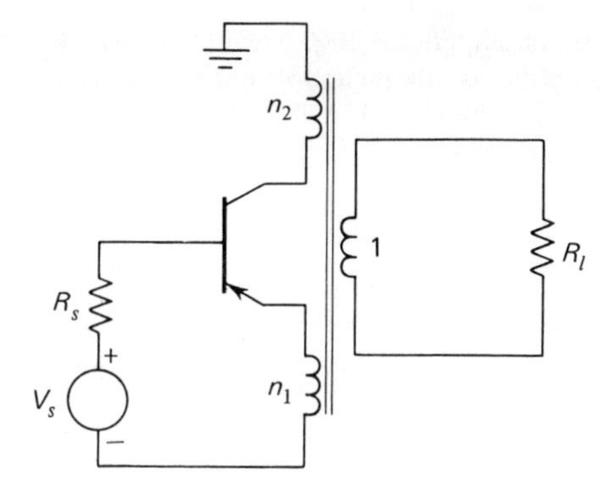

Fig. P11.8.

11.9 Figure P11.9 shows the Schmitt trigger circuit. Assuming that the circuit is operated in its linear region, calculate the critical value of the potential divider ratio m that will cause the closed-loop transmittance V_2/V_1 to become infinite. The two tubes are identical, each having $r_p = 10\,\text{k}\Omega$ and $\mu = 20$.

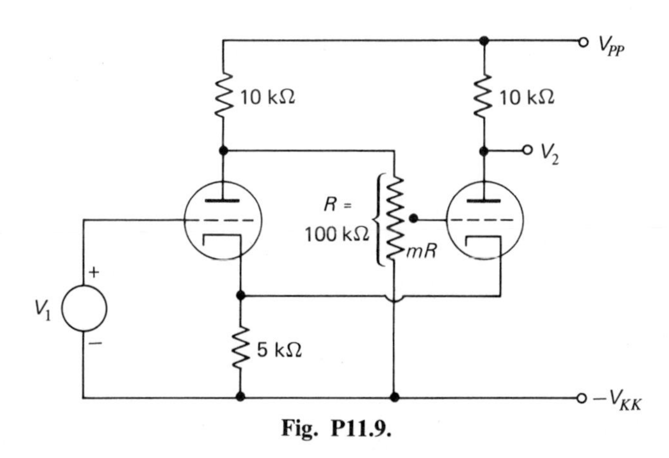

Fig. P11.9.

11.10 A common-emitter stage has a load resistance of 10 kΩ and is driven from a voltage source of resistance 1 kΩ. The transistor has the following set of parameters:

$$h_{ie} = 1\,\text{k}\Omega \qquad h_{re} = 10^{-4}$$

$$h_{fe} = 50 \qquad h_{oe} = 25\,\mu\text{℧}.$$

Determine the return difference of the stage with reference to:

a) the h_{21}-parameter,
b) the y_{21}-parameter,
c) the z_{21}-parameter, and
d) the g_{21}-parameter.

Comment on the results.

11.11 The voltage gain of an amplifier is 1000 when the output port is open-circuited. The input impedance of the amplifier is infinite and its output impedance is 5 kΩ. Negative feedback is applied by the network shown in Fig. P11.11. Evaluate the voltage gain and output impedance of this feedback amplifier.

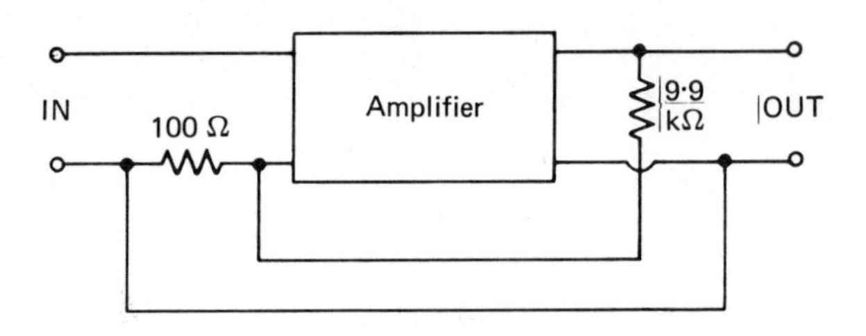

Fig. P11.11.

11.12 Figure P11.12 shows the circuit diagram of a differential amplifier.[16] Determine the voltage V_o developed across the load.

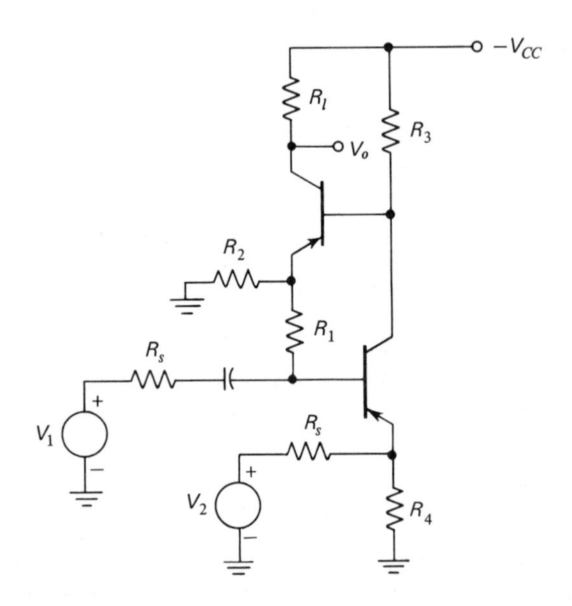

Fig. P11.12.

11.13 Derive the results of Eqs. (4.146)–(4.148) for a common-emitter stage with local series feedback.

11.14 Derive the results of Eqs. (4.153)–(4.155) for a common-emitter stage with local shunt feedback.

11.15 Continuing the analysis of the emitter-to-emitter feedback amplifier considered in Example 11.8, replace the h_{ie}'s of the transistors with equivalent current-controlled voltage sources, and hence evaluate the sensitivity of the closed-loop transmittance to variations in the h_{ie}'s.

CHAPTER 12

THE STABILITY PROBLEM IN FEEDBACK AMPLIFIERS

12.1 INTRODUCTION

In the previous chapter we found that a large amount of negative feedback is required to make the overall gain of a linear amplifier less sensitive to parameter variations, reduce the effects of noise and non-linear distortion, etc. These improvements in performance are, however, achieved at the cost of reduced overall gain. In addition, we are faced with a stability problem in that at some frequency the amplifier may satisfy the conditions that permit it to operate as a phase-shift oscillator. This tendency toward instability becomes progressively more pronounced and more difficult to control as the amount of applied negative feedback in the useful frequency band is increased and as we attempt to extend the feedback over more and more amplifier stages. In the design of a feedback amplifier our task is therefore not only to meet the various performance requirements imposed but also to ensure that the amplifier is stable and remains stable under all possible operating conditions.

The stability of a linear feedback amplifier, like any other electrical system, is completely determined by the location of its natural frequencies in the s-plane, the natural frequencies being defined as roots of the characteristic equation $\Delta(s) = 0$, where $\Delta(s)$ is the circuit determinant. The feedback amplifier is stable if all the roots of $\Delta(s)$ are confined to the left half of the s-plane. Although Routh's criterion would readily enable us to determine the number of roots of $\Delta(s)$ in the right half of the s-plane, if any, its application requires that $\Delta(s)$ be calculated as a function of s. As such, therefore, Routh's criterion is of little use in most practical situations. Furthermore, it does not tell us the degree of stability when the feedback amplifier is stable, nor does it provide us with any information about how to stabilize the amplifier if it is found to be unstable. These limitations are overcome by applying the *Nyquist stability criterion* to the return difference $F(s)$ for some specified element, which is the basis of the frequency-response method of studying the stability problem and its design limitations.

Alternatively, we may consider the stability of a feedback amplifier in terms of the desired transient response and natural frequency locations. It is with this latter approach that we begin our study of the stability problem.

12.2 CONTROL OF NATURAL FREQUENCIES

To evaluate the manner in which the natural frequencies of a feedback amplifier migrate inside the s-plane as the amount of applied feedback is varied, consider the return difference for some specified element W, as defined by:

$$F(s) = 1 + T(s) = \frac{\Delta(s)}{\Delta^0(s)}, \tag{12.1}$$

where $\Delta^0(s)$ is the circuit determinant which results when the element W vanishes. Equation (12.1) reveals that the zeros and poles of $F(s)$ are the same as the zeros of $\Delta(s)$ and those of $\Delta^0(s)$, respectively, unless one or more zeros of $\Delta(s)$ are fortuitously cancelled by zeros of $\Delta^0(s)$. There are, however, good reasons for assuming that this is unlikely to happen in practice. The zeros of $F(s)$ are therefore natural frequencies of the feedback amplifier. Equivalently, the values of s for which the return ratio $T(s)$ is equal to -1 are the natural frequencies of the feedback amplifier.

Throughout this section we shall assume that the mid-band (in our case, the zero-frequency) return ratio, $T(0)$, is positive; that is, at mid-band frequencies the applied feedback is negative. To illustrate the effect of varying $T(0)$ upon the amplifier natural frequencies, consider first the simple case of a feedback amplifier for which the return ratio $T(s)$ has a single pole at $s = \sigma_1$ where σ_1 is a negative real number, so that the pole is located on the negative real axis of the s-plane. Thus

$$T(s) = \frac{T(0)}{1 - s/\sigma_1}. \tag{12.2}$$

The corresponding return difference will therefore have a zero at $s = \sigma_1(1 + T(0))$. As $T(0)$ is increased, this zero moves along the negative real axis of the s-plane further away from the origin, as shown in Fig. 12.1. This means that a single-loop

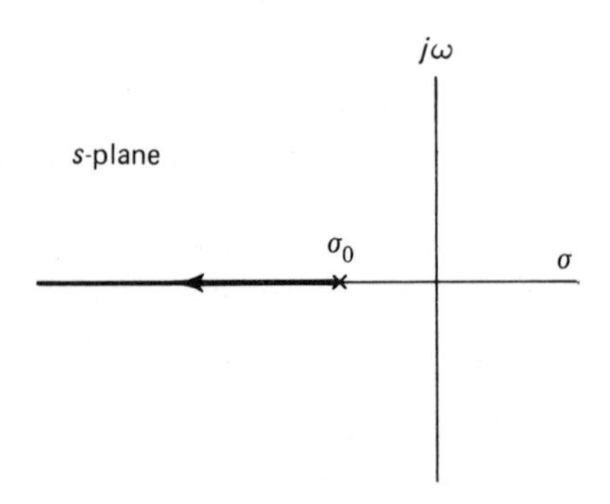

Fig. 12.1. Root-locus diagram for return ratio $T(s)$ with single pole.

feedback amplifier for which the return ratio $T(s)$ has a single pole on the negative real axis, is inherently stable because it is impossible to shift the natural frequency of the amplifier into the right half plane for any positive value of $T(0)$.

Consider next the case of a return ratio $T(s)$ with poles at $s = \sigma_1$ and $s = \sigma_2$, both being located on the negative real axis of the s-plane; that is, σ_1 and σ_2 are negative real numbers. Thus:

$$T(s) = \frac{T(0)}{[1 - (s/\sigma_1)][1 - (s/\sigma_2)]}. \tag{12.3}$$

The natural frequencies of the feedback amplifier are therefore given by the roots of the equation

$$s^2 - (\sigma_1 + \sigma_2)s + \sigma_1\sigma_2(1 + T(0)) = 0, \tag{12.4}$$

which is a quadratic in s, and is obtained by setting $1 + T(s) = 0$. From Eq. (12.4) we see that the natural frequencies s_a and s_b, the zeros of $1 + T(s)$, lie on the negative real axis of the s-plane for $T(0)$ less than $(\sigma_2 - \sigma_1)^2/4\sigma_1\sigma_2$. When $T(0)$ is greater than this value, the amplifier has a complex-conjugate pair of natural frequencies. Figure 12.2 shows the locus traced out by the natural frequencies s_a and s_b as $T(0)$ is increased from zero. Here again we find that the natural frequencies of the feedback amplifier always lie in the left half plane, and as such there is no stability problem.

When, however, the return ratio $T(s)$ has three or more poles, then there is a distinct possibility of the feedback amplifier becoming unstable for large values of $T(0)$. In such cases, the effect of varying $T(0)$ upon the natural frequencies of the system is most conveniently studied from a knowledge of the poles and zeros of $T(s)$, using a graphical technique widely known as the *root-locus method* due to Evans.[1]

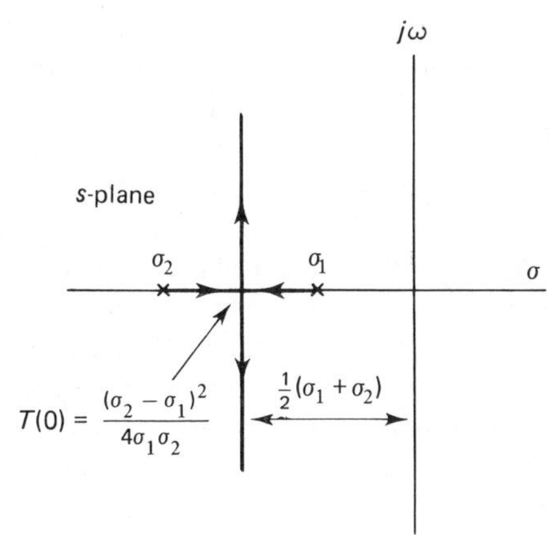

Fig. 12.2. Root-locus diagram for return ratio $T(s)$ with two real poles.

Root-locus Method

The characteristic equation $1 + T(s) = 0$, or equivalently $T(s) = -1$ is satisfied for any value of s for which the magnitude and phase angle of the return ratio $T(s)$ meet the following conditions, respectively,

$$|T(s)| = 1 \tag{12.5}$$

$$\text{ang}\, T(s) = (1 + 2k)\pi, \tag{12.6}$$

where k is any integer, including zero. When the poles and zeros of $T(s)$ are known, $T(s)$ may be written in the form:

$$T(s) = T(0)\frac{\prod_{i=1}^{m} [1 - (s/z_i)]}{\prod_{i=1}^{n} [1 - (s/p_i)]}, \tag{12.7}$$

where p_i and z_i are respectively the poles and zeros of $T(s)$. Therefore, in terms of these poles and zeros, and $T(0)$, we find that the conditions of Eqs. (12.5) and (12.6) are respectively equivalent to:

$$\frac{\prod_{i=1}^{m} |1 - (s/z_i)|}{\prod_{i=1}^{n} |1 - (s/p_i)|} = \frac{1}{T(0)}, \tag{12.8}$$

$$\sum_{i=1}^{m} \text{ang}\left(1 - \frac{s}{z_i}\right) - \sum_{i=1}^{n} \text{ang}\left(1 - \frac{s}{p_i}\right) = (1 + 2k)\pi. \tag{12.9}$$

From Eq. (12.8) we see that as $T(0)$ approaches zero, s approaches p_i, while as $T(0)$ approaches infinity, s approaches z_i. In other words, for $T(0)$ increasing from zero to infinity, the root loci start at the poles of $T(s)$ and terminate upon the zeros of $T(s)$. When $T(s)$ has more finite poles than finite zeros, some of the root loci will terminate at infinity. As for Eq. (12.9), since it is independent of $T(0)$, it applies over all points of the root loci. Thus we may define a root locus as being the locus of those points in the s-plane for which the phase angle of the return ratio $T(s)$ is equal to $(1 + 2k)\pi$, where k is any integer.

In a physical feedback system the order of the denominator polynomial of $T(s)$ is ordinarily greater than that of its numerator polynomial. It follows therefore that the number of root loci is equal to the number of poles of $T(s)$. Also, the root-locus diagram is symmetrical with respect to the real axis of the s-plane, because the poles and zeros of $T(s)$ are either real or else occur in complex-conjugate pairs.

Another interesting property of a root-locus diagram is that for very large values of s, the root loci become asymptotic to straight lines making angles of

$-(1 + 2k)\pi/(n - m)$ radians with the real axis of the s-plane, where n and m are the numbers of poles and zeros of $T(s)$, respectively, and $k = 0, 1, 2, \ldots,$ up to $k = n - m$. Furthermore, the asymptotes meet at a point $s = \sigma'$ located along the real axis and determined by[2]

$$\sigma' = \frac{\sum_{i=1}^{n} p_i - \sum_{i=1}^{m} z_i}{n - m}. \tag{12.10}$$

Example 12.1. *Three-stage feedback amplifier.* To exemplify the root-locus method, consider a feedback amplifier consisting of three stages, with a return ratio

$$T(s) = \frac{T(0)}{(1 + 10s)(1 + s)(1 + 0{\cdot}5s)}.$$

We observe that $T(s)$ has poles at $s = -0{\cdot}1$, $s = -1$, and $s = -2$, and that there are no finite zeros. Therefore, there are three loci moving continuously from the points $s = -0{\cdot}1$, $s = -1$ and $s = -2$ to infinity.

For very large values of s, the asymptotes of the loci make angles of $-60°$, $-180°$, and $-300°$ with the real axis of the s-plane; and from Eq. (12.10) we see that they intersect at $s = -1{\cdot}03$. To start with, all three roots of $1 + T(s) = 0$ are real; then beyond a certain value of $T(0)$, two of the roots become complex. This means that there must be a point at which the equation $1 + T(s) = 0$ has two equal real roots. Let these two roots be denoted by $-a$, and the remaining root by $-b$. Then, we must have

$$(s + a)^2(s + b) = \tfrac{1}{5}[(1 + 10s)(1 + s)(1 + 0{\cdot}5s) + T(0)].$$

Equating coefficients of like powers of s, we get

$$a^2b = \tfrac{1}{5}(1 + T(0))$$

$$2ab + a^2 = 2{\cdot}3$$

$$2a + b = 3{\cdot}1.$$

There are two possible solutions for a, b and $T(0)$; however, since the point of departure from the real axis must lie between the points $s = -0{\cdot}1$ and $s = -1$, we find that only the following solution is acceptable:

$$a = 0{\cdot}483$$

$$b = 2{\cdot}13$$

$$T(0) = 1{\cdot}5.$$

Also, from Routh's criterion we find that when $T(0) = 34{\cdot}7$, two of the root loci intersect the imaginary axis at the points $s = \pm j\,1{\cdot}52$. When $T(0) > 34{\cdot}7$, these two loci enter the right half plane and the amplifier becomes unstable. The above results thus enable us to plot the root locus diagram shown in Fig. 12.3.

In general, the problem of plotting a root-locus diagram is not as simple as this example may indicate; however, the construction of such a diagram is greatly simplified by using a complete set of rules established elsewhere.[2]

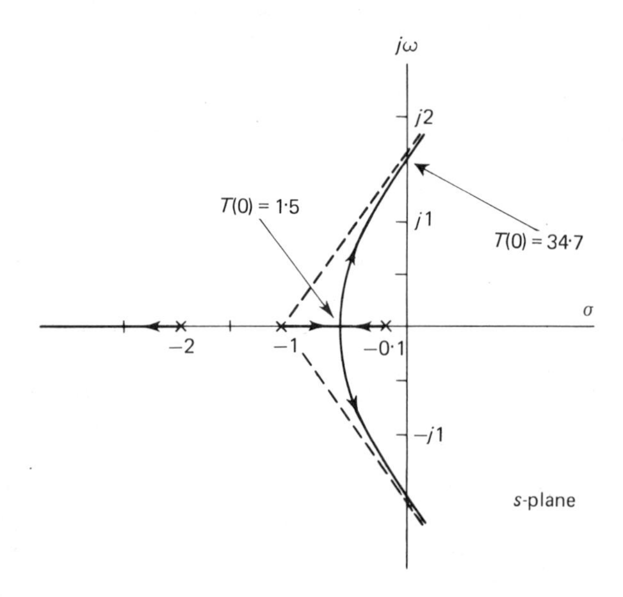

Fig. 12.3. Root-locus diagram for $T(s) = T(0)/[(1 + 10s)(1 + s)(1 + 0{\cdot}5s)]$.

A Stabilization Design Criterion

The root-locus method predicts that, for large values of mid-band return ratio $T(0)$, the natural frequencies of a feedback amplifier or equivalently the poles of the amplifier's closed-loop transmittance move asymptotically along equally spaced radial lines having the general form depicted in Fig. 12.4. This diagram reveals that as $T(0)$ is increased, the natural frequencies move further and further away from one another.

Mulligan[3] has shown that under suitable conditions of separation among the natural frequencies in the complex-frequency plane, the time response of a linear network to a step function excitation can be well approximated at the first maximum and beyond by a constant term plus a *dominant oscillatory mode* in the form of a single exponentially damped sinusoid. In general, although not always, the pair of complex-conjugate natural frequencies located nearest to the imaginary axis in the complex-frequency plane will produce the dominant oscillatory mode in the time response. The approximation to the actual time response given

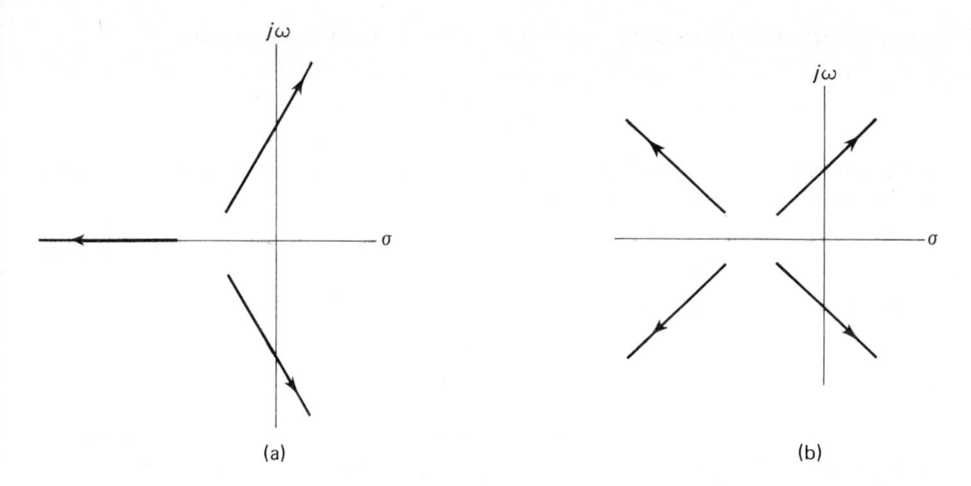

Fig. 12.4. Asymptotic movements of feedback amplifier natural frequencies as mid-band return ratio is increased indefinitely: (a) For an amplifier with three more poles than zeros in $T(s)$; (b) For an amplifier with four more poles than zeros in $T(s)$.

by the dominant oscillatory mode is improved as the separation among the natural frequencies increases. By consideration of the root loci asymptotes obtained from the return ratio $T(s)$ of common vacuum-tube and transistor feedback amplifiers and the corresponding natural-frequency movements which are implied, we find that the dominant oscillatory-mode theory has wide application in the approximation of the transient response of such amplifiers. A design using the dominant oscillatory-mode approximation should, therefore, have considerable utility.[4]

Consider a linear network where the dominant oscillatory mode approximation applies, as indicated by the pole pattern shown in Fig. 12.5. Then, the step

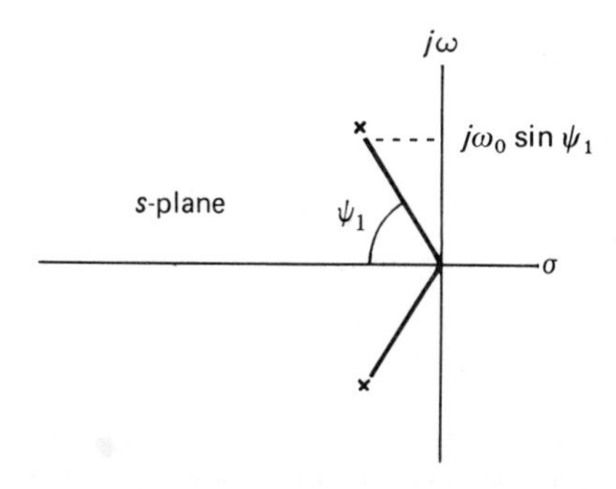

Fig. 12.5. A dominant pair of natural frequencies.

response (normalized to have final value unity) is readily obtained as:

$$h_n^{-1}(t) = 1 - \frac{\varepsilon^{-\omega_0 t \cos \psi_1}}{\sin \psi_1} \sin(\omega_0 t \sin \psi_1 + \psi_1). \tag{12.11}$$

Differentiating $h_n^{-1}(t)$ with respect to time t, and setting the result to zero, we find that the maxima and minima occur when

$$\tan(\omega_0 t \sin \psi_1 + \psi_1) = \tan \psi_1. \tag{12.12}$$

Solving Eq. (12.12) for t, we get

$$t = \frac{k\pi}{\omega_0 \sin \psi_1}, \tag{12.13}$$

where $k = 1, 2, 3, \ldots$. The maxima and minima of $h_n^{-1}(t)$ occur for odd and even integer values of k, respectively. Thus, evaluating the ratio η of the deviations from unity in the step response $h_n^{-1}(t)$ at two successive extreme values, we obtain

$$\eta = \varepsilon^{-\pi \cot \psi_1}. \tag{12.14}$$

Equation (12.14) provides, for feedback amplifiers, a stabilization design criterion based on control of the transient response. For the feedback amplifier to be stable, it is necessary that $\eta < 1$, or that $\psi_1 < \pi/2$. Furthermore, ψ_1 must be so chosen as to ensure that the dominant oscillatory mode in the time response will decay at a certain minimum rate.

A statement of the stabilization problem can now be made in terms of the s-plane. Suppose that a feedback amplifier has been designed to meet various performance requirements and it is found to have two complex natural frequencies in the right half plane, as in Fig. 12.6(a). Also, let the dominant oscillatory mode in the time response be required to decay at a certain minimum rate. That is, we must achieve the condition pictured in Fig. 12.6(b) in which the dominant pair

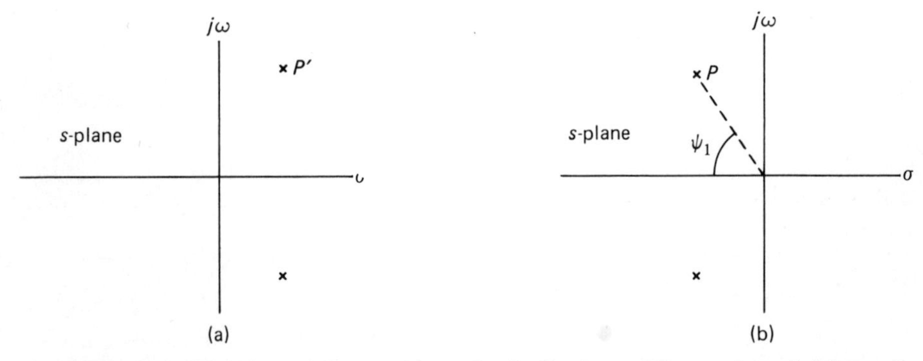

Fig. 12.6. Illustrating the stability problem of a feedback amplifier as that of shifting the position of the complex natural frequency at P' (and its conjugate) in (a) to point P in (b).

of natural frequencies has an angle ψ_1 determined from the prescribed rate of attenuation of the transient response. Then the amplifier design must be altered in such a manner that with the desired mid-band return ratio $T(0)$ unchanged, the point P' in Fig. 12.6(a) moves to P in Fig. 12.6(b). The actual technique employed to effect this change depends on the particular amplifier-design problem.

Example 12.2. *Stabilization with a corrective shunt capacitor.** Consider a three-stage feedback amplifier having a return ratio expressed by

$$T(s) = \frac{T(0)}{(1 - s/\sigma_0)^3}, \tag{12.15}$$

in which $T(0) = 40$. We find that for $T(0) > 8$, the return difference $F(s) = 1 + T(s)$ has two complex roots in the right half plane, so the amplifier is unstable. Assume that it is required to stabilize the amplifier so that $\eta \leq 0{\cdot}2$, that is, the deviation from unity at one local extreme value of the transient response is to be (at least) five times the deviation at the next maximum deviation. Also, it is assumed that the stabilization is effected by the addition of a capacitor in parallel with the input port of one of the active elements, thereby producing the pole pattern shown in Fig. 12.7(a) for the modified return ratio, that is:

$$T(s) = \frac{T(0)}{(1 - s/\sigma_0)^2(1 - s/a\sigma_0)}. \tag{12.16}$$

Figure 12.7(b) shows the root pattern for $1 + T(s) = 0$ which is derived from Eq. (12.16) and meets the condition that point P lies on the radial line with displacement ψ_1 from the negative real axis. By using Eq. (12.14) with $\eta = 0{\cdot}2$, we find that $\psi_1 = 62{\cdot}9°$. In terms of the quantities defined in Fig. 12.7(b), we find that

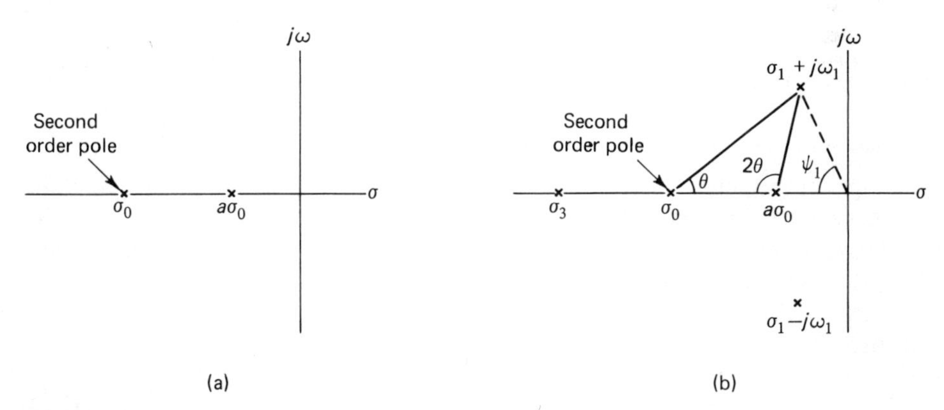

Fig. 12.7. (a) Modified pole pattern of $T(s)$ caused by addition of capacitor to one stage; (b) Desired roots of $1 + T(s) = 0$.

* This example is taken from Ref. 4.

for the conditions

$$|T(s)| = 1$$

and

$$\text{ang } T(s) = -\pi$$

to be satisfied as the root locus crosses the radial line at P, it is necessary that

$$\sin(2\theta - \psi_1)\sin^2(\theta + \psi_1) = \frac{\sin^3 \psi_1}{T(0)}. \tag{12.17}$$

After finding θ using Eq. (12.17), the quantity a is determined from the expression:

$$a = \frac{1}{2T(0)\cos\theta} \frac{\sin^3 \psi_1}{\sin^3(\theta + \psi_1)}. \tag{12.18}$$

The co-ordinates of the natural frequencies of the feedback amplifier are as follows:

$$\sigma_1 = \sigma_0 \frac{\sin\theta \cos\psi_1}{\sin(\theta + \psi_1)}, \tag{12.19}$$

$$\omega_1 = -\sigma_0 \frac{\sin\theta \sin\psi_1}{\sin(\theta + \psi_1)} \tag{12.20}$$

and

$$\sigma_3 = 2\sigma_0\left[1 + \frac{a}{2} - \frac{\sin\theta \cos\psi_1}{\sin(\theta + \psi_1)}\right]. \tag{12.21}$$

For $\psi_1 = 62{\cdot}1°$, and $T(0) = 40$, we thus find that $a = 0{\cdot}0105$, $\sigma_1 = 0{\cdot}242\sigma_0$, $\omega_1 = -0{\cdot}473\,\sigma_0$, and $\sigma_3 = 1{\cdot}526\,\sigma_0$.

Sensitivity of Natural Frequencies to Parameter Variations

Changes in the active as well as passive elements comprising a feedback amplifier cause corresponding changes in the mid-band value $T(0)$ and the poles and zeros of the return ratio $T(s)$. It is therefore of interest to know how sensitive the natural frequencies of the amplifier are to small changes in these elements.

The sensitivity of a natural frequency s_ν, with respect to a specified circuit element W, is defined by[5]

$$S_{s_\nu} = \frac{ds_\nu}{dW/W}, \tag{12.22}$$

where ds_ν is the change in the position of the natural frequency s_ν for a variation dW in the element of design value W. Suppose the return ratio $T(s)$ is expressed in the form of Eq. (12.7), or

$$T(s) = T(0)\frac{N(s)}{D(s)}, \tag{12.23}$$

where $D(s)$ and $N(s)$ are polynomials in s, as defined by

$$N(s) = \prod_{i=1}^{m}\left(1 - \frac{s}{z_i}\right)$$

$$D(s) = \prod_{i=1}^{n}\left(1 - \frac{s}{p_i}\right).$$

The characteristic equation of the feedback amplifier is therefore

$$Q(s, T(0)) = D(s) + T(0)N(s) = 0. \tag{12.24}$$

Let s_v be a zero of $Q(s, T(0))$, that is,

$$Q(s_v, T(0)) = 0. \tag{12.25}$$

Then if $T(0)$ changes to $T(0) + dT(0)$, the natural frequency s_v will correspondingly move to $s_v + ds_v$, so that

$$Q(s_v + ds_v, T(0) + dT(0)) = 0. \tag{12.26}$$

Neglecting second-order terms in ds and $dT(0)$, we obtain

$$ds_v\frac{\partial Q}{\partial s_v} + dT(0)\frac{\partial Q}{\partial T(0)} = 0. \tag{12.27}$$

Hence, use of Eqs. (12.24) and (12.27) gives the sensitivity of the natural frequency s_v with respect to $T(0)$ to be

$$S_{s_v} = \frac{ds_v}{dT(0)/T(0)} = -T(0)\frac{\partial Q/\partial T(0)}{\partial Q/\partial s_v} = -T(0)\frac{N(s_v)}{\partial Q/\partial s_v} = \frac{D(s_v)}{\partial Q/\partial s_v} \tag{12.28}$$

or

$$S_{s_v} = \frac{D(s_v)}{\dfrac{\partial D(s_v)}{\partial s_v} + T(0)\dfrac{\partial N(s_v)}{\partial s_v}}. \tag{12.29}$$

In a similar manner we may evaluate the sensitivity of the natural frequency s_v to small changes in the positions of poles and zeros of the return ratio $T(s)$. Thus if the pole p_i of $T(s)$ changes by a small amount dp_i, the corresponding change in s_v is[6]

$$ds_v = S_{s_v}\frac{dp_i}{s_v - p_i}. \tag{12.30}$$

If, on the other hand, the zero z_i of $T(s)$ changes by a small amount dz_i, we have

$$ds_v = -S_{s_v} \frac{dz_i}{s_v - z_i}. \tag{12.31}$$

The total displacement in the position of the natural frequency s_v caused by simultaneous variations in the mid-band value and the poles and zeros of the return ratio $T(s)$ is therefore

$$ds_v = S_{s_v} \left(\frac{dT(0)}{T(0)} + \sum_{i=1}^{n} \frac{dp_i}{s_v - p_i} - \sum_{i=1}^{m} \frac{dz_i}{s_v - z_i} \right). \tag{12.32}$$

12.3 NYQUIST STABILITY CRITERION[7]

Consider a function $F(s)$ which has a total number of N zeros and P poles inside a closed contour C in the s-plane. In Section 3.3 we showed that as a result of the variable s moving around the contour C once in the counter-clockwise direction, the plot of $F(s)$ will encircle the origin of the F-plane in the counter-clockwise direction a number of times equal to $N - P$. Suppose the function $F(s)$ is the return difference for the control parameter W of some controlled source in the circuit model representing one of the vacuum tubes or transistors in the amplifier. Also, let the contour C be the path which extends along the imaginary axis and folds back around a semicircle of infinite radius, thereby enclosing the entire

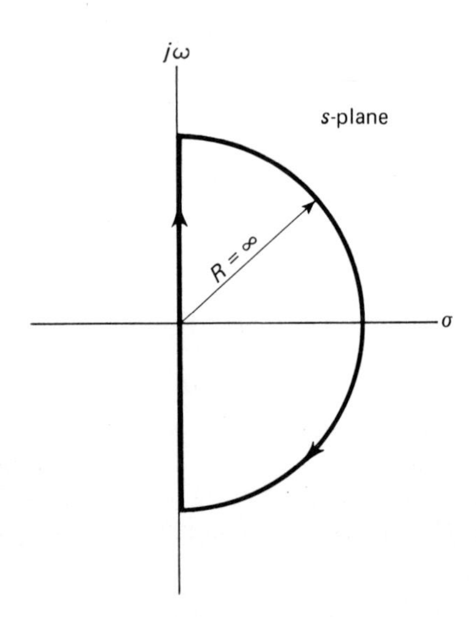

Fig. 12.8. Contour for derivation of Nyquist criterion.

right half of the s-plane, as indicated in Fig. 12.8. In this diagram the contour C is shown traversed in the clockwise direction, so that along the imaginary axis it corresponds to increasing frequency.

From physical considerations it is known that as s approaches infinity the forward transmission through a vacuum tube or transistor is reduced to zero. This is equivalent to saying that the return ratio $T(s)$ will vanish, or the return difference $F(s)$ will approach unity, as s approaches infinity. Therefore, the infinite semicircular path can have no effect on the number of times the plot of $F(s)$ will encircle the origin of the F-plane, so that only the values of $F(s)$ along the imaginary axis are required. Since $F(s) = \Delta(s)/\Delta^0(s)$, it follows that the number of times the plot of $F(s)$ encircles the origin of the F-plane in the clockwise direction while s itself moves from $s = -j\infty$ to $s = j\infty$ along the imaginary axis of the s-plane, is equal to the number of zeros of $\Delta(s)$ in the right half plane diminished by the number of zeros of $\Delta^0(s)$ in the same region. Suppose, however, the feedback amplifier is known to be stable when the element W vanishes, so that the determinant $\Delta^0(s)$ can have no zeros in the right half of the s-plane. In this case the stability or instability of the feedback amplifier can be determined directly from the Nyquist plot of $F(j\omega)$. The return difference $F(s)$ has no zeros in the right half plane, and the amplifier will therefore be stable, if the plot of $F(j\omega)$ does not encircle the origin of the F-plane.

In practice it is found more convenient to work with the return ratio $T(j\omega)$ because it is a directly measurable quantity. The difference between a plot of $F(j\omega)$ and that of $T(j\omega)$ is simply a shift in the position of the imaginary axis. Since $F(j\omega) = 1 + T(j\omega)$, it follows that the origin of the F-plane corresponds to the point $(-1 + j0)$ in the T-plane. We can, therefore, state that if a feedback amplifier is stable when a specified element W vanishes, the necessary and sufficient condition for the feedback amplifier to remain stable when the element W assumes its normal value is that the Nyquist plot of the return ratio $T(j\omega)$ for the element W should not enclose the critical point $(-1 + j0)$ in the T-plane. This is known as the *Nyquist stability criterion.*

When the Nyquist plot of $T(j\omega)$ passes through the point $(-1 + j0)$ the feedback amplifier is critically stable, corresponding to a pair of conjugate natural frequencies on the imaginary axis of the s-plane. This condition provides a useful measure of the theoretical limit of feedback from which we must withdraw to ensure a margin of safety against instability. The nominal value of the return difference in the useful frequency band which just places an amplifier in the critically stable condition, we shall call the *threshold* or *critical feedback.*

The required plotting of $T(j\omega)$ is simplified by observing that the magnitude of $T(j\omega)$ is an even function of ω, and its phase angle is an odd function of ω, so that

$$\begin{aligned} |T(j\omega)| &= |T(-j\omega)| \\ \text{ang } T(j\omega) &= -\text{ang } T(-j\omega). \end{aligned} \tag{12.33}$$

Therefore, it is necessary only to plot $T(j\omega)$ for positive frequencies from $\omega = 0$ to $\omega = \infty$. The remaining plot for negative frequencies from $\omega = -\infty$ to $\omega = 0$ can be inserted simply as the mirror image of this part with respect to the real axis of the T-plane.

Two illustrative Nyquist plots are shown in Fig. 12.9. Clearly, the plot shown in part (a) of this diagram represents a stable system, while that in part (b) represents an unstable system with two natural frequencies in the right half plane as the plot of $T(j\omega)$ encloses the critical point $(-1 + j0)$ twice.

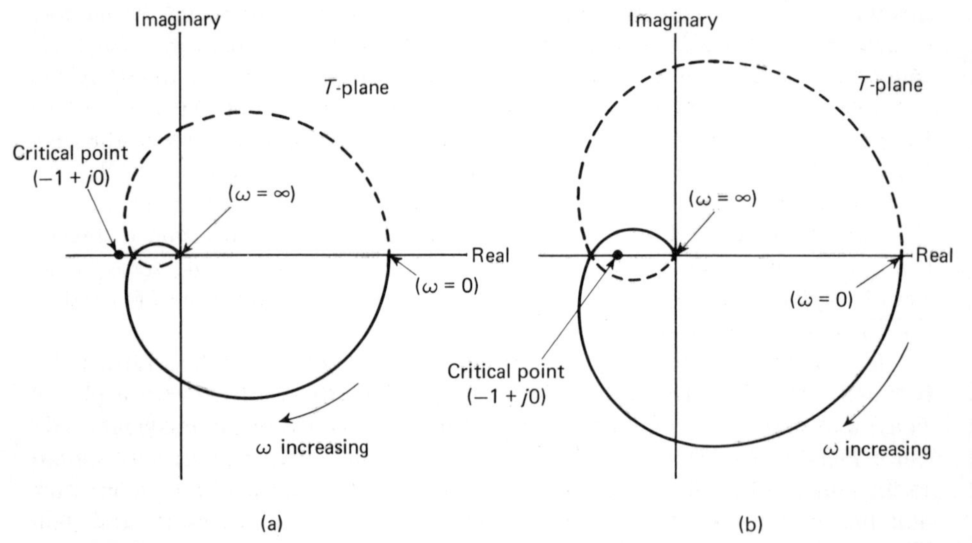

Fig. 12.9. Nyquist plots for: (a) Stable feedback amplifier; (b) Unstable feedback amplier.

Stability Margins

The Nyquist diagram also provides a quantitative measure of the degree of stability of the feedback amplifier. In Fig. 12.10 we see that when the phase angle of the return ratio $T(j\omega)$ equals $-180°$, its magnitude equals $1/T_m$. The quantity $20 \log_{10} T_m$, which is called the *gain margin*, is thus equal to the number of decibels by which the mid-band value $T(0)$ of the return ratio can be increased to make the amplifier critically stable. The frequency ω_p at which $|T(j\omega_p)| = 1/T_m$ is termed the *phase-crossover frequency*. In Fig. 12.10 we also see that when the magnitude of the return ratio $T(j\omega)$ equals unity, its phase angle equals $(-180 + \phi_m)°$. The angle ϕ_m, which is called the *phase margin*, is thus equal to the additional phase lag (at unit magnitude) required to make the amplifier critically stable. The frequency ω_g at which $|T(j\omega_g)| = 1$ and ang $T(j\omega_g) = (-180 + \phi_m)°$, is termed the *gain-crossover frequency*. In an absolutely stable feedback amplifier ω_g is necessarily smaller than ω_p.

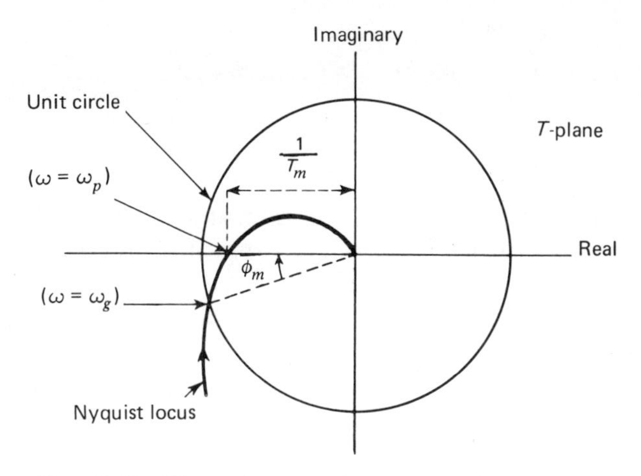

Fig. 12.10. Illustrating the definition of stability margins.

The gain and phase margins are the generally accepted measures of the degree of stability of a feedback amplifier. They are intended to guard against the effect of variations in vacuum-tube or transistor parameters due to supply-voltage changes, temperature changes, sample-to-sample variations, etc. In practice it is found that a gain margin of about 10 dB and a phase margin of about 30° are quite adequate.

Example 12.3. *Three-stage transistor feedback amplifier.* As an illustration, we will investigate the stability performance of the three-stage transistor amplifier of Fig. 12.11 with a single external feedback loop provided by the resistor R_f. This resistor effectively controls the mid-band value of the return ratio. We will assume that all three transistors are identical, having the following set of hybrid-π

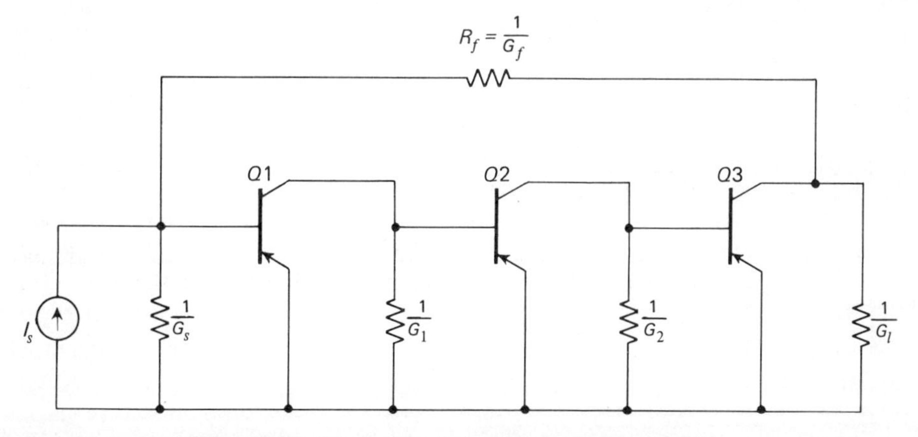

Fig. 12.11. Three-stage transistor feedback amplifier.

parameters:

$$r_{b'} = 100\,\Omega \qquad C_{b'c} = 6{\cdot}4\,\text{pF}$$
$$r_{b'e} = 150\,\Omega \qquad g_m = 182\,\text{mA/volt.}$$
$$C_{b'e} = 200\,\text{pF}$$

The return ratio $T(j\omega)$ for the y_{fe1} of transistor $Q1$ can be determined from Fig. 12.12 where the feedback loop has been opened at the input of the transistor

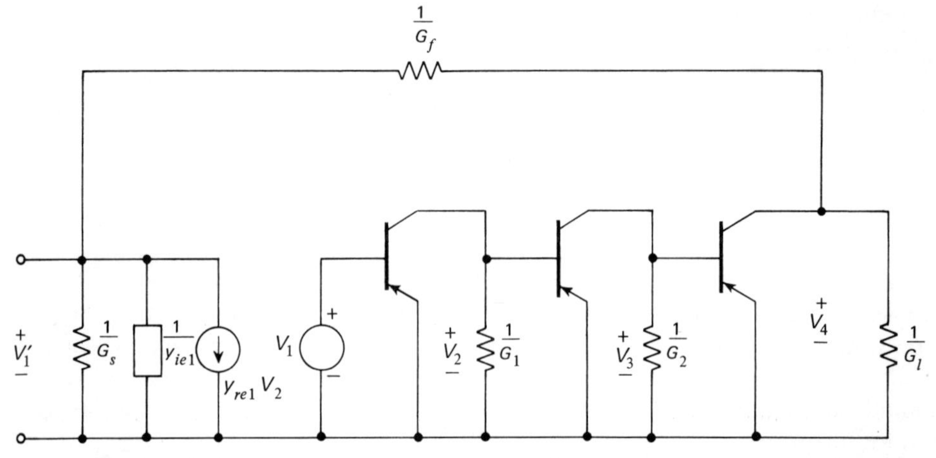

Fig. 12.12. Network for evaluating return difference for y_{fe} of transistor $Q1$.

in question and terminated as required. In terms of the y-parameters of the individual transistors we obtain the signal-flow graph shown in Fig. 12.13 from which we find, by using Mason's direct rule, that

$$T(j\omega) = -\frac{V_1'}{V_1} = -\frac{P_1(1 - L_3) + P_f}{1 - (L_2 + L_3)}$$

where

$$P_1 = \frac{y_{re1}y_{fe1}}{(G_s + y_{ie1})(y_{oe1} + G_1 + y_{ie2})}$$

$$P_f = -\frac{G_f y_{fe1} y_{fe2} y_{fe3}}{(G_s + y_{ie1})(y_{oe1} + G_1 + y_{ie2})(y_{oe2} + G_2 + y_{ie3})(y_{oe3} + G_l)} \qquad (12.34)$$

$$L_2 = \frac{y_{re2}y_{fe2}}{(y_{oe1} + G_1 + y_{ie2})(y_{oe2} + G_2 + y_{ie3})}$$

$$L_3 = \frac{y_{re3}y_{fe3}}{(y_{oe2} + G_2 + y_{ie3})(y_{oe3} + G_l)}.$$

In Fig. 12.13 we have assumed that G_f presents negligible loading at both the input and output ports of the amplifier, that is $G_f \ll |G_s + y_{ie1}|$ and $G_f \ll |G_l + y_{oe3}|$. The common-emitter y-parameters with subscript 1 refer to transistor $Q1$, those with subscript 2 refer to transistor $Q2$, and those with subscript 3 refer to transistor $Q3$.

The path transmittance P_1 represents the effect of the internal feedback associated with transistor $Q1$; when the loop is closed, P_1 assumes the role of a loop transmittance. The path transmittance P_f represents the effect of the externally applied feedback due to G_f. The loop transmittances L_2 and L_3 represent the internal feedbacks associated with transistors $Q2$ and $Q3$, respectively.

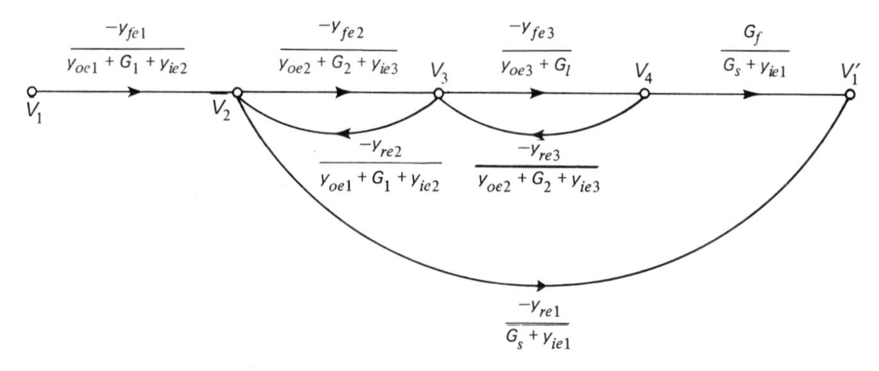

Fig. 12.13. Signal-flow graph for network of Fig. 12.12.

The common-emitter y-parameters are related to the hybrid-π parameters as follows:

$$y_{ie} = \frac{1 + sC_{b'e}r_{b'e}}{r_{b'} + r_{b'e} + sC_{b'e}r_{b'}r_{b'e}}, \qquad y_{fe} = \frac{g_m r_{b'e}}{r_{b'} + r_{b'e} + sC_{b'e}r_{b'}r_{b'e}},$$

$$y_{re} = \frac{-sC_{b'c}r_{b'e}}{r_{b'} + r_{b'e} + sC_{b'e}r_{b'}r_{b'e}}, \qquad y_{oe} = \frac{sC_{b'c}g_m r_{b'e}r_{b'}}{r_{b'} + r_{b'e} + sC_{b'e}r_{b'}r_{b'e}}. \tag{12.35}$$

Therefore, substitution of the given hybrid-π parameter values in Eq. (12.35) leads to the following set of common y-parameters for all three transistors

$$y_{ie} = \frac{4(1 + 3s)}{1 + 1{\cdot}2s}\,\text{m℧} \qquad y_{fe} = \frac{109}{1 + 1{\cdot}2s}\,\text{m℧}$$

$$y_{re} = \frac{-0{\cdot}384s}{1 + 1{\cdot}2s}\,\text{m℧} \qquad y_{oe} = \frac{7s}{1 + 1{\cdot}2s}\,\text{m℧},$$

where s is in units of 10^8 rad/sec. If, in turn, we substitute these results in Eq. (12.34), together with the given values of $G_1 = G_2 = 1\,\text{m℧}$, $G_s = G_l = 20\,\text{m℧}$, and if we assume that the externally applied feedback is large enough to justify neglecting the effect of the internal feedback within transistor $Q1$, we ultimately get:

$$T(s) = \frac{T(0)(1 + s/0{\cdot}833)}{(1 + s/0{\cdot}129)(1 + s/0{\cdot}381)(1 + s/0{\cdot}667)(1 + s/0{\cdot}8)}, \tag{12.36}$$

where

$$T(0) = 110G_f,$$

that is,

$$R_f = \frac{1}{G_f} = \frac{110}{T(0)}\,\text{k}\Omega.$$

In Eq. (12.36) the pole at $s = -0{\cdot}8$ is fairly close to the zero at $s = -0{\cdot}833$; therefore, we may cancel this pole–zero pair, obtaining

$$T(s) \simeq \frac{T(0)}{(1 + s/0{\cdot}129)(1 + s/0{\cdot}381)(1 + s/0{\cdot}667)}. \tag{12.37}$$

For $s = j\omega$ we have plotted in Fig. 12.14 the return ratio of Eq. (12.37) with $T(0) = 10$ (i.e. $R_f = 11\,\text{k}\Omega$) and $T(0) = 30$ (i.e. $R_f = 3{\cdot}7\,\text{k}\Omega$). Since $T(s)$ has no poles (that is, $\Delta^0(s)$ has no zeros) in the right half plane, it follows that for $T(0) = 10$ the feedback amplifier is stable with a gain margin $= 1{\cdot}6$ dB and phase margin $= 10°$, while for $T(0) = 30$ the amplifier is unstable as the Nyquist locus encircles the critical point $(-1 + j0)$.

Bode Diagram

In displaying the Nyquist diagram in polar form there are definite limitations to the ease and speed with which the necessary calculations for a number of different frequencies can be performed. We therefore find that for design purposes, it is generally more desirable to plot the *logarithmic loop gain* and *loop-phase shift* as separate functions of frequency (usually logarithm of frequency) instead of combining them in a single curve. The logarithmic loop gain (or simply loop gain) and loop-phase shift are defined as the real and imaginary parts of the logarithm of the return ratio, as shown by

$$\begin{aligned} \log_\varepsilon T(j\omega) &= \log_\varepsilon |T(j\omega)| + j\,\text{ang}\,T(j\omega) \\ &= A(\omega) + jB(\omega). \end{aligned} \tag{12.38}$$

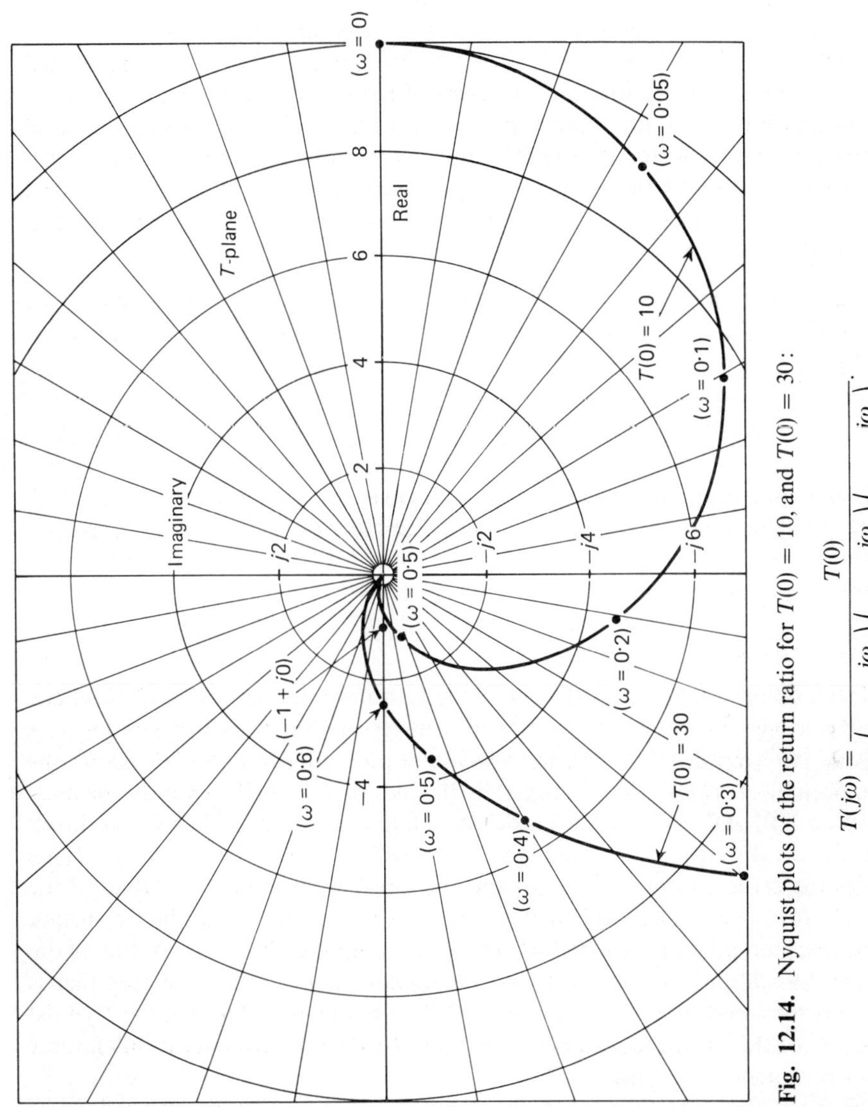

Fig. 12.14. Nyquist plots of the return ratio for $T(0) = 10$, and $T(0) = 30$:

$$T(j\omega) = \frac{T(0)}{\left(1 + \dfrac{j\omega}{0{\cdot}129}\right)\left(1 + \dfrac{j\omega}{0{\cdot}381}\right)\left(1 + \dfrac{j\omega}{0{\cdot}667}\right)}.$$

It is important to observe that when the feedback is nominally negative there is one net phase reversal around the feedback loop in addition to the phase shift $B(\omega)$.

In plotting separate curves for the loop gain and loop-phase shift, the essential features of the Nyquist stability criterion are retained and, in addition, we are in a somewhat better position to visualize whatever modifications may be needed to improve the stability situation. In terms of the separate frequency responses, the critical point $(-1 + j0)$ of the T-plane is represented by the zero-decibel level of loop gain, and by the locus of 180° of loop-phase shift; the frequencies at which these loci are crossed by the respective response curves are the gain-crossover and phase-crossover frequencies. Critical feedback is indicated when the gain and phase crossovers coincide.

Example 12.4. *The three-stage transistor feedback amplifier* (continued). The return ratio of the three-stage transistor feedback amplifier considered earlier is given by

$$T(j\omega) = \frac{T(0)}{(1 + j\omega/0{\cdot}129)(1 + j\omega/0{\cdot}381)(1 + j\omega/0{\cdot}667)}.$$

For $T(0) = 30$, we obtain the total gain and phase characteristics of Fig. 12.15. We see that the amplifier is unstable as the gain-crossover frequency is greater than the phase-crossover frequency. The amplifier is rendered stable by reducing the mid-band value of the loop gain by 8 dB.

Conditional Stability

A special kind of stability is illustrated by the Nyquist diagram of Fig. 12.16 where, for convenience, only the portion of the plot corresponding to positive frequencies is shown. An amplifier having such a Nyquist plot is clearly stable because the critical point $(-1 + j0)$ is not encircled by the plot of $T(j\omega)$. If, however, the mid-band value $T(0)$ of the return ratio is decreased, thereby causing the plot to shrink uniformly at all frequencies, a point will be reached at which the plot of $T(j\omega)$ encircles the critical point $(-1 + j0)$ and the amplifier becomes unstable. If the value of $T(0)$ is decreased still further, and the plot is correspondingly shrunk further, another point is reached at which the amplifier becomes stable again. This form of stability is referred to as *conditional stability* since the amplifier is stable only for a certain range of values of $T(0)$. On the other hand, if the Nyquist diagram is of the absolutely stable type, then $T(0)$ can be decreased indefinitely without producing instability.

The main objection to conditional stability is that a momentary overload of the amplifier will drive some element into saturation and thereby reduce the return ratio to the value where the amplifier becomes unstable. The ensuing oscillations may then build up so rapidly that the amplifier remains saturated, and therefore unstable, as a result of its own oscillations. This effect could also

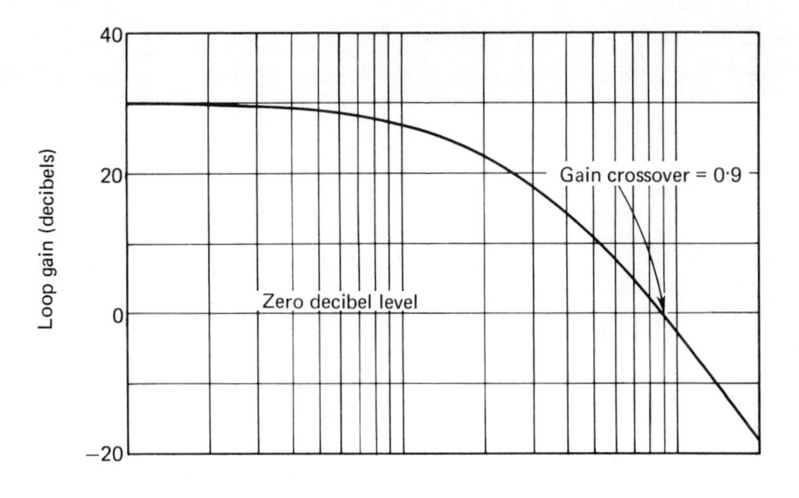

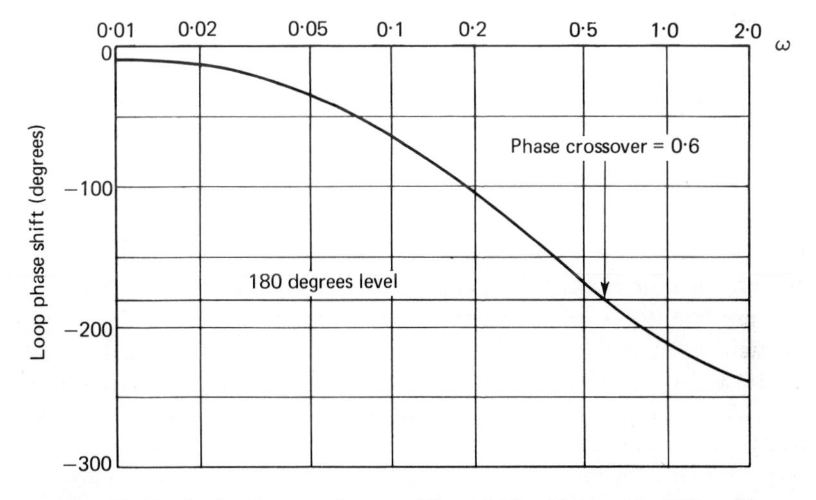

Fig. 12.15. Bode diagram for amplifier of Fig. 12.11 with $T(0) = 30$.

occur during the warming-up period of a vacuum-tube amplifier. The gain of the tubes will be very small until the cathodes are warm. Therefore, as the amplifier is switched on, and the cathode temperatures increase, the Nyquist diagram begins by first being very small and then expands continuously to its final form. If the Nyquist diagram is of the type shown in Fig. 12.16, the amplifier may become unstable as it warms up. In such a case, the ensuing oscillations may grow to an amplitude large enough to saturate the amplifier and prevent it from reaching its final stable state.

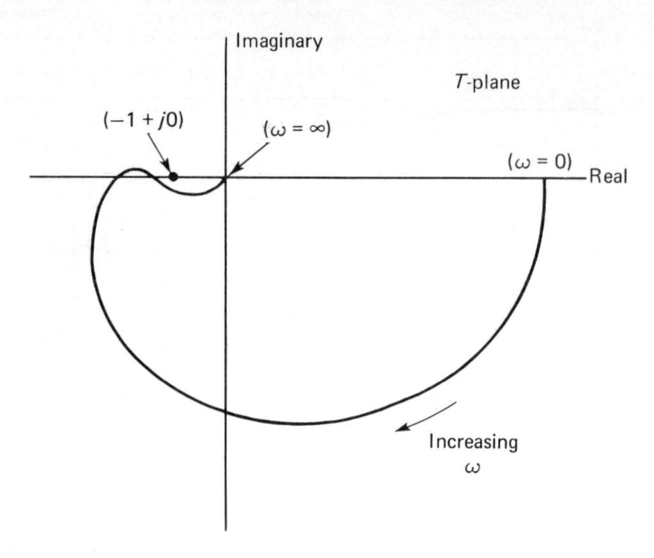

Fig. 12.16. Nyquist plot of a conditionally stable feedback amplifier.

12.4 A GRAPHICAL METHOD FOR EVALUATING THE DOMINANT OSCILLATORY MODE FROM THE NYQUIST DIAGRAM[8]

In Section 12.2 it was explained that in a great majority of feedback amplifiers the transient response to a step function can be closely approximated by a constant plus a dominant oscillatory mode due to a pair of natural frequencies of the amplifier which lie nearest to the imaginary axis of the s-plane. The co-ordinates of this dominant pair of natural frequencies can be evaluated from the Nyquist diagram by applying a graphical technique which uses the frequency scale on the Nyquist plot of $T(j\omega)$ to construct an orthogonal diagram such that the T-plane becomes a conformal transformation of the s-plane. In a conformal transformation any small geometrical figure in the s-plane is transformed into a similar figure in the T-plane with the angles of intersection and the approximate geometrical shapes being preserved as we go from one plane to the other. Accordingly, we find that a set of small equilateral triangles based on the imaginary axis of the s-plane, as in Fig. 12.17(a) is transformed in the T-plane into a corresponding set of curvilinear triangles based on the Nyquist plot. If the triangles in the s-plane are chosen small enough, the corresponding curvilinear triangles in the T-plane may be approximated with straight-sided triangles, as in Fig. 12.17(b).

To determine the dominant oscillatory mode, we first select on the Nyquist plot a number of points which represent equal increments $\delta\omega$ of frequency starting from (or close to) the gain crossover frequency, ω_g. The successive points are joined together by straight lines, and on each line as base an equilateral triangle is constructed. Then the apices of successive triangles are connected together with

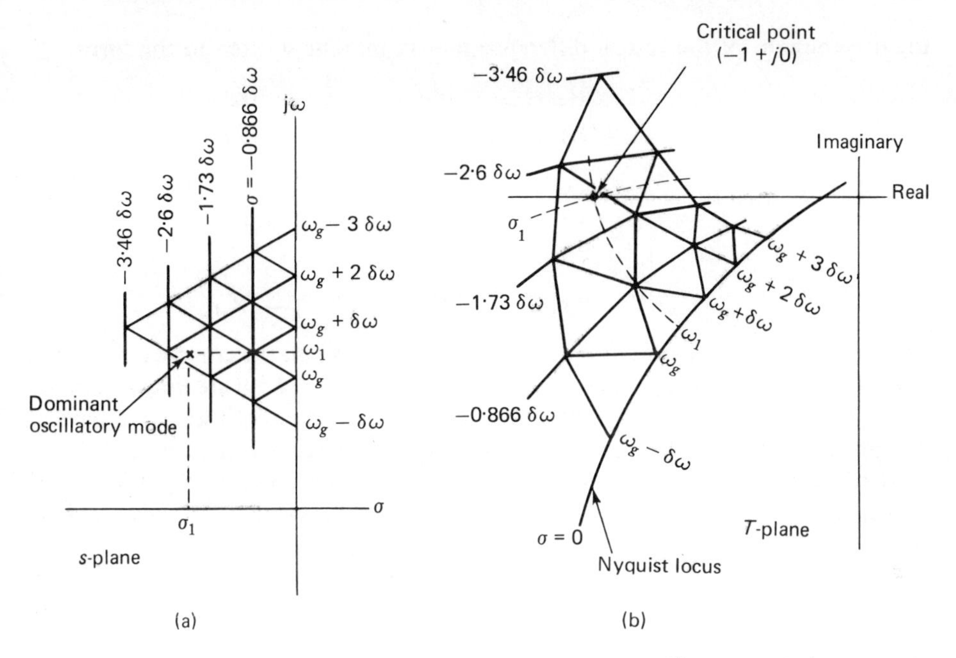

Fig. 12.17. Illustrating the graphical evaluation of dominant oscillatory mode from Nyquist diagram.

a new set of straight lines, and the process is continued till the graphical construction crosses the negative real axis of the T-plane.

The curves drawn through the apices of successive rows of triangles represent loci for which σ changes at increments of $-\delta\omega \sin 60° = -0{\cdot}866\ \delta\omega$. The number of such increments lying between the Nyquist locus and the critical point $(-1 + j0)$ is equal to the real co-ordinate σ_1 of the dominant pair of natural frequencies of the feedback amplifier. To determine the imaginary co-ordinate ω_1 of these natural frequencies, we construct a line to pass through the critical point $(-1 + j0)$ and at right angles to the constant σ-loci. The frequency at which this line intersects the Nyquist locus gives the required value of ω_1. Provided the frequency increment $\delta\omega$ is chosen sufficiently small then, at the expense of increased constructional effort, it is possible to achieve fairly accurate results with this graphical procedure.

12.5 CRITERION OF STABILITY FOR A MULTIPLE-LOOP FEEDBACK AMPLIFIER

Consider a linear multiple-loop feedback amplifier containing n active elements in the form of controlled sources, the control parameters of which are designated as $W_1, W_2, \ldots, W_n$. From the previous discussion in Section 11.5 we find that the determinant of the return-difference matrix may be written in the form

$$\Delta_F(s, W) = \begin{vmatrix} 1 - W_1 t_{11} & -W_2 t_{21} & \cdots & -W_n t_{n1} \\ -W_1 t_{12} & 1 - W_2 t_{22} & \cdots & -W_n t_{n2} \\ \cdot & \cdot & & \cdot \\ \cdot & \cdot & & \cdot \\ \cdot & \cdot & & \cdot \\ -W_1 t_{1n} & -W_2 t_{2n} & \cdots & 1 - W_n t_{nn} \end{vmatrix}. \tag{12.39}$$

The functional dependence of the determinant $\Delta_F(s, W)$ on the complex variable s arises from the fact that, in general, the various t's in Eq. (12.39) are linear, rational functions of s determined by the frequency characteristics of the active devices and passive components used in the interstage and feedback networks.

For mathematical convenience, let the control parameters $W_1, W_2, \ldots, W_n$ be normalized with respect to their normal operating values $W_{10}, W_{20}, \ldots, W_{n0}$, so that

$$a_1 = \frac{W_1}{W_{10}}, \qquad a_2 = \frac{W_2}{W_{20}}, \ldots, a_n = \frac{W_n}{W_{n0}}. \tag{12.40}$$

Then expanding the determinant on the right of Eq. (12.39), we find that Δ_F is a multilinear function of the control parameters or the a's, as shown by

$$\begin{aligned} \Delta_F(s, a) = 1 &+ a_1 T_1 + a_2 T_2 + \cdots + a_n T_n + a_1 a_2 T_{12} + a_1 a_3 T_{13} + \cdots \\ &+ a_1 a_2 a_3 T_{123} + \cdots + a_1 a_2 a_3 \cdots a_n T_{123\ldots n} \end{aligned} \tag{12.41}$$

where

$$\begin{aligned} T_i &= -W_{i0} t_{ii} \qquad \text{for } i = 1, 2, \ldots, n, \\ T_{ij} &= W_{i0} W_{j0} \begin{vmatrix} t_{ii} & i_{ji} \\ t_{ij} & t_{jj} \end{vmatrix} \qquad \text{for } i \neq j, \\ T_{ijk} &= -W_{i0} W_{j0} W_{k0} \begin{vmatrix} t_{ii} & t_{ji} & t_{ki} \\ t_{ij} & t_{jj} & t_{kj} \\ t_{ik} & t_{jk} & t_{kk} \end{vmatrix} \qquad \text{for } i \neq j \neq k, \end{aligned} \tag{12.42}$$

and so on.

It should be mentioned that the Δ_F as expressed in the form of Eq. (12.41) may also be obtained by evaluating the circuit determinant Δ of the network, under normal operating conditions, using the loop or nodal method of analysis, and then dividing it by the value $\Delta^{000\ldots}$ of the circuit determinant which results when all the n active elements are dead. The n parameters a_i, for $i = 1, 2, \ldots, n$ may be regarded as defining a unit n-dimensional Euclidean space which will be designated as the *space of operating points*, $\boldsymbol{a}$. It will be assumed that the range of variations for each of the a's is over the closed interval of real values from zero to unity, with normal operation of the active elements corresponding to $a_i = 1$. However, as a result of aging, deterioration or even failure, some or all the a's will decrease and may even become zero, so that all possible operating points that can exist in the circuit are located on and within the n-dimensional space, $\boldsymbol{a}$.

Equation (12.41) may be regarded as the characteristic equation of the multiple-loop feedback amplifier in that the amplifier is stable for all operating points on and within the n-dimensional space, $\boldsymbol{a}$, if $\Delta_F(s, a)$ does not vanish for any value of s in the right half of the s-plane.

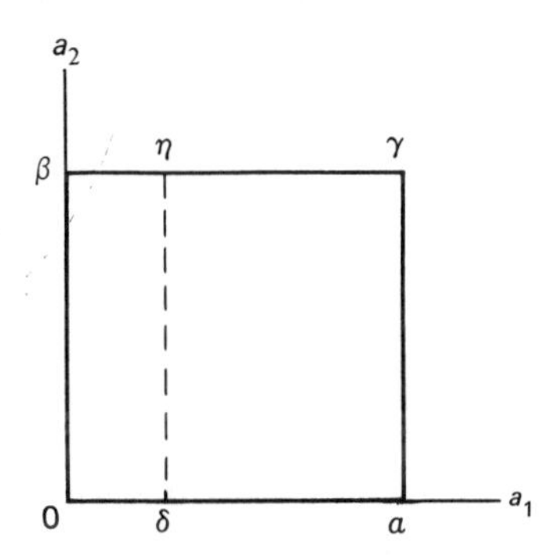

Fig. 12.18. Space of operating points for $n = 2$.

As with the Nyquist criterion, it is necessary only to examine the behaviour of the characteristic equation along the imaginary axis of the s-plane. Furthermore, it will be shown that it is necessary only to examine the stability of the operating points corresponding to the vertices of space $\boldsymbol{a}$ in order to determine the stability of a multiple-loop feedback amplifier.[9,10,11] For convenience, define:

$$T(s, a) = \Delta_F(s, a) - 1. \tag{12.43}$$

The function $T(s, a)$ is useful because it maps the origin in the space of operating points into the origin of the T-plane. Since $T(j\omega, a)$ is a multilinear function of the a's, it follows that for a fixed value of frequency ω any straight-line segment in the space of operating points parallel to an axis of co-ordinates is mapped into a straight-line segment in the T-plane. For example, in a circuit with two active elements the space of operating points is a square with four vertices designated as O, α, β and γ, as shown in Fig. 12.18. For a fixed frequency ω, the pertinent function

$$T(j\omega, a) = a_1 T_1 + a_2 T_2 + a_1 a_2 T_{12} \tag{12.44}$$

maps the four sides of the square $[O, \alpha]$, $[\alpha, \gamma]$, $[O, \beta]$ and $[\beta, \gamma]$ into the convex quadrilateral shown in Fig. 12.19. A quadrilateral is said to be convex if a straight line segment joining any two points on the quadrilateral lies inside it. Consider the segment $[\delta, \eta]$ parallel to the a_2-axis; this segment is mapped into the segment

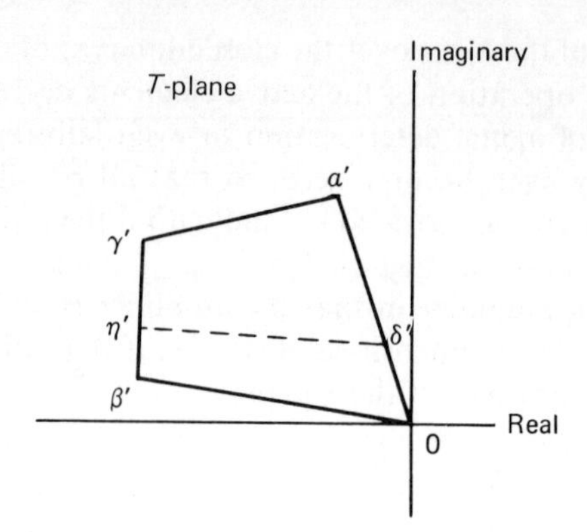

Fig. 12.19. A T-plane image of the space of operating points in Fig. 12.18, for some fixed frequency.

$[\delta', \eta']$ in the T-plane. As the segment $[\delta, \eta]$ sweeps through the whole of the square $\boldsymbol{a}$, starting from the side $[O, \beta]$ and terminating on the $[\alpha, \gamma]$-side, we see that the segment $[\delta', \eta']$ sweeps through the whole quadrilateral $O\alpha'\gamma'\beta'$, or in other words, the images of all operating points in $\boldsymbol{a}$ will lie inside the quadrilateral $O\alpha'\gamma'\beta'$. Therefore, for the particular ω, the mapping of square $\boldsymbol{a}$ by the function $T(j\omega, a)$ requires only the plotting of the images of the vertices of $\boldsymbol{a}$.

If, on the other hand, the function $T(j\omega, a)$ of Eq. (12.44) maps the square $\boldsymbol{a}$ into a quadrilateral which is not convex, as in Fig. 12.20, then some of the points in $\boldsymbol{a}$ will map into points in the T-plane which lie outside the quadrilateral. Thus the segment $[\delta, \eta]$ is shown mapped into $[\delta'', \eta'']$, some points of which are outside the quadrilateral. However, if the vertices α' and β' are connected by a straight line, then the images of all points in $\boldsymbol{a}$ will lie inside the triangle $O\alpha'\beta'$.

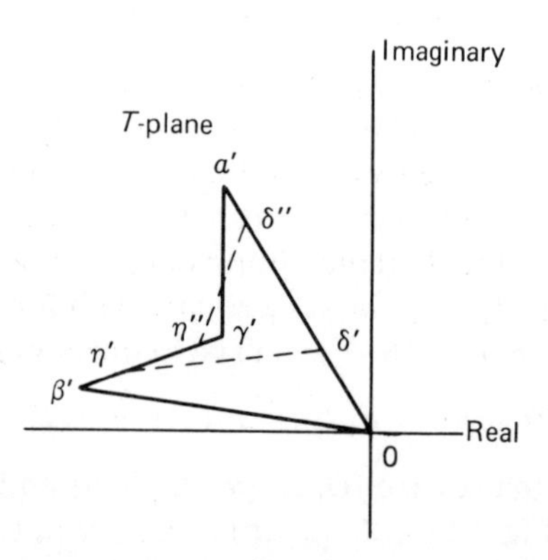

Fig. 12.20. A non-convex mapping of the space of operating points.

We may therefore state that in the general case of a multiple-loop feedback amplifier containing n active elements, the function $T(j\omega, a)$ maps all of the operating points in space $\boldsymbol{a}$ into the *convex hull* of the images of all the vertices of $\boldsymbol{a}$; the convex hull being defined as the smallest convex set it is possible to construct which contains all the images of the vertices of space $\boldsymbol{a}$. For example, in Fig. 12.19 the convex hull of points O, α', β' and γ' is the quadrilateral $O\alpha'\gamma'\beta'$. On the other hand, in Fig. 12.20 the convex hull of points O, α', β' and γ' is the triangle $O\alpha'\beta'$.

As the frequency ω is allowed to vary from $-\infty$ to $+\infty$, the convex hull will move on a pivot about the origin in the T-plane, and the area traced out by the convex hull covers all possible values of $T(j\omega, a)$ as the a's vary independently and continuously over their respective finite, closed intervals from zero to unity. The picture obtained for $T(j\omega, a)$ at negative frequencies is merely the mirror image, about the real axis in the T-plane, of that obtained for positive frequencies. Therefore, as with the Nyquist criterion, it is only necessary to check stability as ω varies from zero to plus infinity.

If the parameters $a_1, a_2, \ldots, a_n$ are each given fixed values, then the function $T(j\omega, a)$ assumes a single, unique value at each frequency, that is, the mapping of $T(j\omega, a)$ is a single point in the T-plane. As the frequency varies from $-\infty$ to $+\infty$, the locus traced out by this point is the well-known Nyquist diagram. The necessary and sufficient condition for the amplifier to be stable is that the Nyquist diagram does not enclose the critical point $(-1 + j0)$ in the T-plane.

When the parameters $a_1, a_2, \ldots, a_n$ are allowed to vary, as is the case under study, then there must be no set of discrete values within their range of variation which would give rise to a Nyquist diagram violating the Nyquist criterion. In other words, the critical point $(-1 + j0)$ in the T-plane must lie outside the area traced out by the convex hull of $T(j\omega, a)$ as it pivots about the origin. We may therefore generalize the Nyquist criterion by stating that a multiple-loop feedback amplifier is stable if the convex hull determined by the images of the vertices of the n-dimensional space of operating points does not enclose the critical point $(-1 + j0)$ in the T-plane, as the frequency varies from zero to infinity.

This is, however, only a sufficient condition for stability, since in the general case there may be points inside and on the convex hull which do not belong to $T(j\omega, a)$. If, therefore, for some ω we find that the mapping of the space of operating points is non-convex and that the convex hull of $T(j\omega, a)$ encloses the critical point $(-1 + j0)$ then we should sketch out the shape of the mapping of $\boldsymbol{a}$ for that ω in order to find out whether or not the point $(-1 + j0)$ does belong to $T(j\omega, a)$, that is, whether or not the circuit is actually unstable. Consider, for example, the case illustrated in Fig. 12.21 which pertains to a circuit with two active elements operating at some frequency ω. We see that the critical point $(-1 + j0)$ is located inside the convex hull, that is, the triangle $O\alpha'\beta'$. However, by constructing line segments parallel to the side $O\beta$ and their images in the T-plane, as indicated in

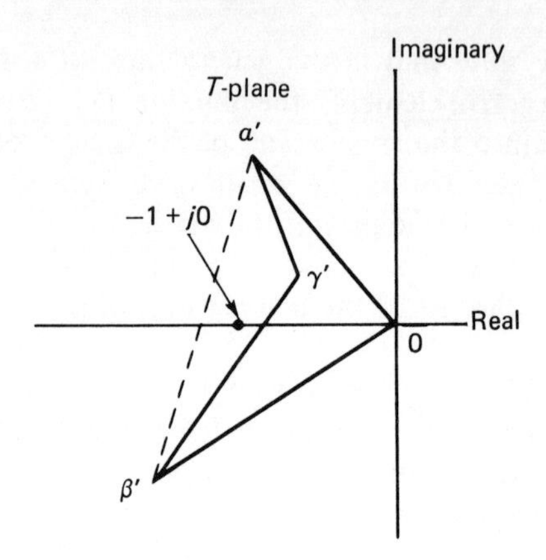

Fig. 12.21. *T*-plane image for some fixed frequency.

Fig. 12.22, we find that the critical point does not belong to $T(j\omega, a)$. We thus need only to trace a few positions of the segment $[\delta', \eta']$ to decide whether or not, for a certain ω, the critical point is included in the mapping of $\boldsymbol{a}$ by $Y(j\omega, a)$. It is of interest to note that in situations of the kind shown in Fig. 12.22, the envelope of the segment $[\delta', \eta']$ as δ moves along the side $O\alpha$ of space $\boldsymbol{a}$ is the arc of a parabola and two straight lines.

An important advantage of the stability criterion developed in this section is that the function $T(j\omega, a)$ can be readily evaluated by noting that each term in the characteristic equation is a return ratio or a product of return ratios for the

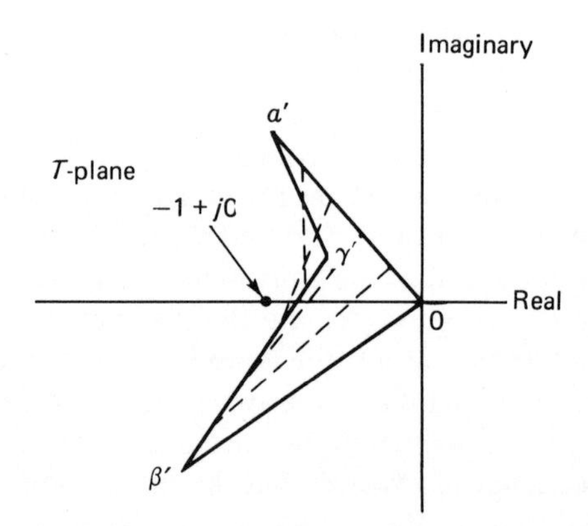

Fig. 12.22. Illustrating the procedure for determining whether or not the critical point $(-1 + j0)$ is included in the non-convex mapping of $\boldsymbol{a}$.

amplifier. Shaping the contours in the T-plane in order to obtain a desired gain and phase margin is therefore equivalent to shaping the return ratios of the individual feedback loops in the amplifier, which in many respects is similar to the problem of shaping the return ratio of a single-loop feedback amplifier.

12.6 STABILIZATION TECHNIQUES

Phase Compensation in the Feedback Path

One of the ways in which the stability performance of a feedback amplifier can be improved is to introduce a phase advance into the open-loop frequency response of the feedback amplifier so as to counteract part of the phase-retarding effect associated with the vacuum tubes or transistors. In the shunt–shunt structure of Fig. 12.11, for example, this may be achieved by adding a capacitor C_f across the resistor R_f in the feedback path.[12] To evaluate the compensating effect of C_f, we may consider the circuit depicted in Fig. 12.23 where the feedback network is shown connected between the collector of the last transistor $Q3$ and the base of the first transistor $Q1$. The resistor R' represents the parallel combination of the source resistance R_s and the low-frequency input resistance of transistor $Q1$. In Fig. 12.23 we see that the reverse transfer impedance for the feedback path is:

$$Z_{12}(s) = \frac{R'R_l(1 + sC_fR_f)}{R' + R_l + R_f + sC_fR_f(R' + R_l)}. \tag{12.45}$$

In the absence of the compensating capacitor C_f, we have

$$Z_{12}(0) = \frac{R'R_l}{R' + R_l + R_f}. \tag{12.46}$$

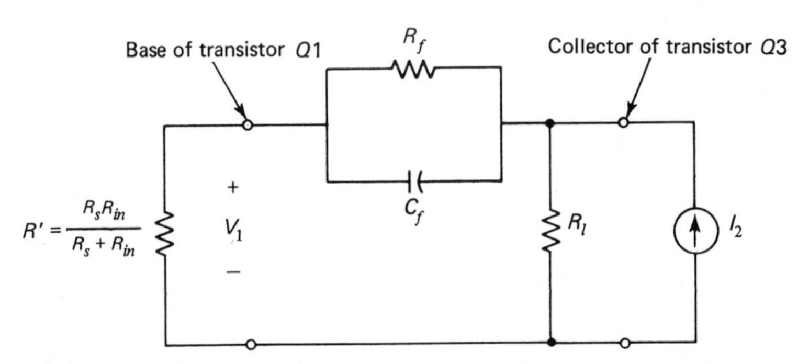

Fig. 12.23. Network for evaluating the effect of inserting a compensating capacitor C_f in the feedback path of the single-loop transistor feedback amplifier of Fig. 12.11.

Hence, the effect of inserting the capacitor C_f in circuit can be expressed by:

$$Z'_{12}(s) = \frac{Z_{12}(s)}{Z_{12}(0)} = \frac{1 + sC_f R_f}{1 + sC_f R_f/m}, \tag{12.47}$$

where

$$m = 1 + \frac{R_f}{R' + R_l}. \tag{12.48}$$

For $s = j\omega$, the gain and phase components of $Z'_{12}(j\omega)$ are as shown plotted in Fig. 12.24 where we see that at all frequencies the phase shift is positive. Therefore, the effect of adding the capacitor C_f in the feedback path is to introduce a phase advance into the loop-phase response of the amplifier. The phase advance

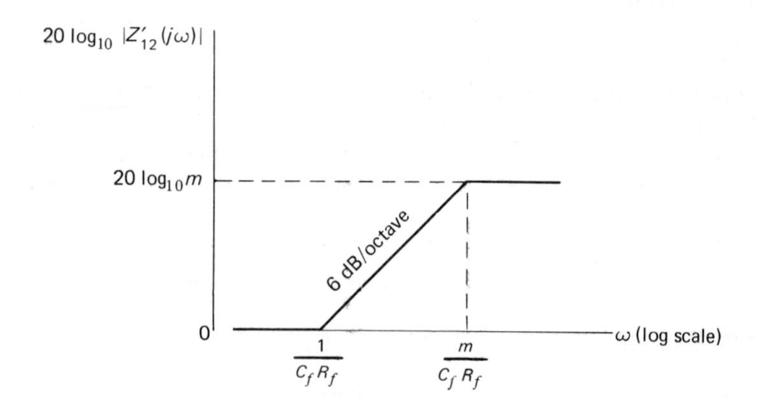

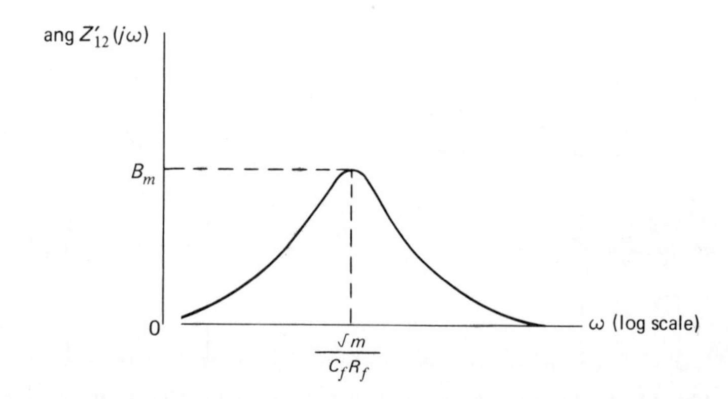

Fig. 12.24. Gain–phase plots of $Z'_{12}(j\omega)$ of Eq. (12.47).

attains its maximum value at the geometric mean of the two corner frequencies $1/C_f R_f$ and $m/C_f R_f$.

Furthermore, at zero and infinite frequencies, respectively, we have $Z'_{12}(0) = 1$ and $Z'_{12}(\infty) = m$. Hence, applying the phase-integral theorem yields (see Eq. 7.34)

$$\int_0^{\infty} [\text{ang } Z'_{12}(j\omega)]\, du = \frac{\pi}{2} \log_\varepsilon m. \tag{12.49}$$

This result shows that the area under the phase characteristic is independent of the compensating capacitor C_f. The effect of varying C_f is merely to shift the gain and phase characteristics along the frequency scale without altering their general shapes. The area under the phase characteristic does, however, depend upon the amount of feedback required at low frequencies in that the smaller the value of R_f, the less the phase advance available for compensation. But a smaller R_f implies larger amounts of feedback, requiring more phase area to make the amplifier stable. It follows, therefore, that as the applied feedback is increased, the available phase advance is reduced and it becomes more difficult to achieve stability with compensation in the feedback path.

There are several ways of evaluating the compensating capacitor C_f. We may, for example, choose C_f so that the loop transmission zero created by insertion of C_f in the feedback path is located at $s = -\omega_p$, where ω_p is the phase-crossover frequency of the uncompensated amplifier, as obtained by calculation or measurement. Then, we have

$$C_f = \frac{1}{\omega_p R_f}. \tag{12.50}$$

Example 12.5. As an illustration, consider the uncompensated transistor feedback amplifier of Example 12.3 for which the loop gain and phase characteristics are shown as curve *I* in Figure 12.25. The phase crossover occurs at $0{\cdot}6 \times 10^8$ rad/sec. Previously, we found that for $T(0) = 30$, the required feedback resistor $R_f = 3{\cdot}7$ kΩ. Hence, from Eq. (12.50) we obtain $C_f = 4{\cdot}5$ pF.

From the values given in Example 12.3 we also have $R_l = 50\ \Omega$ and $R' = 42\ \Omega$; so Eq. (12.48) gives $m = 41$. The effect of adding the capacitor C_f is therefore to modify the loop transmission into the form defined by curves *II* in Fig. 12.25. We see that the inclusion of capacitor C_f has not only stabilized the amplifier for $T(0) = 30$, for which it was previously unstable, but has also enabled us to realize a gain margin of 18 dB and a phase margin of 22°.

Bode's Ideal Loop-gain Characteristic

In the practical design of feedback amplifiers, the requirements are ordinarily for a specified loop gain (preferably constant) throughout the useful frequency band and for specified gain and phase margins to guard against instability.

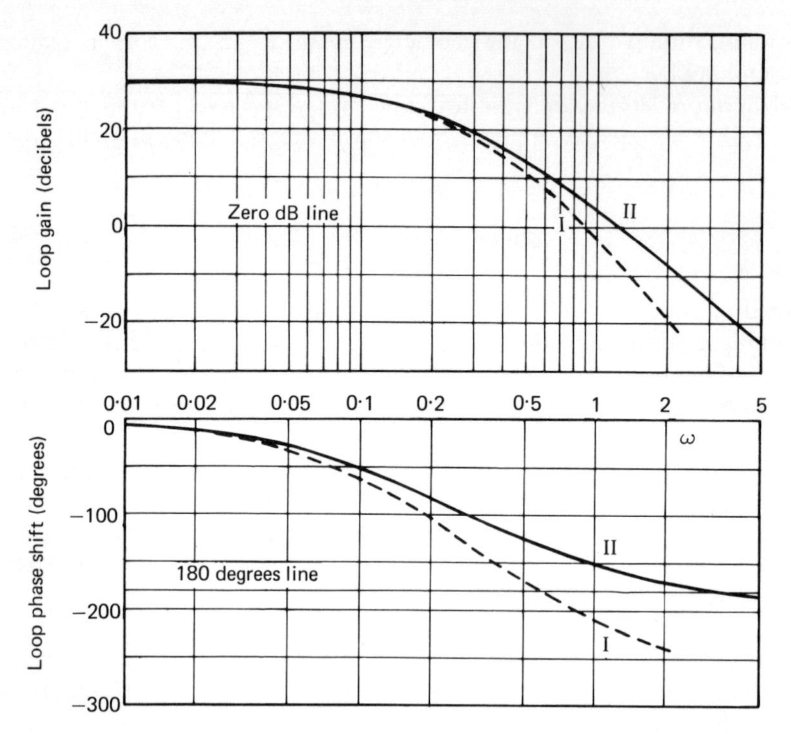

Fig. 12.25. Gain–phase plots of return ratio of uncompensated and compensated feedback amplifier.

These design requirements can all be satisfied by shaping the open-loop frequency response to follow a constant-phase cut-off type of characteristic, so that the loop gain and loop-phase shift have the characteristics shown in Fig. 12.26. The constant phase base is set at an angle of $-k\pi/2$ radians, thereby providing a phase margin of $(1 - k/2)\pi$ radians, where $k < 2$. Also, the flat portion of the gain characteristic, extending up to ω_0, provides a constant loop gain throughout the useful frequency band. The resulting Nyquist plot is as shown in Fig. 12.27.

In practice, however, the open-loop frequency response can be shaped to follow the constant-phase cut-off characteristic only up to a certain frequency, for the following reason. Suppose that for some specified element the return ratio $T(s)$ has m zeros and n poles; then as the frequency ω approaches infinity, we have

$$\lim_{\omega \to \infty} T(j\omega) = K_0(j\omega)^{-(n-m)}, \tag{12.51}$$

where K_0 is a scale factor. On a logarithmic frequency scale, Eq. (12.51) represents a straight line, referred to as the *final asymptote*, with slope $n - m$ in units of 6 dB/octave. The quantity $n - m$ will be represented by l in future discussion.

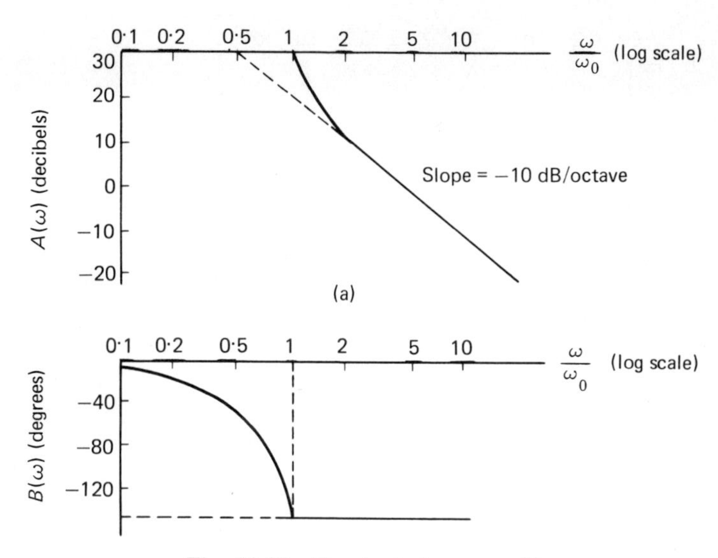

Fig. 12.26. Constant-phase cut off.

Since the logarithmic gain of a common-cathode or common-emitter stage falls off ultimately at the rate of at least -6 dB/octave it is clearly impossible for l to be less than the number of vacuum tubes or transistors making up the internal amplifier component. Furthermore, in negative-feedback amplifiers where there is a likelihood of instability occurring we find that the slope of the final asymptote is greater in absolute value than that of the constant-phase cut-off characteristic, that is, $l > k$. Therefore, the gain component of the constant-phase cut-off can be realized only up to the frequency where it intersects the final asymptote, as indicated in Fig. 12.28(a) for the case of $l = 3$ and $k = \frac{5}{3}$.

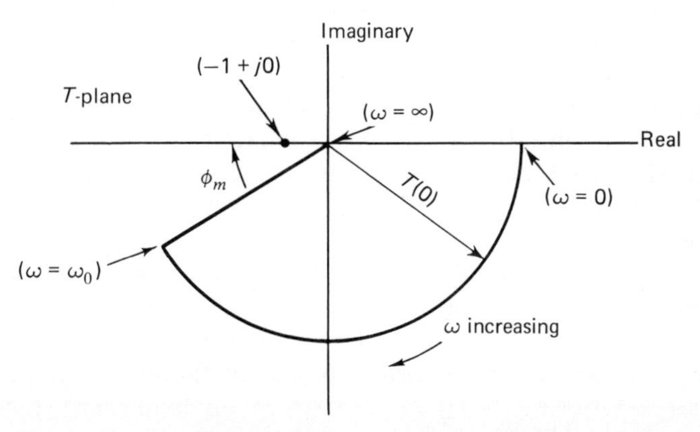

Fig. 12.27. Nyquist diagram of idealized feedback amplifier.

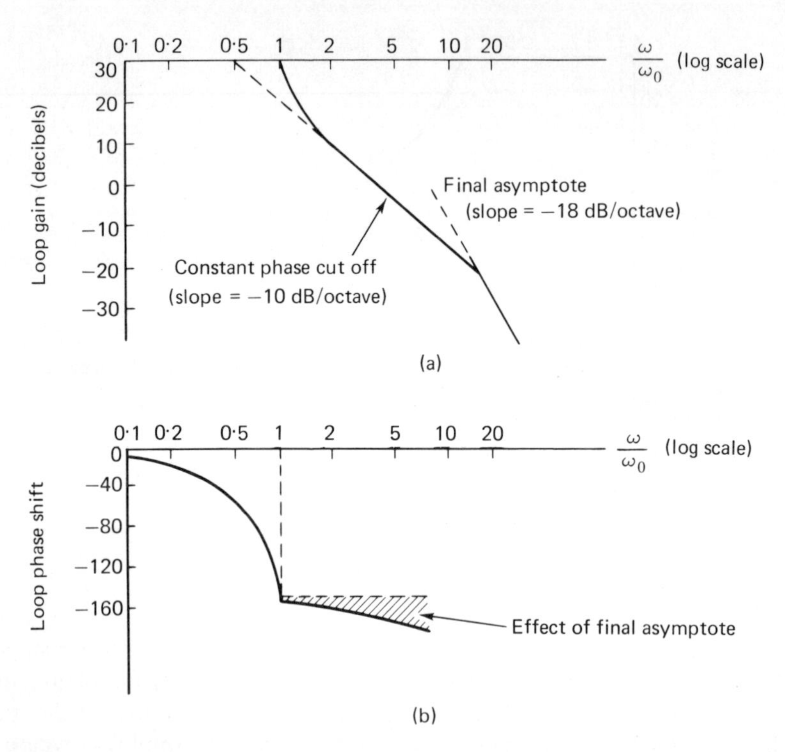

Fig. 12.28. Illustrating the degrading effect of final asymptote upon the loop phase-shift characteristic.

The effect of the final asymptote can be taken into account by adding, to the constant-phase cut-off, a semi-infinite slope equal to $-6(l - k)$ dB/octave and beginning at the intersection frequency. This added semi-infinite slope modifies the total phase shift to the form shown in Fig. 12.28(b), where we see that the much-desired constant-phase feature has been completely destroyed. However, we can counteract this effect, at least within a limited frequency range, by introducing a minor transition region in the form of a horizontal step between the constant-phase cut-off and the final asymptote as shown in Fig. 12.29(a). The horizontal step is located below the zero-decibel line by an amount equal to the specified gain margin. The design becomes complete once the frequency ω_b is determined.

The characteristic of Fig. 12.29(a) can be resolved into the sum of a constant-phase cut-off extending to infinity and two semi-infinite slopes, one of which starts at the frequency ω_b and has a positive slope of $6k$ dB/octave, and the other starts at ω_d and has a negative slope equal to that of the final asymptote, that is, $-6l$ dB/octave. From Eq. (7.48) we see that, at low frequencies, the phase shift of the first semi-infinite asymptote is equal to $2\,k\omega/\pi\omega_b$ radians and that of the

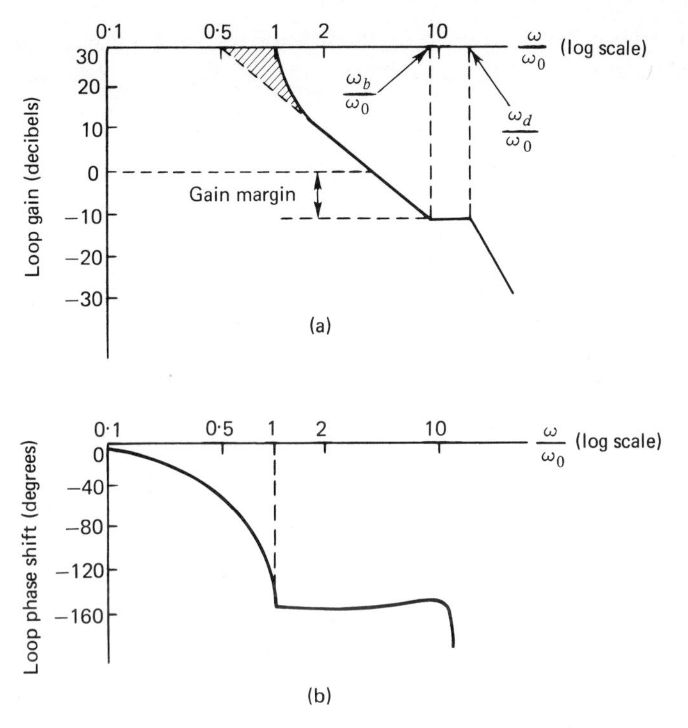

Fig. 12.29. Bode's ideal loop-gain characteristic.

second one is equal to $-2l\omega/\pi\omega_d$ radians. These two phase shifts can be made to cancel one another if we choose

$$\omega_b = \frac{k\omega_d}{l}. \tag{12.52}$$

At low frequencies, we now find that the total phase shift associated with the loop-gain characteristic of Fig. 12.29(a) becomes practically the same as that obtained from a constant-phase cut-off. For the case of $k = \frac{5}{3}$ (i.e., phase margin $= 30°$) and $l = 3$, the complete phase characteristic is as shown in Fig. 12.29(b). The gain characteristic of Fig. 12.29(a) is known as *Bode's ideal loop gain characteristic.*[13]

It is significant to observe that the so-called *fillet*, which is shown shaded in Fig. 12.29(a), results in an increase of $6k$ dB in the loop gain inside the useful band. Furthermore, we see that the cut-off interval extending from the edge of the useful band to the junction of the ideal characteristic with the final asymptote, in octaves, is one more than the feedback in the useful band, expressed as a multiple of 10 dB. We may state, therefore, that the width of the cut-off interval over which effective control of the loop transmission characteristic is necessary is roughly one octave for each 10 dB of feedback in the useful band, plus one additional octave.

In Fig. 12.30(a), curve *I* represents the uncorrected loop-gain characteristic of a three-stage feedback amplifier, while curve *II* represents the required ideal characteristic realizing specified gain and phase margins. The difference indicated by the shaded area is therefore the loss characteristic which should be introduced into the loop, by some corrective network. The effect of such a change may be seen by comparing the corresponding phase characteristics shown by curves *I* and *II* of Fig. 12.30(b). The areas under the two curves are the same, but the insertion of the additional loss characteristic redistributes the total area so that the loop phase remains in magnitude less than 180° minus the required phase margin over a much broader frequency interval.

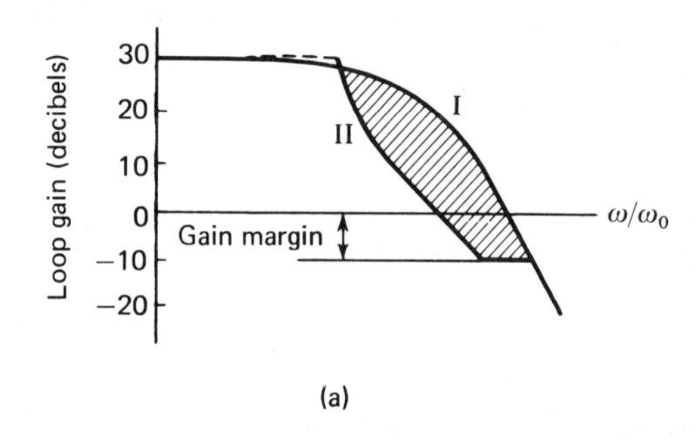

(a)

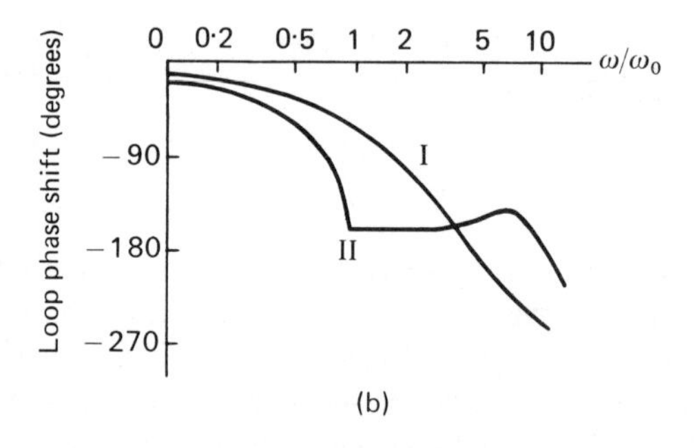

(b)

Fig. 12.30. Shaping of the open-loop frequency response of a three-stage feedback amplifier.

Synthesis of Corrective Network

The precise methods by which Bode's ideal loop-gain characteristic is to be achieved would naturally differ from amplifier to amplifier. However, a technique which is widely used in practice is to add a corrective one-port network of admittance Y_c to one of the inter-stages, as indicated in Fig. 12.31. To evaluate the

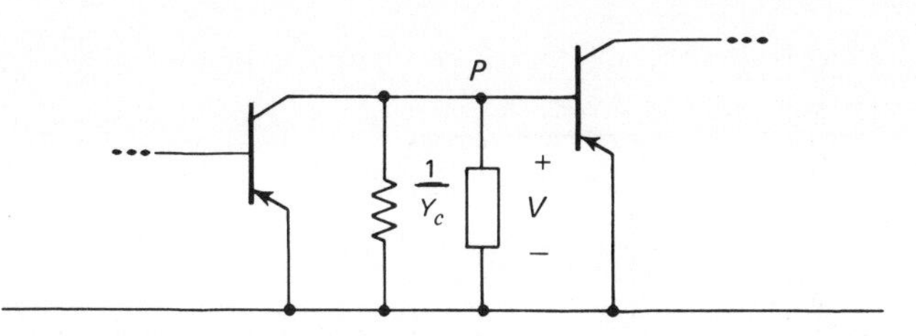

Fig. 12.31. Part of transistor feedback amplifier showing corrective network in position.

effect of inserting this network, we may consider the model shown in Fig. 12.32 where Y_t represents the total admittance measured between the common terminal and interstage point P of the amplifier, with the feedback loop broken and the corrective network removed from circuit. The voltage V developed at the point P is $I/(Y_t + Y_c)$. If, however, the corrective network is removed from circuit we have a voltage $V' = I/Y_t$ developed at the point P. Hence, the insertion-voltage ratio V'/V is given by

$$\varepsilon^{\theta} = \frac{V'}{V} = 1 + \frac{Y_c}{Y_t} \tag{12.53}$$

with $\theta = A + jB$; A is the insertion loss, in nepers, of the corrective one-port network, and B is the phase shift, in radians, associated with the loss A. Solving Eq. (12.53) for the admittance Y_c, we get

$$Y_c = Y_t(\varepsilon^{\theta} - 1). \tag{12.54}$$

Hence, if ε^{θ} is chosen to modify the amplifier corner frequencies in the required way, Y_c may be determined from a knowledge of Y_t and ε^{θ}.

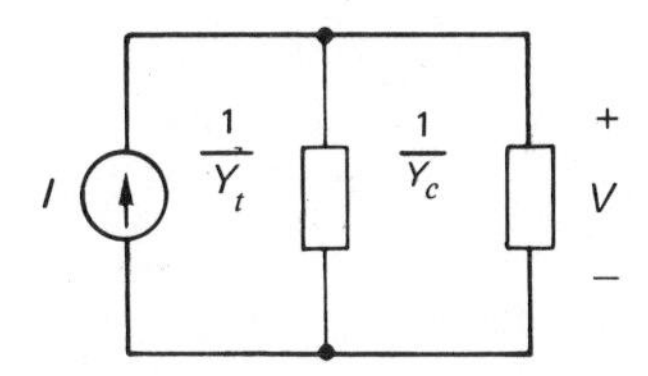

Fig. 12.32. Model for evaluating the insertion effect of corrective network.

In most cases the desired insertion loss characteristic is closely realized by the straight-line approximation of Fig. 12.33 which corresponds to:

$$\varepsilon^{\theta} = K_0 \frac{(s - z_1)(s - z_2)}{(s - p_1)(s - p_2)}. \tag{12.55}$$

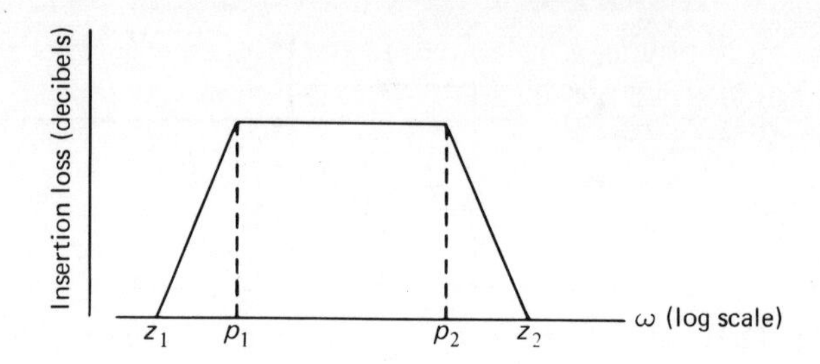

Fig. 12.33. Straight-line approximation of desired insertion loss characteristic.

The zeros z_1, z_2 and poles p_1, p_2 are located on the negative real axis of the s-plane in the manner illustrated in Fig. 12.34.

In Fig. 12.33 we see that the insertion loss A is zero at both zero and infinite frequency; we must have, therefore, that the scale factor K_0 is unity, and

$$z_1 z_2 = p_1 p_2. \tag{12.56}$$

Accordingly, Eq. (12.55) may be written in the following expanded form

$$\varepsilon^{\theta} = \frac{s^2 - (z_1 + p_1 p_2/z_1)s + p_1 p_2}{s^2 - (p_1 + p_2)s + p_1 p_2}, \tag{12.57}$$

from which it follows that

$$\varepsilon^{\theta} - 1 = \frac{s(z_1 - p_1)(p_2 - z_1)}{z_1(s - p_1)(s - p_2)}. \tag{12.58}$$

In a transistor amplifier using common-emitter stages, as in Fig. 12.31, we find, in general, that the admittance Y_t can be closely approximated by a resistance–capacitance type of driving-point admittance,

$$Y_t = G_0 \frac{p_0(s - z_0)}{z_0(s - p_0)}, \tag{12.59}$$

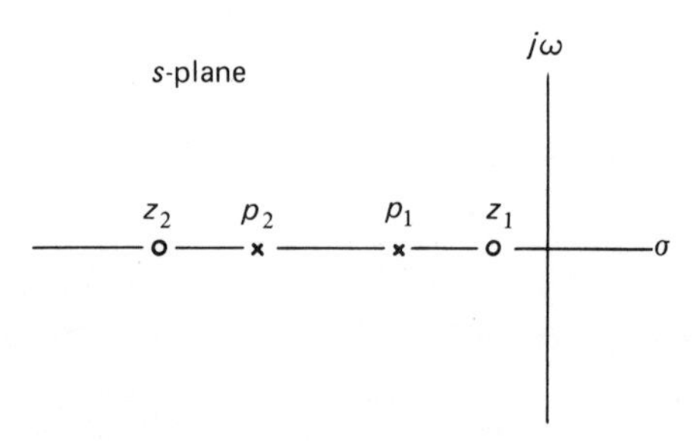

Fig. 12.34. Pole–zero pattern of ε^{θ}.

where the zero z_0 and pole p_0 are located on the negative real axis, with the zero nearest to the origin. The conductance and susceptance components of Y_t are, respectively, given by

$$G_t = G_0 \frac{p_0(\omega^2 + z_0 p_0)}{z_0(\omega^2 + p_0^2)} \tag{12.60}$$

$$B_t = G_0 \frac{\omega p_0(z_0 - p_0)}{z_0(\omega^2 + p_0^2)}. \tag{12.61}$$

We see, therefore, that G_0 and $G_0 p_0/z_0$ are the values of the conductance G_t at zero and infinite frequencies, respectively, and that the conductance G_t increases with frequency in the manner indicated by the approximate straight line characteristic of Fig. 12.35.

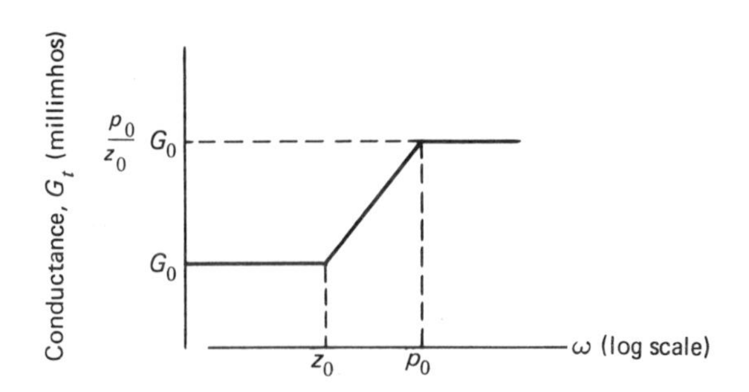

Fig. 12.35. Straight-line approximation of conductance G_t.

Substitution of Eqs. (12.58) and (12.59) in Eq. (12.54) yields

$$Y_c = \frac{G_0 p_0}{z_0 z_1}(z_1 - p_1)(p_2 - z_1)\frac{s(s - z_0)}{(s - p_0)(s - p_1)(s - p_2)}. \tag{12.62}$$

Suppose, however, that the p_1 of ε^θ and the z_0 of Y_t are chosen such that

$$z_0 = p_1. \tag{12.63}$$

Then, a pole–zero cancellation takes place, and accordingly Eq. (12.62) reduces to

$$Y_c = \frac{G_0 p_0}{p_1 z_1}(z_1 - p_1)(p_2 - z_1)\frac{s}{(s - p_0)(s - p_2)}. \tag{12.64}$$

On an impedance basis, we thus get

$$Z_c = \frac{1}{Y_c} = \frac{p_1 z_1}{G_0 p_0(z_1 - p_1)(p_2 - z_1)}\left[s + \frac{p_0 p_2}{s} - (p_0 + p_2)\right]. \tag{12.65}$$

Therefore, the corrective network can be realized by the series *LCR* network of Fig. 12.36 having the following element values

$$L = \frac{p_1 z_1}{G_0 p_0 (z_1 - p_1)(p_2 - z_1)}$$

$$C = \frac{G_0 (z_1 - p_1)(p_2 - z_1)}{p_1 p_2 z_1} \tag{12.66}$$

$$R = \frac{p_1 z_1 (p_0 + p_2)}{G_0 p_0 (p_1 - z_1)(p_2 - z_1)}.$$

In Eqs. (12.66) the elements L, C, and R have positive values and are, therefore, assured of physical realizability.

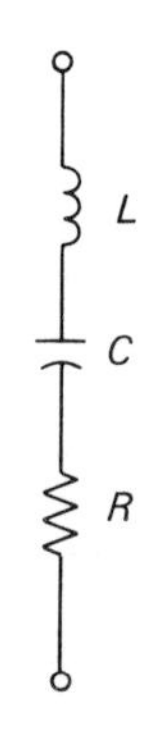

Fig. 12.36. Corrective network.

Equations (12.56) and (12.63) impose two conditions on the possible choice of the two poles and two zeros of the transfer function ε^{θ}. In spite of these restrictions, however, it is still quite possible to choose suitable values for z_1 and p_2 so as to approximate closely to the desired insertion loss characteristic.

Example 12.6. As an illustrative example, consider the synthesis of a corrective network required to introduce the insertion loss characteristic of Fig. 12.37 into the loop. The measured conductance (G_t) and susceptance (B_t) characteristics at the port where the corrective network is to be connected are given in Fig. 12.38. In this diagram we observe that

$$G_0 = 1{\cdot}5 \text{ m℧}.$$

Let Eq. (12.60) be forced to pass through two widely spaced points on the conductance characteristic of Fig. 12.38. For example, two such points are the following

$$\omega = 0{\cdot}5, \qquad G_t = 4{\cdot}5 \text{ m℧}$$

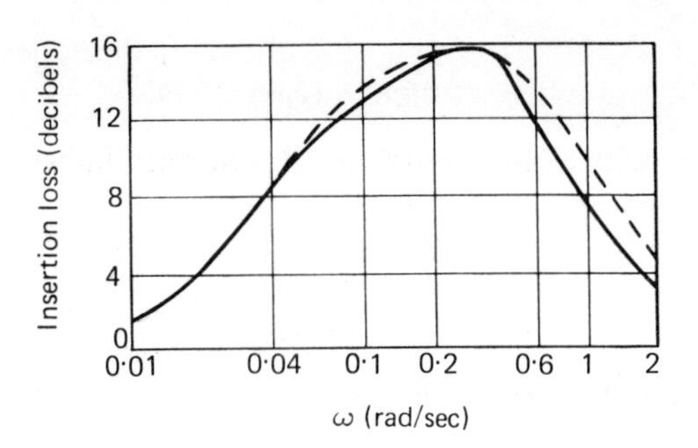

Fig. 12.37. The insertion loss of corrective network (Continuous curve is the actual characteristic; dashed curve is obtained from Eq. 12.55).

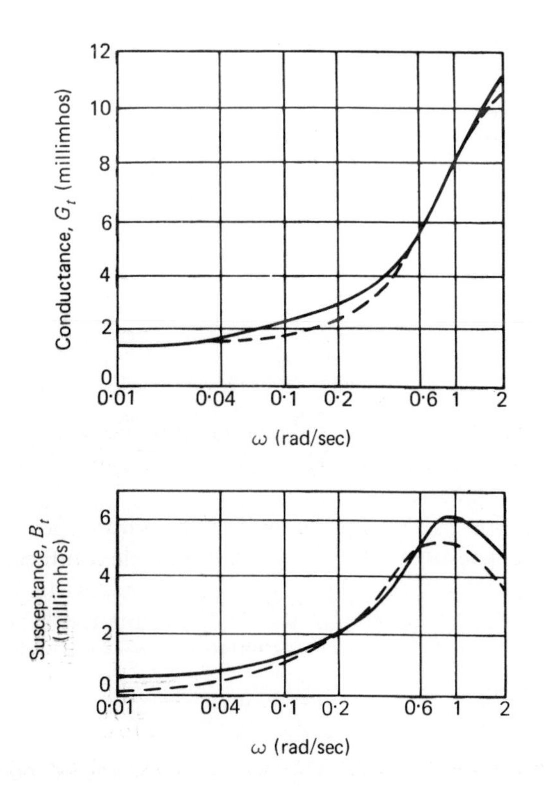

Fig. 12.38. The conductance, G_t, and susceptance, B_t, characteristics at the amplifier interstage point where the corrective network is to be introduced (Continuous curves are the actual characteristics; dashed curves are obtained from Eqs. 12.67 and 12.68).

and

$$\omega = 1{\cdot}25, \qquad G_t = 9 \text{ m℧}$$

where ω is normalized with respect to 10^6 rad/sec. Substituting these values in Eq. (12.60), and solving for z_0 and p_0, we find that

$$z_0 = -0{\cdot}099$$

$$p_0 = -0{\cdot}792.$$

Hence, Eqs. (12.60) and (12.61) give

$$G_t = \frac{1{\cdot}5(1 + 12{\cdot}8\omega^2)}{1 + 1{\cdot}6\omega^2} \text{ m℧}, \tag{12.67}$$

$$B_t = \frac{13{\cdot}3\omega}{1 + 1{\cdot}6\omega^2} \text{ m℧}. \tag{12.68}$$

These are shown plotted as the dashed curves in Fig. 12.38 where sufficiently close agreement with the actual conductance and susceptance characteristics is observed.

Next, we have to determine the poles and zeros of the transfer function ε^{θ} so as to realize the insertion loss characteristic of Fig. 12.37 and, at the same time, satisfy the conditions of Eqs. (12.56) and (12.63). By a process of trial and error it is found the following set of poles and zeros is acceptable:

$$z_1 = -0{\cdot}015$$

$$z_2 = -3{\cdot}31$$

$$p_1 = -0{\cdot}099$$

$$p_2 = -0{\cdot}5.$$

The corresponding insertion loss, calculated from Eq. (12.57) is shown plotted as the dashed curve in Fig. 12.37. Here again close agreement with the required insertion loss characteristic is observed.

Finally, using Eqs. (12.66) and denormalizing with respect to 10^6 rad/sec, we find that the corrective network of Fig. 12.36 has the following element values:

$$L = 30{\cdot}8\ \mu\text{H}$$

$$C = 8200 \text{ pF}$$

$$R = 39{\cdot}6\ \Omega.$$

12.7 DESCRIBING FUNCTION ANALYSIS OF NONLINEAR FEEDBACK CIRCUITS

The Nyquist criterion as previously described is applicable only to linear circuits. It may, however, be extended to testing the stability of a feedback circuit containing a nonlinear element provided that the waveforms in the circuit are essentially sinusoidal in shape. This necessary condition is valid if the nonlinearity is inherently small, or if some kind of low-pass filtering action takes place in the circuit, so as to minimize the harmonic frequencies generated by the nonlinearity. The procedure is based on the principle of equivalent linearization which involves finding a kind of equivalent transfer function for the nonlinear element by considering the fundamental components only. This equivalent function, called the *describing function* for the nonlinear element, is defined as the ratio of the amplitude of the fundamental component of the output signal to the amplitude of the input signal, as expressed by[14]

$$H = \frac{R_1}{E_1}, \tag{12.69}$$

where H is the describing function, E_1 is the amplitude of the sinusoidal input signal $E_1 \sin \omega t$, and R_1 is the amplitude of the fundamental component $R_1 \sin \omega t$ of the output.* The amplitude R_1 is readily determined using a conventional Fourier analysis. Typically, the describing function is dependent upon the amplitude E_1. It would also depend on the frequency ω if the nonlinear device contains any energy storage components; otherwise it is independent of ω.

The stability analysis of a nonlinear feedback circuit begins with a calculation of the describing function for the nonlinear device which is thereafter treated as an element with a transfer function dependent upon the input-signal level. Then using a modification of the Nyquist diagram, as illustrated in the example below, we may study the stability of the circuit as a function of the signal level. It should, however, be emphasized again that the describing function analysis is meaningful and leads to results of useful accuracy only when the circuit is of such a nature that the fundamental is the only significant signal component circulating around the complete feedback loop.

Example 12.7. *RC Phase Shift Oscillator.* Consider the feedback oscillator shown in Fig. 12.39 in which the feedback network consists of a low-pass *RC* ladder. The amplifier component is a common-cathode stage which is assumed to be biased in such a manner that the output-plate voltage is symmetrically limited by the tube being driven into saturation at one end and cut off at the other, as illustrated in the idealized dynamic-transfer characteristic of Fig. 12.40. It is assumed that the feedback network presents negligible loading at the output port of the amplifier.

According to Fig. 12.40, for an input grid excitation of $v_g(t) = V_g \sin \omega t$ in which $V_g \leq V_c$, the plate voltage remains sinusoidal, as in Fig. 12.41(a). If, on the other hand, $V_g > V_c$, then the plate voltage is nonsinusoidal but still periodic, as indicated in Fig. 12.41(b). In this latter case, we find from symmetry considerations that the Fourier series representation of the output signal contains only sine

* Equation (12.69) assumes that the nonlinear device exhibits no hysteresis; otherwise, the describing function has a magnitude as well as phase.

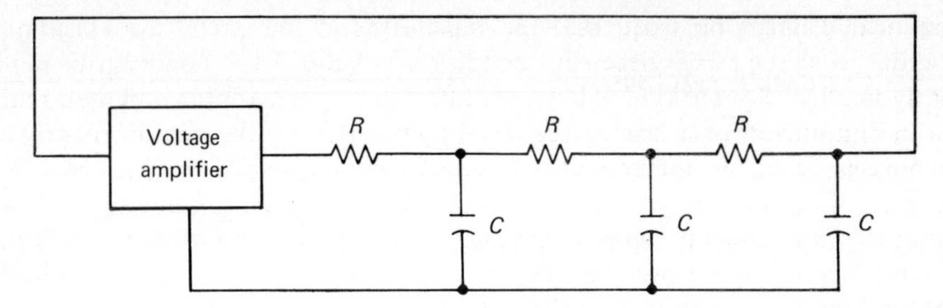

Fig. 12.39. *RC* phase-shift oscillator.

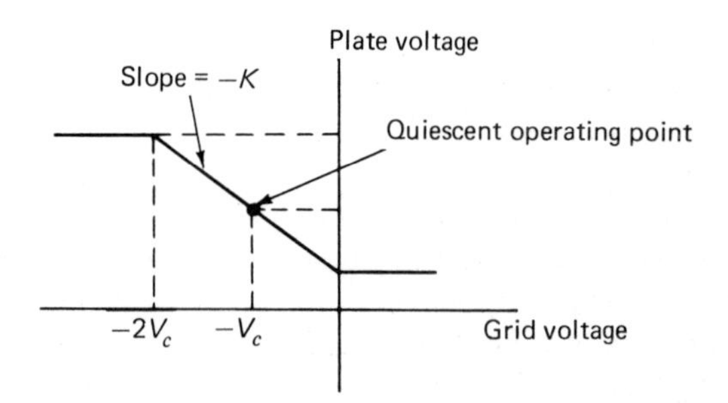

Fig. 12.40. Idealized dynamic transfer characteristic of common-cathode amplifier.

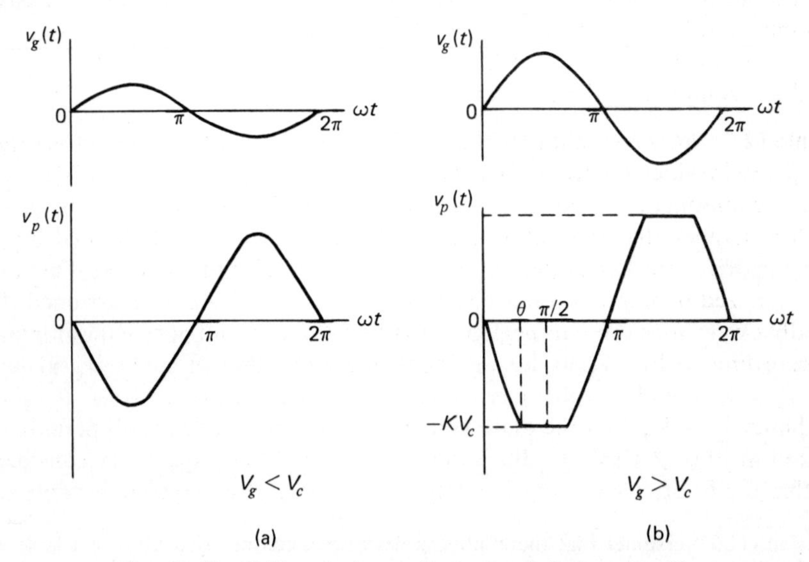

Fig. 12.41. Grid and plate voltage signals for varying V_g.

terms at odd harmonic frequencies, as expressed by

$$v_p(t) = -V_{p1}\sin\omega t - V_{p3}\sin 3\omega t - V_{p5}\sin 5\omega t - \cdots, \tag{12.70}$$

in which, however, only the fundamental component is of direct interest to us. The amplitude of this component is given by the formula

$$V_{p1} = \left|\frac{4}{\pi}\int_0^{\pi/2} v_p(t)\sin\omega t\, d(\omega t)\right|. \tag{12.71}$$

In Fig. 12.41(b) we see that during the first quarter cycle

$$v_p(t) = \begin{cases} -KV_g\sin\omega t & 0 \le \omega t \le \theta \\ \\ -KV_c & \theta \le \omega t \le \dfrac{\pi}{2}, \end{cases} \tag{12.72}$$

where K is the incremental voltage amplification in the linear region of operation, and the angle θ is defined by

$$\theta = \sin^{-1}\left(\frac{V_c}{V_g}\right), \qquad V_g > V_c.$$

Therefore, substituting Eq. (12.72) in (12.71) and carrying out the integration, we obtain,

$$V_{p1} = \frac{KV_g}{\pi}(2\theta + \sin 2\theta). \tag{12.73}$$

For convenience of stability analysis we will model the vacuum-tube amplifier as the cascade combination of a limiter having a describing function $H(V_g)$ dependent upon the amplitude V_g of the input and a linear amplifier having an incremental voltage amplification $-K$, as in the signal-flow graph of Fig. 12.42.

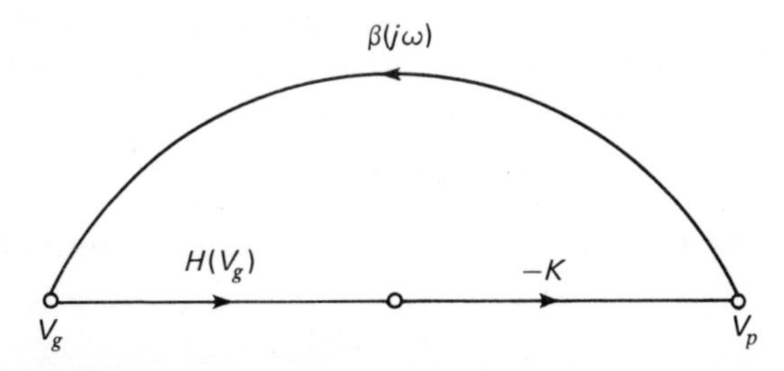

Fig. 12.42. Signal-flow graph representation of *RC* phase-shift oscillator.

The describing function $H(V_g)$ is defined by

$$H(V_g) = \begin{cases} 1 & V_g < V_c \\ \dfrac{1}{\pi}(2\theta + \sin 2\theta) & V_g > V_c. \end{cases} \tag{12.74}$$

The variation of $H(V_g)$ with V_g is shown in Fig. 12.43. The diagram indicates that for V_g greater than the critical level V_c, the effective amplification of the vacuum-tube amplifier decreases with increasing V_g.

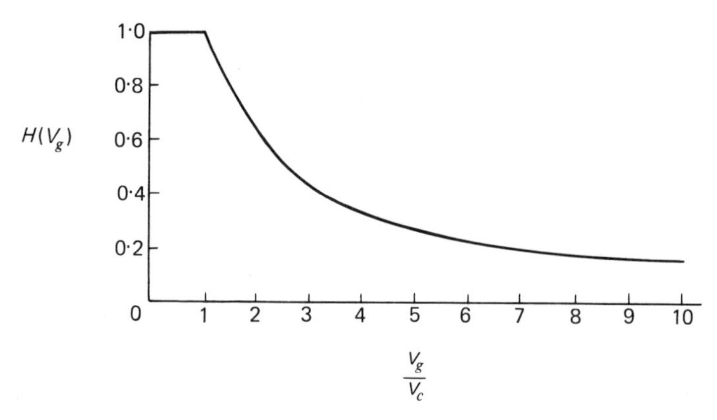

Fig. 12.43. The describing function $H(V_g)$.

In the signal-flow graph of Fig. 12.42 we have also included a branch with a transmittance $\beta(j\omega)$ representing the reverse transmission through the feedback network. The return ratio $T'(j\omega, V_g)$ for K is simply the negative of the loop transmittance $-K\beta(j\omega)H(V_g)$, or

$$T'(j\omega, V_g) = T(j\omega)H(V_g) \tag{12.75}$$

where

$$T(j\omega) = K\beta(j\omega). \tag{12.76}$$

The $T(j\omega)$ of Eq. (12.76) is the return ratio assuming ideal operation of the amplifier without amplitude limiting. The stability of the nonlinear feedback circuit is governed by roots of the characteristic equation,

$$1 + T(j\omega)H(V_g) = 0, \tag{12.77}$$

which can be put into more convenient form for study by expressing it as

$$\frac{1}{T(j\omega)} = -H(V_g). \tag{12.78}$$

Equation (12.78) is the condition for steady-state oscillation.

Evaluating the reverse transmittance $\beta(j\omega)$ for the RC ladder shown in Fig. 12.38 and then multiplying the result by K, we obtain

$$T(j\omega) = \frac{K}{1 - 5\omega^2 R^2 C^2 + j\omega RC(6 - \omega^2 C^2 R^2)}. \tag{12.79}$$

To be specific, suppose $K = 32$ which is slightly greater than the minimum value of 29 required to produce instability. In Fig. 12.44 we have plotted two loci representing the two sides of Eq. (12.78), one of which is the inverse Nyquist diagram,

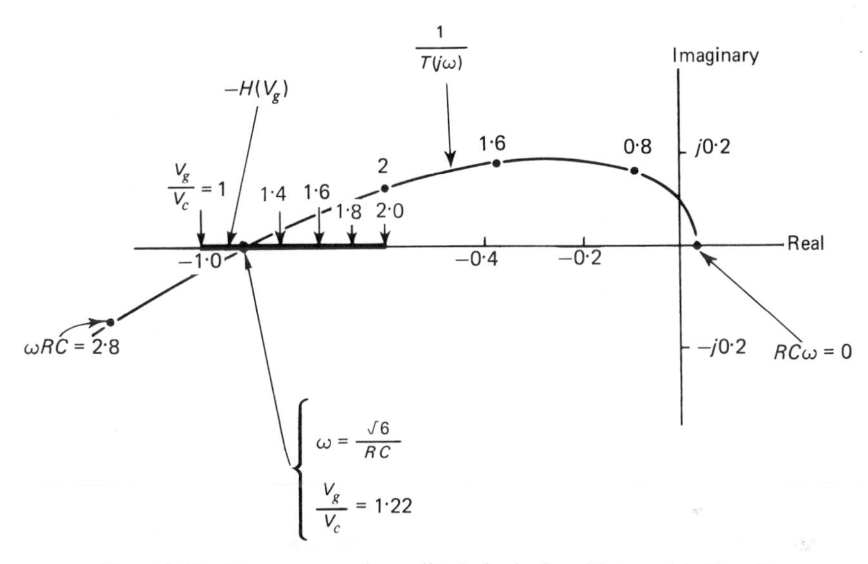

Fig. 12.44. Frequency and amplitude loci of oscillator with $K = 32$.

or *frequency locus*, involving a plot of the imaginary versus the real part of $1/T(j\omega)$; and the other is the *amplitude locus* showing the variation of $-H(V_g)$ with the amplitude V_g of the grid voltage. Along the frequency locus we have marked certain values of the quantity ωRC, while along the amplitude locus we have marked certain values of the ratio V_g/V_c. Equation (12.78) is satisfied when the two loci intersect, and the conditions determined by the intersection point correspond to steady-state oscillations in the nonlinear feedback circuit. In the case of Fig. 12.44 we see therefore that steady-state oscillation occurs with a frequency $\omega = \sqrt{6}/RC$ and an amplitude $V_g = 1{\cdot}22V_c$.

It should, however, be noted that as the voltage amplification K pertaining to the linear region of operation is increased further and further beyond the value of 29, which is the minimum necessary condition for sustained oscillations, the output waveform of the vacuum-tube amplifier approaches more nearly a square wave, becoming more enriched in harmonics and thereby making the describing function approach less applicable.

Concluding Remarks

To conclude, the describing function of a nonlinear device is defined on the basis of the frequency response, postulating two basic assumptions:

a) The input signal to the nonlinear device is a pure sine wave.
b) The effects of second- and higher-order harmonics appearing in the periodic output signal of the nonlinear device may be neglected.

These two assumptions are adequately satisfied in a feedback circuit that contains sufficient low-pass filtering action in the linear components located between the output and input terminals of the nonlinear device inside the loop. The describing-function technique is thus ideally suited to the study of feedback oscillators that are actually designed to produce an almost sinusoidal behaviour. In an oscillator, nonlinearity has the effect of limiting the amplitude of oscillation, which is accomplished by driving the amplifier into saturation for part of the cycle, or alternatively by including an amplitude-controlled resistor (e.g., thermistor) as part of the amplifier or frequency-selective feedback network.

REFERENCES

1. W. R. Evans, "Graphical Analysis of Control Systems," *Trans. A.I.E.E.* **67,** 547 (1948).
2. J. G. Truxal, *Automatic Feedback Control System Synthesis.* McGraw-Hill, 1955.
3. J. H. Mulligan, "The Effect of Pole and Zero Locations on the Transient Response of Linear Dynamic Systems," *Proc. I.R.E.* **37,** 516 (1949).
4. J. H. Mulligan, "Transient Response and the Stabilization of Feedback Amplifiers," *Trans. A.I.E.E.* **78,** part II, 495 (1960).
5. J. G. Truxal and I. M. Horowitz, "Sensitivity Considerations in Active Network Synthesis," *Proc. Second Midwest Symposium on Circuit Theory*, 1956.
6. H. Ur, "Root Locus Properties and Sensitivity Relations in Control Systems," *Trans. I.R.E.* **AC-5,** 57 (1960).
7. H. Nyquist, "Regeneration Theory," *B.S.T.J.* **11,** 126 (1932).
8. J. C. West, *Servomechanisms.* E.U.P., 1953.
9. F. H. Blecher, "Transistor Multiple-loop Feedback Amplifiers," *Proc. National Electronics Conference* **13,** 19 (1957).
10. B. R. Myers, "A Useful Extension of the Nyquist Criterion to Stability Analysis of Multiloop Feedback Amplifiers," *Proc. Fourth Midwest Symposium on Circuit Theory*, J1–J17 (1959).
11. L. A. Zadeh and C. A. Desoer, *Linear System Theory.* McGraw-Hill, 1963.
12. J. Almond and A. R. Boothroyd, "Broadband Transistor Feedback Amplifiers," *Proc. I.E.E.* **103,** 93 (1956).

13. H. W. BODE, *Network Analysis and Feedback Amplifier Design*. Van Nostrand, 1945.
14. R. J. KOCHENBURGER, "A Frequency Response Method for Analysing and Synthesizing Contactor Servomechanisms," *Trans. A.I.E.E.* **69**, part I, 270 (1950).
15. R. A. HADDAD and J. G. TRUXAL, "Sensitivity and Stability in Multiloop Systems," *Trans. I.E.E.E.* **AC-9**, 548 (1964).
16. A. J. COTE and J. B. OAKES, *Linear Vacuum-tube and Transistor Circuits*. McGraw-Hill, 1961.
17. L. STRAUSS, *Wave Generation and Shaping*. McGraw-Hill, 1960.

PROBLEMS

12.1 Determine the natural frequencies of the common-emitter amplifier of Fig. P12.1, using the hybrid-π model to represent the transistor. Investigate the effect of varying the inductor L_c upon the natural frequencies of the circuit. What is the value of L_c required for critically damped transient response?

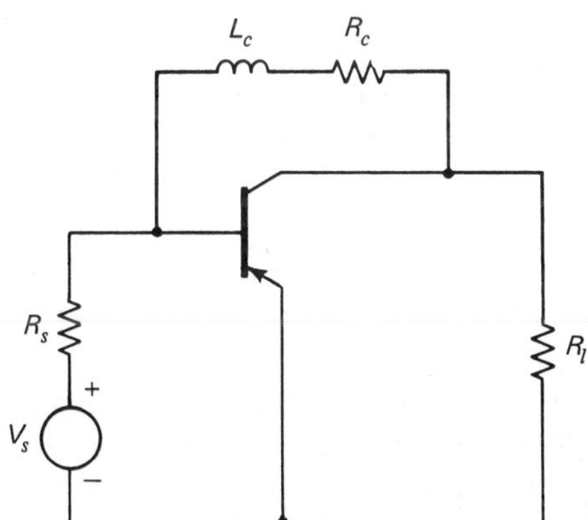

Fig. P12.1.

12.2 A three-stage feedback amplifier has the return ratio

$$T(s) = \frac{T(0)}{(1 - s/\sigma_1)(1 - s/\sigma_2)(1 - s/\sigma_3)}$$

in which σ_1, σ_2 and σ_3 are negative real numbers. Show that the critical value of $T(0)$ which places the amplifier on the verge of instability is given by

$$T_c = (\sigma_1 + \sigma_2 + \sigma_3)\left(\frac{1}{\sigma_1} + \frac{1}{\sigma_2} + \frac{1}{\sigma_3}\right) - 1.$$

For this condition, show that the amplifier has a pair of natural frequencies at $s = \pm j\sqrt{\sigma_1\sigma_2 + \sigma_2\sigma_3 + \sigma_3\sigma_1}$.

12.3 Sketch the root-locus diagram for a negative-feedback amplifier having the return ratio

$$T(s) = T(0)\frac{(1 + s)}{(1 + 5s)^2(1 + s/5)^2}.$$

On the diagram indicate such important data as the asymptotes, points of intersection of the locus with the imaginary axis, and the critical value of $T(0)$ which places the amplifier on the threshold of instability.

For $T(0) = 30$, plot the Nyquist and Bode diagrams, and determine the gain and phase margins of the amplifier.

12.4 In a certain application it is required that a negative-feedback amplifier is not only stable but also that its dominant natural frequencies have a real part more negative than a specified real number σ_0. Show that such a requirement can be investigated by applying the Nyquist criterion to a modified function $T(jx)$ which is obtained from the return ratio $T(j\omega)$ of the amplifier by substituting $j\omega = jx + \sigma_0$. Establish whether or not the dominant natural frequencies have a real part more negative than $-0{\cdot}6$, if the return ratio is as given in Problem 12.3 and $T(0) = 30$.

12.5 Determine the return ratio for the transistor parameter h_{fe} in the RC phase-shift oscillator of Fig. P12.5, using the simplified hybrid model. Assuming that h_{ie} and h_{fe} are both real, and using the Nyquist criterion, show that

a) The minimum value of h_{fe} required for steady-state oscillation is 44·5 corresponding to a minimum value of 0·372 for the ratio $n = R/R_l$.

b) If the conditions in (a) are satisfied, the resulting frequency of oscillation is $\omega = 0{\cdot}242/CR$.

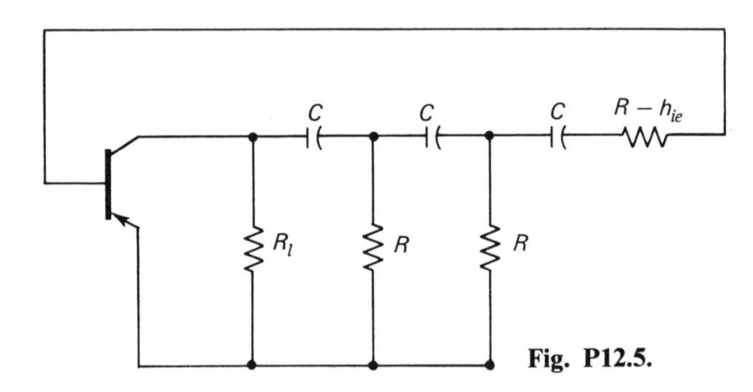

Fig. P12.5.

12.6 The feedback loop of a linear circuit consists of an amplifier and a lossless transmission line of fixed delay τ. The return ratio for the circuit is

$$T(j\omega) = T(0)\frac{\varepsilon^{-j\omega\tau}}{1 + j\omega/\omega_0}.$$

Discuss the effect of the delay factor $\varepsilon^{-j\omega\tau}$ on the Nyquist locus, and show that the circuit oscillates if $T(0) = 1/\cos \omega\tau$.

12.7 The determinant of the return difference matrix, $\Delta_F(s, W)$, of a multiple-loop feedback circuit is said to be complete if every point in the convex hull of $\Delta_F(s, W)$ is also a point of $\Delta_F(s, W)$ itself. Show that the mapping of $\Delta_F(s, W)$ is complete[10] for circuits having the signal-flow graph representations of Fig. P12.7.

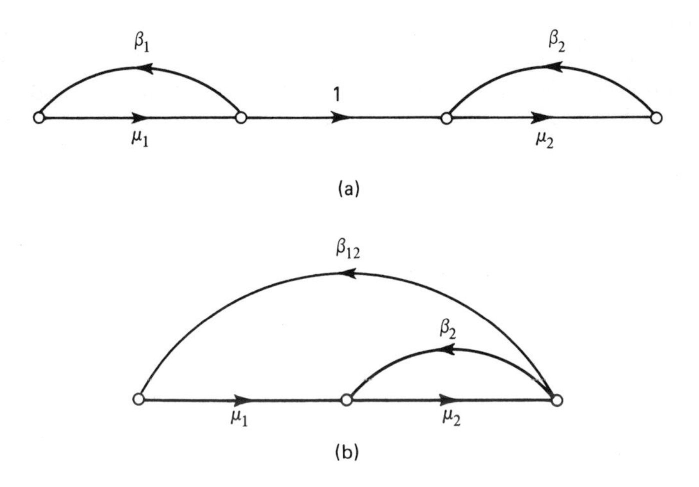

Fig. P12.7.

12.8 Figure P12.8 shows the circuit diagram of the so-called Llewellyn amplifier.[13] Investigate the stability of this multiple-loop feedback circuit, given that $k_1 = 0{\cdot}001$, $k_2 = 0{\cdot}01$, and the admittance $Y = s + 1 + 1/s$. The mutual conductances of the tubes have the following set of normalized values: $g_{m1} = 10$, $g_{m2} = 20$, and $g_{m3} = 40$.

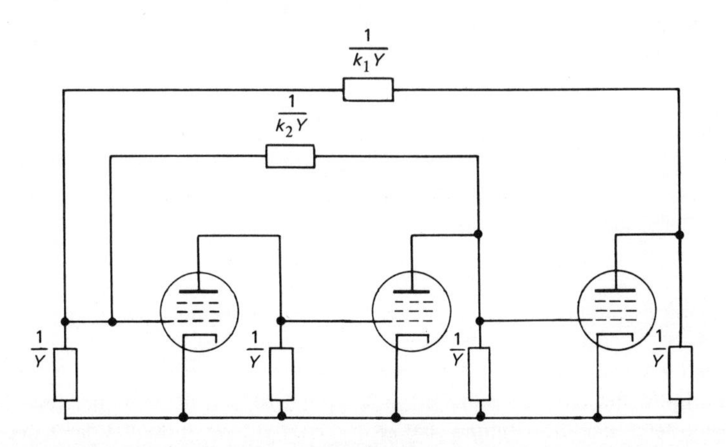

Fig. P12.8.

12.9 A single-loop feedback amplifier involves three stages having cut-off frequencies of 10^6, 2×10^6, and 4×10^6 rad/sec. It is required to have $T(0) = 100$. Construct Bode's ideal loop gain characteristic to realize a gain margin of 10 dB and a phase margin of 30°. Hence, determine the insertion loss characteristic of the required corrective network.

Plot the Nyquist diagram for the corrected amplifier, and then using the graphical technique of Section 12.4, determine approximate values for the co-ordinates of the dominant oscillatory mode.

12.10 The sensitivity function of the closed-loop transmittance of a feedback amplifier with respect to a specified element W is designated by $S(j\omega)$. Show that the critical value, W_c, of this element which places the amplifier on the verge of instability with a pair of natural frequencies at $s = \pm j\omega_c$, is given by either of the following two relations[15]

a)
$$\frac{W_c}{W_0} = 1 + \frac{1}{S(j\omega_c)}$$

which assumes a unity null return difference for W.

b)
$$\frac{W_c}{W_0} = 1 + \frac{1}{S(j\omega_c) - 1}$$

which assumes an infinite null return difference for W.

In both cases, W_0 is the nominal value of the element in question, and the sensitivity $S(j\omega_c)$ is evaluated for $W = W_0$.

How could the above relations be utilized in providing a quantitative measure of circuit stability relative to the element W?

12.11 The cut-off frequency of an amplifier is defined as the frequency at which the gain drops by 3 dB below its midband value. Show that the closed-loop cut-off frequency of a low-pass feedback amplifier which is shaped to follow Bode's ideal loop gain characteristic, is approximately equal to the frequency at which[16]

$$|T(j\omega)| = \frac{1}{\cos\phi_m + (\cos^2\phi_m + 1)^{1/2}},$$

where ϕ_m is the phase margin. Assume that the closed-loop transmittance of the amplifier is expressible in the form

$$K(j\omega) = k\frac{T(j\omega)}{1 + T(j\omega)},$$

where k is independent of frequency. Also, $T(0) \gg 1$ and the gain margin is at least 10 dB.

12.12 Figure 3.13(a) shows a Wien-bridge oscillator in which the amplitude of oscillation is controlled by a thermistor having the relation

$$r = (10 - 14V_2)\,\Omega,$$

where r is the incremental resistance of the thermistor and V_2 is the *RMS* value of the voltage existing across it. Assuming that the amplifier has infinite input impedance, zero output impedance and a voltage gain $K = 1000$, determine the value of resistance R_1 required to produce an output sinusoid that is limited to 30 volts RMS. If the amplifier gain decreases by 50 per cent, determine the resulting change in the amplitude of oscillation. Comment on the performance of the circuit.[17]

CHAPTER 13

SYNTHESIS OF *RC* ACTIVE FILTERS

13.1 INTRODUCTION

Filters using resistors and capacitors only are rather attractive on the grounds that these elements are usually cheaper, more compact and more nearly ideal elements than are inductors. However, because of restrictions imposed on the impedance functions physically realizable with R's and C's only, we find that the natural frequencies of passive RC networks are restricted to lie exclusively on the negative real axis of the s-plane. This, therefore, means that the poles of the transfer function of a passive RC filter can only occur on the negative real axis. The restriction on the location of the poles can be removed if we permit the use of an active element. Then it becomes possible to realize transfer functions having complex-conjugate poles anywhere inside the left half of the s-plane. This in turn enables us to realize highly selective filtering characteristics usually obtained by means of conventional LCR filters. The advantages of eliminating inductors from the filters are particularly apparent at low frequencies when the inductors become rather lossy, bulky and expensive. Also, RC active filters can be conveniently implemented in integrated circuit form.

In this chapter we will study synthesis procedures for two important classes of inductorless filters:

1. *RC–NIC filters* which use a negative-impedance converter (NIC) as the active element.
2. *RC–gyrator filters* in which a gyrator, by virtue of its impedance inverting properties, is used to eliminate the need for an inductor. Although the gyrator is basically a lossless passive element, its practical implementation, however, requires the use of such active devices as transistors. It is on this basis alone that we may justify referring to an RC-gyrator filter as an active filter.

13.2 POSITIVE *RC*–NEGATIVE *RC* CASCADE SYNTHESIS

In a pioneering paper, Linvill[1] proposed an RC active synthesis procedure involving a negative-impedance converter (NIC) and two RC two-port networks connected in cascade, as shown in block diagrammatic form in Fig. 13.1. From Section 4.9 it is recalled that a NIC is an active two-port device which presents at either of its two ports the negative of the impedance connected across the other

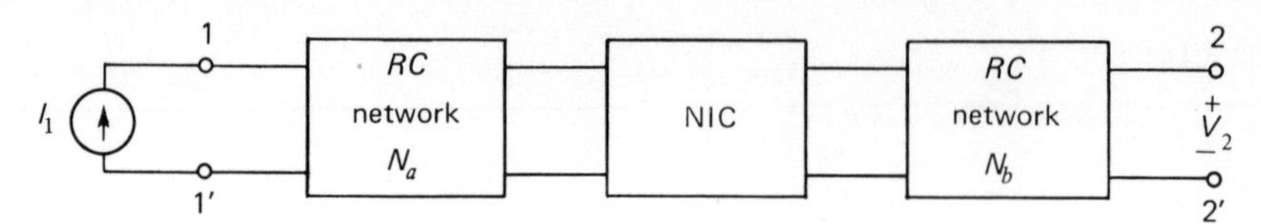

Fig. 13.1. Linvill's *RC*–NIC filter.

port. Assuming that the NIC has a unity conversion ratio, we find by a straightforward analysis that the overall open-circuit transfer impedance of the structure shown in Fig. 13.1 is given by

$$Z_{21}(s) = \left.\frac{V_2}{I_1}\right|_{I_2=0} = \frac{z_{21a}z_{21b}}{z_{22a} - z_{11b}}, \tag{13.1}$$

in which z_{21a} and z_{21b} are the open-circuit transfer impedances of the networks N_a and N_b; z_{22a} and z_{11b} are their driving-point impedances as seen from the ports connected to the NIC. Equation (13.1) pertains to a filter employing a NIC of the voltage-inversion type. The result of using a NIC of the current-inversion type is the same, except for a multiplying factor of -1.

If the driving-point impedances z_{22a} and z_{11b} have, respectively, the same poles as the transfer impedances z_{21a} and z_{21b}, then from Eq. (13.1) it is clear that the poles of the overall transfer impedance $Z_{21}(s)$ are the values of the complex frequency s at which the difference of the two *RC* driving-point impedances z_{22a} and z_{11b} is zero. The zeros of $Z_{21}(s)$ are the values of s at which the product term $z_{21a}z_{21b}$ is zero. It follows therefore that in Fig. 13.1, the natural frequencies of the complete structure are controlled by the NIC, while the zeros of transmission are solely determined by the zeros of the transfer impedances of the two *RC* networks.

Consider a transfer impedance $Z_{21}(s)$ expressed as the ratio of two polynomials in s,

$$Z_{21}(s) = \frac{N(s)}{Q(s)} = K_0 \frac{\prod_{i=1}^{m} (s - s_i')}{\prod_{i=1}^{n} (s - s_i)}, \tag{13.2}$$

in which $m \leq n$ and K_0 is a scale factor. To proceed with the synthesis, we first divide the numerator and denominator polynomials of $Z_{21}(s)$ by an auxiliary polynomial $P(s)$ having n distinct negative real roots $\sigma_1, \sigma_2, \ldots, \sigma_n$, the choice of which is arbitrary as far as the overall transfer impedance of the filter is concerned; of course, none of these roots should coincide with the poles of $Z_{21}(s)$. In certain cases, $P(s)$ may be required to have roots greater than n in number.

We thus obtain

$$Z_{21}(s) = \frac{N(s)/P(s)}{Q(s)/P(s)}, \tag{13.3}$$

where

$$P(s) = \prod_{i=1}^{n} (s - \sigma_i). \tag{13.4}$$

Equations (13.1) and (13.3) can be satisfied by making the identifications

$$z_{21a}z_{21b} = \frac{N(s)}{P(s)} = K_0 \frac{\prod_{i=1}^{m} (s - s_i')}{\prod_{i=1}^{n} (s - \sigma_i)}, \tag{13.5}$$

$$z_{22a} - z_{11b} = \frac{Q(s)}{P(s)} = \frac{\prod_{i=1}^{n} (s - s_i)}{\prod_{i=1}^{n} (s - \sigma_i)}. \tag{13.6}$$

Next, expanding Eq. (13.6) in partial fractions yields,

$$z_{22a} - z_{11b} = 1 + \frac{k_1}{s - \sigma_1} + \frac{k_2}{s - \sigma_2} + \cdots + \frac{k_n}{s - \sigma_n}, \tag{13.7}$$

in which some of the residues will be positive and the remainder negative. However, it is known that any rational function with simple poles on the negative real axis of the s-plane and positive real residues at these poles represents the driving-point impedance of an *RC* network.[2] Hence, in the partial-fraction expansion of $Q(s)/P(s)$, if we group together all terms with positive residues and group together all terms with negative residues, the first group should be associated with z_{22a}, the second group with $-z_{11b}$, that is,

$$z_{22a} = 1 + \sum_{\mu} \frac{k_\mu}{s - \sigma_\mu}, \qquad k_\mu > 0 \tag{13.8}$$

$$-z_{11b} = \sum_{\nu} \frac{k_\nu}{s - \sigma_\nu}, \qquad k_\nu < 0. \tag{13.9}$$

Once the driving-point impedances z_{22a} and z_{11b} are known, we can associate with each of networks N_a and N_b the appropriate number of factors of $N(s)$, thus obtaining the transfer impedances z_{21a} and z_{21b} to within a constant multiplier. We can then determine the *RC* networks N_a and N_b, using established passive-network synthesis techniques.

Design Optimization

As explained previously, the natural frequencies of the filter in Fig. 13.1 which are roots of the characteristic polynomial, $Q(s)$, are determined by the difference of two *RC* driving-point impedances. Accordingly, their locations in the s-plane can be quite sensitive to variations in the conversion ratio of the negative-impedance converter, k, as well as variations in the passive circuit elements. In order to develop the design steps for rendering the filter least sensitive to parameter variations, suppose $Q(s)$ is expressed as the difference of two polynomials, $A(s)$ and $B(s)$, each of which has negative real roots only, as shown by,

$$Q(s) = \prod_i (s - s_i) = A(s) - kB(s). \tag{13.10}$$

For a given $Q(s)$, the roots of both polynomials $A(s)$ and $B(s)$ are determined by the selection of polynomial $P(s)$. Also, in Eq. (13.10) the conversion ratio k has a nominal value of unity. The sensitivity of a root s_i of the characteristic polynomial $Q(s)$ to small changes in k is therefore

$$S_k^{s_i} = \frac{ds_i}{dk/k} = k\frac{B(s_i)}{\dot{Q}(s_i)} = \frac{A(s_i)}{\dot{Q}(s_i)}, \tag{13.11}$$

where

$$\dot{Q}(s_i) = \lim_{s \to s_i} \frac{Q(s)}{s - s_i} = \frac{dQ(s_i)}{ds_i}. \tag{13.12}$$

Equation (13.11) indicates that if $Q(s)$ contains two pairs of complex-conjugate roots s_1, s_1^* and s_2, s_2^*, the term $s_1 - s_2$ appears in the denominators of both $S_k^{s_1}$ and $S_k^{s_2}$. Hence, should the separation between s_1 and s_2 be small, the corresponding sensitivities would become so inordinately large that a practical realization of the filter using a single negative-impedance converter is impossible. Consequently, in practice, we find that when a transfer function has more than one pair of complex poles it is ordinarily preferable to realize the function in sections each of which accounts for a single pair of complex poles, and then cascade these sections through isolation amplifiers such as emitter followers,[3,4] as shown in Fig. 13.2. The second-order transfer function, therefore, is of particular importance and its realization will be studied in some detail.

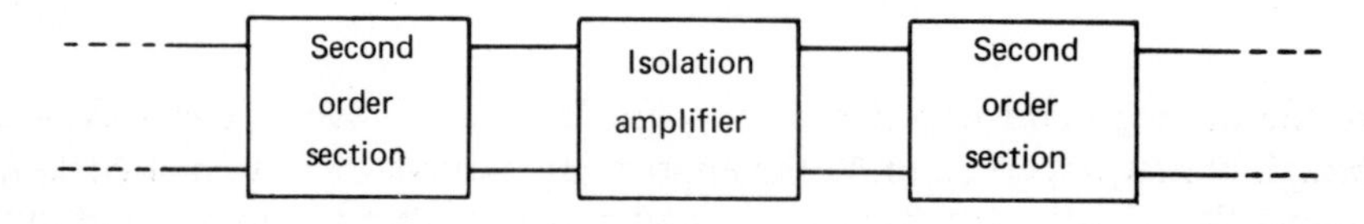

Fig. 13.2. Realization of high-order filters using second-order sections with isolation amplifiers.

Consider the transfer impedance

$$Z_{21}(s) = \frac{N(s)}{Q(s)} = \frac{Z_{21}(0)}{s^2 + 2\zeta s + 1}, \tag{13.13}$$

where s is normalized with respect to the undamped natural frequency of oscillation. When the relative damping ratio ζ is less than unity, the two poles of $Z_{21}(s)$ are complex and located at

$$s_1, s_1^* = -\zeta \pm j\sqrt{1 - \zeta^2}, \tag{13.14}$$

which lie on a semicircle of unit radius in the left half of the s-plane, as in Fig. 13.3. Choosing the auxiliary polynomial $P(s)$ as

$$P(s) = (s - \sigma_1)(s - \sigma_2) \tag{13.15}$$

and expanding $Q(s)/P(s)$ in partial fractions yields

$$\frac{Q(s)}{P(s)} = 1 + \frac{k_1}{s - \sigma_1} + \frac{k_2}{s - \sigma_2} \tag{13.16}$$

where

$$k_1 = \frac{\sigma_1^2 + 2\zeta\sigma_1 + 1}{\sigma_1 - \sigma_2}, \tag{13.17}$$

$$k_2 = \frac{\sigma_2^2 + 2\zeta\sigma_2 + 1}{\sigma_2 - \sigma_1}. \tag{13.18}$$

Assuming that $\sigma_1 > \sigma_2$, we find that the residue k_1 is positive and k_2 is negative.

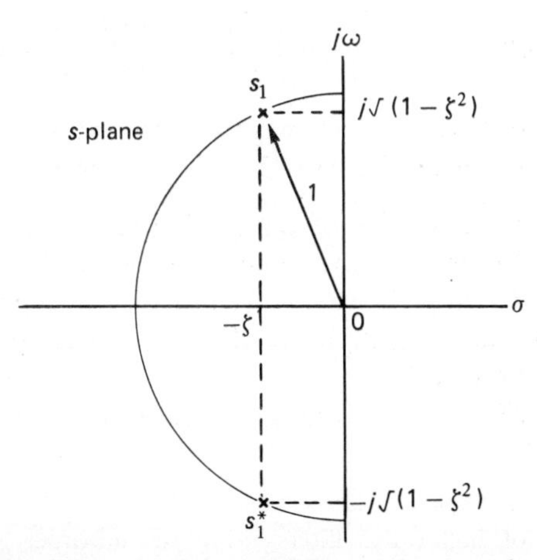

Fig. 13.3. A pair of complex-conjugate poles.

Hence, we may write

$$A(s) = (s - \sigma_1)(s - \sigma_2) + k_1(s - \sigma_2), \tag{13.19}$$

$$B(s) = -k_2(s - \sigma_1). \tag{13.20}$$

Using Eqs. (13.11)–(13.20), we obtain

$$S_k^{s_1} = \frac{\sigma_2^2 + 2\zeta\sigma_2 + 1}{2(\sigma_1 - \sigma_2)}\left(1 + j\frac{\zeta + \sigma_1}{\sqrt{1 - \zeta^2}}\right), \tag{13.21}$$

the magnitude of which is given by

$$|S_k^{s_1}| = \frac{\sigma_2^2 + 2\zeta\sigma_2 + 1}{2(\sigma_1 - \sigma_2)}\left(\frac{\sigma_1^2 + 2\zeta\sigma_1 + 1}{1 - \zeta^2}\right)^{1/2}. \tag{13.22}$$

Both σ_1 and σ_2 are restricted to lie on the negative real axis of the s-plane. When $\sigma_1 = 0$ and $\sigma_2 = -1$, the pole sensitivity $S_k^{s_1}$ attains its minimum magnitude, given by

$$|S_k^{s_1}|_{min} = \left(\frac{1 - \zeta}{1 + \zeta}\right)^{1/2}. \tag{13.23}$$

This may be demonstrated in two steps:

1. Suppose $\sigma_1 = 0$. Then Eq. (13.22) gives

$$|S_k^{s_1}|_{\sigma_1 = 0} = \frac{-1}{2(1 - \zeta^2)^{1/2}}\left(\sigma_2 + 2\zeta + \frac{1}{\sigma_2}\right), \tag{13.24}$$

   which clearly attains the minimum value of Eq. (13.23) when $\sigma_2 = -1$.

2. Suppose next $\sigma_2 = -1$. Then Eq. (13.22) gives

$$|S_k^{s_1}|_{\sigma_2 = -1} = \left(\frac{1 - \zeta}{1 + \zeta}\right)^{1/2}\left[1 - \frac{2\sigma_1(1 - \zeta)}{(1 + \sigma_1)^2}\right]^{1/2}. \tag{13.25}$$

   With σ_1 permitted to have only negative real values, it is clear that the mimimum possible value of Eq. (13.23) occurs when $\sigma_1 = 0$.

The minimum pole sensitivity of Eq. (13.23) is shown plotted against ζ as curve (a) in Fig. 13.4. We see that the smaller the value of ζ, that is, the closer the poles of overall transfer impedance $Z_{21}(s)$ move to the imaginary axis of the s-plane, the greater will the sensitivity of the poles of $Z_{21}(s)$ be to variations in the conversion ratio of the NIC.

Corresponding with the optimum set of conditions, $\sigma_1 = 0$ and $\sigma_2 = -1$, we thus have

$$P(s) = s(s + 1), \tag{13.26}$$

$$Q(s) = (s + 1)^2 - 2(1 - \zeta)s \tag{13.27}$$

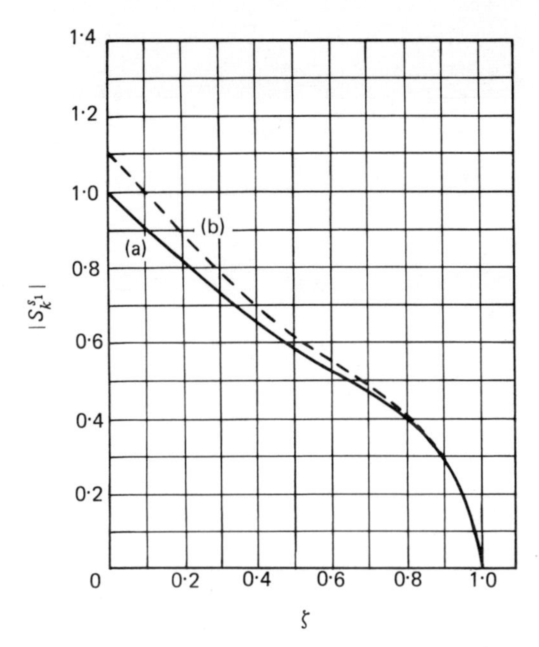

Fig. 13.4. Effect of varying the damping ratio, ζ, upon the pole sensitivity, $|S_k^{s_1}|$, for: (a) $\sigma_1 = 0$ and $\sigma_2 = -1$; (b) $\sigma_1 = -0{\cdot}1$ and $\sigma_2 = -1$.

and

$$\frac{Q(s)}{P(s)} = \frac{s+1}{s} - \frac{2(1-\zeta)}{s+1}. \tag{13.28}$$

The driving-point impedances $z_{22a}(s)$ and $z_{11b}(s)$ are therefore

$$z_{22a} = \frac{s+1}{s}, \tag{13.29}$$

$$z_{11b} = \frac{2(1-\zeta)}{s+1} \tag{13.30}$$

and the associated transfer impedances $z_{21a}(s)$ and $z_{21b}(s)$, respectively, are

$$z_{21a} = \frac{1}{s} \tag{13.31}$$

$$z_{21b} = \frac{2(1-\zeta)}{s+1}, \tag{13.32}$$

where we have assigned to z_{21a} and z_{21b} the same poles as z_{22a} and z_{22b} respectively. We thus obtain the realizations shown in Fig. 13.5 for networks N_a and N_b.

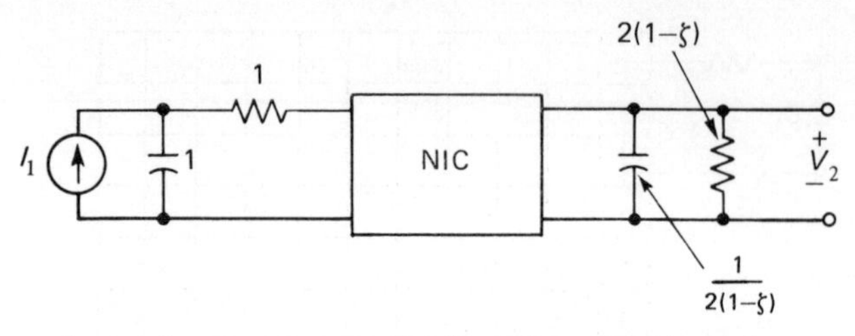

Fig. 13.5. Network realizing $Z_{21}(s) = Z_{21}(0)/(s^2 + 2\zeta s + 1)$ with minimum pole sensitivity.

The resistor connected across the output port can serve the purpose of a load if so required.

However, in the realization of Fig. 13.5 there is no provision for a finite source resistance at the input port. This limitation is overcome by choosing a finite, but relatively small, value for σ_1. Thus let $\sigma_1 = -0{\cdot}1$, say; so that

$$P(s) = (s + 0{\cdot}1)(s + 1), \tag{13.33}$$

where the value of σ_2 is left at -1 as before. Therefore,

$$\begin{aligned} \frac{Q(s)}{P(s)} &= \frac{s^2 + 2\zeta s + 1}{(s + 0{\cdot}1)(s + 1)} \\ &= 1 + \frac{1{\cdot}122 - 0{\cdot}222\zeta}{s + 0{\cdot}1} - \frac{2{\cdot}22(1 - \zeta)}{s + 1}, \end{aligned} \tag{13.34}$$

which, following the established procedure, leads to the realization shown in Fig. 13.6(a) for the filter. We now have a resistance R_1 across the input port which is available to accommodate a definite source resistance. This provision is, however, accompanied by a relatively slight increase in the magnitude of pole sensitivity, $|S_k^{s_1}|$, as shown by curve (b) in Fig. 13.4 which is obtained by plotting Eq. (13.25) for $\sigma_1 = -0{\cdot}1$. By transforming the parallel combination of the current source and resistance R_1 into an equivalent voltage source with series resistance, as in Fig. 13.6(b), we may also realize a transfer-voltage ratio,

$$\begin{aligned} K(s) &= \frac{V_2}{V_1} \\ &= \frac{K(0)}{s^2 + 2\zeta s + 1}. \end{aligned} \tag{13.35}$$

Example 13.1. *Fourth-order low-pass filter.* Consider the design of an *RC* active filter with a low-pass Butterworth characteristic cutting off at the rate of 24 dB/octave. For such a characteristic the poles of transfer function $K(s)$, say, are known to occur on a circle of unit radius,[2] as shown in Fig. 13.7. The zeros of transmission (i.e., zeros of $K(s)$) all fall at infinity.

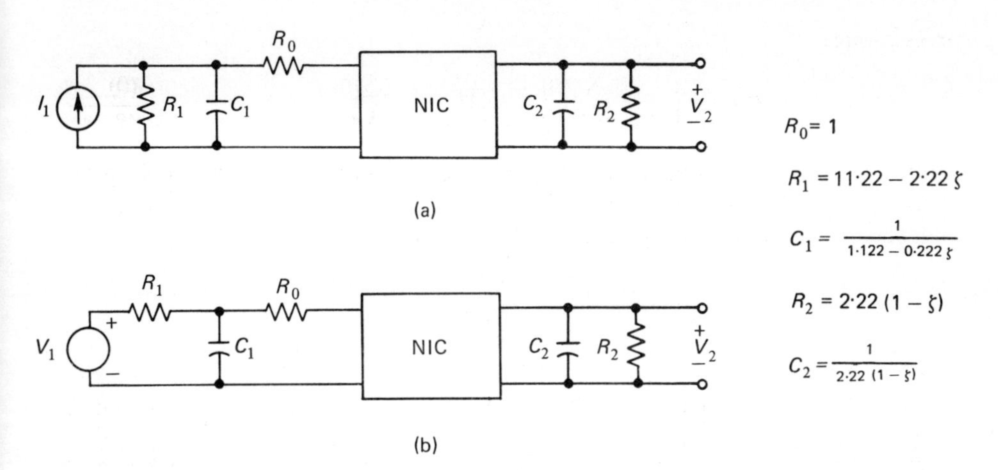

Fig. 13.6. Second-order RC–NIC filter having $\sigma_1 = -0{\cdot}1, \sigma_2 = -1$ with (a) Current excitation; (b) Voltage excitation.

We therefore have

$$K(s) = \frac{K(0)}{(s^2 + 0{\cdot}7654s + 1)(s^2 + 1{\cdot}848s + 1)}.$$

The transfer function $K(s)$ is expressed as a product of two second-order

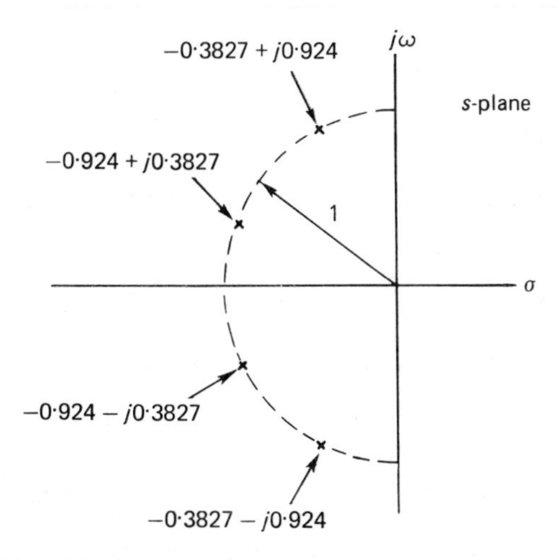

Fig. 13.7. Pole-pattern of fourth-order low-pass filter having Butterworth characteristic.

expressions:

$$K(s) = \left(\frac{K_1(0)}{s^2 + 0{\cdot}7654s + 1}\right)\left(\frac{K_2(0)}{s^2 + 1{\cdot}848s + 1}\right).$$

Using the network of Fig. 13.6(b) to realize each factor, and then coupling the resulting two sections through an isolation amplifier of negligible output impedance, we obtain the complete filter shown in Fig. 13.8.

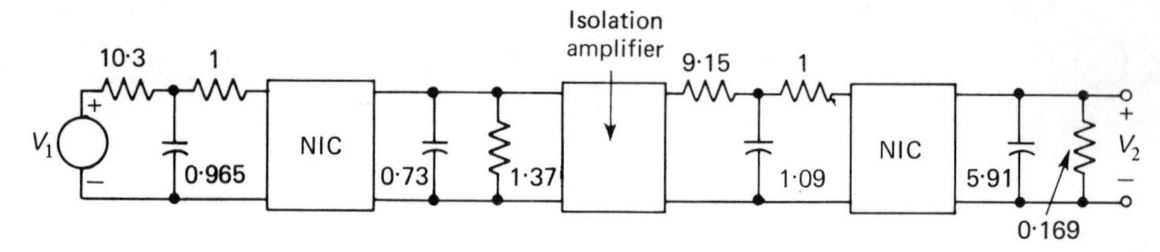

Fig. 13.8. Network realizing $K(s) = K(0)/[(s^2 + 0{\cdot}7654s + 1)(s^2 + 1{\cdot}848s + 1)]$.

13.3 PARALLEL *RC* ACTIVE NETWORK SYNTHESIS

Another synthesis procedure for *RC* active filters was proposed by Yanagisawa,[5] using the structure shown in Fig. 13.9. For this procedure, it is necessary to have the negative-impedance converter of the current-inversion type.* The open-

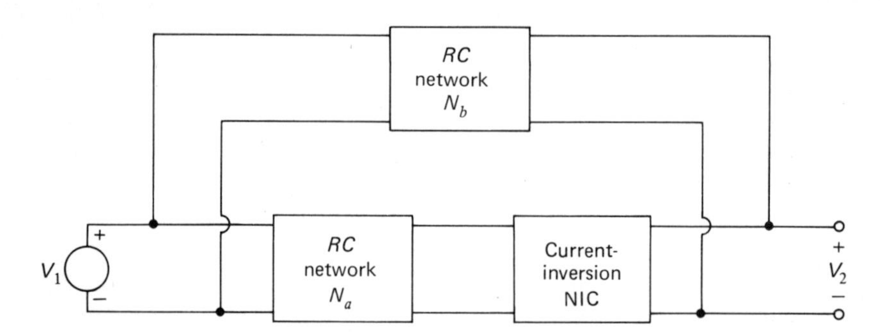

Fig. 13.9. Yanagisawa's *RC*–NIC filter.

* If the converter were of the voltage-inversion type, we would obtain

$$\left.\frac{V_2}{V_1}\right|_{I_2=0} = -\frac{y_{12a} + y_{12b}}{y_{22a} - y_{22b}},$$

in which the numerator is unaffected by the activity of the converter.

circuit transfer-voltage ratio of the network is given by

$$K(s) = \left.\frac{V_2}{V_1}\right|_{I_2=0} = -\frac{y_{12a} - y_{12b}}{y_{22a} - y_{22b}}. \tag{13.36}$$

In contrast to Linvill's *RC* active filter, we see that in Eq. (13.36) the activity of the converter acts on the denominator as well as numerator of the transfer-voltage ratio. In other words, the NIC controls the zeros as well as poles of the transfer-voltage ratio; consequently, this synthesis technique permits realization of complex zeros of transmission using relatively simple *RC* structures for networks N_a and N_b, whereas such a realization by means of Linvill's cascade method requires the use of bridged-*T* or twin-*T* networks.

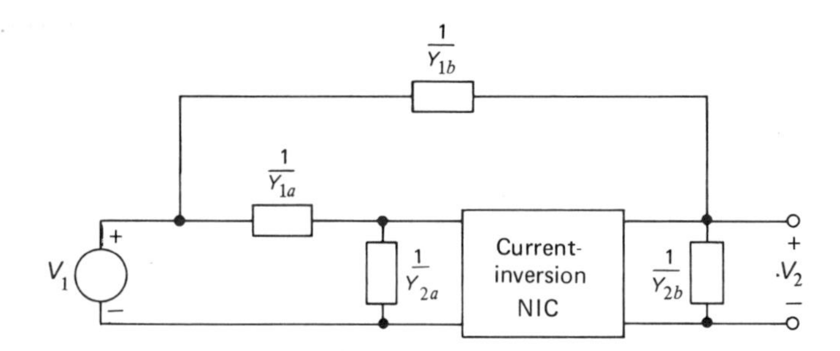

Fig. 13.10. Yanagisawa's active filter using *L*-sections for networks N_a and N_b.

From Eq. (13.36) it is clear that y_{22a} and y_{12a} cannot be synthesized separately. To solve this problem, Yanagisawa proposed the use of inverted *L*-sections for networks N_a and N_b, as in Fig. 13.10. For this circuit we have

$$K(s) = \left.\frac{V_2}{V_1}\right|_{I_2=0} = \frac{Y_{1b} - Y_{1a}}{(Y_{1b} - Y_{1a}) + (Y_{2b} - Y_{2a})}. \tag{13.37}$$

Let $K(s)$ be written as

$$K(s) = \frac{N(s)}{Q(s)} = \frac{N(s)}{N(s) + Q(s) - N(s)}. \tag{13.38}$$

The numerator and denominator of Eq. (13.38) are then divided by a polynomial $P(s)$ with $n - 1$ distinct negative real roots, $\sigma_1, \sigma_2, \ldots, \sigma_{n-1}$, where n is the order of $Q(s)$, that is,

$$P(s) = \prod_{i=1}^{n-1} (s - \sigma_i). \tag{13.39}$$

We therefore obtain

$$Y_{1b} - Y_{1a} = \frac{N(s)}{P(s)} = k_\infty s + k_0 + \sum_{i=1}^{n-1} \frac{k_i s}{s - \sigma_i}, \tag{13.40}$$

$$Y_{2b} - Y_{2a} = \frac{Q(s) - N(s)}{P(s)} = k'_\infty s + k'_0 + \sum_{i=1}^{n-1} \frac{k'_i s}{s - \sigma_i}. \tag{13.41}$$

In Eq. (13.40) the partial-fraction terms with positive residues are realized by Y_{1b}, while those with negative residues are realized by Y_{1a}. Similarly, in Eq. (13.41) terms with positive residues are realized by Y_{2b}, while those with negative residues are realized by Y_{2a}.

It should be noted that the structure of Fig. 13.10 can accommodate any prescribed physical load simply by adding the load admittance to both Y_{2a} and Y_{2b}. Such an addition obviously leaves the complete response unchanged.

Example 13.2. Consider the realization of a *notch filter* characterized by the transfer-voltage ratio:

$$\frac{V_2}{V_1} = \frac{\frac{1}{2}(s^2 + 2)}{s^2 + \frac{1}{6}s + 1}$$

which has zeros of transmission at $s = \pm j\sqrt{2}$. Choosing

$$P(s) = s + 1$$

we obtain

$$Y_{1b} - Y_{1a} = \frac{\frac{1}{2}(s^2 + 2)}{s + 1}.$$

Expanding $(Y_{1b} - Y_{1a})/s$ in partial fractions yields,

$$\frac{Y_{1b} - Y_{1a}}{s} = \frac{\frac{1}{2}(s^2 + 2)}{s(s + 1)} = \frac{1}{2} + \frac{1}{s} - \frac{3}{2(s + 1)}.$$

Therefore,

$$Y_{1b} = \frac{s}{2} + 1, \qquad Y_{1a} = \frac{3s}{2(s + 1)}.$$

Similarly,

$$Y_{2b} - Y_{2a} = \frac{s^2/2 + s/6}{s + 1} = \frac{s}{2} - \frac{s}{3(s + 1)}$$

which gives

$$Y_{2b} = \frac{s}{2}, \qquad Y_{2a} = \frac{s}{3(s + 1)}.$$

The resulting network realization is therefore as shown in Fig. 13.11.

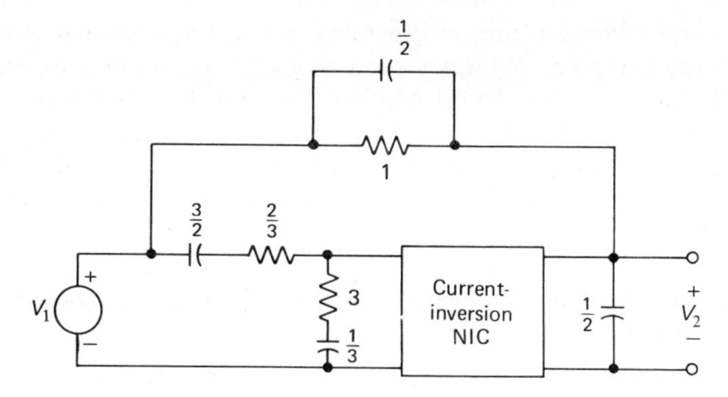

Fig. 13.11. Network realizing $V_2/V_1 = \frac{1}{2}(s^2 + 2)/(s^2 + \frac{1}{6}s + 1)$.

13.4 POSITIVE *RC*–NEGATIVE *RC* DRIVING-POINT SYNTHESIS

The synthesis of driving-point impedances using *RC*-active networks is equal in importance to the synthesis of transfer functions. There are many applications which require the realization of complex driving-point impedances at extremely low frequencies, as, for example, in the design of control systems. Furthermore, driving-point impedances can be used to yield any desired transfer characteristic. Kinariwala[6] has shown that any driving-point impedance function, not necessarily positive real but expressible as the ratio of two polynomials in s, can be realized by a transformerless passive *RC* structure in which is embedded only one active element in the form of a controlled source. The synthesis technique involved is, however, not practical because it requires the use of a large number of resistors. A more practical method of realizing driving-point impedances is to use Kinariwala's *cascade procedure* which involves cascading a passive *RC* network with another passive *RC* network through a NIC, as depicted in Fig. 13.12. In the special, but important case, of second-order driving-point impedances, the structures thus obtained are particularly simple.

Consider an impedance function

$$Z(s) = \frac{N(s)}{Q(s)}, \tag{13.42}$$

where $N(s)$ and $Q(s)$ are polynomials in s, the degrees of which do not differ by

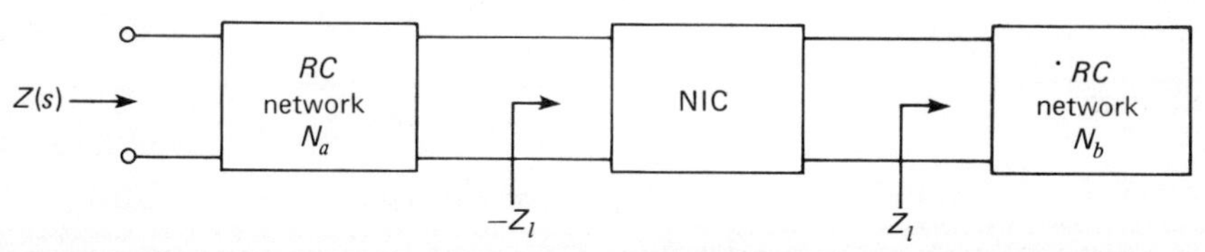

Fig. 13.12. *RC*–NIC structure for realization of driving-point impedance $Z(s)$.

more than one. Without loss of generality, we will assume that $Z(s)$ has only complex zeros and poles. As in the preceding synthesis techniques, the first step is to divide the numerator and denominator of $Z(s)$ by a polynomial $P(s)$ with n distinct negative real roots $\sigma_1, \sigma_2, \ldots, \sigma_n$, where n is equal to or greater than the degree of $N(s)$ or $Q(s)$, whichever is greater. The polynomial $P(s)$ is then expressed as the product of $P_1(s)$ and $P_2(s)$ in such a manner as to satisfy the two conditions:

1. The roots of $P_1(s)$ alternate with the roots of $P_2(s)$ along the negative real axis of the s-plane.
2. The root nearest to the origin of the s-plane belongs to $P_1(s)$.

Consequently, in a partial-fraction expansion of $Q(s)/P_1(s)P_2(s)$ we find that the residues in the factors of $P_1(s)$ are all positive and the residues in the factors of $P_2(s)$ are all negative. The same result is obtained in a partial-fraction expansion of $N(s)/P_1(s)P_2(s)$. We may therefore write

$$Z(s) = \frac{[N_1(s)/P_1(s)] - [N_2(s)/P_2(s)]}{[Q_1(s)/P_1(s)] - [Q_2(s)/P_2(s)]} \tag{13.43}$$

or, equivalently,

$$Z(s) = \frac{N_1(s)}{Q_1(s)} \frac{[N_2(s)/N_1(s)] - [P_2(s)/P_1(s)]}{[Q_2(s)/Q_1(s)] - [P_2(s)/P_1(s)]}. \tag{13.44}$$

When an *RC* two-port network is terminated in a negative impedance at the port 2–2′, say, the impedance measured looking into port 1–1′ is

$$Z(s) = \frac{(z_{11}z_{22} - z_{12}^2) - z_{11}Z_l}{z_{22} - Z_l}, \tag{13.45}$$

where z_{11}, z_{22} and z_{12} are the open-circuit driving-point and transfer impedances characterizing the *RC* two-port network N_a, and $-Z_l$ is the terminating impedance. Through factoring z_{11} out of Eq. (13.45), and noting that

$$y_{22} = \frac{z_{11}}{z_{11}z_{22} - z_{12}^2}, \tag{13.46}$$

we obtain

$$Z(s) = z_{11} \frac{\dfrac{1}{y_{22}} - Z_l}{z_{22} - Z_l}. \tag{13.47}$$

Comparing Eqs. (13.44) and (13.47), we may therefore identify

$$\begin{aligned} z_{11} &= \frac{N_1(s)}{Q_1(s)} \\ z_{22} &= \frac{Q_2(s)}{Q_1(s)} \\ \frac{1}{y_{22}} &= \frac{N_2(s)}{N_1(s)} \\ Z_l &= \frac{P_2(s)}{P_1(s)}. \end{aligned} \tag{13.48}$$

By the choice of roots for $P_1(s)$ and $P_2(s)$, we have $P_2(s)/P_1(s)$ corresponding to an *RC* driving-point impedance, so that the network N_b with input impedance Z_l can be readily synthesized as an *RC* structure. It remains to show that z_{11}, z_{22}, y_{22} represent a two-port *RC* network. Solving Eqs. (13.46) and (13.48) for z_{12} yields

$$z_{12} = \frac{\sqrt{N_1(s)Q_2(s) - N_2(s)Q_1(s)}}{Q_1(s)}. \tag{13.49}$$

Since z_{12} must be a rational function, it follows that $(N_1Q_2 - N_2Q_1)$ must be a full square. Thus, the necessary and sufficient conditions for network N_a to be physically realizable are that [2]

1. z_{11} and z_{22} represent *RC* driving-point impedances,
2. $(N_1Q_2 - N_2Q_1)$ be a full square with a positive leading coefficient, and
3. the residue condition be satisfied at all the poles, including the point $s = \infty$.

The residue condition is

$$k_{11}k_{22} - k_{12}^2 \geq 0, \tag{13.50}$$

where k_{ij} represents the residue of z_{ij} at a given pole. Let σ_i denote a root of polynomial $Q_1(s)$; then, the residues of z_{11} and z_{22} may be written as

$$\begin{aligned} k_{11}^i &= \frac{N_1(\sigma_i)}{\dot{Q}_1(\sigma_i)} \\ k_{22}^i &= \frac{Q_2(\sigma_i)}{\dot{Q}_1(\sigma_i)}, \end{aligned} \tag{13.51}$$

where $\dot{Q}_1(\sigma_i)$ is the first derivative of $Q_1(s)$ evaluated at $s = \sigma_i$. The associated residue of z_{12} is

$$k_{12}^i = \frac{\sqrt{N_1(\sigma_i)Q_2(\sigma_i) - N_2(\sigma_i)Q_1(\sigma_i)}}{\dot{Q}_1(\sigma_i)}. \tag{13.52}$$

However, by assumption, $Q_1(\sigma_i)$ is zero; therefore,

$$k_{11}^i k_{22}^i - (k_{12}^i)^2 = 0. \tag{13.53}$$

That is, the residue condition is satisfied with an equal sign at all the poles. From Eqs. (13.48) and (13.49) we see that the requirements at infinity are also satisfied.

If $Z(s)$ has only complex poles, as previously assumed, then z_{22} is always an *RC* impedance, so that its residues are necessarily positive.[6] Furthermore, if z_{12} is a ratio of two polynomials, the residues at its poles will be real and $(k_{12}^i)^2$ will always be positive. Then, from Eq. (13.53) we see that z_{11} will automatically have positive residues, that is, it must be an *RC* impedance. It follows, therefore, that in order to realize network N_a as an *RC* two-port network we need only make $(N_1Q_2 - N_2Q_1)$ a full square.

This can always be achieved by multiplying the numerator and denominator of Eq. (13.42) by surplus factors. In the case of a driving-point impedance of the second order, however, there exists a simpler procedure for rationalizing z_{12} of network N_a. The procedure involves the addition and subtraction of a positive constant, κ, in the denominator of Eq. (13.43); thus we may write

$$\begin{aligned} \frac{Q_1'(s)}{P_1(s)} &= \frac{Q_1(s)}{P_1(s)} + \kappa \\ \frac{Q_2'(s)}{P_2(s)} &= \frac{Q_2(s)}{P_2(s)} + \kappa. \end{aligned} \tag{13.54}$$

The ratio $Q_2'(s)/Q_1'(s)$ remains on *RC* impedance, independently of κ. By replacing $Q_1(s)$ and $Q_2(s)$ with $Q_1'(s)$ and $Q_2'(s)$, respectively, in Eqs. (13.48), there now exists the possibility of choosing κ so that $(N_1Q_2' - N_2Q_1')$ is a full square.

In some cases the structure of Fig. 13.12 can be used to realize a driving-point function of the form $Z(s)/s$ which has a pole at the origin. This requires the addition and subtraction of a term of the form κ/s, where κ is a positive constant, in the denominator of Eq. (13.43), so that

$$\begin{aligned} \frac{Q_1''(s)}{sP_1(s)} &= \frac{Q_1(s)}{P_1(s)} + \frac{\kappa}{s} \\ \frac{Q_2''(s)}{sP_2(s)} &= \frac{Q_2(s)}{P_2(s)} + \frac{\kappa}{s}. \end{aligned} \tag{13.55}$$

Here, again, the ratio $Q_2''(s)/Q_1''(s)$ remains an *RC* impedance, independently of κ. If in Eqs. (13.48) we replace $Q_1(s)$ and $Q_2(s)$ with $Q_1''(s)/s$ and $Q_2''(s)/s$, respectively, it may be possible to choose κ so that $(N_1Q_2'' - N_2Q_1'')$ is a full square.

Realization of Second-order Transfer Function

Synthesis of driving-point functions can be adapted in several ways to the realization of transfer functions. Two particularly simple situations are of practical importance:

1) In some cases, the input impedance may be chosen equal to

$$Z(s) = \frac{V_o}{I_o} = \frac{N(s)}{Q(s)} \tag{13.56}$$

and synthesized so that the resulting network has a shunt resistance R across the input terminals. Then conversion from current source to voltage source, $V_i = RI_o$, as in Fig. 13.13(a), yields the desired transfer function within a constant multiplier,

$$K(s) = \frac{V_o}{V_i} = \frac{1}{R}\frac{N(s)}{Q(s)}. \tag{13.57}$$

2) In other cases, $Z(s)$ may be chosen so that

$$Z(s) = \frac{V_o}{I_o} = \frac{N(s)}{sQ(s)} \tag{13.58}$$

and synthesized so that the resulting network has a shunt capacitor C across the input terminals. Then, conversion from current source to voltage source, $V_i = I_o/sC$, as in Fig. 13.13(b), yields the desired transfer function within a constant multiplier

$$K(s) = \frac{V_o}{V_i} = C\frac{N(s)}{Q(s)}. \tag{13.59}$$

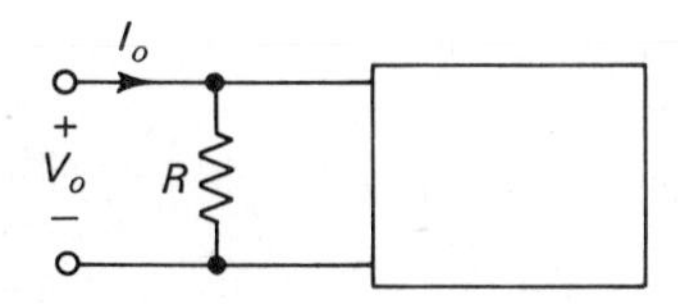

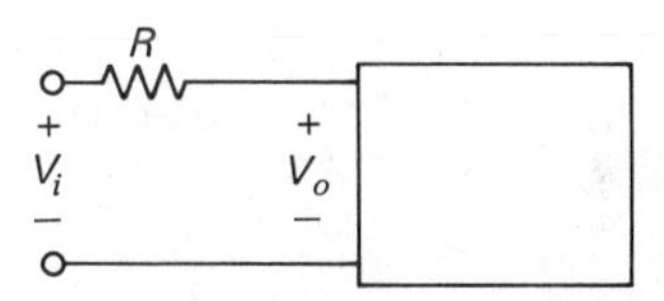

(a)

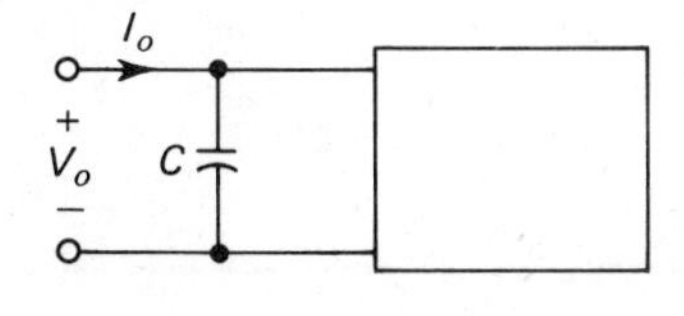

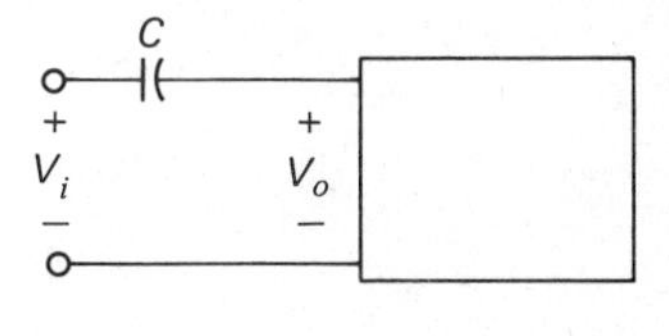

(b)

Fig. 13.13. Adaptation of driving-point synthesis to realize transfer function: (a) $Z(s)$ is realized with shunt resistor across input terminals; (b) $Z(s)$ is realized with shunt capacitor across input terminals.

Consider next the realization of a second-order open-circuit transfer-voltage ratio

$$K(s) = K(\infty)\frac{s^2 + a_1 s + a_0}{s^2 + b_2 s + b_0}, \tag{13.60}$$

where it is assumed that $K(s)$ has complex-conjugate poles and zeros in the left half plane. If b_0 is greater than a_0, then the rationalization of z_{12} is accomplished through the addition and subtraction of an appropriate constant κ in the denominator of Eq. (13.43), and $Z(s)$ proportional to $K(s)$ may be synthesized so that the resulting network has a shunt resistor across its input terminals. The desired transfer function of Eq. (13.60) can then be realized using the structure of Fig. 13.13(a). If, on the other hand, b_0 is less than a_0, the addition and subtraction of κ/s in the denominator of Eq. (13.43) will permit the rationalization of z_{12}, and $Z(s)$ proportional to $K(s)/s$ may be synthesized so that the resulting network has a shunt capacitor across its input terminals. In this case, the transfer function of Eq. (13.60) can be realized using the structure of Fig. 13.13(b).

Example 13.3. Consider the transfer function

$$K(s) = K(\infty)\frac{s^2 + s + 1}{s^2 + s + 2}$$

which pertains to the case $b_0 > a_0$. Choosing

$$P(s) = (s + 1)(s + 2),$$

we obtain

$$\frac{s^2 + s + 1}{(s + 1)(s + 2)} = \frac{s + 2}{s + 1} - \frac{3}{s + 2}$$

$$\frac{s^2 + s + 2}{(s + 1)(s + 2)} = \frac{s + 3}{s + 1} - \frac{4}{s + 2}.$$

Therefore,

$$P_1(s) = s + 1$$

$$P_2(s) = s + 2$$

$$N_1(s) = s + 2$$

$$N_2(s) = 3$$

$$Q_1'(s) = (1 + \kappa)s + (3 + \kappa)$$

$$Q_2'(s) = \kappa s + (4 + 2\kappa).$$

Evaluating $(N_1Q_2' - N_2Q_1')$, we obtain

$$N_1Q_2' - N_2Q_1' = \kappa s^2 + (\kappa + 1)s + (\kappa - 1).$$

For this to be a full square, we require

$$3\kappa^2 - 6\kappa - 1 = 0,$$

that is,

$$\kappa = 2{\cdot}155.$$

The z-parameters of the two-port network N_a are therefore

$$z_{11} = \frac{0{\cdot}317(s + 2)}{s + 1{\cdot}635}$$

$$z_{22} = \frac{0{\cdot}683(s + 3{\cdot}86)}{s + 1{\cdot}635}$$

$$z_{12} = \frac{0{\cdot}465(s + 0{\cdot}731)}{s + 1{\cdot}635}.$$

The need for an ideal transformer in the realization of network N_a can be avoided by ensuring that at $s = \infty$,

$$z_{11}(\infty) = z_{22}(\infty) = z_{12}(\infty).$$

This requirement is achieved simply by multiplying the impedance level on the output side of network N_a by a factor $m = 0{\cdot}464$. That is to say, we multiply z_{22} by m, z_{12} by $\sqrt{m}$, and terminate in $-mZ_l$ instead of $-Z_l$. Such an impedance scaling will, of course, leave the driving-point impedance $Z(s)$ measured at port 1–1′ of the terminated network N_a unchanged.

The scaled network N_a is synthesized using a zero-shifting technique[2] which involves expanding the impedance z_{11} (or mz_{22}) so as to realize the zero of transmission at $s = -0{\cdot}731$ as prescribed by z_{12}. In parts (a) and (b) of Fig. 13.14 we have shown z_{11} and its reciprocal, the admittance $Y_1 = 1/z_{11}$, plotted for values of the variable s along the real axis (σ-axis). We propose to produce a transmission zero at $s = -0{\cdot}731$, and so we must shift the zero of Y_1 into this position so that the inverted remainder will have a pole there. This is achieved by removing a shunt element with an admittance equal to the value of Y_1 at $s = -0{\cdot}731$. At this frequency, we have

$$z_{11}(-0{\cdot}731) = \frac{0{\cdot}317(-0{\cdot}731 + 2)}{-0{\cdot}731 + 1{\cdot}635} = 0{\cdot}445,$$

that is,

$$Y_1(-0{\cdot}731) = \frac{1}{0{\cdot}445} = 2{\cdot}25.$$

The remainder function after subtracting a conductance of 2·25 ℧ from Y_1 is

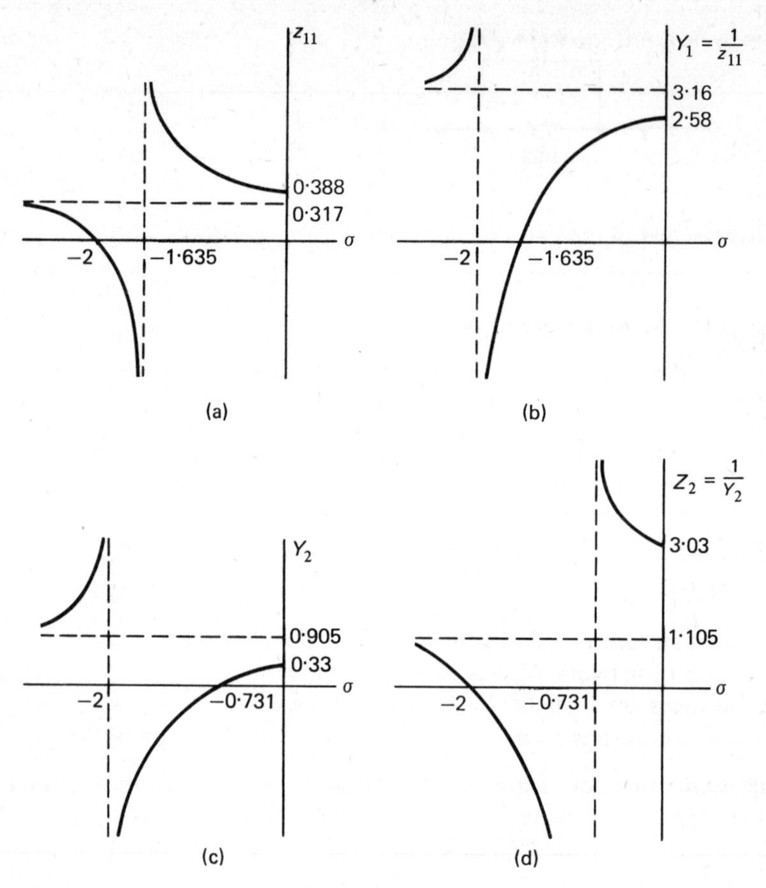

Fig. 13.14. (a) Behaviour of z_{11} along real axis; (b) Behaviour of $Y_1 = 1/z_{11}$ along real axis; (c) Effect of removing part of $Y_1(0)$; (d) Behaviour of $Z_2 = 1/Y_2$ along real axis.

therefore

$$Y_2 = Y_1 - 2{\cdot}25 = \frac{0{\cdot}905(s + 0{\cdot}731)}{s + 2}$$

which, as required, has a zero at $s = -0{\cdot}731$ (see also Fig. 13.14c). The reciprocal function Z_2, shown plotted in Fig. 13.14(d), has a pole at $s = -0{\cdot}731$. A partial-fraction expansion of Z_2 gives,

$$Z_2 = \frac{1}{Y_2} = \frac{1{\cdot}4}{s + 0{\cdot}731} + 1{\cdot}1.$$

Complete removal of the pole at $s = -0{\cdot}731$ as a series arm consisting of a parallel connection of a capacitance of 0·715 F and a resistance of 1·92 Ω leaves a constant

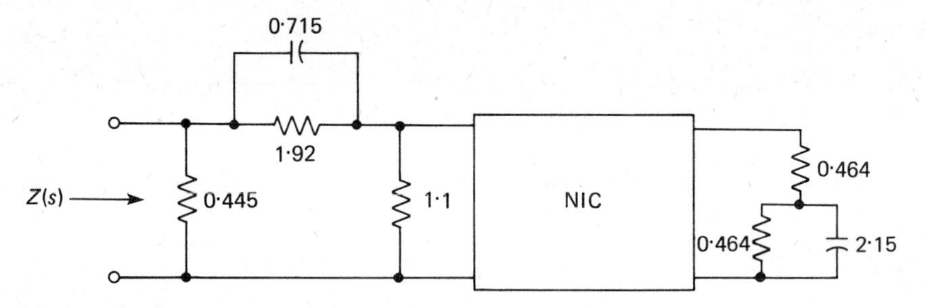

Fig. 13.15. Network realizing the driving-point impedance

$$Z(s) = \frac{s^2 + s + 1}{s^2 + s + 2}.$$

impedance of 1·1 Ω which is connected as a shunt element, as in Fig. 13.15. In this diagram we have also included the *RC* realization of the load impedance, after scaling by the factor $m = 0{\cdot}464$:

$$mZ_l = m\frac{P_2(s)}{P_1(s)} = 0{\cdot}464 + \frac{0{\cdot}464}{s + 1}.$$

The final structure is shown in Fig. 13.16 where only one half of the 0·445 Ω shunt resistor of network N_a has been utilized for the source conversion, so that equal resistances are available for use as source and load resistances. The desired transfer function is thus realized within a constant multiplier $K(\infty) = 1{\cdot}122$.

Example 13.4. Consider next the transfer function

$$K(s) = K(\infty)\frac{s^2 + s + 2}{s^2 + s + 1}$$

which pertains to the case $b_0 < a_0$. Choosing

$$P(s) = s(s + 2),$$

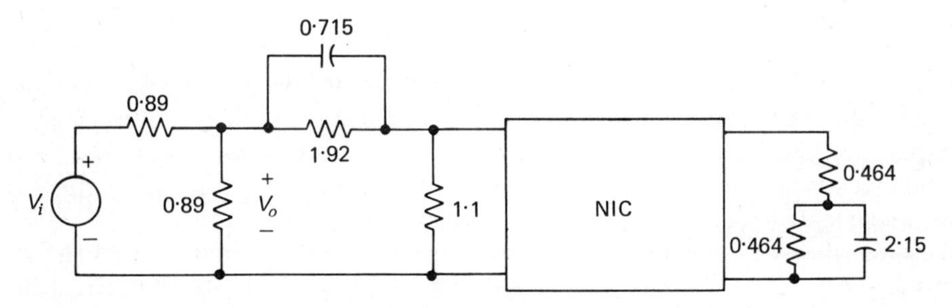

Fig. 13.16. Network realizing the transfer function

$$\frac{V_o}{V_i} = 1{\cdot}122\frac{s^2 + s + 1}{s^2 + s + 2}.$$

we obtain

$$\frac{s^2 + s + 2}{s(s + 2)} = \frac{s + 1}{s} - \frac{2}{s + 2}$$

$$\frac{s^2 + s + 1}{s(s + 2)} = \frac{s + \frac{1}{2}}{s} - \frac{\frac{3}{2}}{s + 2}.$$

We therefore have

$$P_1(s) = s$$

$$P_2(s) = s + 2$$

$$N_1(s) = s + 1$$

$$N_2(s) = 2$$

$$Q_1''(s) = s^2 + (\kappa + \tfrac{1}{2})s$$

$$Q_2''(s) = (\kappa + \tfrac{3}{2})s + 2\kappa.$$

In order to make $(N_1 Q_2'' - N_2 Q_1'')$ a full square, we must have

$$28\kappa^2 - 20\kappa - 1 = 0,$$

that is,

$$\kappa = 0{\cdot}761.$$

The corresponding z-parameters of network N_a are therefore,

$$z_{11} = \frac{s + 1}{s(s + 1{\cdot}261)},$$

$$z_{22} = \frac{2{\cdot}261(s + 0{\cdot}674)}{s(s + 1{\cdot}261)},$$

$$z_{12} = \frac{0{\cdot}511(s + 2{\cdot}42)}{s(s + 1{\cdot}261)}.$$

In this case the need for an ideal transformer is avoided by ensuring that all three z-parameters have the same residue at $s = 0$, which is achieved by multiplying the impedance level on the output side of network N_a by a factor $m = 0{\cdot}657$. Then following a procedure similar to that used in the previous example, we synthesize the scaled network N_a by expanding the impedance z_{11} in such a way as to realize the prescribed transmission zeros at $s = -2{\cdot}42$ and $s = \infty$; the result is shown in Fig. 13.17. This diagram also includes the RC realization of the scaled load impedance

$$mZ_l = 0{\cdot}657\frac{s + 2}{s} = 0{\cdot}657 + \frac{1}{0{\cdot}76s}.$$

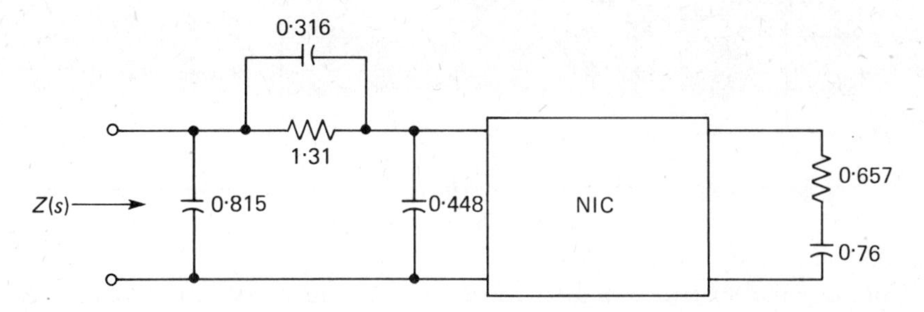

Fig. 13.17. Network realizing the driving-point impedance

$$Z(s) = \frac{s^2 + s + 2}{s^2 + s + 1}.$$

At the port 1–1′, the network of Fig. 13.17 presents a driving-point impedance $Z(s) = (s^2 + s + 2)/[s(s^2 + s + 1)]$. By using part of the 0·815 F shunt capacitor of network N_a to convert from current to voltage source, as in Fig. 13.18, the specified transfer function $K(s)$ is realized within a constant multiplier $K(\infty) = 0{\cdot}408$.

13.5 POSITIVE *RC*–POSITIVE *RL* CASCADE SYNTHESIS

From Section 3.9 we recall that an ideal gyrator is a lossless, passive two-port device that presents at either port an impedance equal to r^2/Z, where Z is the impedance connected at the other port, and r is the gyration resistance (i.e., reciprocal of the gyration conductance, g). Consequently, the impedance measured looking into the input port (let us say) of a gyrator terminated in an *RC* one-port network is equivalent to the driving-point impedance of an *RL* one-port network. The gyrator can, therefore, be particularly useful in the construction of inductorless filters.

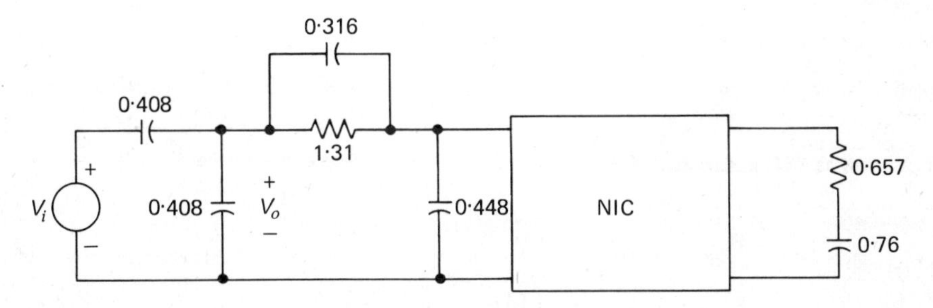

Fig. 13.18. Network realizing the transfer function

$$\frac{V_o}{V_i} = 0{\cdot}408\,\frac{s^2 + s + 2}{s^2 + s + 1}.$$

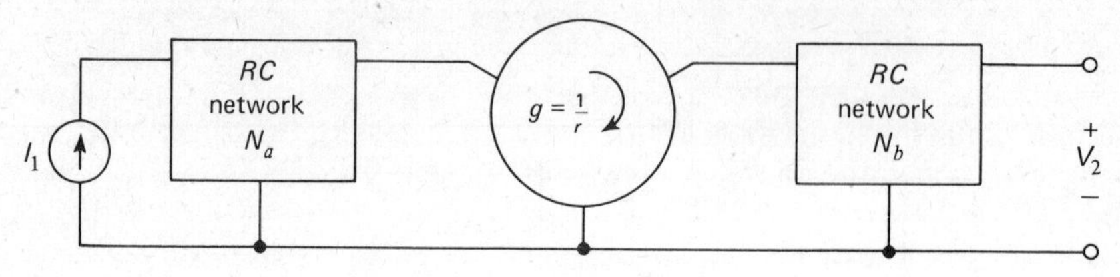

Fig. 13.19. *RC*–gyrator filter.

Consider the structure of Fig. 13.19 where an ideal gyrator is shown connected between two *RC* two-port networks N_a and N_b. The structure is driven from a current source I_1. The overall open-circuit transfer impedance of the network is therefore

$$Z_{21}(s) = \left.\frac{V_2}{I_1}\right|_{I_2=0} = \frac{rz_{21a}z_{21b}/z_{11b}}{z_{22a} + r^2/z_{11b}}. \tag{13.61}$$

Let the specified $Z_{21}(s)$ be denoted as $N(s)/Q(s)$, where $Q(s)$ is of degree n. The realization of $Z_{21}(s)$ begins, first, with dividing both $N(s)$ and $Q(s)$ by an auxiliary polynomial $P(s)$ having n distinct negative real roots $\sigma_1, \sigma_2, \ldots, \sigma_n$, and then expanding $Q(s)/P(s)$ in partial fractions, as shown by

$$\frac{Q(s)}{P(s)} = 1 + \sum_{\mu} \frac{k_\mu}{s - \sigma_\mu} + \sum_{\nu} \frac{k_\nu}{s - \sigma_\nu}, \tag{13.62}$$

where $k_\mu > 0$ and $k_\nu < 0$. In some cases[7] (depending upon the number and positions of the poles of $Z_{21}(s)$, i.e., roots of $Q(s)$) it is possible to add and subtract a constant κ with $0 < \kappa \leq 1$ such that in Eq. (13.63)

$$\frac{Q(s)}{P(s)} = \left(1 - \kappa + \sum_{\mu} \frac{k_\mu}{s - \sigma_\mu}\right) + \left(\kappa + \sum_{\nu} \frac{k_\nu}{s - \sigma_\nu}\right) \tag{13.63}$$

the sum of κ and the terms with negative residues will have a zero frequency value equal to or greater than zero in order to ensure physical realizability of networks N_a and N_b as *RC* networks. This requires that*

$$\kappa \geq \sum_{\nu} \frac{k_\nu}{\sigma_\nu}. \tag{13.64}$$

* Equation (13.64) is satisfied if

$$\sum_i \text{ang}(s_i) \leq \frac{\pi}{2}$$

where

$$Q(s) = \prod_i (s - s_i)(s - s_i^*)$$

and the s_i's lie in the upper half of the s-plane. For a proof, see D. A. Calahan, "Sensitivity minimization in active RC synthesis," *Trans. I.R.E.*, **CT-9**, 38 (1962).

Then, the terms with positive residues minus κ (that is, those in the first parentheses in Eq. 13.63) constitute an *RC* driving-point impedance and can be identified with z_{22a}, while the terms with negative residues plus κ (that is, those in the second parentheses) constitute an *RL* driving-point impedance and can be identified with r^2/z_{11b}.

We are specifically interested in the realization of the second-order low-pass transfer impedance of Eq. (13.13) with minimum pole sensitivity. Using $(s - \sigma_1)(s - \sigma_2)$ as the auxiliary polynomial $P(s)$, and then expanding $Q(s)/P(s)$ in partial fractions, we obtain

$$\frac{Q(s)}{P(s)} = \left(1 - \kappa + \frac{k_1}{s - \sigma_1}\right) + \left(\kappa + \frac{k_2}{s - \sigma_2}\right), \tag{13.65}$$

where $\sigma_1 > \sigma_2$. The residues k_1 and k_2 are defined by Eqs. (13.17) and (13.18), with k_1 positive and k_2 negative. We may therefore write

$$Q(s) = s^2 + 2\zeta s + 1 = (s - s_1)(s - s_1^*) = A(s) + r^2 B(s), \tag{13.66}$$

where

$$A(s) = (1 - \kappa)(s - \sigma_2)\left(s - \sigma_1 + \frac{k_1}{1 - \kappa}\right) \tag{13.67}$$

$$B(s) = \kappa(s - \sigma_1)\left(s - \sigma_2 + \frac{k_2}{\kappa}\right). \tag{13.68}$$

Nominally, the gyration resistance r has a scaled value of unity. The sensitivity of root s_1 with respect to changes in the impedance-inversion factor r^2 is therefore

$$S_{r^2}^{s_1} = \frac{ds_1}{d(r^2)/r^2} = -\frac{r^2 B(s_1)}{s_1 - s_1^*} = \frac{A(s_1)}{s_1 - s_1^*}. \tag{13.69}$$

Hence,

$$|S_{r^2}^{s_1}| = \frac{r^2|B(s_1)|}{2\sqrt{1 - \zeta^2}} = \frac{|A(s_1)|}{2\sqrt{1 - \zeta^2}}. \tag{13.70}$$

The magnitude of the pole sensitivity is thus minimized if $|A(s_1)|$ or $|B(s_1)|$ is minimized. From Eq. (13.67) we see that, for a given σ_2, the minimization of $|A(s_1)|$ requires

$$\frac{\partial}{\partial \sigma_1}\left(-\sigma_1 + \frac{k_1}{1 - \kappa}\right) = 0,$$

which yields

$$1 + 2\zeta\sigma_2 + \sigma_2^2 - \kappa(\sigma_2 - \sigma_1)^2 = 0. \tag{13.71}$$

Next, from Eq. (13.68) we see that, for a given σ_1, the minimization of $|B(s_1)|$

requires

$$\frac{\partial}{\partial\sigma_2}\left(-\sigma_2 + \frac{k_2}{\kappa}\right) = 0,$$

which yields

$$1 + 2\sigma_1(\sigma_2 + \zeta) - \sigma_2^2 + \kappa(\sigma_2 - \sigma_1)^2 = 0. \tag{13.72}$$

Solving Eqs. (13.71) and (13.72) for σ_1 and σ_2, we obtain

$$\sigma_1 = -\zeta + \sqrt{\frac{(1-\kappa)(1-\zeta^2)}{\kappa}}, \tag{13.73}$$

$$\sigma_2 = -\zeta - \sqrt{\frac{\kappa(1-\zeta^2)}{1-\kappa}}. \tag{13.74}$$

With σ_1 restricted to lie on the negative real axis, the lower limit on the permissible value of κ is $1 - \zeta^2$; so that the permissible range for the constant κ is $1 - \zeta^2 \leq \kappa \leq 1$. For a selected value of κ, Eqs. (13.73) and (13.74) define the σ_1 and σ_2 that result in the minimum pole sensitivity,

$$|S_{r2}^{s_1}| = \tfrac{1}{2}\sqrt{1-\zeta^2}, \tag{13.75}$$

which is independent of κ. Equation (13.75) is shown plotted against ζ in Fig. 13.20. Comparison of this diagram with Fig. 13.4 reveals that for transfer functions that contain only one pair of complex poles and zeros only at infinite

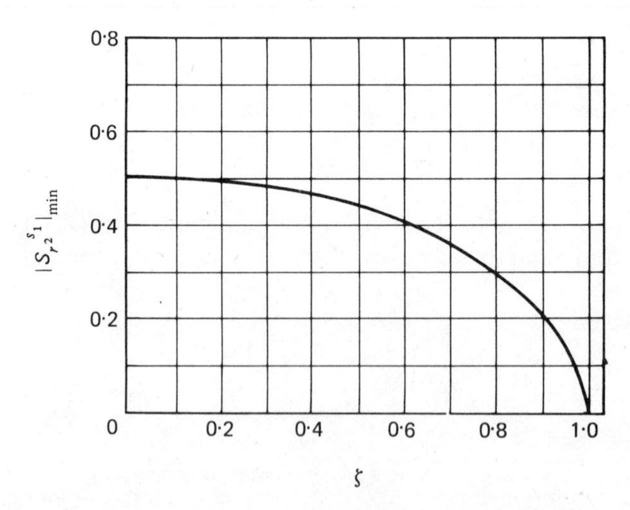

Fig. 13.20. Effect of varying ζ upon the pole sensitivity, $|S_{r2}^{s_1}|_{\min}$.

frequencies, the minimum pole sensitivity with respect to changes in the impedance inversion factor r^2 in the *RC*–gyrator filter is, for all ζ, less than that with respect to changes in the conversion factor k in the *RC*–NIC filter.

The optimum decomposition of polynomial $Q(s)$, pertaining to the set of conditions defined by Eqs. (13.73) and (13.74), can thus be expressed as

$$Q(s) = (1 - \kappa)(s - \sigma_2)^2 + \kappa(s - \sigma_1)^2. \tag{13.76}$$

This decomposition, unlike the *RC*–NIC decomposition of Eq. (13.27), is not unique, since the constant κ can be arbitrarily selected within the permissible range, $1 - \zeta^2 \leq \kappa \leq 1$. For a selected κ, we have

$$\frac{Q(s)}{P(s)} = (1 - \kappa)\frac{s - \sigma_2}{s - \sigma_1} + \kappa\frac{s - \sigma_1}{s - \sigma_2}. \tag{13.77}$$

Hence, for $r = 1$, we may identify network N_a with

$$\begin{aligned} z_{22a} &= (1 - \kappa)\frac{s - \sigma_2}{s - \sigma_1} \\ z_{21a} &= \frac{(1 - \kappa)(\sigma_1 - \sigma_2)}{s - \sigma_1}, \end{aligned} \tag{13.78}$$

and network N_b with

$$\begin{aligned} z_{11b} &= \frac{1}{\kappa}\left(\frac{s - \sigma_2}{s - \sigma_1}\right) \\ z_{21b} &= \frac{\sigma_1 - \sigma_2}{\kappa(s - \sigma_1)}. \end{aligned} \tag{13.79}$$

We thus obtain the realization shown in Fig. 13.21(a). Converting from a current to voltage source as in Fig. 13.21(b), we may also realize a transfer-voltage ratio.

The following two limiting cases are of special interest:

1) When $\kappa = 1 - \zeta^2$, the optimum polynomial decomposition of Eq. (13.76) reduces to

$$Q(s) = \zeta^2\left(s + \frac{1}{\zeta}\right)^2 + (1 - \zeta^2)s^2, \tag{13.80}$$

which, for a given ζ, is unique. Then, the corresponding realizations of networks N_a and N_b take on the forms shown in Fig. 13.22.

2) When $\kappa = 1$, we obtain

$$Q(s) = (s + \zeta)^2 + (1 - \zeta^2) \tag{13.81}$$

which, again, is unique for a prescribed ζ. The corresponding realizations of networks N_a and N_b are as shown in Fig. 13.23. In this second case, it is also

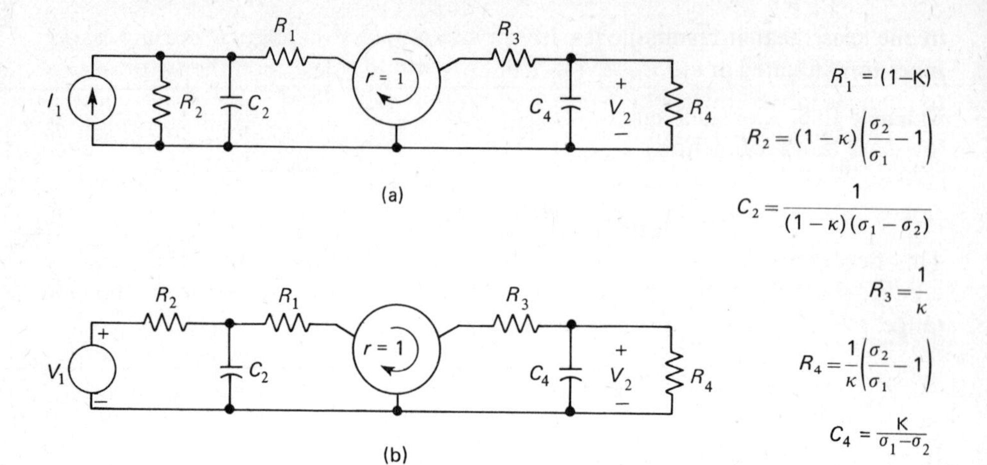

Fig. 13.21. Second-order low-pass *RC* gyrator filters having $1 - \zeta^2 \leqslant \kappa \leqslant 1$ and with (a) Current excitation; (b) Voltage excitation.

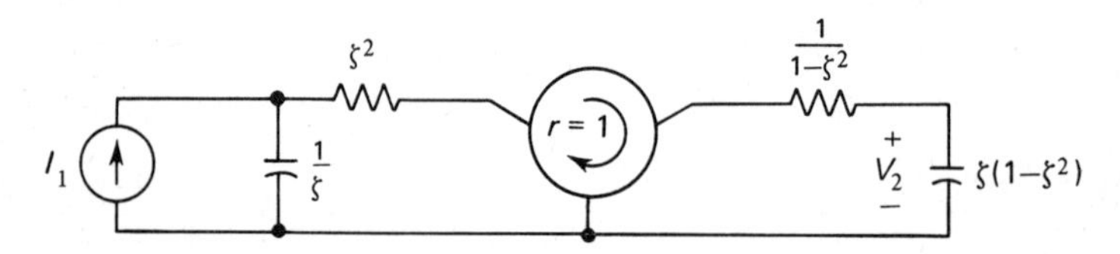

Fig. 13.22. Second-order low-pass *RC* gyrator filter with $\kappa = 1 - \zeta^2$.

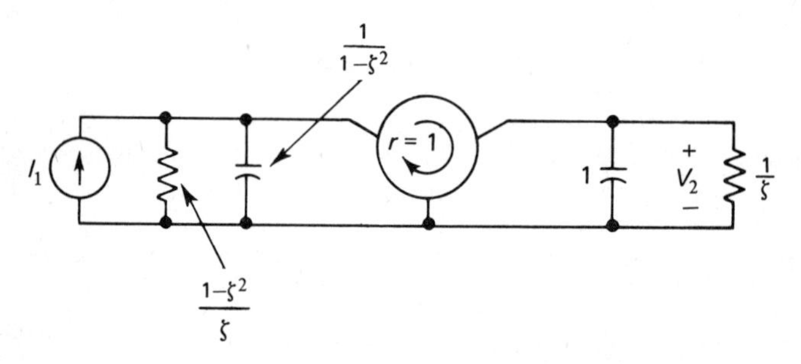

Fig. 13.23. Second-order low-pass *RC* gyrator filter with $\kappa = 1$.

significant to observe that

$$S_{r^2}^{s_1} = -\frac{j}{2}\sqrt{1 - \zeta^2}, \tag{13.82}$$

which shows that the direction of the pole movement for a change in r^2 is parallel to the imaginary axis; this is the optimum direction for a change in pole-position in the sense that it results in the least change in the magnitude of the transfer function evaluated at s_1.

Example 13.5. Consider again the realization, within a constant multiplier, of the low-pass transfer function

$$K(s) = \frac{K(0)}{(s^2 + 0{\cdot}7654s + 1)(s^2 + 1{\cdot}848s + 1)}.$$

Expressing this function as the product of two second-order expressions, with $\zeta_1 = 0{\cdot}3827$ and $\zeta_2 = 0{\cdot}924$, and realizing them separately, we find that for the first section the permissible range for the constant κ is $0{\cdot}854 \leq \kappa_1 \leq 1$, while for the second section the permissible range is $0{\cdot}149 \leq \kappa_2 \leq 1$. Therefore, choosing $\kappa_1 = 0{\cdot}92$ and, $\kappa_2 = 0{\cdot}6$, we obtain the realization shown in Fig. 13.24 for the complete filter. It is assumed that the isolating amplifier has negligible output impedance.

Figure 13.25 shows the realization obtained for the special case of $\kappa = 1$ for both sections of the filter.

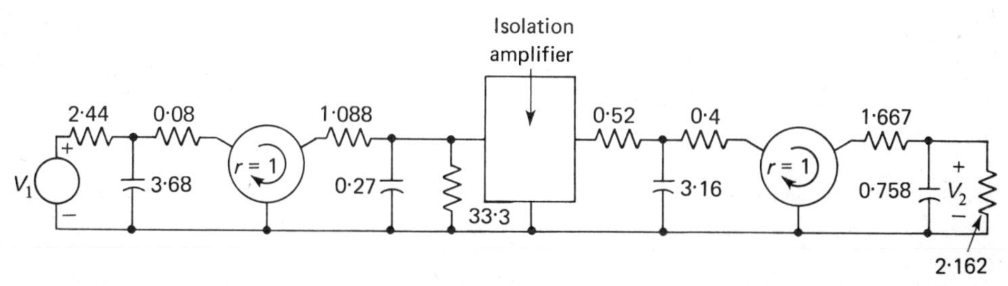

Fig. 13.24. Fourth-order low-pass filter having Butterworth characteristics ($\kappa_1 = 0{\cdot}92$ and $\kappa_2 = 0{\cdot}6$).

13.6 CAPACITOR–GYRATOR ADAPTATIONS OF *LC* LADDER FILTERS

When a flat-passband *LC* ladder filter is designed to operate from a resistive source into a resistive load, with the source delivering its maximum available power into the load at the frequencies of minimum loss over the passband, we find, to a first order of approximation, that at every frequency in the passband and for every component the sensitivity of the loss to component variations is zero.[8] This is easily checked intuitively by noting that when the loss is zero in a reactance network, a component change, either up or down, can only cause the loss to increase. That is to say, in the neighbourhood of the correct value, the curve relating loss to any component value must be quadratic, so that the first derivative of the loss with respect to the component in question must be zero. Indeed, the ability to make high-quality *LC* filters meeting stringent requirements relies heavily on this desensitizing property which occurs at zero loss in a reactance network only when

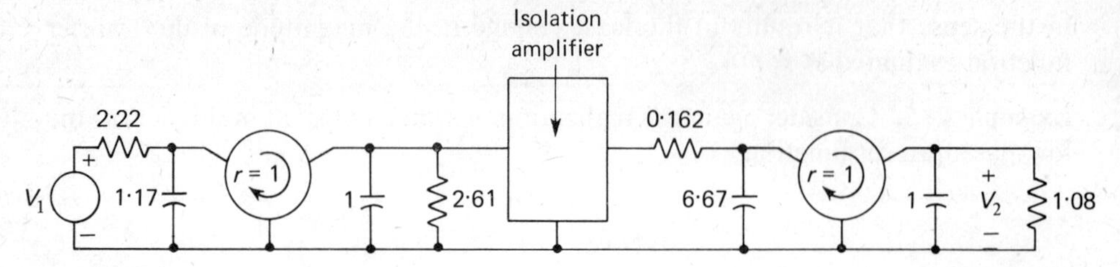

Fig. 13.25. Fourth-order low-pass filter having Butterworth characteristic ($\kappa = 1$ for both sections).

it is doubly loaded. In constructing inductorless filters, we can retain this unique property by designing a conventional doubly loaded *LC* ladder filter to meet the prescribed specification, using established filter theory and design tables, and then simply replace each inductor by a gyrator terminated with an appropriate capacitor.

Example 13.6. To illustrate this procedure, consider Fig. 13.26(a) which gives the element values of a doubly loaded *LC* filter having a maximally flat loss

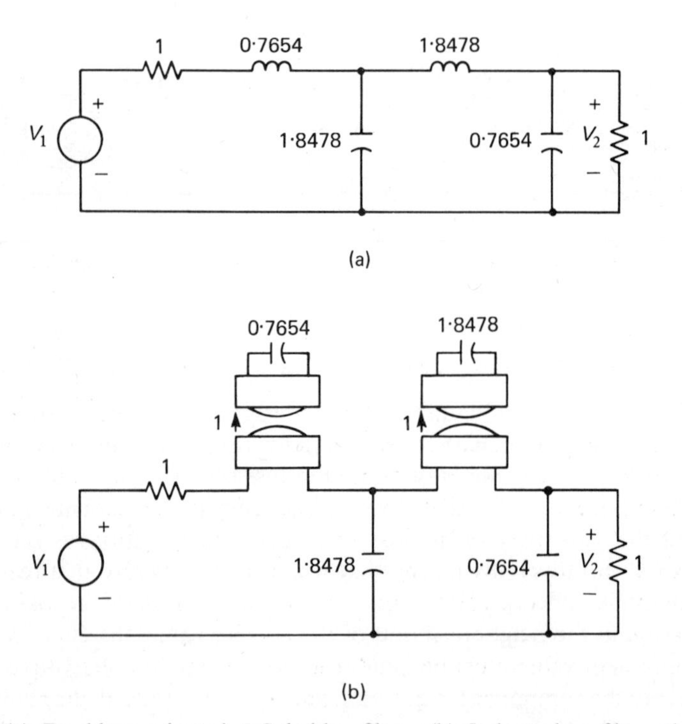

Fig. 13.26. (a) Double-terminated *LC* ladder filter; (b) Inductorless filter obtained by replacing inductors with capacitively terminated gyrators.

characteristic and cutting off at the rate of 24 dB/octave outside the pass band.[9] The corresponding capacitor–gyrator equivalent network is shown in Fig. 13.26(b) where it is assumed that the gyrators are identical and have a scaled gyration resistance of unity.

The direct replacement of an inductor in a series arm of an *LC* ladder filter by a capacitively terminated gyrator necessarily requires the use of a so-called *floating gyrator* with no terminal common to the input and output ports.[10] Such a need can, however, be eliminated, at the expense of an increase in the number of gyrators used, by recognizing that a series inductor at either port of a gyrator is transformed into a shunt capacitor at the other port. Thus, in the case of the low-pass filter of Fig. 13.26(a), by working from the load end toward the source, say, and eliminating the inductors in a step-by-step manner, as illustrated in Fig. 13.27, we ultimately obtain a filter which consists of a cascade of three-terminal gyrators with interstage shunt capacitors accounting for the cut off at higher frequencies. In Fig. 13.27 we have, for convenience, assumed that all the gyrators are identical, with unity gyration resistance.

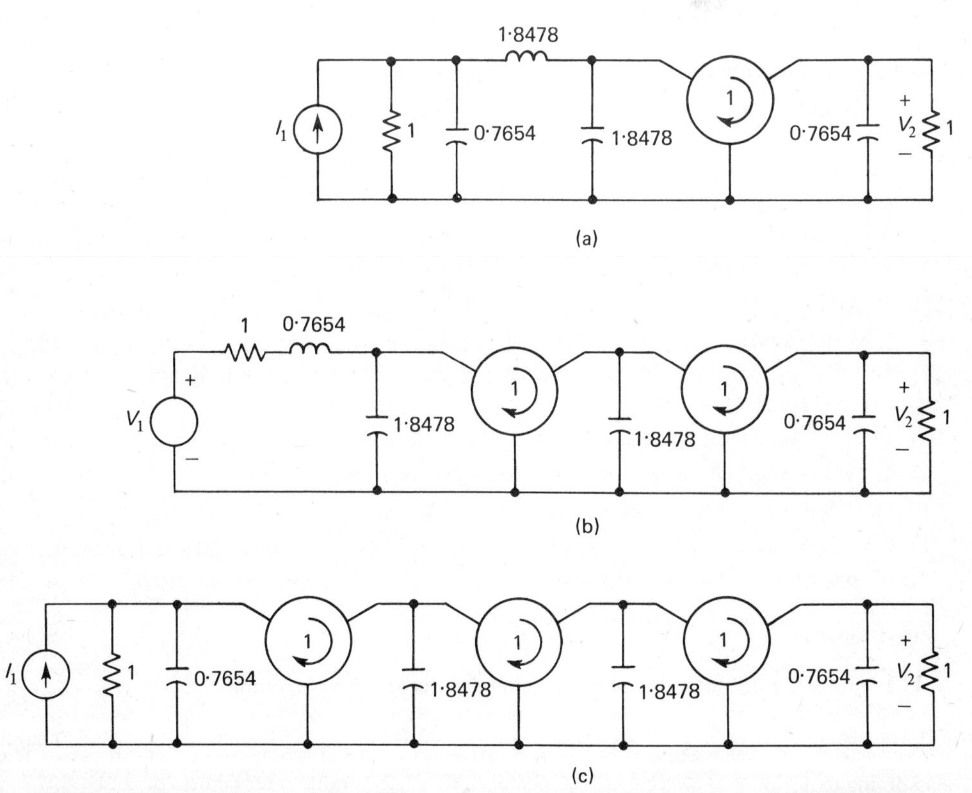

Fig. 13.27. The development of an inductorless filter using three-terminal gyrators.

Concluding Remarks

In our study of *RC*–NIC and *RC*–gyrator filters, the only pole sensitivity considered has been with respect to the conversion factor of the NIC or the impedance inversion factor of the gyrator. It has been shown[11] that if this sensitivity is optimized, then the pole sensitivity of the system to the passive elements is also ordinarily optimized, with a value of the same order of magnitude as the pole sensitivity with respect to the active element. This, therefore, means that in a practical *RC* active filter the NIC or gyrator used must have about the same degree of precision as the passive elements.

The replacement of each inductor in a doubly loaded *LC* ladder filter by an appropriate gyrator–capacitor combination apparently represents the optimum solution to the problem of synthesizing inductorless filters in that a prescribed loss characteristic is realized with zero sensitivity to component tolerances, at frequencies of zero loss. This desensitizing property does not, however, occur in either singly loaded or dissipation-compensated filters.

REFERENCES

1. J. G. LINVILL, "*RC* active filters," *Proc. I.R.E.* **42,** 555 (1954).
2. E. A. GUILLEMIN, *Synthesis of Passive Networks.* Wiley, 1957.
3. F. H. BLECHER, "Application of Synthesis Techniques to Electronic Circuit Design," *I.R.E. International Convention Record* **8,** part 2, 210 (1960).
4. D. J. STOREY and W. J. CULLYER, "Network Synthesis Using Negative-impedance Converters," *Proc. I.E.E.* **111,** 891 (1964).
5. T. YANAGISAWA, "*RC* Active Networks Using Current-inversion Type Negative-impedance Converters," *Trans. I.R.E.* **CT-4,** 140 (1957).
6. B. K. KINARIWALA, "Synthesis of Active *RC* Networks," *B.S.T.J.* **38,** 1269 (1959).
7. I. M. HOROWITZ, "Active *RC* Transfer Function Synthesis by Means of Cascaded *RL* and *RC* Structures," *Polytechnic Institute of Brooklyn*, Research Report R-583-57, PIB-503 (1958).
8. H. J. ORCHARD, "Inductorless Filters," *Electronics Letters* **2,** 224 (1966).
9. L. WEINBERG, *Network Analysis and Synthesis.* McGraw-Hill, 1962.
10. D. F. SHEEHAN, "Gyrator-flotation Circuit," *Electronics Letters* **3,** 39 (1967).
11. I. M. HOROWITZ, "Optimization of Negative-impedance Conversion Methods of Active *RC* Synthesis," *Trans. I.R.E.* **CT-6,** 296 (1959).

PROBLEMS

13.1 Using Linvill's cascade method, synthesize the low-pass function

$$Z_{21}(s) = \frac{Z_{21}(0)}{(s+1)(s^2+s+1)},$$

assuming that

$$P(s) = (s+\tfrac{1}{2})(s+2)(s+4).$$

13.2 Synthesize a fourth-order equi-ripple low-pass filter characterized by

$$\frac{V_2}{V_1} = \frac{K(0)}{(s^2 + 0{\cdot}279s + 0{\cdot}987)(s^2 + 0{\cdot}674s + 0{\cdot}279)}$$

with minimum pole sensitivity, using a cascade of

a) second-order *RC*–NIC sections

b) second-order *RC*–gyrator sections.

13.3 Using Linvill's cascade method, synthesize the high-pass function

$$\frac{V_2}{V_1} = K(\infty)\frac{s^2}{s^2 + \sqrt{2}s + 1}.$$

13.4 Using Kinariwala's procedure, synthesize

a) $$\frac{V_2}{V_1} = K(\infty)\frac{s^2 + 0{\cdot}42}{s^2 + 0{\cdot}023s + 0{\cdot}93}$$

b) $$\frac{V_2}{V_1} = K(\infty)\frac{s^2 + 1{\cdot}3}{s^2 + 0{\cdot}018s + 0{\cdot}6}.$$

13.5 Repeat Problem 13.4 using Yanagisawa's procedure.

13.6 Develop capacitor–gyrator equivalent networks for the band-pass structures shown in Fig. P13.6 using

a) floating gyrators,

b) three-terminal gyrators.

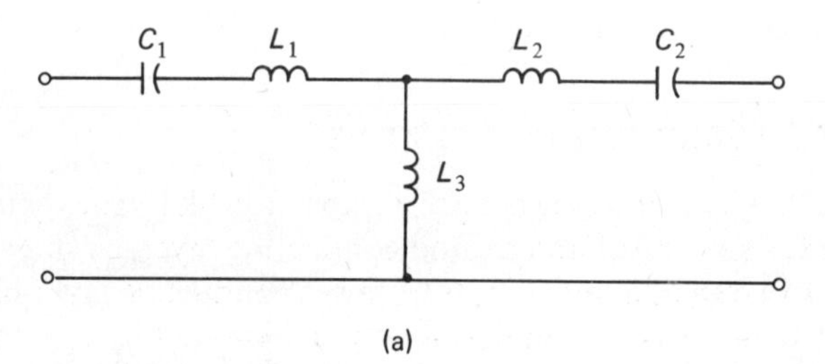

(a)

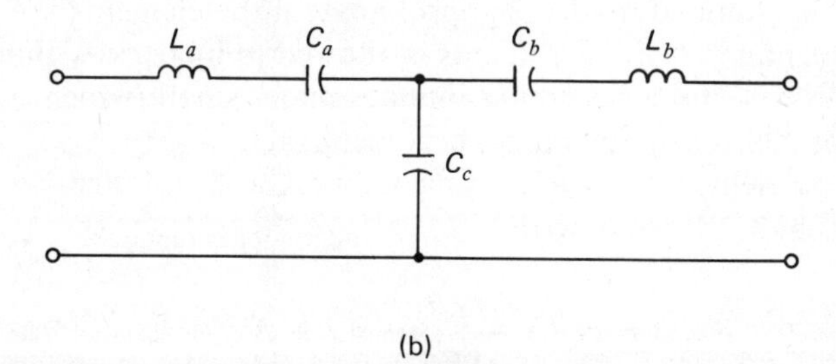

(b)

Fig. P13.6.

CHAPTER 14

DISTRIBUTED *RC* NETWORKS

14.1 INTRODUCTION

In an integrated circuit, resistance and capacitance generally occur in distributed form with the result that certain properties appear which are not realizable in conventional lumped circuits. Thus, for example, a distributed *RC* network can be used to construct a low-pass filter with sharper cut-off than is possible with lumped *RC* networks. A distributed *RC* network can provide more phase shift with less attenuation than is obtainable with lumped *RC* networks. Also, at high frequencies, the use of a distributed *RC* network makes it possible to eliminate the need for an inductance in certain applications, such as the construction of narrow-band tuned amplifiers. In this chapter we shall study the behaviour of uniform distributed and exponentially tapered distributed *RC* networks, and consider some of their applications. Also, we shall develop a technique for adapting the synthesis procedures developed in the previous chapter in order to synthesize distributed *RC* networks with prescribed characteristics.

14.2 UNIFORM DISTRIBUTED *RC* NETWORK

A distributed *RC* structure consists of a capacitor-like sandwich of dielectric layer between a resistive layer and a good conducting layer, as shown in idealized form in Fig. 14.1; we shall assume that the lower conductive layer has zero resistance. Such a structure can be approximately represented by a number of series-resistance and shunt-capacitance elements in a uniform ladder network, as in Fig. 14.2 where r_0 is the resistance per unit length and c_0 is the capacitance per unit length. This lumped model is approximate if the elements are of finite size. If, however, the number of the elements is allowed to increase without limit, each element at the same time becoming infinitesimally small (which is achieved by letting Δx approach zero), the model becomes exact.

Consider an elemental portion of the network at a distance x from the input end, as in Fig. 14.3. We may write

$$i(x, t) = c_0\,\Delta x \frac{\partial v(x, t)}{\partial t} + i(x + \Delta x, t), \tag{14.1}$$

$$v(x, t) = r_0\,\Delta x\; i(x + \Delta x, t) + v(x + \Delta x, t). \tag{14.2}$$

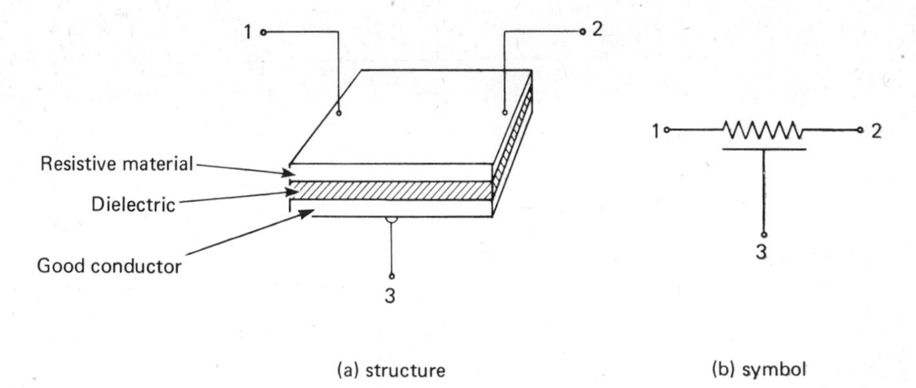

Fig. 14.1. Uniform distributed RC network.

When $\Delta x \to 0$, Eqs. (14.1) and (14.2) take on the form of a pair of partial differential equations, as shown by

$$\frac{\partial i(x, t)}{\partial x} = -c_0 \frac{\partial v(x, t)}{\partial t}, \tag{14.3}$$

$$\frac{\partial v(x, t)}{\partial x} = -r_0 i(x, t). \tag{14.4}$$

Substitution of Eq. (14.4) in (14.3) yields the classical one-dimensional diffusion equation,

$$\frac{\partial^2 v(x, t)}{\partial x^2} = r_0 c_0 \frac{\partial v(x, t)}{\partial t}. \tag{14.5}$$

The current $i(x, t)$ satisfies a similar equation.

Taking the Laplace transforms of both sides of Eq. (14.5), the partial derivatives become total derivatives. Thus, assuming the initial conditions to be zero, and recognizing that differentiation in the time domain corresponds to multiplication by s in the complex-frequency domain, we obtain

$$\frac{d^2 V(x, s)}{dx^2} - s r_0 c_0 V(x, s) = 0. \tag{14.6}$$

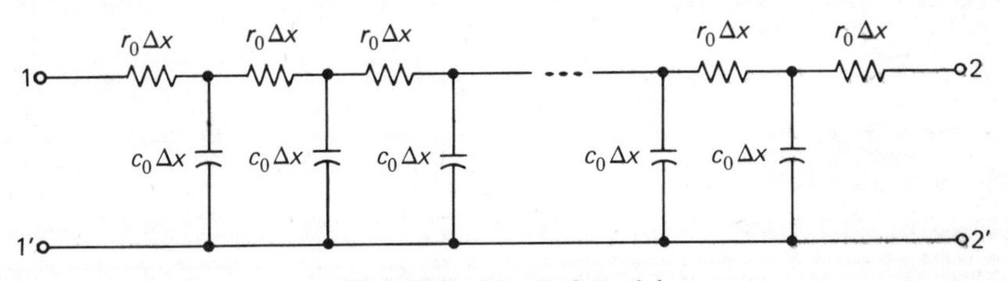

Fig. 14.2. Lumped model.

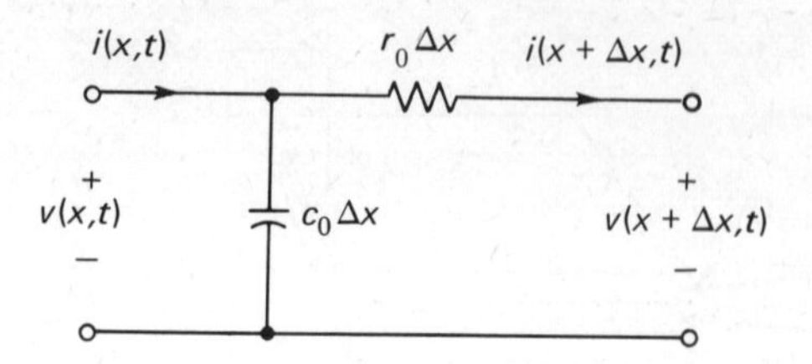

Fig. 14.3. Elemental portion of the lumped model.

The second-order differential equation (14.6) is known to have a general solution of the following form

$$V(x, s) = A_1\varepsilon^{\gamma_0 x} + A_2\varepsilon^{-\gamma_0 x}, \tag{14.7}$$

where

$$\gamma_0 = \sqrt{sr_0c_0} \tag{14.8}$$

and the A's are constants with the dimensions of voltage, to be evaluated by the boundary conditions imposed upon the network. The quantity γ_0 is seen to govern the manner in which $V(x, s)$ varies with x; it is therefore given the name *propagation constant.*

We shall assume the structure to have a length equal to l, and treat it as a two-port network with I_1, V_1 denoting the conditions at port 1–1′, and I_2, V_2 the conditions at port 2–2′, in the usual manner. Thus, evaluating Eq. (14.7) at $x = 0$ and $x = l$, we obtain

$$\begin{aligned} V_1 &= V(0, s) = A_1 + A_2 \\ V_2 &= V(l, s) = A_1\varepsilon^{\gamma_0 l} + A_2\varepsilon^{-\gamma_0 l}. \end{aligned} \tag{14.9}$$

Solving Eqs. (14.9) for A_1 and A_2, we get

$$\begin{aligned} A_1 &= -\frac{\varepsilon^{-\gamma_0 l}V_1}{\varepsilon^{\gamma_0 l} - \varepsilon^{-\gamma_0 l}} + \frac{V_2}{\varepsilon^{\gamma_0 l} - \varepsilon^{-\gamma_0 l}} \\ A_2 &= \frac{\varepsilon^{\gamma_0 l}V_1}{\varepsilon^{\gamma_0 l} - \varepsilon^{-\gamma_0 l}} - \frac{V_2}{\varepsilon^{\gamma_0 l} - \varepsilon^{-\gamma_0 l}}. \end{aligned} \tag{14.10}$$

To determine the current, we first transform Eq. (14.4) into the s-domain, obtaining

$$I(x, s) = -\frac{1}{r_0}\frac{dV(x, s)}{dx}. \tag{14.11}$$

Then, use of Eqs. (14.7) and (14.11) yields

$$I(x, s) = -\frac{1}{Z_0}(A_1\varepsilon^{\gamma_0 x} - A_2\varepsilon^{-\gamma_0 x}), \tag{14.12}$$

where Z_0 is the *characteristic impedance* of the network defined by

$$Z_0 = \sqrt{\frac{r_0}{sc_0}}. \tag{14.13}$$

Evaluating Eq. (14.12) at $x = 0$ and $x = l$, we get

$$\begin{aligned} I_1 &= I(0, s) = -\frac{1}{Z_0}(A_1 - A_2) \\ I_2 &= -I(l, s) = \frac{1}{Z_0}(A_1 \varepsilon^{\gamma_0 l} - A_2 \varepsilon^{-\gamma_0 l}) \end{aligned} \tag{14.14}$$

or, from Eq. (14.10),

$$\begin{bmatrix} I_1 \\ I_2 \end{bmatrix} = \frac{1}{Z_0}\begin{bmatrix} \coth \gamma_0 l & -\operatorname{csch} \gamma_0 l \\ -\operatorname{csch} \gamma_0 l & \coth \gamma_0 l \end{bmatrix} \begin{bmatrix} V_1 \\ V_2 \end{bmatrix}. \tag{14.15}$$

The y-parameters of the network are therefore

$$\begin{aligned} y_{11} = y_{22} &= \frac{\coth \gamma_0 l}{Z_0} = \frac{\cosh \gamma_0 l}{Z_0 \sinh \gamma_0 l} \\ y_{12} = y_{21} &= -\frac{\operatorname{csch} \gamma_0 l}{Z_0} = -\frac{1}{Z_0 \sinh \gamma_0 l}. \end{aligned} \tag{14.16}$$

In terms of the z-parameter representation, we have

$$\begin{bmatrix} V_1 \\ V_2 \end{bmatrix} = Z_0\begin{bmatrix} \coth \gamma_0 l & \operatorname{csch} \gamma_0 l \\ \operatorname{csch} \gamma_0 l & \coth \gamma_0 l \end{bmatrix} \begin{bmatrix} I_1 \\ I_2 \end{bmatrix}, \tag{14.17}$$

with

$$\begin{aligned} z_{11} = z_{22} &= Z_0 \coth \gamma_0 l = Z_0 \frac{\cosh \gamma_0 l}{\sinh \gamma_0 l} \\ z_{12} = z_{21} &= Z_0 \operatorname{csch} \gamma_0 l = \frac{Z_0}{\sinh \gamma_0 l}. \end{aligned} \tag{14.18}$$

Pole–Zero Patterns of the Admittance and Impedance Parameters and their Frequency Characteristics

To determine the pole–zero locations of the two-port parameters of the *RC* distributed network, we may expand Eqs. (14.16) and (14.18) into product form by making use of the relations

$$\begin{aligned} \sinh \theta &= \theta \prod_{k=1}^{\infty} \left(1 + \frac{\theta^2}{k^2 \pi^2}\right) \\ \cosh \theta &= \prod_{k=1}^{\infty} \left(1 + \frac{4\theta^2}{(2k-1)^2 \pi^2}\right). \end{aligned} \tag{14.19}$$

The y-parameters of Eq. (14.16) may thus be expressed in the form

$$
\begin{aligned}
y_{11}(s) = y_{22}(s) &= \frac{1}{R} \prod_{k=1}^{\infty} \frac{\left(1 + \dfrac{4RCs}{(2k-1)^2\pi^2}\right)}{\left(1 + \dfrac{RCs}{k^2\pi^2}\right)} \\
y_{12}(s) = y_{21}(s) &= -\frac{1}{R} \prod_{k=1}^{\infty} \frac{1}{\left(1 + \dfrac{RCs}{k^2\pi^2}\right)},
\end{aligned}
\tag{14.20}
$$

where $R = r_0 l$ is the total resistance of the network and $C = c_0 l$ is the total capacitance. The poles and zeros of the short-circuit driving-point admittances of a distributed RC network are therefore infinite in number, alternating along the negative real axis of the s-plane with a zero as the nearest critical frequency to the origin, as in Fig. 14.4(a). The transfer admittances have the same poles as the driving-point admittances, but no finite zeros, as in Fig. 14.4(b).

To compute the actual frequency dependence of the short-circuit input admittance $y_{11}(j\omega)$ we may use the first line of Eqs. (14.16) for $s = j\omega$, obtaining the characteristics shown in Fig. 14.5. At low frequencies, $\omega \ll 1/RC$, the normalized admittance function $y_{11}(j\omega)R$ approaches unity as a limit, while at high frequencies, $\omega \gg 1/RC$, it approaches the limiting value of $(j\omega RC)^{1/2}$. This means that at high frequencies, the final asymptote of $20 \log_{10}(|y_{11}(j\omega)|R)$ has a slope of

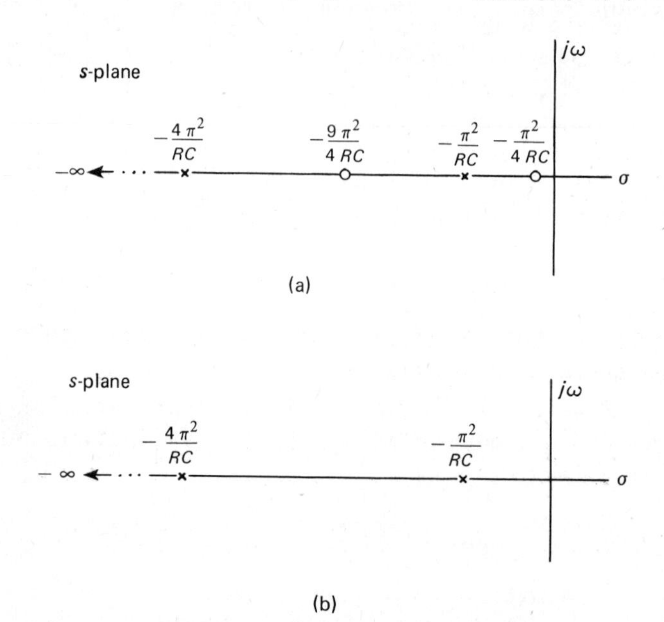

Fig. 14.4. Pole–zero patterns for: (a) $y_{11}(s)$; (b) $y_{12}(s)$.

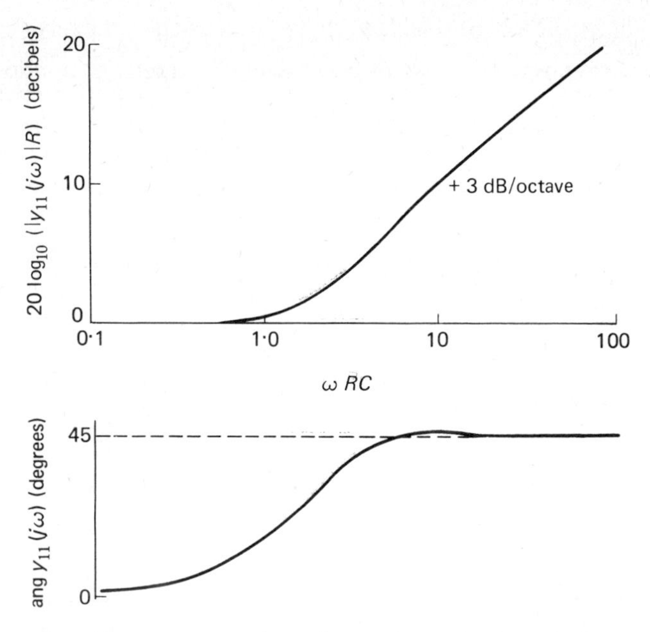

Fig. 14.5. Frequency dependence of magnitude and phase angle of $y_{11}(j\omega)$.

3 decibels per octave, while the phase angle of $y_{11}(j\omega)$ approaches 45 degrees as a limit, both characteristics being precisely half of those for a lumped network consisting of R and C in parallel. It is also apparent that if the high-frequency asymptote of the magnitude characteristic were extended, it would intersect the frequency axis at $\omega = 1/RC$. We may thus approximate the magnitude function, $20 \log_{10}(|y_{11}(j\omega)|R)$, by two line segments, one of which extends along the frequency axis up to $\omega = 1/RC$, and the other extends from this frequency upwards at the rate of 3 decibels per octave. This approximation is simpler and more accurate than that obtained using an infinite succession of the usual RX slopes at ± 6 decibels per octave and with corner frequencies as defined by the first line of Eqs. (14.20).

For the open-circuit impedances of the network, we have

$$
\begin{aligned}
z_{11}(s) = z_{22}(s) &= \frac{1}{sC} \prod_{k=1}^{\infty} \frac{\left(1 + \dfrac{4RCs}{(2k-1)^2\pi^2}\right)}{\left(1 + \dfrac{RCs}{k^2\pi^2}\right)} \\
z_{12}(s) = z_{21}(s) &= \frac{1}{sC} \prod_{k=1}^{\infty} \frac{1}{\left(1 + \dfrac{RCs}{k^2\pi^2}\right)}
\end{aligned}
\tag{14.21}
$$

which corresponds to the pole–zero patterns shown in Fig. 14.6. Here, again, we see that the open-circuit driving-point impedances have an infinite number of poles and zeros, alternating along the negative real axis, with a pole at the origin. Also, the transfer impedances have the same poles as the driving-point impedances but no finite zeros.

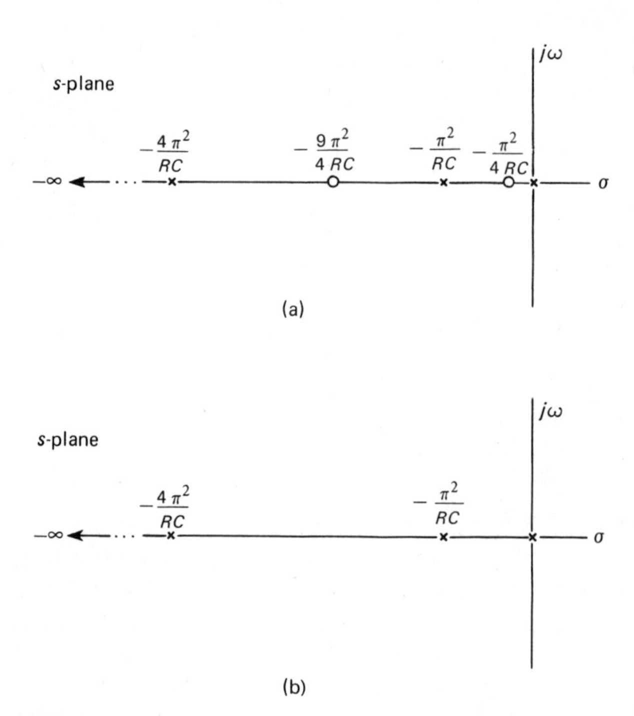

Fig. 14.6. Pole–zero patterns for: (a) $z_{11}(s)$; (b) $z_{12}(s)$.

To compute the actual frequency dependence of the open-circuit input impedance $z_{11}(j\omega)$ we may use the first line of Eqs. (14.18) for $s = j\omega$, obtaining the characteristics shown in Fig. 14.7. The normalized impedance function $z_{11}(j\omega)/R$ approaches the limiting value of $Z_0/\gamma_0 lR$, or $1/j\omega CR$ at low frequencies, and the limiting value of Z_0/R, or, $(j\omega RC)^{-1/2}$ at high frequencies. Thus, at low frequencies the magnitude function, $20 \log_{10}(|z_{11}(j\omega)|/R)$, decreases at the rate of 6 decibels per octave as for a conventional capacitor, while at high frequencies, it decreases at the rate of 3 decibels per octave; the low- and high-frequency asymptotes, if extended, would intersect each other on the frequency axis at $\omega = 1/RC$. We also see that, in the limit, the phase angle of $z_{11}(j\omega)$ approaches -90 degrees at low frequencies, and -45 degrees at high frequencies.

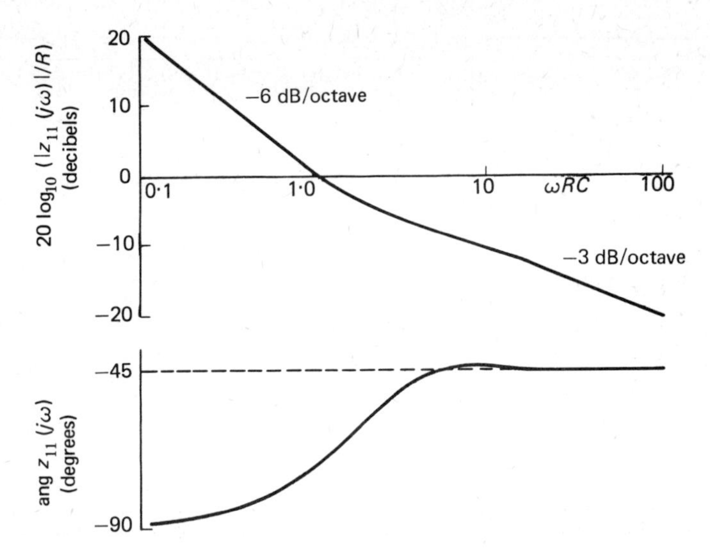

Fig. 14.7. Frequency dependence of magnitude and phase angle of z_{11} $(j\omega)$.

Example 14.1. *Phase-shift oscillator.*[1] A rather useful application of a distributed *RC* network is in the construction of a phase-shift oscillator, as illustrated in Fig. 14.8(a). In the idealized model of Fig. 14.8(b) we have assumed, for simplicity, that the amplifier is unilateral with infinite input impedance and zero output impedance. The return ratio for the voltage gain μ of the amplifier is therefore

$$T(s) = -\frac{\mu y_{12}(s)}{y_{11}(s)}, \tag{14.22}$$

where y_{11} and y_{12} refer to the distributed *RC* network component. Equations (14.16) and (14.22) thus give

$$T(s) = \frac{\mu}{\cosh \gamma_0 l}. \tag{14.23}$$

For $s = j\omega$, we have $\gamma_0 = \sqrt{j\omega r_0 c_0}$, indicating that the propagation constant is a complex number. We may thus write

$$\gamma_0 = \alpha_0 + j\beta_0 = \sqrt{j\omega r_0 c_0}; \tag{14.24}$$

α_0 is called the *attenuation constant*, and β_0 the *phase constant*. Since $\sqrt{j} = (1 + j)/\sqrt{2}$, we may identify

$$\alpha_0 = \beta_0 = \sqrt{\frac{\omega r_0 c_0}{2}}. \tag{14.25}$$

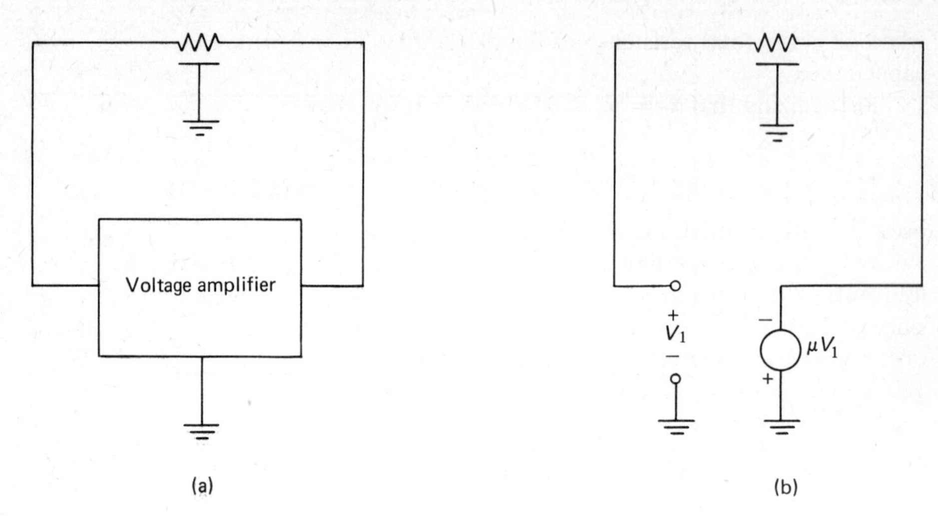

Fig. 14.8. (a) Phase-shift oscillator using distributed *RC* network; (b) Idealized model.

Both α_0 and β_0 are therefore frequency dependent. For Eq. (14.23), we thus have

$$T(j\omega) = \frac{\mu}{\cosh(\alpha_0 + j\beta_0)l}$$

$$= \frac{\mu}{\cosh \alpha_0 l \cos \beta_0 l + j \sinh \alpha_0 l \sin \beta_0 l}. \tag{14.26}$$

For sustained oscillations at some specified frequency ω_0, we must have $T(j\omega_0) = -1$. This requires that the two conditions

$$\sinh \alpha_0 l \sin \beta_0 l = 0, \tag{14.27}$$

$$\frac{\mu}{\cosh \alpha_0 l \cos \beta_0 l} = -1 \tag{14.28}$$

be simultaneously satisfied at $\omega = \omega_0$. Now, for Eq. (14.27) to be satisfied, either $\sinh \alpha_0 l$ or $\sin \beta_0 l$ must be zero. However, since $\sinh \alpha_0 l = 0$ only when $\alpha_0 l = 0$, and $\alpha_0 l$ cannot be zero for any usable case, it follows that $\sin \beta_0 l$ must be zero, which is true for $\beta_0 l = k\pi$ with k as integer. Therefore, at $\omega = \omega_0$, we have

$$\beta_0 l = l\sqrt{\frac{\omega_0 r_0 c_0}{2}} = k\pi. \tag{14.29}$$

Solving for ω_0, we get

$$\omega_0 = \frac{2k^2\pi^2}{r_0 c_0 l^2} = \frac{2k^2\pi^2}{RC}, \tag{14.30}$$

where R is the total resistance of the distributed RC network and C is the total capacitance.

Recognizing that $\alpha_0 = \beta_0$ and substituting Eq. (14.29) in Eq. (14.28), we get

$$\mu = -\cosh k\pi \cos k\pi. \tag{14.31}$$

Equation (14.31) defines the voltage gain the amplifier must provide in order to oscillate; μ is positive for odd-integer values of k and negative for even-integer values of k, corresponding to 180° phase shift and 0° phase shift through the feedback path, respectively. For μ to be a minimum, k must also be a minimum since $\cosh k\pi$ always increases with increasing k. The smallest odd positive value of k is unity, and therefore the frequency for minimum required voltage gain at 180° phase shift through the distributed RC network is

$$\omega_0 = \frac{2\pi^2}{RC} \tag{14.32}$$

and the required gain is

$$\mu = \cosh \pi = 11{\cdot}6. \tag{14.33}$$

These are the equations for oscillation frequency and voltage gain when the uniform distributed RC network is used in a phase-shift oscillator, assuming ideal amplifier conditions as described by the model of Fig. 14.8(b). The requirement $\mu = 11{\cdot}6$ represents a considerable improvement over the more conventional uniform three-section RC phase-shift oscillator for which the minimum μ is required to be 29 for sustained oscillations.

14.3 EXPONENTIALLY TAPERED DISTRIBUTED *RC* NETWORK[2,3]

The desirable characteristics of a distributed RC network can be further enhanced by tapering the structure; Fig. 14.9 illustrates the idealized case of a distributed RC structure with an exponential taper. In such a structure the resistance $r(x)$ per unit length and capacitance $c(x)$ per unit length are both functions of the

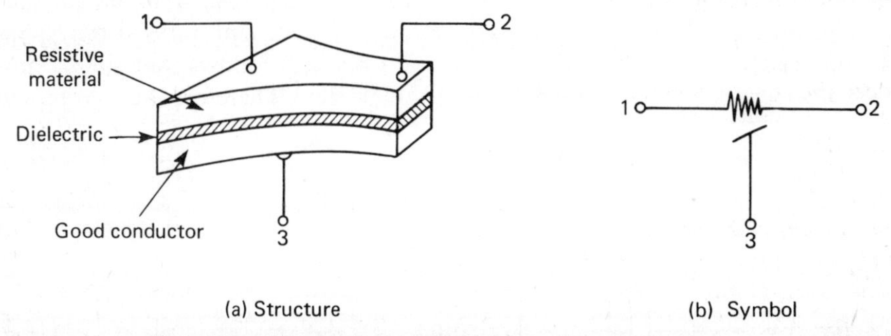

Fig. 14.9. Exponentially tapered distributed RC network.

distance x from the input end, so that, in place of Eqs. (14.3) and (14.4), we may write

$$\frac{\partial i(x, t)}{\partial x} = -c(x)\frac{\partial v(x, t)}{\partial t} \tag{14.34}$$

$$\frac{\partial v(x, t)}{\partial x} = -r(x)i(x, t). \tag{14.35}$$

Substitution of Eq. (14.35) in (14.34) yields

$$\frac{\partial}{\partial x}\left[-\frac{1}{r(x)}\frac{\partial v(x, t)}{\partial x}\right] = -c(x)\frac{\partial v(x, t)}{\partial t}$$

or

$$\frac{\partial^2 v(x, t)}{\partial x^2} - \frac{1}{r(x)}\frac{dr(x)}{dx}\frac{\partial v(x, t)}{\partial x} = r(x)c(x)\frac{\partial v(x, t)}{\partial t}. \tag{14.36}$$

In a distributed *RC* network with an exponential taper, $r(x)$ and $c(x)$ are defined by

$$\begin{aligned} r(x) &= r_0\varepsilon^{mx} \\ c(x) &= c_0\varepsilon^{-mx}, \end{aligned} \tag{14.37}$$

where m is a constant. For such a structure, Eq. (14.36) gives

$$\frac{\partial^2 v(x, t)}{\partial x^2} - m\frac{\partial v(x, t)}{\partial x} = r_0 c_0\frac{\partial v(x, t)}{\partial t}. \tag{14.38}$$

When Eq. (14.38) is transformed into the s-domain, assuming zero initial conditions, we get

$$\frac{d^2 V(x, s)}{dx^2} - m\frac{dV(x, s)}{dx} - sr_0c_0V(x, s) = 0. \tag{14.39}$$

The solution to Eq. (14.39) is obtained as

$$V(x, s) = \varepsilon^{mx/2}(A_1 \sinh \gamma x + A_2 \cosh \gamma x), \tag{14.40}$$

where the A's are arbitrary constants, and γ is defined by

$$\gamma = \sqrt{\frac{m^2}{4} + sr_0c_0}. \tag{14.41}$$

The current is determined by substituting Eq. (14.40) into the transformed version of Eq. (14.35); we thus obtain

$$I(x, s) = -\frac{\varepsilon^{-mx/2}}{r_0}\left[\left(\frac{A_1 m}{2} + A_2\gamma\right)\sinh \gamma x + \left(A_1\gamma + \frac{A_2 m}{2}\right)\cosh \gamma x\right]. \tag{14.42}$$

If, next, we insert the boundary conditions I_1, V_1 and I_2, V_2 for the input and output ports, respectively, proceeding as previously described by Eqs. (14.9)–(14.15), we obtain the y-parameters of an exponentially tapered distributed *RC* network as

$$
\begin{aligned}
y_{11} &= \frac{1}{r_0 \sinh \gamma l}\left(\gamma \cosh \gamma l - \frac{m}{2} \sinh \gamma l\right) \\
y_{12} = y_{21} &= -\frac{\gamma \varepsilon^{-ml/2}}{r_0 \sinh \gamma l} \\
y_{22} &= \frac{\varepsilon^{-ml}}{r_0 \sinh \gamma l}\left(\gamma \cosh \gamma l + \frac{m}{2} \sinh \gamma l\right).
\end{aligned}
\tag{14.43}
$$

The z-parameters of the network are given by

$$
\begin{aligned}
z_{11} &= \frac{1}{sc_0 \sinh \gamma l}\left(\gamma \cosh \gamma l + \frac{m}{2} \sinh \gamma l\right) \\
z_{12} = z_{21} &= \frac{\gamma \varepsilon^{ml/2}}{sc_0 \sinh \gamma l} \\
z_{22} &= \frac{\varepsilon^{ml}}{sc_0 \sinh \gamma l}\left(\gamma \cosh \gamma l - \frac{m}{2} \sinh \gamma l\right).
\end{aligned}
\tag{14.44}
$$

Clearly, in the limit as parameter m approaches zero, we have $\gamma = \gamma_0$, and Eqs. (14.43) and (14.44) reduce to the same form as Eqs. (14.16) and (14.18), respectively, which pertain to a uniform distributed *RC* network.

Transfer Characteristics

Consider a distributed *RC* network with an exponential taper operated as in Fig. 14.10 with terminal 3 common and terminal 2 open. The transfer-voltage ratio of the network, measured in the forward direction, is therefore,

$$
\begin{aligned}
K_1(s) &= \frac{V_{23}}{V_{13}} \\
&= \frac{z_{21}}{z_{11}} \\
&= \frac{\gamma \varepsilon^{ml/2}}{\gamma \cosh \gamma l + \dfrac{m}{2} \sinh \gamma l},
\end{aligned}
\tag{14.45}
$$

where the dependence of propagation constant γ on the complex variable s is defined by Eq. (14.41). The transfer function of Eq. (14.45) represents a low-pass characteristic, as evidenced by Fig. 14.11(a) showing the gain component of $K_1(j\omega)$ plotted for different values of the product ml. To display the magnitude

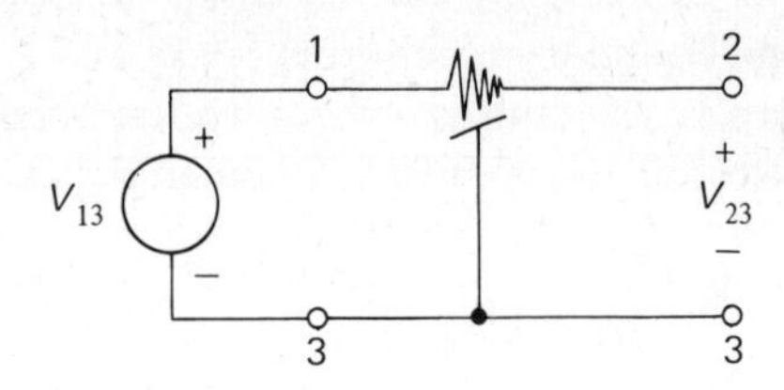

Fig. 14.10. *RC* distributed network operated as low-pass filter.

as well as phase, $K_1(j\omega)$ is given in a polar plot in Fig. 14.11(b). A positive value of m represents a taper in the direction of signal transmission, while a negative m represents a taper in the opposite direction. In Fig. 14.11 we see that for positive m, a tapered structure can produce more phase shift with less attenuation than is obtainable with a uniform structure for which $m = 0$. This means that when a

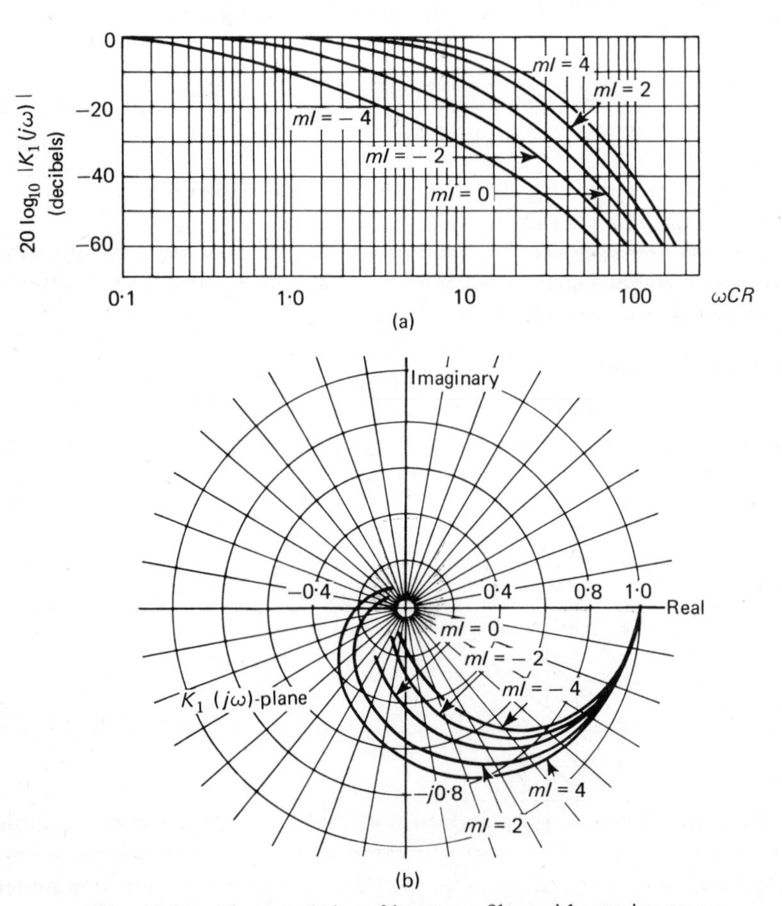

Fig. 14.11. Characteristics of low-pass filter with varying taper.

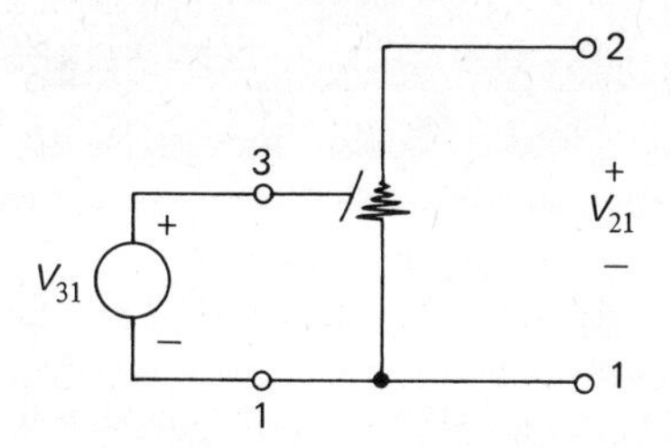

Fig. 14.12. *RC* distributed network operated as high-pass filter.

distributed *RC* network with taper is used in the feedback path of a phase-shift oscillator, it permits a further reduction in the gain requirement on the amplifier.

Consider, next, operation of the distributed structure in the manner shown in Fig. 14.12. For this configuration, the open-circuit transfer-voltage ratio is

$$K_2(s) = \frac{V_{21}}{V_{31}} = \frac{V_{23} + V_{31}}{V_{31}}$$
$$= 1 - K_1(s), \tag{14.46}$$

where $K_1(s)$ is given in Eq. (14.45). The transfer function $K_2(s)$ of Eq. (14.46) represents a high-pass characteristic, as demonstrated by the gain characteristic of Fig. 14.13. In this case, we see that it is possible for $|K_2(j\omega)|$ to be greater than unity for some frequencies. However, since $|K_1(j\omega)|$ is always less than unity (see Fig. 14.11), then

$$|K_2(j\omega)| = |1 - K_1(j\omega)| \leq 1 + |K_1(j\omega)| < 2. \tag{14.47}$$

Equation (14.47) indicates that the open-circuit transfer-voltage ratio of the structure shown in Fig. 14.12 cannot be greater than two.

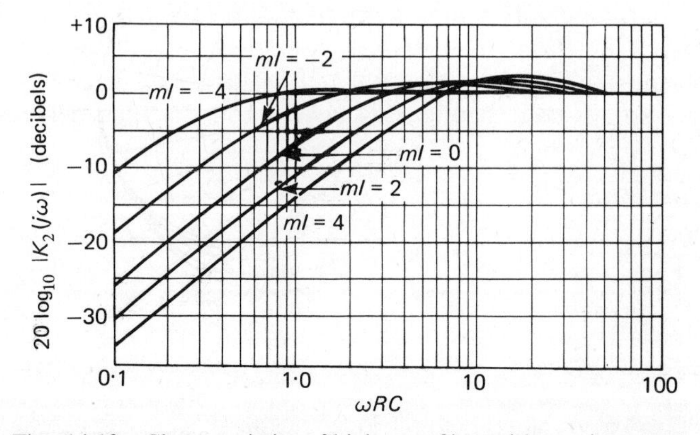

Fig. 14.13. Characteristics of high-pass filter with varying taper.

14.4 SYNTHESIS USING DISTRIBUTED *RC* NETWORKS

To determine the necessary and sufficient conditions for a function to represent the driving-point impedance of a network composed of uniform distributed *RC* sections, we may use a transformation developed by O'Shea.[4] This transformation is rather general, its only limitation being that all the distributed *RC* sections of the complete network must have the same *RC* product. From Eqs. (14.8) and (14.18), the frequency dependence of the z-parameters of a uniform distributed *RC* network can be expressed as follows

$$\mathbf{Z}(s) = \begin{bmatrix} \dfrac{R\coth\sqrt{sRC}}{\sqrt{sRC}} & \dfrac{R\,\mathrm{csch}\sqrt{sRC}}{\sqrt{sRC}} \\ \dfrac{R\,\mathrm{csch}\sqrt{sRC}}{\sqrt{sRC}} & \dfrac{R\coth\sqrt{sRC}}{\sqrt{sRC}} \end{bmatrix}, \tag{14.48}$$

where R is the total resistance of the network and C is the total capacitance. We see that if we multiply Eq. (14.48) by $\sqrt{sRC}\sinh\sqrt{sRC}$ and then set

$$p = \cosh\sqrt{sRC}, \tag{14.49}$$

we get

$$\mathbf{Z}'(p) = \begin{bmatrix} Rp & R \\ R & Rp \end{bmatrix}. \tag{14.50}$$

The elements of the new impedance matrix $\mathbf{Z}'(p)$ are seen to be rational in p. The impedance matrix of Eq. (14.50) can be represented by a lumped network consisting of resistors and inductors, as in Fig. 14.14(b). The presence of negative resistors in Fig. 14.14(b) does not mean that the uniform distributed *RC* structure is not physically realizable; it merely indicates that the transformation is not positive real.

The necessary and sufficient conditions that a function $Z(s)$ be the driving-point impedance function of a network composed of uniform distributed *RC* sections with constant *RC* product are that $Z'(p)$, obtained from $Z(s)$ by applying the transformation as described by Eqs. (14.48) and (14.50), satisfy the following conditions:

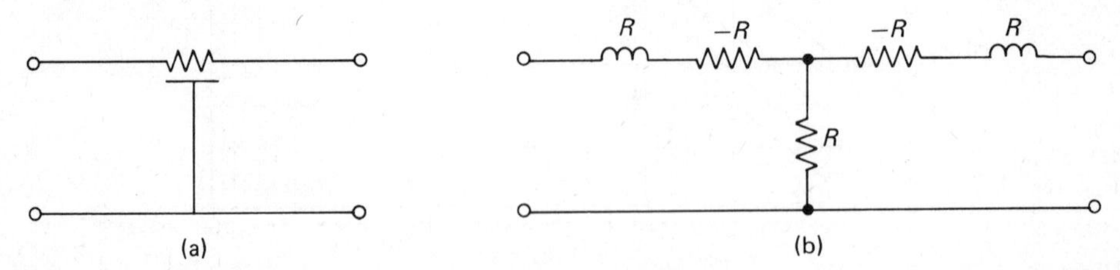

Fig. 14.14. (a) Uniform *RC* distributed network; (b) Equivalent *RL* network in p-domain.

1. $Z'(p)$ is a rational function in p.
2. The poles and zeros of $Z'(p)$ are simple and are interlaced along the real axis of the p-plane.
3. The most negative and most positive of the finite critical points are zeros.
4. All of the poles and zeros are less than or equal to one in magnitude.
5. If

$$Z'(p) = K\frac{\prod_{i=1}^{n+1}(p - p_{oi})}{\prod_{i=1}^{n}(p - p_i)} \tag{14.51}$$

then $K > 0$.

These conditions may be readily proved[4] from a study of the *RL* network shown in Fig. 14.14(b).

Let $Z(s)$ be a driving-point impedance function that satisfies the above set of conditions. From a study of the pole–zero plot of $Z'(p)/(p^2 - 1)$, it is seen that if this function is expanded in partial fractions, all of the coefficients are positive. Therefore we may write

$$\begin{aligned}\frac{Z'(p)}{p^2 - 1} &= \frac{Z'(p)}{(p - 1)(p + 1)} \\ &= \frac{k_1}{p - 1} + \frac{k_2}{p + 1} + \sum_{i=1}^{n} \frac{h_i}{p - p_i},\end{aligned} \tag{14.52}$$

where k_1, k_2 and the h_i's are positive numbers, and $-1 < p_i < +1$. Multiplying both sides of Eq. (14.52) by $p^2 - 1$, we obtain

$$Z'(p) = k_1(p + 1) + k_2(p - 1) + \sum_{i=1}^{n} \frac{h_i(p + 1)(p - 1)}{p - p_i}. \tag{14.53}$$

The terms $k_1(p + 1)$ and $k_2(p - 1)$ can be realized in the p-domain as shown in parts (b) and (d) of Fig. 14.15, respectively. This diagram also gives the corresponding s-domain realizations using uniform distributed *RC* sections. Consider, next, a typical term in the summation of Eq. (14.53); expanding its reciprocal in partial fractions, we get

$$\frac{p - p_i}{h_i(p + 1)(p - 1)} = \frac{(1 + p_i)/2h_i}{p + 1} + \frac{(1 - p_i)/2h_i}{p - 1}. \tag{14.54}$$

The first admittance term of Eq. (14.54) can be realized using the network of Fig. 14.15(b) with $R = 4h_i/(1 + p_i)$, while the second admittance term can be realized using the network of Fig. 14.15(d) with $R = h_i/(1 - p_i)$; the parallel

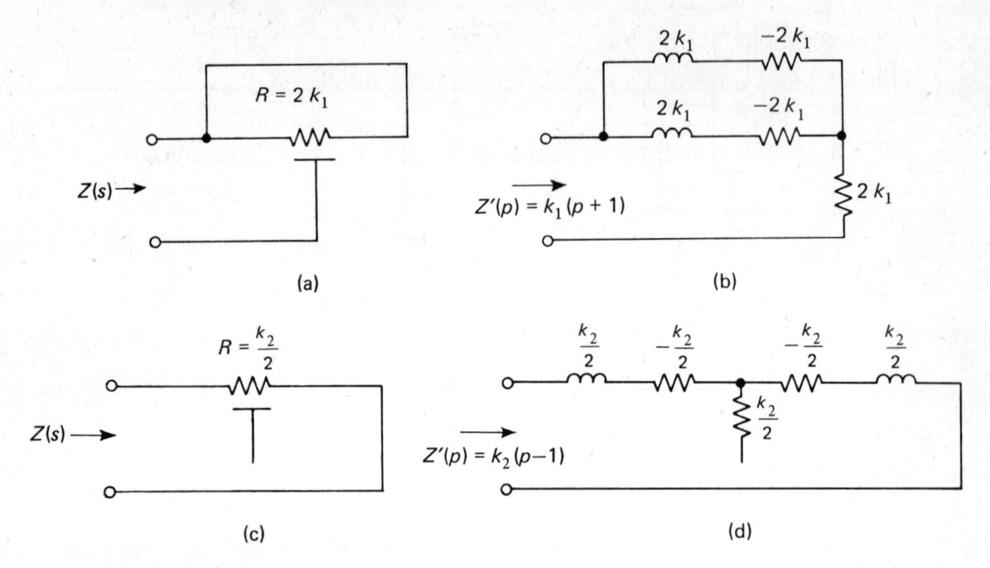

Fig. 14.15. Sections for realizing $Z'(p) = k_1(p + 1)$ and $Z'(p) = k_2(p - 1)$, and corresponding distributed sections.

connection of these two networks, therefore, realizes the admittance function of Eq. (14.54). We have thus demonstrated that an impedance function $Z(s)$ that satisfies the conditions as stated above is completely realizable with uniform distributed *RC* sections.

Example 14.2. As an illustrative example, consider

$$Z(s) = \frac{\tanh \sqrt{s}\,(\tanh^2 \sqrt{s} + 8)}{\sqrt{s}\,(\tanh^2 \sqrt{s} + 3)}.$$

Applying O'Shea's transformation, we get

$$Z'(p) = \frac{9}{4}\,\frac{(p^2 - 1)(p^2 - \frac{1}{9})}{p(p^2 - \frac{1}{4})}.$$

Next, expanding $Z'(p)/(p^2 - 1)$ in partial fractions, and then multiplying both sides by $(p^2 - 1)$, there results

$$Z'(p) = \frac{p^2 - 1}{p} + \frac{\frac{5}{8}(p^2 - 1)}{p - \frac{1}{2}} + \frac{\frac{5}{8}(p^2 - 1)}{p + \frac{1}{2}},$$

which yields the realization shown in Fig. 14.16 for $Z(s)$.

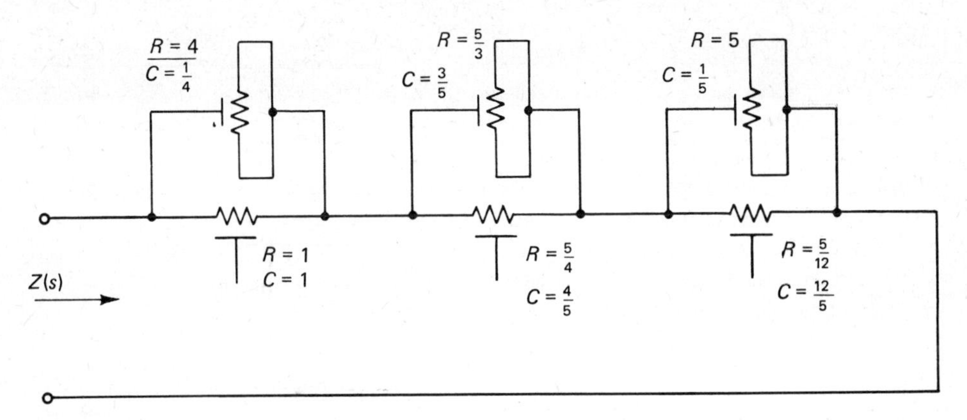

Fig. 14.16. Network realizing the driving-point impedance

$$Z(s) = \frac{\tanh\sqrt{s}(\tanh^2\sqrt{s} + 8)}{\sqrt{s}(\tanh^2\sqrt{s} + 3)}.$$

Distributed *RC* Active Filters

O'Shea's transformation may also be used readily to adapt the various synthesis procedures of Chapter 13 to realize active-distributed *RC* filters.[5,6] For example, consider Yanagisawa's procedure involving the use of a current-inversion NIC which, for convenience, is reproduced in Fig. 14.17. The open-circuit transfer-voltage ratio of this network is given by

$$K(s) = \frac{V_2}{V_1} = \frac{Y_{1b}(s) - Y_{1a}(s)}{[Y_{1b}(s) - Y_{1a}(s)] + [Y_{2b}(s) - Y_{2a}(s)]}. \tag{14.55}$$

Applying O'Shea's transformation to the admittances of Eq. (14.55), we obtain

$$K'(p) = \frac{Y'_{1b}(p) - Y'_{1a}(p)}{[Y'_{1b}(p) - Y'_{1a}(p)] + [Y'_{2b}(p) - Y'_{2a}(p)]}. \tag{14.56}$$

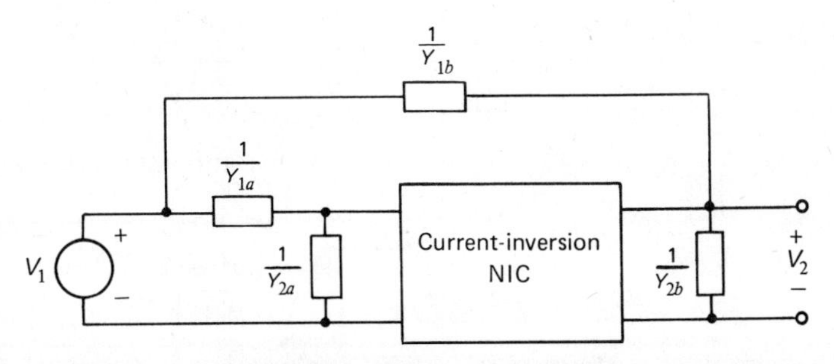

Fig. 14.17. Yanagisawa's structure.

Let

$$K'(p) = \frac{N(p)}{Q(p)}. \tag{14.57}$$

For realizability, the numerator polynomial $N(p)$ must be of no greater degree than the denominator polynomial $Q(p)$. The zeros of $K'(p)$ can lie anywhere in the p-plane but, for stability reasons, the poles of $K'(p)$ must be restricted. With $p = \cosh \sqrt{sRC}$, it can be shown that for stability the p-plane poles must lie within the enclosed curve of Fig. 14.18 which is formed by mapping the portion of the $j\omega$-axis for $-2\pi^2 \leq \omega RC \leq 2\pi^2$ into the p-plane.

Proceeding as in Section 13.3, we may write

$$Y'_{1b}(p) - Y'_{1a}(p) = \frac{N(p)}{P(p)}, \tag{14.58}$$

$$Y'_{2b}(p) - Y'_{2a}(p) = \frac{Q(p) - N(p)}{P(p)}, \tag{14.59}$$

where $P(p)$ is an auxiliary polynomial chosen such that all of its zeros in the p-plane are simple and less than one in magnitude. Also, the degree of $P(p)$ must be at least $n + 1$, with n defining the order of $K'(p)$. Having chosen a suitable $P(p)$, we next expand $N(p)/P(p)$ in partial fractions, allocating the sum of all the terms with positive residues to $Y'_{1b}(p)$ and the remaining terms to $Y'_{1a}(p)$. Similarly, $Y'_{2b}(p)$ and $Y'_{2a}(p)$ are obtained from Eq. (14.59). Finally, these admittances are synthesized as previously described.

Example 14.3. Consider the transfer-voltage ratio

$$K'(p) = \frac{(p-1)(p+3)}{p^2 + p + 1}.$$

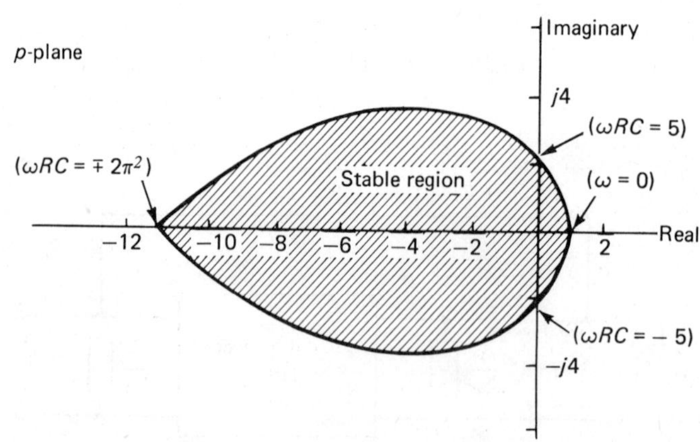

Fig. 14.18. Illustrating the stability restriction imposed upon the p plane poles.

Choosing the auxiliary polynomial as

$$P(p) = p(p-1)(p+1),$$

we get

$$Y'_{1b}(p) - Y'_{1a}(p) = \frac{(p-1)(p+3)}{p(p-1)(p+1)}$$

$$= \frac{3}{p} - \frac{2}{p+1}.$$

We may therefore identify

$$Y'_{1b}(p) = \frac{3}{p}$$

$$Y'_{1a}(p) = \frac{2}{p+1}.$$

Similarly,

$$Y'_{2b}(p) - Y'_{2a}(p) = \frac{-p+4}{p(p-1)(p+1)}$$

$$= \frac{3}{2(p-1)} + \frac{5}{2(p+1)} - \frac{4}{p}.$$

Therefore,

$$Y'_{2b}(p) = \frac{3}{2(p-1)} + \frac{5}{2(p+1)}$$

$$Y'_{2a}(p) = \frac{4}{p}.$$

The distributed *RC* active network may thus be synthesized as shown in Fig. 14.19.

14.5 OTHER APPLICATIONS

a) Notch Filter

The lumped-distributed network of Fig. 14.20 is capable of realizing complex zeros of transmission.[7] The *notch filter* effect is obtained when the transmission zeros are designed to lie on the imaginary axis of the *s*-plane. We shall assume the distributed *RC* network to be uniform. Let

$$a = \frac{R}{R_1}, \tag{14.60}$$

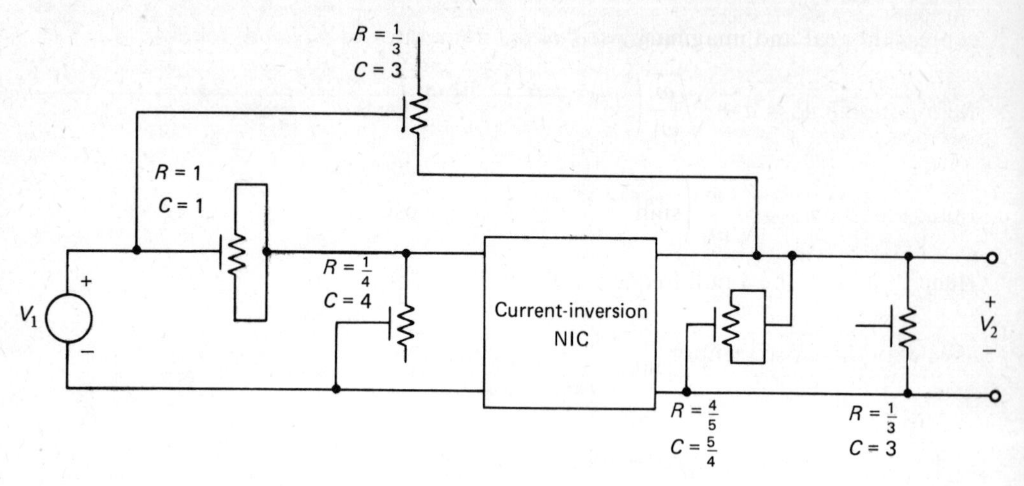

Fig. 14.19. Network realizing $K'(p) = (p - 1)(p + 3)/(p^2 + p + 1)$.

and R and C be the total resistance and total capacitance of the distributed section. Then, the open-circuit transfer-voltage ratio of the structure in Fig. 14.20 is obtained as

$$K(s) = \left.\frac{V_2}{V_1}\right|_{I_2=0} = \frac{a + \sqrt{sRC}\sinh\sqrt{sRC}}{a\cosh\sqrt{sRC} + \sqrt{sRC}\sinh\sqrt{sRC}}. \tag{14.61}$$

For $s = j\omega$, we have

$$K(j\omega) = \frac{a + u\sinh u}{a\cosh u + u\sinh u}, \tag{14.62}$$

where

$$u = (1 + j)\sqrt{\frac{\omega}{\omega_1}}, \tag{14.63}$$

$$\omega_1 = \frac{2}{RC}. \tag{14.64}$$

For a null frequency to be located at ω_0, say, we must have $(1 - \omega^2/\omega_0^2)$ as a factor in the numerator of $K(j\omega)$ in Eq. (14.62). With u defined by Eq. (14.63), we may

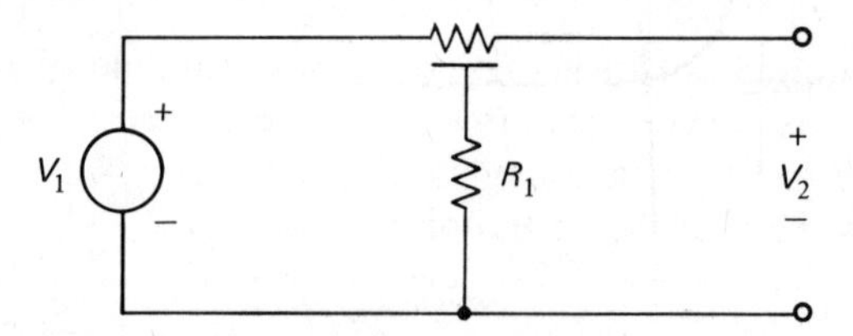

Fig. 14.20. Lumped-distributed notch filter.

express the real and imaginary parts of the numerator of $K(j\omega)$ as follows:

$$\operatorname{Re}[a + u \sinh u] = a + \sqrt{\frac{\omega}{\omega_1}}\left(\sinh\sqrt{\frac{\omega}{\omega_1}}\cos\sqrt{\frac{\omega}{\omega_1}} - \cosh\sqrt{\frac{\omega}{\omega_1}}\sin\sqrt{\frac{\omega}{\omega_1}}\right)$$
$$\operatorname{Im}[a + u \sinh u] = \sqrt{\frac{\omega}{\omega_1}}\left(\sinh\sqrt{\frac{\omega}{\omega_1}}\cos\sqrt{\frac{\omega}{\omega_1}} + \cosh\sqrt{\frac{\omega}{\omega_1}}\sin\sqrt{\frac{\omega}{\omega_1}}\right). \tag{14.65}$$

Hence, for ω_0 to be a null frequency, the two conditions

$$\tan\sqrt{\frac{\omega}{\omega_1}} = -\tanh\sqrt{\frac{\omega}{\omega_1}}$$
$$a = 2\sqrt{\frac{\omega}{\omega_1}}\cosh\sqrt{\frac{\omega}{\omega_1}}\sin\sqrt{\frac{\omega}{\omega_1}} \tag{14.66}$$

must be simultaneously satisfied at $\omega = \omega_0$. Equations (14.66) yield an infinite set of values for a and ω_0. The sets will be denoted a_{0n} and ω_{0n} where n can take on any odd-integer value. As indicated in Fig. 14.21, the integer values of n correspond to the odd-numbered intersections of the $-\tanh\sqrt{\omega/\omega_1}$ curve with the family of curves labelled $\tan\sqrt{\omega/\omega_1}$. The first intersection occurs at

$$\frac{\omega_{01}}{\omega_1} = 5{\cdot}6$$

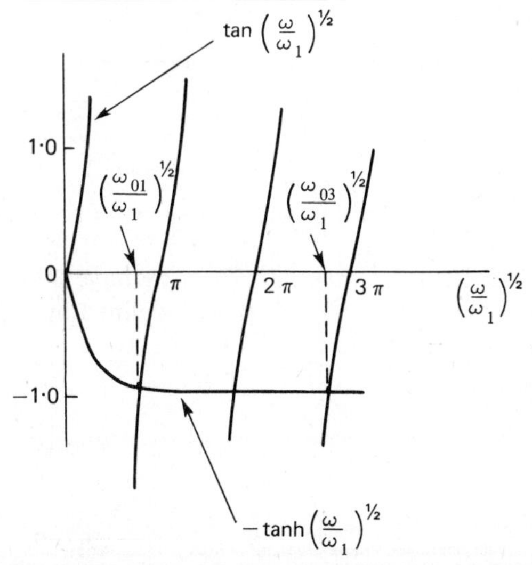

Fig. 14.21. Illustrating the solution of Eqs. (14.66).

and the corresponding value of a_{01} obtained from Eq. (14.66) is

$$a_{01} = 17{\cdot}8.$$

The third intersection and all higher ordered intersections occur very closely at

$$\frac{\omega_{0n}}{\omega_1} = \left(n\pi - \frac{\pi}{4}\right)^2. \tag{14.67}$$

The null frequency thus depends on a and on ω_1; for a_{0n}, the null frequency is ω_{0n}.

For

$$a = 17{\cdot}8,$$

the gain characteristic of the notch filter is as shown plotted in Fig. 14.22.

b) Narrow-band Tuned Amplifier

The lumped-distributed notch filter of Fig. 14.20 may also be used in conjunction with a high-gain amplifier to form a narrow-band tuned amplifier. This is achieved by using the filter as the feedback network of a negative feedback amplifier, as shown in Fig. 14.23(a). In the idealized model of Fig. 14.23(b) it is assumed that the amplifier has infinite input impedance and zero output impedance. The

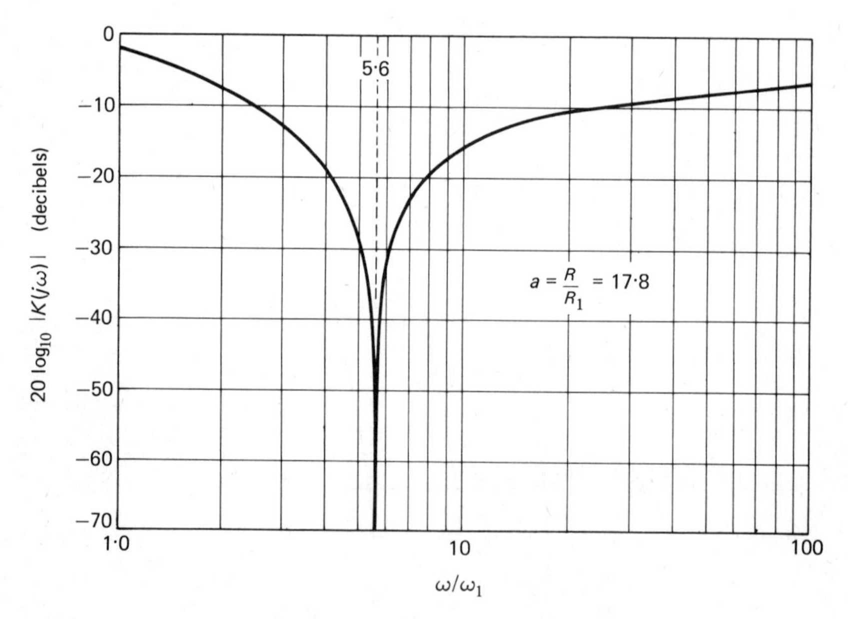

Fig. 14.22. Gain characteristic of notch filter.

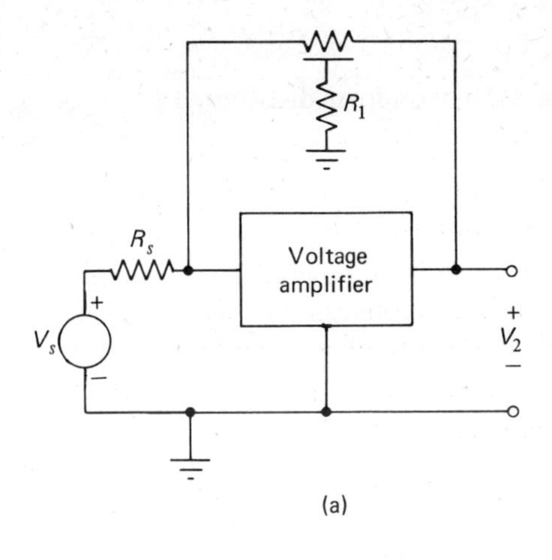

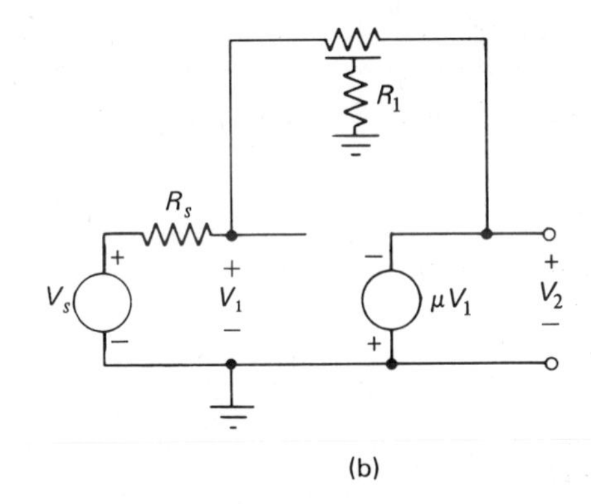

Fig. 14.23. (a) Tuned amplifier using notch filter; (b) Idealized model.

overall transfer-voltage ratio of the feedback amplifier is therefore

$$K(s) = \frac{V_2}{V_s} = \frac{-\mu}{1 + y'_{11}R_s - \mu y'_{12}R_s}, \tag{14.68}$$

where the y-parameters y'_{11} and y'_{12} refer to the filter, and R_s is the source resistance. When the voltage gain μ of the amplifier is high, the overall response closely approaches the reciprocal of the product $y'_{12}R_s$ as shown by

$$K(s) \simeq \frac{1}{y'_{12}R_s}. \tag{14.69}$$

If the filter is adjusted to a_{01}, the Q-factor attainable with the tuned amplifier of Fig. 14.23 is theoretically unlimited, depending only on the amplifier gain μ.

Concluding Remarks

To conclude, a comparison of uniform distributed *RC* networks with exponentially tapered *RC* distributed networks reveals three inherent advantages in tapering:

1. A tapered structure used as a low-pass filter provides a sharper cut-off than is obtainable with a uniform structure.
2. For the same phase shift, a tapered structure produces less attenuation, which, in turn, permits the construction of a phase-shift oscillator with a lower gain requirement on the amplifier component.
3. In the case of a notch filter, the rejection bandwidth, at the same attenuation, is narrower in a tapered structure.

Also, we may use a tapered structure to provide different input and output impedance levels, whereas in a uniform structure both impedance levels are the same.

The effects of a linear taper in distributed *RC* networks have also been studied.[2] It appears that there is not much difference between exponential and linear tapering, but that exponential tapering is somewhat more effective.

REFERENCES

1. R. W. JOHNSON, "Extending the Frequency Range of the Phase-shift Oscillator," *Proc. I.R.E.* **33,** 597 (1945).
2. W. M. KAUFMAN and S. J. GARRETT, "Tapered Distributed Filters," *Trans. I.R.E.* **CT-9,** 329 (1962).
3. B. L. H. WILSON and R. B. WILSON, "Shaping of Distributed *RC* Networks," *Proc. I.R.E.* **49,** 1330 (1961).
4. R. P. O'SHEA, "Synthesis Using Distributed *RC* Networks," *Trans. I.E.E.E.* **CT-12,** 546 (1965).
5. P. M. CHIRLIAN, *Integrated and Active Network Analysis and Synthesis.* Prentice-Hall, 1967.
6. M. S. GHAUSI and J. J. KELLY, *Introduction to Distributed-parameter Networks.* Holt, Rinehart and Winston, 1968.
7. W. M. KAUFMAN, "Theory of a Monolithic Null Device and Some Novel Circuits," *Proc. I.R.E.* **48,** 1540 (1960).

PROBLEMS

14.1 The phase-shift oscillator of Fig. P14.1 employs a non-uniform three-section lumped *RC* network. Show that the circuit oscillates at a frequency

$$\omega = \frac{1}{RC}\sqrt{3 + \frac{2}{n} + \frac{1}{n^2}},$$

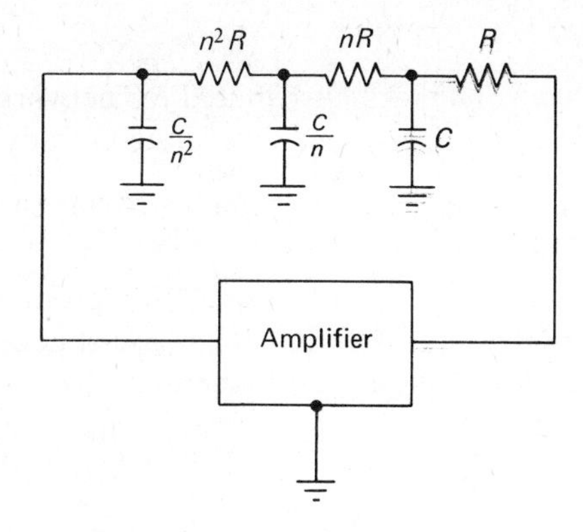

Fig. P14.1.

provided the amplifier has an open-circuit voltage gain

$$\mu \geq 8 + \frac{12}{n} + \frac{7}{n^2} + \frac{2}{n^3}.$$

Assume the amplifier is unilateral, with infinite input impedance and zero output impedance. Compare the performance of this oscillator with a phase-shift oscillator employing a uniform three-section lumped RC network.

14.2 A unit-step voltage is applied across the input port of a uniform distributed RC network with the output port open-circuited. Determine Elmore's delay and rise time for the voltage developed across the output port.

14.3 Determine the open-circuit transfer-voltage ratio of the lumped-distributed network shown in Fig. P14.3. What type of a filter does this structure represent?

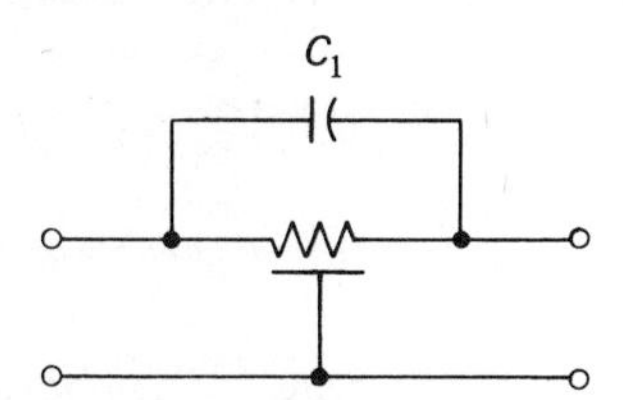

Fig. P14.3.

14.4 Using O'Shea's transformation, synthesize the driving-point impedance function,

$$Z(s) = \frac{\tanh\sqrt{4s}}{\sqrt{s}} + \frac{1}{\sqrt{s}\tanh\sqrt{4s}} + \frac{\tanh\sqrt{4s}}{\sqrt{s}(\tanh^2\sqrt{4s} + 1)}.$$

14.5 Using a current-inversion NIC and uniform distributed RC networks, synthesize the following transfer-voltage ratio specified in the p-domain:

$$K'(p) = \frac{p}{p^2 + \sqrt{2}\,p + 1}.$$

INDEX